STATISTICA™

Volume III: STATISTICS II

1. Nonlinear Estimation 3001
2. Discriminant Function Analysis 3063
3. Reliability and Item Analysis 3107
4. Canonical Correlation 3135
5. Cluster Analysis 3165
6. Factor Analysis 3197
7. Multidimensional Scaling 3235
8. Time Series 3259
9. Log-Linear Analysis 3447
10. Survival Analysis 3479
11. Structural Equation Modeling 3539
12. Megafile Manager 3689
    References 3739
    Index 3765

Volume I:	CONVENTIONS & STATISTICS I
Volume II:	GRAPHICS
Volume III:	STATISTICS II
Volume IV:	INDUSTRIAL STATISTICS
Volume V:	LANGUAGES: BASIC and SCL

USER LICENSE

STATSOFT SINGLE USER LICENSE AGREEMENT

The following constitutes the terms of the License Agreement between a single user (User) of this software package, and the producer of the package, StatSoft, Inc. (called Statsoft hereafter). By opening the package, you (the User) are agreeing to become bound by the terms of this agreement. If you do not agree to the terms of this agreement do not open the package, and contact the StatSoft Customer Service Department (or an authorized StatSoft dealer) in order to obtain an authorization number for the return of the package. This License Agreement pertains also to all third party software included in or distributed with StatSoft products.

License

Unless explicitly stated on the program disks, the enclosed software package is sold to be used on one computer system by one user at a time. This License Agreement explicitly excludes renting or loaning the package. Unless explicitly stated on the program disks, this License Agreement explicitly excludes the use of this package on mulituser systems, networks, or any time sharing systems. (Contact StatSoft concerning Multiuser License Programs.) The user is allowed to make a backup copy for archival purposes and/or to install the software package on a hard disk. However, the software will never be installed on more than one hard disk at a time. The documentation accompanying this software package (or any of its parts) shall not be copied or reproduced in any form.

Disclaimer of Warranty

Although producing error free software is obviously a goal of every software manufacturer, it can never be guaranteed that a software program is actually free of errors. Business and scientific application software is inherently complex (and it can be used with virtually unlimited numbers of data and command settings, producing idiosyncratic operational environments for the software); therefore, the User is cautioned to verify the results of his or her work. This software package is provided "as is" without warranty of any kind. StatSoft and distributors of StatSoft software products make no representation or warranties with respect to the contents of this software package and specifically disclaim any implied warranties or merchantability or fitness for any particular purpose. In no event shall StatSoft be liable for any damages whatsoever arising out of the use of, inability to use, or malfunctioning of this software package. StatSoft does not warrant that this software package will meet the User's requirements or that the operation of the software package will be uninterrupted or error free.

Limited Warranty

If within 30 days from the date when the software package was purchased (i.e., the StatSoft invoice date), the program disks are found to be defective (i.e., they are found to be unreadable by the properly aligned disk drive of the computer system on which the package is intended to run), StatSoft will replace the disks free of charge. After 30 days, the User will be charged for the replacement a nominal disk replacement fee. If within 90 days from the date when the software package was purchased (i.e., invoice date), the software package was found by the User not capable of performing any of its main (i.e., basic) functions described explicitly in promotional materials published by StatSoft, StatSoft will provide the User with replacement disks free of defects, or if the replacement cannot be provided within 90 days from the date when StatSoft was notified by the User about the defect, the User will receive a refund of the purchasing price of the software package.

Updates, Corrections, Improvements

The User has a right to purchase all subsequent updates, new releases, new versions, and modifications of the software package introduced by StatSoft for an update fee or for a reduced price (depending on the scope of the modification). StatSoft is not obligated to inform the User about new updates, improvements, modifications, and/or corrections of errors introduced to its software packages. In no event shall StatSoft be liable for any damages whatsoever arising out of the failure to notify the User about a known defect of the software package.

Copyright © StatSoft, 1995

Chapter 1:
NONLINEAR ESTIMATION

Table of Contents

Introductory Overview 3007
    General Purpose 3007
    Estimating Linear and Nonlinear Models 3007
    Common Nonlinear Regression Models 3008
    Nonlinear Estimation Procedures 3013
    Evaluating the Fit of the Model 3018
    When the Estimation Process Does Not Converge 3020
Program Overview 3023
Examples 3025
    Probit and Logit Models 3025
    Exponential Regression Models 3031
    Other Regression Models 3036
    Specifying Loss Functions 3039
    Concluding Remarks 3041
Dialogs, Options, Statistics 3043
    Startup Panel 3043
    Probit/Logit Regression 3043
    Exponential Regression 3044
    Breakpoint Regression 3045
    User-defined Models 3046
    Estimated Function and Loss Function 3046
    Model Estimation 3047
    Start Values 3050
    Step Sizes 3050
    Parameter Estimation 3051
    Results 3051
Notes 3057
    Syntax for User-specified Regression Models 3057
    Some Useful *STATISTICA BASIC* Programs 3060
Index 3061

The *Detailed Table of Contents* follows on the next page.

1. NONLINEAR ESTIMATION - CONTENTS

Detailed Table of Contents

INTRODUCTORY OVERVIEW **3007**
   General Purpose 3007
   Estimating Linear and Nonlinear Models 3007
   Common Nonlinear Regression Models 3008
      Intrinsically Linear Regression Models 3008
      Intrinsically Nonlinear Regression Models 3009
      General Growth Model 3010
      Models for Binary Responses: Probit and Logit 3010
      General Logistic Regression Model 3011
      Drug Responsiveness and Half-Maximal Response 3012
      Discontinuous Regression Models 3012
   Nonlinear Estimation Procedures 3013
      Least Squares Estimation 3013
      Loss Functions 3013
      Weighted Least Squares 3014
      Maximum Likelihood 3014
      Function Minimization Algorithms 3015
      Start Values, Step Sizes, Convergence Criteria 3016
      Penalty Functions, Constraining Parameters 3016
      Local Minima 3016
      Quasi-Newton Method 3017
      Simplex Procedure 3017
      Hooke-Jeeves Pattern Moves 3017
      Rosenbrock Pattern Search 3017
      Hessian Matrix and Standard Errors 3018
   Evaluating the Fit of the Model 3018
      Proportion of Variance Explained 3018
      Goodness of fit *Chi-square* 3018
      Plot of Observed vs. Predicted Values 3019
      Normal and Half-Normal Probability Plots 3019
      Plot of the Fitted Function 3019
      Variance/Covariance Matrix for Parameters 3020
   When the Estimation Process Does Not Converge 3020

PROGRAM OVERVIEW **3023**
   Overview 3023
   Models 3023
   Loss Functions 3023
   Estimation Procedures 3023

1. NONLINEAR ESTIMATION - CONTENTS

    Output 3023
    Alternative Procedures 3024

EXAMPLES **3025**
    Overview 3025
    Probit and Logit Models 3025
        Example 1: Predicting Success/Failure 3025
        Example 2: Predicting Redemption of Coupons 3029
    Exponential Regression Models 3031
        Example 3: Predicting Recovery from Injury 3031
        Example 4: Comparing Two Learning Curves 3033
        Example 4.1: Estimating Two Different Models 3035
    Other Regression Models 3036
        Example 5: Regression in Pieces 3036
        Example 6: Estimating Drug Responsiveness (Half-Maximal Response) 3038
    Specifying Loss Functions 3039
        Example 7: Weighted Least Squares 3039
    Concluding Remarks 3041

DIALOGS, OPTIONS, STATISTICS **3043**
    Startup Panel 3043
    Probit/Logit Regression 3043
        Input File 3043
        Variables 3044
        Codes 3044
        Missing Data 3044
    Exponential Regression 3044
        Variables 3045
        Missing Data 3045
    Breakpoint Regression 3045
        Variables 3045
        Breakpoint 3045
        Missing Data 3045
    User-defined Models 3046
        Function to be Estimated and Loss Function 3046
        Missing Data 3046
    Estimated Function and Loss Function 3046
        Open 3046
        Save As 3046
        Variables 3046
    Model Estimation 3047
        Estimation Method 3047
        Asymptotic Standard Errors 3048
        Eta for Finite Difference Approximation 3048
        Maximum Number of Iterations 3048

1. NONLINEAR ESTIMATION - CONTENTS

  Convergence Criterion 3049
  Start Values 3049
  Initial Step Size 3049
  Means and Standard Deviations 3049
  Matrix Plot for all Variables 3049
  Box and Whisker Plot for all Variables 3050
 Start Values 3050
  Individual Start Values 3050
  Common Values 3050
  Apply 3050
 Step Sizes 3050
  Individual Step Size Values 3051
  Common Values 3051
  Apply 3051
 Parameter Estimation 3051
 Results 3051
  Parameter Estimates (and Standard Errors) 3052
  Cov./Corrs. of Parameters 3052
  Scale MS-error to 1 3052
  Residual Values 3052
  Predicted Values 3052
  Observed Values 3052
  Classification of Cases and Odds Ratios 3052
  Means and Standard Deviations 3053
  Difference (Previous Model) 3053
  Save Predicted and Residual Values 3053
  Plot Options 3053
  Fitted 2D Function and Observed Values 3053
  Fitted 3D Function and Observed Values 3053
  Distribution of Residuals 3054
  Normal Probability Plot of Residuals 3054
  Half-normal Probability Plot of Residuals 3054
  Predicted vs. Observed Values 3054
  Predicted vs. Residual Values 3055
  Matrix Plot for all Variables 3055
  Box and Whisker Plot for all Variables 3055

NOTES **3057**
 Syntax for User-specified Regression Models 3057
  General Syntax Conventions: Regression Equations 3057
  General Syntax Conventions: Loss Function 3059
  Math Errors 3059
 Some Useful *STATISTICA BASIC* Programs 3060
  Weighted Least Squares Estimation 3060

1. NONLINEAR ESTIMATION - CONTENTS

Two-Stage Least Squares Estimation 3060
Box-Cox and Box-Tidwell Transformations 3060

INDEX **3061**

1. NONLINEAR ESTIMATION - CONTENTS

Chapter 1:
NONLINEAR ESTIMATION

INTRODUCTORY OVERVIEW

General Purpose

In the most general terms, the *Nonlinear Estimation* module will compute the relationship between a set of independent variables and a dependent variable. For example, you may want to compute the relationship between the dose of a drug and its effectiveness, the relationship between training and subsequent performance on a task, the relationship between the price of a house and the time it takes to sell it, etc. You may recognize research issues in these examples that are commonly addressed by such techniques as multiple regression (see *Multiple Regression*, Volume I, Chapter 6) or analysis of variance (see *ANOVA/MANOVA*, Volume I, Chapter 7). In fact, you may think of the *Nonlinear Estimation* module as a *generalization* of those methods. Specifically, multiple regression (and ANOVA) assumes that the relationship between the independent variable(s) and the dependent variable is *linear* in nature. The *Nonlinear Estimation* module leaves it up to you to specify the nature of the relationship. For example, you may specify the dependent variable to be a logarithmic function of the independent variable(s), an exponential function, or a function of some complex ratio of variables, etc.

When allowing for any type of relationship between the independent variables and the dependent variable, two issues arise. First, what types of relationships "make sense," that is, are interpretable in a meaningful manner? Note that the simple linear relationship is very convenient in that it allows you to make such straightforward interpretations as "the more of x (e.g., the higher the price of a house), the more there is of y (the longer it takes to sell it); and given a particular increase in x, a proportional increase in y can be expected." Nonlinear relationships cannot usually be interpreted and verbalized in such a simple manner. The second issue that needs to be addressed is how to exactly compute the relationship, that is, how to arrive at results that allow you to say whether or not there is a nonlinear relationship as predicted.

In the following paragraphs, the nonlinear regression problem will be discussed in a somewhat more formal manner; that is, common terminology will be introduced that will allow you to examine the nature of these techniques more closely, and how they are used to address important questions in various research domains (medicine, social sciences, physics, chemistry, pharmacology, engineering, etc.).

Estimating Linear and Nonlinear Models

Technically speaking, the *Nonlinear Estimation* module is a general fitting procedure that will estimate any kind of relationship between a dependent (or response) variable), and a list of independent variables. In general, all regression models may be stated as:

$$y = F(x_1, x_2, \ldots, x_n)$$

In most general terms, the user is interested in whether and how a dependent variable is related to a list of independent variables; the term $F(x...)$ in the expression above means that y, the dependent or response variable, is a *function* of the x's, that is, the independent variables.

An example of this type of model would be the linear multiple regression model as described in *Multiple Regression* (Volume I, Chapter 6). For this

1. NONLINEAR ESTIMATION - INTRODUCTORY OVERVIEW

model, one assumes the dependent variable to be a *linear* function of the independent variables, that is:

$$y = a + b_1*x_1 + b_2*x_2 + \ldots + b_n*x_n$$

If you are not familiar with multiple linear regression, you may want to read the *Introductory Overview* section to the *Multiple Regression* module (Volume I, Chapter 6) at this point (however, it is not necessary to understand all of the nuances of multiple linear regression techniques in order to understand the methods discussed here).

The *Nonlinear Estimation* module allows you to specify essentially any type of continuous or discontinuous regression model. Some of the most common nonlinear models (such as *probit*, *logit*, *exponential growth*, and *breakpoint regression*) are pre-defined in the *Nonlinear Estimation* module and can simply be requested as dialog options. However, you can also type in any type of regression equation, which *STATISTICA* will then fit to your data (see *User-defined Models*, page 3046). Moreover, you can specify either standard *least squares* estimation, *maximum likelihood* estimation (where appropriate), or, again, define your own "loss function" (see below) by typing in the respective equation.

In general, whenever the simple linear regression model does not appear to adequately represent the relationships between variables, then the nonlinear regression model approach is appropriate.

Common Nonlinear Regression Models

Intrinsically Linear Regression Models

Polynomial regression. A common "nonlinear" model is polynomial regression. The term *nonlinear* is put in quotes here because the nature of this model is actually linear. For example, suppose you measure in a learning experiment subjects'

physiological arousal and their performance on a complex tracking task. Based on the well-known Yerkes-Dodson law, you could expect a curvilinear relationship between arousal and performance; this expectation can be expressed in the regression equation:

$$\texttt{Performance} = a + b_1*\texttt{Arousal} + b_2*\texttt{Arousal}^2$$

In this equation, *a* represents the intercept, and b_1 and b_2 are regression coefficients. The non-linearity of this model is expressed in the term *Arousal*^{2}. However, the *nature* of the model is still linear, except that when estimating it, you would square the measure of arousal. The *Multiple Regression* module *fixed nonlinear* option could also be used to estimate the regression coefficients for this model. These types of models, where you include some transformation of the independent variables in a linear equation, are also referred to as models that are *nonlinear in the variables*.

Models that are nonlinear in the parameters.

To contrast the example above, consider the relationship between a human's age from birth (the *x* variable) and his or her growth rate (the *y* variable). Clearly, the relationship between these two variables in the first year of a person's life (when most growth occurs) is very different than during adulthood (when almost no growth occurs). Thus, the relationship could probably best be expressed in terms of some negative exponential function:

$$\texttt{Growth} = \exp(-b_1*\texttt{Age})$$

If you plotted this relationship for a particular estimate of the regression coefficient (b_1) you would obtain a curve that looks something like this.

NLN - 3008

Copyright © StatSoft, 1995

1. NONLINEAR ESTIMATION - INTRODUCTORY OVERVIEW

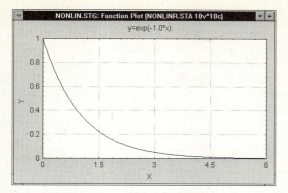

Note that the *nature* of this model is no longer linear, that is, the expression shown above does not simply represent a linear regression model, with some transformation of the independent variable. This type of model is said to be *nonlinear in the parameters*.

Making nonlinear models linear. In general, whenever a regression model can be "made" into a linear model, this is the preferred route to pursue (for estimating the respective model). The linear multiple regression model (see *Multiple Regression*, Volume I, Chapter 6) is very well understood mathematically and, from a pragmatic standpoint, is most easily interpreted. Therefore, returning to the simple exponential regression model of *Growth* as a function of *Age* shown above, you could convert this nonlinear regression equation into a linear one by simply taking the logarithm of both sides of the equations, so that:

```
Log(Growth) = -b₁*Age
```

If you now substitute *log(Growth)* with *y*, you have the standard linear regression model as shown earlier (without the intercept which was ignored here to simplify matters). Thus, you could log-transform the *Growth* rate data (e.g., using the spreadsheet formula transformations, see *General Conventions*, Volume I, Chapter 1) and then use the *Multiple Regression* module to estimate the relationship between *Age* and *Growth*, that is, compute the regression coefficient b_1.

Model adequacy. Of course, by using the "wrong" transformation, one could end up with an inadequate model. Therefore, after "linearizing" a model such as the one shown above, it is particularly important to use the extensive residual statistics in the *Multiple Regression* module.

Intrinsically Nonlinear Regression Models

Some regression models which cannot be transformed into linear ones, can only be estimated via the *Nonlinear Estimation* module. In the growth rate example above, the random error in the dependent variable was purposely "forgotten" about. Of course, the growth rate is affected by many other variables (other than time), and you can expect a considerable amount of random (*residual*) fluctuation about the fitted line. If you add this *error* or residual variability to the model, you can rewrite it as follows:

```
Growth = exp(-b₁*Age) + error
```

Additive error. In this model it is assumed that the error variability is independent of age, that is, that the amount of residual error variability is the same at any age. Because the error term in this model is additive, you can no longer linearize this model by taking the logarithm of both sides. If for a given data set, you were to log-transform variable *Growth* anyway and fit the simple linear model, then you would find that the residuals from the analysis would no longer be evenly distributed over the range of variable *Age*; and thus, the standard linear regression analysis (via *Multiple Regression*, Chapter 6, Volume I) would no longer be appropriate. Therefore, the only way to estimate the parameters for this model is via the *Nonlinear Estimation* module.

Multiplicative error. To "defend" the previous example, in this particular instance it is not likely that the error variability is constant at all ages, that is, that the error is additive. Most likely, there is

NLN - 3009

1. NONLINEAR ESTIMATION - INTRODUCTORY OVERVIEW

more random and unpredictable fluctuation of the growth rate at the earlier ages than the later ages, when growth comes to a virtual standstill anyway. Thus, a more realistic model including the error would be:

`Growth = exp(-b₁*Age) * error`

Put into words, the greater the age, the smaller the term $exp(-b_1*Age)$, and, consequently, the smaller the resultant error variability. If you now take the log of both sides of the equation, the residual error term will become an additive factor in a linear equation, and you can go ahead and estimate b_1 via standard multiple regression.

`Log(Growth) = -b₁*Age + error`

Now consider some regression models (that are nonlinear in their parameters) which cannot be "made into" linear models through simple transformations of the raw data.

General Growth Model

The general growth model, is similar to the example that was previously considered:

`y = b₀ + b₁*exp(b₂*x) + error`

This model is commonly used in studies of any kind of growth (*y*), when the rate of growth at any given point in time (*x*) is proportional to the amount of growth remaining. The parameter b_0 in this model represents the maximum growth value. A typical example where this model would be adequate is when one wants to describe the concentration of a substance (e.g., in water) as a function of time.

Models for Binary Responses: Probit and Logit

It is not uncommon that a dependent or response variable is binary in nature, that is, that it can have only two possible values. For example, patients either do or do not recover from an injury, job applicants either succeed or fail at an employment test, journal subscribers either do or do not renew a subscription, coupons may or may not be returned, etc. In all of these cases, one may be interested in estimating a model that describes the relationship between one or more continuous independent variable(s) to the binary dependent variable.

Using linear regression. Of course, one could use standard multiple regression procedures to compute standard regression coefficients. For example, if one studied the renewal of journal subscriptions, one could create a *y* variable with *1*'s and *0*'s, where *1* indicates that the respective subscriber renewed, and *0* indicates that the subscriber did not renew. However, there is a problem: the *Multiple Regression* module does not "know" that the response variable is binary in nature. Therefore, it will inevitably fit a model that leads to predicted values that are greater than *1* or less than *0*. However, predicted values that are greater than *1* or less than *0* are not valid; thus, the restriction in the range of the binary variable (between *0* and *1*) is ignored if one uses standard multiple regression.

Continuous response functions. The regression problem could be rephrased so that, rather than predicting a binary variable, you are predicting a *continuous* variable that naturally stays within the 0-1 bounds. The two most common regression models that accomplish exactly this are the *logit* and the *probit* regression models.

Logit regression. In the logit regression model, the predicted values for the dependent variable will never be less than (or equal to) *0*, or greater than (or equal to) *1*, regardless of the values of the independent variables. This is accomplished by applying the following regression equation, which actually has some "deeper meaning" as you will see shortly:

$$y = \frac{exp(b_0+b_1*x_1+\ldots+b_n*x_n)}{1+exp(b_0+b_1*x_1+\ldots+b_n*x_n)}$$

One can easily recognize that, regardless of the regression coefficients or the magnitude of the x values, this model will always produce predicted values (predicted y's) in the range of 0 to 1.

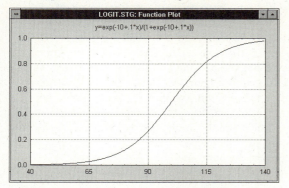

For example, shown above are the expected (predicted) values for a single independent variable with $b_0=-10$ and $b_1=.1$.

The name *logit* stems from the fact that one can easily linearize this model via the *logit* transformation. Suppose one thinks of the binary dependent variable y in terms of an underlying continuous probability p, ranging from 0 to 1. One can then transform that probability p as:

`p' = log_e[p/(1-p)]`

This transformation is referred to as the *logit* or *logistic* transformation. Note that p' can theoretically assume any value between minus and plus infinity. Since the logit transform solves the issue of the 0-1 boundaries for the original dependent variable (probability), you could use those (logit transformed) values in an ordinary linear regression equation. In fact, if you perform the logit transform on both sides of the logit regression equation stated earlier, the standard linear regression model is obtained:

`p' = b_0 + b_1*x_1 + b_2*x_2 + ... + b_n*x_n`

Probit regression. One may consider the binary response variable to be the result of a normally distributed underlying variable that actually ranges from minus infinity to positive infinity. For example, a subscriber to a journal can feel very strongly about not renewing a subscription, be almost undecided, "tend towards" renewing the subscription, or feel very much in favor of renewing the subscription. In any event, all that you (the publisher of the journal) will *see* is the binary response of renewal or failure to renew the subscription. However, if you set up the standard linear regression equation based on the underlying "feeling" or attitude you could write:

`feeling ... = b_0 + b_1*x_1 + ... + b_n*x_n`

which is, of course, the standard regression model. It is reasonable to assume that these feelings are normally distributed, and that the probability p of renewing the subscription is about equal to the relative *space* under the normal curve. Therefore, if you transform each side of the equation so as to reflect normal probabilities, you obtain:

`NP(feeling) = NP(b_0 + b_1*x_1 + ... + b_n*x_n)`

where *NP* stands for *normal probability* (space under the normal curve), as tabulated in practically all statistics texts. The equation shown above is also referred to as the *probit* regression model.

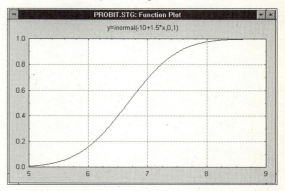

General Logistic Regression Model

The general logistic model can be stated as:

1. NONLINEAR ESTIMATION - INTRODUCTORY OVERVIEW

$$y = b_0/[1+b_1*\exp(b_2*x)]$$

You can think of this model as an extension of the logit or logistic model for binary responses. However, while the logit model restricts the dependent response variable to only two values, this model allows the response to vary within a particular lower and upper limit. For example, suppose you are interested in the population growth of a species that is introduced to a new habitat, as a function of time. The dependent variable would be the number of individuals of that species in the respective habitat. Obviously, there is a lower limit on the dependent variable, since fewer than 0 individuals cannot exist in the habitat; however, there also is most likely an upper limit that will be reached at some point in time.

Drug Responsiveness and Half-Maximal Response

In pharmacology, the following model is often used to describe the effects of different dose levels of a drug:

$$y = b_0 - b_0/[1+(x/b_2)^{b_1}]$$

In this model, x is the dose level (usually in some coded form, so that $x \geq 1$) and y is the responsiveness, in terms of the percent of maximum possible responsiveness. The parameter b_0 then denotes the expected response at the level of dose saturation and b_2 is the concentration that produces a half-maximal response; the parameter b_1 determines the slope of the function.

Discontinuous Regression Models

Piecewise linear regression. It is not uncommon that the *nature* of the relationship between one or more independent variables and a dependent variable changes over the range of the independent variables. For example, suppose you monitor the per-unit manufacturing cost of a particular product as a function of the number of units manufactured (output) per month. In general, the more units per month you produce, the lower is your per-unit cost, and this linear relationship may hold over a wide range of different levels of production output. However, it is conceivable that above a certain point, there is a discontinuity in the relationship between these two variables. For example, the per-unit cost may decrease relatively less quickly when older (less efficient) machines have to be put on-line in order to cope with the larger volume. Suppose that the older machines go on-line when the production output rises above 500 units per month; you may specify a regression model for cost-per-unit as:

$$y = b_0 + b_1*x*(x \leq 500) + b_2*x*(x > 500)$$

In the above formula, y stands for the estimated per-unit cost; x is the output per month. The expressions ($x \leq 500$) and ($x > 500$) denote logical conditions that evaluate to *0* if false, and to *1* if true. Thus, this model specifies a common intercept (b_0), and a slope that is either equal to b_1 (if $x \leq 500$ is true, that is, equal to *1*) or b_2 (if $x > 500$ is true, that is, equal to 1). Instead of *specifying* the point where the discontinuity in the regression line occurs (at *500* units per months in the example above), one could also *estimate* that point. For example, one might have noticed or suspected that there is a discontinuity in the cost-per-unit at one particular point; however, one may not know where that point is. In that case, simply replace the *500* in the equation above with an additional parameter (e.g., b_3). The *Nonlinear Estimation* module would then estimate the point of discontinuity.

One could also adjust the equation above to reflect a "jump" in the regression line. For example, imagine that, after the older machines are put on-line, the per-unit-cost jumps to a higher level and then slowly goes down as volume continues to increase. In that case, simply specify an additional intercept (b_3), so that:

$$y = (b_0+b_1*x)*(x \leq 500) + (b_3+b_2*x)*(x > 500)$$

The *Nonlinear Estimation* module includes a pre-defined breakpoint regression model that can be chosen as a dialog option. However, unlike the model shown above, that option will fit different regression models to different ranges of the dependent *y* variable.

Comparing groups. The method described here to estimate different regression equations in different domains of the independent variable can also be used to distinguish between groups. For example, suppose in the example above there are three different plants; to simplify the example, ignore the breakpoint for now. If you coded the three plants in a grouping variable by using the values *1*, *2*, and *3*, you could simultaneously estimate three different regression equations by specifying:

```
y = (xp=1)*(b10+b11*x) +
    (xp=2)*(b20+b21*x) +
    (xp=3)*(b30+b31*x)
```

In this equation, x_p denotes the grouping variable containing the codes that identify each plant, b_{10}, b_{20}, and b_{30} are the three different intercepts, and b_{11}, b_{21}, and b_{31} refer to the slope parameters (regression coefficients) for each plant. One could compare the fit of this model to the fit of the common regression model (without considering the different plants) in order to determine which model is more appropriate.

Nonlinear Estimation Procedures

Least Squares Estimation

Some of the more common nonlinear regression models have been reviewed above. Now the question arises as to how these models are estimated. If you are familiar with linear regression techniques (as described in *Multiple Regression*, Volume I, Chapter 6) or analysis of variance (ANOVA) techniques (as described in *ANOVA/MANOVA*, Volume I, Chapter 7), then you may be aware of the fact that all of those methods use *least squares* estimation procedures. In the most general terms, least squares estimation is aimed at minimizing the sum of squared deviations of the observed values for the dependent variable from those predicted by the model.

Loss Functions

In standard multiple regression, the regression coefficients are estimated by "finding" those coefficients that minimize the residual variance (sum of squared residuals) around the regression line. Any deviation of an observed score from a predicted score signifies some *loss* in the accuracy of your prediction, for example, due to random noise (error). Therefore, it can be said that the goal of least squares estimation is to minimize a *loss function*; specifically, this loss function is defined as the sum of the squared deviations about the predicted values. When this function is at its minimum, then you get the same parameter estimates (intercept, regression coefficients) as you would in the *Multiple Regression* module; because of the particular loss functions that yielded those estimates, you can call the estimates *least squares estimates*.

Phrased in this manner, there is no reason why you cannot consider other loss functions. For example, rather than minimizing the sum of *squared* deviations, why not minimize the sum of *absolute* deviations? Indeed, this is sometimes useful in order to "de-emphasize" outliers. Relative to all other residuals, a large residual will become *much* larger when squared. However, if one only takes the absolute value of the deviations, then the resulting regression line will most likely be less affected by outliers.

The *Nonlinear Estimation* module contains several function minimization methods that can be used to minimize any kind of loss function. Now, some commonly used loss functions will be reviewed.

Weighted Least Squares

In addition to least squares and absolute deviation regression (see above), weighted least squares estimation is probably the most commonly used technique. Ordinary least squares techniques assume that the residual variance around the regression line is the same across all values of the independent variable(s). Put another way, it is assumed that the error variance in the measurement of each case is identical. Often, this is not a realistic assumption; in particular, violations frequently occur in business, economic, or biological applications. Also, this assumption is clearly violated in the case of the logistic and probit regression models for binary dependent variables (and the *maximum likelihood* loss function is used in the parameter estimation process for those models; see below).

For example, suppose you wanted to study the relationship between the projected cost of construction projects and the actual cost. This may be useful in order to gage the expected cost overruns. In this case it is reasonable to assume that the absolute magnitude (dollar amount) by which the estimates are off is proportional to the size of the project. Thus, you would use a weighted least squares loss function to fit a linear regression model. Specifically, the loss function would be (see, for example, Neter, Wasserman, and Kutner, 1985, page 168):

```
Loss = (Obs-Pred)² * (1/x²)
```

In this equation, the loss function first specifies the standard least squares loss function (*Obs*erved minus *Pred*icted squared; i.e., the squared residual) and then weighs this loss by the inverse of the squared value of the independent variable (*x*) for each case. In the actual estimation, the program will sum up the value of the loss function for each case (e.g., construction project), as specified above, and estimate the parameters that minimize that sum. To return to the example, the larger the project (*x*) the less weight is placed on the deviation from the predicted value (cost). This method will yield more stable estimates of regression parameters (for more details, see Neter, Wasserman, and Kutner, 1985).

Maximum Likelihood

An alternative to the least squares loss function (see above) is to maximize the *likelihood* or *log-likelihood* function (or to minimize the negative *log-likelihood* function). In most general terms, the likelihood function is defined as:

```
L = F(Y,Model)
  = ∏{p[yᵢ,Model Parameters(xᵢ)]}
```

In theory, you can compute the probability (now called *L*, the *likelihood*) of the specific dependent variable values to occur in the sample, given the respective regression model. Provided that all observations are independent of each other, this likelihood is the geometric sum (∏, across $i = 1$ to n cases) of probabilities for each individual observation (*i*) to occur, given the respective model and parameters for the *x* values. (The geometric sum means that one would *multiply* out the individual probabilities across cases.) It is also customary to express this function as a natural logarithm, in which case the geometric sum becomes a regular arithmetic sum (Σ, across $i = 1$ to n cases).

Given the respective model, the larger the likelihood of the model, the larger is the probability of the dependent variable values to occur in the sample. Therefore, the greater the likelihood, the better is the fit of the model to the data. The actual computations for particular models here can become quite complicated because you need to "track" (compute) the probabilities of the *y* values to occur (given the model and the respective *x* values). As it turns out, if all assumptions for standard multiple regression are met (as described in *Multiple Regression*, Volume I, Chapter 6), then the standard least squares estimation method (see above) will yield results identical to the maximum likelihood method. If the assumption of equal error variances across the range of the *x* variable(s) is violated, then the weighted least squares method described earlier will yield maximum likelihood estimates (note that *STATISTICA* includes the *STATISTICA BASIC* example program *Wls.stb* in the *STBASIC* subdirectory, that will compute weighted least squares estimates for multiple linear regression models; see also the *Notes* section, page 3060).

Maximum likelihood and probit/logit models.

The likelihood function has been "worked out" for probit and logit regression models. Specifically, the loss function for these models is computed as the sum of the natural log of the logit or probit likelihood L_1 so that:

$$\log(L_1) = \Sigma[y_i*\log(p_i)+(1-y_i)*\log(1-p_i)]$$

where

$\log(L_1)$ is the natural log of the (logit or probit) likelihood (log-likelihood) for the current model

y_i is the observed value for case *i*

p_i is the expected (predicted or fitted) probability (between 0 and 1)

The log-likelihood of the null model (L_0), that is, the model containing the intercept only (and no regression coefficients) is computed as:

$$\log(L_0) = n_0*[\log(n_0/n)]+n_1*[\log(n_1/n)]$$

where

$\log(L_0)$ is the natural log of the (logit or probit) likelihood of the null model (intercept only)

n_0 is the number of observations with a value of 0 (zero)

n_1 is the number of observations with a value of 1

n is the total number of observations

These formulas are provided here for your reference; maximum likelihood estimation will automatically be used for probit and logit models, so you do not need to type this complex formula into the loss function dialog (see page 3046).

Function Minimization Algorithms

Now that the different regression models and the loss functions that can be used to estimate them have been discussed, the only "mystery" that is left is how to minimize the loss functions (to find the best-fitting set of parameters) and how to estimate the standard errors of the parameter estimates. The *Nonlinear Estimation* module uses one very efficient algorithm (*quasi-Newton*) that approximates the second-order derivatives of the loss function to guide the search for the minimum (i.e., for the best parameter estimates, given the respective loss function). In addition, the *Nonlinear Estimation* module offers several more general function minimization algorithms that follow different search strategies (which do not depend on the second-order derivatives). These strategies are sometimes more effective for estimating loss functions with local minima; therefore, these methods are often particularly useful to find appropriate *start values* for the estimation via the *quasi-Newton* method.

In all cases, *STATISTICA* can compute (if requested by the user) the standard errors of the parameter estimates (see *Estimation Procedure*, page 3047). These standard errors are based on the second-order partial derivatives for the parameters, which are computed via finite difference approximation.

If you are not interested in how the minimization of the loss function is done, only that it *can* be done, you may skip the following paragraphs. However, you may find it useful to know a little about these procedures in case your regression model "refuses" to be fit to the data. In that case, the iterative estimation procedure will fail to converge, producing ever "stranger" (e.g., very large or very small) parameter estimates.

1. NONLINEAR ESTIMATION - INTRODUCTORY OVERVIEW

In the following paragraphs, some general issues involved in unconstrained optimization will first be discussed, and then the methods used in this module will briefly be reviewed. For more detailed discussions of these procedures you may refer to Brent (1973), Gill and Murray (1974), Peressini, Sullivan, and Uhl (1988), and Wilde and Beightler (1967). For specific algorithms, see Dennis and Schnabel (1983), Eason and Fenton (1974), Fletcher (1969), Fletcher and Powell (1963), Fletcher and Reeves (1964), Hooke and Jeeves (1961), Jacoby, Kowalik, and Pizzo (1972), and Nelder and Mead (1964).

Start Values, Step Sizes, Convergence Criteria

A common aspect of all estimation procedures is that they require the user to specify some start values, initial step sizes, and a criterion for convergence (see *Estimation Procedure*, page 3047). All methods will begin with a particular set of initial estimates (*start values*), which will be changed in some systematic manner from iteration to iteration; in the first iteration, the *step size* determines by how much the parameters will be moved. Finally, the *convergence criterion* determines when the iteration process will stop. For example, the process may stop when the improvements in the loss function from iteration to iteration are less than a specific amount. The *Nonlinear Estimation* module has preset defaults for all of these parameters, which are appropriate in most cases.

Penalty Functions, Constraining Parameters

All estimation procedures in the *Nonlinear Estimation* module are unconstrained in nature. That means that the program will move parameters around without any regard for whether or not permissible values result. For example, in the course of logit regression you may get estimated values that are equal to *0.0*, in which case the logarithm cannot be computed (since the log of 0 is undefined). When this happens, the program will assign a *penalty* to the loss function, that is, a very large value. As a result, the various estimation procedures usually move away from the regions that produce those functions. However, in some circumstances, the estimation will "get stuck," and as a result, you would see a very large value of the loss function. This could happen, if, for example, the regression equation involves taking the logarithm of an independent variable which has a value of zero for some cases (in which case the logarithm cannot be computed).

If the user wishes to constrain a procedure in the *Nonlinear Estimation* module, then this constraint must be specified in the loss function as a penalty function (assessment). By doing this, the user may control what permissible values of the parameters to be estimated may be manipulated by the program. For example, if two parameters (*a* and *b*) are to be constrained to be greater than or equal to zero, then one must assess a large penalty to these parameters if this condition is not met. Below is an example of a user-specified regression and loss function, including a penalty assessment designed to "penalize" the parameters *a* and/or *b* if either one is not greater than or equal to zero:

Estimated function:

`v3 = a + b*v1 + c*v1`

Loss function:

`L = (obs-pred)`^{2}`+(a<0)*100000+(b<0)*100000`

Local Minima

The most "treacherous" threat to unconstrained function minimization is *local minima*. For example, a particular loss function may become slightly larger, regardless of how a particular parameter is moved. However, if the parameter were to be moved into a completely different place, the loss function may actually become smaller. You

NLN - 3016

Copyright © StatSoft, 1995

can think of such local minima as local "valleys" or minor "dents" in the loss function. However, in most practical applications, local minima will produce "outrageous" and extremely large or small parameter estimates with very large standard errors. In those cases, specify different start values and try again.

Also note that the *Simplex* method (see below) is particularly "smart" in avoiding such minima; therefore, this method may be particularly suited in order to find appropriate start values for complex functions.

Quasi-Newton Method

As you may remember, the slope of a function at a particular point can be computed as the first-order derivative of the function (at that point). The "slope of the slope" is the second-order derivative, which tells one how fast the slope is changing at the respective point, and in which direction. The *quasi-Newton* method will, at each step, evaluate the function at different points in order to estimate the first-order derivatives and second-order derivatives. It will then use this information to follow a path towards the minimum of the loss function.

Simplex Procedure

This algorithm does not rely on the computation or estimation of the derivatives of the loss function. Instead, at each iteration the function will be evaluated at *m+1* points in the *m* dimensional parameter space. For example, in two dimensions (i.e., when there are two parameters to be estimated), the program will evaluate the function at three points around the current optimum. These three points would define a triangle; in more than two dimensions, the "figure" produced by these points is called a *Simplex*. Intuitively, in two dimensions, three points will allow you to determine "which way to go," that is, in which direction in the two-dimensional space to proceed in order to minimize

the function. The same principle can be applied to the multidimensional parameter space; that is, the *Simplex* will "move" downhill; when the current step sizes become too "crude" to detect a clear downhill direction, (i.e., the *Simplex* is too large), the *Simplex* will "contract" and try again.

An additional strength of this method is that when a minimum appears to have been found, the *Simplex* will again be expanded to a larger size to see whether the respective minimum is a local minimum. Thus, in a way, the *Simplex* moves like a smooth, single-cell organism down the loss function, contracting and expanding as local minima or significant ridges are encountered.

Hooke-Jeeves Pattern Moves

In a sense, this is the simplest of all algorithms. At each iteration, this method first defines a pattern of points by moving each parameter one by one, so as to optimize the current loss function. The entire pattern of points is then shifted or moved to a new location; this new location is determined by extrapolating the line from the old base point in the *m* dimensional parameter space to the new base point. The step sizes in this process are constantly adjusted to "zero in" on the respective optimum. This method is usually quite effective and should be tried if both the *quasi-Newton* and *Simplex* methods (see above) fail to produce reasonable estimates.

Rosenbrock Pattern Search

Where all other methods fail, the *Rosenbrock Pattern Search* method often succeeds. This method will rotate the parameter space and align one axis with a ridge (this method is also called the *method of rotating coordinates*); all other axes will remain orthogonal to this axis. If the loss function is unimodal and has detectable ridges pointing towards the minimum of the function, then this method will proceed with sure-footed accuracy towards the minimum of the function. However, note that this search algorithm may terminate early when there are several constraint boundaries (resulting in the penalty value; see above) that intersect, leading to a discontinuity in the ridges.

Hessian Matrix and Standard Errors

The matrix of second-order (partial) derivatives is also called the Hessian matrix. It turns out that the inverse of the Hessian matrix approximates the variance/covariance matrix of parameter estimates. Intuitively, there *should* be an inverse relationship between the second-order derivative for a parameter and its standard error. If the change of the slope around the minimum of the function is very sharp, then the second-order derivative will be large; however, the parameter estimate will be quite stable in the sense that the minimum with respect to the parameter is clearly identifiable. If the second-order derivative is nearly zero, then the change in the slope around the minimum is zero, meaning that you can practically move the parameter in any direction without greatly affecting the loss function. Thus, the standard error of the parameter will be very large.

The Hessian matrix (and asymptotic standard errors for the parameters) is computed separately via finite difference approximation by selecting the option *Asymptotic Standard Errors* from the *Model Estimation* dialog. This procedure yields very precise asymptotic standard errors for all estimation methods.

Evaluating the Fit of the Model

After estimating the regression parameters, an essential aspect of the analysis is to test the appropriateness of the overall model. For example, if one specified a linear regression model, but the relationship is intrinsically nonlinear (see page 3009), then the parameter estimates (regression coefficients) and the estimated standard errors of those estimates may be significantly "off." Some of the ways to evaluate the appropriateness of a model will now be reviewed.

Proportion of Variance Explained

Regardless of the model, one can always compute the total variance of the dependent variable (total sum of squares, SST), the proportion of variance due to the residual (error sum of squares, SSE), and the proportion of variance due to the regression model (regression sum of squares, SSR = SST - SSE). The ratio of the regression sum of squares to the total sum of squares (SSR/SST) explains the proportion of variance accounted for in the dependent variable (y) by the model; thus, this ratio is equivalent to the *R-square* ($0 \leq R\text{-}square \leq 1$, the *coefficient of determination*). Even when the dependent variable is not normally distributed across cases, this measure may help evaluate how well the model fits the data.

Goodness of fit Chi-square

For probit and logit regression models, the *Nonlinear Estimation* module will use maximum likelihood estimation (i.e., maximize the likelihood function; see *Probit/Logit Regression*, page 3043). As it turns out, one can directly compare the likelihood L_0 for the null model where all slope parameters are zero, with the likelihood L_1 of the fitted model. Specifically, one can compute the *Chi-square* statistic (χ^2) for this comparison as:

$$\chi^2 = -2*[\log(L_0) - \log(L_1)]$$

The degrees of freedom for this *Chi-square* value are equal to the difference in the number of parameters for the null and the fitted model; thus, the degrees of freedom will be equal to the number of independent variables in the logit or probit regression. If the *p*-level associated with this *Chi-square* is significant, then you can say that the estimated model yields a significantly better fit to the data than the null model, that is, that the regression parameters are statistically significant.

Plot of Observed vs. Predicted Values

It is always a good idea to inspect a scatterplot of predicted vs. observed values.

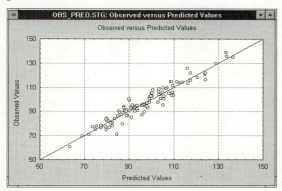

If the model is appropriate for the data, then you would expect the points to roughly follow a straight line; if the model is incorrectly specified, then this plot will indicate a nonlinear pattern.

Normal and Half-Normal Probability Plots

The *normal probability plot* of residuals will give an indication of whether or not the residuals (i.e., errors) are normally distributed. This plot is constructed as follows. First, the residuals (observed minus predicted values) are rank ordered. From these ranks, the z values (i.e., standardized values of the normal distribution) are computed based on the assumption that the data come from a normal distribution. These z values are plotted on the Y-axis in the plot. If the observed residuals (plotted on the X-axis) are normally distributed, then all values should fall onto a straight line.

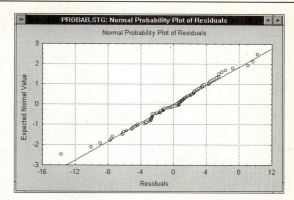

If the residuals are not normally distributed, then they will deviate from the line.

The *half-normal probability plot* is constructed in the same manner as the standard normal probability plot, except that only the positive half of the normal curve is considered. Consequently, only positive normal values will be plotted on the Y-axis. This plot is basically used when one wants to ignore the sign of the residual, that is, when one is mostly interested in the distribution of absolute residuals, regardless of sign.

Plot of the Fitted Function

For models involving two or three variables (one or two predictors) the *Results* dialog (see page 3051) contains options to plot the fitted function using the final parameter estimates; for example:

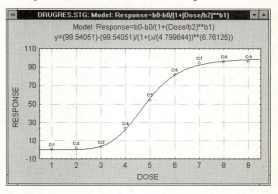

1. NONLINEAR ESTIMATION - INTRODUCTORY OVERVIEW

Here is an example of a 3D plot with two predictor variables:

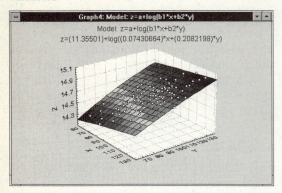

This type of plot represents the most direct visual check of whether or not a model fits the data, and whether there are apparent outliers.

Variance/Covariance Matrix for Parameters

When a model is grossly misspecified or the estimation procedure gets "hung up" in a local minimum, the standard errors for the parameter estimates (which are optionally computed by the program via finite difference approximation) can become very large. This means that regardless of how the parameters were moved around the final values, the resulting loss function did not change much.

Also, the correlations between parameters may become very large, indicating that parameters are very redundant; put another way, when the estimation algorithm moved one parameter away from the final value, then the increase in the loss function could be almost entirely compensated for by moving another parameter. Thus, the effect of those two parameters on the loss function was very redundant.

When the Estimation Process Does Not Converge

In general, follow these steps to estimate the parameters for any nonlinear regression model:

(1) First, accept the default values for the starting parameters, step sizes, and convergence criterion.

(2) If the estimation procedure does not converge, set the convergence criterion to *0.1*, and then experiment with the start values (try different sets of reasonable values).

(3) If the estimation process continues to "go astray," change the step sizes to smaller values (e.g., *0.05*).

(4) The different estimation procedures in the *Nonlinear Estimation* module use very different minimization algorithms (see *Nonlinear Estimation Procedures*, page 3047), each with its own strengths and weaknesses. If problems persist, switch to a different estimation procedure. Try them in the order suggested by the respective dialog, that is, begin with the *Simplex* method. Start with the default settings for all parameters. However, you may want to increase the maximum number of iterations. (*Simplex* usually requires more, but each iteration is faster.) Next use the *Hooke-Jeeves* method, and finally try the *Rosenbrock method*.

(5) Continue to experiment with the starting values and step sizes. For example, *Simplex* sometimes converges more easily when the convergence criterion is set to the minimum (*0.0000001*). If *Simplex* returns a very large loss value (e.g., *1E+37*), it indicates that the estimation got "trapped" with illegal parameters; continue to change the start values and step sizes.

(6) Try combining the methods of estimation. Because the *Simplex*, *Hooke-Jeeves*, and *Rosenbrock* methods are generally less sensitive to local minima, you may use any one of these methods with the *quasi-Newton* method. This is particularly useful if you are not sure about the appropriate start values for the estimation. In that case, the first method may generate initial parameter estimates, that will then be used in subsequent *quasi-Newton* iterations.

Nonlinear regression is not well-suited for exploratory data analysis purposes. Rather, the researcher should bring to the analysis some understanding of the nature of and relationships between variables. Such understanding will suggest appropriate models that can be estimated and start values that are fairly close to the actual parameters. Given such prior understanding of one's data and some experience with the different estimation methods, it is usually not difficult to fit models that "make sense" and satisfactorily reproduce the data.

PROGRAM OVERVIEW

Overview

The *Nonlinear Estimation* module allows the user to fit essentially all types of regression models. These models can be fit using least squares estimation or any user-specified loss function. The user can choose between four very different estimation procedures so that stable parameter estimates can be obtained in practically all cases.

Models

User-specified models. The user can specify any type of regression model by typing in the respective equation into an equation dialog (*Estimated function and loss function* dialog, page 3046). In this manner, the user may, for example, specify exponential growth models, general logistic models, nonlinear drug responsiveness models, etc. The equations may include logical operators; thus, discontinuous (piecewise) regression models and models including indicator variables can also be estimated. These models (including user-defined loss functions) may be saved for later use.

Predefined models. The most common nonlinear regression models are predefined in the *Nonlinear Estimation* module, and can be chosen simply as dialog options. Those regression models include probit and logit regression for binary dependent variables, the exponential regression model, and linear piecewise (breakpoint) regression.

Loss Functions

User-specified loss functions. Parameter estimation in regression amounts to minimizing a particular loss function. By default, the *Nonlinear Estimation* module will use the least squares loss function and produce standard least squares estimates. However, the user may specify any kind of loss function by typing in the respective equation. For example, in this manner, the user can specify absolute deviation regression, weighted least squares estimation, maximum likelihood estimation, etc. Again, these equations may be saved for later use.

Predefined loss functions. For logit and probit regression, the program will compute maximum likelihood parameter estimates.

Estimation Procedures

The user can choose from among four very different and powerful function minimization procedures: *quasi-Newton*, *Simplex*, *Hooke-Jeeves* pattern moves, and *Rosenbrock* pattern search (method of rotating coordinates). The user has control over the start values, step sizes, and convergence criteria used in these methods.

Output

In addition to various descriptive statistics, standard results in the *Nonlinear Estimation* module include the parameter estimates and their standard errors, the variance/covariance matrix of parameter estimates (only if the *quasi-Newton* method was used), the predicted values, residuals, and appropriate measures of goodness of fit (e.g., log-likelihood of estimated/null models and *Chi-square* test of difference, proportion of variance accounted for, etc.). Like all other results in *STATISTICA*, predicted and residual values can be saved to a standard *STATISTICA* data file for further analyses.

Incremental (stepwise) analysis. For probit and logit models, the incremental fit is also automatically computed when adding or deleting parameters from the regression model. This allows the user to explore the data via a stepwise (nonlinear) estimation procedure.

Plots. A wide selection of graphs are available to aid the user. Histograms for the distribution of all selected variables can be produced, as well as

NLN - 3023

histograms of residual values, scatterplots of observed versus predicted values and predicted versus residual values, and normal and half-normal plots of residuals. 2D and 3D plots of the fitted function and input data (including user-defined functions) can also be produced. Also, a large selection of descriptive graphs are available to aid in the interpretation of input data and the output. For example, all multi-way tables can be reviewed in categorized histograms and 3D bivariate histograms.

Alternative Procedures

For standard linear regression models, use the *Multiple Regression* module. If both the dependent and independent variables in a regression model are categorical in nature, use the *Log-Linear* module. Specialized regression models for censored and uncensored survival or failure time analysis are also included in the *Survival Analysis* module [proportional hazard (Cox) with time-dependent covariates, normal, log-normal, and exponential regression model].

STATISTICA BASIC programs. The *STATISTICA* system includes several *STATISTICA BASIC* example programs (in subdirectory *STBASIC*) that are useful for solving certain regression problems (see also the *Notes* section, page 3060). The program *Wls.stb* will perform weighted least squares analysis and compute various alternative R-square statistics (see Kvålseth, 1985). Program *2stls.stb* provides an example for 2-stage least squares regression. Often, when the residuals from an ordinary least squares analysis do not follow the normal distribution, one can apply certain transformations of the independent or dependent variables to correct the problem. Program *Boxcox.stb* will compute the maximum likelihood estimate for *lambda* for the *Box-Cox* transformation of the dependent variable (Box and Cox, 1964; Mason, Gunst, and Hess, 1989; see also page 3060). Program *Boxtid.stb* will compute the parameters for the *Box-Tidwell* transformation for the independent variables in a linear regression (see Mason, Gunst, Hess, 1989).

EXAMPLES

Overview

The following examples are mostly based on demonstration data sets provided in Neter, Wasserman, and Kutner (1985). The discussion of those examples will focus primarily on how to specify models and how to interpret results. Additional guidelines for cases when the estimation procedure fails or for dealing with unusual cases are also provided (see page 3020).

The *logit* and *probit* models for binary dependent variables will first be reviewed; these models are "pre-wired" into the *Nonlinear Estimation* module and can be chosen as dialog options. Numerous examples of how to fit models and use different loss functions specified by the user via the *Estimated function and loss function* dialog will then be reviewed.

Probit and Logit Models

Example 1: Predicting Success/Failure

This example is based on a data set described in Neter, Wasserman, and Kutner (1985, page 357; however, note that those authors fit a linear regression model to the data). Suppose you want to study whether experience helps programmers complete complex programming tasks within a specified amount of time. Twenty-five programmers were selected with different degrees of experience (measured in months). They were then asked to complete a complex programming task within a certain amount of time. The binary dependent variable is the programmers' success or failure in completing the task. These data are recorded in the file *Program.sta*; shown below is a partial listing of this file.

Specifying the analysis. After starting the *Nonlinear Estimation* module, the startup panel will open in which you can open the data file *Program.sta*.

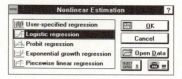

Click on the *Logistic regression* option in the startup panel (see above) and the *Logistic Regression (logit)* dialog will open (see below).

Click on the *Variables* button in this dialog and select the variable *Expernce* as the independent variable and *Success* as the dependent variable. The program will automatically enter the codes for the dependent variable in this dialog. You can also specify the type of missing data deletion (*casewise* deletion of missing data or *mean substitution* of missing data).

Note that the *Nonlinear Estimation* module will always compute maximum likelihood parameter estimates for logit and probit models. Ordinary least squares estimation is based on the assumption of

1. NONLINEAR ESTIMATION - EXAMPLES

constant error (or residual) variance at different values of the independent variables. In the case of binary dependent variables, this assumption is clearly violated, and thus, the maximum likelihood criterion should be used to estimate the parameters of the logistic (and probit) regression model.

Accept the program defaults and click *OK* in this dialog to open the *Model Definition* dialog. In the *Model Definition* dialog, you can select the estimation method as well as specify the convergence criterion, start values, etc. You can also elect to compute separately (via finite difference approximation) the asymptotic standard errors for the parameter estimates. Click on this option to select it for this example.

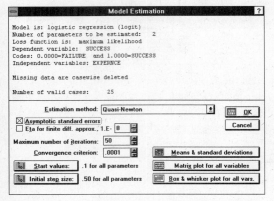

To review the descriptive statistics for all selected variables, click on the *Means and standard deviations* button. As in most other descriptive statistics Scrollsheets in *STATISTICA*, the default graph is the histogram with the normal curve superimposed (click on the Scrollsheet with the right-mouse-button and select the *Quick Stats Graphs - Histogram/Normal* option from the flying menu). Thus you could at this point evaluate the distributions of the variables.

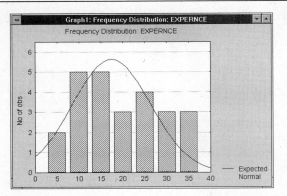

The different estimation procedures in the *Nonlinear Estimation* module are discussed in the *Introductory Overview* section (page 3013). Click on the *Estimation Method* combo box in the *Model Estimation* dialog to see the different options.

A good way to start the analysis is with the default settings in this dialog. As discussed in the *Introductory Overview* section, all estimation procedures require as input start values, initial step sizes, and the convergence criterion (see page 3016). Again, simply accept the defaults in this dialog and click *OK* to estimate the parameters.

Estimating the parameters. The program will display in a window the values of the loss function and the current estimate for each parameter at each iteration. After a few iterations the *quasi-Newton* estimation procedure will converge, that is, arrive at the final parameter estimates.

Reviewing results. Now, click *OK* to proceed to the *Results* dialog.

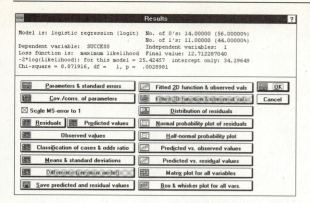

The *Chi-square* value for the difference between the current model and the intercept-only model is highly significant. Thus, one can conclude that experience is related to programmers' success.

Now, review the parameter estimates by clicking on the *Parameter and standard errors* button. As described in the *Introductory Overview* section (page 3018), the standard errors are computed from the finite difference approximation of the Hessian matrix of second-order derivatives. By dividing the estimates by their respective standard errors, you can compute approximate *t*-values, and thus compute the statistical significance levels for each parameter. In the Scrollsheet below, both parameters are significant at the $p<.05$ level.

NONLIN. ESTIMAT.	Model: Logistic regression (logit) N of 0's:14 1's:11 Dep. var: SUCCESS Loss: Max likelihood (MS-err. scaled to 1) Final loss: 12.712287040 Chi²(1)=8.8719 p=.00290	
N=25	Const.B0	EXPERNCE
Estimate	-3.05970	.161486
Std.Err.	1.25959	.064995
t(23)	-2.42912	2.484594
p-level	.02336	.020684

Estimation of standard errors. You may have noticed that the *Scale MS-error to 1* switch on the *Results* dialog shown earlier was checked (selected). In maximum likelihood estimation, it is common practice to rescale the mean square error to 1.0 when computing the estimates for the parameter standard errors (see Jennrich and Moore, 1975, for details).

Interpreting the parameter estimates. In principle, the parameter estimates can be interpreted as in standard linear regression, that is, in terms of an intercept (*Const.B0*) and slope (*Expernce*). Thus, essentially, the results of this study show that the amount of prior experience is significantly related to the successful completion of the programming task. However, as described in the *Introductory Overview* section (page 3010), the parameter estimates pertain to the prediction of *logits* (computed as $log[p/(1-p)]$), not the actual probabilities (*p*) underlying success or failure. Logits will assume values from minus infinity to plus infinity, as the probability *p* moves from *0* to *1*.

Predicted values. Now look at the predicted values, that is, click on the *Predicted values* button in the *Results* dialog. Remember that the logit regression model ensures that the predicted values will never step out of the 0-1 bounds. Thus, you may look at the predicted values as *probabilities*; for example, the predicted probability of success for the second case (*Henry*) is (*.84*).

NONLIN. ESTIMAT.	Predicted Values (pro... SUCCESS Predictd
Frank	.310262
Henry	.835263
Tom	.109996
Beth	.726602
Susan	.461837
Harry	.082130
Paul	.461837
Pete	.245666
Diana	.620812
Louise	.109996

Note that you can save the predicted and residual values for further analysis via the *Save predicted and residual values* button on the *Results* dialog.

Classification of cases. Option *Classification of cases & odds ratio* will bring up the table of correctly and incorrectly classified cases, given the current model (i.e., parameter estimates).

NONLIN. ESTIMAT. Observed	Classification of Cases (program.sta) Odds ratio: 9.7778		
	Pred. FAILURE	Pred. SUCCESS	Percent Correct
FAILURE	11	3	78.57143
SUCCESS	3	8	72.72727

All cases with a predicted value (probability) less than or equal to .5 are classified as *Failure*, those with a predicted value greater than .5 are classified as *Success*. The *Odds ratio* is computed as the ratio of the product of the correctly classified cases over the product of the incorrectly classified cases. Odds ratios that are greater than *1* indicate that the classification is better than what one would expect by pure chance. However, remember that these are *post-hoc* classifications, because the parameters were computed so as to maximize the probability of the observed data (see the description of the *maximum likelihood* loss function in the *Introductory Overview* section, page 3014). Thus, you should not expect to do this well if you applied the current model to classify new (future) observations.

Normal probability plot. The *Introductory Overview* section (page 3019) discusses how normal and half-normal probability plots are constructed. Now click on the *Normal probability plot of residuals* button to bring up this graph.

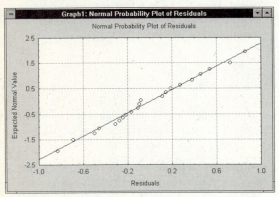

If the residuals (observed minus predicted values) are normally distributed, they will approximately fall onto a straight line in the normal probability plot. Essentially, all points (residuals) in the normal probability plot above are very close to the line expected if the residuals are normally distributed.

Histogram of residuals. Another quick visual way to inspect the distribution of the residuals is via the *Distribution of residuals* button. This resulting plot is shown below:

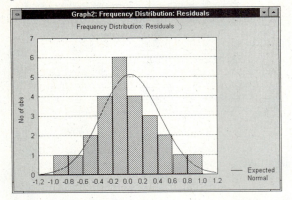

Again, it appears that the residuals are basically normally distributed.

Fitting a probit model. For comparison, now fit the probit regression model to these data. Click on the *Cancel* button in the *Results* dialog and again in the *Logistic Regression (Logit)* dialog to return to the startup panel. Double-click on the *Probit regression* option in the startup panel. In the *Probit Regression* dialog, click on the *Variables* button to select the dependent variable (*Success*) and independent variable (*Expernce*) for the analysis; then click *OK*.

Click *OK* in the *Model Estimation* and *Parameter Estimation* dialogs to go to the *Results* dialog.

Reviewing results. As before, the *Chi-square* value for the comparison of the current model with the intercept-only model is statistically significant.

In fact, the *Chi-square* value is practically the same as before. Thus, the probit regression also leads to the conclusion that programming experience is significantly related to success.

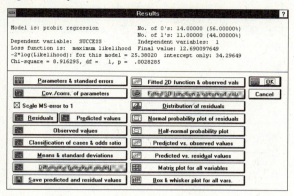

Interpreting the parameter estimates. The parameter estimates for the probit regression can also be interpreted as in standard linear regression. However, the parameter estimates in *probit* regression pertain to the prediction of z values of the normal distribution. If you then take the normal integral of those z values, you end up with values that will never step out of the 0-1 boundaries. The predicted values can again be thought of as probabilities, in this case, as the space under the normal curve associated with the respective predicted z value. Now, look at the *predicted* values.

As you can see in the Scrollsheet above, the predicted values (probabilities of success) for each case under the probit model are very similar to those for the logit model (page 3027). In fact, in most cases, the difference between these two models will be negligible.

Incremental fit with more than one variable.
Before concluding this example, there is an additional feature when fitting probit or logit models that will be discussed. If there are, for example, two independent variables, you can first estimate the model with one variable, and then the model with both variables. In this case, the *Results* dialog *Difference (previous model)* button will no longer be dimmed. This option will compute the difference in the loss function (maximum likelihood) from the previous model (with one independent variable) to the current model (with two independent variables). The *Chi-square* value and *p*-level for the increment in goodness of fit is also reported. In general, you can enter or remove in successive analyses one or more variables; if the resulting model is a subset or superset of the previously fitted model, the incremental goodness of fit will automatically be computed.

Example 2: Predicting Redemption of Coupons

In this example, an alternative way of setting up data files for a probit or logit analysis will be demonstrated. The data set to be analyzed is reported in Neter, Wasserman, and Kutner (1985, page 365) and describes the results of a study of coupon redemption. Specifically, coupons were sent to 1000 randomly selected homes. These coupons differed in their value, that is, with regard to the price reduction offered (either 5, 10, 15, 20, or 30 cents off). Each type of coupon was sent to 200 households. The dependent variable of interest was how many coupons of each type were redeemed. Shown below is a listing of the data file *Coupons.sta*.

1. NONLINEAR ESTIMATION - EXAMPLES

Setup of data file. In this data file, the number of households that did and did not redeem the coupons were recorded. Thus, the dependent variable (likelihood of coupon redemption) really consists of two variables here: Variable *Redeemed* which contains the *codes* to indicate whether or not coupons were redeemed, and variable *Houshlds* which contains the *counts*, that is, information about how many households did or did not redeem the coupon. In a sense, you may think of this data file setup as a crosstabulation table of price reduction by redemption, where variable *Houshlds* contains the frequencies. In this manner, even very large studies can be summarized in a relatively small file. Now, see how this file can be analyzed.

Specifying the analysis. Neter et al. (1985) fit a logit model to these data (using weighted least squares estimation). After opening the *Nonlinear Estimation* module (or after returning to the startup panel), open the data file *Coupons.sta*. Click on the *Logistic regression* option in the startup panel. In the resulting dialog, click on the *Input file contains:* combo box and select *Codes and counts*. Now, click on the *Variables* button and specify *Reductn* as the independent variable, *Redeemed* as the dependent variable, and *Houshlds* as the variable containing the frequency counts.

Click *OK* to go to the *Model Definition* dialog where you can select the *Asymptotic Standard Errors* option (if it is not already selected). Now, click *OK* here and again in the *Parameter Estimation* dialog to open the *Results* dialog.

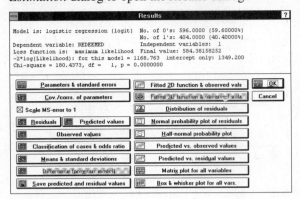

Reviewing results. The *Chi-square* value for the difference between the intercept-only model and the current model is highly significant. Thus, we can conclude that variable *Reductn* is a significant predictor of the likelihood of coupon redemption. Next click on the *Parameter and standard errors* button.

Model: Logistic regression (logit) N of 0's:596 1's:404		
NONLIN. ESTIMAT.	Dep. var: REDEEMED Loss: Max likelihood (MS-err. scaled to 1) Final loss: 584.38158252 Chi²(1)=180.44 p=0.0000	
N=10	Const.B0	REDUCTN
Estimate	-2.1855	.10872
Std.Err.	.1647	.00884
t(8)	-13.2709	12.29312
p-level	.0000	.00000

In this particular example, these parameter estimates are useful for predicting the rate of redemption for coupons offering an intermediate price reduction. For example, suppose you want to estimate the expected proportion of coupons worth 25 cents that will be redeemed. Following the logit model, you could compute:

```
Logit = -2.186 + .109*25 = .539
```

Now remember that logits are computed as $log[p/(1-p)]$; thus, you can compute p as:

```
p = exp(logit)/[1+exp(Logit)]
```

NLN - 3030

Copyright © StatSoft, 1995

= .632

Therefore, you can expect that roughly 63% of the 25-cent coupons would be redeemed.

Exponential Regression Models

Example 3: Predicting Recovery from Injury

This example is also based on a data set reported in Neter, Wasserman, and Kutner (1985, page 469). Suppose a hospital administrator wants to explore the relationship between the chances for long term recovery of severely injured patients and the number of days spent in the hospital. The data file *Patients.sta* contains data for 15 patients; specifically, the file contains information on the number of days that each patient was hospitalized (in variable *Days*) and an index of the prognosis for long-term recovery for each patient (in variable *Prognos*; larger values reflect a better prognosis). Shown below is a listing of this data file.

Case	DAYS	PROGNOS
Patient A	2	54
Patient B	5	50
Patient C	7	45
Patient D	10	37
Patient E	14	35
Patient F	19	25
Patient G	26	20
Patient H	31	16
Patient I	34	18
Patient J	38	13
Patient K	45	8
Patient L	52	11
Patient M	53	8
Patient N	60	4
Patient O	65	6

Neter et al. (1985) fit the following regression model to the data:

$$y = b_0 * \exp(b_1 * x)$$

where *y* denotes the prognosis and *x* represents the number of days that each patient was hospitalized. This model is not offered in the *Nonlinear Estimation* startup panel; therefore, you will have to select the *User-specified regression* option in the startup panel and enter this model directly in the *Estimated function and loss function* dialog.

Specifying the analysis. After starting the *Nonlinear Estimation* module, open the data file *Patients.sta*. Next, double-click on the *User-specified regression* option in the startup panel.

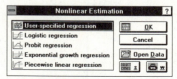

You will now need to specify the regression function. Thus, click on the *Function to be estimated and loss function* button in the *User-Specified Regression Function* dialog to open the *Estimated function and loss function* dialog. As you can see below, you can specify essentially any kind of model via this editor. The syntax for this dialog is reviewed in the *Notes* section of this chapter (see page 3057). The most important rules to remember when entering functions are:

(1) Variables can be referenced by their names or by using the convention *Vxxx* where *xxx* is the number of the variable to be referenced;

(2) All unrecognized names are interpreted as parameters to be estimated by the model.

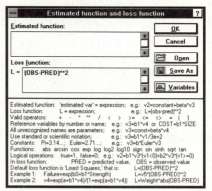

1. NONLINEAR ESTIMATION - EXAMPLES

Shown below is the dialog after typing in the exponential regression equation for this model (with the default loss function in the second edit window).

As you can see, the specification of regression models is rather straightforward. Note that instead of actual variable names, you could have typed in *v1* and *v2*; also, instead of *Param_1* and *Param_2* to refer to the parameters, you could have used *b1* and *b2* or any other name that is not a valid variable name or reserved keyword.

Loss function. The idea of loss functions is reviewed in the *Introductory Overview* section on page 3013. In the loss function equation, the keywords *Pred* and *Obs* refer to the predicted and observed values, respectively. Thus, the default loss function shown in the illustration above specifies the ordinary least squares estimation (predicted minus observed squared). Remember that complex equations can easily be saved in this dialog via the *Save As* button and opened for future use using the *Open* button. Once you click *OK* in this dialog, STATISTICA will check the syntax of the functions, and if it is acceptable it will return to the *User-Specified Regression Function* dialog.

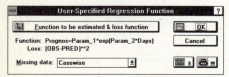

Reviewing results. Now, click *OK* in this dialog to go to the *Model Estimation* dialog.

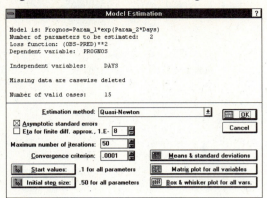

Click on the *Asymptotic standard errors* check box to select it and then click *OK* to accept all the other default selections. Once again, click *OK* in the *Parameter estimation* dialog to go to the *Results* dialog. The *Results* dialog shows that overall, 99% of the variability in the prognosis index from the number of days that patients are hospitalized can be explained.

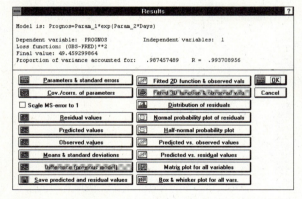

The closeness of the fit of the model to the data is also evident when you plot the fitted function; click on the *Fitted 2D function and observed vals* button to produce the graph.

1. NONLINEAR ESTIMATION - EXAMPLES

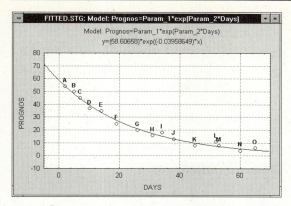

The parameter estimates are shown below (click on the *Parameters and standard errors* button); in order to display the parameters with the greatest precision, the *Increase Column Width* Scrollsheet toolbar button (see Volume I, Chapter 1) was used to increase the width of the columns to the maximum width.

Model: Prognos=Param_1*exp(Param_2*Days) (patients.sta)		
NONLIN. ESTIMAT.	Dep. var: PROGNOS Loss: (OBS-PRED)**2 Final loss: 49.459299864 R=.99371 Variance explained: 98.746%	
N=15	PARAM_1	PARAM_2
Estimate	58.606583524459	-.0395864876778
Std.Err.	1.5051656144375	.00179033179932
t(13)	38.936966777812	-22.11125764106
p-level	.00000000000001	.00000000001067

As you can see, both parameters in this model are highly significant. Click on the *Normal probability plot of residuals* button to evaluate the adequacy of the model fit.

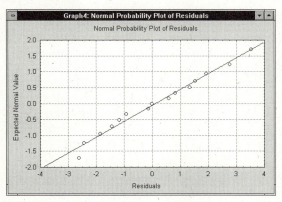

The residuals appear to follow closely the normal distribution; thus, it appears that the exponential model provides an adequate fit to the data.

Example 4: Comparing Two Learning Curves

This example will demonstrate how one can evaluate differences in the regression model across groups. The data set for this example is again based on Neter, Wasserman, and Kutner (1985, page 484) and is contained in the file *Learning.sta*, shown below.

Data: LEARNING.STA 3v Relative efficiency of production		
# LOCATION	WEEKS	EFFICNCY
1 PLANT_A	1	.483
2 PLANT_A	2	.539
3 PLANT_A	3	.618
4 PLANT_A	5	.707
5 PLANT_A	7	.762
6 PLANT_A	10	.815
7 PLANT_A	15	.881
8 PLANT_A	20	.919
9 PLANT_A	30	.964
10 PLANT_A	40	.959
11 PLANT_A	50	.968
12 PLANT_A	60	.971
13 PLANT_A	70	.960
14 PLANT_A	80	.967
15 PLANT_A	90	.975

Suppose you are about to introduce a new electronics product that will be manufactured in two different plants: plant *A* and plant *B*. You may know that plant *B* is a more modern facility; therefore, you would expect that plant to adapt to the new production process more quickly. To study the efficiency of the two plants, the ratio of the per-unit-production cost that would be expected in a modern facility after learning has occurred over the actual per-unit-production cost for selected weeks over a 90-week span was recorded in the variable *Efficncy*.

Neter et al. fit the following model to these data:

$$y = b_0 + b_1 * x_g + b_3 * exp(b_2 * x)$$

In this model, x_g denotes a coding variable that identifies each plant (plant *A = 0*; plant *B = 1*; the same codes are used in *Learning.sta*). If parameter b_1 is significant then it can be concluded that there is a significant constant (or additive) difference between groups.

NLN - 3033

1. NONLINEAR ESTIMATION - EXAMPLES

Specifying the analysis. After starting the *Nonlinear Estimation* module, open the data file *Learning.sta* and click on the *User-specified regression* option in the startup panel. Click on the *Function to be estimated and loss function* button in the *User-Specified Regression Function* dialog to open the *Estimated function and loss function* dialog. Now enter the regression equation shown below and accept the default ordinary least squares loss function.

In this case, the variables have been referenced via the *Vxxx* convention (where *xxx* is a variable number), and *bx* was used to refer to the parameters. Thus, typing in this formula is not that difficult (i.e., lengthy). However, note that you can also easily save the formulas via the *Save As* button and open them at a later time using the *Open* button.

Estimating the parameters. Now, click *OK* to continue on to the *Model Estimation* dialog. It turns out that with the default start values, the only estimation procedure that will "handle" this equation is the *Rosenbrock pattern search* method (or *method for rotating coordinates*; refer to the estimation methods described in the *Introductory Overview* section, page 3013). However, even if it fails to converge completely, it will produce "workable" start values that can be specified for the *quasi-Newton* algorithm. You may try the other algorithms with the default start values. For example, the *quasi-Newton* method will move the first parameter into a completely wrong direction and *Simplex* refuses to converge.

Now select the *Rosenbrock and quasi-Newton* method from the *Estimation method* combo box and also select the *Asymptotic standard errors* option. Click on the *Start values* button to open the *Specify Start Values* dialog. Set all initial parameter estimates to *0* by entering *0* in the *Common Value* edit box and clicking on the *Apply* button.

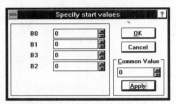

Click *OK* in this dialog and again in the *Model Estimation* dialog to begin the estimation.

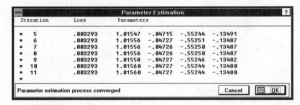

Note that on slower computers, this estimation may take awhile. In that case, instead of choosing the method described here, enter start values that are similar but not identical to the final parameters and then use *quasi-Newton* for estimating the parameters; this will speed up the estimation.

After about 8 iterations, the *Rosenbrock* method will quit and the program will continue with the *quasi-Newton* method, starting where the *Rosenbrock* method left off. After only 11 additional *quasi-Newton* iterations, the estimation process will converge. Shown below is the *Results* dialog:

1. NONLINEAR ESTIMATION - EXAMPLES

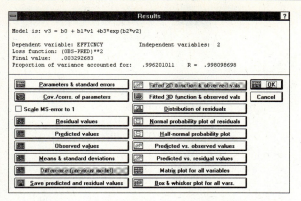

As you can see, over 99% of the variability in the efficiency index can be accounted for. Click on the *Fitted 3D function and observed vals* button to visually check the closeness of the fit of the model to the data.

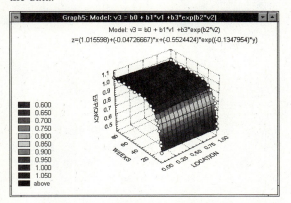

Now, look at the parameter estimates by clicking on the *Parameters and standard error* button in the *Results* dialog.

The crucial parameter b_1 is highly significant ($p<.001$); thus, you can conclude that the two plants do indeed differ in their efficiency. Return to the *Results* dialog and click on the *Normal probability*

plot of residuals button to make sure that the model is appropriate:

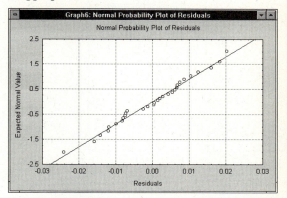

All points are very close to the line denoting the normal distribution; therefore, it can be concluded that the model is quite appropriate.

Example 4.1: Estimating Two Different Models

When specifying regression functions, you can use logical expressions that will evaluate to *0* if they are false, and to *1* if they are true. This example will demonstrate how one can estimate separate regression parameters for each of the plants in Example 4. Specifically, instead of using the b_1 parameter in the model above, use logical operators to estimate the different b_0 parameters directly. After returning to the *User-Specified Regression Function* dialog, click on the *Function to be estimated and loss function* button to open the *Estimated function and loss function* dialog. Specify the following model in this dialog:

1. NONLINEAR ESTIMATION - EXAMPLES

Note how the logical operators are used in this example. If $v1$ (*Location*) is equal to *1*, then the program will estimate b_{0A}, that is, the constant for plant A; if $v1$ is equal to *0* then the program will estimate b_{0B}. Now, proceed as in Example 4: select *0* as the common start value (or to speed things up, you can choose start values that are closer to the final solution) and *Rosenbrock and quasi-Newton* as the estimation method in the *Model Estimation* dialog. The parameter estimates are shown below (click on the *Parameters and standard error* button in the *Results* dialog).

Model: v3=b0A*(v1=1)+b0B*(v1=0)+b3*exp(b2*v2) (learning.sta)				
NONLIN. ESTIMAT.	Dep. var: EFFICNCY Loss: (OBS-PRED)**2 Final loss: .003292683 R=.99810 Variance explained: 99.620%			
N=30	B0A	B0B	B3	B2
Estimate	.9683	1.0156	-.5524	-.1348
Std.Err.	.0037	.0037	.0083	.0045
t(26)	261.9571	274.7439	-66.8346	-29.7037
p-level	0.0000	0.0000	.0000	.0000

The estimates are as expected based on the previous analysis, that is, the constant for plant A is *1.0156-.04727*1=.9683*, and for plant B it is *1.0156-.04727*0=1.0156*.

Other Regression Models

Example 5: Regression in Pieces

The logical operators demonstrated in the previous example can also be used in order to specify different regression models for different regions of the independent variable(s), that is, to estimate piecewise regression models (see the *Introductory Overview*, page 3012). This example is also based on a data set reported in Neter, Wasserman, and Kutner (1985, page 348). Specifically, the data set pertains to a production process in which the per-unit cost is related to the lot size. Shown below is the data file *Lotsize.sta*.

Data: LOTSIZE.STA	
Unit cost of production as	
1 COST	2 LOT_SIZE
1 2.570	650.000
2 4.400	340.000
3 4.520	400.000
4 1.390	800.000
5 4.750	300.000
6 3.550	570.000
7 2.490	720.000
8 3.770	480.000

Supposedly, for lots greater than 500, the relationship between the variables changes; Neter et al. (1985) fit a linear model that allows for different slopes for lots of sizes less than or equal to 500, and lots greater than 500. Specifically, Neter, et al. fit the following model:

$$y = b_0 + b_1 * x + b_2 * (x-500) * (x>500)$$

Again, this model has a logical expression ($x>500$) that serves as a multiplier: If the expression is true, it will evaluate to *1*; if it is false, it will evaluate to *0*. Therefore, this equation actually represents two models. For values of x that are less than or equal to 500 ($x>500$ is false, i.e., equal to *0*):

$$y = b_0 + b_1 * x$$

For values of x that are greater than 500 [i.e., when ($x>500$) is equal to *1*], the equation is:

NLN - 3036

Copyright © StatSoft, 1995

1. NONLINEAR ESTIMATION - EXAMPLES

$y = b_0 + b_1*x + b_2*(x-500)$

If you multiply out this equation, you can see that for *x* values greater than 500, the slope is equal to b_1+b_2, and the intercept is equal to $(b_0-500*b_2)$.

Specifying the model. Shown below is how this model can be specified in the *Estimated function and loss function* dialog.

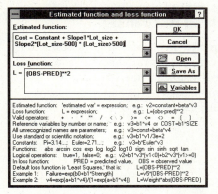

As you can see, essentially you type in the equation in a straightforward manner. Note that in the illustration above, the variable names were used to denote variables, and the names *Constant*, *Slope1*, and *Slope2* were used to denote the parameters. Remember that all unknown names will be interpreted by the program as model parameters. Also recall that you can record such long formulas via the *Save As* button in this dialog. They can later be opened via the *Open* button. You could also have used the *Vxxx* convention to refer to variables (where *xxx* is the variable number) and named the parameters *a*, *b*, and *c*. This would have saved you some typing but makes the formula less readable.

Estimating the model. Now, proceed as before, that is, accept all defaults in the *Parameter Estimation* dialog and click on the *OK* button until the *Results* dialog appears.

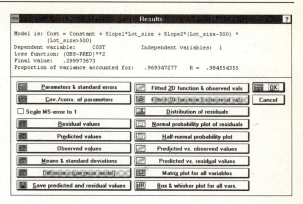

Note that the *quasi-Newton* estimation procedure with the default start values will converge after 11 iterations. Click on the *Parameters and standard errors* button to display the parameter estimates in a Scrollsheet.

The significance level for parameter *Slope1* is *.045*; however, the other parameter is much less significant (*.153*), which may suggest that the initial (*a priori*) breakpoint (*500*) was adequate (i.e., the best regression models for the two ranges of *Lot-size* are different). In the next step, this will be more directly verified by estimating the value of the breakpoint itself.

Estimating the breakpoint itself. Logical expressions, such as the one used in this model, can also contain parameters (rather than constants only). For example, if you did not know where the breakpoint is in your model, you could specify:

1. NONLINEAR ESTIMATION - EXAMPLES

Here, the parameter *Breakpnt* was added to the model. Thus, in this manner you can estimate the breakpoint.

To estimate the above model, it is a good idea to set the start value for the *Breakpnt* parameter to about *500* (click on the *Start values* button in the *Model Estimation* dialog). If you start the estimation with the default start values (*0.1*), then the *quasi-Newton* estimation will move the breakpoint to negative values. Since there are no negative values in the data file, this is essentially like fitting a straight line without a breakpoint.

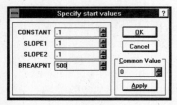

The resulting parameter estimates (including the breakpoint) for this model are shown in the Scrollsheet below.

NONLIN. ESTIMAT.	Dep. var: COST Loss: (OBS-PRED)**2 Final loss: .257947376 R=.98673 Variance explained: 97.364%			
N=8	CONSTANT	SLOPE1	SLOPE2	BREAKPNT
Estimate	6.121232	-.00462	-.00400	570.0000
Std.Err.	.654538	.00159	.00227	95.7179
t(4)	9.351986	-2.89982	-1.75706	5.9550
p-level	.000728	.04413	.15374	.0040

Option "piecewise linear regression." You may wonder why the piecewise linear regression model that is available as a dialog option on the *Nonlinear Estimation* startup panel was not used. That model allows the user to specify or estimate breakpoints for the range of the *dependent* or *y* variable. Thus, that model was not applicable to this case.

Example 6: Estimating Drug Responsiveness (Half-Maximal Response)

A common problem in pharmacological research is to estimate an organism's responsiveness to a drug. A model (also discussed in the *Introductory Overview* section, page 3012) that can be used to estimate the expected response at the point of saturation, and the concentration of the drug that produces a half-maximal response is:

$$y = b_0 - b_0/[1+(x/b_2)^{b_1}]$$

In this model, b_0 is the expected response at saturation, b_2 is the concentration for a half-maximal response, and b_1 determines the slope of the function. Usually, the *x* values (concentration levels) are coded so that they are always larger than *1.0*, and the *y* values are expressed as proportions of the maximum possible responsiveness to the drug. Neter, Wasserman, and Kutner (1985, page 489) present the following example data set:

DOSE	RESPONSE
1	.500
2	2.300
3	3.400
4	24.000
5	54.700
6	82.100
7	94.800
8	96.200
9	96.400

These data are contained in the file *Drugres.sta*. Open the data file and select the *User-specified regression* option in the startup panel, then click *OK*. Shown below is how the model is specified in the *Estimated function and loss function* dialog (click on the *Function to be estimated and loss function*

NLN - 3038

1. NONLINEAR ESTIMATION - EXAMPLES

button in the *User-Specified Regression Function* dialog).

Estimating the model. Now, click *OK* in this dialog and again in the *User-Specified Regression* dialog to open the *Model Definition* dialog. To estimate this model, click on the *Start value* button and set the start value for b_0 to *100*; the other start values can remain at the default value of *0.1*. If you simply use the default start values in this case in conjunction with the *quasi-Newton* method, the model will converge but the model fit will clearly be "off" (zero percent of variance explained, normal probability plot shows clear deviations of residuals form normal). Click *OK* in this and the *Parameter Estimation* dialog to go to the *Results* dialog.

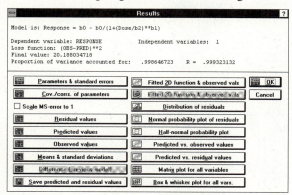

Now, click on the *Fitted 2D function and observed vals* button to see how well the specified function fits the observed data. The fit is almost perfect.

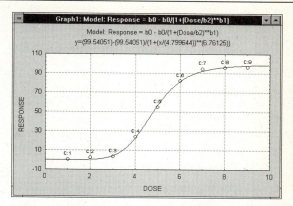

Click on the *Parameters and standard errors* button to obtain the final parameter estimates (see below).

Model: Response = b0 - b0/(1+(Dose/b2)**b1) (drugres.sta)			
NONLIN. ESTIMAT.	Dep. var: RESPONSE Loss: (OBS-PRED)**2 Final loss: 20.188034718 R=.99932 Variance explained: 99.865%		
N=9	B0	B2	B1
Estimate	99.54051	4.79964	6.76125
Std.Err.	1.57190	.05045	.42802
t(6)	63.32503	95.13118	15.79652
p-level	.00000	.00000	.00000

In this example, the half-maximal response can be expected at a (coded) dose level of *4.80*.

Specifying Loss Functions

Example 7: Weighted Least Squares

Most regression models can be estimated via least squares methods, that is, by using as the loss function in the estimation procedure the sum of squared deviations of the observed values from the predicted values. Least squares is the default loss function for user-specified regression models. However, as discussed in the *Introductory Overview* section (page 3014), there are instances where weighting of the squared residuals might be in order.

For example, the computation of the standard errors of regression weights in linear regression rests on the assumption that the residuals are distributed evenly

1. NONLINEAR ESTIMATION - EXAMPLES

around the regression line over the entire range of the independent variables. When this assumption is violated, one should use weighted least squares instead.

For example, suppose a construction company wants to estimate the relationship between the size of a bid and the cost of preparing the bid. Intuitively, it makes sense to assume that the larger the project, the greater will be the residual variability of the cost about the estimated regression line.

Shown below is a scatterplot of a data set reported in Neter, Wasserman, and Kutner (1985, page 169; see the data file *Bid_prep.sta*, below). Note that the size of the bid (on the horizontal axis) is scaled in terms of millions of dollars, and the preparation cost is scaled in terms of thousands of dollars. As you can see, the variability of the residuals about the regression line tends to be larger for larger bids.

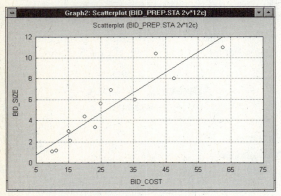

Without going into details (see Neter et al., 1985, page 167), intuitively, it makes some sense to weight the residuals by the inverse of the squared *x* values; what this would accomplish is that about an equal emphasis is placed on the larger and the smaller bids in the estimation, and the resultant estimates may become more stable. This estimation method is also called weighted least squares estimation.

Specifying the equation and loss function.
Now, specify the weighted least squares. After

starting the *Nonlinear Estimation* module, open the data file *Bid_prep.sta*.

Select the *User-specified regression* option in the *Nonlinear Estimation* startup panel and click on the *Function to be estimated and loss function* button. In the *Estimated function and loss function* dialog, specify the regular linear regression equation (*bid_cost = intercpt + slope*bid_size*) and the following loss function:

The first part of the loss function (i.e., *(OBS-PRED)**2*) specifies standard least squares. However, for each case, the loss is weighted by the inverse of the squared bid size (i.e., multiplied by *(1/bid_size**2)*).

Estimation and results. You can estimate this model without changing any of the estimation default values in the *Model Estimation* dialog. Shown below are the parameter estimates and their standard errors (accessible by clicking on the

Parameter and standard errors button in the *Results* dialog):

The approximate standard errors for the intercept and slope parameters are *.965* and *.404*, respectively. If you re-analyze these data with the standard least squares loss function (or via the *Multiple Regression* module), you would obtain standard errors of *3.252* and *.5285*, respectively. Thus, you can conclude that the parameter estimates based on the weighted least squares estimation are more stable (less subject to random sampling variation).

Weighted least squares via *STATISTICA* BASIC.

STATISTICA includes the *STATISTICA BASIC* example program *Wls.stb* (in the *STBASIC* subdirectory), that will also compute weighted least squares estimates for multiple linear regression models. To replicate the results from this example via that program, first add a variable to *Bid_prep.sta* and compute the values of that new variable (e.g., called *Weight*) as *1/bid_size**2*; that is, compute the same weight that you specified to modify the ordinary least squares *Loss function* in the *Estimated function and loss function* dialog shown earlier. Because program *Wls.stb* uses matrix algebra expressions (e.g., see Neter, Wassermann, and Kutner, 1985) to estimate the weighted least squares parameters, rather than the iterative procedure used in this example, for large data files program *Wls.stb* is more efficient, and often much faster.

Concluding Remarks

Of course, all examples in this section were textbook examples, which usually have the tendency to work well. In actual research applications, you will surely come across data sets that seem to "actively reject" the model chosen for them. Please refer to the *Introductory Overview* section (page 3020) for a review of some strategies that you may follow if a model cannot be fit to the data.

In general, it is always a good idea to begin with the simplest model possible. Also, the nonlinear regression techniques presented here can be considered hypothesis testing procedures, thus, they should generally not be used for exploratory data analyses. Rather, the researcher should bring to the data a prior understanding of the underlying mechanisms that relate the variables of interest.

1. NONLINEAR ESTIMATION - DIALOGS AND OPTIONS

DIALOGS, OPTIONS, STATISTICS

Startup Panel

Select the desired regression model from this dialog. When you select the *User-specified regression* (default) option, you will be able to type in a formula to specify the relationship between the independent variables and the dependent variable.

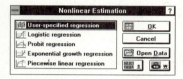

For details concerning the different nonlinear models, see the *Introductory Overview* (page 3008).

Probit/Logit Regression

When you click on either the *Logistic Regression* or the *Probit Regression* options from the startup panel (see above), the *Probit* or *Logit Regression* dialogs will come up.

Loss Function. To estimate the parameters of logit and probit models, the program will maximize the likelihood for the data (i.e., compute *maximum likelihood* parameter estimates; see also page 3013). To use other loss functions (e.g., weighted least squares), choose option *User-specified regression* (see also page 3046); you can then specify the model by typing in the model equation, and the loss function by specifying the equation that describes the quantity that is to be minimized during the parameter estimation process.

Input File

There are two ways to set up a data file for a probit or logit regression analysis. First, a data file may contain raw data, with a coding variable as the dependent variable (containing the two codes that indicate the two levels of the dichotomous dependent variable).

If the data are arranged in this manner, then set this combo box to *Codes and no counts* and the *Probit* and *Logit Regression* dialogs will appear as follows:

Alternatively, one may enter frequencies directly into the file. This way of arranging the data file is often more efficient when the data contain failure (or survival) rates as a function of time.

For example, Cox (1970) reported the following data on tests of objects for failures after certain times. Variable *Failure* is an indicator variable (0 = no failure, 1 = failure), and variable *Count* represents the respective number of objects that did or did not fail in the respective time interval:

```
Time   Failure   Count
----   -------   -----
  7       0        55
 14       0       155
 14       1         2
 27       0       152
 27       1         7
 51       0        13
 51       1         3
```

NLN - 3043

1. NONLINEAR ESTIMATION - DIALOGS AND OPTIONS

Thus, at time *7* there were no failures among *55* objects. At time *14*, there were *2* failures (coded *1*) and *155* non-failures (coded *0*, for a total of *157* objects). At time *27* there were *7* failures and *152* non-failures, etc.

To analyze this data set, one would set the *Input file* combo box to *Codes and counts* and then click on the *Variables* button to specify the desired variables.

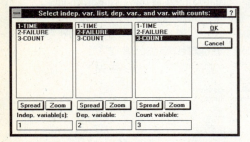

In the *Specify Variables* dialog, specify *Time* as the independent variable (in the first list), *Failure* as the dependent variable (in the second list), and variable *Count* as the variable containing the counts (in the third list). The *Probit* and *Logit Regression* dialogs would appear as follows (the *Probit* dialog is shown):

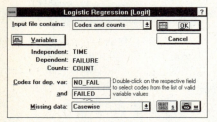

Note that the total frequencies reported in the *Results* dialog will be the actual frequencies, that is, taking into account the counting variable.

Variables

This option will bring up the standard variable selection dialog in which you can select one dependent variable and a set of independent or predictor variables. In addition, if the *Input file* combo box is set to *Codes and counts*, then

STATISTICA expects you to specify the *Count* variable containing the frequency counts (see the example above).

Codes

Logit and probit regression models, in a sense, predict probabilities underlying the dichotomous dependent variable, and these methods will produce predicted (expected) values in the range between *0* and *1* (for details, see page 3010). If the dependent variable is not coded in this way, that is, as *0* and *1*, then the user must specify the respective codes in the *Codes* edit boxes. As the data are read, the dependent variable will then be transformed so that all values that match the first code become *0* (zero), and all values that match the second code become *1*.

Missing Data

Missing data can either be deleted casewise or substituted by the means for the respective variables. If *Casewise* deletion of missing data is selected, a case will be deleted from the analysis if it has missing data for any of the selected variables. If *Mean substitution* is specified, a missing data point will be replaced by the mean for the respective variable.

Exponential Regression

STATISTICA will estimate, using least squares, the model:

$$y = c + \exp(b_0 + b_1 * x_1 + b_2 * x_2 + \ldots + b_m * x_m)$$

where

c, b_i are parameters (for the *m* independent variables)

This model is commonly used in order to study the growth of populations.

1. NONLINEAR ESTIMATION - DIALOGS AND OPTIONS

Variables

When you click on this button, the standard variable selection dialog will open in which you will be able to select the independent or predictor variables, and a dependent variable.

Missing Data

Missing data can either be deleted casewise or substituted by the means for the respective variables. For more information on these types of missing data deletion, see page 3044.

Breakpoint Regression

When you select the *Piecewise Linear Regression* option from the *Nonlinear Regression* startup panel, *STATISTICA* will estimate, using least squares, the model:

$$y = (b_{01}+b_{11}*x_1+...+b_{m1}*x_m)*(y \leq b_n)$$
$$+ (b_{02}+b_{12}*x_1+...+b_{m2}*x_m)*(y > b_n)$$

Thus, the program will estimate two separate linear regression equations; one for the y values that are less than or equal to the breakpoint (b_0) and one for the y values that are greater than the breakpoint. For a general description of these types of models, see the *Introductory Overview*, page 3012; for an example of this type of model, see *Example 5* in the *Examples* section, page 3036.

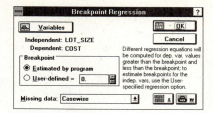

To estimate models with breakpoints (discontinuities) in the independent variables, choose option *User-specified regression* from the startup panel (see also page 3046), and then use logical operators in the model equation to define the breakpoints (see page 3057 for a summary of the syntax rules for defining user-specified models).

Variables

When you click on this button, the standard variable selection dialog will open in which you will be able to select the independent or predictor variables, and a dependent variable.

Breakpoint

By default, *STATISTICA* will estimate the breakpoint (starting at the mean for the dependent variable; the *Estimated by Program* radio button is selected). You can specify a breakpoint for the regression by checking the *User-defined* radio button (in which case, the breakpoint value that you enter in the edit box will be treated as a constant in the estimation procedure, rather than as a parameter to be estimated).

Missing Data

Missing data can either be deleted casewise or substituted by the means for the respective variables. For more information on these types of missing data deletion, see page 3044.

User-defined Models

You can specify a regression model and loss function by first clicking on the *User-specified Regression* option in the *Nonlinear Regression* startup panel to open the *User-Specified Regression Function* dialog.

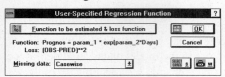

Function to be Estimated and Loss Function

When you click on this button, the *Estimated function and loss function* dialog will open (see below) in which you can specify the regression model and the loss function. Different regression models and loss functions are described in the *Introductory Overview* (page 3008 and 3013, respectively).

Missing Data

Missing data can either be deleted casewise or substituted by the means for the respective variables. For more information on these types of missing data deletion, see page 3044.

Estimated Function and Loss Function

Enter in this dialog, a user-defined regression model and/or loss function for the analysis.

Shown below the edit fields in this dialog is a summary of the syntax. The general conventions for the equations and syntax reference are described in the *Notes* section of this chapter (page 3057; see also the *Electronic Manual* accessible by pressing F1).

Open

You can use a previously saved regression model and loss function in this dialog by clicking on the *Open* button and selecting the name of the file with the equations.

Save As

You can save the equation specifying a model and the loss function using the *Save As* button in this dialog. The file will be saved in a standard text format, so that it can also be edited via other editors (e.g., Notepad); it consists of two lines of text: The model and the loss function.

Variables

Click on this button to open the *Variables* dialog in which you can view the variable names in the data file. You can also highlight a variable in this dialog and look at its values or long variable name.

Model Estimation

Once you have specified the regression model, the *Model Estimation* dialog will open.

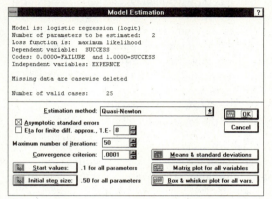

Here, you can specify the estimation method, number of iterations, convergence criterion, and other options. Also given in this dialog is a summary of the model and parameters established so far in the analysis.

Summary box. The summary box lists your choices for the regression model, i.e., number of parameters, loss function, dependent variable, codes, independent variables, missing data deletion, and the number of valid cases. You can review those choices here and if they are not appropriate, click on the *Cancel* button to return to the previous dialog in order to make any corrections.

Estimation Method

There are four different estimation procedures available in *Nonlinear Estimation* which can also be combined. This combo box allows you to select a particular estimation procedure. *Nonlinear Estimation* provides four estimation methods (listed below). These methods, their strengths and weaknesses, are described in the *Introductory Overview* (page 3017).

The section *When the Estimation Process Does Not Converge* (page 3020) contains several practical guidelines that you may follow (involving the choice of the estimation method) if the estimation procedure fails to converge.

Quasi-Newton. For most applications, the default *quasi-Newton* method will yield the best performance, that is, it is the fastest method to converge. In this method the second-order (partial) derivatives of the loss function are asymptotically estimated, and used to determine the movement of parameters from iteration to iteration. To the extent that the second-order derivatives of the loss function are meaningful (and they usually are), this procedure is more efficient than any of the others. For more information, see page 3017.

The following procedures do not estimate the second-order derivatives of the loss function but rather use various geometrical approaches to function minimization. They have the general advantage of being more "robust," that is, they are less likely to converge on local minima, and are less sensitive to "bad" (i.e., grossly inadequate) start values.

Simplex. This algorithm does not rely on the computation or estimation of the derivatives of the loss function. Instead, at each iteration the function will be evaluated at $m+1$ points in the m dimensional parameter space. For more information, see page 3017.

Hooke-Jeeves pattern moves. At each iteration, this method first defines a pattern of points by moving each parameter one by one, so as to optimize the current loss function. The entire pattern of points is then shifted or moved to a new location; this new location is determined by extrapolating the line from the old base point in the m dimensional parameter space to the new base point.

The step sizes in this process are constantly adjusted to "zero in" on the respective optimum. This method is usually quite effective, and should be tried if both the *quasi-Newton* and *Simplex* methods (see above) fail to produce reasonable estimates.

Rosenbrock pattern search. This method will rotate the parameter space and align one axis with a ridge (this method is also called the *method of rotating coordinates*); all other axes will remain orthogonal to this axis.

If the loss function is unimodal and has detectable ridges pointing towards the minimum of the function, then this method will proceed with sure-footed accuracy towards the minimum of the function. However, note that this search algorithm may terminate early when there are several constraint boundaries (resulting in the penalty value; see above) that intersect, leading to a discontinuity in the ridges.

Choosing combinations of methods. Because the *Simplex*, *Hooke-Jeeves*, and *Rosenbrock* methods are generally less sensitive to local minima, you may use any one of these methods with the *quasi-Newton* method. This is particularly useful if you are not sure about the appropriate start values for the estimation. In that case, the first method may generate initial parameter estimates that will then be used in subsequent *quasi-Newton* iterations.

Asymptotic Standard Errors

Set this check box if you want to compute the standard errors for the parameter estimates (and the variance/covariance matrix of parameter estimates). These standard errors are computed via finite difference approximation of the second-order partial derivatives (i.e., the Hessian matrix; refer to the *Introductory Overview*, page 3018, for details). Note that the *Parameter estimates and standard errors* option in the *Results* dialog is only available if this check box is set.

Eta for Finite Difference Approximation

This option is only available if *Asymptotic standard errors* were requested (see above). As described in the overview section (page 3015), the standard errors for the parameter estimates are computed via finite differencing. Specifically, the matrix of second-order partial derivatives is approximated. In order to obtain accurate estimates for the derivatives, some *a priori* knowledge is necessary of the reliability of the loss value.

This reliability can be expressed as parameter η (*Eta*) so that $\eta = 10^{-DIGITS}$, where *Digits* is the number of reliable base-10 digits computed from the loss function. By default (i.e., when this check box is not set), η will be estimated automatically by the program (by checking the "responsiveness" of the loss function to small changes in the parameter values). However, in some cases, when the magnitudes of the first order partial derivatives for two or more parameters are very different, the default estimation of η may not be optimal. In that case, you may enter a user-defined constant; specifically, the integer value entered into the edit field to the right of this check box will be interpreted as the *Digits* in η.

In practice, experiment with this parameter (start with the default value of 10^{-8}) in cases when the parameter estimation converges with reasonable values, but when requesting the Scrollsheet with parameter values and their standard errors (from the *Results* dialog), the message appears: *Matrix ill-conditioned; cannot compute standard errors*.

Maximum Number of Iterations

The estimation of parameters in nonlinear regression is an iterative procedure. At each iteration, the program evaluates whether the fit of the model (to the data) has improved from the previous iteration. This edit box allows you to specify the maximum number of iterations that are to be performed.

Convergence Criterion

This edit box allows you to change the convergence criterion value (by default, *0.0001*). The exact meaning of this parameter depends, among other things, on the estimation method that is chosen. Refer to Fletcher (1972) for details about the *quasi-Newton* method; refer to O'Neill (1971) or Nelder and Mead (1965) for a discussion of the *Simplex* procedure; refer to Fletcher and Reeves (1964), and Hooke and Jeeves (1961) for details concerning the *Hooke-Jeeves* method and the *Rosenbrock* pattern method.

The section *When the Estimation Process Does Not Converge* (page 3020) contains several practical guidelines that you may follow (involving the choice of the estimation method) if the estimation procedure fails to converge.

Start Values

The start values are used in the first iteration of each estimation method to determine the initial value of the loss function (see also, page 3016). This option allows you to change the start values (by default, *0.1*).

Clicking on this button will bring up the *Specify Start Values* dialog (see below) in which you can enter the individual start values for each parameter or one common value for all parameters.

Initial Step Size

This option will allow you to change the default step size (*0.5* for *quasi-Newton*, *1.0* for *Simplex* and *Rosenbrock*, *2.0* for *Hooke-Jeeves*). Clicking on this button will bring up the *Specify Initial Step Sizes* dialog (see below) in which you can enter the individual step size values for each parameter or one common step size for all parameters.

The step size values are used during the initial iterations to "scale" the problem, that is, to determine by how much to move each parameter. The exact impact of these values on the estimation depends, among other things, on the estimation method that is chosen.

Means and Standard Deviations

This option will bring up a Scrollsheet with the means and standard deviations for all selected variables (i.e., including the variable containing counts, if one was chosen). The default graph for this Scrollsheet is the histogram.

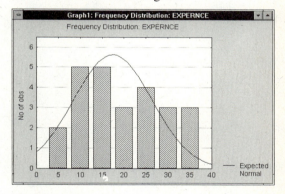

The histograms will show the observed frequency distribution for the respective variable, and the expected frequencies for the normal distribution.

Matrix Plot for all Variables

This option will bring up a scatterplot matrix for all variables selected for the analysis.

1. NONLINEAR ESTIMATION - DIALOGS AND OPTIONS

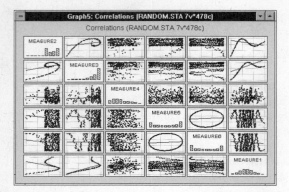

Box and Whisker Plot for all Variables

This option allows the user to produce a box and whisker plot for the variables selected for the analysis.

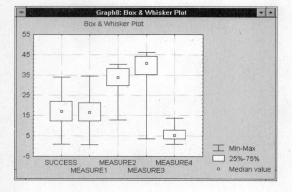

Start Values

The start values are used in the first iteration of each estimation method to determine the initial value of the loss function (see the *Introductory Overview* section, page 3016, for details). In this dialog, specify the desired start values.

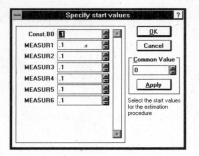

Individual Start Values

This box lists the parameters currently in the model. You can change the start values for each of these parameters individually in the appropriate edit boxes or specify a common start value (see below).

Common Values

Specify a start value that will be used for each of the parameters.

Apply

Once you have specified a common start value, you can click on the *Apply* button to copy it to the individual start values edit boxes.

Step Sizes

The step size values are used during the initial iterations to "scale" the problem, that is, to determine by how much to move each parameter. The exact impact of these values on the estimation depends, among other things, on the estimation method that is chosen. Specify here, the desired step size values.

1. NONLINEAR ESTIMATION - DIALOGS AND OPTIONS

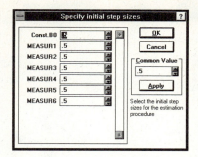

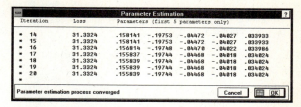

Clicking *OK* in this dialog will open the *Results* dialog (see below).

Individual Step Size Values

This box lists the parameters in the current model. You can change the step size values for each of these parameters individually in the appropriate edit boxes or specify a common step size value.

Common Values

Specify a step size that will be used for each of the parameters.

Apply

Once you have specified a common step size value, you can click on the *Apply* button to copy it to the individual step size values edit boxes.

Parameter Estimation

When you click *OK* in the *Model Estimation* dialog, STATISTICA will begin the iterative estimation process and display the resulting iterations, loss values, and parameter estimates in the *Parameter Estimation* dialog.

Results

The *Results* dialog is the final dialog in the *Nonlinear Regression* process. In this dialog, you can selectively review the results of the regression as well as other options (described below). Note that for this model, the *Difference (previous model)* and *Fitted 2D function and observed vals* buttons are dimmed.

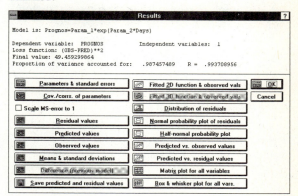

This dialog also summarizes the choices that you have made in previous dialogs.

Summary box. The summary box lists your choices for the regression analysis, i.e., the regression model, dependent variable, loss function, proportion of variance accounted for by the model, *R*, number of *0*'s and *1*'s, independent variables and final value.

NLN - 3051

1. NONLINEAR ESTIMATION - DIALOGS AND OPTIONS

Parameter Estimates (and Standard Errors)

This option will bring up a Scrollsheet with the parameter estimates for the current model. Summary statistics for the current analysis will be displayed in the headers of the Scrollsheet. If the standard errors were requested on the previous dialog (see *Estimation Procedure*, page 3047), then the Scrollsheet will also show the (asymptotic) standard errors of the parameter estimates, the respective *t*-values (parameters divided by standard errors), and the associated *p*-levels.

Cov./Corrs. of Parameters

This option is only available if you have selected the *Asymptotic Standard Errors* option in the *Estimation Procedure* dialog (page 3047). Clicking this button will first bring up a Scrollsheet with the (asymptotic) variance/covariance matrix for the parameter estimates, and then a Scrollsheet with the correlations between parameter estimates.

Scale MS-error to 1

For maximum likelihood estimates, it is recommended to rescale the mean square error to *1*. The resulting standard deviations for the parameter estimates are then the usual information theory standard errors (Jennrich and Moore, 1975). When using maximum likelihood estimation for *probit* or *logit* models, this check box will automatically be set.

Residual Values

Clicking on this button will bring up a Scrollsheet with the residual values (i.e., observed values minus predicted values) for each case.

Predicted Values

Clicking on this button will bring up a Scrollsheet with the predicted values for each case.

Observed Values

Clicking on this button will bring up a Scrollsheet with the observed values for each case.

Classification of Cases and Odds Ratios

This option is only available when fitting logit and probit regression models. After clicking on this button, a Scrollsheet will be displayed with the observed and predicted classifications, and the so-called odds ratio.

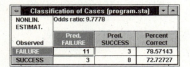

A case is assigned to (classified as belonging to) the first group if the predicted value is less than or equal to 0.5, and it is assigned to the second group if the respective predicted value is greater than 0.5. The odds ratio is computed as (f11*f22)/(f12*f21); where fij. stands for the respective frequencies in the 2 by 2 table. The percentages of correctly classified cases are also reported.

Odds ratios that are greater than *1* indicate that the classification is better than what one would expect by pure chance. However, remember that these are *post-hoc* classifications, because the parameters were computed so as to maximize the probability of the observed data (see the description of the *maximum likelihood* loss function in the *Introductory Overview* section, page 3014). Thus, you should not expect to do this well if you applied the current model to classify new (future) observations.

1. NONLINEAR ESTIMATION - DIALOGS AND OPTIONS

Means and Standard Deviations

Clicking on this button will bring up a Scrollsheet with the means and standard deviations for all selected variables (i.e., including the variable containing counts, if one was chosen).

Difference (Previous Model)

This option is only available if *logit* or *probit* regression was selected in the startup panel, and when the current model is hierarchically related to the previous model that was estimated. "Hierarchically related" means that the current model is identical to the previous model with the exception of an addition or deletion of one or more independent variables. In that case, comparisons between the goodness of fit for the two models are meaningful and can be reviewed by clicking on this button.

For logit and probit models the program will compute maximum likelihood parameter estimates. For those models, this Scrollsheet will also show the incremental maximum likelihood *Chi-square* value for the difference between the two hierarchically related models, and the respective degrees of freedom and *p*-level.

Save Predicted and Residual Values

Predicted and residual scores (as well as all variables in the current data file) can be saved together in a data file for further analysis.

Plot Options

After clicking on any of the graphics buttons, the respective graph will be displayed

Fitted 2D Function and Observed Values

This plot allows a 2-dimensional visual examination (i.e., qualitative evaluation) of the fit of the data to the model.

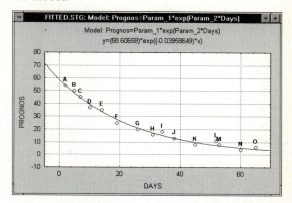

It is useful for identification of outliers, which can then be marked in the *Graph Data Editor* using the *Brushing Tool* (see Volume II).

Fitted 3D Function and Observed Values

This plot allows a 3-dimensional visual examination (i.e., qualitative evaluation) of the fit of the data to the model. Once again, it is useful for identification of outliers.

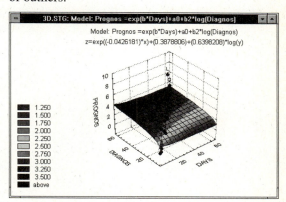

Distribution of Residuals

Click on this button to display a plot of the frequency distribution (histogram) of the residuals, overlaid with the normal curve (the expected frequencies for the normal distribution).

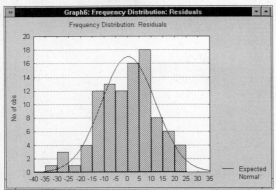

Normal Probability Plot of Residuals

Normal probability plots provide a quick way to visually inspect to what extent the pattern of residuals follows a normal distribution. If the residuals are not normally distributed, they will deviate from the line.

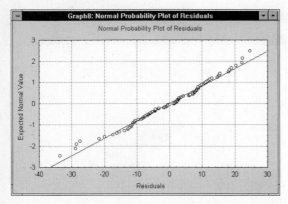

Outliers may also become evident in this plot. If there is a general lack of fit, and the data seem to form a clear pattern (e.g., an *S* shape) around the line, then the dependent variable may have to be transformed in some way (e.g., a log transformation to "pull in" the tail of the distribution, etc.; e.g., use the spreadsheet formulas, see *General Conventions*, Volume I, Chapter 1).

Half-normal Probability Plot of Residuals

When you click on this button, a half-normal probability plot will be constructed.

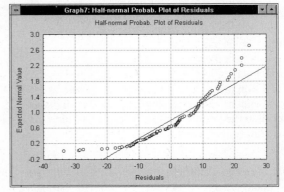

The half-normal probability plot is constructed in the same way as the standard normal probability plot, except that only the positive half of the normal curve is considered. Consequently, only positive normal values will be plotted on the *Y*-axis. This plot is used when one wants to ignore the sign of the residual, that is, when one is mostly interested in the distribution of absolute residuals, regardless of the sign.

Predicted vs. Observed Values

Choose this option when you want a plot of the predicted values vs. the observed values.

1. NONLINEAR ESTIMATION - DIALOGS AND OPTIONS

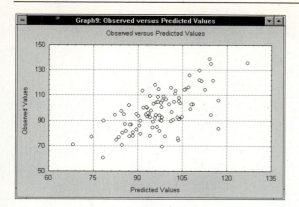

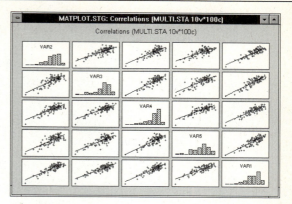

Predicted vs. Residual Values

Choose this option when you want a plot of the predicted values vs. the residual values.

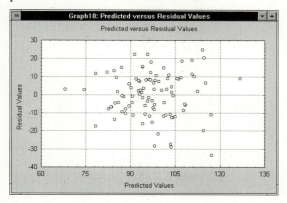

Matrix Plot for all Variables

Select this option to produce a visual (scatterplot) representation of the correlations between several variables in a matrix format.

When you click on this button, the *Select Variables* dialog will open in which you can select the variables to be plotted against each other (one scatterplot is generated for each pair of variables) in a square matrix of scatterplots.

Box and Whisker Plot for all Variables

The box plot resulting from this option represents the central tendencies and variability's of the selected variables. As measures of central tendency, you can choose between plotting the means and medians and their respective measures of variability; standard deviations or standard errors (associated with the means), and ranges (associated with the medians). Choose one of four types of box plots to represent the distribution of the values of the currently highlighted variable in the *Box and Whisker Plot Type* dialog.

NLN - 3055

Copyright © StatSoft, 1995

1. NONLINEAR ESTIMATION - DIALOGS AND OPTIONS

NOTES

Syntax for User-specified Regression Models

The syntax for specifying regression models and loss functions is straightforward, because it follows the standard notation for mathematical expressions. Also, the syntax conventions are the same as those used in the spreadsheet (data transformation) formulas described in the *General Conventions* chapter (see Volume I, Chapter 1). Thus, for the most current list of all available functions and operators, review the description of those transformation facilities in the *Electronic Manual* (e.g., press F1 in any module of *STATISTICA*).

The fastest way to familiarize yourself with the syntax is to review the examples provided in the *Examples* section of this chapter. Note that you can save the equation specifying a model and loss function using the *Save As* button (see below). The file will be saved in a standard text format, so that it can also be edited via other editors (e.g., via the *Text/output* window, or the Notepad); it consists of two lines of text; the model and the loss function (the default loss function is $(PRED-OBS)^2$).

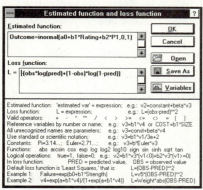

Shown below the edit fields in this dialog is a summary of the syntax. The general conventions for the equations and syntax reference are described below. An on-line reference is available in the *Electronic Manual* accessible by pressing F1 in this dialog.

General Syntax Conventions: Regression Equations

All regression models have the same form. In general:

```
dependent variable = expression including
                     independent variables
```

On the left side of the equation specify the dependent variable; on the right side specify the expression including independent variables and the parameters to be estimated.

Variables. Refer to variables either by their numbers (e.g., = $v1 - v2$) or their names (e.g., = *Retail - Cost*). Note that you can type variable names in either upper or lower case letters (i.e., *GENDER* is equivalent to *gender*).

Parameters. All names that are not recognized by the program as variable names or valid reserved keywords are interpreted to be parameters.

Logical expressions. Equations may contain logical expressions that involve constants, variables, parameters, or any mixture of the three. Logical expressions evaluate to *0* if they are *false*, and to *1* if they are *true*; thus models containing conditional expressions can be evaluated (see *Example 5*, page 3036, for an example of a model with a logical expression). The section on spreadsheet formulas in the *General Conventions* chapter (Volume I, Chapter 1) explains and illustrates this logic in detail.

Operators. Use parentheses to specify complex logical conditions (see above) or change the default precedence of the arithmetic and logical operators shown below (successive operators take precedence over preceding operators).

1. NONLINEAR ESTIMATION - NOTES

`=, <, >,` `<>, <=, >=`	-	relational operators
`+`	-	addition
`-`	-	subtraction
`*`	-	multiplication
`/`	-	division
`or`	-	inclusive disjunction
`div`	-	integer division
`and`	-	conjunction
`**, ^`	-	exponentiation
`not`	-	negation

To reiterate, this list is sorted in terms of computational precedence, from lowest to highest. For example, consider the expression:

`y=a+b1*x+b2*x*((x>500) and (x<1000))`

This equations specifies the standard linear regression model, plus an additional parameter $b2$ that applies only to cases with values of x in the range: $500<x<1000$; because the relational operators "<" and ">" are evaluated *after* the conjunction operator "*and*," and the parentheses around each condition ($x>500$, $x<1000$) are necessary here.

Constants. Constants can be entered as integers, floating point numbers, or in scientific notation. The following special constants can also be used.

`Euler`	-	e (2.71...)
`Pi`	-	π (3.14...)

Functions. All standard arithmetic functions as shown in the bottom part of the *Estimated function and loss function* dialog are supported:

`Abs(x)`	-	absolute value of x
`ArcSin(x)`	-	arc sine of x
`Cos(x)`	-	cosine of x
`Log(x)`	-	natural logarithm of x (base *Euler*)
`Log10(x)`	-	common logarithm of x (base *10*)
`Log2(x)`	-	binary logarithm of x (base *2*)
`Sign(x)`	-	returns the sign of x ($x<0 \rightarrow -1$; $x=0 \rightarrow 0$; $x>0 \rightarrow 1$)
`Sin(x)`	-	sine of x
`SinH(x)`	-	hyperbolic sine of x
`Sqrt(x)`	-	square root of x
`Tan(x)`	-	tangent of x

Distributions and their integrals. All distribution functions supported in *STATISTICA* can be used in custom regression models. For example:

`y = inormal(b0 + b1 * x, 0, 1);`

where *b0* and *b1* are the regression parameters to be estimated, and *inormal* stands for the normal integral, for the standard normal distribution ($\mu = 0$, $\sigma = 1$). (You may recognize this model as the *probit* model described in the *Introductory Overview*, page 3011).

The distributions (and their integrals) supported in these expressions are listed below.

Distri- bution	Density or Probability Function	Distribution Function	Inverse Distribution Function
Beta	beta(x,v,ω)	ibeta(x,v,ω)	vbeta(x,v,ω)
Binomial	binom(x,p,n)	ibinom(x,p,n)	
Cauchy	cauchy(x,η,θ)	icauchy(x,η,θ)	vcauchy(x,η,θ)
Chi-square	chi2(x,v)	ichi2(x,v)	vchi2(x,v)
Exponential	expon(x,λ)	iexpon(x,λ)	vexpon(x,λ)
Extreme	extreme(x,a,b)	iextreme(x,a,b)	vextreme(x,a,b)
F	F(x,v,ω)	iF(x,v,ω)	vF(x,v,ω)
Gamma	gamma(x,c)	igamma(x,c)	vgamma(x,c)
Geometric	geom(x,p)	igeom(x,p)	

Laplace	laplace(x,a,b)	ilaplace(x,a,b)	vlaplace(x,a,b)
Logistic	logis(x,a,b)	ilogis(x,a,b)	vlogis(x,a,b)
Lognormal	lognorm(x,μ,σ)	ilognorm(x,μ,σ)	vlognorm(x,μ,σ)
Normal	normal(x,μ,σ)	inormal(x,μ,σ)	vnormal(x,μ,σ)
Pareto	pareto(x,c)	ipareto(x,c)	vpareto(x,c)
Poisson	poisson(x,λ)	ipoisson(x,λ)	
Rayleigh	rayleigh(x,b)	irayleigh(x,b)	vrayleigh(x,b)
Student's	student(x,df)	istudent(x,df)	vstudent(x,df)
Weibull	weibull(x,b,c,θ)	iweibull(x,b,c,θ)	vweibull(x,b,c,θ)

For more information on these distributions, see the *Electronic Manual*.

Missing data. If the value of any variable used in the formula is missing (in the current case), then the expression evaluates to missing data (for the current case).

Rescaling/recoding of data. If any of the variables in the model need to be transformed or rescaled prior to the estimation of parameters, use the spreadsheet formulas (see Volume I, Chapter 1). For more extensive data transformations, use *STATISTICA BASIC* (a programming language available from the *Analysis* pull down menu which supports a variety of special facilities for statistical data management and transformations; see Volume I, Chapter 2).

For data recoding, use the *Recode Values of Variable* dialog accessible from the spreadsheet.

General Syntax Conventions: Loss Function

The specification of the loss function is also quite straightforward. In the estimation procedure, the specified value will be computed for each case in the data file; the estimation procedure will then attempt to find parameters that minimize the sum of those values across all cases.

In general, all rules apply as outlined for the specification of the regression equation for the model. In addition, the two keywords *Pred* and *Obs* (upper- or lower-case) are available to allow you to refer to the predicted and observed values, respectively, for the dependent variable. For example, the default least squares loss function can be specified as:

```
L = (Obs - Pred)**2
```

Constraining parameters (penalty function). When specifying regression functions, you can assess a penalty function in the loss function in order to constrain parameter values (see also page 3016). The following example will demonstrate how one can constrain the parameters b_0 and b_1 to be greater than or equal to zero.

Step 1: In the *Estimated function and loss function* dialog, specify the regression model:

$$v_3 = b_0 + b_1*v_1 + b_3*\exp(b_1*v_2)$$

Step 2: Now, specify the desired loss function with a penalty assessment as follows:

```
L = (obs-pred)²+(b₀<0)*100000+(b₁<0)*100000
```

Note how the logical operators are used in this loss function. If either b_0 or b_1 is less than zero, then the program adds a large number, thus assessing a penalty to the loss function. As a result, the various estimation procedures move away from the regions that produce those functions.

Math Errors

If a math error such as division by zero or square root of a negative number is detected during the evaluation of an equation, then the program will automatically assign a very large number to the loss function. This way of dealing with invalid parameter estimates is also referred to as assigning a *penalty function* (see above; see also *Introductory Overview*, page 3016). In most cases, the penalty

1. NONLINEAR ESTIMATION - NOTES

function will "entice" the parameters back into the realm of valid values.

Some Useful STATISTICA BASIC Programs

The *STATISTICA* system includes several *STATISTICA BASIC* example programs (in subdirectory *STBASIC*) that are useful for solving certain regression problems.

Weighted Least Squares Estimation

The program *Wls.stb* will perform weighted least squares analyses and compute various alternative R-square statistics (see Kvålseth, 1985). Instead of the iterative derivative-free estimation method used in *Example 7* (page 3039), program *Wls.stb* uses matrix algebra expressions (e.g., see Neter, Wassermann, and Kutner, 1985) to estimate the weighted least squares parameters. Thus, for large data files program *Wls.stb* is more efficient, and often much faster.

Two-Stage Least Squares Estimation

Program *2stls.stb* provides an example for 2-stage least squares regression. That program can easily be expanded or tailored to, for example, perform iteratively reweighted least squares analysis.

Box-Cox and Box-Tidwell Transformations

Often, when the residuals from an ordinary least squares analysis do not follow the normal distribution, one can apply certain transformations of the independent or dependent variables to correct the problem. A family of simple power transformations for positive response values *y* can be expressed as:

$$z = y^\lambda, \quad \lambda \neq 0$$
$$ = \text{natural log (y)}, \quad \lambda = 0$$

Note that for $\lambda = 1$ (*lambda =1*) no transformation is performed; for $\lambda = -1$ the result of the transformation is the reciprocal of *y*, for $\lambda = \frac{1}{2}$ the result of the transformation is the square root of y. Thus, various common transformations of the dependent variable can be expressed in this general form. Program *Boxcox.stb* will compute the maximum likelihood estimate for *lambda* for the *Box-Cox* transformation of the dependent variable (Box and Cox, 1964; see also Mason, Gunst, and Hess, 1989), and allows you to construct graphs of the error sums of squares as a function of different values for *lambda*.

The same type of power-transformation can be applied to the predictor (independent) variables in a linear regression problem. Program *Boxtid.stb* will compute the *lambda* parameters for this so-called *Box-Tidwell* transformation for the independent variables (see Mason, Gunst, Hess, 1989).

INDEX

A

Absolute value of x, 3058
Additive error, 3009
Arc sine of x, 3058
Asymptotic standard errors, 3018

B

BASIC (STATISTICA) programs,
　useful for nonlinear regression,
　3024, 3060
Binary logarithm of x (base 2), 3058
Box-Cox and Box-Tidwell
　transformations, 3060
Breakpoint regression, 3012, 3036,
　3045

C

Classification of cases, from
　logit/probit regression results,
　3052
Common logarithm of x (base 10),
　3058
Comparing groups, 3013
Constraining parameters, 3016, 3059
Continuous response functions, 3010
Convergence criteria, 3016
Cosine of x, 3058
Crossproduct ratios, for logit/probit
　regression results, 3052

D

Discontinuous regression models
　breakpoint regression, 3012, 3036
　comparing groups, 3013
　piecewise linear regression, 3012
Distributions and their integrals, in
　user-defined regression models,
　3058
Drug responsiveness and half-maximal
　response, 3012

E

Error
　additive, 3009
　multiplicative, 3009
Estimated function and loss function,
　3046
Estimating linear models, 3007
Estimating nonlinear models, 3007
Estimation method
　quasi-Newton, 3047
Eta for finite difference approximation,
　3048
Exponential regression models, 3031,
　3044

F

Finite difference approximation of
　standard errors, 3020
Function minimization algorithms,
　3015

G

General growth model, 3010
General logistic regression model, 3011
General syntax conventions
　loss functions, 3059
　regression equations, 3057
Goodness of fit Chi-square, 3018

H

Half-maximal response, 3012
Hessian matrix, 3018
Histograms
　2D histograms, 3026, 3028
Hooke-Jeeves pattern moves, 3017
Hyperbolic sine of x, 3058

I

Incremental fit, 3029
Intrinsically linear regression models,
　3008
Intrinsically nonlinear regression
　models, 3009

L

Least squares estimation, 3013
Local minima, 3016

Logistic transformation, 3011
Logit and probit models, 3010, 3015,
　3025, 3043
Loss functions, 3013, 3023, 3032,
　3039, 3059

M

Math errors, 3059
Matrix plots
　scatterplots, 3049
Maximum likelihood estimation, 3014,
　3015
Method of rotating coordinates, 3017
Model estimation, 3047
Models for binary responses
　probit and logit, 3010
Multiplicative error, 3009

N

Natural logarithm of x (base Euler),
　3058
Nonlinear estimation
　BASIC (STATISTICA) programs,
　　3024, 3060
　Box-Cox and Box-Tidwell
　　transformations, 3060
　breakpoint regression, 3012,
　　3036, 3045
　classification of cases, 3052
　comparing groups, 3013
　comparing two learning curves,
　　3033
　compound operators, 3058
　constants, 3058
　constraining parameters, 3016,
　　3059
　convergence criteria, 3016
　crossproduct ratios, 3052
　distributions and their integrals,
　　3058
　estimating linear and nonlinear
　　models, 3007
　estimating two different models,
　　3035
　eta for finite difference
　　approximation, 3048
　evaluating the fit of the model,
　　3018
　exponential regression models,
　　3031, 3044
　finite difference approximation of
　　standard errors, 3020

1. NONLINEAR ESTIMATION - INDEX

Nonlinear estimation (continued)
    function minimization algorithms, 3015
    functions, 3058
    general growth model, 3010
    general logistic regression model, 3011
    half-maximal response, 3012, 3038
    Hooke-Jeeves pattern moves, 3017
    incremental fit with more than one variable, 3029
    intrinsically linear regression models, 3008
    intrinsically nonlinear regression models, 3009
    least squares estimation, 3013
    local minima, 3016
    logical expressions, 3057
    logical operators, 3058
    logistic transformation, 3011
    logit and probit models, 3010, 3043
    logit models, 3010, 3025, 3030
    loss functions, 3013, 3032, 3039
    making nonlinear models linear, 3009
    math errors, 3059
    maximum likelihood estimation, 3014, 3028
    maximum likelihood estimation and probit/logit models, 3015
    method of rotating coordinates, 3034
    models that are nonlinear in the parameters, 3008
    normal distribution function, 3058
    odds ratios, 3052
    operators, 3057
    overview, 3007
    parameter estimates, 3052
    parameters, 3057
    penalty functions, 3016, 3059
    piecewise linear regression, 3036, 3038, 3045
    polynomial regression, 3008
    precedence, of computations in user-specified models, 3058
    predicting recovery from injury, 3031

Nonlinear estimation (continued)
    predicting redemption of coupons, 3029
    predicting success/failure, 3025
    probit models, 3010
    proportion of variance explained, 3018
    quasi-Newton method, 3017
    regression in pieces, 3012, 3036, 3045
    Rosenbrock pattern search, 3017, 3034
    simplex procedure, 3017
    standard errors, 3052
    start values, 3016
    STATISTICA BASIC programs, 3024, 3060
    step sizes, 3016
    stepwise analysis, 3023, 3029
    syntax for user-specified regression models, 3057
    two-stage least squares, 3060
    user-defined models, 3046
    weighted least squares, 3014, 3015, 3024, 3060
Nonlinear in the parameters, 3008
Nonlinear in the variables, 3008

O

Odds ratios, for logit/probit regression results, 3052

P

Parameter estimation, 3051, 3052
Penalty functions, 3016, 3059
Piecewise linear regression, 3012, 3036, 3038, 3045
Plot of fitted 2D function and observed values, 3032, 3039, 3053
Plot of fitted 3D function and observed values, 3053
Plot of observed vs. predicted values, 3019
Polynomial regression, 3008
Predefined loss functions, 3023
Predicting success/failure, 3025
Probability plots
    half-normal probability plots, 3019
    normal probability plots, 3019, 3028

Probit and logit models, 3011, 3025, 3043

Q

Quasi-Newton method, 3017, 3047

R

Regression in pieces, 3012, 3036, 3045
Return the sign of x, 3058
Rosenbrock pattern search, 3017, 3034

S

Scatterplots
    matrix scatterplots, 3049
Simplex procedure, 3017
Sine of x, 3058
Square root of x, 3058
Standard errors, 3048, 3052
Start values, 3016, 3050
STATISTICA BASIC programs, useful for nonlinear regression, 3024, 3060
Step sizes, 3016, 3050
Stepwise analysis, 3023, 3029
Syntax for user-specified regression models, 3057

T

Tangent of x, 3058
Two-stage least squares, 3060

U

User-specified loss functions, 3023
User-specified regression models, 3023, 3031, 3046, 3057

V

Variance/covariance matrix for parameters, 3019, 3020

W

Weighted least squares, 3014, 3015, 3024, 3039, 3060
When the estimation process does not converge, 3020

Chapter 2:
DISCRIMINANT FUNCTION ANALYSIS

Table of Contents

Introductory Overview 3067
    General Purpose 3067
    Computational Approach 3067
    Stepwise Discriminant Analysis 3068
    Interpreting a Two-Group Discriminant Function 3069
    Discriminant Functions for Multiple Groups 3069
    Assumptions 3071
    Classification 3072
Program Overview 3077
Example 3079
    Overview 3079
    Specifying the Analysis 3079
    Reviewing the Results of Discriminant Analysis 3084
    Classification 3088
    Summary 3090
Dialogs, Options, Statistics 3091
    Startup Panel 3091
    Specifying the Analysis 3091
    Descriptive Statistics 3093
    Results and Classification 3096
    Canonical Analysis 3100
    A priori Probabilities 3101
Technical Notes 3103
Index 3105

Detailed Table of Contents

INTRODUCTORY OVERVIEW **3067**
    General Purpose 3067
    Computational Approach 3067
    Stepwise Discriminant Analysis 3068
    Interpreting a Two-Group Discriminant Function 3069

2. DISCRIMINANT FUNCTION ANALYSIS - CONTENTS

    Discriminant Functions for Multiple Groups 3069
    Assumptions 3071
    Classification 3072

PROGRAM OVERVIEW 3077
    Overview 3077
    Descriptive Statistics and Plots 3077
    Output Statistics 3077
    Classification 3077
    Alternative Procedures 3078

EXAMPLE 3079
    Overview 3079
    Specifying the Analysis 3079
        Data File 3079
        Startup Panel 3079
        Reviewing Descriptive Statistics 3080
        Specifying Discriminant Function Analysis 3082
    Reviewing the Results of Discriminant Analysis 3084
        Results at Step 0 3084
        Results at Step 2 3085
        Results at Step 4 (Final Step) 3085
        Canonical Analysis 3086
    Classification 3088
        Classification Functions 3088
        A priori Probabilities 3088
        Classification Matrix 3089
        Classification of Cases 3089
    Summary 3090

DIALOGS, OPTIONS, STATISTICS 3091
    Startup Panel 3091
        Variables 3091
        Codes for Grouping Variable 3091
        Missing Data 3091
    Specifying the Analysis 3091
        Variables 3091
        Method 3091
        Tolerance Value 3092
        Stepwise Discriminant Function Analysis 3092
        Review Correlations, Statistics, and Graphs for Groups 3093
    Descriptive Statistics 3093
        Pooled Within-Groups Covariances and Correlations 3093
        Total Covariances and Correlations 3093
        Graph (Scatterplot Matrix) 3094

2. DISCRIMINANT FUNCTION ANALYSIS - CONTENTS

  Means and Number of Cases 3094
  Box and Whisker Plot 3094
  Standard Deviations 3095
  Categorized Histogram (by Group) 3095
  Box and Whisker Plot (by Group) 3095
  Categorized Scatterplot (by Group) 3095
  Categorized Probability Plot (by Group) 3096
 Results and Classification 3096
  Variables in the Model 3096
  Variables not in the Model 3097
  Distances Between Groups 3097
  Summary of Stepwise Analysis 3097
  Canonical Analysis 3097
  Classification Functions 3097
  Classification Matrix 3098
  Classification of Cases 3098
  Squared Mahalanobis Distances 3098
  Posterior Probabilities 3098
  Save Classifications, Distances, Probabilities 3099
  A priori Classification Probabilities 3099
  Next 3099
 Canonical Analysis 3100
  Chi-square Tests of Successive Roots 3100
  Coefficients for Canonical Variables 3100
  Factor Structure 3100
  Means for Canonical Variables 3100
  Canonical Scores for Each Case 3101
  Save Canonical Scores 3101
  Histogram of Canonical Scores for Selected Group 3101
  Combined Histogram for all Groups 3101
  Scatterplot of Canonical Scores 3101
 A priori Probabilities 3101

TECHNICAL NOTES **3103**
 Computational Method 3103
 Formulas 3103

INDEX **3105**

2. DISCRIMINANT FUNCTION ANALYSIS - CONTENTS

Chapter 2:
DISCRIMINANT FUNCTION ANALYSIS

INTRODUCTORY OVERVIEW

General Purpose

Discriminant function analysis is used to determine which variables discriminate between two or more naturally occurring (or *a priori* defined) groups. For example, an educational researcher may want to investigate which variables discriminate between high school graduates who decide (1) to go to college, (2) to attend a trade or professional school, or (3) to seek no further training or education. For that purpose the researcher could collect data on numerous variables prior to students' graduation. After graduation, most students will naturally fall into one of the three categories. *Discriminant Analysis* could then be used to determine which variable(s) are the best predictors of students' subsequent educational choice. A medical researcher may record different variables relating to patients' backgrounds in order to learn which variables best predict whether a patient is likely to recover completely (group 1), partially (group 2), or not at all (group 3). A biologist could record different characteristics of similar types (groups) of flowers, and then perform a discriminant function analysis to determine the set of characteristics that allows for the best discrimination between the types.

Computational Approach

Computationally, discriminant function analysis is very similar to analysis of variance (see *ANOVA-/MANOVA*, Volume I). Consider the following simple example. Suppose you measure height in a random sample of 50 males and 50 females. Females are, on the average, not as tall as males, and this difference will be reflected in the difference in means (for the variable *Height*). Therefore, the variable *Height* allows you to discriminate between males and females with a better than chance probability: If a person is tall, then he is likely to be a male; if a person is short, then she is likely to be a female.

You can generalize this reasoning to groups and variables that are less "trivial." For example, suppose you have two groups of high school graduates: Those who choose to attend college after graduation and those who do not. You could have measured students' stated intention to continue on to college one year prior to graduation. If the means for the two groups (those who actually went to college and those who did not) are different, then you can say that intention to attend college as stated one year prior to graduation allows one to discriminate between those who are and are not college-bound (and this information may be used by career counselors to provide the appropriate guidance to the respective students). To summarize the discussion so far, the basic idea underlying discriminant function analysis is to determine whether groups differ with regard to the mean of a variable, and then to use that variable to predict group membership (e.g., of new cases).

Analysis of variance. Stated in this manner, the discriminant function problem can be rephrased as a one-way analysis of variance (ANOVA) problem. Specifically, one can ask whether or not two or more groups are *significantly different* from each other with respect to the mean of a particular variable. To learn more about how one can test for the statistical significance of differences between means in different groups, you may want to read the *Introductory Overview* section to *ANOVA/MANOVA* (Volume I, Chapter 7). However, it should be clear that if the means for a variable are significantly different in different groups, then you can say that this variable discriminates between the groups.

In the case of a single variable, the final significance test of whether or not a variable discriminates between groups is the F test. As described in *Elementary Concepts* (Volume I) and *ANOVA/MANOVA* (Volume I), F is essentially computed as the ratio of the between-groups variance in the data over the pooled (average) within-group variance. If the between-group variance is significantly larger, then there must be significant differences between means.

Multiple variables. Usually, one includes several variables in a study in order to see which of them contribute to the discrimination between groups. In that case, you have a matrix of total variances and covariances; likewise, you have a matrix of pooled within-group variances and covariances. You can compare those two matrices via multivariate F tests in order to determine whether or not there are any significant differences (with regard to all variables) between groups.

This procedure is identical to multivariate analysis of variance or *MANOVA*. As in *MANOVA*, one could first perform the multivariate test and, if statistically significant, proceed to see which of the variables have significantly different means across the groups. Thus, even though the computations with multiple variables are more complex, the principal reasoning still applies, namely, that you are looking for variables that discriminate between groups, as is evident in observed mean differences. In fact, you may perform discriminant function analysis with the *ANOVA/MANOVA* module; however, different types of statistics are customarily computed and interpreted in discriminant analysis (as described later).

Stepwise Discriminant Analysis

Probably the most common application of discriminant function analysis is to include many measures in the study in order to determine the ones that best discriminate between groups. For example, an educational researcher interested in predicting high school graduates' choices for further education would probably include as many measures of personality, achievement motivation, academic performance, etc. as possible in order to learn which measures offer the best prediction.

Model. Put another way, you want to build a "model" of how you can best predict to which group a case belongs. In the following discussion the term "in the model" will be used in order to refer to variables that are included in the prediction of group membership, and variables will be referred to as being "not in the model" if they are not included.

Forward stepwise analysis. In stepwise discriminant function analysis, *STATISTICA* will "build" a model of discrimination step-by-step. Specifically, at each step *STATISTICA* will review all variables and evaluate which one will contribute most to the discrimination between groups. That variable will then be included in the model, and *STATISTICA* will proceed to the next step.

Backward stepwise analysis. One can also step backwards; in that case *STATISTICA* will first include all variables in the model and then, at each step, eliminate the variable that contributes least to the prediction of group membership. Thus, as the result of a successful discriminant function analysis, one would only keep the "important" variables in the model, that is, those variables that contribute the most to the discrimination between groups.

F to enter, F to remove. The stepwise procedure is "guided" by the respective F to enter and F to remove values. The F value for a variable indicates its statistical significance in the discrimination between groups, that is, it is a measure of the extent to which a variable makes a unique contribution to the prediction of group membership. If you are familiar with stepwise multiple regression procedures (see *Multiple Regression*, Volume I, Chapter 6), then you may interpret the *F* to enter-/remove values in the same way as in stepwise regression.

In general, STATISTICA will continue to choose variables to be included in the model, as long as the respective *F* values for those variables are larger than the user-specified *F* to enter; STATISTICA will exclude (remove) variables from the model if their significance is less than the user-specified *F* to remove.

Capitalizing on chance. A common misinterpretation of the results of stepwise discriminant analysis is to take statistical significance levels at face value. When STATISTICA decides which variable to include or exclude in the next step of the analysis, it will actually compute the significance of the contribution of *each* variable under consideration. Therefore, by nature, the stepwise procedures will capitalize on chance because they "pick and choose" the variables to be included in the model so as to yield maximum discrimination. Thus, when using the stepwise approach, the researcher should be aware that the significance levels do not reflect the true *alpha* error rate, that is, the probability of erroneously rejecting H_0 (the null hypothesis that there is no discrimination between groups).

Interpreting a Two-Group Discriminant Function

In the two-group case, discriminant function analysis can also be thought of as (and is analogous to) multiple regression (see *Multiple Regression*, Volume I, Chapter 6; the two-group discriminant analysis is also called *Fisher linear discriminant analysis* after Fisher, 1936; computationally all of these approaches are analogous). If you code the two groups in the analysis as *1* and *2*, and use that variable as the dependent variable in a multiple regression analysis, then you would get results that are analogous to those you would obtain via the *Discriminant Analysis* module. In general, in the two-group case you fit a linear equation of the type:

$$\text{Group} = a + b_1 \ast x_1 + b_2 \ast x_2 + \ldots + b_m \ast x_m$$

where *a* is a constant and b_1 through b_m are regression coefficients. The interpretation of the results of a two-group problem is straightforward and closely follows the logic of multiple regression: Those variables with the largest (standardized or *beta*) regression coefficients are the ones that contribute most to the prediction of group membership.

Discriminant Functions for Multiple Groups

When there are more than two groups, then you can estimate more than one discriminant function like the one presented above. For example, when there are three groups, you could estimate (1) a function for discriminating between group 1 and groups 2 and 3 combined, and (2) another function for discriminating between group 2 and group 3. For example, you could have one function that discriminates between those high school graduates that go to college and those who do not (but rather get a job or go to a professional or trade school), and a second function to discriminate between those graduates that go to a professional or trade school versus those who get a job. The *b* coefficients in those discriminant functions could then be interpreted as before.

Canonical analysis. When actually performing a multiple group discriminant analysis, you do not have to specify how to combine groups so as to form different discriminant functions. Rather,

2. DISCRIMINANT FUNCTION ANALYSIS - INTRODUCTORY OVERVIEW

STATISTICA will automatically determine some optimal combination of variables so that the first function provides the most overall discrimination between groups, the second provides second-most, and so on. Moreover, the functions will be independent or *orthogonal*, that is, their contributions to the discrimination between groups will not overlap. Computationally, STATISTICA will perform a *canonical correlation* analysis (see also *Canonical Correlation*, Chapter 4) that will determine the successive canonical *roots* and functions. The maximum number of functions that STATISTICA will compute will be equal to the number of groups minus one, or the number of variables in the analysis, whichever is smaller.

Interpreting the discriminant functions. As stated before, you will get *b* (and standardized *beta*) coefficients for each variable in each discriminant (now also called *canonical*) function, and they can be interpreted as usual: the larger the standardized coefficient, the greater is the contribution of the respective variable to the discrimination between groups. (Note that you could also interpret the *structure coefficients*; see below.) However, these coefficients do not tell you between which of the groups the respective functions discriminate. You can identify the nature of the discrimination for each discriminant (canonical) function by looking at the means for the functions across groups. Shown below is a Scrollsheet with the means for two canonical functions (also called *canonical variables*) for three groups.

The results shown below pertain to a data set reported by Fisher (1936). Specifically, that data set contains measures of the lengths and widths of sepals and petals for 50 flowers of three types of irises: *Setosa*, *Versicol*, and *Virginic*. The analysis of this data set is further discussed in the *Example* section of this chapter (starting on page 3079). Shown in the illustration below are the means for the first and second discriminant function (*Root 1* and *Root 2*, respectively) for each group, that is, type of iris.

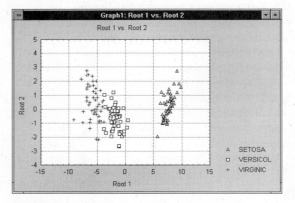

The first function seems to discriminate mostly between *Setosa* which shows a large positive mean, versus *Versicol* and *Virginic* which both show somewhat negative means. The second function seems to distinguish between *Versicol* versus *Setosa* and *Virginic* combined. However, the means on the second function are much less distinct.

Plot of discriminant functions. You can also visualize how these two functions discriminate between groups by plotting the individual scores for the two discriminant functions.

Again, function *1* seems to discriminate mostly between groups *Setosa*, and *Virginic* and *Versicol* combined. In the vertical direction (*Root 2*), the slight trend of *Versicol* points to fall below the center line (*0*) is apparent, but minor as one could have expected from the means for the two functions.

Factor structure matrix. Another way to determine which variables "mark" or define a particular discriminant function is to look at the factor structure. The factor structure coefficients are the correlations between the variables in the model and the discriminant functions; if you are familiar with factor analysis (see *Factor Analysis*, Chapter 6), you may think of these correlations as factor *loadings* of the variables on each discriminant function.

Some authors have argued that these structure coefficients should be used when interpreting the substantive "meaning" of discriminant functions. The reasons given by those authors are that (1) supposedly the structure coefficients are more stable, and (2) they allow for the interpretation of factors (discriminant functions) in the manner that is analogous to factor analysis. However, subsequent Monte Carlo research (Barcikowski and Stevens, 1975; Huberty, 1975) has shown that the discriminant function coefficients and the structure coefficients are about equally unstable, unless the n is fairly large (e.g., if there are 20 times more cases than there are variables). The most important thing to remember is that the discriminant function coefficients denote the unique (partial) contribution of each variable to the discriminant function(s), while the structure coefficients denote the simple correlations between the variables and the function(s). If one wants to assign substantive "meaningful" labels to the discriminant functions (akin to the interpretation of factors in factor analysis), then the structure coefficients should be used (interpreted); if one wants to learn what is each variable's unique contribution to the discriminant function, then use the discriminant function coefficients (weights).

Significance of discriminant functions. One can test the number of roots that add *significantly* to the discrimination between groups. Only those found to be statistically significant should be used for interpretation; non-significant functions (roots) should be ignored.

Summary. To summarize, when interpreting multiple discriminant functions, which arise from analyses with more than two groups and more than one variable, one would first test the different functions for statistical significance, and only consider the significant functions for further examination. Next, you would look at the standardized *beta* coefficients for each variable, for each significant function. The larger the standardized *beta* coefficient, the larger is the respective variable's unique contribution to the discrimination specified by the respective discriminant function. In order to derive substantive "meaningful" labels for the discriminant functions, one can also examine the factor structure matrix with the correlations between the variables and the discriminant functions. Finally, you would look at the means for the significant discriminant functions in order to determine between which groups the respective functions seem to discriminate.

Assumptions

As mentioned earlier, discriminant function analysis is computationally very similar to *MANOVA*, and all assumptions for *MANOVA* mentioned in the *ANOVA/MANOVA* chapter (Volume I) apply. In fact, you may use the wide range of diagnostics and statistical tests of assumption that are available in *ANOVA/MANOVA* to examine your data for the discriminant analysis (to avoid unnecessary duplications, the extensive set of facilities provided in *ANOVA/MANOVA* is not repeated in *Discriminant Analysis*).

Normal distribution. It is assumed that the data (for the variables) represent a sample from a multivariate normal distribution. Note that the *Discriminant Analysis* module makes it very simple to produce histograms of frequency distributions from within Scrollsheets (with a single keystroke). Thus, the user can examine whether or not variables are normally distributed. However, note that violations of the normality assumption are usually not "fatal," meaning that the resultant significance tests, etc., are still "trustworthy." The *ANOVA-/MANOVA* module provides specific tests for normality.

Homogeneity of variances/covariances. It is assumed that the variance/covariance matrices of variables are homogeneous across groups. Again,

2. DISCRIMINANT FUNCTION ANALYSIS - INTRODUCTORY OVERVIEW

minor deviations are not that important; however, before accepting final conclusions for an important study it is probably a good idea to review the within-groups variances and correlation matrices. In particular, the scatterplot matrix that can be produced from the *Descriptive Statistics* dialog (page 3093) can be very useful for this purpose. When in doubt, try re-running the analyses excluding one or two groups that are of less interest. If the overall results (interpretations) hold up, you probably do not have a problem. You may also use the numerous tests and facilities in the *ANOVA/MANOVA* module to examine whether or not this assumption is violated in your data. However, as mentioned in the *ANOVA/MANOVA* chapter (Volume I), the multivariate Box *M* test for homogeneity of variances/covariances is particularly sensitive to deviations from multivariate normality and should not be taken too "seriously."

Correlations between means and variances.
The major "real" threat to the validity of significance tests occurs when the means for variables across groups are correlated with the variances (or standard deviations). Intuitively, if there is large variability in a group with particularly high means on some variables, then those high means are not reliable. However, the overall significance tests are based on pooled variances, that is, the average variance across all groups. Thus, the significance tests of the relatively larger means (with the large variances) would be based on the relatively smaller pooled variances, erroneously resulting in statistical significance. In practice, this pattern may occur if one group in the study contains a few extreme outliers, which have a large impact on the means and also increase the variability. To guard against this problem, inspect the descriptive statistics, that is, the means and standard deviations or variances for such a correlation. The *ANOVA/MANOVA* module also allows you to plot the means and variances (or standard deviations) in a scatterplot.

The matrix ill-conditioning problem. Another assumption of discriminant function analysis is that the variables that are used to discriminate between groups are not completely redundant. As part of the computations involved in discriminant analysis, STATISTICA will invert the variance/covariance matrix of the variables in the model. If any one of the variables is completely redundant with the other variables, then the matrix is said to be *ill-conditioned*, and it cannot be inverted. For example, if a variable is the sum of three other variables that are also in the model, then the matrix is ill-conditioned.

Tolerance values. In order to guard against matrix ill-conditioning, STATISTICA will constantly check the so-called *tolerance value* for each variable. This value is also routinely displayed when you ask to review the summary statistics for variables that are in the model, and those that are not in the model. This tolerance value is computed as *1 minus R-square* of the respective variable with all other variables included in the current model. Thus, it is the proportion of variance that is unique to the respective variable. You may also refer to *Multiple Regression* (Volume I) to learn more about multiple regression and the interpretation of the tolerance value. In general, when a variable is almost completely redundant (and, therefore, the matrix ill-conditioning problem is likely to occur), the tolerance value for that variable will approach 0. The default value in the *Discriminant Analysis* module for the minimum acceptable tolerance is *0.01*. STATISTICA will issue a matrix ill-conditioning message when the tolerance for any variable falls below that value, that is, if any variable is more than 99% redundant (the user may change this default value).

Classification

Another major purpose to which discriminant analysis is applied is the issue of predictive classification of cases. Once a model has been finalized and the discriminant functions have been derived, how well can they *predict* to which group a particular case belongs?

A priori and post hoc predictions. Before going into the details of different estimation procedures, it is important to make sure that this difference is clear. Obviously, if you estimate, based on some data set, the discriminant functions that best discriminate between groups, and then use the *same* data to evaluate how accurate your prediction is, then you are very much capitalizing on chance. In general, one will *always* get a worse classification when predicting cases that were not used for the estimation of the discriminant function. Put another way, *post hoc* predictions are always better than *a priori* predictions. (The trouble with predicting the future *a priori* is that one does not know what will happen; it is much easier to find ways to predict what already has happened.) Therefore, one should never base one's confidence regarding the correct classification of future observations on the same data set from which the discriminant functions were derived; rather, if one wants to classify cases predictively, it is necessary to collect new data to "try out" (validate) the utility of the discriminant functions.

Classification functions. The *Discriminant Analysis* module will automatically compute the classification functions. These are not to be confused with the discriminant functions. The classification functions can be used to determine to which group each case most likely belongs. There are as many classification functions as there are groups. Here are the classification functions for the three types of irises from the example discussed earlier (and later in the *Example* section, starting on page 3079).

DISCRIM. ANALYSIS	SETOSA p=.33333	VERSICOL p=.33333	VIRGINIC p=.33333
SEPALLEN	23.5442	15.6982	12.446
SEPALWID	23.5879	7.0725	3.685
PETALLEN	-16.4306	5.2115	12.767
PETALWID	-17.3984	6.4342	21.079
Constant	-86.3085	-72.8526	-104.368

Classification Functions; grouping: IRISTYPE (irisdat.sta)

Classification scores. Each function allows you to compute *classification scores* for each case for each group, by applying the formula:

$$S_i = c_i + w_{i1}*x_1 + w_{i2}*x_2 + \ldots + w_{im}*x_m$$

In this formula, the subscript i denotes the respective group; the subscripts $1, 2, \ldots, m$ denote the m variables; c_i is a constant for the i'th group, w_{ij} is the weight for the j'th variable in the computation of the classification score for the i'th group; x_i is the observed value for the respective case for the j'th variable. S_i is the resultant classification score.

For example, to compute the classification score for the first type of iris (*Setosa*) from the results shown in the Scrollsheet above, you would use:

$$S_1 = -86.3 + 23.5*x_1 + 23.6*x_2 - 16.4*x_3 - 17.4*x_4$$

where the x_i's denote the observed values for the four variables for the respective case. Thus, you can use the classification functions to directly compute classification scores for some new observations (for example, these functions can be specified in the spreadsheet as the formulas for computing new variables; as new cases are added to the file, the classification scores are then automatically computed).

Classification of cases. Once you have computed the classification scores for a case, it is easy to decide how to classify the case: in general, the case is classified as belonging to the group for which it has the highest classification score (unless the *a priori* classification probabilities are widely disparate; see below). Thus, if you were to study high school students' post-graduation career-/educational choices (e.g., attending college, attending a professional or trade school, or getting a job) based on several variables assessed one year prior to graduation, you could use the classification functions to predict what each student is most likely to do after graduation. However, you would also like to know the *probability* that the student will make the predicted choice. Those probabilities are called *posterior* probabilities, and they can also be

2. DISCRIMINANT FUNCTION ANALYSIS - INTRODUCTORY OVERVIEW

computed. However, to understand how those probabilities are derived, first consider the so-called *Mahalanobis distances*.

Mahalanobis distances. You may have read about these distances in other parts of the manual (e.g., in *Multiple Regression*, Volume 1). In general, the Mahalanobis distance is a measure of distance between two points in the space defined by two or more correlated variables.

For example, if there are two variables that are uncorrelated, then you could plot points (cases) in a standard two-dimensional scatterplot; the Mahalanobis distances between the points would then be identical to the Euclidean distance; that is, the distance as measured, for example, by a ruler. If there are three uncorrelated variables, you could also simply use a ruler (in a 3D plot) to determine the distances between points. If there are more than 3 variables, you cannot represent the distances in a plot any more. Also, when the variables are correlated, then the axes in the plots can be thought of as being *non-orthogonal* that is, they would not be positioned at right angles to each other. In those cases, the simple Euclidean distance is not an appropriate measure, while the Mahalanobis distance will adequately account for the correlations.

Mahalanobis distances and classification. For each group in our sample, you can determine the location of the point that represents the means for all variables in the multivariate space defined by the variables in the model. These points are called group *centroids*. For each case you can then compute the Mahalanobis distances (of the respective case) from each of the group centroids. Again, you would classify the case as belonging to the group to which it is closest, that is, where the Mahalanobis distance is smallest.

Posterior classification probabilities. Using the Mahalanobis distances to do the classification, you can now derive probabilities. The probability that a case belongs to a particular group is basically proportional to the Mahalanobis distance from that group centroid (it is not exactly proportional because a multivariate normal distribution around each centroid is assumed). Because the location of each case is computed from your prior knowledge of the values for that case on the variables in the model, these probabilities are called *posterior* probabilities. In summary, the *posterior* probability is the probability, based on your knowledge of the values of other variables, that the respective case belongs to a particular group. Of course, the *Discriminant Analysis* module will automatically compute those probabilities for all cases (or for selected cases only for validation studies).

A priori classification probabilities. There is one additional factor that needs to be considered when classifying cases. Sometimes, you know ahead of time that there are more observations in one group than in any other; thus, the *a priori* probability that a case belongs to that group is higher.

For example, if you know ahead of time that 60% of the graduates from your high school usually go to college (20% go to a professional school, and another 20% get a job), then you should adjust your prediction accordingly: *a priori*, and all other things being equal, it is more likely that a student will attend college than choose either of the other two options. The *Discriminant Analysis* module will allow you to specify different *a priori* probabilities, which will then be used to adjust the classification of cases (and the computation of *posterior* probabilities) accordingly.

In practice, the researcher needs to ask him or herself whether the unequal number of cases in different groups in the sample is a reflection of the true distribution in the population, or whether it is only the (random) result of the sampling procedure. In the former case, you would set the *a priori* probabilities to be proportional to the sizes of the groups in the sample; in the latter case you would specify the *a priori* probabilities as being equal in each group. The specification of different *a priori*

2. DISCRIMINANT FUNCTION ANALYSIS - INTRODUCTORY OVERVIEW

probabilities can greatly affect the accuracy of the prediction.

Summary of the prediction. A common result that one looks at in order to determine how well the current classification functions predict group membership of cases is the *classification matrix*. The classification matrix shows the number of cases that were correctly classified (on the diagonal of the matrix) and those that were misclassified.

DISCRIM. ANALYSIS Group	Percent Correct	SETOSA p=.33333	VERSICOL p=.33333	VIRGINIC p=.33333
SETOSA	100.0000	50	0	0
VERSICOL	96.0000	0	48	2
VIRGINIC	98.0000	0	1	49
Total	98.0000	50	49	51

Rows: Observed classifications
Columns: Predicted classifications

Shown above is the classification matrix from the iris flowers example. This Scrollsheet shows the number of cases that were correctly classified (on the diagonal of the matrix) and those that were misclassified. Also shown in the first row of each column header are the *a priori* classification probabilities. As you can see, 98% of all cases are classified correctly in this case; 100% of the flowers of type *Setosa* are correctly classified, 96% of the type *Versicol* are correctly classified, and 98% of the type *Virginic* are correctly classified.

Another word of caution. To reiterate, *post hoc* predicting of what has happened in the past is not that difficult. It is not uncommon to obtain very good classification if one uses the same cases from which the classification functions were computed. In order to get an idea of how well the current classification functions "perform," one must classify (*a priori*) *different* cases, that is, cases that were not used to estimate the classification functions.

In the *Discriminant Analysis* module, you can flexibly use the selection conditions (double-click on the *Sel:OFF* field on the status bar) to include or exclude cases from the computations; thus, the classification matrix can be computed for "old" cases as well as "new" cases. Only the classification of new cases allows one to assess the predictive validity of the classification functions; the classification of old cases only provides a useful diagnostic tool to identify outliers or areas where the classification function seems to be less adequate.

2. DISCRIMINANT FUNCTION ANALYSIS - INTRODUCTORY OVERVIEW

PROGRAM OVERVIEW

Overview

The *Discriminant Analysis* module is a full implementation of stepwise multiple discriminant function analysis. The program will perform forward stepwise or backward stepwise analyses, or enter user-specified blocks of variables into the model. In addition to the numerous graphics and statistics describing the discriminant functions, the program also provides a wide range of options and statistics for the classification of "old" or "new" cases (for validation).

Descriptive Statistics and Plots

The user can compute the full range of descriptive statistics for each variable and group. As the default plot of the descriptive statistics Scrollsheet, the user can produce the histogram of the distribution and thus provide a quick visual check for violations of the normality assumption. The user can also review scatterplots of any pair of variables in the analysis. When requested from within the Scrollsheet of the overall correlations (or variances/covariances), then the respective plot will be constructed as a standard scatterplot; when requested from within the Scrollsheet of *pooled* within-groups correlations, this plot will show the within-groups deviations from the respective means (and thus reflect the pooled within-groups correlation). A large selection of descriptive graphs are also available to aid in the interpretation of input data and the output. For example, a selection of categorized graphs are available which automatically generate multiple sub-graphs (one for each group in the current design); bivariate distributions of variables can be reviewed in categorized histograms and 3D bivariate histograms.

Output Statistics

For each variable, the program will compute the respective Wilks' *lambda*, partial *lambda*, F to enter (or remove), the p-level, the tolerance value, and the R-square value of the respective variable with all other variables in the model. The program will perform a full canonical analysis and report the raw and cumulative eigenvalues (roots) for all functions and their statistical significance. The user can review the raw and standardized discriminant (canonical) function coefficients, the structure coefficient matrix (of factor loadings), the means for the discriminant functions, and the discriminant scores for each case. Like all other output in *STATISTICA*, these scores may also be saved in a *STATISTICA* data file for further analysis.

Plots. The user can produce histograms of the canonical scores within each group, as well as for all groups combined. To aid in the interpretation of discriminant functions, the program will also produce special scatterplots for pairs of canonical variables in which the group membership of individual cases is visibly marked.

Classification

The *Discriminant Analysis* module will compute the standard classification functions for each group. The classification of cases can be reviewed in terms of Mahalanobis distances (of cases from respective group centroids), *posterior* probabilities, or actual classifications. All of these values can be saved in a *STATISTICA* data file for further analyses. The summary classification matrix of the number and percent of correctly classified cases can also be displayed.

Specifying *a priori* probabilities. The user has several options to specify the *a priori* classification probabilities. Classifications can be based on equal *a priori* probabilities for each group, *a priori* probabilities that are proportional to the sample

DIS - 3077

2. DISCRIMINANT FUNCTION ANALYSIS - PROGRAM OVERVIEW

sizes, or they can be based on user-specified *a priori* probabilities.

Selecting cases for classification. The user can specify selection conditions to include or exclude selected cases from the classification. This option is useful when one wants to exclude cases that were used to estimate the classification functions. This procedure is common when one wants to validate the classification functions in a new sample.

Alternative Procedures

The nature of the computations involved in discriminant function analysis is very similar to that of multivariate analysis of variance. In fact, many of the statistics reported by the *Discriminant Analysis* module are also available in the *ANOVA/MANOVA* module; therefore, the *ANOVA/MANOVA* module should be used for factorial discriminant function analysis. The *ANOVA/MANOVA* module also includes a comprehensive list of statistics for testing various assumptions of discriminant function analysis (tests of homogeneity of variances and covariances, test of normality, scatterplot of correlations between means and standard deviations, etc.). If there are only two groups in the analysis, then the *Multiple Regression* module will provide the same basic results as the *Discriminant Analysis* module. However, in that case, it may be more appropriate to use the *probit* and *logit* regression procedures in the *Nonlinear Estimation* module (Chapter 1).

In order to test complex hypotheses about relationships between variables in different groups, you can use the *Structural Equation Modeling (SEPATH)* module of *STATISTICA* (see Chapter 11). In that module you can fit hypothesized models about the relationships between variables to correlation (covariance) matrices in one or more groups. The *SEPATH* module allows you to constrain parameters (to equality) across groups, and to analyze moment matrices, so you can test hypotheses about differences in means.

Finally, you may want to refer to *Cluster Analysis* (Chapter 5), which contains procedures that are also commonly used for classification purposes. However, note that cluster analysis techniques are mostly used when the number of groups and the group membership of cases are unknown. In those cases, the *Cluster Analysis* module may detect apparent clusters of cases or variables.

2. DISCRIMINANT FUNCTION ANALYSIS - EXAMPLE

EXAMPLE

Overview

The following example is based on a classic example data set reported by Fisher (1936). It reports the lengths and widths of sepals and petals of three types of irises (*Setosa*, *Versicol*, and *Virginic*). The purpose of the analysis is to learn how one can discriminate between the three types of flowers, based on the four measures of width and length of petals and sepals. In principle, all discriminant analyses address similar questions. If you are an educational researcher, you may substitute "type of flower" with "type of drop-out," and the variables (measures of sepal/petal widths and lengths) with "grades in four key courses." If you are a social scientist, you may study variables which predict people's choices of careers. In a personnel selection study, you may be interested in variables that discriminate between employees who will perform above average and will later be promoted, employees who do an adequate job, and employees who are unacceptable. Thus, even though the present example falls into the domain of biology, the general procedures shown here are generally applicable.

Specifying the Analysis

Data File

The data set for this analysis is contained in the file *Irisdat.sta*. A partial listing of this file is shown below. The first two variables in this file (*Sepallen*, *Sepalwid*) pertain to the length and width of sepals; the next two variables (*Petallen*, *Petalwid*) pertain to the length and width of petals. The last variable in this file is a grouping or coding variable that identifies to which type of iris each flower belongs

(*Setosa*, *Versicol*, and *Virginic*). In all, there are 150 flowers in this sample, 50 of each type.

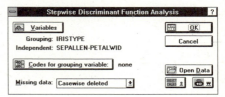

Startup Panel

After starting the *Discriminant Analysis* module, open the data file *Irisdat.sta*. Next, you will need to specify the grouping variable (variable *Iristype*) and the continuous *independent* variables that will be used in order to discriminate between iris types; Click on the *Variables* button in the startup panel to specify these variables

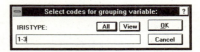

Specify *Iristype* as the grouping variable and the remaining variables (*Sepallen*, *Sepalwid*, *Petallen*, *Petalwid*) as the independent variables. Next, you need to specify the codes that were used in the grouping variable to identify to which group each case belongs. Click on the *Codes for grouping variables* button and either enter *1-3*, click on the *All* button, or use the asterisk (*) convention to select all codes.

Alternatively, you can click *OK* in the startup panel and *STATISTICA* will automatically search the

2. DISCRIMINANT FUNCTION ANALYSIS - EXAMPLE

grouping variable(s) and select all codes for those variables.

Deletion of missing data. This particular data file does not contain any missing data. However, if there are missing data in the file, you may either choose to ignore cases with missing data (set the *Missing data* combo box to *casewise deleted*) or to substitute missing data by the respective means (set the switch to *substituted by means*).

Reviewing Descriptive Statistics

Now click *OK* to begin the analysis. A dialog will open that will allow you to define the discriminant analysis and to review the descriptive statistics.

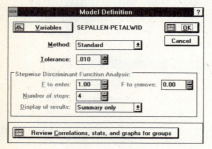

Before specifying the discriminant function analysis, click on the *Review correlations, stats, and graphs for groups* button to look at the distribution of some of the variables and their intercorrelations.

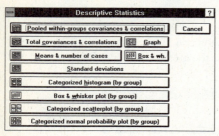

First, look at the means. Click on the *Means and numbers of cases* button and a Scrollsheet with the means and valid *n* for each group and for all groups combined will appear.

Continue...	SEPALLEN	SEPALWID	PETALLEN	PETALWID	Valid N
SETOSA	5.006000	3.428000	1.462000	.246000	50
VERSICOL	5.936000	2.770000	4.260000	1.326000	50
VIRGINIC	6.588000	2.974000	5.552000	2.026000	50
All Grps	5.843333	3.057333	3.758000	1.199333	150

Producing a histogram from a Scrollsheet.

In order to produce a histogram of the frequency distribution for a variable, first click on the desired Scrollsheet column to highlight it. You can either produce the histogram for all groups combined or for selected groups only.

For example, to produce the histogram for variable *Sepalwid* for type *Versicol* only, move the cursor to the second column and the second row of the above Scrollsheet. Then, click with the right-mouse-button to open the flying menu and select the *Quick Stats Graphs* option. Now, select the default graph option - *Histogram/Normal* to produce the following graph.

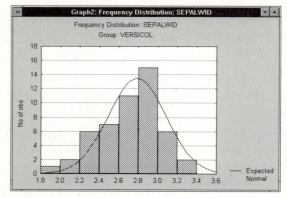

Many other options to graphically view the data are available in the *Descriptive Statistics* dialog. These options are described below.

Box and whisker plot. Click on the *Box & wh.* button (next to the *Means and numbers of cases* button) to produce a box and whisker plot of the independent variables. This plot is useful to summarize the distribution of the variables by three components:

(1) A central line to indicate central tendency or location (i.e., mean or median);

(2) A box to indicate variability around this central tendency (i.e., quartiles, standard errors, or standard deviations);

(3) Whiskers around the box to indicate the range of the variable [i.e., ranges, standard deviations, 1.96 times the standard deviations (95% normal prediction interval for individual observations around the mean), or 1.96 times the standard errors of the means (95% confidence interval)].

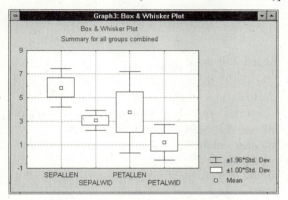

You can also view the distribution of the variables within each level of the grouping variable by clicking on the *Box & whisker plot (by group)* button and selecting the variable *Petallen*. Then, in the following dialog, select the *Mean/SD/1.96*SD* type of box and whisker plot.

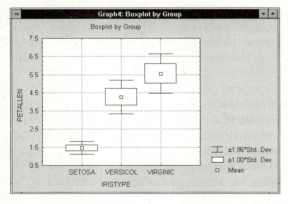

Categorized histograms. You can graphically display histograms of a variable as categorized by a grouping variable when you click on the *Categorized histogram (by group)* button on the *Descriptive Statistics* dialog. When you click on this button, you will be able to select a variable from a list of the previously selected independent variables. For this example, select the variable *Sepalwid*. The histograms as categorized by the grouping variable selected in the startup panel are shown below.

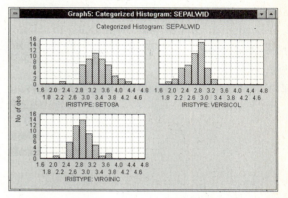

As you can see, this variable is basically normally distributed within each group (type of flower). In some cases, you may want to view one of these histograms in a separate graph using the *Quick Stats Graphs* option from the right-mouse-button flying menu.

Scatterplots. Now return to the *Descriptive Statistics* dialog. Another type of graph of interest would be the scatterplots of correlations between variables included in the analysis. To graphically view the correlations between variables together in a matrix scatterplot, click on the *Graph* button next to the *Total covariances & correlations* button.

2. DISCRIMINANT FUNCTION ANALYSIS - EXAMPLE

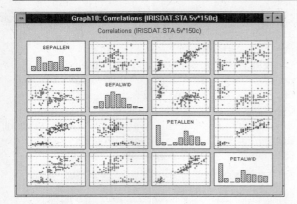

Now, look at the scatterplot for variables *Sepallen* and *Petallen*.

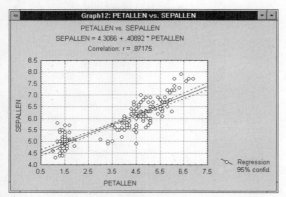

It appears that there are two "clouds" of points in this plot. Perhaps the points in the lower left-hand corner of this plot all belong to one iris type. If so, then there is good "hope" for this discriminant analysis. However, if not, then the possibility that the underlying distribution for these two variables is not bivariate normal, but rather multi-modal with more than one "peak," would have to be considered. To explore this possibility, return to the *Descriptive Statistics* dialog and click on the *Pooled within-groups covariances & correlations* button. In the *Pooled within-groups correlations* Scrollsheet, place the cursor in the cell containing the correlation associated with the variables *Petallen* and *Sepallen*. Click with the right-mouse-button and select the *Quick Stats Graphs - Scatterplot/Conf* option.

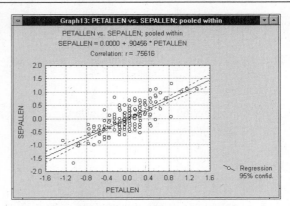

This scatterplot shows the correlation between variables *Sepallen* and *Petallen within* groups. Specifically, it shows the correlations between *deviations* for each variable from the respective group mean. In a sense, the differences between group means from the first scatterplot were "taken out" and now there only appears to be one cloud of points. Thus, it can be concluded that the assumption of a bivariate normal distribution within each group is probably not violated for this particular pair of variables.

Specifying Discriminant Function Analysis

Now, return to the primary goal of the analysis; click *Cancel* in the *Descriptive Statistics* dialog in order to return to the *Model Definition* dialog. A stepwise analysis will be performed in order to see what happens at each step of the discriminant analysis. Click on the *Method* combo box and select the *forward stepwise* option. In this setting, the program will enter variables into the discriminant function model one by one, always choosing the variable that makes the most significant contribution to the discrimination.

2. DISCRIMINANT FUNCTION ANALYSIS - EXAMPLE

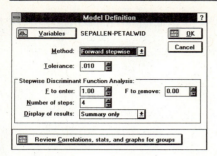

Stop rules. *STATISTICA* will keep "stepping" until one of four things happen. The program will terminate the stepwise procedure when:

(1) All variables have been entered or removed, or

(2) The maximum number of steps has been reached, as specified by the *Number of steps* option, or

(3) No other variable that is not in the model has an *F* value greater than the *F to enter* that is specified in this dialog and when no other variable in the model has an *F* value that is smaller than the *F to remove* specified in this dialog, or

(4) Any variable after the next step would have a tolerance value that is smaller than that specified under the *Tolerance* option.

F to enter/remove. When stepping forward, the program will select the variable for inclusion that makes the most significant unique (additional) contribution to the discrimination between groups; that is, the program will choose the variable with the largest *F* value (greater than the respective user-specified *F to enter* value). When stepping backward, the program will select the variable for exclusion that is least significant, that is, the variable with the smallest *F* value (less than the respective user-specified *F to remove* value). Therefore, if you want to enter all variables in a forward stepwise analysis, set the *F to enter* value as small as possible (and the *F to remove* to *0*).

If you want to remove all variables from a model, one by one, set *F to enter* to a very large value (e.g., *9999.*), and also set *F to remove* to a very large value that is only marginally smaller than the *F to enter* value (e.g., *9998.*). Remember that the *F to enter* value must always be set to a larger value than the *F to remove* value.

Tolerance. The meaning of the *Tolerance* value was introduced in the *Introductory Overview* section (page 3072). In short, at each step the program will compute the multiple correlation (*R-square*) for each variable with all other variables that are currently included in the model. The *tolerance* value of a variable is then computed as *1 minus R-square*. Thus, the *tolerance* value is a measure of the redundancy of a variable.

For example, if a variable that is about to enter into the model has a tolerance value of *.01*, then this variable can be considered to be 99% redundant with the variables already included. At one point, when one or more variables become too redundant, the variance/covariance matrix of variables included in the model can no longer be inverted, and the discriminant function analysis cannot be performed.

It is generally recommended that you leave the *Tolerance* setting at its default value of *0.01*. If a variable is included in the model that is more than 99% redundant with other variables, then its practical contribution to the improvement of the discriminatory power is dubious. More importantly, if you set the tolerance to a much smaller value, round-off errors may result, leading to unstable estimates of parameters.

Starting the analysis. After reviewing the different options on this dialog, you may proceed in the usual manner, that is, do not change any of the default settings for now. However, in order to view the results of the analyses at each step, change the *Display of results* combo box to *At each step*.

2. DISCRIMINANT FUNCTION ANALYSIS - EXAMPLE

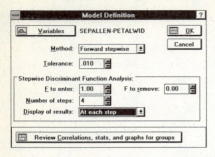

Now, click *OK* to begin the discriminant analysis.

Reviewing the Results of Discriminant Analysis

Results at Step 0

First, the *Results* at step 0 will be displayed. *Step 0* means that no variable has yet been included into the model.

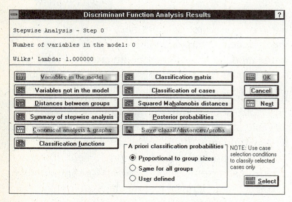

Because no variable has been entered yet, most options on this dialog are not yet available (i.e., they are dimmed). However, you can review the variables *not* in the equation via the *Variables not in the model* button.

DISCRIM. ANALYSIS N=150	Df for all F-tests: 2,147					
	Wilks' Lambda	Partial Lambda	F to enter	p-level	Toler.	1-Toler. (R-Sqr.)
SEPALLEN	.381294	.381294	119.264	0.000000	1.000000	0.00
SEPALWID	.599217	.599217	49.160	.000000	1.000000	0.00
PETALLEN	.058628	.058628	1180.161	0.000000	1.000000	0.00
PETALWID	.071117	.071117	960.007	0.000000	1.000000	0.00

Wilks' *lambda*. In general, Wilks' *lambda* is the standard statistic that is used to denote the statistical significance of the discriminatory power of the current model. Its value will range from *1.0* (no discriminatory power) to *0.0* (perfect discriminatory power). Each value in the first column of the Scrollsheet shown above denotes the Wilks' *lambda after* the respective variable is entered into the model.

Partial Wilks' *lambda*. This is the Wilks' *lambda* for the unique contribution of the respective variable to the discrimination between groups. In a sense, one can look at this value as the equivalent to the partial correlation coefficients reported in *Multiple Regression* (Volume I, Chapter 6). Because a *lambda* of *0.0* denotes perfect discriminatory power, the lower the value in this column, the greater is the unique discriminatory power of the respective variable. Because no variable has been entered into the model yet, the partial Wilks' *lambda* at step 0 is equal to the Wilks' *lambda* after the variable is entered, that is, the values reported in the first column of the Scrollsheet.

***F* to enter and *p*-level.** Wilks' *lambda* can be converted to a standard *F* value (see the *Technical Notes* section, page 3103), and you can compute the corresponding *p*-levels for each *F*. However, as discussed in the *Introductory Overview* section (page 3068), one should generally not take these *p*-levels at face value. One is always capitalizing on chance when including several variables in an analysis without having any *a priori* hypotheses about them, and choosing to interpret only those that happen to be "significant" is not appropriate.

In short, there is a big difference between predicting *a priori* a significant effect for a particular variable and then finding that variable to be significant, as compared to choosing from among 100 variables in the analysis the one that happens to be significant. Without going into details, in purely practical terms, in the latter case, it is not very likely that you would find the same variable to be significant if you were

2. DISCRIMINANT FUNCTION ANALYSIS - EXAMPLE

to replicate the study. When reporting the results of a discriminant function analysis, one should be careful not to leave the impression as if only the significant variables were chosen in the first place (for some theoretical reasons), when, in fact, they were chosen because they happened to "work."

Looking at the Scrollsheet above, you can see that the largest *F to enter* is shown for variable *Petallen*. Thus, that variable will be entered into the model at the next (first) step.

Tolerance and *R-square*. The tolerance value was discussed earlier in this section (refer also to the *Introductory Overview*, page 3072); to reiterate, it is defined as *1 minus R-square* of the respective variable with all other variables in the model, and this value gives an indication of the redundancy of the respective variable. Since no other variables have been chosen yet, all *R-squares* are equal to *1.0*.

Results at Step 2

Now, click on the *Next* button to go to the next step. Step 1 will not be discussed here, so click on the *Next* button again to go to Step 2 (the model with 2 variables). The *Results* dialog will look like this:

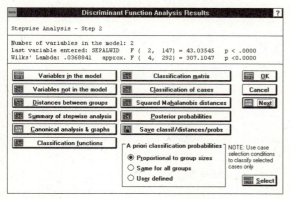

Overall, the discrimination between types of irises is highly significant (Wilks' *lambda = .037*; $F = 307.1$, $p<0.0001$). Now look at the independent contributions to the prediction for each variable in the model.

Variables in the model. Click on the *Variables in the model* button to bring up the Scrollsheet of results for the variables currently in the model. As you can see, both variables are highly significant.

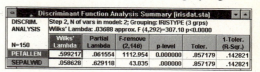

Variables not in the model. Now click on the *Continue* button in the Scrollsheet to return to the *Results* dialog. Click on the *Variables not in the model* button to bring up a Scrollsheet with the same statistics that were reviewed earlier.

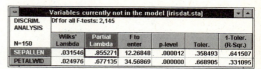

As you can see, both variables that are not yet in the model have *F to enter* values that are larger than *1*; thus, you know that the stepping will continue and that the next variable that will enter into the model is the variable *Petalwid*.

Results at Step 4 (Final Step)

Once again, click on the *Next* button in the *Results* dialog to go to the next step in the analysis.

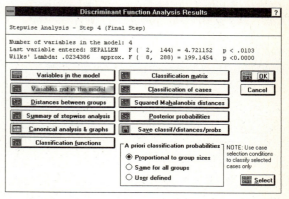

DIS - 3085

2. DISCRIMINANT FUNCTION ANALYSIS - EXAMPLE

Step 3 will not be reviewed here, so click on the *Next* button again to go to the final step in the analysis - Step 4 (shown above). Now, click on the *Variables in the model* button to review the independent contributions for each variable to the overall discrimination between types of irises.

DISCRIM. ANALYSIS N=150	Discriminant Function Analysis Summary (irisdat.sta) Step 4, N of vars in model: 4; Grouping: IRISTYPE (3 grps) Wilks' Lambda: .02344 approx. F (8,288)=199.15 p<0.0000					
	Wilks' Lambda	Partial Lambda	F-remove (2,144)	p-level	Toler.	1-Toler. (R-Sqr.)
PETALLEN	.035025	.669206	35.59018	.000000	.365126	.634874
SEPALWID	.030580	.766480	21.93593	.000000	.608859	.391141
PETALWID	.031546	.743001	24.90433	.000000	.649314	.350686
SEPALLEN	.024976	.938464	4.72115	.010329	.347993	.652007

The partial Wilks' *lambda* indicates that variable *Petallen* contributes most, variable *Petalwid* second most, variable *Sepalwid* third most, and variable *Sepallen* contributes least to the overall discrimination. (Remember that the smaller the Wilks' *lambda*, the greater is the contribution to the overall discrimination.) Thus, you may conclude at this point that the measures of the petals are the major variables that allow you to discriminate between different types of irises. To learn more about the nature of the discrimination, you need to perform a canonical analysis. Thus, click on the *Continue* button to return to the *Results* dialog.

Canonical Analysis

Next, compute the actual discriminant functions to see how the four variables discriminate between the different groups (types of irises). Click on the *Canonical analysis and graphs* button to perform the canonical analysis and open the *Canonical Analysis* dialog.

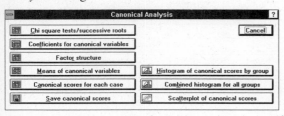

As discussed in the *Introductory Overview* section (page 3069), the program will compute different independent (orthogonal) discriminant functions. Each successive discriminant function will contribute less to the overall discriminatory power. The maximum number of functions that is estimated is either equal to the number of variables or the number of groups minus one, whichever number is smaller. In this case, two discriminant functions will be estimated.

Significance of roots. First, determine whether both discriminant functions (roots) are statistically significant. Click on the *Chi-square test/successive roots* button and the following Scrollsheet will be displayed.

DISCRIM. ANALYSIS	Chi-Square Tests with Successive Roots Removed (irisdat.sta)					
	Eigenvalue	Canonicl R	Wilks' Lambda	Chi-Sqr.	df	p-level
0	32.19193	.984821	.023439	546.1153	8	0.000000
1	.28539	.471197	.777973	36.5297	3	.000000

In general, this Scrollsheet reports a step-down test of all canonical roots. The first line always contains the significance test for all roots; the second line reports the significance of the remaining roots, after removing the first root, and so on. Thus, this Scrollsheet tells you how many canonical roots (discriminant functions) to interpret. In this example, both discriminant (or canonical) functions are statistically significant. Thus, you will have to come up with two separate conclusions (interpretations) of how the measures of sepals and petals allow you to discriminate between iris types.

Discriminant function coefficients. Click on the *Coefficients for canonical variables* button in the *Results* dialog. Two Scrollsheets will be produced, one for the *Raw Coefficients* and one for the *Standardized Coefficients*. Now, look at the Scrollsheet for *Raw Coefficients*.

DISCRIM. ANALYSIS	Raw Coefficients (irisdat.sta) for Canonical Variables	
Variable	Root 1	Root 2
PETALLEN	-2.20121	-.93192
SEPALWID	1.53447	2.16452
PETALWID	-2.81046	2.83919
SEPALLEN	.82938	.02410
Constant	2.10511	-6.66147
Eigenval	32.19193	.28539
Cum.Prop	.99121	1.00000

Raw here means that the coefficients can be used in conjunction with the observed data to compute (raw) discriminant function scores.

The standardized coefficients are the ones that are customarily used for interpretation, because they pertain to the standardized variables and therefore refer to comparable scales.

Standardized Coefficients (irisdat.sta)		
DISCRIM. ANALYSIS	for Canonical Variables	
Variable	Root 1	Root 2
PETALLEN	-.94726	-.401038
SEPALWID	.52124	.735261
PETALWID	-.57516	.581040
SEPALLEN	.42695	.012408
Eigenval	32.19193	.285391
Cum.Prop	.99121	1.000000

The first discriminant function is weighted most heavily by the length and width of petals (variable *Petallen* and *Petalwid*, respectively). The other two variables also contribute to this function. The second function seems to be marked mostly by variables *Sepalwid*, and to a lesser extent by *Petalwid* and *Petallen*.

Eigenvalues. Also shown in the Scrollsheet above are the eigenvalues (roots) for each discriminant function and the cumulative proportion of explained variance accounted for by each function. As you can see, the first function accounts for over 99% of the explained variance; that is, 99% of all discriminatory power is explained by this function. Thus, this first function is clearly the most "important" one.

Factor structure coefficients. These coefficients (which can be viewed via the *Factor structure* button on the *Canonical Analysis Results* dialog) represent the correlations between the variables and the discriminant functions and are commonly used in order to interpret the "meaning" of discriminant functions (see also the discussion on page 3070 in the *Introductory Overview* section).

In educational or psychological research it is sometimes desired to attach meaningful labels to functions (e.g., "extroversion," "achievement motivation"), using the same reasoning as in factor analysis (see *Factor Analysis*, Chapter 6). In those cases, the interpretation of factors should be based on the factor structure coefficients. However, such meaningful labels for these functions will not be considered for this example.

Factor Structure Matrix (irisdat.sta)		
DISCRIM. ANALYSIS	Correlations Variables - Canonical Roots (Pooled-within-groups correlations)	
Variable	Root 1	Root 2
PETALLEN	-.706065	.167701
SEPALWID	.119011	.863681
PETALWID	-.633178	.737242
SEPALLEN	-.222596	.310812

Means of canonical variables. You now know how the variables participate in the discrimination between different types of irises. The next question is to determine the nature of the discrimination for each canonical root. The first step to answer this question is to look at the canonical means. Click on the *Means of canonical variables* button in the *Canonical Analysis Results* dialog.

Means of Canonical Variables (irisdat.sta)		
DISCRIM. ANALYSIS	Root 1	Root 2
SETOSA	7.60760	.215133
VERSICOL	-1.82505	-.727900
VIRGINIC	-5.78255	.512767

Apparently, the first discriminant function discriminates mostly between the type *Setosa* and the other iris types. The canonical mean for *Setosa* is quite different from that of the other groups. The second discriminant function seems to distinguish mostly between type *Versicol* and the other iris types; however, as one would expect based on the review of the eigenvalues earlier, the magnitude of the discrimination is much smaller.

Scatterplot of canonical scores. A quick way of visualizing these results is to produce a scatterplot for the two discriminant functions. Return to the *Canonical Analysis Results* dialog and click on the *Scatterplot of canonical scores* button to plot the unstandardized scores for *Root 1* vs. *Root 2*.

This plot (see below) confirms the interpretation so far. Clearly, the flowers of type *Setosa* are plotted much further to the right in the scatterplot. Thus, the first discriminant function mostly discriminates between that type of iris and the two others. The second function seems to provide *some* discrimination between the flowers of type *Versicol* (which mostly show negative values for the second canonical function) and the others (which have mostly positive values). However, the discrimination is not nearly as clear as that provided by the first canonical function (root).

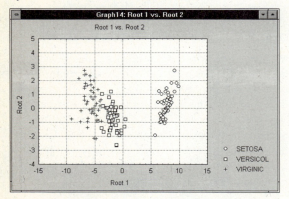

Summary. To summarize the findings so far, it appears that the most significant and clear discrimination is possible for flowers of type *Setosa* by the first discriminant function. This function is marked by negative coefficients for the width and length of petals and positive weights for the width and length of sepals. Thus, the longer and wider the petals, and the shorter and smaller the sepals, the *less* likely it is that the flower is of iris type *Setosa* (remember that in the scatterplot of the canonical functions, the flowers of type *Setosa* were plotted to the right, that is, they were distinguished by *high* values on this function).

Classification

Now, return to the *Discriminant Function Analysis Results* dialog (click on the *Cancel* button in the *Canonical Analysis Results* dialog) and turn to the problem of classification. As discussed in the *Introductory Overview* section (page 3072), one goal of a discriminant function analysis is to enable the researcher to classify cases. Now, see how well the current discriminant functions classify the flowers.

Classification Functions

First look at the classification functions. As described in the *Introductory Overview* section, these are not to be confused with the discriminant functions. Rather, the classification functions are computed for each group and can be used directly to classify cases. You would classify a case into the group for which it has the highest classification score. Click on the *Classification functions* button to see those functions.

Classification Functions; grouping: IRISTYPE (irisdat.sta)			
DISCRIM. ANALYSIS	SETOSA p=.33333	VERSICOL p=.33333	VIRGINIC p=.33333
PETALLEN	-16.4306	5.2115	12.767
SEPALWID	23.5879	7.0725	3.685
PETALWID	-17.3984	6.4342	21.079
SEPALLEN	23.5442	15.6982	12.446
Constant	-86.3085	-72.8526	-104.368

You could use these functions in the *Data Management* module to define the transformations for three new variables. As you would then enter new cases, the program would automatically compute the classification scores for each group.

A priori Probabilities

As discussed in the *Introductory Overview* (page 3072), you can specify different *a priori* probabilities for each group (select *User-defined* in the *a priori classification probabilities* box of the *Results* dialog). These are the probabilities that a case belongs to a respective group, without using any knowledge of the values for the variables in the model. For example, you may know *a priori* that there are more flowers of type *Versicol* in the world, and therefore, the *a priori* probability of a flower to belong to that group is higher than that for any other group. *A priori* probabilities can greatly affect the accuracy of the classification. You can also compute the results for selected cases only (click on the *Select* button). This is particularly useful if you want to validate the discriminant function analysis results with new additional data. However, for now, simply accept the default selection of *Proportional to group sizes*.

Classification Matrix

Now, click on the *Classification matrix* button. In the resulting Scrollsheet (see below), the second line in each column header indicates the *a priori* classification probabilities.

DISCRIM. ANALYSIS	Rows: Observed classifications Columns: Predicted classifications			
Group	Percent Correct	SETOSA p=.33333	VERSICOL p=.33333	VIRGINIC p=.33333
SETOSA	100.0000	50	0	0
VERSICOL	96.0000	0	48	2
VIRGINIC	98.0000	0	1	49
Total	98.0000	50	49	51

Because there were exactly 50 flowers of each type, and you chose those probabilities to be proportional to the sample sizes, the *a priori* probabilities are equal to *1/3* for each group. In the first column of the Scrollsheet, you see the percent of cases that are correctly classified in each group by the current classification functions. The remaining columns show the number of cases that are misclassified in each group, and how they are misclassified.

***A priori* versus *post hoc* classification.** As discussed in the *Introductory Overview* section, when classifying cases from which the discriminant functions were computed, you usually obtain a fairly good discrimination (although usually not as good as in this example). However, one should only look at those classifications as a diagnostic tool for identifying areas of strengths and weaknesses in the current classification functions, because these classifications are not *a priori predictions* but rather *post hoc classifications*. Only if one classifies different (new) cases can one interpret this table in terms of predictive discriminatory power. Thus, it would be unjustified to claim that you can successfully predict the type of iris in 98 percent of all cases, based on only four measurements. Because you capitalized on chance, you could expect much less accuracy if you were to classify new cases (flowers).

Classification of Cases

Mahalanobis distances and *posterior* probabilities. Now, return again to the *Results* dialog. As described in the *Introductory Overview* section, cases are classified into the group to which they are closest. The Mahalanobis distance is a measure of the distance that can be used in the multivariate space defined by the variables in the model. You can compute the distance between each case and the center of each group (i.e., the group *centroid*, defined by the respective group means for each variable). The closer the case is to a group centroid, the more confidence you can have that it belongs to that group. Mahalanobis distances can be computed via the *Squared Mahalanobis distances* button.

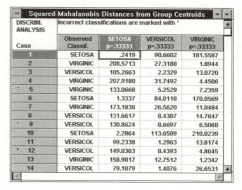

You can also directly compute the probability that a case belongs to a particular group. This is a conditional probability, that is, it is contingent on your knowledge of the values for the variables in the model. Thus, these probabilities are called *posterior* probabilities. You can request those probabilities via the *Posterior probabilities* button. Note that as in the case of the classification matrix, you can

2. DISCRIMINANT FUNCTION ANALYSIS - EXAMPLE

select cases to be classified, and you can specify different *a priori* probabilities.

Actual classifications. Shown below are the actual classifications of cases (flowers; *Classification of cases* button).

The classifications are ordered into a first, second, and third choice. The column under the header *1* contains the first classification choice, that is, the group for which the respective case had the highest *posterior* probability. The rows marked by the asterisk (*) are cases that are misclassified. Again, in this example, the classification accuracy is very high, even considering the fact that these are all *post hoc* classifications. Such accuracy is rarely attained in research in the social sciences.

Summary

This example illustrates the basic ideas of discriminant function analysis. In general, in many cases where there are naturally occurring groups that you would like to be able to discriminate, this technique is appropriate. However, as stated at various points in the preceding discussion, if correct predictive classification is the goal of the research, then at least two studies must be conducted: one in order to build the classification functions and another to validate them.

DIALOGS, OPTIONS, STATISTICS

Startup Panel

When you first open the *Discriminant Analysis* module, the *Stepwise Discriminant Function Analysis* dialog (startup panel) will open.

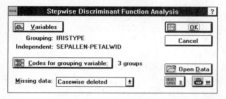

Here, you will be able to specify the variables for the analysis and the type of missing data deletion (see below).

Variables

Choosing this option will bring up a standard variable selection window. *STATISTICA* expects (1) that you specify a grouping variable containing codes (text values) that uniquely identify to which group each case belongs, and (2) that you specify an independent variable list of continuous variables that will later be used in the discriminant function analysis. Note that you can later "deselect" independent variables. Therefore, specify at this point all variables that you might want to use later.

Codes for Grouping Variable

This option will bring up a standard codes selection window in which you can specify the codes that are used in the data file to identify to which group each case belongs.

Missing Data

The *Discriminant Analysis* module allows the user to ignore cases with missing data on any variable (missing data are *casewise deleted*), or to substitute missing data by the overall means for the respective variables (missing data are *substituted by means*). This combo box allows the user to choose between these two methods of dealing with missing data.

Specifying the Analysis

The *Model Definition* dialog allows you to specify the type of analysis and other options.

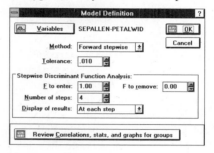

Variables

With this option the user can select or de-select variables for the analysis from among those selected on the startup panel. After clicking on this button the standard variable selection dialog will appear.

Method

This combo box has three settings: standard, forward stepwise, and backward stepwise.

Standard. If the standard method is selected, then all selected variables will be simultaneously entered into the model.

Forward stepwise. If forward stepwise is selected, then *STATISTICA* will move variables into the model in successive steps; at each step the variable with the largest F value (greater than the respective user-specified *F to enter* value) will be chosen for inclusion in the model. The stepping will terminate when no other variable has an *F* value that is greater than that specified under the option *F to enter* (see below).

Backward stepwise. If backward stepwise is selected, *STATISTICA* will first move all selected variables into the equation and then remove variables one by one; at each step, the variable with the smallest *F* value (less than the respective user-specified *F to remove* value) will be removed from the model. The stepping will terminate when no other variable in the model has an *F* value smaller than that specified under the option *F to remove* (see below).

Tolerance Value

The *tolerance* value is discussed in the *Introductory Overview* (page 3072). In short, the tolerance value of a variable is computed as the *1 minus R-square* of that variable with all other variables in the model. Thus, if a variable has a tolerance value that is smaller than the default value of *0.01*, then that variable is more than 99% redundant with the other variables already in the model. If *STATISTICA* detects a variable with a tolerance value smaller than that specified in this option, it will issue an error message. It is not recommended to set this value much smaller than the default value. The inclusion of highly redundant variables in the analysis may later lead to potentially serious round-off errors in the matrix inversion or canonical analysis, due to matrix *ill-conditioning*.

Stepwise Discriminant Function Analysis

This set of options is only available if you select one of the two stepwise methods above.

F to enter/remove. In general, in stepwise discriminant function analysis, variables will be entered into the model if their respective *F* value is larger than the *F to enter* specified in this option; variables will be removed from the model if their respective *F* value is smaller than the *F to remove* specified in this option. Note that the *F to enter* value must always be larger than the *F to remove* value. If, in a *forward stepwise* analysis, you want to enter all variables, set the *F to enter* value to a very small number (e.g., *0.0001*) and the *F to remove* value to *0.0*. If, in a *backward stepwise* analysis, you want to remove all variables from a model, set the *F to enter* value to a very large number (e.g., *9999.*) and the *F to remove* value to a marginally smaller value of equal magnitude (e.g., *9998.*).

Number of steps. The number specified in this option will determine the maximum number of steps that will be performed. Note that this option takes precedence over the *F to enter/remove* values. Stepping will terminate after the maximum number of steps has been reached, regardless of whether or not additional variables would qualify for inclusion in or exclusion from the model based on their *F* values.

Display of results. If this combo box is set to *summary only*, then *STATISTICA* will perform all steps of the stepwise analysis and only display the full *Results* dialog after the last step. However, that dialog will include an option to review the major summary statistics for the stepwise procedure. Specifically, the *Summary of stepwise analysis* option on the *Results* dialog will display for each step in the analysis the respective *F* to enter or remove and associated *p*-level, the number of variables in the model, and the overall Wilks' *lambda* with the associated *F* value and significance level. If this combo box is set to *at each step*, then *STATISTICA* will display the full *Results* dialog at each step (starting with step 0).

Review Correlations, Statistics, and Graphs for Groups

This option will bring up the *Descriptive Statistics* dialog (see below) that will allow the user to review the means, standard deviations, and the variances-/covariances or correlations for all variables. Means and standard deviations can be reviewed for all groups combined or by groups; the correlations can be computed by collapsing across groups, or by pooling the within-groups correlations.

Histograms of the frequency distributions of the variables can also be produced for all groups combined or by groups; scatterplots can be produced for the overall correlations (collapsed across groups) as well as for the pooled within-groups correlations.

Descriptive Statistics

When you click on the *Review correlations, stats, and graphs for groups* button in the *Model Definition* dialog, the *Descriptive Statistics* dialog will open.

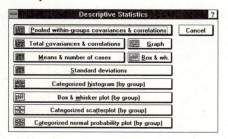

Pooled Within-Groups Covariances & Correlations

This option will first bring up a Scrollsheet with the pooled within-groups covariances for all variables, and another Scrollsheet with the pooled within-groups correlations.

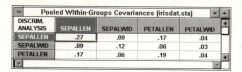

The default graph for these Scrollsheets is the scatterplot, which will be computed to reflect the pooled within-groups correlations.

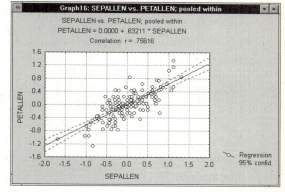

Specifically, the scatterplot will show the individual values in terms of their deviations from the respective group mean. In this plot, the variable on the horizontal axis actually represents deviations of the respective cases from the means of the variable in the respective groups to which the cases belong; in short, the two axes in the plot represent within-groups deviation scores (from the respective means).

Total Covariances and Correlations

This option will first bring up a Scrollsheet with the overall covariances and another Scrollsheet with the overall correlations. These covariances and correlations are computed in the usual manner, that is, collapsed across all groups. The default graph in

2. DISCRIMINANT FUNCTION ANALYSIS - DIALOGS AND OPTIONS

these Scrollsheets is again the scatterplot which is constructed in the usual manner.

Graph (Scatterplot Matrix)

This option will bring up a scatterplot matrix for the independent variables selected for the analysis (see page 3080).

Means and Number of Cases

This option will bring up a Scrollsheet with the means for each variable for each group (see page 3080); the last column of the Scrollsheet will contain the number of cases in each group, and the last row of the Scrollsheet will contain the overall means (collapsed across all groups, i.e., the *weighted means*). The default graph for this Scrollsheet is the histogram, which can be produced for all cases or for selected groups only. In all histograms, the normal distribution will be superimposed over the observed distribution.

Histogram for selected groups. To produce a histogram for a selected group, first move the highlight (click with the mouse) in the Scrollsheet over the desired variable (column) and the desired group (row).

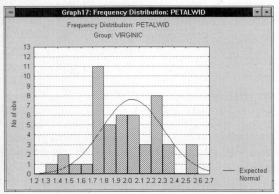

Then click with the right-mouse-button on that cell and select *Quick Stats Graphs* from the flying menu.

In the *Quick Stats Graph* menu, select the *Histogram/Normal* option to produce this graph.

Histogram for all groups combined. To produce a histogram for all groups combined, move the highlight (click with the mouse) in the Scrollsheet over the desired variable (column) and into the last row (labeled *All Groups*). Then click with the right-mouse-button on that cell and select *Quick Stats Graphs* from the flying menu. In the *Quick Stats Graph* menu, select the *Histogram/Normal* option to produce this graph.

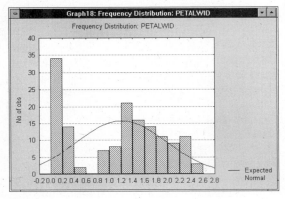

Box and Whisker Plot

With this option the user can produce a box and whisker plot for the selected variables (see page 3080). Note that this is a graph of each selected variable, collapsed across all groups. In order to graphically compare groups, choose the *Box & whisker plot (by group)* option or the *Categorized histogram* option (see below).

Median and quartiles. After selecting this option, and after selecting a variable for the plot, the *Box-Whisker Type* dialog will come up. If you select to compute for the box-whisker plot the median and the quartiles, those values will be computed according to the setting of the *Percentiles* option on the *STATISTICA Defaults: General* dialog (choose option *General* on the *Options* pull-down menu).

For computational details, refer to the description of this dialog in the *General Conventions* chapter in Volume I.

Standard Deviations

This option will bring up a Scrollsheet of the standard deviations. As in the Scrollsheet of means, the total number of cases in each group will be displayed in the last column of the Scrollsheet, and the total standard deviations for each variable will be displayed in the last row of the Scrollsheet. The default graph for this Scrollsheet is also the histogram which works in the same way as described under option *Means & number of cases*, above.

Categorized Histogram (by group)

This option allows the user to produce a categorized histogram for a selected variable.

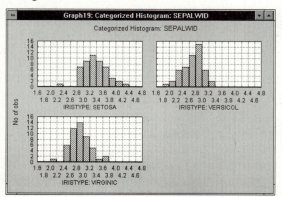

This graph shows the distribution of the variable in the different groups.

Box and Whisker Plot (by group)

This option allows the user to produce a box and whisker plot for a selected variable.

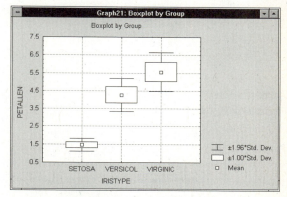

This graph shows the mean or median, and different measures of dispersion for the variable (e.g., standard deviation, quartile range) in the different groups (see page 3080). The specific type of the box and whisker plot can be selected in an intermediate dialog. For computational details concerning the median and the quartiles refer to the description of the *STATISTICA Defaults: General* dialog in the *General Conventions* chapter (Volume I).

Categorized Scatterplot (by group)

This option allows the user to produce a categorized scatterplot for two selected variables, by group.

DIS - 3095

2. DISCRIMINANT FUNCTION ANALYSIS - DIALOGS AND OPTIONS

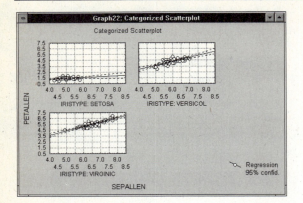

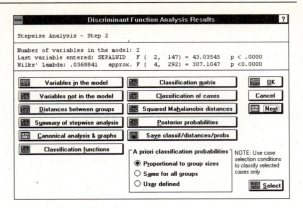

Categorized Probability Plot (by group)

This option allows the user to produce a categorized probability plot for a selected variable, by group.

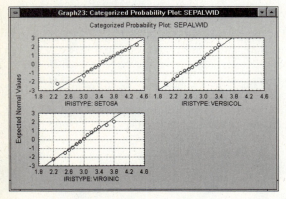

The normal probability plot provides a quick check of whether or not the respective variable is normally distributed in the groups. If so, the points in the plot will approximate a straight line.

Results and Classification

The *Discriminant Function Analysis Results* dialog offers the following options:

Variables in the Model

This option will first bring up a Scrollsheet with the following summary statistics for all variables currently in the model.

Wilks' *lambda*. This is the Wilks' *lambda* for the overall model that will result after removing the respective variable. Remember that Wilks' *lambda* can assume values in the range of *0* (perfect discrimination) to *1* (no discrimination).

Partial *lambda*. This is the Wilks' *lambda* associated with the unique contribution of the respective variable to the discriminatory power of the model.

F to remove. This is the *F* value associated with the respective partial Wilks' *lambda*.

p-level. This is the *p*-level associated with the respective *F to remove*. Note that in stepwise analyses, these *p*-levels should be interpreted with caution (see page 3069).

Tolerance. The tolerance value of a variable is computed as *1-R square* of the respective variable with all other variables in the model (in this Scrollsheet). Thus, the tolerance is a measure of the respective variable's redundancy. For example, a tolerance value of *.10* means that the variable is 90% redundant with the other variables in the model.

DIS - 3096

Copyright © StatSoft, 1995

1-Tolerance (*R-square*). This is the *R-square* value of the respective variable with all other variables in the model (in this Scrollsheet).

Variables not in the Model

This option will bring up a Scrollsheet with the summary statistics for all variables that are currently not in the model. The statistics displayed in this Scrollsheet are essentially identical to those described above (see *Variables in the Model*, above for a description of these options).

Distances Between Groups

This option will bring up a Scrollsheet with the squared Mahalanobis distances between the group centroids. The Mahalanobis distance is similar to the standard Euclidean distance measure, except that it takes into account the correlations between variables. The larger the differences in this Scrollsheet, the farther are the respective groups apart from each other, and the more discriminatory power does our current model possess for discriminating between the respective two groups.

Together with the Scrollsheet of Mahalanobis distances, two other Scrollsheets will be displayed: one with the *F* values associated with the respective distances and another with the respective *p*-levels. Again, those *p*-levels should be interpreted with caution, unless one brings to the analysis strong *a priori* hypotheses concerning which pairs of groups should show particularly large (and significant) distances.

Summary of Stepwise Analysis

This option is only available if a stepwise discriminant function analysis was performed. After clicking on this button, a Scrollsheet will come up which summarizes for each step:

(1) Which variable was entered or removed (indicated by the letters *E* or *R* in parentheses after the respective variable name in the first column of the Scrollsheet);

(2) The step number;

(3) The *F* to enter or remove;

(4) The respective degrees of freedom for that *F* value;

(5) The *p*-level for that *F* value;

(6) The number of variables in the model after the respective step;

(7) The overall value of Wilks' *lambda* after the respective step;

(8) The *F* value associated with the *lambda*;

(9) The degrees of freedom for that *F* value;

(10) The *p*-level for that *F* value.

Canonical Analysis

A canonical analysis can only be performed if at least three groups were specified for the analysis and if there are at least two variables in the model. In that case, *STATISTICA* will perform a complete canonical analysis, and compute the discriminant functions. These and related values can be reviewed from a separate results dialog (*Canonical Analysis Results*, see page 3100).

Classification Functions

This option will bring up the classification functions. Classification functions are computed for each group and can be used directly to classify cases. A case would be classified into the group for which it has the highest classification score. You could use these functions in the *Data Management* module to define the transformations for three new variables (one for each group). As you would then enter new cases, *STATISTICA* would automatically compute the classification scores for each group.

Classification Matrix

The classification matrix contains information about the number and percent of correctly classified cases in each group. Note that you can use the standard case selection conditions (click on the *Sel* field on the status bar) to classify only cases that were not used to compute the current classification functions. Also, the computations for the classification of cases will be based on *a priori* classification probabilities that are either:

(1) The same for all groups,

(2) Proportional to the respective group sizes, or

(3) User-defined (see *a priori Classification Probabilities*, below).

If *User-defined a priori* probabilities are selected, then, before the computations begin, you will be prompted to specify the *a priori* classification probabilities for each group (see page 3101).

Classification of Cases

After clicking on this button a Scrollsheet will be displayed with the classification for each selected case. The classifications are ordered into a first, second, and third choice. The column under the header *1* contains the first classification choice, that is, the group for which the respective case had the highest *posterior* probability (see option *Posterior Probabilities* below). The rows marked by the asterisk (*) are cases that are misclassified.

Squared Mahalanobis Distances

After clicking on this option a Scrollsheet with the squared Mahalanobis distances of each case from each group centroid will be displayed. These distances are similar to the squared Euclidean distances of the respective case from the centroids for each group (the point defined by the means for all variables in the respective group). However, unlike the Euclidean distance, the Mahalanobis distance takes into account the intercorrelations between the variables in the model (which define the multivariate space). A case will generally be classified into the group that it is closest to, unless widely disparate *a priori* probabilities lead to very different *posterior* classification probabilities. Asterisks (*) in the first column of the Scrollsheet will mark misclassified cases. The default custom graph (from the right-mouse-button flying menu) for this Scrollsheet is the icon plot.

This plot allows you to quickly compare the Mahalanobis distances of each case from the different groups.

Posterior Probabilities

Given the Mahalanobis distances of a case from the different group centroids, you can compute the respective *posterior* classification probabilities for each group. In general, the further away a case is from a group centroid, the less likely it is that the case belongs to that group. The *posterior* probabilities are determined by the Mahalanobis distances and the *a priori* probabilities (see below). A case will be classified into the group for which it has the highest *posterior* classification probability. Misclassified cases will be marked in the Scrollsheet by asterisks.

The default custom graph (from the right-mouse-button flying menu) for this Scrollsheet is the icon plot.

```
Graph27: From: Posterior Probabilities (syrup.sta)
        From: Posterior Probabilities (syrup.sta)
        Incorrect classifications are marked with *

    *  1    2    3    4  * 5    6    7    8  * 9
      10   11   12   13   14   15   16 * 17   18
      19 * 20   21   22   23   24   25   26 * 27
    * 28   29 * 30   31   32   33 * 34   35   36
    * 37   38   39   40   41   42   43   44 * 45
    * 46   47   48 * 49   50   51   52 * 53   54
LEGEND (clockwise):  Low    p=.33333, Medium   p=.33333, High   p=.33333,
```

This plot allows you to quickly compare the *posterior* probabilities for each case, for each group.

Save Classifications, Distances, Probabilities

After clicking on this button, a pop-up dialog will be displayed that allows you to save the actual classifications for each case, the squared Mahalanobis distances, or the *posterior* probabilities.

A priori Classification Probabilities

There are three choices available here: *proportional to group sizes*, *same for all groups*, and *user-*

defined. The *a priori* probabilities specify how likely it is, without using any prior knowledge of the values for the variables in the model, that a case will fall into one of the groups. For example, in an educational study of high school drop-outs, it may happen that overall there are fewer drop-outs than students who stay (i.e., there are different *base rates*); thus, the *a priori* probability that a student drops out is lower than that a student remains in school. The *a priori* probabilities can greatly affect the accuracy of the classification. If differential base rates are not of interest for the study, or if one knows that there are about an equal number of cases in each group, then one could set the *a priori* probabilities to be the *same for all groups*. If the differential base rates are reflected in the sample sizes (as they would be, if the sample is a probability sample), then set the *a priori* probabilities to be *proportional to group sizes*.

Finally, if you have specific knowledge about the base rates (for example, based on previous research), then set the *a priori* probabilities to *user-defined*. In that case, after you subsequently request classification of cases, a dialog window will appear allowing you to specify the *a priori* probabilities for each group (see page 3101). If those probabilities do not add up to *1*, STATISTICA will automatically adjust them proportionately.

Next

This button will only appear if you are in the process of performing a stepwise discriminant analysis and requested to display results *At each step* (on the *Model Definition* dialog, page 3091). Clicking the *Next* button will bring up the results for the next step in the stepwise analysis.

Canonical Analysis

When you click on the *Canonical Analysis* button in the *Discriminant Function Analysis Results* dialog, the *Canonical Analysis* dialog will open.

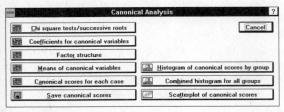

Chi-square Tests of Successive Roots

This option will first bring up a Scrollsheet with a step-down test for canonical roots (or discriminant functions). The first row in that Scrollsheet contains the test of significance for all roots combined. The second row contains the significance of the remaining roots, after removing the first root, and so on. Thus, this Scrollsheet allows you to evaluate how many significant roots to interpret.

Coefficients for Canonical Variables

Raw discriminant function coefficients. This option will first bring up a Scrollsheet with the raw discriminant (canonical) function coefficients. These are the coefficients that can be used to compute the raw canonical scores for each case for each discriminant function. Also included in this Scrollsheet will be the eigenvalues for each discriminant function and the cumulative proportion of (common) variance extracted by each discriminant function.

Standardized discriminant function coefficients. A second Scrollsheet that will be displayed reports the standardized discriminant function coefficients. These are the coefficients that are based on standardized variables, and thus, these coefficients pertain to equal scales of measurement. Therefore, these coefficients may be compared in order to determine the magnitudes and directions of the (unique) contributions of the variables to each canonical function.

Factor Structure

This option will bring up a Scrollsheet with the pooled within-groups correlations of variables with the respective discriminant (canonical) functions. If you are familiar with factor analysis (see *Factor Analysis*, Chapter 6), you may think of these correlations as the factor loadings of the respective variables on the discriminant functions.

Some authors have argued that to interpret the "meaning" of the discriminant functions, one should use these structure coefficients rather than the standardized discriminant function coefficients. Refer to the *Introductory Overview* section (page 3070) for a discussion of this argument. The most important thing to remember is that the discriminant function *coefficients* denote the unique (partial) contribution of each variable to the discriminant functions, while the structure coefficients denote the simple correlations between the variables and the functions; therefore, the structure coefficients are usually more appropriate for substantive interpretations of functions.

Means for Canonical Variables

This option will bring up the means for the discriminant functions. These means allow one to determine the groups that are best identified (discriminated) by each discriminant function.

2. **DISCRIMINANT FUNCTION ANALYSIS** - DIALOGS AND OPTIONS

Canonical Scores for Each Case

This option will allow the user to review the discriminant function scores for each case.

Save Canonical Scores

This option allows the user to save the canonical scores in a *STATISTICA* data file.

Histogram of Canonical Scores for Selected Group

This option allows the user to produce a histogram of the frequency distribution of the canonical scores for selected groups.

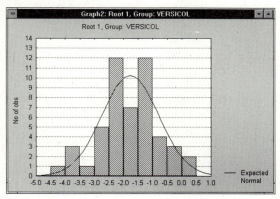

Combined Histogram for all Groups

This option allows the user to produce a histogram of the frequency distribution of the canonical scores for all groups combined.

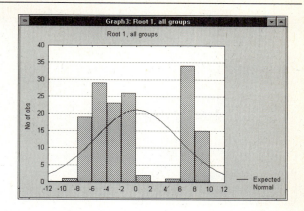

Scatterplot of Canonical Scores

This option allows the user to produce a scatterplot of the canonical scores for pairs of discriminant functions (canonical roots). This plot is very useful for determining how each discriminant function contributes to the discrimination between groups.

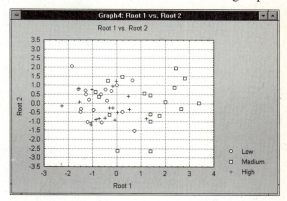

This option is not available if only one canonical (discriminant) function was extracted.

A *priori* Probabilities

Specify the *a priori* classification probabilities for each group in this dialog. You can enter a *common value* for the probabilities that will be used in each group (e.g., *.333*) and then click on the *Apply* button

DIS - 3101

Copyright © StatSoft, 1995

to apply that value to the individual groups, or you can specify individual probabilities for each group by typing in the values in the appropriate edit fields. The probabilities should add up to *1.0*; if they do not, then *STATISTICA* will proportionately adjust them accordingly.

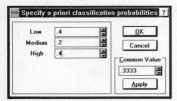

The *a priori* probabilities specify how likely it is, without using any prior knowledge of the values for the variables in the model, that a case will fall into one of the groups. For example, in an educational study of high school drop-outs, it may happen that overall there are fewer drop-outs than students who stay (i.e., there are different *base rates*); thus, the *a priori* probability that a student drops out is lower than *a priori* probability that a student remains in school. The *a priori* probabilities can greatly affect the accuracy of the classification. If differential base rates are not of interest for the study, or if one knows that there are about an equal number of cases in each group, then one could set the *a priori* probabilities to be the *same for all groups*. If the differential base rates are reflected in the sample sizes (as they would be if the sample is a probability sample), then set the *a priori* probabilities to be *proportional to group sizes*. Finally, if you have specific knowledge about the base rates (for example, based on previous research), then set the *a priori* probabilities to *user-defined*.

TECHNICAL NOTES

Computational Method

The Discriminant Function Analysis module uses the standard formulas for computing the discriminant functions, performing the canonical analysis, and for computing the classification statistics. A step-by-step discussion of the computations can be found in Jennrich (1977), who also describes the computation of the group classification function coefficients and the posterior classification probabilities.

Formulas

The formulas for computing the Mahalanobis distances and classification of cases, given a-priori probabilities, are described in Lindeman, Merenda, and Gold (1980, Chapter 6). Detailed description of computational formulas and examples are also presented in, for example, Cooley and Lohnes (1971, Chapters 9 and 10) and Pedhazur (1973, Chapters 17 and 18).

Wilks' *lambda*. The Wilks' *lambda* statistic for the overall discrimination is computed as the ratio of the determinant (*det*) of the within-groups variance-/covariance matrix over the determinant of the total variance covariance matrix:

`Wilks' λ = det(W)/det(T)`

The *F* approximation to Wilks' *lambda* is computed following Rao (1951).

Partial *lambda*. The partial *lambda* is computed as the multiplicative increment in *lambda* that results from adding the respective variable:

`partial λ = λ(after)/λ(before)`

Put another way, the partial *lambda* is the ratio of Wilks' *lambda* after adding the respective variable over the Wilks' *lambda* before adding the variable.

The corresponding *F* statistic (see Rao, 1965, page 470) is computed as:

`F=[(n-q-p)/(q-1)]*[(1-partial λ)/partial λ]`

where

`n`	is the number of cases
`q`	is the number of groups
`p`	is the number of variables
`partial l`	is the partial *lambda*

DIS - 3103

INDEX

A

A priori
    classification probabilities, 3074, 3088, 3099, 3101
    predictions, 3073
    versus post hoc classification, 3089
Analysis of variance, 3067
Assumptions
    correlations between means and variances, 3072
    homogeneity of variances/covariances, 3071
    matrix ill-conditioning, 3072, 3092
    normality, 3071
    tolerance values, 3072

B

Box and whisker plots
    by group, 3095
    for all variables, 3080, 3094

C

Canonical analysis, 3069, 3086, 3097, 3100
Canonical scores for each case, 3101
Centroids, 3074, 3089
Chi-square tests
    successive roots, 3086, 3100
Classification
    a priori classification probabilities, 3074, 3088, 3099, 3101
    a priori versus post hoc classification, 3073, 3089
    classification functions, 3073, 3088, 3097
    classification matrix, 3089, 3098
    classification of cases, 3073, 3089, 3098
    classification scores, 3073
    Mahalanobis distances, 3074, 3089

Classification (continued)
    posterior classification probabilities, 3074, 3089, 3098
Classification functions, 3073, 3088, 3097
Classification of cases, 3073, 3089, 3098
Classification scores, 3073
Coefficients for canonical variables, 3100
Correlations between means and variances, 3072

D

Discriminant analysis
    assumptions, 3071
    backward stepwise analysis, 3068
    canonical analysis, 3069, 3086, 3097
    canonical scores for each case, 3101
    centroids, 3074, 3089
    classification, 3072, 3088
    coefficients for canonical variables, 3100
    computational approach, 3067
    descriptive statistics, 3080
    discriminant functions for multiple groups, 3069
    eigenvalues, 3087
    example, 3079
    F to enter, F to remove, 3068, 3083, 3091
    factor structure coefficients, 3070, 3087, 3100
    Fisher linear discriminant analysis, 3069
    forward stepwise analysis, 3068
    group centroids, 3074, 3089
    homogeneity of variances/covariances, 3071
    Mahalanobis distances, 3074, 3089
    means of canonical variables, 3087, 3100
    multiple variables, 3068
    overview, 3067
    partial Wilks' lambda, 3084, 3103
    raw discriminant function coefficients, 3100
    reviewing results, 3084
    R-square, 3085, 3097

Discriminant analysis (continued)
    specifying the analysis, 3079
    specifying the discriminant analysis, 3082
    squared Mahalanobis distances, 3089, 3097, 3098
    standardized discriminant function coefficients, 3100
    startup panel, 3091
    stepwise discriminant analysis, 3068
    stop rules, 3083
    tolerance values, 3083, 3085, 3092, 3096
    two-group discriminant function, 3069
    Wilks' lambda, 3084, 3103
Discriminant function coefficients, 3086
Discriminant functions for multiple groups, 3069

E

Eigenvalues, 3087

F

F to enter, F to remove, 3068, 3083, 3091
Factor structure coefficients, 3070, 3087, 3100
Fisher linear discriminant analysis, 3069

G

Group centroids, 3074, 3089

H

Histograms
    2D histograms, 3080
    categorized histograms, 3081, 3095
    combined histogram for all groups, 3094, 3101
    histogram for selected groups, 3094
    histogram of canonical scores for selected group, 3101
Homogeneity of variances/covariances, 3071

2. DISCRIMINANT FUNCTION ANALYSIS - INDEX

I

Icon plot, 3098
Interpreting the discriminant functions, 3070

M

Mahalanobis distances, 3074, 3089
Matrix ill-conditioning, 3072, 3092
Means of canonical variables, 3087, 3100
Missing data deletion, 3080

N

Normal distribution, 3071

O

One-way analysis of variance, 3067

P

Partial Wilks' lambda, 3084, 3096, 3103
Pooled within-groups covariances and correlations, 3093
Post hoc predictions, 3073
Posterior classification probabilities, 3074, 3089, 3098
Probability plots
    categorized normal probability plot, 3096

R

Raw discriminant function coefficients, 3100
R-square, 3085, 3097

S

Scatterplots
    categorized scatterplot, 3095
    matrix scatterplots, 3081, 3094
    scatterplot of canonical scores, 3087, 3101
Select cases for classification, 3078
Significance of discriminant functions, 3071
Significance of roots, 3086

Squared Mahalanobis distances, 3089, 3097, 3098
Standardized discriminant function coefficients, 3100
Stepwise discriminant analysis
    backward stepwise analysis, 3068, 3092
    capitalizing on chance, 3069
    F to enter, F to remove, 3068, 3091, 3092
    forward stepwise analysis, 3068, 3091
Stop rules, 3083

T

Tolerance values, 3072, 3083, 3085, 3092, 3096
Two-group discriminant function, 3069

W

Wilks' lambda, 3084, 3096, 3103

DIS - 3106

Copyright © StatSoft, 1995

Chapter 3:
RELIABILITY AND ITEM ANALYSIS

Table of Contents

Introductory Overview	3111
Program Overview	3117
Examples	3119
Example 1: Evaluating the Reliability of Items in a Questionnaire	3119
Example 2: Split-Half Reliability	3123
Dialogs, Options, Statistics	3125
Startup Panel	3125
Review Descriptive Statistics	3126
Results	3127
Split-Half Reliability	3130
Technical Notes	3131
Index	3133

Detailed Table of Contents

INTRODUCTORY OVERVIEW	**3111**
General Purpose	3111
Basic Ideas	3111
Classical Testing Model	3112
Reliability	3112
Sum Scales	3112
Cronbach's *Alpha*	3113
Split-Half Reliability	3113
Correction for Attenuation	3113
Designing a Reliable Scale	3114
Final Remarks	3115
PROGRAM OVERVIEW	**3117**
Methods	3117
Data Files	3117
Results	3117
Alternative Procedures	3117

3. RELIABILITY AND ITEM ANALYSIS - CONTENTS

EXAMPLES 3119
- Example 1: Evaluating the Reliability of Items in a Questionnaire 3119
  - Overview and Data File 3119
  - Starting the Analysis 3119
  - Correlation Matrix 3119
  - Means and Standard Deviations 3120
  - Reviewing Results 3121
- Example 2: Split-Half Reliability 3123

DIALOGS, OPTIONS, STATISTICS 3125
- Startup Panel 3125
  - Variables 3125
  - Split-Half Reliability 3125
  - Input File 3125
  - Missing Data Deletion 3125
  - Correlation Matrix 3125
  - Codes for Dichotomized Variables 3126
  - Compute Multiple Regression of the Items with the Scale 3126
  - Batch Processing/Printing 3126
- Review Descriptive Statistics 3126
  - Means and Standard 3126
  - Deviations 3126
  - Correlations 3127
  - Covariances 3127
  - Save Correlation Matrix 3127
  - Box and Whisker Plot 3127
  - Matrix Plot of Correlations 3127
  - The SD Check Box 3127
- Results 3127
  - Item-Total Statistics 3128
  - Split-Half Reliability 3128
  - Analysis of Variance 3128
  - Correlation Matrix 3128
  - Matrix Plot 3128
  - Means and Standard Deviations 3128
  - Box and Whisker Plot 3129
  - Attenuation Correction 3129
  - *What If More Items* Option 3129
  - *How Many More Items* option 3129
- Split-Half Reliability 3130
  - Summary 3130

TECHNICAL NOTES 3131
- Computations 3131
- Standardized *Alpha* 3131

REL - 3108

Copyright © StatSoft, 1995

3. RELIABILITY AND ITEM ANALYSIS - CONTENTS

Guttman Split-Half Reliability 3131
Items with Zero Variances 3131

INDEX **3133**

3. RELIABILITY AND ITEM ANALYSIS - CONTENTS

Chapter 3:
RELIABILITY AND ITEM ANALYSIS

INTRODUCTORY OVERVIEW

General Purpose

In many areas of research, the precise measurement of hypothesized processes or variables (theoretical *constructs*) poses a challenge by itself. For example, in psychology, the precise measurement of personality variables or attitudes is usually a necessary first step before any theories of personality or attitudes can be considered. In general, in all social sciences, unreliable measurements of people's beliefs or intentions will obviously hamper efforts to predict their behavior. The issue of precision of measurement will also come up in applied research, whenever variables are difficult to observe. For example, reliable measurement of employee performance is usually a difficult task; yet, it is obviously a necessary precursor to any performance-based compensation system.

In all of these cases, the *Reliability and Item Analysis* module may by used to construct reliable measurement scales, to improve existing scales, and to evaluate the reliability of scales already in use. Specifically, the *Reliability and Item Analysis* module will aid in the design and evaluation of *sum scales*, that is, scales that are made up of multiple individual measurements (e.g., different items, repeated measurements, different measurement devices, etc.). The program will compute numerous statistics that allow the user to build and evaluate scales following the so-called *classical testing theory* model.

The assessment of scale reliability is based on the correlations between the individual items or measurements that make up the scale, relative to the variances of the items. If you are not familiar with the *correlation coefficient* or the *variance* statistic, it is recommended that you review the respective discussions provided in the *Introductory Overview* section of *Basic Statistics and Tables* (Volume I).

The classical testing theory model of scale construction has a long history, and there are many textbooks available on the subject. For additional detailed discussions, you may refer to, for example, Carmines and Zeller (1980), De Gruitjer and Van Der Kamp (1976), Kline (1979, 1986), or Thorndyke and Hagen (1977). A widely acclaimed "classic" in this area, with an emphasis on psychological and educational testing, is Nunally (1970). A short summary of the standard techniques and statistics that are commonly used will be presented in the following paragraphs.

Testing hypotheses about relationships between items and tests. Note that *STATISTICA* includes the general linear structural equation modeling procedure *SEPATH* (see Chapter 11). In that module you can test specific hypotheses about the relationship between sets of items or different tests (e.g., test whether two sets of items measure the same construct, analyze multi-trait multi-method matrices, etc.). For additional details and illustrations, refer to the *Examples* section of the *SEPATH* chapter (Chapter 11; see also Bollen, 1979; Jöreskog and Sörbom, 1979).

Basic Ideas

Suppose you want to construct a questionnaire to measure people's prejudices against foreign-made cars. You could start out by generating a number of items such as: "Foreign cars lack personality," "Foreign cars all look the same," etc. You could then submit those questionnaire items to a group of subjects (for example, people who have never owned a foreign-made car). You could ask the

3. RELIABILITY AND ITEM ANALYSIS - INTRODUCTORY OVERVIEW

subjects to indicate their agreement with these statements on 9-point scales, anchored at *1=disagree* and *9=agree*.

True scores and error. Consider more closely what is meant by precise measurement in this case. You hypothesize that there is such a thing (theoretical construct) as "prejudice against foreign cars," and that each item "taps into" this concept to some extent. Therefore, you may say that a subject's response to a particular item reflects two aspects: first, the response reflects the prejudice against foreign cars, and second, it will reflect some esoteric aspect of the respective question.

For example, consider the item "Foreign cars all look the same." A subject's agreement or disagreement with that statement will partially depend on his or her general prejudices, and partially on some other aspects of the question or person. For example, the subject may have a friend who just bought a very different-looking foreign car.

Classical Testing Model

To summarize, each measurement (response to an item) reflects, to some extent the true score for the intended concept (prejudice against foreign cars), and to some extent esoteric, random error. This can be expressed in an equation as:

`X = tau + error`

In this equation, *X* refers to the respective actual measurement, that is, the subject's response to a particular item; *tau* is commonly used to refer to the *true score*, and *error* refers to the random error component in the measurement.

Reliability

In this context the definition of *reliability* is straightforward: A measurement is reliable if it reflects mostly true score, relative to the error. For example, an item such as "Red foreign cars are particularly ugly" would likely provide an unreliable measurement of prejudices against foreign-made cars. This is because there probably are ample individual differences concerning the likes and dislikes of colors. Thus, this item would "capture" not only a person's prejudice but also his or her color preference. Therefore, the proportion of true score (for prejudice) in the subjects' response to that item would be relatively small.

Measures of reliability. From the above discussion, one can easily infer a measure or statistic to describe the reliability of an item or scale. Specifically, an *index of reliability* may be defined in terms of the proportion of true score variability that is captured across subjects or respondents, relative to the total observed variability. In equation form, you can say:

`Reliability`$= \sigma^2_{\text{(true score)}} / \sigma^2_{\text{(total observed)}}$

Sum Scales

What will happen when you sum up several more-or-less reliable items designed to measure prejudice against foreign-made cars? Suppose the items were written so as to cover a wide range of possible prejudices against foreign-made cars. If the error component in subjects' responses to each question is truly random, then you may expect that the different components will cancel each other out across items. In slightly more technical terms, the expected value or mean of the error component across items will be zero. The true score component remains the same when summing across items. Therefore, the more items are added, the more true score (relative to the error score) will be reflected in the sum scale.

Number of items and reliability. This conclusion describes a basic principle of test design. Namely, the more items there are in a scale designed to measure a particular concept, the more reliable will the measurement (sum scale) be. Perhaps a somewhat more practical example will further clarify this point. Suppose you want to measure the height

REL - 3112

Copyright © StatSoft, 1995

of 10 persons, using only a crude stick as the measurement device. Note that you are not interested in this example in the absolute correctness of measurement (i.e., in inches or centimeters), but rather in the ability to distinguish reliably between the 10 individuals in terms of their height. If you measure each person only once in terms of multiples of lengths of your crude measurement stick, the resultant measurement may not be very reliable. However, if you measure each person 100 times, and then take the average of those 100 measurements as the summary of the respective person's height, then you will be able to make very precise and reliable distinctions between people (based solely on the crude measurement stick).

Now look at some of the common statistics that are used to estimate the reliability of a sum scale.

Cronbach's *Alpha*

To return to the prejudice example, if there are several subjects who respond to the items, then you can compute the variance for each item and the variance for the sum scale. The variance of the sum scale will be smaller than the sum of item variances if the items measure the *same* variability between subjects, that is, if they measure some true score. Technically, the variance of the sum of two items is equal to the sum of the two variances *minus* (two times) the covariance, that is, the amount of true score variance common to the two items.

You can estimate the proportion of true score variance that is captured by the items by comparing the sum of item variances with the variance of the sum scale. Specifically, you can compute:

$$\alpha = [k/(k-1)] * [1 - \Sigma(s_i^2)/s_{sum}^2]$$

This is the formula for the most common index of reliability, namely, Cronbach's coefficient *alpha* (α). In this formula, the s_i^2's denote the variances for the k individual items; s_{sum}^2 denotes the variance for the sum of all items. If there is no true score but only error in the items (which is esoteric and unique, and, therefore, uncorrelated across subjects), then the variance of the sum will be the same as the sum of variances of the individual items. Therefore, coefficient *alpha* will be equal to zero. If all items are perfectly reliable and measure the same thing (true score), then coefficient *alpha* is equal to *1*. (Specifically, $1 - \Sigma(s_i^2)/s_{sum}^2$ will become equal to $(k-1)/k$; if you multiply this by $k/(k-1)$, you obtain 1.)

Alternative terminology. Cronbach's *alpha*, when computed for binary (e.g., true/false) items, is identical to the so-called *Kuder-Richardson-20* formula of reliability for sum scales. In either case, because the reliability is actually estimated from the consistency of all items in the sum scales, the reliability coefficient computed in this manner is also referred to as the *internal-consistency reliability*.

Split-Half Reliability

An alternative way of computing the reliability of a sum scale is to divide it in some random manner into two halves. If the sum scale is perfectly reliable, then it would be expected that the two halves are perfectly correlated (i.e., *r = 1.0*). Less than perfect reliability will lead to less than perfect correlations. You can estimate the reliability of the sum scale via the *Spearman-Brown split-half* coefficient:

$$r_{sb} = 2r_{xy}/(1 + r_{xy})$$

In this formula, r_{sb} is the split-half reliability coefficient, and r_{xy} represents the correlation between the two halves of the scale.

Correction for Attenuation

Now consider some of the consequences of less than perfect reliability. Suppose you use your scale of prejudice against foreign-made cars to predict some other criterion, such as subsequent actual purchase of a car. If your scale correlates with such a criterion, it would raise your confidence in the

validity of the scale, that is, that it really measures prejudices against foreign-made cars and not something completely different. In actual test design, the *validation* of a scale is a lengthy process that requires the researcher to correlate the scale with various external criteria that, in theory, should be related to the concept that is supposedly being measured by the scale.

How will validity be affected by less than perfect scale reliability? The random error portion of the scale is unlikely to correlate with some external criterion. Therefore, if the proportion of true score in a scale is only 60% (that is, the reliability is only .60), then the correlation between the scale and the criterion variable will be *attenuated*; that is, it will be smaller than the actual correlation of true scores. In fact, the validity of a scale is always limited by its reliability.

Given the reliability of the two measures in a correlation (i.e., the scale and the criterion variable), you can estimate the actual correlation of true scores in both measures. Put another way, you can *correct* the correlation for *attenuation*:

$$r_{xy,corrected} = r_{xy}/\sqrt{(r_{xx} * r_{yy})}$$

In this formula, $r_{xy,corrected}$ stands for the corrected correlation coefficient; that is, it is the estimate of the correlation between the true scores in the two measures x and y. The term r_{xy} denotes the uncorrected correlation, and r_{xx} and r_{yy} denote the reliability of measures (scales) x and y. The *Reliability* module provides an option to compute the attenuation correction based on user-specified values or based on actual raw data (in which case the reliabilities of the two measures are estimated from the data).

Designing a Reliable Scale

After the discussion so far, it should be clear that the more reliable a scale, the better (e.g., more valid) the scale. As mentioned earlier, one way to make a sum scale more valid is by adding items. The *Reliability and Item Analysis* module includes options that allow the user to compute how many items would have to be added in order to achieve a particular reliability, or how reliable the scale would be if a certain number of items were added. However, in practice, the number of items on a questionnaire is usually limited by various other factors (e.g., respondents get tired, overall space is limited, etc.). Now, returning to the prejudice example, the steps that one would generally follow in order to design the scale so that it will be reliable are outlined below:

Step 1: Generating items. The first step is to write the items. This is essentially a creative process where the researcher makes up as many items as possible that seem to relate to prejudices against foreign-made cars. In theory, one should "sample items" from the domain defined by the concept. In practice, for example in marketing research, *focus groups* are often utilized to illuminate as many aspects of the concept as possible. For example, you could ask a small group of highly committed American car buyers to express their general thoughts and feelings about foreign-made cars. In educational and psychological testing, one commonly looks at other similar questionnaires at this stage of the scale design, again, in order to gain as wide a perspective on the concept as possible.

Step 2: Choosing items of optimum difficulty. In the first draft of your prejudice questionnaire, include as many items as possible (note that the *Reliability* module will handle up to 300 items in a single scale). Now, administer this questionnaire to an initial sample of typical respondents, and examine the results for each item. First, you would look at various characteristics of the items, for example, in order to identify *floor* or *ceiling* effects. If all respondents agree or disagree with an item, then it obviously does not help to discriminate between respondents, and thus, it is useless for the design of a reliable scale. In test construction, the proportion of respondents who agree or disagree with an item, or who answer a test item correctly, is often referred to as the *item difficulty*. In essence, you would look at the item means and standard deviations and eliminate those items that show extreme means, and zero or nearly zero variances.

Step 3: Choosing internally consistent items.
Remember that a reliable scale is made up of items that proportionately measure mostly true score; in this example, you would like to select items that mostly measure prejudice against foreign-made cars and few esoteric aspects you consider random error. To do so, look at the following Scrollsheet:

RELIABL. ANALYSIS variable	Summary for scale: Mean=46.1100 Std.Dv.=8.26444 Valid N:100 Cronbach alpha: .794313 Standardized alpha: .800491 Average inter-item corr.: .297818					
	Mean if deleted	Var. if deleted	StDv. if deleted	Itm-Totl Correl.	Squared Multp. R	Alpha if deleted
ITEM1	41.61000	51.93790	7.206795	.656298	.507160	.752243
ITEM2	41.37000	53.79310	7.334378	.666111	.533015	.754692
ITEM3	41.41000	54.86190	7.406882	.549226	.363895	.766778
ITEM4	41.63000	56.57310	7.521509	.470852	.305573	.776015
ITEM5	41.52000	64.16961	8.010593	.054609	.057399	.824907
ITEM6	41.56000	62.68640	7.917474	.118561	.045653	.817907
ITEM7	41.46000	54.02840	7.350401	.587637	.443563	.762033
ITEM8	41.33000	53.32110	7.302130	.609204	.446298	.758992
ITEM9	41.44000	55.06640	7.420674	.502529	.328149	.772013
ITEM10	41.66000	53.78440	7.333785	.572875	.410561	.763314

Shown above are the results for 10 items that are discussed in greater detail in the *Example* section (page 3119). Of most interest are the three right-most columns in this Scrollsheet. They show the correlation between the respective item and the total sum score (without the respective item; *Itm-Totl Correl.*), the squared multiple correlation between the respective item and all others (*Squared Multp. R*), and the internal consistency of the scale (coefficient *alpha*) if the respective item would be deleted (*Alpha if deleted*).

Clearly, items 5 and 6 "stick out," in that they are not consistent with the rest of the scale. Their correlations with the sum scale are *.05* and *.12*, respectively, while all other items correlate at *.45* or better. In the right-most column (*Alpha if deleted*), you can see that the reliability of the scale would be about *.82* if either of the two items were to be deleted. Thus, you would probably delete the two items from this scale.

Step 4: Returning to Step 1.
After deleting all items that are not consistent with the scale, you may not be left with enough items to make up an overall reliable scale (remember that the fewer items, the less reliable the scale). In practice, one often goes through several rounds of generating items and eliminating items, until one arrives at a final set that makes up a reliable scale.

Final Remarks

The *Reliability* module was designed to make the construction of sum scales as efficient as possible. *STATISTICA* will handle a large number of items (e.g., 300 items in a single scale), allows the user to identify and delete "bad" items efficiently, provides various options for estimating the number of items that should be added to the scale, and allows the user to examine the correlation of the scale with other measures after correcting for attenuation (validity). The *Example* section (page 3119) will provide additional details concerning the utility of the various options and statistics provided in the program.

Tetrachoric correlations.
In educational and psychological testing, it is common to use *yes/no* type items, that is, to prompt the respondent to answer either yes or no to a question. An alternative to the regular correlation coefficient in that case is the *tetrachoric* correlation coefficient. Usually, the tetrachoric correlation coefficient is larger than the standard correlation coefficient; therefore, Nunally (1970, page 102) discourages the use of this coefficient for estimating reliability. However, it is still used (e.g., in modeling), and the *Reliability and Item Analysis* module offers different ways to compute this coefficient.

3. RELIABILITY AND ITEM ANALYSIS - INTRODUCTORY OVERVIEW

PROGRAM OVERVIEW

Methods

This module includes a comprehensive selection of procedures for the development of surveys, questionnaires, or tests. The program is specifically designed to handle very large item "pools;" scales with up to 300 items can be processed in a single run. The user can calculate reliability statistics for all items in a scale, interactively select subsets, or obtain comparisons between subsets of items via the *split-half* (or *split-part*) method. When interactively deleting items, the reliability of the resulting scale is computed instantly without processing the data file again.

Tetrachoric correlations. By default, the program will base all computations on the standard correlations between the items in the analysis. For modeling purposes, an option is provided to base the calculations on tetrachoric correlation coefficients. These coefficients can be computed based on the quick cosine *pi* formula or calculated via a more precise iterative estimation procedure (Gaussian quadrature and Newton-Raphson iterations).

Data Files

The *Reliability and Item Analysis* module will accept raw data files as well as correlation matrices (either created by other modules or entered directly into the spreadsheet; see *Matrix File Format*, Volume I). Correlation matrices can also be saved via this module in the *STATISTICA* matrix file format. Missing data may be substituted by the respective means.

Results

The output includes correlation matrices and descriptive statistics for items and the sum scale (including the skewness and kurtosis of the sum). In addition, the program will compute Cronbach's *alpha*, the standardized *alpha*, the average inter-item correlation, the complete analysis of variance table for the scale, and the complete set of item-total statistics (e.g., *alpha* if the item is deleted; scale mean, variance, and standard deviation if item is deleted; item-total correlations, and squared multiple item-total R's).

Split-half (part) statistics include Cronbach's *alpha* and descriptive statistics for each half, the correlation between halves, the correlation corrected for *attenuation*, and the split-half reliability (Spearman-Brown and Guttman). A number of interactive "what if" procedures are provided to aid in the development of scales.

For example, the user can calculate the expected (hypothetical) reliability after adding a particular number of items to the scale, or the user can compute the number of items that would have to be added to the scale in order to achieve a particular reliability. Also, the user can estimate the correlation corrected for attenuation between the current scale and another measure (given the reliability of the current scale).

Alternative Procedures

The reliability/item analysis statistics computed in the *Reliability and Item Analysis* module are highly specialized, and there are no similar techniques available in other modules. However, the construction of measurement scales often involves factor analysis (*Factor Analysis*, Chapter 6) to identify underlying dimensions in a pool of items, multiple regression (*Multiple Regression*, Volume I) and discriminant analysis (*Discriminant Analysis*, Chapter 2) during the validation stage of the scale construction.

Testing hypotheses about relationships between items and tests. To test specific

hypotheses about the relationship between sets of items or different tests (e.g., test whether two sets of items measure the same construct, analyze multi-trait multi-method matrices, etc.), use the *Structural Equation Modeling (SEPATH)* procedure (see Chapter 11).

3. RELIABILITY AND ITEM ANALYSIS - EXAMPLES

EXAMPLES

Example 1: Evaluating the Reliability of Items in a Questionnaire

Overview and Data File

The following example is based on a (fictitious) data set consisting of 10 items and 100 cases. To continue the example described in the *Introductory Overview* (page 3111), suppose you wanted to design a questionnaire to measure people's prejudices against foreign-made cars. You have already gone through several rounds of designing and selecting questionnaire items (as explained on page 3114 in the *Introductory Overview*), and the current study represents one of the final steps. The ten final items under consideration are displayed in the Scrollsheet below (click on the *Variables - All Specs* option on the *Edit* pull-down menu).

	Name	MD Code	Format	Long Name (label, formula or link)
1	ITEM1	-9999	3.0	Foreign cars all look the same
2	ITEM2	-9999	3.0	One should generally "buy American"
3	ITEM3	-9999	3.0	Foreign cars do not provide enough space
4	ITEM4	-9999	3.0	American cars are just as well built
5	ITEM5	-9999	3.0	Foreign cars are generally too expensive
6	ITEM6	-9999	3.0	Foreign cars use "foreign technology"
7	ITEM7	-9999	3.0	Foreign cars lack personality
8	ITEM8	-9999	3.0	Foreign cars are not of better quality
9	ITEM9	-9999	3.0	Foreign cars lack "spirit"
10	ITEM10	-9999	3.0	Foreign cars are no better than American

These items were accompanied by 9-point scales, with the anchors *1=disagree* and *9=agree* at the two ends of each scale. The questionnaire containing the items was administered to a sample of 100 individuals. Their responses were recorded in the data file *10items.sta*.

Starting the Analysis

After starting the *Reliability and Item Analysis* module, the startup panel will open in which you can open the file *10items.sta*. Click on the *Variables* button in the startup panel and select all 10 variables (items) in the file.

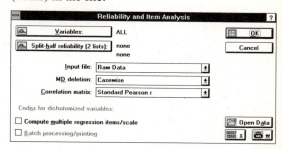

Now, click *OK* in the startup panel and a dialog will appear for reviewing descriptive statistics and the correlation matrix of items.

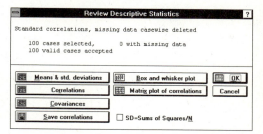

Correlation Matrix

First look at the correlation matrix; click on the *Correlations* button to open the Scrollsheet with correlations between all of the previously selected variables (see below).

Continue...	ITEM 1	ITEM 2	ITEM 3	ITEM 4	ITEM 5	ITEM 6	ITEM 7	ITEM 8	ITEM 9	ITEM 10
ITEM1	1.00	.58	.49	.43	.04	.14	.54	.38	.40	.50
ITEM2	.58	1.00	.46	.36	-.03	.12	.57	.55	.44	.47
ITEM3	.49	.46	1.00	.36	.11	.05	.47	.38	.35	.29
ITEM4	.43	.36	.36	1.00	.06	-.04	.27	.35	.42	.37
ITEM5	.04	-.03	.11	.06	1.00	.03	.01	.13	-.04	.01
ITEM6	.14	.12	.05	-.04	.03	1.00	.14	.10	.07	.09
ITEM7	.54	.57	.47	.27	.01	.14	1.00	.44	.29	.43
ITEM8	.38	.55	.38	.35	.13	.10	.44	1.00	.41	.51
ITEM9	.40	.44	.35	.42	-.04	.07	.29	.41	1.00	.41
ITEM10	.50	.47	.29	.37	.01	.09	.43	.51	.41	1.00

By and large, most correlations between items seem to be positive and substantial. A few exceptions seem to exist in the columns for *Item5* and *Item6*.

3. RELIABILITY AND ITEM ANALYSIS - EXAMPLES

Plots. These correlations can be graphically represented via a matrix of scatterplots (click on the *Matrix plot of correlations* button) or a scatterplot produced from the Scrollsheet. As in most other Scrollsheets of correlation matrices, the default graph for this Scrollsheet is the scatterplot of the respective two highlighted variables. For example, to obtain the scatterplot depicting the correlation between *Item1* and *Item2*, click with the right-mouse-button on that correlation in the Scrollsheet and select the *Quick Stats Graphs - Scatterplot/Conf* option on the flying menu. This scatterplot is shown below.

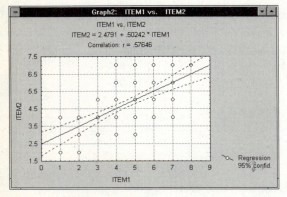

When you click on the *Matrix plot of correlations* button, you can select the variables to be plotted from a list of variables selected in the startup panel.

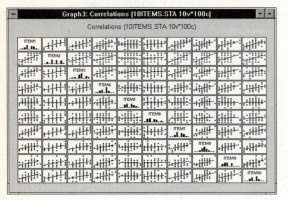

The matrix scatterplot will display the individual scatterplots for the correlations on the off-diagonal with histograms of the selected variables on the diagonal of the matrix plot. This plot is useful in determining which sets of correlations may contain outliers. You can then use the *Zoom-in* button on the graphics window toolbar to "blow-up" the specific scatterplot and determine which cases are the outliers. After reviewing the matrix of scatterplots, one may wish to examine a few of those scatterplots in order to get a closer view and detect specific outliers among the cases. Such outliers may greatly bias the computation of the correlation coefficient, and thus the estimation of the scale's reliability. In this case, no outliers are evident.

Means and Standard Deviations

Now, click on the *Means & std. deviations* button to display the Scrollsheet of means and standard deviations for the variables selected in the startup panel.

Means and Standard Deviations (10items.sta)		
Continue...	mean	st. dev.
ITEM1	4.500000	1.445998
ITEM2	4.740000	1.260271
ITEM3	4.700000	1.352140
ITEM4	4.480000	1.321768
ITEM5	4.590000	1.477747
ITEM6	4.550000	1.479660
ITEM7	4.650000	1.366075
ITEM8	4.780000	1.396822
ITEM9	4.670000	1.421729
ITEM10	4.450000	1.416889

The default graph for the Scrollsheet of means and standard deviations is the histogram of the frequency distribution for the respective variable. To produce this graph, click on the Scrollsheet with the right-mouse-button and select the *Quick Stats Graphs - Histogram/Normal* option on the flying menu. These histograms indicate the expected normal distribution.

3. RELIABILITY AND ITEM ANALYSIS - EXAMPLES

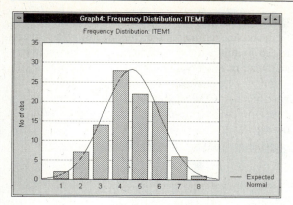

One should routinely examine these histograms in order to make sure that the variable of interest is normally distributed. In particular, it may happen that the distribution of responses is *multi-modal*, that is, that there are, for example, two "peaks" in the observed distribution. Such peaks may indicate that the population is not homogeneous with regard to the concept that you are trying to measure. Under these conditions, the correlations between items may become inflated, and consequently the reliability estimate for the scale may become inflated. As a result, you could end up with a scale that distinguishes well between, for example, two groups of people (the two "peaks"), but not at all between people in each group.

Now, click on the *Box and whisker plot* button in the *Review Descriptive Statistics* dialog and select all variables in the data set.

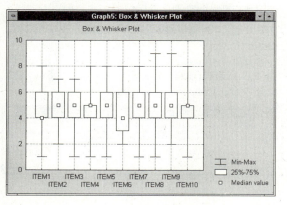

A window will open in which you can select from four types of box and whisker plots (for this example, select *median/quart./range*). Box and whisker plots are useful to determine if the distribution of a variable is symmetrical. If the distribution is not symmetrical, then you may want to view the histogram for the respective variable. These plots display the central tendency (e.g., median) and variability (e.g., quartile and range) of the selected variables.

Reviewing Results

Now, click *OK* in the *Review Descriptive Statistics* dialog to proceed to the *Results* dialog.

The statistics reported on the *Results* dialog (see below) summarize the characteristics of the sum scale. The values for the *skewness* and *kurtosis* of the sum scale are both close to zero; therefore, it can be concluded that the values for the sum scale are more-or-less normally distributed in the sample (see also the *Introductory Overview* section to *Basic Statistics and Tables*, Volume I, or *Nonparametrics and Distributions*, Volume I, for details concerning *skewness* and *kurtosis*).

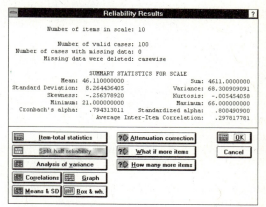

Looking at the summary section of the *Results* dialog, overall, the internal consistency reliability (*Cronbach's alpha*) for the sum is estimated at .79. The *standardized alpha* reported here is the reliability that would result if you were to use the

3. RELIABILITY AND ITEM ANALYSIS - EXAMPLES

standardized (*z* transformed) values for the items in the computation of Cronbach's *alpha* (see also the *Technical Notes* section, page 3131). The magnitude of the Cronbach's *alpha* value is "not bad" for a sum scale of only 10 items. As described in the *Introductory Overview* section (page 3113), you can interpret this value to indicate that about 70% of the variability in the sum score is *true score* variability, that is, true variability between respondents concerning the (prejudice) concept common in all items.

Item-total statistics. Now look at whether and how you can further improve the reliability of the scale. To do so, click on the *Item-total statistics* button.

RELIABL. ANALYSIS variable	Summary for scale: Mean=46.1100 Std.Dv.=8.26444 Valid N:100 Cronbach alpha: .794313 Standardized alpha: .800491 Average inter-item corr.: .297818				
	Mean if deleted	Var. if deleted	StDv. if deleted	Itm-Totl Correl.	Alpha if deleted
ITEM1	41.61000	51.93790	7.206795	.656298	.752243
ITEM2	41.37000	53.79310	7.334378	.666111	.754692
ITEM3	41.41000	54.86190	7.406882	.549226	.766778
ITEM4	41.63000	56.57310	7.521509	.470852	.776015
ITEM5	41.52000	64.16961	8.010593	.054609	.824907
ITEM6	41.56000	62.68640	7.917474	.118561	.817907
ITEM7	41.46000	54.02840	7.350401	.587637	.762033
ITEM8	41.33000	53.32110	7.302130	.609204	.758992
ITEM9	41.44000	55.06640	7.420674	.502529	.772013
ITEM10	41.66000	53.78440	7.333785	.572875	.763314

As described in the *Introductory Overview* section of this chapter (page 3114), the two right-most columns in this Scrollsheet are of particular importance (to obtain the *Squared Multp. R* column, click on the *Compute multiple regression items/scale option* in the startup panel). The correlations between the items and the sums score (without the item) are shown in the next-to-the-last column (*Itm-Totl Correl.*). The last column (*Alpha if deleted*) shows the resultant Cronbach's *alpha* value if the respective item were to be deleted.

It seems that both *Item5* ("Foreign cars are generally too expensive") and *Item6* ("Foreign cars use foreign technology") show a lower correlation with the sum scale than any of the other items. Deleting either one of the items would result in a reliability of .82.

Deleting items. Now to delete these two items from the scale, click on the *Cancel* button to return to the startup panel. Click on the *Variables* button and specify as the variables for the analysis, items *1-4* and *7-10*, that is, delete items *5* and *6* from the variable list. Click *OK* in this and the *Review Descriptive Statistics* dialog to go to the *Results* dialog.

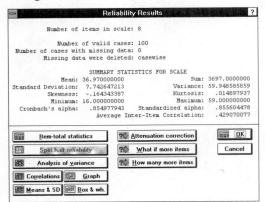

Reviewing new results. As you can see in the summary section of this dialog, the reliability of the scale has improved; Cronbach's *alpha* is now equal to *.85*. If you click on the *Item-total statistics* button again, you will see that there are no other variables that should be deleted, that is, whose deletion would further increase the overall scale reliability. Remember that in general, the fewer items there are in the sum scale, the smaller will the reliability of the scale become (see page 3112).

"What if..." analyses. How can the scale be further improved? Click on the *How many more items* button and a dialog will appear that will allow you to specify a desired target reliability. Now, enter *.9* and click on the *Compute* button.

REL - 3122

Copyright © StatSoft, 1995

3. RELIABILITY AND ITEM ANALYSIS - EXAMPLES

In order to obtain a reliability for the sum scale that is equal to .9, you would have to add *4* items of the same average "quality" (reliability) as the other items in the scale.

Now click *Cancel* to return to the *Results* dialog and then click on the *What if more items* button to open the following window.

Here, you can enter the number of items that you are considering to add; the program will then compute the expected reliability. For example, enter *5* as the number of items that you are considering to add and click on the *Compute* button.

As you can see, if you were to add *5* more items (of the same "quality" as the other items) to the scale, the resultant reliability for the sum scale would increase to about *.91*.

Attenuation correction. Another useful option on this dialog is the *Attenuation correction* option. As described in the *Introductory Overview* section (page 3113), the correlation between a scale and some other variable or scale is limited by the reliability of the respective scales or variables. For example, suppose in a validation study you were to obtain a correlation of $r = .4$ between the current 8-item scale with another scale designed to measure *xenophobia* (fear of strangers or foreigners). Also, assume that the xenophobia scale has an estimated reliability of *alpha = .70*.

The attenuation correction is generally useful for estimating the true similarity of or correlation between two concepts, which are both measured

with imperfect scales. Thus, it is often computed in the course of validation studies (where a scale is correlated with other scales measuring similar concepts).

Now click on the *Attenuation correction* button and specify those values (first the correlation, that is, *.4*, then the reliability of the xenophobia scale, that is, *.7*), and then click on the *Compute* button.

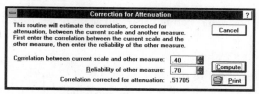

The correlation between these two concepts, corrected for attenuation due to unreliability, is equal to *.52*. This is the correlation that one could expect if both scales were perfectly reliable, that is, if they would only measure the true scores of the respective concepts.

Example 2:
Split-Half Reliability

Suppose you want to use the questionnaire to assess the success of a new advertisement campaign. In that case you may want to divide the 8 item scale into two halves that can be used before and after respondents have been exposed to the new commercials. You can split the scale "in the middle," that is, make up one sum scale of items *1* through *4*, and the other scale of items *7* through *10*.

Specifying the analysis. In the *Reliability and Item Analysis* startup panel, click on the *Split-half reliability* button and select variables *Item1-Item4* in the first list and variables *Item7-Item10* in the second list.

3. RELIABILITY AND ITEM ANALYSIS - EXAMPLES

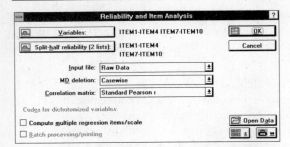

Click *OK* in the startup panel and again in the *Review Descriptive Statistics* dialog to go to the *Results* dialog.

Reviewing the results. Now, the *Split-half reliability* button will not be dimmed in the *Results* dialog (see below).

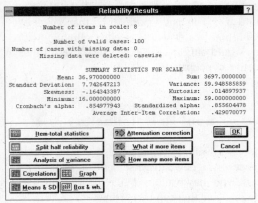

Click on this button to display the *Split-half reliability* results dialog (for more information on the options available in this dialog, see page 3130).

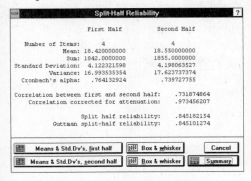

As you can see in the summary section of this dialog, the estimate of reliability based on the *split-half reliability* formula (see page 3113 in the *Introductory Overview*) is practically identical to the internal consistency estimate (Cronbach's *alpha*). The *Guttman split-half reliability* reported on this dialog can be interpreted as the Cronbach's *alpha* that would result if you treat the two halves of the scales as two single items of a two-item overall scale. Again, this estimate is very similar to the other reliability coefficients.

Finally, it appears that the means and standard deviations for the two halves of the scale are also comparable.

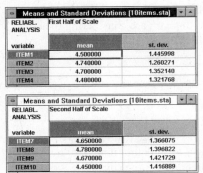

Therefore, you may use these two halves as equivalent scales to measure prejudice against foreign-made cars.

3. RELIABILITY AND ITEM ANALYSIS - DIALOGS AND OPTIONS

DIALOGS, OPTIONS, STATISTICS

Startup Panel

The following options are available on the *Reliability and Item Analysis* startup panel. Once you have made your selections, click *OK* to continue the analysis.

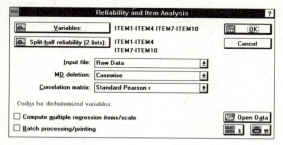

Variables

In order to perform an item analysis and to compute the reliability of a scale, first select the variables (items) via the *Variables* button.

Split-Half Reliability

This button remains dimmed until after the variables (items) have been selected, above. Once the variables have been selected for the analysis, you can divide the items into two variable (item) lists (the two lists do not have to contain all the items that were originally selected, see also page 3113). For each item list, the program will calculate the reliability, the correlation between the two lists (i.e., their sums), the correlation corrected for attenuation, as well as the split-half reliability.

Input File

Raw data files or correlation matrices may be read as input into the *Reliability and Item Analysis* module.

Missing Data Deletion

Missing data can be deleted casewise, or substituted by means.

Correlation Matrix

The following types of correlations are available in this option:

No correlations. In order to speed up calculations, the calculation of the correlation matrix can be suppressed by selecting *No*. Note that, if more than 100 variables (items) are selected, the correlation matrix will not be computed in this dialog and this option will automatically default to *No*. However, as in all modules of *STATISTICA*, you can compute an overall correlation matrix for more variables (e.g., 1,000 or more, depending on the amount of memory available in your computer) via the *Quick Basic Stats - Correlation matrices* option from the *Analysis* pull-down menu, or by pressing the *Quick Basic Stats* toolbar button (see the *General Conventions* chapter in Volume I).

Pearson r. If you select this option, *STATISTICA* will compute the standard Pearson product-moment correlation coefficients, and the reliability analysis will be based on those correlations.

Tetrachoric r (quick cos p approx). In this setting, the analysis will be based on tetrachoric correlations, computed via the *cosine pi* formula (applicable to dichotomous data) that yields up to 4 digits of precision, as long as the proportions (i.e., the number of *0*'s and *1*'s) are not extreme (e.g., 95% *0*'s and 5% *1*'s).

Tetrachoric r (iterative approx). In this setting, the analysis will be based on tetrachoric correlations, computed using a more time-consuming but also more precise procedure based on (Gaussian quadrature and Newton-Raphson iterations (see Kirk, 1973). This iterative procedure is considerably slower than the above tetrachoric correlation procedure. Tetrachoric correlations are not recommended for most standard analyses (see page 3115).

Codes for Dichotomized Vars

This option is only available if tetrachoric correlations are selected above. The tetrachoric correlation coefficient assumes that the data in the selected variables (items) consist of codes that indicate which one of two responses the respective respondent gave for the respective item. You may enter two integer values in the range from -999 to +999 to indicate the (integer) values used in the selected items to indicate the two responses.

Note that the program always assumes that the first code that is entered represents the lower value, and the second code the higher value of the dichotomized variable, and the directions (signs) of the tetrachoric correlations are determined accordingly.

Compute Multiple Regression Items/Scale

The multiple correlation (between each item and the sum of all others in the scale) may be calculated by selecting this check box; this may increase processing time if the scale contains many items (i.e., if many items were selected for the analysis).

Batch Processing-/Printing

This option is only available if *printer*, *disk file*, and/or *Text/output Window* is currently selected in the *Page/Output Setup* dialog (click on the *Output* field on the status bar to open this dialog; see Volume I). The selection of the *Batch Processing/Printing* option affects the way in which the reliability output is printed. The default setting of the option is *interactive processing* (i.e., *Batch Processing/Printing* is de-selected). In this mode you can determine (interactively) which Scrollsheets to print. In the *Batch Processing/Printing* mode, the resulting Scrollsheets will automatically be printed, saved to the printer file, and/or sent to the *Text/output Window* (whichever you have selected in the *Page/Output Setup* dialog).

Review Descriptive Statistics

This intermediate dialog (available if you selected one of the three types of correlations in the *Reliability and Item Analysis* startup panel) in reliability analysis will allow you to review specific descriptive statistics on the previously selected variables (items).

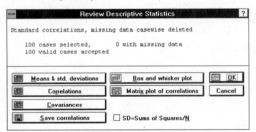

Means and Standard Deviations

This option will bring up a Scrollsheet with means and standard deviations for all items in the scale. By default, the standard deviation is computed as the square root of the sum of squared deviations about the mean (SS) divided by $n - 1$ (an unbiased estimate of σ^2). You can use the *SD = Sums of Squares/n* option (see below) to compute the standard deviation as the SS divided by n.

The default graph for this Scrollsheet is a histogram (frequency distribution) of the chosen variable, with the normal distribution superimposed (see page 3120). This plot can be very useful as a quick check of whether the assumption of normality has been violated (on which the computation of the correlation coefficients is based; for more information, see *Basic Statistics and Tables*, Volume I).

Correlations

This option will bring up a Scrollsheet with correlations (either tetrachoric correlations or Pearson-product moment correlations, whichever was selected in the startup panel). The default graph for this Scrollsheet is a scatterplot of the respective correlation coefficients (see page 3120). The plot will indicate the regression line as well as the 95% confidence limits for the regression line.

Covariances

This option will bring up a Scrollsheet with the covariance matrix for all items in the scale. The default graph for this Scrollsheet is a scatterplot of the selected variables (see page 3120). The plot will indicate the regression line and the 95% confidence limits for the regression line.

Save Correlation Matrix

When you click on this button, the *Save Matrix As* dialog will open in which you can save the correlation matrix as a standard *STATISTICA* matrix file.

Box and Whisker Plot

This option allows you to produce a box and whisker plot (showing medians and quartiles, means and standard deviations, etc.) for the selected variables (items).

Median and quartiles. After selecting this option, and after selecting a variable for the plot, the *Box-Whisker Type* dialog will come up.

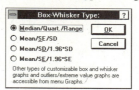

If you select to compute for the box-whisker plot the median and the quartiles, those values will be computed according to the setting of the *Percentiles* option on the *STATISTICA Defaults: General* dialog (choose option *General* on the *Options* pull-down menu). For computational details, refer to the description of this dialog in the *General Conventions* chapter in Volume I.

Matrix Plot of Correlations

This option will bring up a matrix scatterplot for the selected variables.

The SD Check Box

There are two ways in which standard deviations can be computed: as the sums of (deviation) squares divided by n (the valid number of cases) or $n-1$. If divided by $n-1$, the resulting standard deviations are estimates of the standard deviations in the population (*sigma*); if divided by n, the resulting standard deviations are descriptions of the sample only. The setting of this switch will not affect the subsequent computations of the reliability coefficients.

Results

This dialog allows the user to review the key results of the analysis; the interpretation of the available statistics are discussed in the *Introductory Overview*.

3. RELIABILITY AND ITEM ANALYSIS - DIALOGS AND OPTIONS

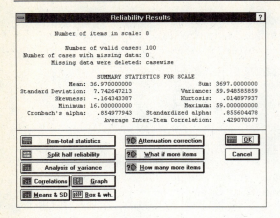

Item-Total Statistics

After clicking on this button, a Scrollsheet with various statistics regarding the relationship between each item and the scale (without the respective item) will be displayed: the mean of the scale without the respective item, the variance of the scale without the respective item, the standard deviation of the scale without the respective item, the correlation between the respective item and the scale (without the item), and Cronbach's *alpha* for the scale without the respective item. In addition, if multiple regression statistics were requested from the startup panel, the multiple *R*-squared of each item with the scale without the respective item will be displayed. Summary statistics for the analysis (e.g., Cronbach's *alpha*, scale mean, etc.) are shown in the header of the Scrollsheet.

Split-Half Reliability

This option is only available if *split-half reliability* was requested from the startup panel (i.e., two variable lists were specified, see page 3125). After clicking on this button, the program will display Cronbach's *alpha* for each half, the correlation between the two halves, the correlation corrected for attenuation, the split-half reliability, and the Guttman split-half reliability.

Analysis of Variance

Clicking on this button will bring up a Scrollsheet with a complete analysis of variance table for the items in the scale. This analysis is equivalent to a one-way repeated measures ANOVA, where each item represents one level of the repeated measures factor (refer to the *Introductory Overview* section of *ANOVA/MANOVA*, Volume I, for more details concerning repeated measures ANOVA).

Correlation Matrix

This option is only available if a correlation matrix was requested from the startup panel (the default selection). After clicking on this button, a Scrollsheet with the correlation matrix will be displayed.

The default plot for this Scrollsheet is the scatterplot of the respective two variables that are highlighted in the Scrollsheet (see page 3120).

Matrix Plot

This option allows you to produce a scatterplot matrix for the selected variables (see page 3120).

Means and Standard Deviations

After clicking on the *Means & SD* button, a Scrollsheet with the means and standard deviations for all items will be displayed. The standard deviations will be computed in the manner indicated by the *SD=sums of squares/n* check box on the *Review Descriptive Statistics* dialog (see page 3127). Namely, they will either be computed as the square roots of the sums of deviation squares divided by *n-1* or *n*. If divided by *n-1*, the resulting standard deviations are estimates of the standard deviations in the population (*sigma*); if divided by *n*, the resulting standard deviations are descriptions of

---

REL - 3128

Copyright © StatSoft, 1995

the sample only. The default graph for this Scrollsheet is a histogram (frequency distribution) of the chosen variable, with the normal distribution superimposed (see page 3120). This plot can be very useful as a quick check of whether the assumption of normality has been violated (on which the computation of the correlation coefficients is based, see *Basic Statistics and Tables*, Volume I).

Box and Whisker Plot

This option allows you to produce a box and whisker plot (showing medians and quartiles, means and standard deviations, etc.) for the selected variables (items, see page 3120). For computational details concerning the median and the quartiles refer to the description of the *STATISTICA Defaults: General* dialog in the *General Conventions* chapter (Volume I).

Attenuation Correction

With this option, the user may estimate the correlation, corrected for attenuation (due to less than perfect reliability of measures), between the current test (given its current reliability) with a criterion (given its user-specified reliability). You can enter the value of a correlation between the current test and another measure, and the reliability of the other measure in the *Correction for Attenuation* dialog.

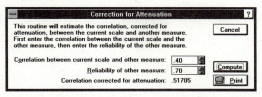

The program will then compute the corrected correlation. For more information on the correction for attenuation, see the *Introductory Overview* section (page 3113).

What If More Items

This option estimates the reliability of the scale if a particular number of items are added. It is assumed that the intercorrelations among the new items are equivalent to that of the old items. When you click on this option, the following dialog will open in which you can enter the number of items to be added to the scale. The program will then compute the reliability of the scale that may be expected if that many items were added.

For more information on the number of items and reliability, see the *Introductory Overview* section (page 3112).

How Many More Items

This option will estimate the number of items which would have to be added to the scale in order to attain a user-specified reliability. When you click on this option, the following dialog will open in which you can enter the desired scale reliability.

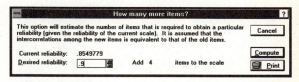

The program will then compute the number of items that would have to be added to attain that reliability. The computations will assume that the items that will be added are of approximately equal "quality" to the other items in the scale, that is, that they will be of comparable reliability (see also the *Introductory Overview* section, page 3112).

Split-Half Reliability

This dialog displays a summary of the descriptive statistics and reliabilities for the two halves of the scale, as specified on the startup panel (page 1167).

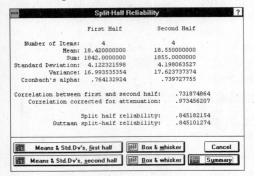

You can display the descriptive statistics for each of the items in each half of the scale by clicking on the respective *Means & standard deviations* buttons (for the first or second half of the items). You can also create box and whisker plots for the selected items in each half of the scale by clicking on the appropriate *Box and whisker plot* button. For example, based on the data file *10items.sta*, the box and whisker plot for the first half is shown below:

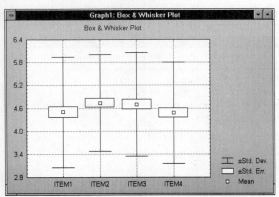

The box and whisker plot for the second half is shown below:

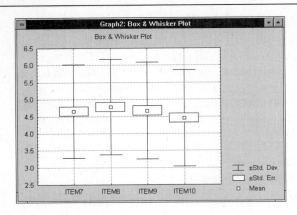

Summary

This option will bring up a Scrollsheet with the summary statistics for the split-half analysis, as well as statistics for each half of the scale, and (below the thick line) the lists of items included in each half of the scale.

TECHNICAL NOTES

Computations

The standard formulas from classical testing theory are used to compute Cronbach's *alpha*, and for the attenuation correction (see Nunally, 1970).

Standardized *Alpha*

The *standardized alpha* may be interpreted as the reliability that would result if all values for each item were standardized (*z* transformed) before computing Cronbach's *alpha*. The computational formula is:

$$\alpha = k * r_{avrg} / [1 + (k-1) * r_{avrg}]$$

where

k is the number of items in the scale

r_{avrg} is the average inter-item correlation

Guttman Split-Half Reliability

The Guttman split-half coefficient is similar to the Spearman-Brown split-half coefficient, but does not assume equal reliabilities or equal variances of the two halves. It is computed as:

$$r_G = 2 * (s_t^2 - s_{t1}^2 - s_{t2}^2) / s_t^2$$

where

r_G is the Guttman split-half coefficient

s_t^2 is the variance of the total scale (both halves)

s_{t1}^2 is the variance of the first half

s_{t2}^2 is the variance of the second half

Items with Zero Variances

If an item has zero variance, a correlation matrix cannot be computed (a warning will be issued by the program). In that case, either eliminate the item(s) with zero variances, or choose not to compute the correlation matrix. If the latter option is chosen, the item(s) with zero variances will be included in all subsequent computations.

Note that if no correlation matrix is requested, then the following formula is used to compute Cronbach's *alpha*:

$$\alpha = [k/(k-1)] * (1 - \sum s_i^2 / s_t^2)$$

where

k is the number of items in the scale

$\sum s_i^2$ denotes the sum of item variances

s_t^2 is the variance of the scale

If items with zero variances are included, they will not affect the variances but only the factor *[k/(k-1)]*.

3. RELIABILITY AND ITEM ANALYSIS - INDEX

INDEX

A

Attenuation correction, 3113, 3123, 3128, 3129

B

Batch processing/printing, 3126
Box and whisker plots
 for all variables, 3121, 3127, 3129

C

Correction for attenuation, 3113, 3123, 3129
Correlation matrix, 3119, 3125, 3127, 3128
Cronbach's alpha, 3113, 3131

D

Deleting items from a scale, 3115, 3122
Designing a reliable scale, 3114

G

Gaussian quadrature, computation of tetrachoric correlations, 3117, 3125
Guttman split-half reliability, 3124, 3131

H

Histogram
 2D histogram, 3120

I

Index of reliability, 3112
Internal-consistency reliability, 3113
Item difficulty, 3115
Item-total statistics, 3122, 3128

K

Kuder-Richardson-20 formula, 3113

M

Measures of reliability, 3112
Missing data deletion, 3117, 3125
Multiple regression of items with scale, 3122, 3126

N

Newton-Raphson iterations, computation of tetrachoric correlations, 3117, 3125
Number of items and reliability, 3112

P

Pearson r correlation coefficients, 3125

R

Reliability and item analysis
 analysis of variance, 3128
 attenuation correction, 3113, 3123, 3129
 classical testing model, 3112
 codes for dichotomized variables, 3126
 correction of attenuation, 3113, 3123, 3129
 Cronbach's alpha, 3113
 deleting items from the scale, 3115, 3122
 descriptive statistics, 3119, 3126
 designing a reliable scale, 3114
 example, 3119
 Gaussian quadrature, computation of tetrachoric correlations, 3117, 3125
 Guttman split-half reliability, 3124, 3131
 how many more items, 3129
 index of reliability, 3112
 internal-consistency reliability, 3113
 items with zero variances, 3131
 item-total statistics, 3122, 3128
 Kuder-Richardson-20 formula, 3113
 measures of reliability, 3112
 missing data deletion, 3117, 3125

Reliability and item analysis (cont.)
 multiple regression of items with scale, 3122, 3126
 Newton-Raphson iterations, computation of tetrachoric correlations, 3117, 3125
 number of items and reliability, 3112
 overview, 3111
 reliability, 3112
 Spearman-Brown split-half coefficient, 3113
 split-half reliability, 3113, 3123, 3125, 3128, 3130
 standardized alpha, 3131
 startup panel, 3125
 sum scales, 3112
 tetrachoric correlations, 3115
 tetrachoric r (iterative approx), 3125
 tetrachoric r (quick cos p approx), 3125
 true scores and error, 3112
 what if more items option, 3123, 3129
 what if... analyses, 3122
 zero variances, items with, 3131

S

Scatterplots
 2D scatterplot, 3120
 matrix scatterplots, 3120, 3127, 3128
Spearman-Brown split-half coefficient, 3113
Split-half reliability, 3113, 3123, 3125, 3128, 3130
Standardized alpha, 3131
Sum scales, 3111, 3112

T

Tetrachoric correlations, 3115, 3125
 iterative approximation, 3125
 quick cos p approximation, 3125
Theoretical construct, 3112

W

What if more items option, 3123

Chapter 4:
CANONICAL CORRELATION

Table of Contents

Introductory Overview 3139
    General Purpose of Canonical Correlation 3139
    General Ideas 3139
    Computational Methods and Results 3141
    Assumptions 3144
Program Overview 3145
Example 3147
    Overview 3147
    Specifying the Canonical Analysis 3149
    Reviewing Results 3149
Dialogs, Options, Statistics 3155
    Startup Panel 3155
    Review Descriptive Statistics 3155
    Define Model 3157
    Results 3157
Technical Notes 3161
Index 3163

Detailed Table of Contents

INTRODUCTORY OVERVIEW **3139**
    General Purpose of Canonical Correlation 3139
    General Ideas 3139
        Sum Scores 3139
        Using a Weighted Sum 3140
        Determining the Weights 3140
        Canonical Roots/Variates 3140
        Number of Roots 3140
        Extraction of Roots 3140
    Computational Methods and Results 3141
        Eigenvalues 3141
        Canonical Correlations 3141

4. CANONICAL CORRELATION - CONTENTS

Significance of Roots	3141
Canonical Weights	3142
Canonical Scores	3142
Factor Structure	3142
Factor Structure versus Canonical Weights	3142
Variance Extracted	3143
Redundancy	3143
Assumptions	3144
Distributions	3144
Sample Sizes	3144
Outliers	3144
Matrix Ill-Conditioning	3144

PROGRAM OVERVIEW 3145
Overview 3145
Canonical Scores 3145
Plots 3145
Data Files 3145
Alternative Procedures 3145

EXAMPLE 3147
Overview 3147
    Purpose of Analysis 3147
    Initial Computations 3147
Specifying the Canonical Analysis 3149
Reviewing Results 3149
    Canonical R 3149
    Overall Redundancy 3149
    Testing the Significance of Canonical Roots 3150
    Factor Structure and Redundancy 3150
    Canonical Scores 3151
    Conclusion 3152

DIALOGS, OPTIONS, STATISTICS 3155
Startup Panel 3155
    Variables 3155
    Input File 3155
    MD Deletion 3155
    Review Descriptive Statistics, Correlation Matrix 3155
Review Descriptive Statistics 3155
    Means & SD 3155
    Correlations 3156
    Covariances 3156
    Save Correlations 3156
    Box and Whisker Plots 3156

Matrix Plot of Correlations	3156
The SD Check Box	3156
Define Model	3157
Variables for Canonical Analysis	3157
Batch Processing/Printing	3157
Means and Standard Deviations	3157
Box and Whisker Plot	3157
Correlation Matrix	3157
Matrix Plot	3157
Results	3157
Correlations Within and Between Sets	3158
Eigenvalues	3158
Canonical Weights for Left and Right Set	3158
Factor Structures and Redundancies	3158
Chi-square Tests of Successive Roots	3158
Sequential Significance Test	3158
Save Canonical Scores	3159
Summary	3159
Plot of Eigenvalues	3159
Plot of Canonical Correlations	3159
Scatterplot of Canonical Correlations	3159
TECHNICAL NOTES	**3161**
INDEX	**3163**

4. CANONICAL CORRELATION - CONTENTS

Chapter 4:
CANONICAL CORRELATION

INTRODUCTORY OVERVIEW

General Purpose of Canonical Correlation

Several *STATISTICA* modules will compute measures of correlation to express the relationship between two or more variables. For example, the standard Pearson product moment correlation coefficient (r) measures the extent to which two variables are related; the *Nonparametrics and Distributions* module (see Volume I) offers various measures of relationships that are based on the similarity of ranks in two variables; the *Multiple Regression* module (see Volume I) allows the user to assess the relationship between a dependent variable and a *set* of independent variables. The *Canonical Correlation* module is an additional procedure for assessing the relationship between variables. Specifically, this module allows the user to investigate the relationship between two *sets* of variables.

For example, an educational researcher may want to compute the (simultaneous) relationship between three measures of scholastic ability with five measures of success in school. A sociologist may want to investigate the relationship between two predictors of social mobility based on interviews, with actual subsequent social mobility as measured by four different indicators. A medical researcher may want to study the relationship of various risk factors to the development of a group of symptoms. In all of these cases, the researcher is interested in the relationship between two *sets* of variables, and the *Canonical Correlation* module would be the appropriate method of analysis. The following paragraphs will briefly introduce the major concepts and statistics in canonical correlation analysis. It is assumed that you are familiar with the correlation coefficient as described in *Basic Statistics and Tables* (Volume I) and the basic ideas of multiple regression as described in the *Multiple Regression* chapter (Volume I).

General Ideas

Suppose you conduct a study in which you measure satisfaction at work with three questionnaire items, and satisfaction in various other domains with an additional seven items. The general question that you may want to answer is how satisfaction at work relates to the satisfaction in those other domains.

Sum Scores

One approach is to simply add up the responses to the work satisfaction items and to correlate that sum with the responses to all other satisfaction items. If the correlation between the two sums is statistically significant, then you could conclude that work satisfaction is related to satisfaction in other domains. In a way this is a rather "crude" conclusion. You still know nothing about the particular domains of satisfaction that are related to work satisfaction. In fact, you could have potentially *lost* important information by simply adding up items. For example, suppose there were two items, one measuring satisfaction with one's relationship with the spouse, and the other measuring satisfaction with one's financial situation. Adding the two together is, obviously, like adding "apples to oranges." Doing so implies that a person who is dissatisfied with her finances but happy with her spouse is comparable overall to a person who is satisfied financially but not happy in the relationship with her spouse. By simply correlating two sums, one might lose important information in the process, and possibly "destroy" important relationships between variables by adding "apples to oranges."

Using a Weighted Sum

It seems reasonable to correlate some kind of a weighted sum instead, so that the "structure" of the variables in the two sets is reflected in the weights. For example, if satisfaction with one's spouse is only marginally related to work satisfaction, but financial satisfaction is strongly related to work satisfaction, then a smaller weight could be assigned to the first item and a greater weight to the second item.

If you have two sets of variables, the first one containing p variables and the second one containing q variables, then you would like to correlate the weighted sums for the two sets of variables (i.e., the linear combination of the p variables with the linear combination of the q variables).

Determining the Weights

You have now formulated the general type of "model equations" for canonical correlation. The only problem that remains is how to determine the weights for the two sets of variables. It seems to make little sense to assign weights so that the two weighted sums do not correlate with each other. A reasonable approach to take is to impose the condition that the two weighted sums shall correlate maximally with each other. This is exactly what the *Canonical Correlation* module will do when performing a canonical analysis based on the overall correlation matrix of all variables.

Canonical Roots/Variates

In the terminology of canonical correlation analysis, the weighted sums define a pair of *canonical variates* or variables; the squared correlation between the two canonical variates is also called the *canonical root*. You can think of those canonical variates (weighted sums) as describing some underlying "latent" variables. For example, if, for a set of diverse satisfaction items, you were to obtain a weighted sum marked by large weights for all items having to do with work, you could conclude that the respective canonical variate measures satisfaction with work.

Number of Roots

So far the discussion implied that only one pair of canonical variates (weighted sums) can be extracted from the two sets of variables. However, suppose that you had among your work satisfaction items particular questions regarding satisfaction with pay and questions pertaining to satisfaction of one's social relationships with other employees. It is possible that the pay satisfaction items correlate with financial satisfaction, and that the social relationship satisfaction items correlate with the reported satisfaction with one's spouse. If so, you should derive two additional weighted sums to reflect this "complexity" in the structure of satisfaction. In fact, the computations involved in canonical correlation analysis will lead to more than one set of weighted sums. To be precise, the number of roots extracted by the program will be equal to the minimum number of variables in either set. For example, if you have three work satisfaction items and seven general satisfaction items, then the program will extract exactly three canonical roots.

Extraction of Roots

As mentioned before, the program will generally extract roots so that the resulting correlation between the canonical variates is maximal. When extracting more than one root, each successive pair of canonical variates will explain a *unique* additional proportion of variability in the two sets of variables. Successively extracted pairs of canonical variates will be uncorrelated with each other, and account for less and less variability.

Computational Methods and Results

Some of the computational issues involved in canonical correlation and the major results that are commonly reported will now be reviewed.

Eigenvalues

When extracting the canonical roots, *STATISTICA* will compute the *eigenvalues*. These can be interpreted as the proportion of variance accounted for by the correlation between the respective canonical variates. Note that the proportion here is computed relative to the variance of the canonical variates, that is, of the weighted sum scores of the two sets of variables; the eigenvalues do *not* tell how much variability is explained in either set of variables. The program will compute as many eigenvalues as there are canonical roots, that is, as many as the minimum number of variables in either of the two sets.

Successive eigenvalues will be of smaller and smaller size. The program will first compute the weights that maximize the correlation of the two sum scores. After this first root has been extracted, the program will find the weights that produce the second largest correlation between sum scores, subject to the constraint that the next set of sum scores does not correlate with the previous one, and so on.

Canonical Correlations

If the square root of the eigenvalues is taken, then the resulting numbers can be interpreted as correlation coefficients (see also *Basic Statistics and Tables*, Volume I). Because the correlations pertain to the canonical variates, they are called *canonical correlations*. Like the eigenvalues, the correlations between successively extracted canonical variates are smaller and smaller. Therefore, as an overall index of the canonical correlation between two sets of variables, it is customary to report the largest correlation, that is, the one for the first root. However, the other canonical variates can also be correlated in a meaningful and interpretable manner (see below).

Significance of Roots

The significance test of the canonical correlations is straightforward in principle. Simply stated, the different canonical correlations are tested, one by one, beginning with the largest one. Only those roots that are statistically significant are then retained for subsequent interpretation. Actually, the nature of the significance test is somewhat different. The program will first evaluate the significance of all roots combined, then of the roots remaining after removing the first root, the second root, etc. For example:

CANONICL ANALYSIS	Canonicl R	Canonicl R-sqr.	Chi-sqr.	df	p	Lambda Prime
0	.884705	.782703	153.5785	21	.000000	.193486
1	.271080	.073484	10.8516	12	.541689	.890422
2	.197374	.038957	3.7153	5	.591097	.961043

In this example, when 0 roots are removed, the results are highly significant (*Chi-square = 153.58, p<.000001*); after the first root is removed, the remaining roots are not significant (*Chi-square = 10.85, p<.54*). Therefore, in this case, the user would conclude that only the first root is statistically significant and should be retained for further examination (see below).

Some authors have criticized this sequential testing procedure for the significance of canonical roots (e.g., Harris, 1976). However, this procedure was "rehabilitated" in a subsequent Monte Carlo study by Mendoza, Markos, and Gonter (1978).

In short, the results of that study showed that this testing procedure will detect strong canonical correlations most of the time, even with samples of relatively small size (e.g., *n = 50*). Weaker canonical correlations (e.g., *R = .3*) require larger sample sizes (*n > 200*) to be detected at least 50% of the time. Note that canonical correlations of small magnitude are often of little practical value, as they account for very little actual variability in the data. This issue, as well as the sample size issue, will be discussed shortly.

Canonical Weights

After determining the number of significant canonical roots, the question arises as to how to interpret each (significant) root. Remember that each root actually represents two weighted sums, one for each set of variables. One way to interpret the "meaning" of each canonical root would be to look at the weights for each set. These weights are called the *canonical weights*.

In general, the larger the weight (i.e., the absolute value of the weight), the greater is the respective variable's unique positive or negative contribution to the sum. To facilitate comparisons between weights, the canonical weights are usually reported for the standardized variables, that is, for the *z* transformed variables with a mean of *0* and a standard deviation of *1*.

If you are familiar with multiple regression (see *Multiple Regression*, Volume I), you may interpret the canonical weights in the same manner as you would interpret the *beta* weights in a multiple regression equation. In a sense, they represent the *partial correlations* of the variables with the respective canonical root. If you are familiar with factor analysis (see *Factor Analysis*, Chapter 6), you may interpret the canonical weights in the same manner as you would interpret the *factor score coefficients*. To summarize, the canonical weights allow the user to understand the "make-up" of each canonical root, that is, it lets the user see how each variable in each set uniquely contributes to the respective weighted sum (canonical variate).

Canonical Scores

Canonical weights can also be used to compute actual values of the canonical variates; that is, you can simply use the weights to compute the respective sums. Again, remember that the canonical weights are customarily reported for the standardized (*z* transformed) variables. The *Canonical Correlation* module will automatically compute the canonical scores for you, which then can be saved for further analyses with other modules.

Factor Structure

Another way of interpreting the canonical roots is to look at the simple correlations between the canonical variates (or *factors*) and the variables in each set. These correlations are also called canonical factor *loadings*. The logic here is that variables that are highly correlated with a canonical variate have more in common with it. Therefore, you should weigh them more heavily when deriving a meaningful interpretation of the respective canonical variate. This method of interpreting canonical variates is identical to the manner in which factors are interpreted in factor analysis (*Factor Analysis*, Chapter 6).

Factor Structure versus Canonical Weights

Sometimes, the canonical weights for a variable are nearly zero, but the respective loading for the variable is very high. The opposite pattern of results may also occur. At first, such a finding may seem contradictory; however, remember that the canonical weights pertain to the *unique* contribution of each variable, while the canonical factor loadings represent simple overall correlations.

For example, suppose you included in your satisfaction survey two items which measured basically the same thing, namely: (1) "Are you satisfied with your supervisors?" and (2) "Are you satisfied with your bosses?" Obviously, these items are very redundant. When the program computes the weights for the weighted sums (canonical variates) in each set so that they correlate maximally, it only "needs" to include one of the items to capture the essence of what they measure. Once a large weight is assigned to the first item, the contribution of the second item is redundant; consequently, it will receive a zero or negligibly small canonical weight. Nevertheless, if you then look at the simple correlations between the respective sum score with the two items (i.e., the factor *loadings*), those may be substantial for *both*.

To reiterate, the canonical weights pertain to the *unique contributions* of the respective variables with a particular weighted sum or canonical variate; the canonical factor loadings pertain to the *overall correlation* of the respective variables with the canonical variate.

Variance Extracted

As discussed earlier, the canonical correlation coefficient refers to the correlation between the weighted sums of the two sets of variables. It tells nothing about how much variability (variance) each canonical root explains in the *variables*. However, you can infer the proportion of variance extracted from each set of variables by a particular root by looking at the canonical factor loadings. Remember that those loadings represent correlations between the canonical variates and the variables in the respective set.

If you square those correlations, the resulting numbers reflect the *proportion* of variance accounted for in each variable (see also the *Introductory Overview* section to *Basic Statistics and Tables*, Volume I, for additional details). For each root, you can take the average of those proportions across variables to get an indication of how much variability is explained, on the average, by the respective canonical variate in that set of variables. Put another way, you can compute in this manner the average proportion of *variance extracted* by each root.

Redundancy

The canonical correlations can be squared to compute the proportion of variance shared by the sum scores (canonical variates) in each set. If you multiply this proportion by the proportion of variance extracted, you arrive at a measure of *redundancy*, that is, of how redundant one set of variables is, given the other set of variables. In equation form, you may express the redundancy as:

$$\text{Redundancy}_{left} = [\Sigma(\text{loadings}_{left}^2)/p]*R_c^2$$

$$\text{Redundancy}_{right} = [\Sigma(\text{loadings}_{right}^2)/q]*R_c^2$$

In these equations, p denotes the number of variables in the first (*left*) set of variables, and q denotes the number of variables in the second (*right*) set; R_c^2 is the respective squared canonical correlation.

Note that you can compute the redundancy of the first (*left*) set of variables given the second (*right*) set, and the redundancy of the second (*right*) set of variables, given the first (*left*) set. Because successively extracted canonical roots are uncorrelated, you could sum up the redundancies across all (or only the first significant) roots to arrive at a single index of redundancy (as proposed by Stewart and Love, 1968).

Practical significance. The measure of redundancy is also useful for assessing the *practical* significance of canonical roots. With large sample sizes (see below), canonical correlations of magnitude $R = .30$ may become statistically significant (see above). If you square this coefficient (*R-square* = .09) and use it in the redundancy formula shown above, it becomes clear that such canonical roots account for only very little variability in the variables. Of course, the final assessment of what does and does not constitute a finding of practical significance is subjective by nature. However, to maintain a realistic appraisal of how much actual variance (in the variables) is accounted for by a canonical root, it is important to always keep in mind the redundancy measure, that is, how much of the actual variability in one set of variables is explained by the other.

Assumptions

The following discussion provides only a list of the most important assumptions of canonical correlation analysis, and the major threats to the reliability and validity of results.

Distributions

The tests of significance of the canonical correlations is based on the assumption that the distributions of the variables in the population (from which the sample was drawn) are multivariate normal. As in most programs in *STATISTICA*, the *Canonical Correlation* module allows the user to graph the variables in the analysis, that is, to produce frequency histograms with the normal curve superimposed, and scatterplots. Little is known about the effects of violations of the multivariate normality assumption. However, with a sufficiently large sample size (see below) the results from canonical correlation analysis are usually quite robust.

Sample Sizes

Stevens (1986) provides a very thorough discussion of the sample sizes that should be used in order to obtain reliable results. As mentioned earlier, if there are strong canonical correlations in the data (e.g., $R > .7$), then even relatively small samples (e.g., $n = 50$) will detect them most of the time. However, in order to arrive at reliable estimates of the canonical factor loadings (for interpretation), Stevens recommends that there should be at least 20 times as many cases as variables in the analysis, if one wants to interpret the most significant canonical root only. To arrive at reliable estimates for two canonical roots, Barcikowski and Stevens (1975) recommend, based on a Monte Carlo study, to include 40 to 60 times as many cases as variables.

Outliers

Outliers can greatly affect the magnitudes of correlation coefficients (see *Basic Statistics and Tables*, Volume I). Since canonical correlation analysis is based on (computed from) correlation coefficients, they can also seriously affect the canonical correlations. Of course, the larger the sample size, the smaller is the impact of one or two outliers. However, it is a good idea to examine the various scatterplots available in the *Canonical Correlation* module to detect possible outliers. Note that scatterplots can be produced not only for variables, but also for canonical variates.

Matrix Ill-Conditioning

One assumption is that the variables in the two sets should not be completely redundant. For example, if you included the *same* variable twice in one of the sets, then it is not clear how to assign different weights to each of them. Computationally, such complete redundancies will "upset" the canonical correlation analysis. When there are perfect correlations in the correlation matrix, or if any of the multiple correlations between one variable and the others is perfect ($R = 1.0$), then the correlation matrix cannot be inverted, and the computations for the canonical analysis cannot be performed. Such correlation matrices are said to be *ill-conditioned*.

Once again, this assumption appears trivial on the surface; however, it often is "almost" violated when the analysis includes very many highly redundant measures, as is often the case when analyzing questionnaire responses. In extreme cases, the program will "refuse" to perform the analysis, and issue a respective error message ("matrix ill-conditioned...").

PROGRAM OVERVIEW

Overview

This module offers a comprehensive implementation of canonical analysis procedures. The results reported by the program include the canonical correlation coefficients, the variances extracted, and redundancy coefficients (for each root and for all roots combined). The canonical roots are tested for statistical significance via the standard sequential testing procedure.

Canonical Scores

The program will compute the canonical scores for each canonical root for each case. Like all other results produced by *STATISTICA*, those values can be saved in a standard *STATISTICA* data file for further analyses with other modules.

Plots

The program allows the user to produce (from within Scrollsheets via the right-mouse-button flying menu, *Quick Stats Graphs* option) histograms of the frequency distributions of variables in the analyses. Those histograms will show the normal distribution curve superimposed over the observed frequencies. Matrix scatterplots can be produced for the correlations between variables; scatterplots can also be produced for the correlations between the canonical variates in each set. Box and whisker plots can also be produced, allowing for a quick comparison of the distributions of the analyzed variables. Additionally, the user can produce line graphs of eigenvalues and canonical correlations, as well as scatterplots of canonical correlations. A large selection of descriptive graphs are also available to aid in the interpretation of input data and the output.

Data Files

The *Canonical Correlation* module will accept raw data files as well as correlation matrices (created by other modules or entered into the spreadsheet; see *Matrix File Format*, Volume I); correlation matrices can also be saved via this module in the *STATISTICA* matrix file format.

Alternative Procedures

The *Canonical Correlation* module is unique in that it is the only statistical procedure which will assess the relationships between two sets of continuous variables. However, if one of the sets represents a dummy-coded categorical variable, then you may use the *Discriminant Analysis* module or the *ANOVA/MANOVA* module instead. If you are primarily interested in the *structure* of variables, then the *Factor Analysis* module may be an appropriate alternative. If one of the lists consists of only a single (dependent) variable, then the *Multiple Regression* module may be used, which in that case will produce results identical to those of the *Canonical Correlation* module.

Confirmatory models. If you have specific hypotheses about the relationships between the different variables in your data, you can test those hypotheses via the *Structural Equation Modeling (SEPATH)* procedure (see Chapter 11). With *SEPATH* you can fit (confirmatory) factor models, path models, models with intercepts, etc. Also, models can be fit to multiple samples, and parameter estimates can be constrained (to be equal) across samples.

EXAMPLE

Overview

The following example is based on a fictitious data set describing a study of life satisfaction. This data set is also analyzed in the *Example* section of the *Factor Analysis* chapter (Chapter 6).

Suppose that a questionnaire was administered to a random sample of 100 adults. The questionnaire contained 10 items that were designed to measure satisfaction at work, satisfaction with hobbies (leisure time satisfaction), satisfaction at home, and general satisfaction in other areas of life. Responses to all questions were recorded via computer, and scaled so that the mean for all items was approximately 100.

The results were entered into the data file *Factor.sta* (see the partial listing shown below).

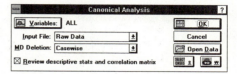

Purpose of Analysis

Suppose that you want to learn about the relationship of work satisfaction to satisfaction in other domains. Conceptually, you will treat the work satisfaction items as the explanatory or independent variables, and the other satisfaction items as the dependent variables.

Initial Computations

After opening the *Canonical Correlation* module, open the data file *Factor.sta*. Click on the *Variables* button in the startup panel and select all variables in this file.

Canonical correlation analysis is based on the correlation matrix of variables. Therefore, the first step of the analysis is to compute that matrix (unless a matrix input file is specified via the *Input* combo box). Note that you can later select variables for the analysis (for the two sets) from among those that are specified at this point.

Set the *Review desc. stats, corr. matrix* check box in the startup panel in order to compute the detailed descriptive statistics (i.e., means, correlations, covariances) for the variables in the current analysis. Now, click *OK* in the startup panel to open the *Review Descriptive Statistics* dialog.

Plots. In order to visualize the distribution of the variables, two types of plots are available from this dialog: *Box & whisker plot of vars* and *Matrix plot of correlations*.

Click on the *Box & whisker plot of vars* button and select all variables in the data set. A window will

4. CANONICAL CORRELATION - EXAMPLE

open in which you can select from four types of box and whisker plots (for this example, select *median/quart./range*). The central tendency (e.g., median) and variability (e.g., quartile and range) of the selected variables are displayed in these plots (note that the specific method for computing the quartiles and the median can be configured in the *STATISTICA Defaults: General* dialog; refer to the description of that dialog in the *General Conventions* chapter in Volume I for details).

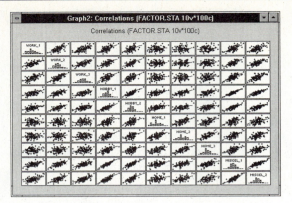

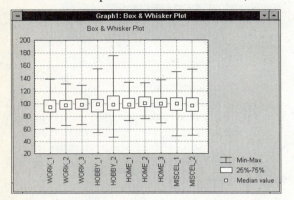

Box and whisker plots are useful to determine if the distribution of a variable is symmetrical. If the distribution is not symmetrical, then you may want to view the histogram for the respective variable (see below).

Now, click on the *Matrix plot of correlations* button to display a matrix of scatterplots between selected variables. These plots (as well as scatterplots from Scrollsheets via the *Quick Stats Graphs* options) should be examined for outliers, which may greatly bias the computation of the correlation coefficients, and thus the canonical analysis (see the discussion in the *Introductory Overview* section, page 3144).

As in most other modules, the default plot for the Scrollsheet of means and standard deviations is the histogram of the distribution of the respective variable. This histogram will show the normal curve superimposed over the observed distribution, to provide a visual check for any violations of the normality assumption. For example, to produce the histogram for variable *Work_1*, click with the right-mouse-button on the mean for variable *Work_1* and then select the *Quick Stats Graphs - Histogram/Normal* option in the flying menu.

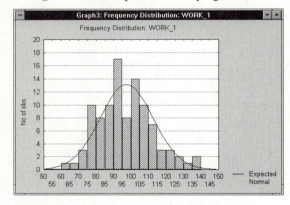

The distribution of this variable (of subjects' responses to the first item) follows the normal distribution. Thus, there is little reason to suspect that this variable violates the normality assumption.

Specifying the Canonical Analysis

To proceed with the canonical correlation analysis, click *OK* to exit the *Review Descriptive Statistics* dialog and open the *Model Definition* dialog.

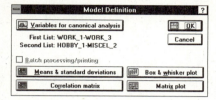

Specifying variables. To specify the two variable lists, click on the *Variables for canonical analysis* button to open the standard two variable selection window. Select the work satisfaction items (i.e., variables *Work_1*, *Work_2*, and *Work_3*) into the first list, and the remaining satisfaction items (i.e., variables *Hobby_1*, *Hobby_2*, *Home_1*, *Home_2*, *Home_3*, *Miscel_1*, *Miscel_2*) into the second list.

Note that the designation of *first* and *second* list is arbitrary here, that is, you could also select the work satisfaction items into the second list and the remaining satisfaction items into the first list. In a sense, canonical analysis is completely "symmetrical," that is, it will compute the same statistics (loadings, weights, etc.) for the variables in both lists.

This dialog offers some other options (means and standard deviations, correlations, and graphic options) which are discussed on page 3157.

Reviewing Results

After specifying the two lists of variables, you are now ready to begin the analysis; thus click *OK*. After a few moments, the *Canonical Analysis Results* dialog will be displayed.

4. CANONICAL CORRELATION - EXAMPLE

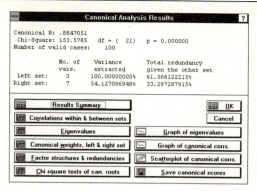

The nature of most of the statistics reported here in the results summary are reviewed in the *Introductory Overview* section of this chapter (page 3141). Therefore, the focus in this section will be on the interpretation of the results.

Click on the *Results Summary* button to bring up the canonical analysis results Scrollsheet.

Canonical R

The overall canonical R is fairly substantial (.88), and highly significant ($p<.001$). Remember that the canonical R reported here pertains to the first and most significant canonical root. Thus, this value can be interpreted as the simple correlation between the weighted sum scores in each set, with the weights pertaining to the first (and most significant) canonical root.

Overall Redundancy

The values in the rows labeled *Variance extracted* and *Total redundancy* give an indication of the

magnitude of the overall correlations between the two sets of variables, relative to the variance of the variables. This is different from the canonical *R-square*, because that statistic expresses the proportion of variance accounted for in the canonical variates (see also page 3143 in the *Introductory Overview* for further details).

Variance extracted. The values in the *Variance extracted* row indicate the average amount of variance extracted from the variables in the respective set by *all canonical roots*. Thus, all three roots extract *100%* of the variance from the left set, that is, the three work satisfaction items, and *54%* of the variance in the right set. Note that one of these values will always be 100% because the program extracts as many roots as the minimum number of variables in either set. Thus, for one set of variables, there are as many independent sums (canonical variates) as there are variables. Intuitively, it should be clear that, for example, three independent sum scores derived from three variables will explain 100% of all variability.

Redundancy. The computation of the *Total redundancy* is explained in the *Introductory Overview* section of this chapter (page 3143). These values can be interpreted such that, based on *all* canonical roots, given the right set of variables (the seven non-work related satisfaction items), you can account for, on the average, *61.6%* of the variance in the variables in the left set (the satisfaction items). Likewise, you can account for *33.3%* of the variance in the non-work related satisfaction items, given the work related satisfaction items. These results suggest a fairly strong overall relationship between the items in the two sets.

Testing the Significance of Canonical Roots

Now, check whether all three canonical roots are significant. Keep in mind that the canonical *R* reported in this Scrollsheet represents only the first root, that is, the strongest and most significant canonical correlation. To test the significance of all canonical roots, click on the *Chi-square tests of can. roots* button. These results are shown below:

CANONICL ANALYSIS	Canonicl R	Canonicl R-sqr.	Chi-sqr.	df	p	Lambda Prime
0	.884705	.782703	153.5785	21	.000000	.193486
1	.271080	.073484	10.8516	12	.541689	.890422
2	.197374	.038957	3.7153	5	.591097	.961043

The maximum number of roots that can be extracted is equal to the smallest number of variables in either set. Since three work satisfaction items were selected into the first set, the program would be expected to extract three canonical roots.

The sequential significance test works as follows (see also, page 3141). First, look at all three canonical variables together, that is, without any *roots removed*. That test is highly significant. Next, the first (and, as you know, most significant) root is "removed" and the statistical significance of the remaining two roots is determined. That test (in the second row of the Scrollsheet) is not significant. You can stop at this point and conclude that only the first canonical root is statistically significant, and it should be examined further. If the second test were also statistically significant, you would then proceed to the third line of the Scrollsheet to see whether the remaining third canonical root is also significant.

Factor Structure and Redundancy

You now know that you should consider further only the first canonical root. How can this root be interpreted, that is, how is it correlated with the variables in the two sets? As discussed in the *Introductory Overview* (page 3142), the interpretation of canonical "factors" follows a similar logic to that in factor analysis (*Factor Analysis*, Chapter 6). Specifically, you can compute the correlations between the items in each set with the respective canonical root or variable (remember that the canonical variable in each set is "created" as the weighted sum of the variables). Those

4. CANONICAL CORRELATION - EXAMPLE

correlations are also called canonical factor *loadings* or *structure coefficients*. You can compute those values (as well as the variance extracted for each set) via the *Factor structures & redundancies* button. Click on this button to view these Scrollsheets.

Factor structure in the left set.
First, examine the loadings for the left set.

Factor Structure, left set (factor.sta)			
CANONICAL ANALYSIS	Root 1	Root 2	Root 3
WORK_1	.796461	-.575348	.186075
WORK_2	.952643	.100725	-.286926
WORK_3	.875390	.217192	.431880

Remember that only the first canonical root is statistically significant, and it is the only one that should be interpreted. As you can see, all three work satisfaction items show substantial loadings on the first canonical factor, that is, they correlate highly with that factor.

As a measure of redundancy, the average amount of variance accounted for in each item by the first root could be computed. To do so, you could sum up the squared canonical factor loadings and divide them by *3* (the number of variables in this set). This Scrollsheet is shown below (click on the *Factor structures & redundancies* button in the *Results* dialog).

Variance Extracted (Proportions), left set (factor.sta)		
CANONICL ANALYSIS	Variance extractd	Reddncy.
Root 1	.769395	.602208
Root 2	.129448	.009512
Root 3	.101157	.003941

As you can see, the first canonical root extracts an average of about 77% of the variance from the work satisfaction items. If you multiply that value with the proportion of shared variance between the canonical variates in the two sets (i.e., with *R-square*), then the number in the second column of the Scrollsheet (*redundancy*) is obtained. Thus, given the variables in the right set (the non-work related satisfaction items), you can account for about 60% of the variance in the work related satisfaction items, based on the first canonical root.

Factor structure in right set.
Apparently, the first canonical root or factor in this Scrollsheet is marked by high loadings on the leisure satisfaction items (*Hobby_1* and *Hobby_2*).

Factor Structure, right set (factor.sta)			
CANONICAL ANALYSIS	Root 1	Root 2	Root 3
HOBBY_1	.823914	-.099577	.055376
HOBBY_2	.808859	.277944	-.149747
HOME_1	.186554	-.017177	.236525
HOME_2	.238228	.253170	.456321
HOME_3	.278220	.414321	.173396
MISCEL_1	.861283	.048749	.328388
MISCEL_2	.820764	.245223	.241010

The loadings are much lower for the home related satisfaction items. Therefore, you can conclude that the significant canonical correlation between the variables in the two sets (based on the first root) is probably the result of a relationship between work satisfaction, and leisure time and general satisfaction. If you consider work satisfaction as the explanatory variable, you could say that work satisfaction affects leisure time and general satisfaction, but not (or much less so) satisfaction with home life.

The Scrollsheet with the redundancies for the right set of variables is shown below.

Variance Extracted (Proportions), right set (factor.sta)		
CANONICAL ANALYSIS	Variance extractd	Reddncy.
Root 1	.416787	.326221
Root 2	.055104	.004049
Root 3	.069379	.002703

As you can see, the first canonical root accounts for an average of roughly 42% of variance in the variables in the right set; given the work satisfaction items, you can account for about 33% of the variance in the other satisfaction items, based on the first canonical root. Note that these numbers are "pulled down" by the relative lack of correlations between this canonical variate and the home satisfaction items.

Canonical Scores

Remember that the canonical variates represent weighted sums of the variables in each set. You can review those weights in separate Scrollsheets by

4. CANONICAL CORRELATION - EXAMPLE

clicking on the *Canonical weights, left and right set* button.

CANONICL ANALYSIS	Canonical Weights, left set (factor.sta)		
	Root 1	Root 2	Root 3
WORK_1	.217021	-1.35890	.24350
WORK_2	.592485	.37139	-1.38769
WORK_3	.300124	.83222	1.28861

CANONICL ANALYSIS	Canonical Weights, right set (factor.sta)		
	Root 1	Root 2	Root 3
HOBBY_1	.249966	-1.63580	-1.29069
HOBBY_2	.221850	1.00467	-1.09591
HOME_1	-.156475	-.43401	.30966
HOME_2	-.447900	.32580	.45678
HOME_3	-.066126	.69965	-.31653
MISCEL_1	.615556	-.36114	1.73651
MISCEL_2	.290859	.79337	.45774

The weights above pertain to the standardized (*z* transformed) variables in the two sets. You may use those weights to compute scores for the canonical variates. The scores computed from the data in the current data file may be saved via the *Save canonical scores* button.

Plotting canonical scores. Now, plot the canonical scores for the variables in the left set against the scores for the variables in the right set. Click on the *Scatterplot of canonical corr.* button to open the *Scatterplot of Canonical Correlation* dialog. To produce a scatterplot for the first (and only significant) canonical variate, select *Root 1* in the left set and *Root 1* in the right set.

Now, click *OK* to produce the desired scatterplot. (Note that a linear regression line was added to the plot via the *Plot Layout* dialog options.)

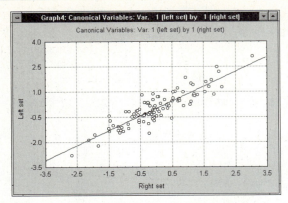

There are no outliers apparent in this plot, nor do the residuals around the regression line indicate any non-linear trend (e.g., by forming a *U* or *S* around the regression line). Therefore, you can be satisfied that no major violations of a main assumption of canonical correlation analysis are evident.

Clusters of cases. Another interesting aspect of this plot is whether or not there is any evidence of clustering of cases. Such clustering may happen if the sample is somehow heterogeneous in nature. For example, suppose respondents from two very different industries who are working under very different conditions were included in the sample. It is conceivable that the canonical correlation represented by the plot above could then be the result of the fact that one group of respondents is generally more satisfied with their work and leisure time (and life in general). If so, this would be reflected in this plot by two distinct clusters of points: one at the low ends of both axes and one at the high ends. However, in this example, there is no evidence of any natural grouping of this kind, and therefore you do not have to be concerned.

Conclusion

It can be concluded from the analysis of this (fictitious) data set that satisfaction at work affects leisure time satisfaction and general satisfaction. Satisfaction with home life did not seem to be affected. In practice, before generalizing these conclusions, you should replicate the study. Specifically, you should ensure that the canonical factor structure that led to the interpretation of the first canonical root is reliable (i.e., replicable).

4. CANONICAL CORRELATION - EXAMPLE

DIALOGS, OPTIONS, STATISTICS

Startup Panel

In this dialog, you can select all variables that are to be included in the analysis.

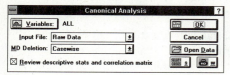

Variables

Clicking on this button will bring up the standard variable selection window in which you can select the variables that will be used later in the canonical analysis. Only variables that are selected at this point will be available later for the specification of the canonical correlation. A correlation matrix will be calculated for all variables that are selected in this dialog.

Input File

This combo box has two settings: *raw data* and *correlation matrix*.

Raw data file. If you set the combo box to *raw data*, then the program expects a standard raw data file as input.

Matrix file. If you set the combo box to *matrix*, then you will need to select a correlation matrix file as the input file. Correlation matrix files can be created from within various other modules (e.g., *Basic Statistics and Tables*, *Factor Analysis*, *Multiple Regression*, etc.), or they can be created directly via *Data Management* (see *Matrix File Format*, Volume I).

MD Deletion

This combo box is only active if a raw data file has been specified (see above). Missing data can be deleted casewise or substituted by means (mean substitution).

Review Desc. Stats, Corr. Matrix

Select this check box in order to open the *Review Descriptive Statistics* dialog after you leave the startup panel. In the *Review Descriptive Statistics* dialog, you can review detailed descriptive statistics for the variables selected above and save the correlation matrix. Click *OK* in the *Review Descriptive Statistics* dialog in order to continue the analysis and open the *Model Definition* dialog.

Note that the *Means & standard deviations* button in the *Model Definition* dialog will bring up a Scrollsheet with the means and standard deviations of the variables selected in the startup panel.

Review Descriptive Statistics

This dialog allows you to compute detailed descriptive statistics for the variables in the current analysis.

Means & SD

This option will bring up a Scrollsheet with the descriptive statistics for the variables in the analysis. If a raw data file is analyzed, then the default statistical graph for that Scrollsheet is the histogram with fitted normal distribution.

Correlations

This option will bring up a Scrollsheet with the current correlation matrix. If a raw data file is analyzed, then the default statistical graph for that Scrollsheet is the scatterplot with the fitted regression line.

Covariances

This option will bring up a Scrollsheet with the current covariance matrix. If a raw data file is analyzed, then the default statistical graph for that Scrollsheet is the scatterplot with the fitted regression line.

Save Correlations

This option allows you to save the current correlation matrix in *STATISTICA Matrix File Format* (see *Data Management*, Volume I). The correlation matrix can then be directly read in subsequent canonical analyses (or analyses via *Multiple Regression*, *Factor Analysis*, *Reliability and Item Analysis*, *Cluster Analysis*, *Multi-dimensional Scaling*, etc.).

Box and Whisker Plots

This option allows you to produce box and whisker plots for the variables in the current analysis (see page 3147). The box and whisker plot will summarize each variable by three components:

(1) A central line to indicate central tendency or location;

(2) A box to indicate variability around this central tendency;

(3) Whiskers around the box to indicate the range of the variable.

After clicking on this button you can choose to plot for each variable:

(1) Medians (central line), quartiles (box), and ranges (whiskers); note that the specific method that is used to compute these values can be configured in the *STATISTICA Defaults: General* dialog (see the *General Conventions* chapter in Volume I);

(2) Means, standard errors of the means, and standard deviations;

(3) Means, standard deviations, and 1.96 times the standard deviations (95% normal confidence interval for individual observations around the mean);

(4) Means, standard errors of the means, and 1.96 times the standard errors of the means (95% normal confidence interval for means).

Many other types of box and whisker and range plots are available via the *Stats 2D Graphs* option in the *Graphs* pull-down menu.

Matrix Plot of Correlations

This option will bring up a matrix scatterplot for the variables selected for the analysis (see page 3148).

The SD Check Box

There are two ways in which standard deviations can be computed: as the sums of (deviation) squares divided by n (the valid number of cases), or $n-1$.

(1) If divided by $n-1$, the resulting standard deviations are estimates of the standard deviations in the population (*sigma*).

(2) If divided by n, then the resulting standard deviations are descriptions of the sample only.

The setting of this switch will not affect the subsequent computations for the canonical analysis.

Define Model

This dialog will allow you to specify the variables for the canonical analysis as well as other options described below.

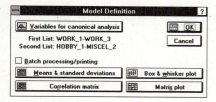

Variables for Canonical Analysis

Clicking on this option will bring up the standard two variable list selection window.

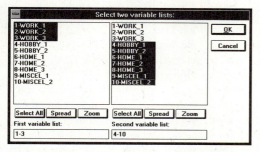

Select the two sets of variables for the canonical correlation analysis in this dialog.

Batch Processing/ Printing

This check box is only available if *printer*, *disk file* output, or *Text/output Window* was selected in the *Page/Output Setup* dialog (see Volume I). When this box is checked, then the entire output from the analysis will be printed without requiring any user input.

Means and Standard Deviations

Clicking on this button will bring up a Scrollsheet with the means and standard deviations for all variables (selected from the startup panel). The default graph for this Scrollsheet is the histogram of the frequency distribution for the respective variable (see page 3148). That plot will also indicate the normal distribution, so it may serve as a quick visual check for any violations of the normality assumption.

Box and Whisker Plot

This option allows you to produce a box and whisker plot (showing medians and quartiles, means and standard deviations, etc.) for selected variables (see page 3156).

Correlation Matrix

This option will bring up a Scrollsheet with the correlation matrix for all variables (selected from the startup panel). The default plot for this Scrollsheet is the scatterplot of the respective two variables that are highlighted.

Matrix Plot

This option allows you to produce a scatterplot matrix for selected variables.

Results

The canonical correlation reported in the summary section at the top of this dialog denotes the correlation between the first and most significant canonical variates in each set (the first canonical root that is extracted). The interpretation of the values reported in the columns labeled *variance extracted* and *total redundancy* are explained under the option *Factor structures and redundancies* below and on page 3149.

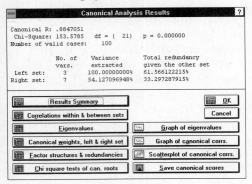

Correlations Within and Between Sets

This button will bring up Scrollsheets with the correlations within each set of variables, and between sets.

Eigenvalues

This button will bring up a Scrollsheet with the eigenvalues for the canonical roots. Note that the square roots of the eigenvalues are the canonical correlation coefficients.

Canonical Weights for Left and Right Set

This button will bring up a Scrollsheet with the canonical weights for each set. These weights can be used to compute the canonical scores for each canonical root, for each set of variables.

Note, that the weights pertain to the standardized (i.e., z transformed) variables in each set. The canonical weights may also be used in the interpretation of the canonical roots.

Factor Structures and Redundancies

This button will bring up Scrollsheets with the canonical factor loadings and redundancies (variance extracted) for both sets of variables. The *canonical factor loadings* can be interpreted as in factor analysis (see *Factor Analysis*, Chapter 6), that is, they represent the correlations between the variables with the respective canonical variates. The *variance extracted* is computed by summing up the squared canonical factor loadings across variables in a set for a particular canonical root and then dividing that sum by the number of variables in the set. The resulting proportion can be interpreted as the average proportion of variance accounted for in the respective variables by the respective root. The total proportion of variance extracted that is reported at the top of the results dialog can be interpreted as the average proportion of variance accounted for by *all* canonical roots. *Redundancies* are computed by multiplying the proportion of variance extracted by the squared canonical correlation. The *redundancies* for a particular root can be interpreted as the average proportion of variance accounted for in the respective set of variables by that root, given the variables in the other set. The *total redundancy* reported at the top of the results dialog is the sum of the redundancies across *all* roots.

Chi-square Tests of Successive Roots

This button will bring up a Scrollsheet reporting, for each canonical root, the canonical R, R-*square*, *Chi-square*, degrees of freedom, p-level, and *lambda* value.

Sequential Significance Test

The results displayed in this Scrollsheet can be used to decide how many canonical roots are statistically significant and should be examined further (that is, interpreted). The so-called *sequential significance test* works as follows (see also *Introductory Overview*, page 3141).

(1) First, look at all canonical roots together, that is, without any *roots removed*. That test (*Chi-square* and *p*-level) is reported in the first row of the Scrollsheet. If that test is significant, then conclude that at least one canonical root is statistically significant.

(2) Next, the first (and most significant) root is "removed," and the statistical significance of the remaining roots is determined. That test is reported in the second row of the Scrollsheet. If significant, conclude that at least two canonical roots are statistically significant. If not, stop and only retain the first root for interpretation.

(3) If the second root is also significant, then that root is "removed," and the statistical significances of each of the remaining roots are determined, and so on.

Save Canonical Scores

This option is only available if a raw data file was specified in the *Canonical Analysis* startup panel. After clicking on this button, the user will be prompted for an output file name. A variable selection window will then be displayed allowing the user to select the variables from the input file that are to be saved together with the canonical scores.

Summary

This option will bring up a Scrollsheet with the summary statistics for the current analysis.

Plot of Eigenvalues

After choosing this option, a line graph of eigenvalues will be displayed (see page 3141).

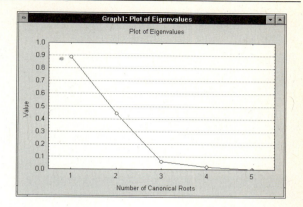

Plot of Canonical Correlations

After clicking on this button, a line graph of canonical correlations (square roots of eigenvalues) will be displayed (see page 3141).

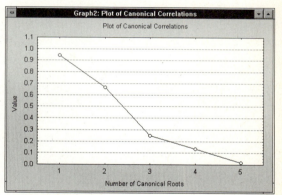

Scatterplot of Canonical Correlations

This option is only available if raw data are analyzed (i.e., *raw data file* was specified in the *Canonical Correlation* startup panel, page 3155). When you click on the *Scatterplot of canonical corr* button, the *Scatterplot of Canonical Correlations* dialog will

4. CANONICAL CORRELATION - DIALOGS AND OPTIONS

open in which you can select the canonical variables to plot.

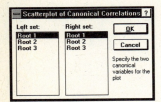

For example, you can produce the following graph:

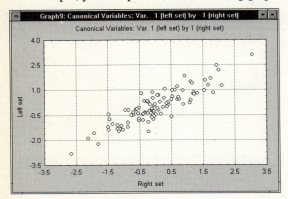

TECHNICAL NOTES

To maximize the precision of computations, a modified Gauss-Jordan algorithm for matrix inversion is used and eigenvectors are extracted from the Householder reduced matrix via a QL algorithm with implicit shifts.

The standard formulas are used for computing the canonical R, canonical weights, canonical loadings and significance levels (see, for example, Bolch and Huang, 1974; Cooley and Lohnes, 1971; Enslein, Ralston, and Wilf, 1977; Finn, 1974; Longley, 1984; Tatsuoka, 1971; Wherry, 1984; Younger, 1985).

4. CANONICAL CORRELATION - TECHNICAL NOTES

INDEX

A

Assumptions
    normality, 3144
    sample size, 3144

B

Box and whisker plots for all variables, 3145, 3147, 3156

C

Canonical correlation
    assumptions, 3144
    canonical R, 3149
    canonical roots, 3140, 3150
    canonical scores, 3142, 3151
    canonical variates, 3140
    canonical weights (coefficients), 3142, 3158
    clustering of cases, 3152
    correlations, 3141, 3159
    descriptive statistics, 3155
    eigenvalues, 3141, 3158, 3159
    example, 3147
    factor loadings, 3142, 3151
    factor structure, 3142
    factor structure vs. canonical weights, 3142
    factor structures and redundancy, 3150, 3158
    input data, 3145
    matrix file input, 3145
    matrix ill-conditioning, 3144
    missing data, 3155
    outliers, 3144
    overview, 3139
    redundancy, 3143, 3149, 3150
    saving canonical scores, 3145, 3159
    sequential significance test, 3141, 3150, 3158
    specifying the canonical analysis, 3149
    startup panel, 3155
    structure coefficients, 3151
    sum scores, 3139

Canonical correlation (continued)
    variance extracted, 3143, 3150
Canonical R, 3149
Canonical roots, 3140
Canonical scores, 3142, 3151
Canonical variates, 3140
Canonical weights (coefficients), 3142, 3158
Chi-square tests
    test of successive roots, 3158
Clustering of cases, 3152

E

Eigenvalues, 3141, 3158, 3159

F

Factor loadings, 3151
Factor structure, 3142
Factor structure vs. canonical weights, 3142
Factor structures and redundancy, 3150, 3158

G

Gauss-Jordan algorithm, matrix inversion, 3161

H

Histograms
    2D histograms, 3148
Householder reduced matrix, 3161

I

Input data, 3145

M

Matrix file input, 3145
Matrix ill-conditioning, 3144
Missing data, 3155

P

Proportion of variability explained, 3140, 3141, 3143, 3151

Q

QL algorithm with implicit shifts, 3161

R

Redundancy, 3143, 3149, 3150, 3158

S

Sample Size, 3144
Saving canonical scores, 3145, 3159
Scatterplots
    matrix scatterplots, 3148
    scatterplot of canonical correlations, 3159
Sequential significance test, 3141, 3150, 3158
Structure coefficients, 3151
Sum scores, 3139

V

Variance extracted, 3143, 3150

Chapter 5:
CLUSTER ANALYSIS

Table of Contents

Introductory Overview 3169
    Overview 3169
    Joining (Tree Clustering) 3169
    Two-way Joining 3172
    K-means Clustering 3173
Program Overview 3175
Examples 3177
    Overview 3177
    Example 1: Joining (Tree Clustering) 3177
    Example 2: K-means Clustering 3181
Dialogs, Options, Statistics 3185
    Cluster Analysis - Startup Panel 3185
    Joining (Tree Clustering) Dialog 3185
    K-means Clustering Dialog 3186
    Two-way Joining (Block Clustering) Dialog 3188
    Joining (Tree Clustering) - Results Dialog 3188
    K-means Clustering - Results Dialog 3190
    Two-way Joining (Block Clustering) - Results Dialog 3192
Index 3195

Detailed Table of Contents

INTRODUCTORY OVERVIEW 3169
    Overview 3169
        Statistical Significance Testing 3169
        Areas of Application 3169
    Joining (Tree Clustering) 3169
        General Logic 3169
        Hierarchical Tree 3170
        Distance Measures 3170
        Amalgamation or Linkage Rules 3171
    Two-way Joining 3172
    K-means Clustering 3173

5. CLUSTER ANALYSIS - CONTENTS

PROGRAM OVERVIEW **3175**
    Methods 3175
    Distance Measures 3175
    Amalgamation/ Linkage Rules 3175
    Output 3175
    Alternative Procedures 3175

EXAMPLES **3177**
    Overview 3177
        Data File 3177
        Scale of Measurement 3177
        Purpose of the Analysis 3177
    Example 1: Joining (Tree Clustering) 3177
        Specifying the Analysis 3177
        Results 3178
        Identifying Clusters 3180
    Example 2: K-means Clustering 3181
        Specifying the Analysis 3181
        Results 3181

DIALOGS, OPTIONS, STATISTICS **3185**
    Cluster Analysis - Startup Panel 3185
    Joining (Tree Clustering) Dialog 3185
        Variables 3185
        Input Data 3185
        Cluster Cases or Variables 3185
        Amalgamation (linkage) Rules 3185
        Distance Measures 3186
        Missing Data Deletion 3186
        Batch Processing and Printing 3186
    K-means Clustering Dialog 3186
        Variables 3186
        Cluster Cases or Variables 3187
        Number of Clusters 3187
        Maximum Number of Iterations 3187
        Missing Data Deletion 3187
        Initial Cluster Centers 3187
        Batch Processing and Printing 3188
    Two-way Joining (Block Clustering) Dialog 3188
        Variables 3188
        Threshold Value 3188
        Missing Data Deletion 3188
        Batch Processing and Printing 3188
    Joining (Tree Clustering) - Results Dialog 3188
        Hierarchical Tree Plot Options 3189

CLU - 3166

5. CLUSTER ANALYSIS - CONTENTS

    Amalgamation Schedule 3190
    Graph of Amalgamation Schedule 3190
    Distance Matrix 3190
    Descriptive Statistics 3190
    Save Distance Matrix 3190
  K-means Clustering - Results Dialog 3190
    Analysis of Variance 3190
    Means of Each Cluster & Distances 3191
    Graph of Means 3191
    Descriptive Statistics for Each Cluster 3191
    Cluster Means and Euclidean Distances 3191
    Save Classifications & Distances 3192
  Two-way Joining (Block Clustering) - Results Dialog 3192
    Descriptive Statistics for Cases (Rows) 3192
    Descriptive Statistics for Variables (Columns) 3192
    Reordered Data Matrix 3192
    Two-way Joining Graph 3193

INDEX **3195**

5. CLUSTER ANALYSIS - CONTENTS

Chapter 5:
CLUSTER ANALYSIS

INTRODUCTORY OVERVIEW

Overview

The term *cluster analysis* actually encompasses a number of different classification algorithms. A general question facing researchers in many areas of inquiry is how to *organize* observed data into meaningful structures, that is, to develop taxonomies. For example, biologists have to organize the different species of animals before a meaningful description of the differences between animals is possible. According to the modern system employed in biology, man belongs to the primates, the mammals, the amniotes, the vertebrates, and the animals.

Note how in this classification, the higher the level of aggregation, the less similar are the members in the respective class. Man has more in common with all other primates (e.g., apes) than it does with the more "distant" members of the mammals (e.g., dogs), etc. The general categories of cluster analysis methods [*joining* (*tree clustering*), *two-way joining* (*block clustering*), and *k-means clustering*] will be reviewed in the following sections.

Statistical Significance Testing

Note that the above discussions refer to clustering *algorithms* and do not mention anything about statistical significance testing. In fact, cluster analysis is not as much a typical statistical test as it is a "collection" of different algorithms that "put objects into clusters." The point here is that, unlike many other statistical procedures, cluster analysis methods are mostly used when you do *not* have any *a priori* hypotheses, but are still in the exploratory phase of your research. In a sense, cluster analysis finds the "most significant solution possible." Therefore, statistical significance testing is really not appropriate here, even in cases when *p*-levels are reported (as in *k-means clustering*).

Areas of Application

Clustering techniques have been applied to a wide variety of research problems. Hartigan (1975) provides an excellent summary of the many published studies reporting the results of cluster analyses.

For example, in the field of medicine, clustering diseases, cures for diseases, or symptoms of diseases can lead to very useful taxonomies. In the field of psychiatry, the correct diagnosis of clusters of symptoms such as paranoia, schizophrenia, etc. is essential for successful therapy. In archeology, researchers have attempted to establish taxonomies of stone tools, funeral objects, etc. by applying cluster analytic techniques.

In general, whenever one needs to classify a "mountain" of information into manageable, meaningful piles, cluster analysis is of great utility.

Joining (Tree Clustering)

General Logic

The above examples illustrate the goal of the joining or *tree clustering* algorithm. The purpose of this algorithm is to join together objects (e.g., animals) into successively larger clusters, using some measure of similarity or distance. A typical result of this type of clustering is the hierarchical tree.

5. CLUSTER ANALYSIS - INTRODUCTORY OVERVIEW

Hierarchical Tree

Consider the *Horizontal Hierarchical Tree Plot* shown below. The plot begins with each object in a class by itself (on the left of the plot). Now imagine that, in very small steps, you "relax" your criterion as to what is and is not unique. Put another way, you lower your threshold regarding the decision when to declare two or more objects to be members of the same cluster.

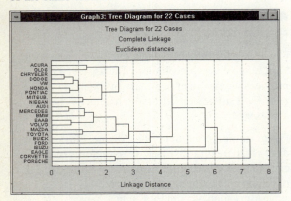

As a result, you *link* more and more objects together and aggregate (*amalgamate*) larger and larger clusters of increasingly dissimilar elements. Finally, in the last step, all objects are joined together. In these plots, the horizontal axis denotes the linkage distance (in *Vertical Icicle Plots*, the vertical axis denotes the linkage distance). Thus, for each node in the graph (where a new cluster is formed) you can read off the criterion distance at which the respective elements were linked together into a new single cluster. When the data contain a clear "structure" in terms of clusters of objects that are similar to each other, then this structure will often be reflected in the hierarchical tree as distinct branches. As the result of a successful analysis with the joining method, one is able to detect clusters (branches) and interpret those branches.

Distance Measures

The joining or tree clustering method uses the dissimilarities or distances between objects when forming the clusters. These distances can be based on a single dimension or multiple dimensions.

For example, if you were to cluster fast foods, you could take into account the number of calories they contain, their price, subjective ratings of taste, etc. The most straightforward way of computing istances between objects in a multi-dimensional space is to compute Euclidean distances. If you had a two- or three-dimensional space, then this measure is the actual geometric distance between objects in the space (i.e., as if measured with a ruler).

However, the joining algorithm does not "care" whether the distances that are "fed" to it are actual real distances or some other derived measure of distance that is more meaningful to the researcher; and it is up to the researcher to select the right method for his/her specific application. The *Cluster Analysis* module will compute various types of distance measures, or the user can compute a matrix of distances independently and use it directly in the procedure.

Euclidean distance. This is probably the most common type of distance chosen. It simply is the geometric distance in the multidimensional space. It is computed as:

`distance(x,y)` $= [\Sigma_i (x_i - y_i)^2]^{\frac{1}{2}}$

Squared Euclidean distance. One may want to square the standard Euclidean distance in order to place progressively greater weight on objects that are further apart. This distance is computed as:

`distance(x,y)` $= \Sigma_i (x_i - y_i)^2$

City-block (Manhattan) distance. This distance is simply the average difference across dimensions. In most cases, this distance measure yields results similar to the simple Euclidean distance. However, note that in this measure, the effect of single large differences (outliers) is dampened (since they are not squared). The city-block distance is computed as:

$$\text{distance}(x,y) = \Sigma_i |x_i - y_i|$$

Chebychev distance. This distance measure may be appropriate in cases when one wants to define two objects as "different" if they are different on any one of the dimensions. The Chebychev distance is computed as:

$$\text{distance}(x,y) = \text{Maximum} |x_i - y_i|$$

Power distance. Sometimes one may want to increase or decrease the progressive weight that is placed on dimensions on which the respective objects are very different. This can be accomplished via the *power distance*. The power distance is computed as:

$$\text{distance}(x,y) = (\Sigma_i |x_i - y_i|^p)^{1/r}$$

where *r* and *p* are user-defined parameters.

A few example calculations may demonstrate how this measure "behaves." Parameter *p* controls the progressive weight that is placed on differences on individual dimensions, parameter *r* controls the progressive weight that is placed on larger differences between objects. If *r* and *p* are equal to 2, then this distance is equal to the Euclidean distance.

Percent disagreement. This measure is particularly useful if the data for the dimensions included in the analysis are categorical in nature. This distance is computed as:

$$\text{distance}(x,y) = (\text{Number of } x_i \neq y_i)/i$$

Amalgamation or Linkage Rules

At the first step, when each object represents its own cluster, the distances between those objects are defined by the chosen distance measure. However, once several objects have been linked together, how do you determine the distances between those new clusters? In other words, a linkage or amalgamation rule is needed to determine when two clusters are sufficiently similar to be linked together. There are various possibilities: for example, you could link two clusters together when *any* two objects in the two clusters are closer together than the respective linkage distance. Put another way, you use the "nearest neighbors" across clusters to determine the distances between clusters; this method is called *single linkage*. This rule produces "stringy" types of clusters, that is, clusters "chained together" by only single objects that happen to be close together.

Alternatively, you may use the neighbors across clusters that are furthest away from each other; this method is called *complete linkage*. There are numerous other linkage rules such as these that have been proposed, and the *Cluster Analysis* module offers a wide choice of them.

Single linkage (nearest neighbor). As described above, in this method the distance between two clusters is determined by the distance of the two closest objects (nearest neighbors) in the different clusters. This rule will, in a sense, *string* objects together to form clusters, and the resulting clusters tend to represent long "chains."

Complete linkage (furthest neighbor). In this method the distances between clusters are determined by the greatest distance between any two objects in the different clusters (i.e., by the "furthest neighbors"). This method usually performs quite well in cases when the objects actually form naturally distinct "clumps." If the clusters tend to be somehow elongated or of a "chain" type nature, then this method is inappropriate.

Unweighted pair-group average. In this method the distance between two clusters is calculated as the average distance between all pairs of objects in the two different clusters. This method is also very efficient when the objects form naturally distinct "clumps;" however, it performs equally well with elongated, "chain" type clusters. Note that in their book, Sneath and Sokal (1973) introduced the abbreviation UPGMA to refer to this method as *unweighted pair-group method using arithmetic averages*.

Weighted pair-group average. This method is identical to the unweighted pair-group average method, except that in the computations, the size of the respective clusters (i.e., the number of objects contained in them) is used as a weight. Thus, this method (rather than the previous method) should be used when the cluster sizes are suspected to be greatly uneven. Note that in their book, Sneath and Sokal (1973) introduced the abbreviation WPGMA to refer to this method as *weighted pair-group method using arithmetic averages*.

Unweighted pair-group centroid. The *centroid* of a cluster is the average point in the multidimensional space defined by the dimensions. In a sense, it is the *center of gravity* for the respective cluster. In this method the distance between two clusters is determined as the difference between centroids. Sneath and Sokal (1973) use the abbreviation UPGMC to refer to this method as *unweighted pair-group method using the centroid average*.

Weighted pair-group centroid (median). This method is identical to the previous one, except that weighting is introduced into the computations to take into consideration differences in cluster sizes (i.e., the number of objects contained in them). Thus, when there are (or one suspects there to be) considerable differences in cluster sizes, this method is preferable to the previous one. Sneath and Sokal (1973) use the abbreviation WPGMC to refer to this method as *weighted pair-group method using the centroid average*.

Ward's method. This method is distinct from all other methods because it uses an analysis of variance approach to evaluate the distances between clusters. In short, this method attempts to minimize the Sum of Squares (SS) of any two (hypothetical) clusters that can be formed at each step. Refer to Ward (1963) for details concerning this method. In general, this method is regarded as very efficient, however, it tends to create clusters of small size.

Two-way Joining

Previously, this method has been discussed in terms of "objects" that are to be clustered (see *Joining (Tree Clustering)*, page 3169). In all other types of analyses in *STATISTICA* the research question of interest is usually expressed in terms of cases (observations) or variables. It turns out that the clustering of both may yield useful results.

For example, imagine a study where a medical researcher has gathered data on different measures of physical fitness (variables) for a sample of heart patients (cases). The researcher may want to cluster cases (patients) to detect clusters of patients with similar syndromes. At the same time, the researcher may want to cluster variables (fitness measures) to detect clusters of measures that appear to tap similar physical abilities. In the *Cluster Analysis* module you can choose to cluster cases as well as variables.

Given the discussion in the paragraph above concerning whether to cluster cases or variables, one may wonder why not cluster both simultaneously? The *Cluster Analysis* module contains a two-way joining procedure to do exactly that. Two-way joining is useful in (the relatively rare) circumstances when one expects that both cases and variables will simultaneously contribute to the uncovering of meaningful patterns of clusters.

For example, returning to the illustration above, the medical researcher may want to identify clusters of patients that are similar with regard to particular clusters of similar measures of physical fitness. The difficulty with interpreting these results may arise from the fact that the similarities between different clusters may pertain to (or be caused by) somewhat different subsets of variables. Thus, the resulting structure (clusters) is by nature not homogeneous. This may seem a bit confusing at first, and indeed, compared to the other clustering methods described

(see *Joining (Tree Clustering)*, page 3169, and *K-means Clustering,* below), two-way joining is probably the least-used method. However, some researchers believe that this method offers a powerful exploratory data analysis tool (for more information, you may want to refer to the detailed description of this method in Hartigan, 1975).

K-means Clustering

General logic. This method of clustering is very different from the *Joining (Tree Clustering)* and *Two-way Joining* methods. Suppose that you already have hypotheses concerning the number of clusters in your cases or variables. You may want to "tell" the computer to form exactly 3 clusters that are to be as distinct as possible. This is the type of research question that can be addressed by the k-means clustering algorithm. In general, the k-means method will produce exactly k different clusters of greatest possible distinction.

Example. In the physical fitness example (see *Two-way Joining*, above), the medical researcher may have a "hunch" from clinical experience that her heart patients basically fall into three different categories with regard to physical fitness. She might wonder whether this intuition can be quantified, that is, whether a k-means cluster analysis of the physical fitness measures would indeed produce the three clusters of patients as expected. If so, the means on the different measures of physical fitness for each cluster would represent a quantitative way of expressing the researcher's hypothesis or intuition (i.e., patients in cluster 1 are high on measure 1, low on measure 2, etc.).

Computations. Computationally, you may think of this method as analysis of variance (ANOVA) "in reverse." The program will start with k random clusters, and then move objects between those clusters with the goal to:

(1) Minimize variability within clusters, and

(2) Maximize variability between clusters.

This is analogous to "ANOVA in reverse" in the sense that the significance test in ANOVA evaluates the between-group variability against the within-group variability when computing the significance test for the hypothesis that the means in the groups are different from each other. In k-means clustering, the program tries to move objects (e.g., cases) in and out of groups (clusters) to get the most significant ANOVA results. Because, among other results, the ANOVA results are part of the standard output from a k-means clustering analysis, you may want to refer to the *ANOVA/MANOVA* chapter (Volume I) to learn more about that method.

Interpretation of results. Usually, as the result of a k-means clustering analysis, you would examine the means for each cluster on each dimension to assess how distinct the k clusters are. Ideally, you would obtain very different means for most, if not all dimensions, used in the analysis. The magnitude of the F values from the analysis of variance performed on each dimension is another indication of how well the respective dimension discriminates between clusters.

5. CLUSTER ANALYSIS - INTRODUCTORY OVERVIEW

PROGRAM OVERVIEW

Methods

This module includes implementations of three clustering methods: *k-means clustering* (see pages 3173, 3181, 3186, 3190), *hierarchical clustering* (joining, tree-clustering; see pages 3169, 3177, 3185, 3188), and *two-way joining* (see pages 3172, 3188, 3192). The program can process data from either raw data files or matrices of distance measures (e.g., correlation matrices).

Distance Measures

The user can cluster cases, variables, or both based on a wide variety of distance measures: *Euclidean*, *Squared Euclidean*, *Manhattan* (city-block), *Chebychev*, *power distances*, *percent disagreement*, *correlation coefficients* (1-Pearson r). Refer to page 3170 for a description of each of these measures.

Amalgamation/ Linkage Rules

Available amalgamation/linkage rules for hierarchical (tree) clustering are single linkage, complete linkage, weighted and unweighted group averages, weighted and unweighted centroid, and Ward's method. Refer to page 3171 for descriptions of these rules.

Output

In addition to the standard cluster analysis output, a comprehensive set of descriptive statistics and extended diagnostics is available (e.g., the complete amalgamation schedule with cohesion levels in hierarchical clustering, the ANOVA table in k-means clustering). Graphics options in the *Cluster Analysis* module include tree diagrams (*horizontal hierarchical plots* and *vertical icicle plots*), two-way joining plots (contour plots), plots of means in k-means clustering, and graphs of amalgamation schedules, as well as a large selection of descriptive graphs (available to aid in the interpretation of the output).

Alternative Procedures

Alternative procedures for determining structure, and for explanatory multivariate data analysis are factor analysis (*Factor Analysis*, Chapter 6), structural equation modeling and confirmatory factor analysis (*SEPATH*, Chapter 11), multidimensional scaling (*Multidimensional Scaling*, Chapter 7), and discriminant function analysis (*Discriminant Analysis*, Chapter 2).

Cluster analysis vs. factor analysis. Factor analysis assumes that the data represent measurements on an interval scale and are distributed as (multivariate) normal variables. Factor analysis uses the correlations between variables (i.e., the similarities in variation of variables across cases) to identify underlying dimensions. These dimensions can then be thought of as "underlying" (latent) variables that correlate with observed measures. Cluster analytic techniques, specifically the joining (tree clustering) method, can be applied regardless of how the distance measures were computed, and there are no stringent requirements regarding underlying distributions or scale of measurement (as long as the distance measurements are valid). Also, clusters are purely descriptive in nature and provide only a *taxonomy* of objects in the analysis; no inferences regarding latent variables are implied by this method (however, note that one may argue that principal components analysis, unlike factor analysis, is also a merely *descriptive* data reduction method).

Cluster analysis vs. multidimensional scaling (MDS). MDS is similar to factor analysis regarding the interpretation of the results. The goal of MDS is to establish a space of fewer dimensions than there are objects in the analysis, based on distances (similarities) between objects. However, like cluster analysis techniques, MDS is also a classification *algorithm* in the sense that it will iteratively move objects around to produce their best representation (i.e., of the distances) in the *k* dimensional space. As in factor analysis, the resulting dimensions can be thought of as underlying (latent) variables that contain the common "essence" of all variables.

Cluster analysis vs. discriminant function analysis. Discriminant function analysis seems to bear some resemblance to k-means clustering; however, it is quite different in nature and interpretation. In discriminant function analysis you know beforehand to which group (cluster) each case belongs. The goal of the analysis is to determine which variables best discriminate between those observed groups. In k-means clustering you do *not* know *a priori* to which cluster each case belongs, nor are you sure of the nature (or sometimes even the number) of clusters that exist in the data. Thus, the goal of k-means clustering is to establish whether and how objects (e.g., cases) fall into groups (clusters), while discriminant function analysis assumes *a priori* knowledge of the number of groups and the group membership of each case (however, this particular distinction becomes blurred when one uses discriminant function analyses strictly to *classify* unclassified cases based on a previously established discriminant function).

EXAMPLES

Overview

Data File

The following example is based on a sample of different automobiles. Specifically, one particular model was randomly chosen from among those offered by the respective manufacturer. The following data for each car were then recorded:

- The approximate price of the car (variable *Price*),
- The acceleration of the car (0 to 60 in seconds; variable *Acceler*),
- The braking performance of the car (braking distance from 80 mph to complete standstill; variable *Braking*),
- An index of road holding capability (variable *Handling*), and
- The gas-mileage of the car (miles per gallon; variable *Mileage*).

Scale of Measurement

All clustering algorithms at one point need to assess the distances between clusters or objects, and obviously, when computing distances, you need to decide on a scale. Because the different measures included here used entirely different types of scales (e.g., number of seconds, thousands of dollars, etc.), the data were standardized (via the *Standardize Variables* option in the *Data Management* module; see Volume I) so that each variable has a mean of 0 and a standard deviation of 1. It is very important that the dimensions (variables in this example) that are used to compute the distances between objects (cars in this example) are of comparable magnitude; otherwise, the analysis will be biased and rely most heavily on the dimension that has the greatest range of values.

	1 PRICE	2 ACCELE	3 BRAKIN	4 HANDLI	5 MILAGE
Acura	-.521	.477	-.007	.382	2.079
Audi	.866	.208	.319	-.091	-.677
BMW	.496	-.802	.192	-.091	-.154
Buick	-.614	1.689	.933	-.210	-.154
Corvette	1.235	-1.811	-.494	.973	-.677
Chrysler	-.614	.073	.427	-.210	-.154
Dodge	-.706	-.196	.481	.145	-.154
Eagle	-.614	1.218	-4.199	-.210	-.677
Ford	-.706	-1.542	.987	.145	-1.724
Honda	-.429	.410	-.007	.027	.369
Isuzu	-.798	.410	-.061	-4.230	1.067
Mazda	.126	.679	-.133	.500	-1.724
Mercedes	1.051	.006	.120	-.091	-.154
Mitsub.	-.614	-1.003	.084	.382	.718
Nissan	-.429	.073	-.007	.263	.997
Olds	-.614	-.734	.409	.382	2.114

The standardized data for this example are contained in the file *Cars.sta*.

Purpose of the Analysis

Given these data, can a taxonomy for the automobiles included in the study be developed? In other words, do these automobiles form "natural" clusters that can be labeled in a meaningful manner? First, perform a joining analysis (tree clustering, hierarchical clustering) on this data.

Example 1: Joining (Tree Clustering)

Specifying the Analysis

After opening the *Cluster Analysis* module, open the data file *Cars.sta*. Select *Joining (tree clustering)* and click *OK* in the *Cluster Method* dialog. Next, click on the *Variables* button in the *Cluster Analysis: Joining (Tree Clustering)* startup panel and select all of the variables. Now, you want to cluster the automobile (cases) based on the different performance indices (variables). However, the default setting of the *Cluster* combo box is *Variables (columns)*; so you need to change this setting. Remember from the introduction (page 3172) that, depending on the research question at hand, one may cluster cases in some instances and variables in others. For example, you could be interested in whether or not the car performance measures (variables) form natural clusters. However, in this instance, you would like to know whether the cars (cases) form clusters, and, therefore, you need to reset the combo box to show *Cases (rows)*. Also, set the *Amalgamation (linkage) rules* combo box to *Complete linkage*; this will be discussed shortly.

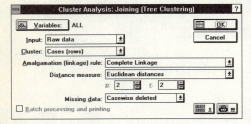

Distance measures. Remember that the tree clustering method will successively link together objects of increasing dissimilarity or distance. There are various ways to compute distances, and they are explained in the *Introductory Overview* section (page 3170). The most straightforward way to compute a distance is to consider the *k* variables as dimensions that make up a *k*-dimensional space. If there were three variables, then they would form a 3-dimensional space. The Euclidean distance in that case would be the same as if you were to measure the distance with a ruler. Accept this default measure (*Euclidean distance*) for this example.

Amalgamation (Linkage) rule. The other issue of some ambiguity in tree clustering is exactly how to determine the distances between *clusters*. Should you use the closest neighbors in different clusters, the furthest neighbors, or some aggregate measure? As it turns out, all of these methods (and more) have been proposed (see page 3171). The default method --*single linkage*-- is the "nearest neighbor" rule.

Thus, as you proceed to form larger and larger clusters of less and less similar objects (cars), the distance between any two clusters is determined by the closest objects in those two clusters. Intuitively, it may occur to you that this will likely result in "stringy" clusters, that is, the program will chain together clusters based on the particular location of single elements. Alternatively, you could have chosen to use the *complete linkage* rule. In that case, the distance between two clusters is determined by the distance of the furthest neighbors. This will result in more "lumpy" clusters. As it turns out for this data, the single linkage rule does in fact produce rather "stringy" and undistinguished clusters.

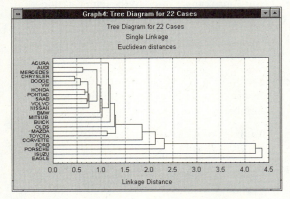

So, for this analysis, the *complete linkage* method was chosen as the *Amalgamation (linkage) rule*.

Results

Now, click *OK* to begin the analysis.

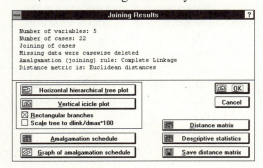

The tree clustering method is an iterative procedure, and after all objects have been joined, the *Joining Results* dialog will be displayed (see above).

The tree diagram. The most important result to consider in a tree clustering analysis is the *hierarchical tree*. The *Cluster Analysis* module offers two types of tree diagrams with two types of branches. For the standard style of tree diagram, select the *Rectangular branches* option and click on the *Horizontal hierarchical tree plot* button.

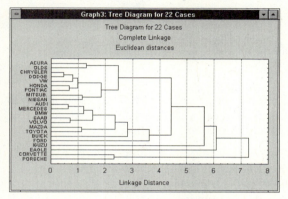

You can also display the tree diagram in a vertical style by clicking on the *Vertical icicle plot* button.

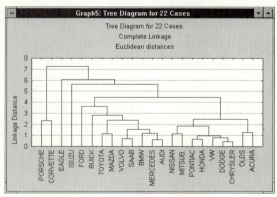

The branches of both types of tree diagrams can be set to rectangular (see the above plots) or diagonal (see below).

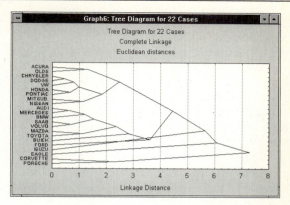

To produce a tree diagram with diagonal branches, de-select the *Rectangular branches* option. The diagonal format may increase the readability of the diagram for solutions with "balanced" joining structures.

You can elect to scale the tree plot to a standardized scale with the *Scale tree to dlink/dmax*100* option. When you select this option, the horizontal axis (or vertical axis for vertical icicle plots) will be scaled in percentages, specifically, as *dlink/dmax*100*. Thus, it represents the percentage of the range from the maximum to the minimum distance in the data. If this option is de-selected, then the scale will be based on the previously selected distance measure.

The tree diagram below illustrates the use of the standardized scale (compare to the horizontal tree plot with rectangular branches, above).

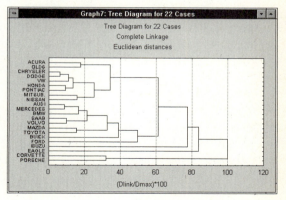

5. CLUSTER ANALYSIS - EXAMPLES

At first, tree diagrams may seem a bit confusing; however, once they are carefully examined, they will be less confusing. The diagram begins on the left in horizontal tree diagrams (or on the bottom in vertical icicle plots) with each car in its own cluster. As you move to the right (or up in vertical icicle plots), cars that are "close together" are joined to form clusters. Each node in the above diagrams represents the joining of two or more clusters; the locations of the nodes on the horizontal (or vertical) axis represent the distances at which the respective clusters were joined.

Identifying Clusters

For this discussion, consider only horizontal hierarchical tree diagrams (see the tree diagram with the standardized scale above), and begin at the top of the diagram. Apparently, first there is a cluster consisting of only *Acura* and *Olds*; next there is a group (i.e., cluster) of seven cars: *Chrysler*, *Dodge*, *VW*, *Honda*, *Pontiac*, *Mitsubishi*, and *Nissan*. As it turns out, in this sample the entry level models (more or less) of these brands were chosen. Thus, you may want to call this cluster the "economy sedan" cluster.

The first two cars, *Acura* and *Olds*, join this cluster at the approximate linkage distance of 32; after that (to the right), this branch of the tree extends out to 60. Thus, these two cars could also be considered as members of the *economy sedan* cluster. Moving down the plot, a cluster starting with *Audi* extends to *Ford*, perhaps all the way to *Eagle*. These cars (i.e., the particular models chosen for the sample) more or less represent high-priced, luxury sedans; thus, this cluster can be identified as the *"luxury"* sedan cluster.

Finally, at the bottom of the plot there are the *Corvette* and *Porsche* that are joined at the linkage distance of approximately 30. You can label this cluster for yourself...

Amalgamation schedule. A non-graphical presentation of these results is the amalgamation schedule (click on the *Amalgamation schedule* button).

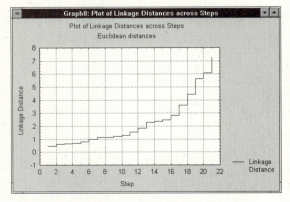

This Scrollsheet lists the objects (cars that are joined together at the respective linkage distances (in the leftmost column of the Scrollsheet).

Graph of amalgamation schedule. Clicking on the *Graph of amalgamation schedule* button will bring up a line graph of the linkage distances at successive clustering steps.

This graph can be very useful by suggesting a cutoff for the tree diagram. Remember that in the tree diagram, as you move to the right (increase the linkage distances), larger and larger clusters are formed of greater and greater within-cluster diversity. If this plot shows a clear plateau, then it means that many clusters were formed at essentially the same linkage distance. That distance may be the optimal cut-off when deciding how many clusters to retain (and interpret).

Example 2:
K-means Clustering

This example will illustrate one other method of clustering: k-means clustering. As described in the *Introductory Overview* (page 3173), the goal of the k-means algorithm is to find the optimum "partition" for dividing a number of objects into *k* clusters. This procedure will move objects around from cluster to cluster with the goal of minimizing the within-cluster variance and maximizing the between-cluster variance. In the previous example, you identified three clusters in the car data set (*Cars.sta*). Now see what kind of solution k-means clustering will suggest for 3 clusters.

Specifying the Analysis

Return to the *Cluster Method* dialog (click *Cancel* in each dialog until you return to *Cluster Method*) and select the *K-means clustering* method to open the *Cluster Analysis: K-means Clustering* dialog.

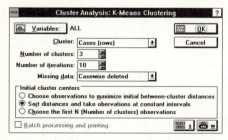

As before, you may cluster cases or variables; in this case, set the *Cluster* combo box to *Cases (rows)* in order to cluster cases (cars). To look at the results for 3 clusters, change the *number of clusters* to 3.

Initial cluster centers. The options in this box control the way in which the initial cluster centers

are computed. The results from the k-means clustering method depend to some extent on the initial configuration (i.e., cluster means or centers). This is particularly the case when there are many small clusters (with few objects) that are clearly distinct. For more information concerning the options in this box refer to the *Dialogs, Options, Statistics* section (page 3187).

For this example, choose the default method *Sort distances and take observations at constant intervals,* and then click *OK* to begin the analysis.

Results

When the analysis is complete, the *K-means Clustering Results* dialog will be displayed.

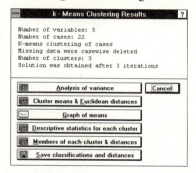

Analysis of variance. In the introduction, k-means clustering was referred to as "analysis of variance in reverse." In an analysis of variance, the between-groups variance is compared to the within-groups variance to decide whether the means for a particular variable are significantly different between groups.

CLUSTER ANALYSIS	Between SS	df	Within SS	df	F	signif. p
PRICE	9.08159	2	11.91841	19	7.23881	.004602
ACCELER	6.74790	2	14.25210	19	4.49794	.025163
BRAKING	10.11892	2	10.88108	19	8.83457	.001938
HANDLING	10.87750	2	10.12250	19	10.20857	.000975
MILAGE	7.99118	2	13.00882	19	5.83575	.010573

Even though significance testing would not be proper in this case (you are very much capitalizing on chance), you may nevertheless look at the

5. CLUSTER ANALYSIS - EXAMPLES

analysis of variance results, comparing the means for each dimension (i.e., performance measure) between groups (clusters of cars). Click on the *Analysis of Variance* button to display the above Scrollsheet. Judging from the magnitude (and significance levels) of the *F* values, variables *Handling*, *Braking*, and *Price* are the major criteria for assigning objects to clusters.

Identification of clusters. Now, see how the program assigned cars to clusters using these criteria. To view the members of each cluster, click on the *Members of each cluster & distances* button to produce a cascade of Scrollsheets (one for each cluster). Cluster 1 consists of *Acura*, *Buick*, *Chrysler*, *Dodge*, *Honda*, *Mitsubishi*, *Nissan*, *Olds*, *Pontiac*, *Saab*, *Toyota*, *VW*, and *Volvo*.

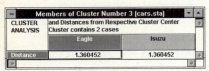

The next Scrollsheet contains the members of Cluster 2:

The second cluster contains *Audi*, *BMW*, *Corvette*, *Ford*, *Mazda*, *Mercedes*, and *Porsche*. The final cluster is given in the third Scrollsheet, below.

Cluster 3 consists of *Eagle* and *Isuzu*.

These results do not entirely match the clusters found in the previous analysis. However, the distinction between *economy sedan* vs. *high luxury sedan* still seems tenable. The *Eagle* and *Isuzu* were probably moved into their own category because they did not "fit" anywhere else, and because any other split between cars did not improve the solution (i.e., increase between-groups sums of squares).

Descriptive statistics for each cluster.
Another way of identifying the nature of each cluster is to examine the means for each cluster for each dimension. You can either display descriptive statistics separately for each cluster (click on the *Descriptive statistics for each cluster* button), display the means for all clusters and the distances (Euclidean and squared Euclidean, see below) between clusters in separate Scrollsheets (click on the *Cluster means & Euclidean distances* button), or plot those means (click on the *Graph of means* button). Usually, the graph provides the best summary (see below).

Looking at the lines for the *economy sedan* cluster (Cluster 1) as compared to the *luxury sedan* cluster (Cluster 2) in the graph below, it is found that, indeed, the cars in the latter cluster are:

(1) More expensive,

(2) Have slower acceleration (probably because of greater weight),

(3) Require about the same braking distances,

(4) Are about equal in handling, and

(5) Get lower gas mileage.

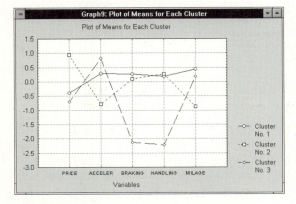

5. CLUSTER ANALYSIS - EXAMPLES

The most distinguishing feature of the cars in the third cluster (*Eagle* and *Isuzu*) in this plot appears to be their shorter braking distances and their poorer handling.

Cluster distances. Another useful result to examine are the Euclidean distances between clusters (click on the *Cluster means & Euclidean distances* button). These distances (Euclidean and squared Euclidean) are computed from the cluster means on each dimension.

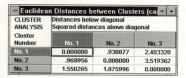

Note that clusters 1 and 2 are relatively close together (Euclidean distance = *2.17*) relative to the distance of cluster 3 from clusters 1 and 2.

5. CLUSTER ANALYSIS - EXAMPLES

DIALOGS, OPTIONS, STATISTICS

Cluster Analysis - Startup Panel

Select the desired clustering method in the *Clustering Method* dialog (startup panel). Refer to the *Introductory Overview* (page 3169) for brief descriptions of the different clustering techniques.

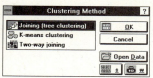

Briefly, the *joining* technique will create tree diagrams of clusters. The *Two-way joining* method will simultaneously cluster cases and variables. The *k-means* clustering method will move objects around among a user-specified number of clusters with the goal of minimizing within-cluster variability while maximizing between-cluster variability.

Joining (Tree Clustering) Dialog

When you click on the *Joining (Tree Clustering)* option in the *Clustering Method* startup panel, the *Cluster Analysis: Joining (Tree Clustering)* dialog will open.

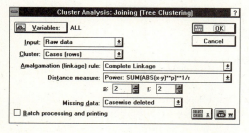

Variables

Choosing this option will bring up a standard variable selection window. Note that the variables that are selected here will be interpreted by the program as *dimensions* if the *Cluster* combo box is set to *cases*; they will be interpreted as *objects* if that combo box is set to *variables*.

Input Data

This combo box allows you to select either *Raw Data* or a matrix (select *Distance Matrix*) for input into the cluster analysis. The input matrix may either be a correlation matrix or a distance (dissimilarity) matrix with numbers indicating the distances or dissimilarities between objects. STATISTICA will automatically determine the contents of the matrix (i.e., whether it contains correlations or dissimilarities, see *Matrix File Format*, Volume I). If the input matrix is a correlation matrix (which indicates the similarity and closeness between objects), it is converted to distances before the analysis begins; specifically, all correlations are transformed as *1-Pearson r*.

Cluster Cases or Variables

This combo box allows the user to select clustering of *Variables (columns)* or clustering of *Cases (rows)*.

Amalgamation (linkage) Rules

One of the main parameters that guides the joining (tree-clustering) process is the linkage rule, that is, the rule that determines when two clusters are to be joined (linked or amalgamated). There are 7 different amalgamation rules available: *Single Linkage*, *Complete Linkage*, *Unweighted Pair-group Average*, *Weighted Pair-group Average*, *Unweighted Pair-group Centroid*, *Weighted Pair-group Centroid (median)*, and *Ward's Method*. The

5. CLUSTER ANALYSIS - DIALOGS AND OPTIONS

default rule is single linkage (also called the "method of the nearest neighbors;" see also the *Introductory Overview* section, page 3171).

Distance Measures

The joining algorithm starts by first computing a matrix of distances between the objects that are to be clustered. There are 6 different distance measures that can be computed from *Raw Data* (see above): *Squared Euclidean Distances*, *Euclidean Distances*, *City-block (Manhattan) Distances*, *Chebychev Distances*, *Power Distances*, *Percent Disagreement*, and *1-Pearson r*. For definitions of each of these distances refer to the *Introductory Overview* (page 3170).

Power distances. If *Power Distances* are selected, then you can specify parameters p and r for the power distance in the edit boxes below this combo box.

Distance matrix. If *Distance Matrix* was selected as input (see above), then the combo box will automatically be set to *Dissimilarities from matrix*. If the input matrix is a correlation matrix, then the correlations (which denote the degree of *similarity*) will be transformed to dissimilarities ($1-r$).

Missing Data Deletion

This combo box allows you to select one of two ways of treating missing data.

Casewise deletion of missing data. Cases will be deleted from the analysis if they have missing data for any variable selected for the analysis.

Substitution by means. Missing data are substituted by the means for the respective dimensions. Note that the substitution by means is performed *after* the program has determined what are the (logical) variables (i.e., dimensions) and what are the (logical) cases (i.e., objects) for the respective analysis.

For example, when joining (tree clustering) cases, logically, the variables become cases and the cases become variables. In that instance, a distance matrix would be computed for cases, treating the variables as dimensions. When the program computes the means for the substitution of missing data, it will compute the means for each case, across variables.

Batch Processing and Printing

This option is only available if *printer*, disk *file*, and/or *Text/output Window* was selected as the output destination in the *Page/Output Setup* dialog. If checked, STATISTICA will automatically perform the entire analysis, print the results (either to the printer, disk file, and/or *Text/output Window*, whichever was selected in the *Page/Output Setup* dialog, see Volume I), and then return to this dialog.

K-means Clustering Dialog

When you click on the *K-means Clustering* option in the *Clustering Method* startup panel, the *Cluster Analysis: K-means clustering* dialog will open.

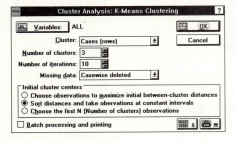

Variables

Choosing this option will bring up a standard variable selection window. Note that the variables that are selected here will be interpreted by the program as *dimensions* if the *Cluster* combo box is

5. CLUSTER ANALYSIS - DIALOGS AND OPTIONS

set to *cases*; they will be interpreted as *objects* if that switch is set to *variables*.

Cluster Cases or Variables

This combo box allows the user to select clustering of *Variables (columns)* or clustering of *Cases (rows)*.

Number of Clusters

The purpose of the k-means clustering procedure is to classify objects into a user-specified number of clusters. The algorithm will move objects into different clusters with the goal of minimizing the within-cluster variability while maximizing the between-cluster variability. For a further discussion of this method, refer also to the *Introductory Overview* (page 3173).

This edit box allows the user to enter the desired number of clusters, which must be greater than *1* and less than the number of objects (i.e., cases or variables depending on the setting of the *Cluster* combo box, see above).

Maximum Number of Iterations

K-means clustering is an iterative procedure; in each iteration objects are moved into different clusters. This edit box allows the user to specify the maximum number of iterations that are to be performed. The algorithm implemented in *Cluster Analysis* is very efficient, and the default setting (10 iterations) usually does not have to be changed.

Missing Data Deletion

This combo box allows you to select one of two ways of treating missing data: *Casewise deleted* or *Substituted by means*. See page 3186 for more information about these methods.

Initial Cluster Centers

These options control the way in which the initial cluster centers are computed. Note that the results from the k-means clustering method depend to some extent on the initial configuration (i.e., cluster means or centers). This is particularly the case when there are many small clusters (with few objects) that are clearly distinct.

Maximize between-cluster distances. This option will take observations or objects as the initial cluster centers; the choice of the object follows rules to maximize the initial cluster distances. Specifically, (1) the program will select the first N (number of clusters) cases to be the respective cluster centers; (2) subsequent cases will replace previous cluster centers if their smallest distance to any of the cluster centers is larger than the smallest distance between clusters; if this is not the case, then (3) subsequent cases will replace initial cluster centers if their smallest distance from a cluster center is larger the distance of that cluster center from any other cluster center. The effect of this selection procedure is to maximize the initial distances between clusters. Note that this procedure may yield clusters with single observations if there are clear outliers in the data.

Sort distances and take observations at constant intervals. If this option is chosen, the distances between all objects will first be sorted, and then objects at constant intervals will be chosen as initial cluster centers.

Choose the first N (number of clusters) clusters observations. This option will take the first N (number of clusters) observations to be the initial cluster centers. Thus, this option provides the user full control over the choice of the initial configuration. This is often useful if one brings a priori expectations regarding the nature of the clusters to the analysis. In that case, move the cases that you want to choose as the initial cluster centers to the beginning of the file.

Batch Processing and Printing

This option is only available if *printer*, disk *file*, and/or *Text/output Window* was selected as the output destination in the *Page/Output Setup* dialog. If checked, STATISTICA will automatically perform the entire analysis, print the results (either to the printer, disk file, and/or *Text/output Window*, whichever was selected in the *Page/Output Setup* dialog), and then return to this dialog.

Two-way Joining (Block Clustering) Dialog

When you click on the *Two-way Joining* option in the startup panel, the *Cluster Analysis: Joining (Tree Clustering)* dialog will open.

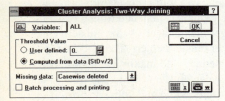

Variables

Choosing this option will bring up a standard variable selection window from which you can select the variables for the analysis.

Threshold Value

The 2-way joining algorithm attempts to simultaneously cluster cases and variables. The *threshold* parameter determines when the algorithm will consider two numbers in the data matrix to be equal, and thus satisfactorily classified in the same cluster. If this value is very large (relative to the numbers in the data matrix), then only one cluster will be formed; if it is very small, then each data point will represent a cluster by itself. The default value (the overall standard deviation divided by 2) is recommended for most cases.

Missing Data Deletion

This combo box allows you to select one of two ways of treating missing data: *Casewise deleted* or *Substituted by means*. See page 3186 for more information about these methods.

Batch Processing and Printing

This option is only available if *printer*, disk *file*, and/or *Text/output Window* was selected as the output destination in the *Page/Output Setup* dialog. If checked, STATISTICA will automatically perform the entire analysis, print the results (either to the printer, disk file, and/or *Text/output Window*, see Volume I), and then return to this dialog.

Joining (Tree Clustering) - Results Dialog

This dialog provides options to review the results of the *Joining (Tree Clustering)* method.

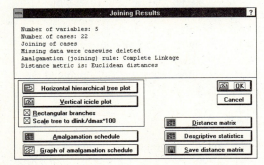

Hierarchical Tree Plot Options

Refer to the *Introductory Overview* (page 3170) for a brief discussion of how to interpret the tree diagram.

Horizontal hierarchical tree plot. This button will bring up a horizontal tree diagram summarizing the successive clustering of objects.

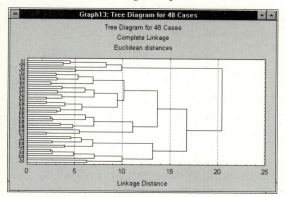

Vertical icicle plot. Click on this button to bring up a vertical tree diagram (the vertical axis denotes the linkage distance).

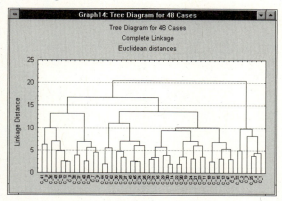

Rectangular branches. You can choose to display both vertical and horizontal tree plots with either rectangular branches (select this option; see above) or diagonal branches (de-select this option; see below). The latter format may increase readability of the diagram for solutions with "balanced" joining structures.

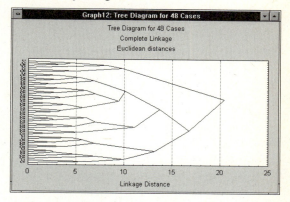

Scale tree to dlink/dmax*100. When you select this option, the tree plot will be scaled to a standardized scale (i.e., dlink/dmax*100).

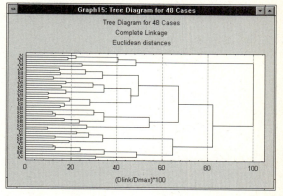

If de-selected, the scale will be based on the linkage distance selected in the startup panel (page 3170).

Note that depending on the current default scale value font-size setting, the item (case or variable) labels in the tree plot may overlap or the display of some (e.g., every other or third) overlapping labels may be suppressed (see *Filters*, Volume II); in such cases, click on the labels in the graph and reduce the font size.

5. CLUSTER ANALYSIS - DIALOGS AND OPTIONS

Amalgamation Schedule

Clicking on this button will bring up a Scrollsheet with the amalgamation (linkage) schedule. The first column of the Scrollsheet will contain the linkage distances at which the respective clusters (as indicated in the respective rows) were formed, and each row contains the names of the objects (cases or variables) that comprise the respective cluster.

Graph of Amalgamation Schedule

This option will bring up a line graph of the linkage distances across consecutive steps of the linking process.

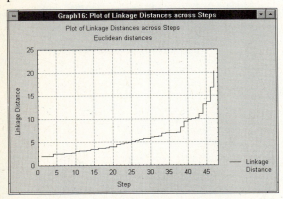

This graph is useful for identifying plateaus where many clusters are formed at approximately the same linkage distance. This may indicate a natural "discontinuity" in terms of distances between the observed objects.

Distance Matrix

This option will bring up a standard Scrollsheet with the matrix of distances. This matrix can be saved using the *Save Distance Matrix* option, below.

Descriptive Statistics

This option will bring up a standard Scrollsheet with means and standard deviations for each object included in the cluster analysis (i.e., for each case or variable, depending on the setting of the *Cluster* combo box that was chosen on the startup panel, page 3185).

Save Distance Matrix

This option allows the user to save the current distance matrix (see above) as a standard *STATISTICA* matrix file (see *Matrix File Format*, Volume I). This file can be accessed later in the *Cluster Analysis* or *Multidimensional Scaling* modules.

K-means Clustering - Results Dialog

The *K-means Clustering Results* dialog provides several options to review the results of the *K-means Clustering* method of the *Cluster Analysis* module.

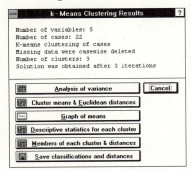

Analysis of Variance

The goal of k-means clustering is to classify objects (cases or variables, depending on the setting of the *Cluster* combo box on the startup panel, page 3186)

CLU - 3190

Copyright © StatSoft, 1995

into a user-specified number of clusters. To evaluate the appropriateness of the classification, one can compare the within-cluster variability (small if the classification is good) with the between-cluster variability (large if the classification is good), that is, perform a standard between-groups analysis of variance for each dimension (case or variable). Clicking on this button will bring up a standard Scrollsheet with those ANOVA's.

Note that although the *F*-ratios and *p*-levels are given in the table, measures of statistical significance should be interpreted with caution since their meanings are not the same as in an actual ANOVA of experimental data (see page 3173). In short, these are not *a priori* tests, and you capitalize on chance by arranging the most statistically significant ANOVA's possible (see Hartigan, 1975, for a more detailed discussion of this point).

Cluster Means and Euclidean Distances

Two Scrollsheets will be displayed with this option:

(1) A Scrollsheet with the means for each cluster for each dimension, and

(2) A Scrollsheet with the Euclidean distances (below the diagonal) and squared Euclidean distances (above the diagonal) between "cluster centers."

Specifically, this matrix shows the Euclidean distances between clusters, computed from the respective cluster means on the dimensions used for the classification.

Distances. The distances between two objects or cluster centers *i* and *j* are computed as:

$$D_{i,j} = \text{Square Root } (\Sigma \ ((x_i - x_j)^2 / ND))$$

where the summation is over the ND dimensions in the current analysis.

Graph of Means

This option produces a line graph of the means across clusters.

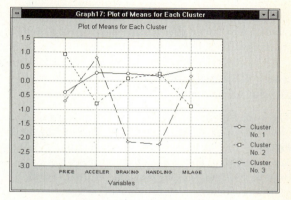

This type of plot is very useful for visually summarizing the differences in means between clusters.

Descriptive Statistics for Each Cluster

This option will bring up Scrollsheets, one for each cluster, with the descriptive statistics.

Members of Each Cluster and Distances

Via this option, the user can display the (Euclidean) distances of the objects (cases or variables, depending on the setting of the *Cluster* combo box on the startup panel, page 3186) from their respective cluster center (mean). This allows one to identify potential "bad" cluster members, that is, objects that are very distant from the cluster center, yet apparently do not belong to any other cluster (i.e., they are even further away from the centers of alternative clusters). Those distances will be displayed in standard Scrollsheets, each Scrollsheet showing the results for one cluster.

Save Classifications & Distances

The k-means method features an option which allows the user to save classification information and distances for clustered objects. After clicking on the *Save classifications and distances* button, the program will ask you for the name of the file into which the classification results will be saved.

If you are clustering variables, then the data file will contain one case for each variable clustered and the original names of each clustered variable will be used as case names in the new file. Three variables will be created and named: *Variable* [the id number of each clustered item (variable) as it appeared in the original input data set], *Cluster* [the cluster number to which each item (variable) was classified], and *Distance* [the distance of that item (variable) from its respective cluster center].

If you are clustering cases, the program will give you an option to save any of the variables (i.e., their raw values) from the current data file along with the cluster information. If the original input data file contained case names, then they will also be used to label cases in the new file. In addition to the (optional) variables selected from the original data file, the new data file will contain three variables, named *Case_No* [the consecutive id number of each clustered item (case) as it appeared in the original input data set], *Cluster* [the cluster number to which each item (case) was classified], and *Distance* [the distance of that item (case) from its respective cluster center].

Two-way Joining (Block Clustering) - Results Dialog

The *Two-way Joining Results* dialog provides several options to review the results of the *Two-way Joining* method of the *Cluster Analysis* module.

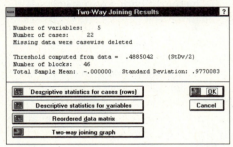

Descriptive Statistics for Cases (Rows)

This option will bring up a standard Scrollsheet with means and standard deviations for each case (i.e., row of the data matrix) across variables (columns of the data matrix).

Descriptive Statistics for Variables (Columns)

Clicking on this button will bring up a standard Scrollsheet with means and standard deviations for each variable (i.e., column of the data matrix) across cases (rows of the data matrix).

Reordered Data Matrix

This option will bring up a standard Scrollsheet with the complete data matrix, reordered according to the results of the two-way joining computations. Note that the goal of this procedure is to simultaneously cluster cases and variables, that is, to move similar data points together as closely as possible by reordering the columns and rows of the data matrix. Refer to Hartigan (1975) for a detailed description of this procedure and its applications.

3D box plot of reordered data matrix. The custom graph for this Scrollsheet is the 3D box plot, which allows you to visually inspect the pattern of similar and dissimilar data points.

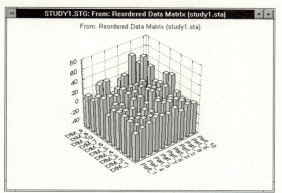

When the graph is difficult to review (e.g., one data series covers another), use the *Animated Stratification* (or "X-ray" review) tool; refer to the *Electronic Manual* or *Quick Reference* Manual for details. Sometimes changing the plot type to *Ribbon* (in the *General Layout* dialog, accessible by double-clicking on the outside background) increases the readability of the display.

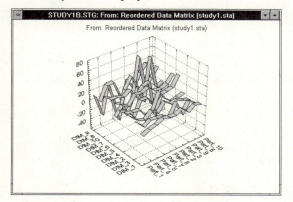

Two-way Joining Graph

This option will bring up a graphical representation of the results of the *Two-way joining* procedure. Specifically, the reordered data matrix is represented in a discrete contour plot.

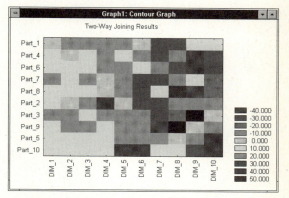

To adjust the color palette, double-click on any color symbol in the legend (see Volume II, for details).

5. CLUSTER ANALYSIS - INDEX

INDEX

A

Amalgamation or linkage rules
    complete linkage (furthest neighbor), 3171
    joining (tree clustering), 3178, 3185
    overview, 3171, 3175, 3178, 3185
    single linkage (nearest neighbor), 3171
    unweighted pair-group average, 3171
    unweighted pair-group centroid, 3172
    Ward's method, 3172
    weighted pair-group average, 3172
    weighted pair-group centroid, 3172

Amalgamation schedule, 3180, 3190
Animated stratification, review of 3D box plot, 3193

B

Batch printing cluster analyses, 3186, 3188
Box plot, 3D (of reordered data matrix), 3193

C

Casewise deletion of missing data, 3186
Chebychev distance, 3171
City-block (Manhattan) distance, 3170
Classifications and distances, 3192
Cluster analysis
    amalgamation rules, 3171, 3175, 3178, 3185
    amalgamation schedule, 3180, 3190
    analysis of variance, 3181, 3190
    animated stratification, review of 3D box plot, 3193
    areas of application, 3169
    batch printing, 3186, 3188
    Chebychev distance, 3171

Cluster analysis (continued)
    city-block (Manhattan) distance, 3170
    cluster cases, 3185, 3187
    cluster distances, 3183
    cluster variables, 3185, 3187
    complete linkage (furthest neighbor), 3171
    descriptive statistics for each cluster, 3182
    diagonal branches, 3189
    distance matrix, 3186, 3190
    distance measures, 3178, 3186
    Euclidean distance, 3170
    example, 3177
    graph of amalgamation schedule, 3180, 3190
    hierarchical tree plot, 3170, 3179, 3189
    horizontal hierarchical tree plot, 3179, 3189
    icicle plot, 3170, 3189
    identification of clusters, 3180, 3182
    initial cluster centers, for k-means clustering, 3181, 3187
    joining (tree clustering), 3169, 3177, 3185, 3188
    K-means clustering, 3173, 3181, 3186, 3190
    linkage rules, 3171, 3175, 3178, 3185
    Manhattan (city-block) distance, 3170
    maximum number of iterations, 3187
    number of clusters, 3187
    overview, 3169
    percent disagreement, 3171
    power distances, 3171, 3186
    rectangular branches, 3189
    scale tree to dlink/dmax*100, 3189
    single linkage (nearest neighbor), 3171
    squared Euclidean distance, 3170
    startup panel, 3185
    statistical significance testing, 3169
    threshold value, 3188
    tree diagram, 3170, 3179, 3189
    two-way joining (block clustering), 3172, 3188, 3192

Cluster analysis (continued)
    unweighted pair-group average, 3171
    unweighted pair-group centroid, 3172
    vertical icicle tree plot, 3179, 3189
    Ward's method, 3172
    weighted pair-group average, 3172
    weighted pair-group centroid, 3172
    X-ray review tool, for 3D box plot, 3193

Cluster analysis vs. discriminant function analysis, 3176
Cluster analysis vs. factor analysis, 3175
Cluster analysis vs. multidimensional scaling (MDS), 3175
Cluster distances, 3183
Complete linkage (furthest neighbor), 3171
Contour plot, 3193

D

Descriptive statistics for each cluster, 3182
Distance matrix, 3186
Distance measure
    Chebychev distance, 3171
    city-block (Manhattan) distance, 3170
    Euclidean distance, 3170
    Manhattan (city-block) distance, 3170
    overview, 3170
    percent disagreement, 3171
    power distances, 3171
    squared Euclidean distance, 3170

E

Euclidean distance, 3170

G

Graph of amalgamation schedule, 3180, 3190

CLU - 3195

5. CLUSTER ANALYSIS - INDEX

H

Hierarchical tree plot, 3170, 3179, 3189
Horizontal hierarchical tree plot, 3189

I

Icicle plot, 3170, 3189
Identification of clusters, 3180, 3182
Initial cluster centers, for k-means clustering, 3181, 3187

J

Joining (tree clustering), 3169, 3177, 3185, 3188

K

K-means clustering, 3173, 3181, 3186, 3190

L

Linkage rules, 3171, 3175, 3178, 3185

M

Manhattan (city-block) distance, 3170
Missing data deletion
    casewise deletion of missing data, 3186
    substitution by means, 3186

P

Percent disagreement, 3171
Power distances, 3171, 3186

R

Reordered data matrix, 3192

S

Scale tree to dlink/dmax*100, 3179
Single linkage (nearest neighbor), 3171
Squared Euclidean distance, 3170
Substitution of missing data by means, 3186

T

Threshold value, 3188
Tree diagram, 3170, 3179, 3189
    diagonal branches, 3179
    hierarchical tree, 3179
    horizontal hierarchical tree plot, 3179
    icicle plot, 3170, 3189
    rectangular branches, 3179
    scale tree to dlink/dmax*100, 3179
    vertical icicle tree plot, 3179, 3189
Two-way joining (block clustering), 3172, 3188, 3192
Two-way joining graph, 3193

U

Unweighted pair-group average, 3171
Unweighted pair-group centroid, 3172

V

Vertical icicle tree plot, 3179, 3189

W

Ward's method, 3172
Weighted pair-group average, 3172
Weighted pair-group centroid (median), 3172

X

X-ray review tool, for 3D box plot, 3193

Chapter 6:

PRINCIPAL COMPONENTS AND FACTOR ANALYSIS

Table of Contents

Introductory Overview 3201
    General Purpose 3201
    Factor Analysis as a Data Reduction Method 3201
    Factor Analysis as a Classification Method 3205
    Miscellaneous Other Issues and Statistics 3208
Program Overview 3209
Example 3211
Dialogs, Options, Statistics 3217
    Startup Panel 3217
    Factor Extraction 3217
    Descriptive Statistics 3219
    Multiple Regression 3222
    Results 3224
    Factor Rotation 3227
    3D Plot of Loadings 3229
    2D Plot of Loadings 3230
Technical Notes 3231
Index 3233

Detailed Table of Contents

INTRODUCTORY OVERVIEW **3201**
    General Purpose 3201
    Factor Analysis as a Data Reduction Method 3201
        Combining Two Variables into a Single Factor 3202
        Principal Components Analysis 3202
        Extracting Principal Components 3202
        Generalizing to the Case of Multiple Variables 3202
        How Many Factors to Extract? 3203
        Reviewing the Results of a Principal Components Analysis 3203
        Eigenvalues 3203
        Eigenvalues and the Number-of-Factors Problem 3204

6. FACTOR ANALYSIS - CONTENTS

    Principal Factors Analysis 3204
  Factor Analysis as a Classification Method 3205
    Factor Loadings 3206
    Rotating the Factor Structure 3206
    Interpreting the Factor Structure 3207
    Oblique Factors 3207
    Hierarchical Factor Analysis 3207
    Confirmatory Factor Analysis 3208
  Miscellaneous Other Issues and Statistics 3208
    Factor Scores 3208
    Reproduced and Residual Correlations 3208
    Matrix Ill-Conditioning 3208

PROGRAM OVERVIEW 3209
  Methods 3209
  Output 3209
  Matrix Input 3209
  Alternative Procedures 3209

EXAMPLE 3211
  Overview 3211
  Specifying the Analysis 3211
  Define Method of Factor Extraction 3211
  Review Descriptive Statistics 3211
  Extraction Method 3212
  Reviewing Results 3212
  Deciding on the Number of Factors 3213
  Factor Loadings 3213
  Rotating the Factor Solution 3214
  Reproduced and Residual Correlation Matrix 3215
  The "Secret" to the Perfect Example 3216
  Miscellaneous Other Results 3216
  Final Comment 3216

DIALOGS, OPTIONS, STATISTICS 3217
  Startup Panel 3217
    Variables 3217
    Input File 3217
    Missing Data Deletion 3217
  Factor Extraction 3217
    Review Correlations/Means/SD 3217
    Perform Multiple Regression 3217
    Extraction Methods 3218
    Maximum Number of Factors 3218
    Minimum Eigenvalue 3219

6. FACTOR ANALYSIS - CONTENTS

    Minimum Change in Communality 3219
    Maximum Number of Iterations 3219
Descriptive Statistics 3219
    Means & SD 3219
    Correlations 3219
    Scatterplot Matrix 3219
    Covariances 3220
    Save Correlations 3220
    Summary Statistics for Pairwise Deletion of Missing Data 3220
    Pairwise Means 3220
    Pairwise Standard Deviations 3220
    Pairwise *n* 3220
    Box and Whisker Plots 3220
    Histograms 3221
    Normal Probability Plots 3221
    3D Bivariate Distribution Plot 3221
    2D Scatterplots 3222
    3D Scatterplots 3222
    Surface Plots 3222
Multiple Regression 3222
    Variables 3223
    Tolerance 3223
    Review Corrs/Means/SD 3223
    Regression Coefficients 3223
    Partial Correlations 3223
    Save Predicted Values/Residuals 3224
    Alpha Level for Highlighting 3224
Results 3224
    Eigenvalues 3224
    Communalities 3224
    Goodness of Fit Test 3225
    Scree Plot 3225
    Reproduced/Residual Correlations 3225
    Factor Rotation 3225
    Factor Loadings 3225
    Plot of Loadings, 2D & 3D 3226
    Hierarchical Analysis of Oblique Factors 3226
    Factor Score Coefficients 3226
    Factor Scores 3226
    Save Factor Scores 3227
    Review Correlations/Means/SD 3227
    Perform Multiple Regression 3227
Factor Rotation 3227
    Varimax Raw 3227

6. FACTOR ANALYSIS - CONTENTS

    Varimax Normalized 3228
    Biquartimax Raw 3228
    Biquartimax Normalized 3228
    Quartimax Raw 3228
    Quartimax Normalized 3228
    Equamax Raw 3228
    Equamax Normalized 3228
    Oblique Factors 3228
  3D Plot of Loadings 3229
  2D Plot of Loadings 3230

TECHNICAL NOTES **3231**
  Eigenvalues 3231
  Matrix Ill-Conditioning and the Modified Correlation Matrix 3231
  Analyzing Covariance or Moment Matrices 3231

INDEX **3233**

Chapter 6:
PRINCIPAL COMPONENTS AND FACTOR ANALYSIS

INTRODUCTORY OVERVIEW

General Purpose

The main applications of factor analytic techniques are:

(1) to *reduce* the number of variables, and

(2) to *detect structure* in the relationships between variables, that is, to *classify variables*.

Therefore, factor analysis is applied as a data reduction or structure detection method. The topics listed below will describe the principles of factor analysis, and how it can be applied towards these two purposes.

It is assumed that you are familiar with the basic logic of statistical reasoning as described in *Elementary Concepts* (Volume I). It is also assumed that you are familiar with the concepts of variance and correlation (e.g., as described in *Basic Statistics and Tables*, Volume I).

There are many excellent books on factor analysis. For example, a hands-on how-to approach can be found in Stevens (1986); more detailed technical descriptions are provided in Cooley and Lohnes (1971); Harman (1976); Kim and Mueller, (1978a, 1978b); Lawley and Maxwell (1971); Lindeman, Merenda, and Gold (1980); Morrison (1967); or Mulaik (1972). The interpretation of secondary factors in hierarchical factor analysis, as an alternative to traditional oblique rotational strategies, is explained in detail by Wherry (1984).

Confirmatory factor analysis. *STATISTICA* also includes the general *Structural Equation Modeling (SEPATH)* module (see Chapter 11). The procedures available in that module allow you to test specific hypotheses about the factor structure for a set of variables (confirmatory factor analysis), in one or several samples (e.g., you can compare factor structures across samples). The *Examples* section of the *SEPATH* chapter (Chapter 11) discusses several examples of such analyses.

Factor Analysis as a Data Reduction Method

Suppose you conducted a (rather "silly") study in which you measured the height of 100 people in inches and centimeters. Thus, you would have two variables that measure height. If in future studies, you want to research, for example, the effect of different nutritional food supplements on height, would you continue to use *both* measures? Probably not; height is one characteristic of a person, regardless of how it is measured.

Now, extrapolate from this "silly" study to something that one might actually do as a researcher. Suppose you want to measure people's satisfaction with their lives. You design a satisfaction questionnaire with various items; among other things you ask your subjects how satisfied they are with their hobbies (item 1) and how intensely they are pursuing a hobby (item 2). Subjects' responses are measured via a computer so that average responses (e.g., satisfaction) are assigned a value of 100, while below or above average responses are assigned smaller and larger values, respectively. Shown below is a scatterplot summarizing the results for the two items.

6. FACTOR ANALYSIS - INTRODUCTORY OVERVIEW

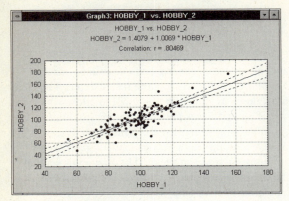

In this plot each respondent is represented by a point. As you can see, the responses to the two items are highly correlated with each other. (If you are not familiar with the correlation coefficient, it is recommend that you read the description in *Basic Statistics and Tables*, Volume I.) Given a high correlation between the two items, you can conclude that they are quite redundant.

Combining Two Variables into a Single Factor

One can summarize the correlation between two variables in a scatterplot. A regression line can then be fitted that represents the "best" summary of the linear relationship between the variables. If you could define a variable that would approximate the regression line in such a plot, then that variable would capture most of the "essence" of the two items. Subjects' single scores on that new factor, represented by the regression line, could then be used in future data analyses to represent that essence of the two items. In a sense, you have reduced the two variables to one factor. Note that the new factor is actually a linear combination of the two variables.

Principal Components Analysis

The example described above, combining two correlated variables into one factor, illustrates the basic idea of factor analysis, or of principal components analysis to be precise (this distinction will be discussed later). If the two-variable example is extended to multiple variables, then the computations become more involved, but the basic principle of expressing two or more variables by a single factor remains the same.

Extracting Principal Components

Details about the computational aspects of principal components analysis can be found in the references provided at the beginning of this section. Basically, the extraction of principal components amounts to a *variance maximizing* (*varimax*) *rotation* of the original variable space. For example, in a scatterplot you can think of the regression line as the original *X*-axis, rotated so that it approximates the regression line. This type of rotation is called *variance maximizing* because the criterion for (goal of) the rotation is to maximize the variance (variability) of the "new" variable (factor), while minimizing the variance around the new variable (see *Rotational Strategies*, page 3206).

Generalizing to the Case of Multiple Variables

When there are more than two variables, you can think of them as defining a "space," just as two variables defined a plane. Thus, when you have three variables, you could plot a three-dimensional scatterplot analogous to that shown above.

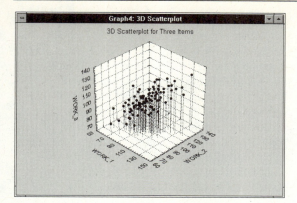

Again, you could fit a line through the data. With more than three variables it becomes impossible to illustrate the points in a scatterplot; however, the logic of rotating the axes so as to maximize the variance of the new factor remains the same.

Multiple orthogonal factors. After you have found the line on which the variance is maximal, there remains some variability around this line. In principal components analysis, after the first factor has been *extracted*, that is, after the first line has been drawn through the data, another line is defined that maximizes the remaining variability, and so on. In this manner, consecutive factors are extracted. Because each consecutive factor is defined to maximize the variability that is not captured by the preceding factor, consecutive factors are independent of each other. Put another way, consecutive factors are uncorrelated or *orthogonal* to each other.

How Many Factors to Extract?

Remember that, so far, principal components analysis is being considered as a data reduction method, that is, as a method for reducing the number of variables. The question then is, how many factors do you want to extract? Note that as you extract consecutive factors, they account for less and less variability. The decision of when to stop extracting factors basically depends on the point when there is only very little "random" variability left. The nature of this decision is arbitrary; however, various guidelines have been developed, and they will be reviewed in the context of the *Eigenvalues and the Number-of-Factors Problem* (see below).

Reviewing the Results of a Principal Components Analysis

Now look at some of the standard results from a principal components analysis. To reiterate, you are extracting factors that account for less and less variance. To simplify matters, one usually starts with the correlation matrix, where the variances of all variables are equal to *1.0*. Therefore, the total variance in that matrix is equal to the number of variables. For example, if you have 10 variables each with a variance of *1*, then the total variability that can potentially be extracted is equal to 10 times 1. Suppose that in the satisfaction study introduced earlier you included 10 items to measure different aspects of satisfaction at home and at work. The variance accounted for by successive factors would be summarized as follows:

FACTOR ANALYSIS	Extraction: Principal components			
Value	Eigenval	% total Variance	Cumul. Eigenval	Cumul. %
1	6.118369	61.18369	6.11837	61.1837
2	1.800682	18.00682	7.91905	79.1905
3	.472888	4.72888	8.39194	83.9194
4	.407996	4.07996	8.79993	87.9993
5	.317222	3.17222	9.11716	91.1716
6	.293300	2.93300	9.41046	94.1046
7	.195808	1.95808	9.60626	96.0626
8	.170431	1.70431	9.77670	97.7670
9	.137970	1.37970	9.91467	99.1467
10	.085334	.85334	10.00000	100.0000

Eigenvalues

In the second column (*Eigenval*) in the Scrollsheet above, you find the variance on the new factors that were successively extracted. The variances extracted by the factors are called the *eigenvalues*. This name derives from the computational issues involved (solving the so-called *Eigenvalue* problem, see also the *Technical Notes* section, page 3231).

6. FACTOR ANALYSIS - INTRODUCTORY OVERVIEW

The third column contains the cumulative variance extracted. These values are expressed as a percent of the total variance (in this example, *10*). As you can see, factor 1 (Value 1) accounts for 61 percent of the variance, factor 2 (Value 2) for 18 percent, and so on. As expected, the sum of the eigenvalues is equal to the number of variables.

Eigenvalues and the Number-of-Factors Problem

Now that you have a measure of how much variance each successive factor extracts, you can return to the question of how many factors to retain. As mentioned earlier, by its nature this is an arbitrary decision. However, there are some guidelines that are commonly used, and that, in practice, seem to yield the best results.

The Kaiser criterion. First, you can retain only factors with eigenvalues greater than *1*. In essence this is like saying that, unless a factor extracts at least as much as the equivalent of one original variable, you drop it. This criterion was proposed by Kaiser (1960) and is probably the one most widely used. In the example above, using this criterion, you would retain 2 factors (principal components).

The scree test. The *scree* test is a graphical method first proposed by Cattell (1966).

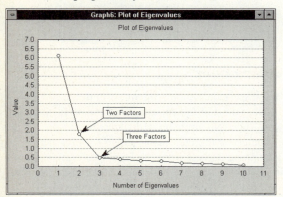

You can plot the eigenvalues shown in the Scrollsheet earlier in a simple line plot (click on the *Eigenval* column with the right-mouse-button and select the *Custom 2D Graphs* option). Cattell suggests finding the place where the smooth decrease of eigenvalues appears to level off to the right of the plot. To the right of this point, presumably, one finds only "factorial scree" -- "scree" is the geological term referring to the debris which collects on the lower part of a rocky slope. According to this criterion, you would probably retain 2 or 3 factors in this example.

Which criterion to use. Both criteria have been studied in detail (Browne, 1968; Cattell and Jaspers, 1967; Hakstian, Rogers, and Cattell, 1982; Linn, 1968; Tucker, Koopman and Linn, 1969). Theoretically, one can evaluate those criteria by generating random data based on a particular number of factors. One can then see whether the number of factors is accurately detected by those criteria. Using this general technique, the first method (*Kaiser criterion*) sometimes retains too many factors, while the second technique (*scree test*) sometimes retains too few; however, both do quite well under normal conditions, that is, when there are relatively few factors and many cases.

In practice, an additional important aspect is the extent to which a solution is interpretable. Therefore, one usually examines several solutions with more or fewer factors and chooses the one that makes the best "sense." This issue will be discussed in the context of factor rotations below.

Principal Factors Analysis

Before you continue to examine the different aspects of the typical output from a principal components analysis, principal factors analysis will be introduced. Return to the satisfaction questionnaire example in order to conceive another "mental model" for factor analysis. You can think of subjects' responses as being dependent on two

components. First, there are some underlying common factors, such as the "satisfaction-with-hobbies" factor you looked at before. Each item measures some part of this common aspect of satisfaction. Second, each item also captures a unique aspect of satisfaction that is not addressed by any other item.

Communalities. If this model is correct, then you should not expect that the factors will extract all variance from the items; rather, only that proportion which is due to the common factors and shared by several items. In the language of factor analysis, the proportion of variance of a particular item that is due to common factors (shared with other items) is called *communality*. Therefore, an additional task facing the user when applying this model is to estimate the communalities for each variable, that is, the proportion of variance that each item has in common with other items. The proportion of variance that is unique to each item is then the respective item's total variance minus the communality.

A common starting point is to use the squared multiple correlation of an item with all other items as an estimate of the communality (refer to *Multiple Regression*, Volume I, for details about multiple regression). Some authors have suggested various iterative "post-solution improvements" to the initial multiple regression communality estimate; for example, the so-called MINRES method (minimum residual factor method; Harman and Jones, 1966) will try various modifications to the factor loadings with the goal of minimizing the residual (unexplained) sums of squares. Refer to the description of the *Extraction* dialog (page 3217) for the different methods available in the *Factor Analysis* module.

Principal factors vs. principal components. The defining characteristic then that distinguishes between the two factor analytic models is that in principal components analysis you assume that *all* variability in an item should be used in the analysis, while in principal factors analysis you only use the variability in an item that is common to the other items. A detailed discussion of the pros and cons of each approach is beyond the scope of this introduction (refer to the general references provided on page 3201).

In most cases, these two methods usually yield very similar results. However, principal components analysis is often preferred as a method for data reduction, while principal factors analysis is often preferred when the goal of the analysis is to detect structure (see *Factor Analysis as a Classification Method* below).

Factor Analysis as a Classification Method

Now, return to the interpretation of the standard results from a factor analysis. The term *factor analysis* will henceforth be used generically to encompass both principal components and principal factors analysis. Assume that you are at the point in your analysis where you basically know how many factors to extract. You may now want to know the meaning of the factors, that is, whether and how you can interpret them in a meaningful manner.

To illustrate how this can be accomplished, work "backwards," that is, begin with a meaningful structure and then see how it is reflected in the results of a factor analysis. Now, return to the satisfaction example; shown below is the correlation matrix for items pertaining to satisfaction at work and satisfaction at home.

FACTOR ANALYSIS	Correlations (factor.sta) Casewise deletion of MD N=100					
Variable	WORK_1	WORK_2	WORK_3	HOME_1	HOME_2	HOME_3
WORK_1	1.00	.65	.65	.14	.15	.14
WORK_2	.65	1.00	.73	.14	.18	.24
WORK_3	.65	.73	1.00	.16	.24	.25
HOME_1	.14	.14	.16	1.00	.66	.59
HOME_2	.15	.18	.24	.66	1.00	.73
HOME_3	.14	.24	.25	.59	.73	1.00

The work satisfaction items are highly inter-correlated amongst themselves, and the home satisfaction items are highly inter-correlated amongst themselves. The correlations across these two types of items (work satisfaction items with home satisfaction items) is comparatively small. It thus seems that there are two relatively independent factors reflected in the correlation matrix, one related to satisfaction at work, and the other related to satisfaction at home.

Factor Loadings

Now, perform a principal components analysis and look at the two-factor solution.

FACTOR ANALYSIS	Factor Loadings (Unrotated) (factor.sta) Extraction: Principal components	
Variable	Factor 1	Factor 2
WORK_1	.654384	.564143
WORK_2	.715256	.541444
WORK_3	.741688	.508212
HOME_1	.634120	-.563123
HOME_2	.706267	-.572658
HOME_3	.707446	-.525602
Expl.Var	2.891313	1.791000
Prp.Totl	.481885	.298500

Specifically, look at the correlations between the variables and the two factors (or "new" variables), as they are extracted by default; these correlations are also called factor *loadings*. Apparently, the first factor is generally more highly correlated with the variables than the second factor. This is to be expected because, as previously described, these factors are extracted successively and will account for less and less variance overall.

Rotating the Factor Structure

You could plot the factor loadings shown above in a scatterplot. In that plot, each variable is represented as a point.

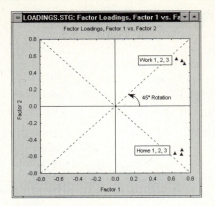

In this plot you could rotate the axes in any direction without changing the *relative* locations of the points to each other; however, the actual coordinates of the points, that is, the factor loadings, would of course change.

In this example, if you produce the plot it will be evident that if you were to rotate the axes about the origin by about 45 degrees you might attain a clear pattern of loadings identifying the work satisfaction items and the home satisfaction items.

Rotational strategies. There are various rotational strategies that have been proposed. The goal of all of these strategies is to obtain a clear pattern of loadings, that is, factors that are somehow clearly marked by high loadings for some variables and low loadings for others. This general pattern is also sometimes referred to as *simple structure* (a more formalized definition can be found in most standard textbooks). Typical rotational strategies are *varimax*, *quartimax*, and *Equamax*; these are described in greater detail in the context of the *Rotation* dialog (page 3227).

The idea of the varimax rotation has been described before (see *Extracting Principal Components*, page 3202), and it can be applied to this problem as well. As before, you want to find a rotation that maximizes the variance on the new axes; put another way, you want to obtain a pattern of loadings on each factor that is as diverse as possible, lending itself to easier interpretation. Below is the table of rotated factor loadings.

FACTOR ANALYSIS Variable	Factor Loadings (Varimax normalized) Extraction: Principal components	
	Factor 1	Factor 2
WORK_1	.862443	.051643
WORK_2	.890267	.110351
WORK_3	.886055	.152603
HOME_1	.062145	.845786
HOME_2	.107230	.902913
HOME_3	.140876	.869995
Expl.Var	2.356684	2.325629
Prp.Totl	.392781	.387605

Interpreting the Factor Structure

Now the pattern is much clearer. As expected, the first factor is marked by high loadings on the work satisfaction items, the second factor is marked by high loadings on the home satisfaction items. You would thus conclude that satisfaction, as measured by your questionnaire, is composed of those two aspects; hence you have arrived at a *classification* of the variables.

Oblique Factors

Some authors (e.g., Harman, 1976; Jennrich and Sampson, 1966; Clarkson and Jennrich, 1988) have discussed in some detail the concept of *oblique* (non-orthogonal) factors, in order to achieve a more interpretable simple structure. Specifically, computational strategies have been developed to rotate factors so as to best represent "clusters" of variables, without the constraint of orthogonality (that is, independence) of factors.

However, the oblique factors produced by such rotations are often not easily interpreted. To return to the example discussed above, suppose you had included in the satisfaction questionnaire above, four items that measured other, "miscellaneous" types of satisfaction. Assume that people's responses to those items were affected about equally by their satisfaction at home (*Factor 1*) and at work (*Factor 2*). An oblique rotation will likely produce two correlated factors with less-than-obvious meaning, that is, with many cross-loadings.

Hierarchical Factor Analysis

Instead of computing loadings for often difficult to interpret oblique factors, the *Factor Analysis* module in *STATISTICA* uses a strategy first proposed by Thompson (1951) and Schmidt and Leiman (1957), which has been elaborated and popularized in the detailed discussions by Wherry (1959, 1975, 1984). In this strategy, *STATISTICA* first identifies clusters of items and rotates axes through those clusters; next, the correlations between those (oblique) factors are computed, and that correlation matrix of oblique factors is further factor-analyzed to yield a set of orthogonal factors that divide the variability in the items into that due to shared or common variance (secondary factors), and unique variance due to the clusters of similar variables (items) in the analysis (primary factors).

FACTOR ANALYSIS Factor	Secondary & Primary (Unique) Factor Loadings		
	Second. 1	Primary 1	Primary 2
WORK_1	.483178	.649499	-.187074
WORK_2	.570953	.687056	-.140627
WORK_3	.565624	.656790	-.115461
HOBBY_1	.776013	.439010	.303672
HOBBY_2	.714183	.455157	.228351
HOME_1	.535812	-.117278	.630076
HOME_2	.615403	-.079910	.668880
HOME_3	.586405	-.065512	.626730
MISCEL_1	.780488	.466823	.280141
MISCEL_2	.734854	.464779	.238512

To return to the example, such a hierarchical analysis might yield the factor loadings shown above. Careful examination of these loadings would lead to the following conclusions:

(1) There is a general (secondary) satisfaction factor that likely affects all types of satisfaction measured by the 10 items;

(2) There appear to be two primary unique areas of satisfaction that can best be described as satisfaction with work and satisfaction with home life.

Wherry (1984) discusses in great detail examples of such hierarchical analyses, and how meaningful and interpretable secondary factors can be derived.

Confirmatory Factor Analysis

Over the past 15 years, so-called confirmatory methods have become increasingly popular (e.g., see Jöreskog and Sörbom,1979). In general, one can specify *a priori* a pattern of factor loadings for a particular number of orthogonal or oblique factors, and then test whether the observed correlation matrix can be reproduced given these specifications. Confirmatory factor analyses can be performed via *STATISTICA*'s general *Structural Equation Modeling (SEPATH)* module (see Chapter 11). Note that the *Examples* section of the *SEPATH* chapter (Chapter 11) discusses several examples of such analyses.

Miscellaneous Other Issues and Statistics

Factor Scores

You can estimate the actual values of individual cases (observations) for the factors. In the *Factor Analysis* module of *STATISTICA*, these scores may be displayed in a Scrollsheet or *Text/output Window*, and also saved to a disk file (whichever you choose in the *Page/Output Setup* dialog, see *General Conventions*, Volume I).

These factor scores are particularly useful when one wants to perform further analyses involving the factors that one has identified in the factor analysis.

Reproduced and Residual Correlations

An additional check for the appropriateness of the respective number of factors that were extracted is to compute the correlation matrix that would result if those were indeed the only factors. That matrix is called the *reproduced* correlation matrix. To see how this matrix deviates from the observed correlation matrix, one can compute the difference between the two; that matrix is called the matrix of *residual* correlations. The residual matrix may point to "misfits," that is, to particular correlation coefficients that cannot be reproduced appropriately by the current number of factors.

Matrix Ill-Conditioning

Computationally, in order to perform a factor analysis, *STATISTICA* needs to invert the correlation matrix. If there are variables in this correlation matrix that are 100% redundant, then the inverse of the matrix cannot be computed.

For example, if a variable is the sum of two other variables selected for the analysis, then the correlation matrix of those variables cannot be inverted, and the factor analysis basically cannot be performed. In practice this happens when you are attempting to factor analyze a set of highly inter-correlated variables, as sometimes occurs, for example, in correlational research with questionnaires.

The *Factor Analysis* module will detect matrix ill-conditioning and issue a respective warning. Then *STATISTICA* will artificially lower all correlations in the correlation matrix by adding a small constant to the diagonal of the matrix, and then re-standardize it. This procedure will usually yield a matrix that can now be inverted and thus factor-analyzed; moreover, the factor patterns should not be affected by this procedure. However, note that the resulting estimates are not exact.

PROGRAM OVERVIEW

Methods

The *Factor Analysis* module contains a wide range of statistics and options, providing a comprehensive implementation of factor analytic techniques. The user may specify a raw data file or a correlation matrix as input for the analyses. The module is especially optimized to handle very large factor analysis problems, and matrices with up to 300 variables (90,000 correlations) can be analyzed (note that even more variables can be processed using the supplementary *STATISTICA BASIC* program *Factoran.stb*, in the *STBASIC* subdirectory). The *Factor Analysis* module will perform principal components and principal factors (common and hierarchical-oblique) analysis. Communalities for principal factors analysis can be computed via multiple regression or several different iterative procedures (*minimum residual* method, *centroid* method, *principal axis* method); maximum likelihood factor analysis is also available. In addition, facilities for the hierarchical factoring of oblique factors are provided in the program. In order to allow the user to explore redundancies among the variables in the analyses, the module also includes a complete implementation of multiple regression statistics, and will fit regression equations with up to 300 variables (note that even more variables can be processed using the supplementary *STATISTICA BASIC* program *Factoran.stb*, in the *STBASIC* subdirectory).

Output

The output from the *Factor Analysis* module contains a comprehensive selection of statistics including factor loadings, factor score coefficients, regular and squared eigenvalues, cumulative eigenvalues, etc.

Predicted and residual correlations. After a factor solution has been determined, the user can recalculate the correlation matrix and evaluate the fit of the factor model by inspecting the residual correlation matrix (or residual variance/covariance matrix).

Factor rotations. Available factor rotations include varimax, biquartimax, quartimax, and equamax; all rotations can be performed based on normalized and raw loadings. Relationships between oblique factors can also be analyzed.

Factor scores. Regression estimates for factor scores can be computed and saved in a data file together with the variables from the input file.

Graphs. A wide variety of specialized graphics options are available to aid in the interpretation of results. For example, raw or rotated factor loadings can be displayed in 2D or 3D scatterplots with labeled data points (variable names), eigenvalues can be plotted in a *scree* plot, variables can be plotted in various ways (2D and 3D) to assess the assumption of multivariate normality.

Matrix Input

If a correlation matrix is chosen as input into the *Factor Analysis* module, the submatrix of selected variables will be extracted from the chosen correlation matrix. Several *STATISTICA* modules will create matrix files (*Basic Statistics and Tables*, *Multiple Regression*, *Factor Analysis*, *Canonical Correlation*, and *Reliability*), but the user can also enter correlations directly into the spreadsheet (see *Matrix File Format*, Volume I).

Alternative Procedures

Confirmatory factor analyses can be performed via the *Structural Equation Modeling (SEPATH)* module (see Chapter 11). Also, *STATISTICA* includes the example *STATISTICA BASIC* program *Factoran.stb* (in the *STBASIC* subdirectory), which

will compute a principal components analysis based on either the correlation matrix, covariance matrix, or moment matrix (for more than 300 variables). Other classification procedures that allow one to detect structure in the relationships between variables are available in the *Cluster Analysis* (see Chapter 5), *Multidimensional Scaling* (see Chapter 7), and *Canonical Analysis* (see Chapter 4) modules.

6. FACTOR ANALYSIS - EXAMPLE

EXAMPLE

Overview

The following example is based on a fictitious data set describing a study of life satisfaction. Suppose that a questionnaire was administered to a random sample of 100 adults. The questionnaire contained 10 items that were designed to measure satisfaction at work, satisfaction with hobbies, satisfaction at home, and general satisfaction in other areas of life. Responses to all questions were recorded via computer and scaled so that the mean for all items was approximately 100.

The results for all respondents were entered into the data file *Factor.sta*; below is a listing of the variables in that file (to obtain this listing, click on the *Var - All Specs* option in the spreadsheet *Edit* pull-down menu or the respective toolbar button).

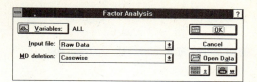

Purpose of analysis. The goal is to learn more about the relationships between satisfaction in the different domains. Specifically, it was desired to learn about the number of factors "behind" these different domains of satisfaction, and their meaning.

Specifying the Analysis

After starting the *Factor Analysis* module, open the data file *Factor.sta*. Click on the *Variables* button in the startup panel (see below) and select all 10 variables in that file.

Other options. In order to perform a standard factor analysis, this is all that you need to specify on this dialog. To give a brief overview of other options available in the startup panel, you can select a *correlation matrix* as the input file via the *Input file* combo box. Via the *MD deletion* combo box, you can choose either *casewise* or *pairwise* deletion, or *mean substitution* of missing data.

Define Method of Factor Extraction

Now, click *OK* to continue to the next dialog titled *Define Method of Factor Extraction*. In this dialog, you will be able to review descriptive statistics, perform a multiple regression analysis, select the extraction method for the factor analysis, select the maximum number of factors and the minimum eigenvalue, and other options related to specific extraction methods.

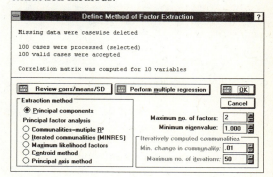

Review Descriptive Statistics

Now, click on the *Review correlations/means/SD* button in this dialog to open the *Review Descriptive*

6. FACTOR ANALYSIS - EXAMPLE

Statistics dialog (the *Define Method of Factor Extraction* dialog will be discussed shortly).

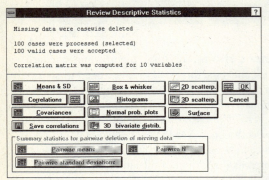

Here, you can review the descriptive statistics graphically or in Scrollsheets.

Computing correlation matrix. Click on the *Correlations* button to display the Scrollsheet of correlations.

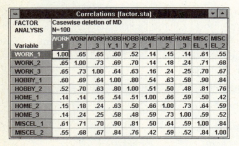

All correlations in the above Scrollsheet are positive; some correlations are of substantial magnitude. For example, variables *Hobby_1* and *Miscel_1* are correlated at the level of *.90*. Some correlations (for example the ones between work satisfaction and home satisfaction) seem comparatively small. So, it looks like there is some clear structure in this matrix.

Extraction Method

Now, click on the *Cancel* button in the *Review Descriptive Statistics* dialog to return to the *Define Method of Factor Extraction* dialog. You can choose from several extraction methods (see *Factor Extraction*, page 3217, for a description of each method and the *Introductory Overview* for a description of *Principal Components*, page 3202, and *Principal Factors*, page 3204). For this example, accept the default extraction method of *Principal Components* and change the *Maximum no. of factors* option to *10* (the maximum number of factors in this example) and the *Minimum eigenvalue* option to *0* (the minimum value for this option).

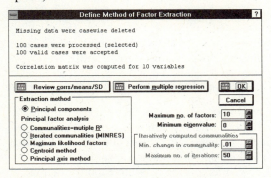

Click *OK* in this dialog to continue the analysis.

Reviewing Results

You can interactively review the results of the factor analysis in the *Factor Analysis Results* dialog.

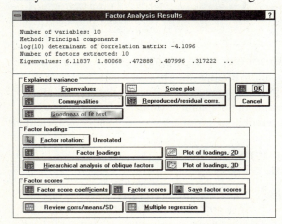

Reviewing the eigenvalues. The meaning of eigenvalues and how they help the user decide how

FAC - 3212

Copyright © StatSoft, 1995

many factors to retain (interpret) was given in the *Introductory Overview* (page 3203). Now, click on the *Eigenvalues* button to display the Scrollsheet of eigenvalues, percent of total variance, cumulative eigenvalues, and cumulative percent.

FACTOR ANALYSIS	Eigenvalues (factor.sta) Extraction: Principal components			
Value	Eigenval	% total Variance	Cumul. Eigenval	Cumul. %
1	6.118369	61.18369	6.11837	61.1837
2	1.800682	18.00682	7.91905	79.1905
3	.472888	4.72888	8.39194	83.9194
4	.407996	4.07996	8.79993	87.9993
5	.317222	3.17222	9.11716	91.1716
6	.293300	2.93300	9.41046	94.1046
7	.195808	1.95808	9.60626	96.0626
8	.170431	1.70431	9.77670	97.7670
9	.137970	1.37970	9.91467	99.1467
10	.085334	.85334	10.00000	100.0000

As you can see, the eigenvalue for the first factor is equal to *6.118369*; the proportion of variance accounted for by the first factor is approximately *61.2%*. Note that these values happen to be easily comparable here because there are 10 variables in the analysis, and thus the sum of all eigenvalues is equal to 10. The second factor accounts for about 18% of the variance. The remaining eigenvalues each account for less than .5 of the variance accounted for.

Deciding on the Number of Factors

The *Introductory Overview* section briefly described how these eigenvalues can be used to decide how many factors to retain, that is, to interpret. According to the *Kaiser* criterion (Kaiser, 1960), you would retain factors with an eigenvalue greater than *1*; based on the eigenvalues in the above Scrollsheet, that criterion would lead you to choose 2 factors.

Scree test. Now, to produce a line graph of the eigenvalues in order to perform Cattell's scree test (Cattell, 1966), click on the *Scree plot* button. The graph below has been "enhanced" to clarify the test. Cattell suggests, based on Monte Carlo studies, that the point where the continuous drop in eigenvalues levels off suggests the cutoff, where only random "noise" is being extracted by additional factors. In the plot below, that point could be at factor 2 or factor 3 (as indicated by the arrows). Therefore, try both solutions and see which one will yield the most interpretable factor pattern.

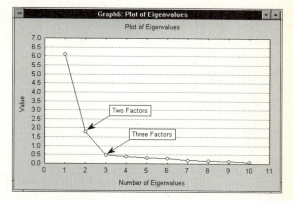

Now, examine the factor loadings.

Factor Loadings

As described in the *Introductory Overview* section (page 3206), factor loadings can be interpreted as the correlations between the factors and the variables. Thus, they represent the most important information on which the interpretation of factors is based.

First look at the (unrotated) factor loadings for all 10 factors. Click on the *Factor Loadings* button to display the Scrollsheet of loadings.

FACTOR ANALYSIS	Factor Loadings (Unrotated) (factor.sta) Extraction: Principal components					
Variable	Factor 1	Factor 2	Factor 3	Factor 4	Factor 9	Factor 10
WORK_1	.652601	-.514217	-.301687	-.439108	.080008	-.003894
WORK_2	.756976	-.494770	.078826	.211795	.103633	-.012210
WORK_3	.745706	-.456680	.104749	-.030826	-.017932	-.038980
HOBBY_1	.941630	.021835	-.012653	-.001861	-.243305	-.171990
HOBBY_2	.875615	-.051643	-.099675	.324541	.088684	-.017996
HOME_1	.576062	.604977	-.490999	.114927	.004027	.019576
HOME_2	.671289	.617962	.125776	-.159963	.145372	-.048318
HOME_3	.641532	.573925	.268572	-.152709	.006890	-.000902
MISCEL_1	.951516	-.013513	.050164	-.026706	-.156713	.223847
MISCEL_2	.900333	-.048154	.151805	.034832	.087690	.030324
Expl.Var	6.118369	1.800682	.472888	.407996	.137970	.085334
Prp.Totl	.611837	.180068	.047289	.040800	.013797	.008533

6. FACTOR ANALYSIS - EXAMPLE

Remember that factors are extracted so that successive factors account for less and less variance (see page 3203, *Introductory Overview*). Therefore, it is not surprising to see that the first factor shows most of the highest loadings. Also note that the sign of the factor loadings only counts insofar as variables with opposite loadings on the same factor relate to that factor in opposite ways. However, you could multiply all loadings in a column by -1, reverse all signs, and the results would not be affected in any way.

Rotating the Factor Solution

As described in the *Introductory Overview* section (page 3206), the actual orientation of the factors in the factorial space is arbitrary, and any rotation of factors will reproduce the correlations as well as any other rotation. This being the case, it seems natural to rotate the factor solution to yield a factor structure that is simplest to interpret; in fact, the formal term *simple structure* was coined and defined by Thurstone (1947) to basically describe the condition when factors are marked by high loadings for some variables, low loadings for others, and when there are few high cross-loadings, that is, few variables with substantial loadings on more than one factor. The most standard computational method of rotation to bring about simple structure is the *varimax* rotation (Kaiser, 1958); others that have been proposed are *quartimax*, *biquartimax*, and *equamax* (see Harman, 1967). These methods are described in greater detail in the *Menus, Options, Statistics* section of this chapter (see the *Factor Rotation* dialog, page 3227).

Specifying a rotation. First, consider the number of factors that you want to rotate, that is, retain and interpret. It was previously decided that two is most likely the appropriate number of factors; however, based on the results of the Scree plot, it was also decided to look at the three factor solution. Start with three factors. Click on the *Cancel* button to return to the *Define Method of Factor Extraction* dialog and change the *Maximum no. of factors* from *10* to *3*, then click *OK* to continue with the analysis.

Now, perform a varimax rotation. In the *Factor Analysis Results* dialog, click on the *Factor Rotation* button to open the *Factor Rotation* dialog.

In this dialog, click on the *Varimax raw* button to perform a varimax rotation and display a Scrollsheet with the resulting factor loadings.

FACTOR ANALYSIS	Extraction: Principal components		
Variable	Factor 1	Factor 2	Factor 3
WORK_1	.839579	-.157384	.227287
WORK_2	.898615	.118837	-.048899
WORK_3	.865608	.151923	-.057003
HOBBY_1	.731038	.501711	.318078
HOBBY_2	.726495	.371499	.336895
HOME_1	.099696	.426704	.864238
HOME_2	.148303	.823540	.384856
HOME_3	.147422	.857607	.236350
MISCEL_1	.758585	.518368	.252834
MISCEL_2	.736270	.524719	.136162
Expl.Var	4.495100	2.591518	1.305320
Prp.Totl	.449510	.259152	.130532

Reviewing the three-factor rotated solution. In the Scrollsheet above, substantial loadings on the first factor appear for all but the home-related items. Factor 2 shows fairly substantial factor loadings for all but the work-related satisfaction items. Factor 3 only has one substantial loading for variable *Home_1*. The fact that only one variable shows a high loading on the third factor should make one wonder whether one cannot do just as well without it (the third factor).

Reviewing the two-factor rotated solution. Once again, click on the *Cancel* button in the *Factor Analysis Results* dialog to return to the *Define Method of Factor Extraction* dialog. Change the

FAC - 3214

Copyright © StatSoft, 1995

Maximum no. of factors from *3* to *2* and click *OK* to continue to the *Results* dialog. Click on the *Factor rotation* button and again select *Varimax raw*.

Variable	Factor 1	Factor 2
WORK_1	.830623	-.019320
WORK_2	.902408	.058905
WORK_3	.870524	.082595
HOBBY_1	.739857	.582885
HOBBY_2	.731191	.484489
HOME_1	.097371	.829676
HOME_2	.165722	.897242
HOME_3	.168370	.844159
MISCEL_1	.768988	.560555
MISCEL_2	.748861	.502121
Expl.Var	4.561544	3.357507
Prp.Totl	.456154	.335751

Factor 1 shows the highest loadings for the items pertaining to work-related satisfaction. The smallest loadings on that factor are for home-related satisfaction items. The other loadings fall in-between. Factor 2 shows the highest loadings for the home-related satisfaction items, lowest loadings for work-related satisfaction items, and loadings in-between for the other items.

Interpreting the two-factor rotated solution.
Does this pattern lend itself to an easy interpretation? It looks like the two factors are best identified as the work satisfaction factor (factor 1) and the home satisfaction factor (factor 2). Satisfaction with hobbies and miscellaneous other aspects of life seem to be related to both factors. This pattern makes some sense in that satisfaction at work and at home may be independent from each other in this sample, but both contribute to leisure time (hobby) satisfaction and satisfaction with other aspects of life.

Plot of the two-factor rotated solution.
Click on the *Plot of loadings, 2D* button in the *Factor Analysis Results* dialog in order to produce a scatterplot of the two factors. The graph below simply shows the two loadings for each variable. Note that this scatterplot nicely illustrates the two independent factors and the 4 variables (*Hobby_1*, *Hobby_2*, *Miscel_1*, *Miscel_2*) with the cross-loadings.

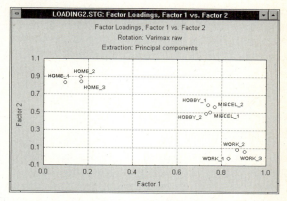

Now, see how well you can reproduce the observed correlation matrix from the two-factor solution.

Reproduced and Residual Correlation Matrix

Click on the *Reproduced/residual corrs.* button to bring up two Scrollsheets with the reproduced correlation matrix and the residual correlations (observed minus reproduced correlations).

Variable	WORK_1	WORK_2	WORK_3	HOBB Y_1	HOBB Y_2	HOME_1	HOME_2	HOME_3	MISC EL_1	MISC EL_2
WORK_1	.31	-.10	-.07	-.01	-.08	.08	.02	.01	-.02	-.06
WORK_2	-.10	.18	-.06	-.01	.01	.01	-.02	.03	-.02	-.02
WORK_3	-.07	-.06	.24	-.06	-.05	.01	.02	.04	-.02	-.02
HOBBY_1	-.01	-.01	-.06	.11	-.02	-.02	-.01	-.03	.01	-.00
HOBBY_2	-.08	.01	-.05	-.02	.23	-.03	-.06	-.05	-.02	-.04
HOME_1	.08	.01	.01	-.02	-.03	.30	-.10	-.13	-.04	-.06
HOME_2	.02	-.02	.02	-.01	-.06	-.10	.17	-.05	.01	.02
HOME_3	.01	.03	.04	-.03	-.05	-.13	-.05	.26	-.05	-.03
MISCEL_1	-.02	-.02	-.02	.01	-.02	-.04	.01	-.02	.09	-.02
MISCEL_2	-.06	-.02	-.02	-.00	-.04	-.06	.02	-.03	-.02	.19

The entries in the *Residual Correlations* Scrollsheet can be interpreted as the "amount" of correlation that cannot be accounted for with the two factor solution. Of course, the diagonal elements in the matrix contain the standard deviation that cannot be accounted for, which is equal to the square root of one minus the respective communalities for two factors (remember that the communality of a variable is the variance that can be explained by the respective number of factors). If you review this matrix carefully you will see that there are virtually no residual correlations left that are greater than *0.1* or less than *-0.1* (actually, a few are about of that magnitude). Add to that the fact that the first two factors accounted for 79% of the total variance (see cumulative % eigenvalues displayed on page 3212).

The "Secret" to the Perfect Example

The example you have reviewed does indeed provide a nearly perfect two-factor solution. It accounts for most of the variance, allows for ready interpretation, and reproduces the correlation matrix with only minor disturbances (remaining residual correlations). Of course, nature rarely affords one such simplicity, and, indeed, this fictitious data set was generated via the normal random number generator accessible in the spreadsheet formulas (see *General Conventions*, Volume I). Specifically, two orthogonal (independent) factors were "planted" into the data, from which the correlations between variables were generated. The factor analysis example retrieved those two factors as intended (i.e., the work satisfaction factor and the home satisfaction factor); thus, had nature planted the two factors, you would have learned something about the underlying or *latent* structure of nature.

Miscellaneous Other Results

Before concluding this example, brief comments on some other results will be made.

Communalities. To view the communalities for the current solution, that is, current numbers of factors, click on the *Communalities* button in the *Factor Analysis Results* dialog. Remember that the communality of a variable is the portion that can be reproduced from the respective number of factors; the rotation of the factor space has no bearing on the communalities. Very low communalities for one or two variables (out of many in the analysis) may indicate that those variables are not well accounted for by the respective factor model.

Factor score coefficients. The factor score coefficients can be used to compute factor scores. These coefficients represent the weights that are used when computing factor scores from the variables. The coefficient matrix itself is usually of little interest; however, factor scores are useful if one wants to perform further analyses on the factors. To view these coefficients, click on the *Factor score coefficients* button in the *Results* dialog.

Factor scores. Factor scores (values) can be thought of as the actual values for each respondent on the underlying factors that you discovered. The *Factor scores* button on the *Factor Analysis Results* dialog allows you to compute factor scores. These scores can be saved via the *Save factor scores* button and used later in other data analyses.

Final Comment

Factor analysis is a not a simple procedure. Anyone who is routinely using factor analysis with many (e.g., 50 or more) variables has seen a wide variety of "pathological behaviors" such as negative eigenvalues, un-interpretable solutions, ill-conditioned matrices, and such. If you are interested in using factor analysis in order to detect structure or meaningful factors in large numbers of variables, it is recommended that you carefully study a textbook on the subject (such as Harman, 1968, or any of the others mentioned at the beginning of the this chapter). Also, because many crucial decisions in factor analysis are by nature subjective (number of factors, rotational method, interpreting loadings), be prepared for the fact that experience is required before you feel comfortable making those judgments. The *Factor Analysis* module was specifically designed to make it easy for the user to switch interactively between different numbers of factors, rotations, etc., so that different solutions can be tried and compared.

6. FACTOR ANALYSIS - DIALOGS AND OPTIONS

DIALOGS, OPTIONS, STATISTICS

Startup Panel

In this dialog, you can select all variables that are to be included in the analysis.

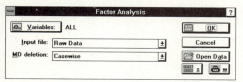

Note that the *Factor Analysis* module can analyze extremely large problems (up to 300 variables, correlation matrices with up to 90,000 elements; see also the note concerning *Matrix ill-conditioning* in the *Technical Notes* section, page 3231.)

Variables

Click on this button to bring up the standard variable selection window in which you can select the variables to be included in the factor analysis.

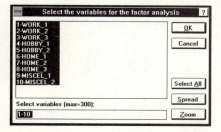

Input File

Input is either *raw data* or a *correlation matrix* previously created via the *Factor Analysis* module or another *STATISTICA* module (e.g., *Basic Statistics and Tables*, *Multiple Regression*, *Canonical Analysis*, *Reliability*, or directly via *Data Management*; see also *Matrix File Format*, Volume I).

Missing Data Deletion

This combo box is only active if a raw data file has been specified (see above). Missing data can be deleted *casewise*, *pairwise*, or *substituted by means*.

Factor Extraction

The choices on this dialog will determine (1) the manner in which factors are extracted, and (2) how many factors will be extracted. Both issues are discussed in greater detail in the *Introductory Overview* section.

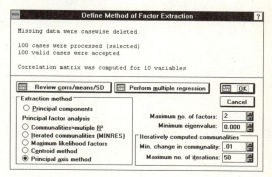

Review Correlations/ Means/SD

This button will bring up a dialog for reviewing various descriptive statistics and graphs (see page 3219). You can also use this option in order to save the current in a matrix file.

Perform Multiple Regression

This button will bring up a dialog for performing *multiple regression analyses* with the currently selected variables (see page 3222). Regression equations with up to 299 independent variables can be evaluated.

Note that extensive options for multiple regression are available in the *Multiple Regression* module of *STATISTICA* (Volume I).

Extraction Methods

The choice of radio buttons in this box will determine the procedure that is used for extracting the factors from the correlation matrix.

Principal components. The first option, *Principal components*, will factor the original correlation matrix (with *1*'s on the diagonal, i.e., for communalities); all other options below will result in principal factor analyses. This distinction is also discussed in the *Introductory Overview* section.

Communalities = multiple *R-square*. If this radio button is set, then, prior to factoring, the diagonal of the correlation matrix (communalities) will be computed as the multiple *R-square* of the respective variable with all other variables. This is a common default method for estimating the communalities for principal factor analysis.

Iterated communalities (MINRES). This method was originally proposed by Harman and Jones (1966). First, multiple *R-square* estimates are used for the communalities. After the initial extraction of factors, the method adjusts the loadings over several iterations and evaluates the goodness of fit of the resulting solution in terms of the residual sums of squares.

When the iterative computation of communalities is in progress, this window will be displayed.

You can click on the *Cancel* button to interrupt the computations. (See also the note concerning *Matrix ill-conditioning* in the *Technical Notes* section, page 3231.)

Maximum likelihood factors. This method of factoring was first proposed by Lawley (1940), and it is discussed in detail in Harman (1976). Unlike in the other methods, it is assumed that the underlying number of factors is known (that is, as set in the *Maximum number of factors* edit box; see below). *STATISTICA* will then estimate the loadings and communalities that maximize the probability of the observed correlation matrix to occur (hence the term *maximum likelihood*). On the *Results* dialog, a *Chi-square* test of the goodness of fit is available (see page 3225).

Centroid method. This method of factoring a correlation matrix was originally developed by Thurstone (1931), and represents a geometrical approach to factor analysis. It is the least "modern" method for factor analysis. Refer to Harman (1976) or Wherry (1984) for additional details.

Principal axis method. In this method, in each iteration, the eigenvalues are computed from current communalities; next the communalities are recomputed based on the extracted eigenvalues and eigenvectors. The new communalities are then placed in the diagonal of the correlation matrix, and the next iteration begins. Iterations will continue until either (1) the *Maximum number of iterations* are exceeded, or (2) the *Min. change in communalities* is less than that specified in the respective edit box (see below).

Maximum No. of Factors

This number will determine how many factors will be extracted. Note that this edit field works in conjunction with the *Minimum eigenvalue* field below; that is, *STATISTICA* will extract either as many factors as requested or as many as have eigenvalues greater than that specified in the *Minimum eigenvalue* field, whichever criterion yields fewer numbers of factors.

Minimum Eigenvalue

This number will determine how many factors are extracted, that is, *STATISTICA* will extract as many factors as there are eigenvalues greater than the number specified in this field. Note that this edit field works in conjunction with the *Maximum no. of factors* field above.

Min. Change in Communality

This option is only available if *Principal axis* factoring or *Centroid* factoring are requested. Those procedures will recompute the communalities in successive iterations until some criterion is met. Specifically, iterations will stop when the change in the successively computed communalities is less than the value specified in this edit field, or when the *Maximum number of iterations* (see below) have been exceeded.

Maximum Number of Iterations

This option is also only available if *Principal axis* factoring or *Centroid* factoring are requested. Those procedures will recompute the communalities in successive iterations; iterations will stop when the change in the successively computed communalities is less than the value specified in the *Min. change in communality* field, or when the *Maximum number of iterations* have been exceeded.

Descriptive Statistics

This dialog allows you to compute detailed descriptive statistics for the variables in the current analysis. Note that two important underlying assumption of factor analysis are

(1) That the distributions of variables follow the normal distribution, and

(2) That the correlations between variables are linear.

The graphics options on this dialog allow the user to visually test both of those assumptions. (Note that the graphics options are not available if a correlation matrix input file is currently being analyzed. See also the note concerning *Matrix ill-conditioning* in the *Technical Notes* section, page 3231.)

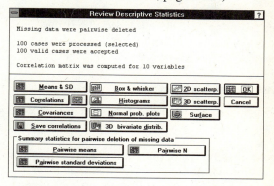

Means & SD

This option will bring up a Scrollsheet with the descriptive statistics for the variables in the analysis. If a raw data file is analyzed, then the default statistical graph for that Scrollsheet is the histogram with fitted normal distribution.

Correlations

This option will bring up a Scrollsheet with the current correlation matrix. If a raw data file is analyzed, then the default statistical graph for that Scrollsheet is the scatterplot with the fitted regression line. If matrix ill-conditioning was detected during the factoring process, then this matrix will have been slightly modified (see the *Technical Notes* section, page 3231).

Scatterplot Matrix

This option is only available if a raw data file is being analyzed. This button will bring up a scatterplot matrix for the variables in the current analysis.

Covariances

This option will bring up a Scrollsheet with the current covariance matrix. If a raw data file is analyzed, the default statistical graph for that Scrollsheet is the scatterplot with the fitted regression line. If matrix ill-conditioning was detected during the factoring process, then this matrix will have been slightly modified (see the *Technical Notes* section, page 3231).

Save Correlations

This option allows you to save the current correlation matrix in *STATISTICA Matrix File Format* (see Volume I). The correlation matrix can then be directly read in subsequent factor analyses (or analyses via *Multiple Regression*, *Canonical Correlation*, *Reliability and Item Analysis*, *Cluster Analysis*, *Multidimensional Scaling,* etc.).

Summary Statistics for Pairwise Deletion of Missing Data

These options are only available if *Pairwise deletion of missing data* was selected on the startup panel (page 3217), and if this dialog was called immediately after reading the data (these various matrices are no longer needed in the actual factor analysis computations, and thus are discarded when the factor extraction process begins). Note that in *pairwise* deletion of missing data, the different correlations in the correlation matrix can be based on unequal numbers of (different) cases. Thus, for each correlation coefficient there can be:

(1) A unique *n* (valid number of cases),

(2) Unique means for the two respective variables, and

(3) Unique standard deviations for the two variables.

If large discrepancies are evident for the values for different correlations (pairs of variables), then the different correlations may have been computed from different actual cases. Before reaching final conclusions from the factor analysis, it is advisable to repeat the analysis either with *casewise* deletion of missing data or *mean substitution*.

Pairwise Means

This option will bring up a Scrollsheet with the pairwise means. The Scrollsheet will show the means for the row variables, when they are paired (correlated) with the respective column variables.

Pairwise Standard Deviations

This option will bring up a Scrollsheet with the pairwise standard deviations. The Scrollsheet will show the standard deviations for the row variables, when they are paired (correlated) with the respective column variables.

Pairwise *n*

This option will bring up a Scrollsheet with the pairwise *n* (valid number of cases).

Box and Whisker Plots

This option allows you to produce box and whisker plots for the variables in the current analysis. The box and whisker plot will summarize each variable by three components:

(1) A central line to indicate central tendency or location,

(2) A box to indicate variability around this central tendency, and

(3) Whiskers around the box to indicate the range of the variable.

After clicking on this button you can choose to plot for each variable:

(1) Medians (central line), quartiles (box), and ranges (whiskers); note that the specific method that is used to compute medians and quartiles can be configured in the *STATISTICA Defaults: General* dialog (see the *General Conventions* chapter in Volume I);

(2) Means, standard errors of the means, and standard deviations,

(3) Means, standard deviations, and 1.96 times the standard deviations (95% normal confidence interval for individual observations around the mean), or

(4) Means standard errors of the means, and 1.96 times the standard errors of the means (95% normal confidence interval for means).

Histograms

This option allows you to produce histograms of the distributions of variables in the current analysis. The plots will also include the expected normal distribution curves.

Normal Probability Plots

This option will produce a cascade of normal probability plots for the selected variables. The standard normal probability plot is constructed as follows. First, the deviations from the respective means (*residuals*) are rank ordered. From these ranks, z values (i.e., standardized values of the normal distribution) can be computed *based on the assumption* that the data come from a normal distribution. These z values are plotted on the *Y*-axis in the plot. If the observed residuals (plotted on the *X*-axis) are normally distributed, then all values should fall onto a straight line.

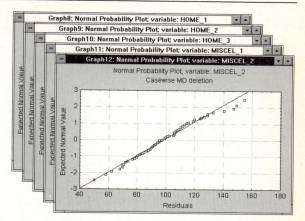

If the residuals are not normally distributed, then they will deviate from the line. Outliers may also become evident in this plot.

3D Bivariate Distribution Plot

This option will produce a cascade of 3D histograms for selected pairs of variables, one plot per pair.

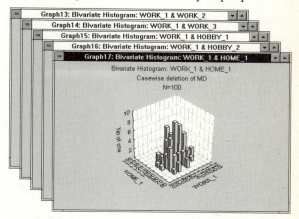

After clicking on this button, the user is first prompted to select two lists of variables (from among those originally selected via the *Variables* option). A cascade of 3D histograms will be produced, one for each variable in the first list with each variable in the second list.

6. FACTOR ANALYSIS - DIALOGS AND OPTIONS

2D Scatterplots

This option will produce a cascade of scatterplots for selected pairs of variables, one plot per variable pair (i.e., correlation).

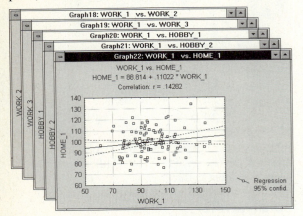

Each plot will also show the regression line and the standard error of the predicted values (as error lines).

3D Scatterplots

This option will produce a 3D scatterplot for a selected triplet of variables.

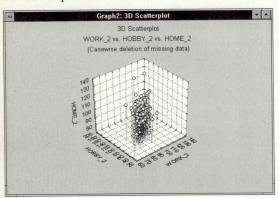

Surface Plots

This option will produce a 3D scatterplot for a selected triplet of variables and fit a quadratic surface to the data.

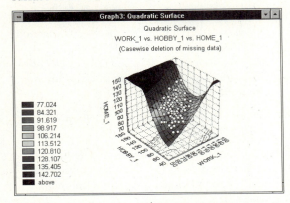

You can switch to other fitting/smoothing methods in the *General Layout* dialog accessible by double-clicking anywhere on the background of the graph (see Volume II, *Graphics*, for details). In this context, a surface may help interpret the pattern of data in *XYZ* scatterplots (please refer to Volume II for more information).

Multiple Regression

This dialog allows you to perform multiple regression analyses for the variables in the current analysis.

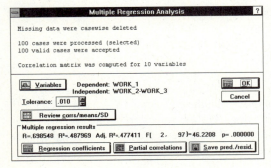

Note that *STATISTICA* includes a designated *Multiple Regression* module (see Volume I), which contains numerous additional options and extensive residual analysis facilities. (See also the note concerning *Matrix ill-conditioning* in the *Technical Notes* section, page 3231.)

Variables

This option will bring up a standard two-variable lists selection dialog in which you can select the independent variables and the dependent variable (to analyze more than 300 variables in a single regression equation, use the supplementary *STATISTICA BASIC* example program *Regressn.stb* in the *STBASIC* subdirectory).

Tolerance

The tolerance of a variable is defined as 1 minus the squared multiple correlation of this variable with all other independent variables in the regression equation. Therefore, the smaller the tolerance of a variable, the more redundant is its contribution to the regression (i.e., it is redundant with the contribution of other independent variables).

If the tolerance of any of the variables in the regression equation is equal to zero (or very close to zero), then the regression equation cannot be evaluated (the matrix is said to be ill-conditioned, and it cannot be inverted, see the *Technical Notes* section, page 3231).

This option allows you to specify the minimum tolerance that is considered acceptable by *STATISTICA*. The minimum value that can be specified here is $1.00E-25$ (i.e., a number with 24 zeros past the decimal point). However, it is not recommended to reset this switch to such an extremely low value.

If the tolerance of a variable entered into the regression equation is less than the default tolerance value (.01) it means that this variable is 99% redundant with (identical to) the variables already in the equation.

Forcing very redundant variables into the regression equation is not only questionable in terms of relevance of results, but the resultant estimates (regression coefficients) will become increasingly unreliable.

Review Corrs/ Means/SD

This option will bring up the *Descriptive Statistics* dialog (see page 3219).

Regression Coefficients

This option will bring up a Scrollsheet with the standardized (*beta*) and nonstandardized (*B*) regression coefficients (weights), their standard error, and statistical significance. The summary statistics for the regression analysis (e.g., *R*, *R-square*, etc.) will be shown in the headers of the Scrollsheet.

Partial Correlations

Clicking on this button will open Scrollsheets with:

(1) the *beta in* (standard regression coefficient for the respective variable if it were to enter into the regression equation as an independent variable),

(2) the *partial* correlation (between the respective variable and the dependent variable, after controlling for all other independent variables in the equation),

(3) the *semi-partial* (part) *correlation* (the correlation between the *unadjusted* dependent variable with the respective variable after controlling for all independent variables in the equation),

(4) the *tolerance* for the respective variable (defined as 1 minus the squared multiple

correlation between the respective variable and all independent variables in the regression equation),

(5) the *minimum tolerance* (the smallest tolerance among all independent variables in the equation if the respective variable were to be entered as an additional independent variable),

(6) the *t* value associated with these statistics for the respective variable,

(7) the statistical significance of the *t* value.

These statistics will first be displayed separately for variables not currently in the regression equation (if any), and for the variables *in* the regression equation.

Save Pred/ Residuals

This option is only available if a raw data file is analyzed (not a matrix file). This option allows you to save the predicted and residual values for the current regression equation.

Alpha Level for Highlighting

The default value for *alpha* is *.05*. This value is used to determine the significance of a specific effect. If the *p*-value is less than or equal to the designated value for *alpha*, then it is concluded that the effect is significant and it is highlighted in the window.

In this window (accessible from the *Options* button on the Scrollsheet toolbar), set the desired *alpha* level (.0001<α<.5) that is to be used for highlighting significant results in the Scrollsheet.

Results

The following options are available from the *Factor Analysis Results* dialog. (See also the note concerning *Matrix ill-conditioning* in the *Technical Notes* section, page 3231.)

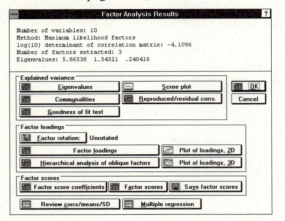

Eigenvalues

This option will bring up a Scrollsheet with the eigenvalues. The relative (percent) and cumulative eigenvalues are also reported. The interpretation of eigenvalues is described in the *Introductory Overview* section (page 3203). In general, the eigenvalues reflect the amount of common variance accounted for by the respective number of factors.

Communalities

This option will bring up a Scrollsheet with the communalities, computed from the respective number of factors extracted in the current analysis. Again, refer to the *Introductory Overview* section (page 3205) for a discussion of communalities; in general, the communalities can be interpreted as the proportion of variance accounted for in the respective variables by the current number of factors.

Goodness of Fit Test

This option is only available if *Maximum Likelihood* factoring was selected in the *Define Method of Factor Extraction* dialog. This option will bring up a Scrollsheet with the *U* test (Lawley, 1940) of fit for the current number of factors. In general, when the variables in the analysis come from a multivariate normal distribution, then the distribution of covariances follows the Wishart distribution (Wishart, 1928). Based on these assumptions, a *Chi-square* test can be constructed testing whether all residual correlations are equal to zero, that is, whether or not the residual correlation matrix is a diagonal matrix. If the test is statistically significant, it can be concluded that the residual correlation matrix is significantly different from the diagonal matrix, and therefore, that significant correlations between variables are still not accounted for. Also, *Chi-square* tests of incremental models (with increasing numbers of factors) can be performed, allowing for statistical significance testing of the number of factors "contained" in the correlation matrix. However, note that the sensitivity of the *Chi-square* test is greatly affected by the sample size (*n*); refer to Harman (1976), Mulaik (1972), or Wherry (1984) for additional details.

Scree Plot

This option will bring up the so-called *scree plot* (Cattell, 1966). Specifically, the successive eigenvalues will be shown in a simple line plot. Cattell suggests finding the place where the smooth decrease of eigenvalues appears to level off to the right of the plot. To the right of this point, presumably, one finds only "factorial scree" -- "scree" is the geological term referring to the debris which collects on the lower part of a rocky slope. Thus, no more than the number of factors to the left of this point should be extracted. Refer to *Introductory Overview* section (page 3204) for a more detailed discussion of the number-of-factors issue in factor analysis.

Reproduced/Residual Correlations

This option will bring up Scrollsheets with the reproduced and residual (observed minus reproduced) correlation matrices. You can enter the cut-off value for the residual correlations that are to be highlighted in the Scrollsheet in this window.

The *Minimum Residual for Highlighting* window is accessible from the *Options* button on the Scrollsheet toolbar.

Factor Rotation

Pressing this button will bring up a dialog allowing the selection of the method of rotation. As discussed in the *Introductory Overview* section, the rotational orientation of axes in factor analysis is more or less arbitrary. However, numerous rotational strategies have been proposed to choose an orientation of axes that is most interpretable (i.e., approximates *simple structure*). Several such strategies are available in the *Factor Analysis* module (refer to the *Factor Rotation* dialog, page 3227). The current orientation of axes is indicated to the right of this button.

Factor Loadings

This option will bring up a Scrollsheet with the current factor loadings, that is, rotated in the manner indicated next to the *Factor rotation* button. The default statistical plot for this Scrollsheet is the 2D or 3D scatterplot of factor loadings. Factor loadings can be interpreted as correlations between the respective variables and factors; thus, they represent the most important information for the interpretation of factors. Refer to *Introductory Overview* section (page 3206) for additional details.

Initially, factor loadings greater than *0.7* are highlighted in the Scrollsheet. You can enter the cut-off value for factor loadings that are to be highlighted in the Scrollsheet in this window (accessible from the *Options* button on the Scrollsheet toolbar).

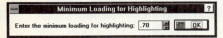

Factor loadings can be interpreted as correlations between the respective variables and factors; thus, they represent the most important information for the interpretation of factors. Refer to the *Introductory Overview* section (page 3206) for additional details.

Plot of Loadings, 2D & 3D

These options will produce 2D or 3D scatterplots of the current factor loadings; a visual inspection of loadings often suggests a clearer interpretation of factors.

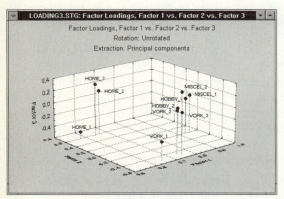

Hierarchical Analysis of Oblique Factors

This option is only available if more than one factor was extracted. In that case, a hierarchical factor analysis (see Thompson, 1951; Schmid and Leiman, 1957; Wherry, 1959, 1975, 1984) can be performed.

In short, *STATISTICA* will first identify clusters of marker variables (with high unique factor loadings, and low cross-loadings) and rotate axes through those clusters. If no such clear clusters of "marker variables" can be identified for each oblique factor, the hierarchical analysis will not be performed.

Next, the correlations between those (oblique) factors is computed, and that correlation matrix of oblique factors is further factor-analyzed to yield a set of orthogonal factors that divide the variability in the items into that due to shared or common variance (secondary factors), and unique variance due to the clusters of similar variables (items) in the analysis (primary factors). After clicking on this button, Scrollsheets with the complete results will be automatically generated.

Initially, factor loadings greater than *0.7* are highlighted in the Scrollsheet. However, this value can be changed via the *Minimum Loading for Highlighting* window (see *Factor Loading*, above).

Factor Score Coefficients

This option will bring up a Scrollsheet with the factor score coefficients. For principal components analysis, exact coefficients can be computed from the (rotated or unrotated) factor loadings. For principal factors analysis (with estimated communalities), regression estimates of the factor score coefficients are computed; see Harman (1976, page 368).

Factor Scores

This option is only available if a raw data file is currently being analyzed. This option will bring up a Scrollsheet with factor scores, based on the factor score coefficients computed as described above. If *casewise* or *pairwise* deletion of missing data was selected, then factor scores can only be computed for complete cases (without missing data on any of the variables in the current analysis); if *mean substitution* was selected, then missing data will be substituted by the respective variable means in the computations (since factor scores are computed from normal *z* values, missing data will in effect be substituted by zero).

Save Factor Scores

This option is only available if a raw data file is currently being analyzed. This option allows you to save the factor scores, which are computed as described above.

Review Correlations/ Means/SD

This button will bring up a dialog for reviewing various descriptive statistics and graphs. This option is also used in order to save the current correlation matrix in a matrix file.

Perform Multiple Regression

This button will bring up a dialog for performing *multiple regression analyses* with the currently selected variables. Regression equations with up to 299 independent variables can be evaluated. Note that extensive options for multiple regression are available in the *Multiple Regression* module of *STATISTICA* (Volume I).

Factor Rotation

This dialog allows you to select the method of factor rotation. There are various rotational strategies that have been proposed. The goal of all of these strategies is to obtain a clear pattern of loadings, that is, factors that are somehow clearly marked by high loadings for some variables and low loadings for others. This general pattern is also sometimes referred to as *simple structure* (a more formalized definition can be found in most standard textbooks). Typical rotational strategies are *varimax*, *biquartimax, quartimax*, and *equamax*.

Some authors (e.g., Harman, 1976; Jennrich and Sampson, 1966; Clarkson and Jennrich, 1988) have discussed in some detail the concept of *oblique* (non-orthogonal) factors, in order to achieve more interpretable simple structure. Specifically, computer (algorithmic) strategies have been developed to rotate factors so as to best represent "clusters" of variables, without the constraint of orthogonality of factors. However, the oblique factors produced by such rotations are often not easily interpreted. Use the *Hierarchical factor analysis of oblique factors* option instead in order to identify (correlated, oblique) clusters of variables. Note that you can also use the *Structural Equation Modeling (SEPATH)* module (see Chapter 11) to test the adequacy (goodness of fit) of specific orthogonal or oblique factor solutions.

Varimax Raw

This option will perform a varimax rotation of the factor loadings. This rotation is aimed at maximizing the variances of the squared raw factor loadings across variables for each factor; this is equivalent to maximizing the variances in the *columns* of the matrix of squared raw factor loadings.

Varimax Normalized

This option will perform a varimax rotation of the *normalized factor loadings* (raw factor loadings divided by the square roots of the respective communalities). This rotation is aimed at maximizing the variances of the squared normalized factor loadings across variables for each factor; this is equivalent to maximizing the variances in the *columns* of the matrix of squared normalized factor loadings. Rather than *Varimax Raw*, this is the method that is most commonly used and is referred to as simply *varimax* rotation.

Biquartimax Raw

This option will perform a biquartimax rotation of the raw factor loadings. This rotation can be considered to be an "even mixture" of the varimax and quartimax rotation. Specifically, it is aimed at *simultaneously* maximizing the sum of variances of the squared raw factor loadings across factors *and* maximizing the sum of variances of the squared raw factor loadings across variables; this is equivalent to simultaneously maximizing the variances in the *rows and columns* of the matrix of squared raw factor loadings.

Biquartimax Normalized

This rotation is equivalent to the biquartimax raw rotation described above, except that it is performed on normalized (standardized) factor loadings.

Quartimax Raw

This option will perform a quartimax rotation of the (raw) factor loadings. This rotation is aimed at maximizing the variances of the squared (raw) factor loadings across factors for each variable; this is equivalent to maximizing the variances in the *rows* of the matrix of squared raw factor loadings.

Quartimax Normalized

This option will perform a quartimax rotation of the normalized factor loadings, that is, the raw factor loadings divided by the respective communalities. This rotation is aimed at maximizing the variances of the squared normalized factor loadings across factors for each variable; this is equivalent to maximizing the variances in the *rows* of the matrix of squared normalized factor loadings. This is the method that is commonly referred to as *quartimax* rotation.

Equamax Raw

This option will perform an equamax rotation of the raw factor loadings. This rotation can be considered to be a "weighted mixture" of the varimax and quartimax rotation. Specifically, it is aimed at *simultaneously* maximizing the sum of variances of the squared raw factor loadings across factors *and* maximizing the sum of variances of the squared raw factor loadings across variables; this is equivalent to simultaneously maximizing the variances in the *rows and columns* of the matrix of squared raw factor loadings. However, unlike the biquartimax rotation, the relative weight assigned to the varimax criterion in the rotation is equal to the number of factors divided by 2.

Equamax Normalized

This option will perform an equamax rotation, as described above (*Equamax raw*); however, this rotation will be performed on the *normalized factor loadings*.

Oblique Factors

In order to analyze relationships between oblique factors, use the option *Hierarchical Analysis of Oblique Factors* from the previous (*Factor Analysis Results*) dialog.

3D Plot of Loadings

From this dialog, select the variables for the 3D scatterplots of factor loadings.

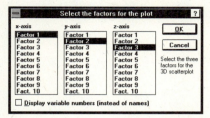

The points in the resulting scatterplot represent individual variables and are labeled with their respective names (de-select the *Display variable numbers* option) or numbers (select the *Display Variable numbers* option).

If the labels overlap. If the labels overlap, try to rotate the graph (via the *Perspective and Rotation* Graph Window toolbar button) or double-click on any data point label to access the *Data Point Labels* dialog (see below) and then click on the *Font* button in order to adjust (reduce) the size of font used to label the points.

You can also use the *Display Filters* option from the Graphics *View* pull-down menu in order to select the degree of filtering (removal of overlapping points or text) in the current graph (the default setting, *Level 2*, suppresses the display of only those labels which overlap to a large degree).

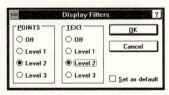

The *Display Filters* facility is described in detail in Volume II.

Finally, you may also adjust the location of specific labels one at a time: double-click on any point label to bring up the *Data Point Labels* dialog.

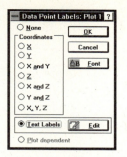

Now, click on the *Edit* button in this dialog to open the *Edit Text Labels* dialog.

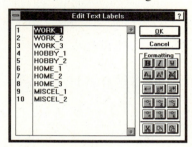

In this dialog, highlight the label to be adjusted and cut it (by pressing CTRL+X); press *OK* to return to the plot. Now, paste (CTRL+V) the label into the graph, and then position the label as desired.

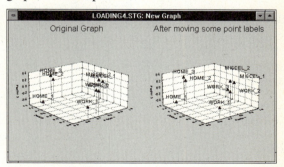

You may need to adjust the font and other attributes of the label if the current defaults of the custom text are different from the defaults set for the data labels, but you will need to make this adjustment only once - the next repositioned label will use the previously set attributes.

2D Plot of Loadings

From this dialog, select the variables for the 2D scatterplot of factor loadings.

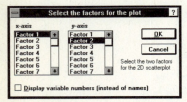

The points in the resulting scatterplot represent individual variables and are labeled with their respective names (de-select the *Display variable numbers* option) or numbers (select the *Display Variable numbers* option).

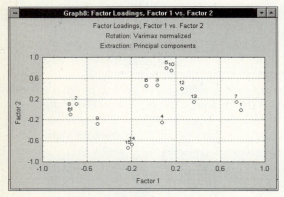

If the labels overlap, use one of the methods described above.

TECHNICAL NOTES

Eigenvalues

At the "heart" of factor analysis is the eigenvalue problem that is solved in this program via the Householder method; see, for example, Golub and Van Loan (1983), Jacobs (1977), or Ralston and Wilf (1967, Vol. II). Eigenvalues are calculated via least squares procedures. The sum of the eigenvalues is equal to the sum of the diagonal elements of the matrix (correlations or covariances) that is analyzed, that is:

$$\Sigma \lambda_j = \text{trace}/S/ = \Sigma s_{ij}$$

where

λ_j is the j'th eigenvalue
$/S/$ is the variance/covariance matrix or correlation matrix
s_{ij} are the diagonal elements of the variance-/covariance matrix or the correlation matrix

Matrix Ill-Conditioning and the Modified Correlation Matrix

The *Introductory Overview* section discusses the issue of matrix ill-conditioning. If during the factoring process you received the message that the correlation matrix cannot be inverted, then the correlation matrix will be modified (to allow inversion). Specifically, a small constant is added to the diagonal elements of the correlation matrix until the determinant of that matrix is greater than *1.E-50*. All subsequent computations will be performed based on the modified (slightly lowered) correlations; to restore the original exact correlation matrix, click *Cancel* to return to the startup panel and then click *OK* to reread the data file.

Analyzing Covariance or Moment Matrices

STATISTICA includes the example *STATISTICA BASIC* program *Factoran.stb* (in the *STBASIC* subdirectory). That program will compute a principal components analysis based on either the correlation matrix, covariance matrix, or moment matrix.

6. FACTOR ANALYSIS - TECHNICAL NOTES

6. FACTOR ANALYSIS - INDEX

INDEX

A

Alpha level for highlighting, 3224

B

Biquartimax normalized, 3228
Biquartimax raw, 3228
Box and whisker plots
    for all variables, 3220

C

Casewise deletion of missing data, 3217
Centroid method, 3218
Chi-square goodness of fit test, 3218, 3225
Combining two variables into a single factor, 3202
Communalities, 3205, 3216, 3224
    computed as multiple R-square, 3218
Confirmatory factor analysis, 3201, 3208
Correlation matrix
    input file, 3211, 3217
    matrix ill-conditioning and the modified correlation matrix, 3231
    reproduced and residual, 3215, 3225
    reviewing, 3219
    saving, 3220
Covariance matrix, factor analysis of, 3231

D

Data reduction method, 3201
Descriptive statistics, 3211, 3219

E

Eigenvalues, 3203, 3212, 3224, 3231
Equamax normalized, 3228
Equamax raw, 3228
Extracting principal components, 3202

Extraction method
    centroid method, 3218
    iterated communalities (MINRES method), 3218
    maximum likelihood factors, 3218
    maximum number of factors, 3218
    maximum number of iterations, 3219
    minimum change in communality, 3219
    minimum eigenvalue, 3219
    multiple R-square, computing communalities as, 3218
    principal axis method, 3218
    principal components, 3218

F

Factor analysis
    2D plot of loadings, 3226, 3230
    3D plot of loadings, 3226, 3229
    alternative procedures, 3210
    assumptions, 3219
    centroid method, 3218
    classification method, 3205
    combining two variables into a single factor, 3202
    communalities, 3205, 3216, 3224
    communalities computed as R-square, 3218
    confirmatory factor analysis, 3201, 3208
    covariance matrix, factor analysis of, 3231
    data reduction method, 3201
    deciding on the number of factors, 3213
    descriptive statistics, 3219
    eigenvalues, 3203, 3212, 3224, 3231
    example, 3211
    extraction method, 3212, 3218
    factor extraction, 3217
    factor loadings, 3206, 3213, 3225
    factor rotation, 3225
    factor score coefficients, 3216, 3226
    factor scores, 3208, 3216, 3226
    generalizing to the case of multiple variables, 3202
    hierarchical analysis of oblique factors, 3207, 3226

Factor analysis (continued)
    how many factors to extract, 3203
    interpreting the factor structure, 3207
    iterated communalities, 3218
    Kaiser criterion, 3204
    matrix ill-conditioning, 3208, 3231
    matrix input, 3209
    maximum likelihood factors, 3218
    MINRES method, 3205, 3218
    moment matrix, factor analysis of, 3231
    multiple orthogonal factors, 3203
    multiple regression, 3222
    oblique factors, 3207, 3226, 3228
    partial correlation, 3223
    principal axis method, 3218
    principal factors analysis, 3204
    principal factors vs. principal components, 3205
    reproduced and residual correlations, 3208, 3215, 3225
    rotating the factor structure, 3206, 3214
    rotational strategies, 3206, 3227
    scree test, 3213, 3225
    startup panel, 3217
    tolerance, 3223
Factor extraction, 3217
Factor loadings, 3206, 3213, 3225
Factor rotation
    biquartimax normalized, 3228
    biquartimax raw, 3228
    equamax normalized, 3228
    equamax raw, 3228
    quartimax normalized, 3228
    quartimax raw, 3228
    varimax normalized, 3228
    varimax raw, 3227
Factor score coefficients, 3216, 3226
Factor scores, 3208, 3216, 3226

G

Goodness of fit Chi-square, 3218, 3225

H

Hierarchical factor analysis, 3207, 3226
Histograms
    2D histograms, 3221

FAC - 3233

Copyright © StatSoft, 1995

6. FACTOR ANALYSIS - INDEX

Histograms (continued)
    3D bivariate histograms, 3221

I

Iterated communalities, 3218

K

Kaiser criterion, 3204

M

Matrix ill-conditioning, 3208, 3231
Maximum likelihood factors, 3218
Mean substitution of missing data, 3217
MINRES method, 3205, 3218
Missing data deletion
    casewise deletion of missing data, 3217
    mean substitution of missing data, 3217
    pairwise deletion of missing data, 3217, 3220
Moment matrix, factor analysis of, 3231
Multiple orthogonal factors, 3203
Multiple regression analysis, 3217, 3222

O

Oblique factors, 3207, 3228
Orthogonal factors, 3203
    and oblique factors, 3207, 3226

P

Pairwise deletion of missing data, 3217, 3220
Pairwise means, 3220
Pairwise n, 3220
Pairwise standard deviations, 3220
Partial correlation, 3223
Principal axis method, 3218
Principal components analysis
    extracting principal components, 3202
    extraction method, 3218
    overview, 3202

Principal components analysis (cont.)
    reviewing the results of a principal components analysis, 3203
Principal factors analysis, 3204
Principal factors vs. principal components, 3205
Probability plots
    normal probability plots, 3221

Q

Quartimax normalized, 3228
Quartimax raw, 3228

R

Reproduced and residual correlations, 3208, 3215, 3225
Rotating the factor structure, 3206, 3214
Rotational strategies, 3206, 3227

S

Scatterplots
    2D scatterplots, 3222
    3D scatterplot, 3222
    3D scatterplot with fitted quadratic surface, 3222
    matrix scatterplot, 3219
Scree test, 3204, 3213, 3225
Simple structure, 3206, 3214
Surface plot, 3222

T

Tolerance, 3223

V

Variance maximizing (varimax) rotation, 3202
Varimax normalized, 3228
Varimax raw, 3227

FAC - 3234

Copyright © StatSoft, 1995

Chapter 7:
MULTIDIMENSIONAL SCALING

Table of Contents

Introductory Overview 3237
Program Overview 3243
Example 3245
Dialogs, Options, Statistics 3249
    Startup Panel 3249
    Estimation 3250
    Results 3251
Technical Notes 3255
Index 3257

Detailed Table of Contents

INTRODUCTORY OVERVIEW 3237
    General Purpose 3237
    Logic of Multidimensional Scaling 3237
    Computational Approach 3237
    How Many Dimensions to Specify? 3238
    Interpreting the Dimensions 3239
    Applications 3240
    Multidimensional Scaling and Factor Analysis 3241

PROGRAM OVERVIEW 3243
    Overview 3243
    Alternative Procedures 3243

EXAMPLE 3245
    Overview and Data File 3245
    Specifying the Analysis 3245
    Performing the Analysis 3245
    Results 3246
    Continuing the Analysis 3247

DIALOGS, OPTIONS, STATISTICS 3249

7. MULTIDIMENSIONAL SCALING - CONTENTS

Startup Panel 3249
    Variables 3249
    Number of Dimensions 3249
    Starting Configuration 3249
    Epsilon 3250
    Number of Iterations 3250
Estimation 3250
    D-stars and D-hats 3250
Results 3251
    Final Configuration 3251
    D-hat Values 3251
    D-star Values 3251
    Distance Matrix 3251
    Summary 3251
    Save Final Configuration 3251
    Graph Final Configuration - 2D, 3D 3252
    Graph D-hat vs. Distances 3253
    Graph D-star vs. Distances 3253
    Shepard Diagram 3253

TECHNICAL NOTES **3255**
    Computational Approach 3255
    D-stars and D-hats 3255

INDEX **3257**

Chapter 7: MULTIDIMENSIONAL SCALING

INTRODUCTORY OVERVIEW

General Purpose

Multidimensional scaling (MDS) can be considered to be an alternative to factor analysis (see *Factor Analysis*, Chapter 6). In general, the goal of the analysis is to detect meaningful underlying dimensions that allow the researcher to explain observed similarities or dissimilarities (distances) between the investigated objects. In factor analysis, the similarities between objects (e.g., variables) are expressed in the correlation matrix. With MDS one may analyze any kind of similarity or dissimilarity matrix, in addition to correlation matrices.

Logic of Multidimensional Scaling

The following simple example may demonstrate the logic of a multidimensional scaling analysis. Suppose you take a matrix of distances between major US cities from a map. You then analyze this matrix, specifying that you want to reproduce the distances based on two dimensions. As a result of the MDS analysis, you would most likely obtain a two-dimensional representation of the locations of the cities, that is, you would basically obtain a two-dimensional map.

In general then, MDS attempts to arrange "objects" (major cities in this example) in a space with a particular number of dimensions (two dimensions in this example) so as to reproduce the observed distances. As a result, you can "explain" the distances in terms of underlying dimensions; in this example, you could explain the distances in terms of the two geographical dimensions: north/south and east/west.

Orientation of axes

As in factor analysis, the actual orientation of axes in the final solution is arbitrary. To return to the example, you could rotate the map in any way you want, and the distances between cities would remain the same. Thus, the final orientation of axes in the plane or space is mostly the result of a subjective decision by the researcher, who will choose an orientation that can be most easily explained. In this example, you could have chosen an orientation of axes other than north/south and east/west; however, that orientation is most convenient because it "makes the most sense" (i.e., it is easily interpretable).

Computational Approach

MDS is not so much an exact procedure as rather a way to "rearrange" objects in an efficient manner, so as to arrive at a configuration that best approximates the observed distances. The program actually moves objects around in the space defined by the requested number of dimensions and checks how well the distances between objects can be reproduced by the new configuration. In more technical terms, the program uses a function minimization algorithm that evaluates different configurations with the goal of maximizing the goodness of fit (or minimizing "lack of fit").

Measures of goodness of fit: Stress

The most common measure that is used to evaluate how well (or poorly) a particular configuration reproduces the observed distance matrix is the stress measure. The raw stress value *Phi* of a configuration is defined by:

$$\text{Phi} = \Sigma[d_{ij} - f(\delta_{ij})]^2$$

7. MULTIDIMENSIONAL SCALING - INTRODUCTORY OVERVIEW

In this formula, d_{ij} stands for the reproduced distances, given the respective number of dimensions, and δ_{ij} (*delta$_{ij}$*) stands for the input data (i.e., observed distances). The expression $f(\delta_{ij})$ indicates a *nonmetric*, monotone transformation of the observed input data (distances). Thus, the program will attempt to reproduce the general rank-ordering of distances between the objects in the analysis.

There are several similar related measures that are commonly used; however, most of them amount to the computation of the sum of squared deviations of the observed distances (or some monotone transformation of those distances) from the reproduced distances. Thus, the smaller the stress value, the better is the fit of the reproduced distance matrix to the observed distance matrix.

Shepard Diagram. One can plot the reproduced distances for a particular number of dimensions against the observed input data (distances). This scatterplot is referred to as a Shepard diagram. Shown below is such a diagram.

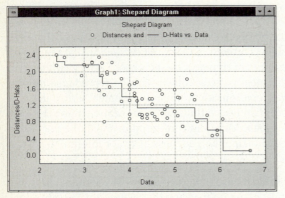

In the example plot above, the input data were actually similarity ratings (see the *Example* section, page 3245); thus, it shows the reproduced distances plotted on the vertical (*Y*) axis versus the original similarities plotted on the horizontal (*X*) axis (hence the generally negative slope). This plot also shows a step-function. This line represents the so-called *D-hat* values, that is, the result of the monotone transformation $f(\delta_{ij})$ of the input data. If all reproduced distances fall onto this step-line, then the rank-ordering of distances (or similarities) would be perfectly reproduced by the respective solution (dimensional model). Deviations from the step-line indicate lack of fit.

How Many Dimensions to Specify?

If you are familiar with factor analysis, you will be quite aware of this issue. If you are not familiar with factor analysis, you may want to read the *Introductory Overview* section of the *Factor Analysis* chapter (Chapter 6); however, this is not necessary in order to understand the following discussion.

In general, the more dimensions you use in order to reproduce the distance matrix, the better is the fit of the reproduced matrix to the observed matrix (i.e., the smaller is the stress value). In fact, if you use as many dimensions as there are variables, then you can perfectly reproduce the observed distance matrix. Of course, your goal is to *reduce* the observed complexity of nature, that is, to explain the distance matrix in terms of fewer underlying dimensions.

To return to the example of distances between cities, once you have a two-dimensional map it is much easier to visualize the location of the cities and navigate between them, as compared to relying on the distance matrix only.

Sources of misfit. Consider for a moment why fewer factors may produce a worse representation of a distance matrix than would more factors. Imagine the three cities *A*, *B*, and *C*, and the three cities *D*, *E*, and *F*; shown below are their distances from each other.

	A	B	C		D	E	F
A	0			D	0		
B	90	0		E	90	0	
C	90	90	0	F	180	90	0

In the first matrix, all cities are exactly 90 miles apart from each other; in the second matrix, cities *D* and *F* are 180 miles apart. Now, can the three cities (objects) be arranged on one dimension (line)? Indeed, cities *D*, *E*, and *F* can be arranged on one dimension:

D —[90 miles]— E —[90 miles]— F

D is 90 miles away from *E*, and *E* is 90 miles away from *F*; thus, *D* is 90+90=180 miles away from *F*.

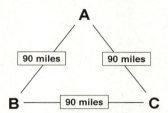

If you try to do the same thing with cities *A*, *B*, and *C*, you will see that there is no way to arrange the three cities on one line so that the distances can be reproduced. However, those cities can be arranged in two dimensions, in the shape of a triangle. Arranging the three cities in this manner, you can perfectly reproduce the distances between them. Without going into too much detail, this small example illustrates how a particular distance matrix implies a particular number of dimensions. Of course, "real" data are never this "clean," and contain a lot of noise, that is, random variability that contributes to the differences between the reproduced and observed matrix.

Scree test. A common way to decide how many dimensions to use is to plot the stress value against different numbers of dimensions. This test was first proposed by Cattell (1966) in the context of the number-of-factors problem in factor analysis (see *Factor Analysis*, Chapter 6); Kruskal and Wish (1978; pages 53-60) discuss the application of this plot to MDS. Shown below is a typical plot.

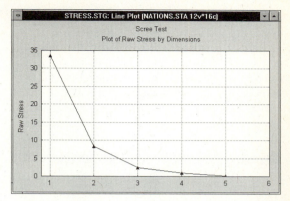

Cattell suggests to find the place where the smooth decrease of stress values (eigenvalues in factor analysis) appears to level off to the right of the plot. To the right of this point one finds, presumably, only "factorial scree" -- "scree" is the geological term referring to the debris which collects on the lower part of a rocky slope. According to this criterion one would probably choose 2 dimensions, given the example plot shown above.

Interpretability of configuration. A second criterion for deciding how many dimensions to interpret is the clarity of the final configuration. Sometimes, as in our example of distances between cities, the resultant dimensions are easily interpreted. At other times, the points in the plot form a sort of "random cloud," and there is no straightforward and easy way to interpret the dimensions.

In the latter case one should try to include more or fewer dimensions and examine the resultant final configurations. Often, more interpretable solutions emerge. However, if the data points in the plot do not follow any pattern, and if the stress plot does not show any clear "elbow," then the data are most likely random "noise."

Interpreting the Dimensions

The interpretation of dimensions usually represents the final step of the analysis. As mentioned earlier, the actual orientations of the axes from the MDS analysis are arbitrary, and can be rotated in any direction. A first step is to produce scatterplots of the objects in the different two-dimensional planes. Shown below is such a plot that summarizes the two-dimensional solution from a study of college students' perception of 12 countries (Kruskal and Wish, 1978, page 30; see also the *Example* section of this chapter, page 3245).

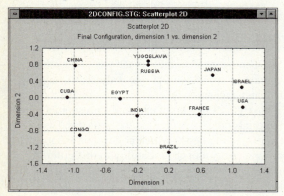

Kruskal and Wish (1978) rotated this solution by approximately 45 degrees, and, based on their general knowledge of the characteristics of those countries, interpreted the rotated dimensions as *developed vs. underdeveloped*, and *pro-western vs. pro-communist* (remember that this study was conducted in the mid 1970's).

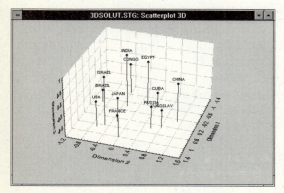

Three-dimensional solutions can also be illustrated graphically, however, their interpretation is somewhat more complex. Note that in addition to "meaningful dimensions," one should also look for clusters of points or particular patterns and configurations (such as circles, manifolds, etc.). For a detailed discussion of how to interpret final configurations, see Borg and Lingoes (1987), Borg and Shye (in press), or Gutman, (1968).

Use of multiple regression techniques. An analytical way of interpreting dimensions (described in Kruskal and Wish, 1978) is to use multiple regression techniques to regress some meaningful variables on the coordinates for the different dimensions. Note that this can easily be done via the *Multiple Regression* module (see Volume I).

Applications

The "beauty" of MDS is that you can analyze any kind of distance or similarity matrix. These similarities can represent people's ratings of similarities between objects, the percent agreement between judges, the number of times a subjects fails to discriminate between stimuli, etc.

For example, MDS methods used to be very popular in psychological research on person-perception where similarities between trait descriptors were analyzed to uncover the underlying dimensionality of people's perceptions of traits (see, for example Rosenberg, 1977). They are also very popular in marketing research, in order to detect the number and nature of dimensions underlying the perceptions of different brands or products (for detailed descriptions of various examples see, for example, Green and Carmone, 1970).

In general, MDS methods allow the researcher to ask relatively unobtrusive questions ("how similar is brand A to brand B") and to derive underlying dimensions from those questions without the respondents ever knowing what is the researcher's real interest.

Multidimensional Scaling and Factor Analysis

Even though there are similarities in the type of research questions to which these two procedures can be applied, MDS and factor analysis are fundamentally different methods. Factor analysis requires that the underlying data are distributed as multivariate normal, and that the relationships are linear. MDS imposes no such restrictions. As long as the rank-ordering of distances (or similarities) in the matrix is meaningful, MDS can be used. In terms of resultant differences, factor analysis tends to extract more factors (dimensions) than MDS; as a result, MDS often yields more readily interpretable solutions. Most importantly, however, MDS can be applied to any kind of distances or similarities, while factor analysis requires that a correlation matrix first be computed. MDS can be based on the subjects' direct assessment of similarities between stimuli, while factor analysis requires subjects to rate those stimuli on some list of attributes (for which the factor analysis is performed). In summary, MDS methods are potentially applicable to a wider variety of research problems because the input data (distance measures) can be obtained in any number of ways (for different examples, refer to the references provided at the beginning of this section).

7. MULTIDIMENSIONAL SCALING - INTRODUCTORY OVERVIEW

PROGRAM OVERVIEW

Overview

The *Multidimensional Scaling* procedure is a full implementation of (nonmetric) multidimensional scaling. The user can analyze matrices of similarities, dissimilarities, or correlations with up to 90 objects (e.g., variables). Up to 9 dimensions can be specified. The starting configuration can be computed by the program (via principal components analysis) or specified by the user. Moreover, in successive analyses, the program will always retain the previous solution in memory, and the user has the option to specify that solution as the starting configuration for the next run.

The program employs an iterative procedure to minimize the stress value (Form 1; see Kruskal 1964) and the coefficient of alienation (Guttman, 1968). The user can monitor the iterations and inspect the changes in these values.

Graphs. A wide selection of specialized graphics options are available to aid in the interpretation of results. The final configuration can be reviewed via Scrollsheets or scatterplots. The goodness of fit can be evaluated via Shepard diagrams, and plots of *D-hats* (see Kruskal, 1964) and *D-stars* (Guttman, 1968). The final configuration can be plotted in 2D or 3D scatterplots (with labeled points) and saved in a standard *STATISTICA* data file. These files may later be used in additional analyses or to produce specialized two- and three-dimensional graphs.

Alternative Procedures

The major alternative to *Multidimensional Scaling* (which analyzes both spatial and distance models) is *Factor Analysis* (which analyzes spatial models). The *Factor Analysis* module (see Chapter 6) offers a full implementation of principal components and principal factors analysis (including maximum likelihood factor analysis). To test hypotheses about specific patterns of factor loadings (confirmatory factor analysis), use the *Structural Equation Modeling (SEPATH)* module (see Chapter 11). If the major goal is to classify objects (i.e., analyze distance models), then the different classification procedures in the *Cluster Analysis* module may present appropriate alternatives to MDS. Also, the *Cluster Analysis* module allows the user to generate distance (dissimilarity) matrices based on a wide variety of distance measures (Euclidean, Manhattan, Chebychev, percent disagreement, etc.); those matrices can be used as input to the *Multidimensional Scaling* module.

7. MULTIDIMENSIONAL SCALING - EXAMPLE

EXAMPLE

Overview and Data File

The present example will be based on the data file *Nations.sta*. These data are discussed in Kruskal and Wish (1978, page 30). The data file contains the mean similarity ratings of 18 students for 12 countries. The countries were Brazil, Congo, Cuba, Egypt, France, India, Israel, Japan, Mainland China, Russia, USA, and Yugoslavia. A partial listing of this similarity matrix is shown below.

	1 BRAZIL	2 CONGO	3 CUBA	10 RUSSIA	11 USA	12 YUGOSLAV
BRAZIL	0.00	4.83	5.28	3.06	5.39	3.17
CONGO	4.83	0.00	4.56	3.39	2.39	3.50
CUBA	5.28	4.56	0.00	5.44	3.17	5.11
EGYPT	3.44	5.00	5.17	4.39	3.33	4.28
FRANCE	4.72	4.00	4.11	5.06	5.94	4.72
INDIA	4.50	4.83	4.00	4.50	4.28	4.00
ISRAEL	3.83	3.33	3.61	4.17	5.94	4.44
JAPAN	3.50	3.39	2.94	4.61	6.06	4.28
CHINA	2.39	4.00	5.50	5.72	2.56	5.06
RUSSIA	3.06	3.39	5.44	0.00	5.00	6.67
USA	5.39	2.39	3.17	5.00	0.00	3.56
YUGOSLAV	3.17	3.50	5.11	6.67	3.56	0.00
Means	0.00	0.00	0.00	0.00	0.00	0.00
Std.Dev.	0.00	0.00	0.00	0.00	0.00	0.00
No.Cases	18.00					
Matrix	2.00					

Note that you can produce a similarity matrix file by entering the distances into a new spreadsheet following the matrix data format conventions (as described in *Matrix File Format*, see *Data Management*, Volume I).

Specifying the Analysis

After starting the *Multidimensional Scaling* module, select the matrix file *Nations.sta*. Then bring up the startup panel, click on the *Variables* button, and select all variables (i.e., objects or nations) for the analysis.

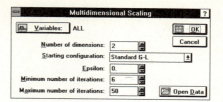

The program assumes that you want to calculate a two-dimensional solution for this similarity matrix, and that the initial solution is to be estimated via principal components analysis (see page 3249). Alternatively, you may also specify the initial configuration by selecting a *STATISTICA* raw data file with the initial coordinates.

Click *OK* to accept these defaults. First, the initial (starting) configuration will be computed, and the coordinates will be displayed in a Scrollsheet behind the *Parameter Estimation* window (for later review).

Starting Configuration [nations.sta]		
MULTIDIM SCALING	(Guttman-Lingoes)	
	DIM. 1	DIM. 2
BRAZIL	-.085995	-.714599
CONGO	-.711952	-.421219
CUBA	-.296830	.121739
EGYPT	-.171395	.073384
FRANCE	.134139	-.056573
INDIA	-.083048	-.113595
ISRAEL	.505443	-.047624
JAPAN	.451928	.081312
CHINA	-.330921	.673819
RUSSIA	.067029	.266788
USA	.516756	-.239711
YUGOSLAV	.004847	.376278

Performing the Analysis

The iterative algorithm for finding an optimum configuration proceeds in two stages: First, the program will use a method known as *steepest descent*. The respective number of steepest descent iterations is listed under the first column (labeled *iter. s:*) in the *Parameter Estimation* window.

After each iteration under steepest descent, the program will perform up to 5 additional iterations to "fine-tune" the configuration (see the *Technical Notes* section, page 3255, for details). The respective numbers of these iterations are listed under the second column in the *Parameter Estimation* window (labeled *iter. t:*). In addition, the stress value (Kruskal, 1964) and coefficient of alienation (Guttman, 1968) are calculated and displayed at each step (see also, page 3237 in the *Introductory Overview* section and page 3255 in the *Technical Notes* section). A detailed discussion of this iterative procedure can be found in Shiffman, Reynolds, and Young (1981, pages 366-370).

iter. s: t:	[dim=2] cosin	step	D-star raw stress	D-star alienation	D-hat raw stress	d-hat stress
15 1	.968	.485			5.310299	.1920341
16 1	.887	.404			5.304969	.1919377
17 1	.694	.230			5.300331	.1918537
18 1	.711	.197			5.295306	.1917628
19 1	.908	.322			5.284995	.1915760
20 1	.960	.456			5.264813	.1912098
21 1	.898	.411			5.242468	.1908036
22 1	.700	.234			5.228747	.1905538
23 1	.662	.180			5.217969	.1903573
24 1	.877	.281			5.201594	.1900584
25 1	.956	.429			5.182857	.1897157
26 1	.561	.188			5.178150	.1896296
27 1	.665	.168			5.175625	.1895833
28 1	.950	.354			5.172096	.1895187
29 1	.968	.487			5.169822	.1894770
30 1	.698	.247			5.169298	.1894674
30 *			8.287207	.2381639	5.169298	.1894674
13 1	.697	.184			5.320363	.1922159
14 1	.938	.349			5.315846	.1921343

Estimation procedure converged

After the program has determined the best two-dimensional configuration, it will display the final stress value; click *OK* to go to the *Results* dialog.

Results

In general, you may examine the results in Scrollsheets or graphs via the options available in the *Results* dialog.

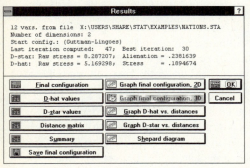

First, examine the table of actual distances and estimated distances.

Reproduced and observed distances. To evaluate the fit of the two-dimensional solution, click on the *Summary* button in the *Results* dialog.

MULTIDIM SCALING	Final Configuration [nations.sta] D-star: Raw stress = 8.287207; Alienation = .2381639 D-hat: Raw stress = 5.169298; Stress = .1894674		
	Distance	D-star	D-hat
D(12,10)	.093247	.093247	.093247
D(11, 8)	.849364	.456247	.591286
D(11, 5)	.578455	.462164	.591286
D(11, 7)	.481080	.481080	.591286
D(6, 4)	.456247	.578455	.591286
D(10, 9)	.938324	.671464	.865152
D(9, 3)	.791981	.786270	.865152
D(10, 3)	1.312809	.791981	1.133368
D(11, 1)	1.450855	.842240	1.133368
D(3, 1)	1.817224	.849364	1.133368
D(4, 3)	.671464	.861000	1.133368
D(12, 3)	1.362808	.872578	1.133368
D(10, 5)	1.373663	.874489	1.133368
D(12, 9)	.933320	.886758	1.133368

There are four columns in this Scrollsheet. The columns labeled *D-hat* and *D-star* contain the monotone transformations of the input data (see the *Introductory Overview*, page 3237): *D-stars* are *rank images* calculated according to Guttman (1968); *D-hats* are monotone regression estimates calculated according to Kruskal (1964).

The rows in the Scrollsheet, each representing one distance as specified in the similarity matrix, are sorted according to the size of *D-star* or *D-hat*. The second column of the Scrollsheet contains the reproduced distances from the current configuration. If the fit of the current model (i.e., the current number of dimensions) is very good, then the order of reproduced distances should be approximately the same as that for the transformed input data (i.e., *D-star* or *D-hat* values). Out-of-order elements indicate lack of fit. The first column of the Scrollsheet references the elements of the original input matrix as $D(X,Y)$, where X is the respective row in the input matrix, and Y is the respective column.

For example, $D(2,1)$ would be the element in the second row and the first column of the input matrix (i.e., in our example, the comparison between *Congo* and *Brazil*). It appears that, by and large, the order of distances was approximately reproduced by the two-dimensional solution.

Shepard diagram. Now examine the Shepard plot. As described in the *Introductory Overview* section (see page 3238), this plot is a scatterplot of the observed input data (similarities or dissimilarities) against the reproduced distances. The plot will also show the *D-hat* values, that is, the monotonically transformed input data, as a step function. To produce the plot, click on the *Continue* button in the Scrollsheet to return to the *Results* dialog, and then click on the *Shepard Diagram* button.

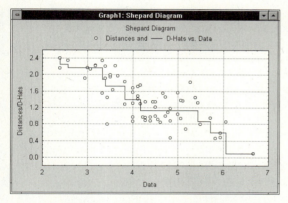

Most points in this plot are clustered around the step-line. Thus, one may conclude for now that this two-dimensional configuration is adequate for describing the similarities between countries.

Interpreting the configuration. In order to interpret this solution, you can display the configuration of nations in the two-dimensional space. Return to the *Results* dialog and then click on the *Graph final configuration - 2D* button. The *Select Dimensions for Scatterplot* intermediate dialog will open in which you can select the dimensions for the 2D scatterplot.

As described in the *Introductory Overview* section (page 3237), the actual orientation of axes in multidimensional scaling is arbitrary (just as in

Factor Analysis). Thus, one may rotate the configuration in order to achieve a more interpretable solution. Kruskal and Wish (1978) used a program called KYST (which uses a slightly different algorithm for multidimensional scaling) in order to analyze the present data, and they obtained a very similar solution. They then rotated their solution by approximately 45 degrees, and interpreted the rotated dimensions as *developed vs. underdeveloped*, and *pro-western vs. pro-communist*. Looking at the plot below (rotated by 45 degrees), this interpretation seems to hold quite well (remember that this study was conducted in the 1970's).

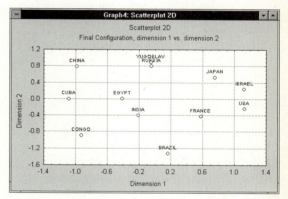

In general, in addition to "meaningful dimensions," one should also look for clusters of points or particular patterns and configurations (such as circles, manifolds, etc.). For a detailed discussion of how to interpret final configurations, see Borg and Lingoes (1987), Borg and Shye (in press), or Gutman, (1968).

Continuing the Analysis

Now click on the *Cancel* button in the *Results* dialog to return to the *Multidimensional Scaling* startup panel.

7. MULTIDIMENSIONAL SCALING - EXAMPLE

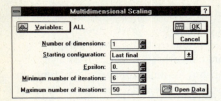

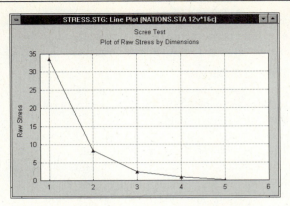

Note that now the default settings on the startup panel are different than when the program was first started. *Multidimensional Scaling* will remember the configuration from the previous analysis (unless you specify a new data file, or if you select new cases). Also, the default number of dimensions is now *1*. You could now click *OK* to compute the one-dimensional solution, using the configuration for the first dimension from the previous analysis as the starting configuration. In this manner, one may efficiently evaluate several consecutive solutions, starting with several dimensions (and working one's way down to the one-dimensional solution).

Scree test: Plotting the stress values. This example began with the two-dimensional solution. Actually, if one is unsure about the dimensionality underlying the matrix, one should plot the stress values for consecutive numbers of dimensions. Then find the place where the smooth decrease of stress values appears to level off to the right of the plot. To the right of this point, presumably, one finds only "factorial scree" -- "scree" is the geological term referring to the debris which collects on the lower part of a rocky slope (see, for example, Kruskal and Wish, 1978, pages 53-56, for a discussion of this plot). Shown below is this plot which was created from the Scrollsheet stress values for consecutive dimensions (1 through 6) for the present data.

Based on the plot above, the two-dimensional solution would, indeed, have been chosen. Perhaps you also would have looked at the three-dimensional solution. You can be the judge of whether the three-dimensional solution is more meaningful than the two-dimensional one. Shown below is a 3D scatterplot of the solution when 3 dimensions were specified in the startup panel. To obtain this graph, click on the *Graph of final configuration - 3D* button (this button is dimmed if 1 or 2 dimensions were specified in the startup panel; if more than 3 dimensions were specified, you will be prompted to select the dimensions to plot in the 3D graph when you click on this button).

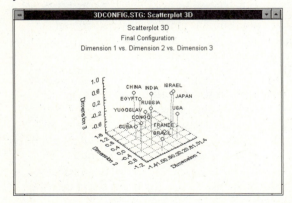

MDS - 3248

Copyright © StatSoft, 1995

DIALOGS, OPTIONS, STATISTICS

Startup Panel

Multidimensional Scaling is the only statistical module in *STATISTICA* that cannot accept raw data files as the main input file. Instead, it requires as input a matrix file (see page 3245), containing either correlations, similarities, or dissimilarities (distances).

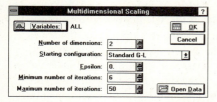

STATISTICA matrix files can be created directly (i.e., entered into the spreadsheet following the *STATISTICA* matrix file format, see Volume I), or they can be produced by other modules of *STATISTICA*. For example, a wide variety of distance matrices can be computed from raw data and saved via *Cluster Analysis*. Also, *STATISTICA* matrix files with correlation matrices can be produced in every module using the *Quick Basic Stats* procedure (accessible from every toolbar).

Variables

This button will bring up the standard variable selection window in which you can select the variables (i.e., items) for the analysis.

Number of Dimensions

In this edit box, the user can enter the number of dimensions for the multidimensional scaling analysis. The default value is *2*. The maximum number of dimensions that can be specified is *9* or the number of variables (objects) minus one, whichever is smaller. Note that in consecutive analyses, by accepting the default settings the user can systematically evaluate solutions for *n* dimensions, *n-1* dimensions, *n-2* dimensions, etc. For example, you may first specify 5 dimensions, and then, after returning from the *Results* dialog, *STATISTICA* will by default compute a solution for 4 dimensions, and so on.

Starting Configuration

This combo box allows the user to select the starting configuration for the current analysis; available options are *Standard Guttman-Lingoes*, *Last final*, or *From a STATISTICA file*.

Standard Guttman-Lingoes. When first entering *Multidimensional Scaling*, the default starting configuration is *Standard Guttman-Lingoes*. This procedure amounts to a principal components analysis, and in most instances it will provide an adequate starting configuration for the iterative fitting procedure. (For a more detailed description of principal components analysis, refer to the *Introductory Overview* section of *Factor Analysis*, Chapter 6.)

Last final. After returning from the *Results* dialog, the default starting configuration is *Last final*. In this setting, *STATISTICA* will use as the starting configuration the configuration of the previously computed model. For example, suppose that you have just computed a three-dimensional solution for a similarity matrix and are now returning to the startup panel. If you next want to calculate the two-dimensional solution (and you do not change the setting of the *Starting configuration* combo box) the program will automatically use the previous configuration for the first two dimensions as the starting configuration for the next analysis. Of course, if you select new variables (*Variables*), or if you increase the number of dimensions, the *Last final* option is not available.

7. MULTIDIMENSIONAL SCALING - DIALOGS AND OPTIONS

From a STATISTICA file. If the *Starting configuration* combo box is set to *From a STATISTICA file*, the user can specify a STATISTICA raw data file with the coordinates for the starting configuration. This file may be created via *Data Management*, or it can be a configuration file saved from the *Multidimensional Scaling Results* dialog (see page 3251) during a previous run. The file containing the starting configuration must have as many cases as the current matrix file, and it must have as many variables as there are dimensions specified for the current run.

Epsilon

All distances that are less than *epsilon* will be considered to be 0 (zero) by the program. Thus, this parameter specifies the smallest distance that is to be considered "important" or significant by the program. The smaller the value, the more iterations will be performed and the more "precise" is the solution. Change this value only if the iterative fitting procedure cannot converge (i.e., produce a solution) even after very many iterations.

Number of Iterations

This option allows you to specify the minimum and maximum number of iterations that are to be performed during the iterative fitting procedure. The minimum number of iterations is 6.

Estimation

Multidimensional Scaling is an implementation of nonmetric multidimensional scaling. After determining the starting configuration (e.g., via principal components analysis, see *Factor Analysis*, Chapter 6) the program will begin iterations under *steepest descent* (see, for example, Schiffman, Reynolds, and Young, 1981).

The goal of these iterations is to minimize the raw stress (or raw *Phi*) and the coefficient of alienation (see Guttman, 1968). The raw stress is defined as:

$$\text{Phi} = \Sigma[d_{ij} - f(\delta_{ij})]^2$$

where

- d_{ij} are the reproduced distances; given the current number of dimensions,
- $f(\delta_{ij})$ represents the monotone transformation of the observed input data δ_{ij} (*delta* $_{ij}$)

The coefficient of alienation K is defined as:

$$K = \{1 - [\Sigma d_{ij} * f(\delta_{ij})]^2 / \Sigma d_{ij}^2 * \Sigma [f(\delta_{ij})]^2\}^{\frac{1}{2}}$$

In general, STATISTICA will attempt to minimize the differences between the reproduced distances and a monotone transformation of the input data; that is, the program will attempt to reproduce the rank-ordering of the input distances or similarities (hence also the name *nonmetric* multidimensional scaling).

Note that under steepest descent, the fitted values are calculated via the *rank-image permutation* procedure (see Guttman, 1968; or Schiffman, Reynolds, and Young, 1981, pages 368-369). After each iteration under steepest descent the program will perform up to five iterations using the *monotone regression transformation* procedure (see Kruskal, 1964; or Schiffman, Reynolds, and Young, 1981, pages 367-368). This procedure is aimed at minimizing the standardized stress (*S*):

$$S = [\Sigma (d_{ij} - f(\delta_{ij}))^2 / \Sigma (d_{ij}^2)]^{\frac{1}{2}}$$

Before final convergence, STATISTICA will perform several monotone regression transformation iterations.

D-stars and D-hats

D-stars. D-stars are calculated via a procedure known as *rank-image permutation* (see Guttman, 1968; or Schiffman, Reynolds, and Young, 1981,

MDS - 3250

Copyright © StatSoft, 1995

pages 368-369). In general, this procedure attempts to reproduce the *rank order* of differences in the similarity or dissimilarity matrix.

D-hats. D-hats are calculated via a procedure referred to as *monotone regression transformation* (see Kruskal, 1964; or Schiffman, Reynolds, and Young, 1981, pages 367-368).

In this procedure, the program attempts to determine the best monotone (regression) transformation to reproduce the similarities (or dissimilarities) in the input matrix.

Results

When you click *OK* in the *MDS - Estimation* window, the *Results* dialog will open. The upper part of the dialog contains a summary of results; the lower part includes a selection of output options.

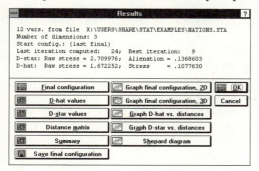

Final Configuration

This option will bring up a Scrollsheet with the coordinates of the final configuration.

D-hat Values

This option will bring up a Scrollsheet with the transformed input data values calculated according to the monotone regression procedure (*D-hat*, see Kruskal, 1964; see also the *Introductory Overview*).

D-star Values

This option will bring up a Scrollsheet with the transformed input data values, calculated according to Guttman's rank image procedure (*D-star*; Guttman, 1968).

Distance Matrix

This option will bring up a Scrollsheet with the matrix of reproduced distances, given the current number of dimensions and configuration of points.

Summary

This option will display the reproduced distances along with the *D-hat* and *D-star* values. The elements in the Scrollsheet are sorted by the transformed input data values (*D-hat, D-star*), and each element is referenced by its matrix subscript. For example, the element denoted as *D(4,2)* refers to the element in the fourth row and the second column of the similarity (or dissimilarity) matrix.

Save Final Configuration

The coordinates for the final configuration can be saved in a standard *STATISTICA* data file. The number of variables in the new data file will always be equal to the number of dimensions specified for the current multidimensional scaling analysis. These files can then be accessed by other *STATISTICA* modules.

For example, you may specify the configuration stored in this file as the starting configuration for subsequent analyses. You may also combine this file (via *Data Management*) with another data file containing additional variables. Multiple regression techniques (regressing these auxiliary variables on the coordinates for the different dimensions) are often used to aid in the interpretation of dimensions. These files can be used to produce specialized 2D and 3D graphs of the respective solution.

7. MULTIDIMENSIONAL SCALING - DIALOGS AND OPTIONS

Graph Final Configuration - 2D, 3D

These options allow the user to plot the final configuration of variables (objects). Note that a variety of specialized 2D and 3D graphs can also be produced from a data file containing the configuration as saved via the option *Save final configuration* (above).

2D plot of final configuration. When you click on the *Graph final configuration, 2D* button in the *Results* dialog, an intermediate dialog will open, allowing you to select the two dimensions to be plotted in the 2D scatterplot.

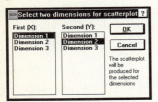

For example, the following graph can help interpret the meaning of the dimensions. If the labels overlap, use one of the methods described below.

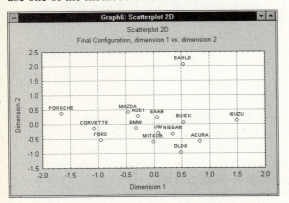

3D plot of final configuration. When you click on *Graph final configuration, 3D* button in the *Results* dialog, an intermediate dialog will open allowing you to select the three dimensions to be plotted in the 3D scatterplot.

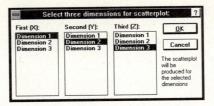

For example, the following graph can help interpret the meaning of the dimensions.

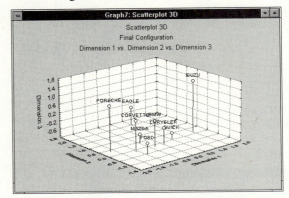

If the labels overlap. If the labels overlap, try to rotate the graph (via the *Perspective and Rotation* Graph Window toolbar button) or double-click on any data point label to access the *Data Point Labels* dialog (see below) and then click on the *Font* button in order to adjust (reduce) the size of font used to label the points.

You can also use the *Display Filters* option from the Graphics *View* pull-down menu in order to select the degree of filtering (removal of overlapping points or text) in the current graph (the default setting, *Level 2*, suppresses the display of only those labels which overlap to a large degree).

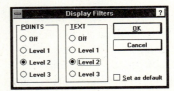

The *Display Filters* facility is described in detail in Volume II.

MDS - 3252

Copyright © StatSoft, 1995

Finally, you may also adjust the location of specific labels one at a time: double-click on any point label to bring up the *Data Point Labels* dialog.

Now, click on the *Edit* button in this dialog to open the *Edit Text Labels* dialog.

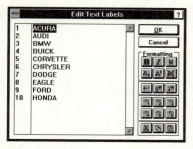

In this dialog, highlight the label to be adjusted and cut it (by pressing CTRL+X); press *OK* to return to the plot. Now, paste (CTRL+V) the label into the graph, then position the label as desired.

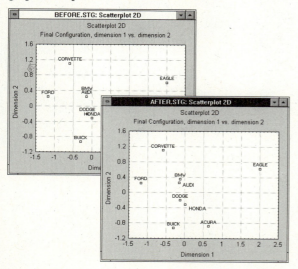

You may need to adjust the font and other attributes of the label if the current defaults of the custom text are different from the defaults set for the data labels, but you will need to make this adjustment only once - the next repositioned label will use the previously set attributes.

Graph D-hat vs. Distances

With this option, the user can produce a plot of the transformed input data values (*D-hat*) against the reproduced distances. The more closely the points in the plot cluster around the diagonal, the better is the fit of the respective model.

Graph D-star vs. Distances

With this option, the user can produce a plot of the transformed input data values (*D-star*) against the reproduced distances. The more closely the points in the plot cluster around the diagonal, the better is the fit of the respective model.

Shepard Diagram

After clicking on this option, a scatterplot will be displayed where each point represents a distance - data point pair (note that the original data are plotted against the horizontal *X* axis in the plot). In addition, the *D-hat* values are indicated in the graph by a step function. As described in the *Introductory Overview* section (page 3237), *STATISTICA* performs nonmetric multidimensional scaling, and the *D-hat* values are computed by a monotone (non-linear) transformation of the original data (dissimilarities). Thus, the step function will be monotonically increasing or decreasing. The closer the fit of the step function to the data points plotted in the graph, the better the fit of the respective model, that is, the better is the reproduction of the distances (or, specifically, the rank order of distances) in the input data, given the respective number of dimensions.

7. MULTIDIMENSIONAL SCALING - DIALOGS AND OPTIONS

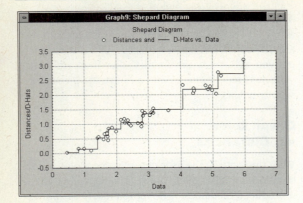

TECHNICAL NOTES

The *Multidimensional Scaling* module is an implementation of non-metric multidimensional scaling. Comprehensive introductions to the computational approach used in this module can be found in Borg and Lingoes (1987), Borg and Shye (in press), Guttman (1968), Schiffman, Reynolds, and Young (1981), Young and Hamer (1987), and in Kruskal and Wish (1978).

Computational Approach

As the starting configuration, principal components of the similarity/dissimilarity matrix are computed. The program will then begin iterations under *steepest descent* (see, for example, Schiffman, Reynolds, and Young, 1981). The goal of these iterations is to minimize the raw stress (or raw *Phi*) and the coefficient of alienation (see Guttman, 1968). The raw stress is defined as:

$$\text{Phi} = \Sigma[d_{ij} - f(\delta_{ij})]^2$$

where

d_{ij} are the reproduced distances; given the current number of dimensions,

$f(\delta_{ij})$ represents the monotone transformation of the observed input data δ_{ij} (*delta $_{ij}$*)

The coefficient of alienation *K* is defined as:

$$K = \{1 - [\Sigma d_{ij} * f(\delta_{ij})]^2 / \Sigma d_{ij}^2 * \Sigma [f(\delta_{ij})]^2\}^{\frac{1}{2}}$$

In general, *STATISTICA* will attempt to minimize the differences between the reproduced distances and a monotone transformation of the input data; that is, the program will attempt to reproduce the rank-ordering of the input distances or similarities (hence also the name *nonmetric* multidimensional scaling).

Note that under steepest descent the fitted values are calculated via the *rank-image permutation* procedure (see Guttman, 1968; or Schiffman, Reynolds, and Young, 1981, pages 368-369). After each iteration under steepest descent, the program will perform up to five iterations using the *monotone regression transformation* procedure (see Kruskal, 1964; or Schiffman, Reynolds, and Young, 1981, pages 367-368). This procedure is aimed at minimizing the standardized stress (*S*):

$$S = [\Sigma(d_{ij} - f(\delta_{ij}))^2 / \Sigma(d_{ij}^2)]^{\frac{1}{2}}$$

Before final convergence, the program will perform several monotone regression transformation iterations.

D-stars and D-hats

D-stars are calculated via a procedure known as *rank-image permutation* (see Guttman, 1968; or Schiffman, Reynolds, and Young, 1981, pages 368-369). In general, this procedure attempts to reproduce the *rank order* of differences in the similarity or dissimilarity matrix. *D-hats* are calculated via a procedure referred to as *monotone regression transformation* (see Kruskal, 1964; Schiffman, Reynolds, and Young, 1981, pages 367-368). In this procedure, the program attempts to determine the best monotone (regression) transformation to reproduce the similarities (or dissimilarities) in the input matrix.

INDEX

C
Coefficient of alienation, 3250, 3255

D
D-hats, 3250, 3251, 3255
Distance matrix, 3240, 3251
D-stars, 3250, 3251, 3255

E
Epsilon, 3250
Estimation, 3250

G
Graphs
    D-hat vs. distances, 3253
    D-star vs. distances, 3253
    scree test, 3239, 3248
    Shepard diagram, 3238, 3253
    three-dimensional plot of final configuration, 3240, 3252
    two-dimensional plot of final configuration, 3240, 3252
Guttman-Lingoes method, 3249

M
MDS (see multidimensional scaling), 3237
Measures of goodness of fit
    stress, 3237
Misfit (sources of), 3238
Monotone regression transformation, 3250, 3251, 3255
Multidimensional scaling
    coefficient of alienation, 3250
    computational approach, 3237
    D-hats, 3250, 3251, 3255
    distance matrix, 3251
    D-stars, 3250, 3251, 3255
    epsilon, 3250
    estimation, 3250
    example, 3245
    graph D-hat vs. distances, 3253
    graph D-star vs. distances, 3253

Multidimensional scaling (continued)
    how many dimensions to specify, 3238
    interpreting the dimensions, 3239, 3247
    logic of MDS, 3237
    MDS and factor analysis, 3241
    misfit (sources of), 3238
    monotone regression transformation, 3250, 3251, 3255
    orientation of axes in MDS, 3237
    overview, 3237
    rank-image permutation, 3250, 3255
    raw stress, 3246, 3250, 3255
    reproduced and observed distances, 3246
    scree test, 3239, 3248
    Shepard diagram, 3238, 3247, 3253
    standardized stress, 3250
    starting configurations, 3249
    startup panel, 3249
    steepest descent method, 3245, 3255

O
Orientation of axes in MDS, 3237
Overlapping labels, 3252

R
Rank-image permutation, 3250, 3255
Raw stress, 3250, 3255
Reproduced and observed distances, 3246

S
Scree test, 3239, 3248
Shepard diagram, 3247
Similarity matrix, 3240
Standardized stress, 3250
Starting configuration
    from a *STATISTICA* file, 3250
    last final, 3249
    standard Guttman-Lingoes, 3249
Steepest descent method, 3245, 3255
Stress measure, 3237, 3246, 3248

T
Three-dimensional plot of final configuration, 3240, 3252
Two-dimensional plot of final configuration, 3240, 3252

Chapter 8:
TIME SERIES ANALYSIS

Table of Contents

Introductory Overview 3269
Program Overview 3309
General Conventions and Opening Dialog 3313
    Active Work Area 3313
    Missing Data 3314
    Display/Plot Options 3315
Transformations of Variables 3319
    Example 3319
    Dialogs, Options, Statistics 3329
ARIMA 3341
    Examples 3341
    Dialogs, Options, Statistics 3357
Seasonal and Non-Seasonal Exponential Smoothing 3371
    Example 3371
    Dialogs, Options, Statistics 3377
Seasonal Decomposition (Census Method I) 3391
    Example 3391
    Dialogs, Options, Statistics 3395
X-11 Seasonal Decomposition (Census Method II) 3401
    Example 3401
    Dialogs, Options, Statistics 3405
Distributed Lags Analysis 3419
    Example 3419
    Dialogs, Options, Statistics 3421
Spectrum (Fourier) Analysis 3425
    Example 3425
    Dialogs, Options, Statistics 3429
Index 3441

Detailed
Table of Contents

INTRODUCTORY OVERVIEW **3269**
    General Overview 3269

8. TIME SERIES ANALYSIS - CONTENTS

Two Main Goals	3269
Overview	3269
Identifying Patterns in Time Series Data	3270
Systematic Pattern and Random Noise	3270
Two General Aspects of Time Series Patterns	3270
Trend Analysis	3270
Smoothing	3271
Fitting a Function	3271
Detecting Serial Correlations and Seasonality via Autocorrelation Analysis	3271
Autocorrelation Correlogram	3272
Examining Correlograms	3272
Partial Autocorrelations	3272
Removing Serial Dependency	3273
ARIMA	3273
Two Common Processes	3274
The ARIMA Model	3274
Model Identification	3275
Parameter Estimation	3276
Evaluation of the Model	3277
Interrupted Time Series ARIMA	3278
Abrupt Permanent Impact	3278
Gradual Permanent Impact	3278
Abrupt Temporary Impact	3279
Exponential Smoothing	3279
Simple Exponential Smoothing	3280
Choosing the Best Value for Parameter α (*alpha*)	3281
Indices of Lack of Fit (Error)	3281
Seasonal and Non-Seasonal Models with or without Trend	3282
Classical Seasonal Decomposition (Census Method I)	3284
General Model	3284
Additive and Multiplicative Seasonality	3285
Additive and Multiplicative Trend-cycle	3285
Computations	3286
X-11 Census Method II Seasonal Adjustment	3287
Seasonal Adjustment: Basic Ideas and Terms	3287
The Census II Method	3289
Result Tables Computed by the X-11 Method	3290
Specific Description of all Result Tables Computed by the X-11 Method	3291
Distributed Lags Analysis	3298
General Purpose	3298
General Model	3299
Almon Distributed Lag	3299
Spectrum and Cross-Spectrum (Fourier) Analysis	3300
Frequency and Period	3300

8. TIME SERIES ANALYSIS - CONTENTS

The General Structural Model	3301
A Simple Example	3302
Periodogram	3302
The Problem of Leakage	3303
Padding the Time Series	3303
Tapering	3303
Data Windows and Spectral Density Estimates	3303
Preparing the Data for Analysis	3304
Results when no Periodicity in the Series Exists	3304
Fast Fourier Transforms (FFT)	3305
Computation of FFT	3305
Cross-Spectrum Analysis	3306
The Cross-periodogram, Cross-density, Quadrature-density, and Cross-amplitude	3307
Squared Coherency, Gain, and Phase Shift	3307
How the Example Data were Created	3307

PROGRAM OVERVIEW 3309

Analysis of Multiple Series, Capacity	3309
Transformations, Modeling, Plots	3309
Autocorrelation Analysis	3309
ARIMA	3310
Seasonal and Non-Seasonal Exponential Smoothing	3310
Classical Seasonal Decomposition (Census Method I)	3311
X-11 Monthly and Quarterly Seasonal Decomposition and Adjustment (Census Method II)	3311
Polynomial Distributed Lag Time Series Models	3311
Spectrum (Fourier) and Cross-Spectrum Analysis	3311
Related Procedures	3312

GENERAL CONVENTIONS AND OPENING DIALOG 3313

Active Work Area	3313
Highlighted Variable	3313
Naming Conventions	3313
Editing Variable Names	3313
Number of Backups per Variable (Series)	3313
Locking Variables (Series)	3314
Deleting Variables (Series) from the Current Work Area	3314
Saving Variables (Series) in the Current Work Area	3314
Missing Data	3314
Overall Mean	3314
Interpolation from Adjacent Points	3314
Mean of n Adjacent Points	3315
Median of n Adjacent Points	3315
Predicted Values from Linear Trend Regression	3315
Display/Plot Options	3315
Label Data Points with Case Numbers, Case Names, Dates, or Consecutive Integers	3315

8. TIME SERIES ANALYSIS - CONTENTS

Scale *X*-Axis in Plots Manually (Min, Step)	3316

TRANSFORMATIONS OF VARIABLES ... 3319

EXAMPLE ... 3319
- General Conventions and Options ... 3319
  - Transformations and the Active Work Area ... 3319
  - Missing Data ... 3320
  - Reviewing the Time Series ... 3321
  - Plotting Two Series with Different Scales ... 3322
  - Transforming Time Series ... 3323
  - The Updated Active Work Area ... 3323
  - Further Processing of the Transformed Series ... 3324
  - Multiple Successive Transformations ... 3324
  - Saving the Series in the Active Work Area ... 3325
- Autocorrelation Analysis ... 3325
  - Plotting the Autocorrelation Function ... 3326

DIALOGS, OPTIONS, STATISTICS ... 3329
- Transformation of Variables Dialog ... 3329
  - Number of Backups per Variable ... 3329
  - Delete ... 3329
  - Save Variables ... 3329
  - Review and Plot Variables ... 3329
  - Plot Variables (Series) After Each Transformation ... 3329
  - Display/Plot Subset of Cases Only ... 3330
  - Review/Plot Highlighted Variable ... 3330
  - Review/Plot Multiple Variables ... 3330
  - Plot Two Var. Lists with Different Scales ... 3330
  - Label Data Points with Case Numbers, Case Names, Dates, or Consecutive Integers ... 3330
  - Scale *X*-Axis in Plots Manually (Min, Step) ... 3331
  - Autocorrelations ... 3331
  - *Alpha* for Highlighting ... 3332
  - White Noise Standard Errors ... 3332
  - Partial Autocorrelations ... 3332
  - Crosscorrelations ... 3332
  - Number of Lags ... 3333
  - 2D Scatterplot ... 3333
  - 3D Scatterplot ... 3333
  - Label Points in Scatterplots ... 3333
  - Histogram ... 3333
  - Descriptive Statistics ... 3333
  - Normal, Detrended, and Half-Normal Probability Plots ... 3333
- General Transformations ... 3334
  - Simple Transformations: $x=f(x)$ ... 3334

8. TIME SERIES ANALYSIS - CONTENTS

    Smoothing 3335
    Two-Series Transformations 3336
    Shift Relative Starting Point of Series 3336
    Filtering and Other Techniques 3336
  Transformations for Spectrum Analysis 3337
    Tapering 3337
    Smoothing Window 3337
    Real & Imaginary Part 3338
    Inverse Fourier Transform 3339

ARIMA 3341

EXAMPLES 3341
  Example 1: Single Series ARIMA 3341
    Identification Phase 3341
    ARIMA Integrated Transformations 3344
    ARIMA Specifications Dialog 3345
    Results 3346
    Analysis of Residuals 3347
    Further Analyses 3348
  Example 2: Interrupted ARIMA 3349
    Model Identification 3349
    Specifying the ARIMA Model 3352
    Preliminary Results 3352
    Specifying Interrupted ARIMA 3353
    Reviewing the Final Results 3354

DIALOGS, OPTIONS, STATISTICS 3357
  Single Series ARIMA 3357
    ARIMA Model Parameters 3357
    Transform Variable (Series) Prior to Analysis 3358
    Estimation of Maximum Likelihood 3358
    Review and Plot Variables 3359
    Autocorrelations 3360
    Histogram 3361
    Descriptive Statistics 3361
    Normal, Detrended, and Half-Normal Probability Plots 3361
  Parameter Estimation Options 3361
    Maximum Number of Iterations 3362
    Convergence Criterion 3362
    Maximum Number of Iterations for Backcasting 3362
    Start Values for Parameter Estimates 3362
  Parameter Estimation Procedure 3363
    If the Parameter Estimation Procedure Fails to Converge 3363
  Interrupted Time Series 3363

TIM - 3263

8. TIME SERIES ANALYSIS - CONTENTS

- Specify Times and Types of Interventions 3364
- Review Types of Impact Patterns 3364
- ARIMA Results 3365
  - Parameter Estimates 3366
  - Print Results 3366
  - Parameter Covariances and Correlations 3367
  - Forecasting 3367
  - Plots of Residuals 3368
  - Review and Plot Variables 3368
  - Autocorrelations of Residuals 3369

SEASONAL AND NON-SEASONAL EXPONENTIAL SMOOTHING 3371

EXAMPLE 3371
- Choosing a Model 3371
- Simple Exponential Smoothing 3372
- Exponential Smoothing with Linear Trend 3373
- Triple Exponential Smoothing: Winters' Method 3374
- Parameter Grid Search 3374
- Automatic Parameter Search 3375
- Final Results 3376

DIALOGS, OPTIONS, STATISTICS 3377
- Exponential Smoothing Specifications 3377
  - Exponential Smoothing Models 3377
  - Model 3382
  - Make Summary Plot for Each Smooth 3384
  - Add Pred./Errors to Work Area 3384
  - Grid Search 3384
  - Automatic Search for Best Parameters 3384
  - Review and Plot Variables 3385
  - Autocorrelations 3386
  - Histogram 3386
  - Descriptive Statistics 3386
  - Normal, Detrended, and Half-Normal Probability Plots 3387
  - Other Transformations and Plots 3387
- Parameter Grid Search 3387
- Automatic Parameter Search 3387
  - Max. Number of Iterations 3388
  - Convergence Criterion 3388
  - Unconstrained Parameter Estimation 3388
  - Lack of Fit Indicator 3388
  - Parameter Start Values 3389

SEASONAL DECOMPOSITION (CENSUS METHOD I) 3391

8. TIME SERIES ANALYSIS - CONTENTS

- **EXAMPLE** 3391
  - Data File 3391
  - Determining the Seasonal Model 3391
  - Seasonal Decomposition 3392
- **DIALOGS, OPTIONS, STATISTICS** 3395
  - Seasonal Decomposition (Census I) 3395
    - Seasonal Model 3395
    - On *OK* Append Components to Active Work Area 3396
    - Other Transformations and Plots 3396
    - Histogram 3397
    - Descriptive Statistics 3397
    - Normal, Detrended, and Half-Normal Probability Plots 3397
    - Review and Plot Variables 3397
    - Autocorrelations 3398

X-11 SEASONAL DECOMPOSITION (CENSUS METHOD II) 3401

- **EXAMPLE** 3401
  - Data File 3401
  - Determining the Seasonal Model 3401
  - Seasonal Decomposition 3402
- **DIALOGS, OPTIONS, STATISTICS** 3405
  - X-11 Monthly Seasonal Decomposition 3405
    - Other Transf. & Plots 3405
    - Seasonal Adjustment 3405
    - Dates (Start of Series) 3406
    - Summary Measures (Input is Seasonally Adjusted) 3406
    - Printout Tables (Detail) 3406
    - Charts 3406
    - Batch Print All Tables/Charts without Interruption 3407
    - Prior Daily Weights 3407
    - Include a Length of Month Allowance 3407
    - Trading-Day Regression & Adjustment of Series 3407
    - Prior Monthly Adjustment Factors 3408
    - *Sigma* Limits for Graduating Extreme Values 3408
    - Moving Averages for Seasonal Factor Curves 3408
    - Henderson Curve Moving Average for Variable Trend-Cycle 3409
    - Adjust Trend-Cycle for Strikes 3410
  - X-11 Quarterly Seasonal Adjustment 3410
    - Other Transf. & Plots 3410
    - Seasonal Adjustment 3411
    - Dates (Start of Series) 3411
    - Summary Measures (Input is Seasonally Adjusted) 3411

TIM - 3265

Printout Tables (Detail)	3411
Charts	3411
Batch Print All Tables/Charts without Interruption	3412
Sigma Limits for Graduating Extreme Values	3412
Adjust Trend-Cycle for Strikes	3412
Review and Plot Variables	3412
Autocorrelations	3413
Histogram	3414
Descriptive Statistics	3414
Normal, Detrended, and Half-Normal Probability Plots	3414
Moving Averages for Seasonal Factor Curves	3415
Selecting Specific Output Tables and/or Charts	3415
Standard Printout Detail (*S*)	3415
Standard Printout Detail for Charts (*S*)	3416
Long Printout Detail (*L*)	3416
Full Printout Detail (*F*)	3417
All Printout Detail for Charts (*A*)	3417

DISTRIBUTED LAGS ANALYSIS 3419

EXAMPLE 3419
- Overview and Data File 3419
- Specifying the Analysis 3419
- Reviewing Results 3420
- Almon Distributed Lag 3420

DIALOGS, OPTIONS, STATISTICS 3421
- Distributed Lags Analysis 3421
    - Independent Variable 3421
    - Lag Length 3421
    - Method 3421
    - Review and Plot Variables 3421
    - Autocorrelations 3422
    - Histogram 3423
    - Descriptive Statistics 3423
    - Normal, Detrended, and Half-Normal Probability Plots 3423
    - Other Transformations and Plots 3423

SPECTRUM (FOURIER) ANALYSIS 3425

EXAMPLE 3425
- Overview and Data File 3425
- Specifying the Analysis 3425
- Results 3426

DIALOGS, OPTIONS, STATISTICS 3429

8. TIME SERIES ANALYSIS - CONTENTS

Spectrum (Fourier) Analysis 3429
    Single Series Fourier Analysis 3429
    Two Series Fourier Analysis 3429
    Transformation of Input Series 3429
    Padding of Input Series 3430
    Histogram 3431
    Descriptive Statistics 3431
    Normal, Detrended, and Half-Normal Probability Plots 3431
    Review and Plot Variables 3431
    Autocorrelations 3432
    Other Transformations and Plots 3433
Single Variable Spectrum Analysis Results 3433
    Summary 3433
    Examine Subset of Periodogram 3434
    n Largest Values 3434
    Data Windows for Spectral Density Estimates 3434
    Append to Work Area on *Cancel* 3435
    Periodogram and Spectral Density Plots 3435
    Histogram of Periodogram Values 3436
    Spectral Density 3437
Cross-Spectrum Analysis Results 3437
    Summary 3437
    Examine Subset of Periodogram 3439
    n Largest Values 3439
    Data Windows for Spectral Density Estimates 3439
    Append to Work Area on *Cancel* 3440
    Periodogram and Spectral Density Plots 3440

INDEX **3441**

8. TIME SERIES ANALYSIS - CONTENTS

Chapter 8:
TIME SERIES ANALYSIS

INTRODUCTORY OVERVIEW

General Overview

The following topics will review techniques that are useful for analyzing time series data, that is, sequences of measurements that follow non-random orders. Unlike the analyses of random samples of observations that are discussed in the context of most other statistics, the analysis of time series is based on the assumption that successive values in the data file represent consecutive measurements taken at equally-spaced time intervals.

Detailed discussions of the methods described in this section can be found in Anderson (1976), Box and Jenkins (1976), Kendall (1984), Kendall and Ord (1990), Montgomery, Johnson, and Gardiner (1990), Pankratz (1983), Shumway (1988), Vandaele (1983), Walker (1991), and Wei (1989). Additional references for specific procedures are provided later in this section.

Two Main Goals

There are two main goals of time series analysis:

(a) Identifying the nature of the phenomenon represented by the sequence of observations, and

(b) Forecasting (predicting future values of the time series variable).

Both of these goals require that the pattern of observed time series data is identified and more or less formally described. Once the pattern is established, you can interpret and integrate it with other data (i.e., use it in your theory of the investigated phenomenon, e.g., seasonal commodity prices). Regardless of the depth of your understanding and the validity of your interpretation (theory) of the phenomenon, you can extrapolate the identified pattern to predict future events.

Overview

The *Time Series* module of *STATISTICA* provides all major techniques for analyzing time series data in a fully integrated environment: An extensive set of transformations and smoothing procedures, which are most commonly used to reveal systematic patterns that are "hidden" in the series, are available in this module. Additional aids for that purpose are the simple and partial autocorrelation and cross-correlation functions and plots. Several techniques are available for separating systematic (trend and/or seasonal) variability in the time series from random fluctuations (e.g., classical seasonal decomposition, X-11 seasonal decomposition). Other techniques allow the user to formulate and test specific hypotheses about the nature and origin of systematic patterns. Specifically, the so-called ARIMA method allows you to test whether and how current observations are influenced by prior observations or specific discrete events (interventions; the latter method is also called interrupted time series analysis). Another very powerful method for generating forecasts is the exponential smoothing technique, which is built on the simple assumption that each observation is the result of a weighted average of prior observations and some random "shock." Finally, there are spectral decomposition techniques that can be used to determine whether there are recurring patterns over time, and the frequencies at which they occur.

All of these methods are fully integrated, that is, the result of one type of analysis can immediately be used in a follow-up with another technique. Of course, as in all other modules of *STATISTICA*, the *Time Series* module offers an extensive set of

8. TIME SERIES ANALYSIS - INTRODUCTORY OVERVIEW

graphics options which are particularly useful here because systematic patterns in time series data are sometimes most easily detected by simple visual inspections of plots.

On the following pages some basic principles and terminology pertaining to the analysis of time series data will first be introduced, along with some basic types of common transformations and smoothing operations. Then, the different techniques described above will be discussed.

Identifying Patterns in Time Series Data

Systematic Pattern and Random Noise

As in most other analyses, in time series analysis it is assumed that the data consist of a systematic pattern (usually a set of identifiable components) and random noise (error) which usually makes the pattern difficult to identify. Most time series analysis techniques involve some form of filtering out the noise in order to make the pattern more salient.

Two General Aspects of Time Series Patterns

Most time series patterns can be described in terms of two basic classes of components: trend and seasonality. The former represents a general systematic linear or nonlinear component that may change over time. The latter refers to a recurring pattern in the data that repeats at systematic intervals over time. Those two general classes of time series components often coexist in real-life data.

For example, sales of a company can rapidly grow over years, but they still follow consistent seasonal patterns (e.g., as much as 25% of yearly sales each year are made in December, whereas only 4% in August).

This general pattern is well illustrated in the "classic" *Series G* data set (Box and Jenkins, 1976, page 531) representing monthly international airline passenger totals (measured in thousands) in twelve consecutive years from 1949 to 1960 (see the example data file *Series_g.sta*).

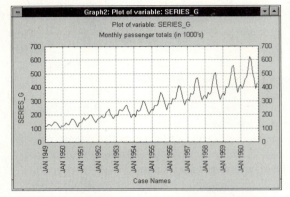

The plot of the monthly airline passenger totals shows a clear, almost linear trend, indicating that the airline industry enjoyed a steady growth over the years (approximately 4 times more passengers traveled in 1960 than in 1949). At the same time, the monthly figures follow an almost identical pattern each year (e.g., more people travel during holidays than during any other time of the year). This example data file also illustrates a very common general type of pattern in time series data, where the amplitude of the seasonal changes increases with the overall trend (i.e., the variance is correlated with the mean over the segments of the series). This pattern which is called *multiplicative seasonality* indicates that the relative amplitude of seasonal changes is constant over time, thus it is related to the trend.

Trend Analysis

There are no proven "automatic" techniques to identify trend components in the time series data; however, as long as the trend is monotonous (consistently increasing or decreasing), that part of

the data analysis is typically not very difficult. If the time series data contain considerable error, then the first step in the process of trend identification is smoothing.

Smoothing

Smoothing always involves some form of local averaging of data such that the nonsystematic components of individual observations cancel each other out. The most common technique is *moving average* smoothing which replaces each element of the series by either the simple or weighted average of n adjacent elements, where n is the width of the smoothing "window" (see Box and Jenkins, 1976; Velleman and Hoaglin, 1981). Medians can be used instead of means. The main advantage of median smoothing as compared to moving average smoothing is that its results are less biased by outliers (within the smoothing window). Thus, if there are outliers in the data (e.g., due to measurement errors), median smoothing typically produces smoother or at least more "reliable" curves than moving average smoothing based on the same window width. The main disadvantage of median smoothing is that in the absence of clear outliers it may produce more "jagged" curves than moving average smoothing and it does not allow for weighting.

All of these techniques are included among the interactive transformations in the *Time Series* module. In the relatively less common cases (in time series data), when the measurement error is very large, the *distance weighted least squares smoothing* or *negative exponentially weighted smoothing* techniques described in the chapters on *Graphics* (Volume II) can be used. All those methods will filter out the noise and convert the data into a smooth curve that is relatively unbiased by outliers (see the respective sections on each of those methods for more details). Series with relatively few and systematically distributed points can be smoothed with *bicubic splines* (also described in the chapters on *Graphics*, Volume II).

Fitting a Function

Many monotonous time series data can be adequately approximated by a linear function. If there is a clear monotonous nonlinear component, the data first need to be transformed to remove the nonlinearity. Usually a logarithmic, exponential, or (less often) polynomial function can be used to transform the data. There are several ways to do this in *STATISTICA*. You can experiment with transformations of unlimited complexity using spreadsheet formulas (or *STATISTICA BASIC*, see Volume V), and later submit the transformed series to linear regression (either via the *Multiple Regression* or the *Time Series* module) and generate forecasts (in *Multiple Regression;* see Volume I). Nonlinear functions of practically unlimited complexity including piecewise estimations with break points (where different functions can be simultaneously fitted to different ranges of the series) can be performed with the *Nonlinear Estimation* module (see Chapter 1). Finally, the *Graphics* options include general purpose curve fitting procedures that can be used to fit polynomial functions (of user-specified order), logarithmic functions (with user-specified bases), exponential and other functions (see Volume II).

Detecting Serial Correlations and Seasonality via Autocorrelation Analysis

Serial and seasonal dependency (seasonality) are other general components of the time series pattern. The concept was illustrated in the example of the airline passengers data above (page 3270). One can easily see that each observation is most similar (closest) to the adjacent observation; in addition, there is the recurring seasonal pattern, that is, each observation is also similar to the same observation a year earlier. In general, serial dependency can formally be defined as correlational dependency of order k between each i'th element of the series and the $(i-k)$'th element (Kendall, 1976), and it can be measured by autocorrelation (i.e., a correlation between the two terms); k is usually called the *lag*. If the measurement error is not too large, seasonality can be visually identified in the series as a pattern that repeats every k elements.

Autocorrelation Correlogram

Seasonal patterns of time series can be examined via correlograms. The correlogram (autocorrelogram) displays graphically and numerically the autocorrelation function (*ACF*), that is, serial correlation coefficients (and their standard errors) for consecutive lags in a specified range of lags (e.g., 1 through 30). Ranges of two standard errors for each lag are usually marked in correlograms, but typically the size of the autocorrelation is of more interest than its reliability because one is usually interested only in very strong (and thus highly significant) autocorrelations.

Examining Correlograms

While examining correlograms one should keep in mind that autocorrelations for consecutive lags are formally dependent. Consider the following example. If the first element is closely related to the second, and the second to the third, then the first element must also be somewhat related to the third one, etc. This implies that the pattern of serial dependencies can change considerably after removing the first order autocorrelation, that is, after differencing the series with a lag of *1*. Also, differencing removes the trend component which typically "overwhelms" all other autocorrelations. If there is, for example, a steady linear trend, as in the airline passenger example, then each observation is to a large extent a (positive) linear function of the preceding observation. Shown below are the autocorrelograms for the airline passenger example

shown earlier (page 3270), both before and after differencing the series.

Before differencing:

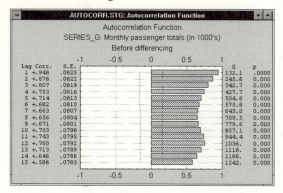

After non-seasonal differencing (with lag=*1*):

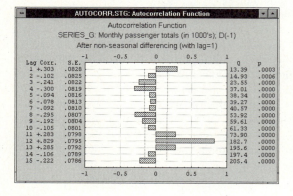

Partial Autocorrelations

Another useful method to examine serial dependencies is to examine the partial autocorrelation function (*PACF*) - an extension of autocorrelation, where the dependence on the intermediate elements (those *within* the lag) is removed. In other words the partial autocorrelation is similar to autocorrelation, except that when calculating it, the (auto) correlations with all the elements within the lag are partialled out (Box and Jenkins, 1976; see also McDowall, McCleary, Meidinger, and Hay, 1980). At the lag of *1* (when there are no intermediate elements within the lag), the partial autocorrelation is equivalent to autocorrelation. In a sense, the partial autocorrelation provides a "cleaner" picture of serial dependencies for individual lags (not confounded by other serial dependencies). Shown below are the partial autocorrelograms for the airline passenger example shown earlier, both before and after differencing the series.

Before differencing:

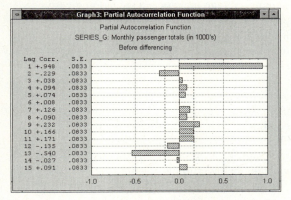

After non-seasonal differencing (with lag=*1*):

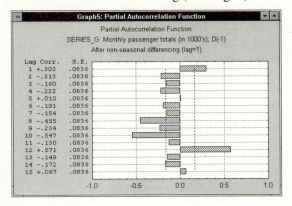

Removing Serial Dependency

As mentioned above, serial dependency for a particular lag of *k* can be removed by differencing the series, that is converting each *i*'th element of the series into its difference from the *(i-k)*'th element. There are two major reasons for such transformations.

First, one can identify the hidden nature of seasonal dependencies in the series. Remember that, as mentioned earlier, autocorrelations for consecutive lags are interdependent. Therefore, removing some of the autocorrelations will change other autocorrelations, that is, it may eliminate them or it may make some other seasonalities more apparent.

The other reason for removing seasonal dependencies is to make the series stationary which is necessary for ARIMA and other techniques (see below).

For more information on simple autocorrelations (introduced in this section) and other autocorrelations, see the *Examples* and *Dialogs, Options, Statistics* sections of this chapter; see also Anderson (1976), Box and Jenkins (1976), Kendall (1984), Pankratz (1983), and Vandaele (1983).

ARIMA

In real-life research and practice, there are often no obvious patterns in the data, individual observations involve considerable error, and you not only want to uncover the hidden patterns in the data, but also generate forecasts. The ARIMA methodology developed by Box and Jenkins (1976) allows you to do just that. This method has gained enormous popularity in many areas, and research practice confirms its power and flexibility (Hoff, 1983; Pankratz, 1983; Vandaele, 1983). However, because of its power and flexibility, ARIMA is a complex technique; it is not easy to use, it requires a great deal of experience, and although it often produces satisfactory results, those results depend on the researcher's level of expertise (Bails and Peppers, 1982). The following sections will introduce the basic ideas of this methodology. For those interested in a brief, applications-oriented (non-mathematical) introduction to ARIMA methods, we recommend McDowall, McCleary, Meidinger, and Hay (1980).

Two Common Processes

Autoregressive process. Most time series consist of elements that are serially dependent in the sense that one can estimate a coefficient or a set of coefficients that describe consecutive elements of the series from specific, time-lagged (previous) elements. This can be summarized in the equation:

$$x_t = \xi + \phi_1 * x_{(t-1)} + \phi_2 * x_{(t-2)} + \ldots + \varepsilon$$

where

ξ is a constant (intercept), and

ϕ_1, ϕ_2 are the autoregressive model parameters

Put into words, each observation is made up of a random error component (random shock, ε) and a linear combination of prior observations.

Stationarity requirement. Note that an autoregressive process will only be stable if the parameters are within a certain range; for example, if there is only one autoregressive parameter, then it must fall within the interval of $-1 < \phi_1 < +1$. Otherwise, past effects would accumulate and the values of successive x_t's would move towards infinity, that is, the series would not be *stationary*. If there is more than one autoregressive parameter, similar (general) restrictions on the parameter values can be defined (e.g., see Box and Jenkins, 1976; Montgomery, 1990). The *Time Series* program will automatically check whether the stationarity requirement is met.

Moving average process. Independent from the autoregressive process, each element in the series can also be affected by the past error (or random shock) that cannot be accounted for by the autoregressive component, that is:

$$x_t = \mu + \varepsilon_t - \theta_1 * \varepsilon_{(t-1)} - \theta_2 * \varepsilon_{(t-2)} - \ldots$$

where

μ is a constant, and

θ_1, θ_2 are the moving average model parameters

Put into words, each observation is made up of a random error component (random shock, ε) and a linear combination of prior random shocks.

Invertibility requirement. Without going into too much detail, there is a "duality" between the moving average process and the autoregressive process (e.g., see Box and Jenkins, 1976; Montgomery, Johnson, and Gardiner, 1990); that is, the moving average equation above can be rewritten (*inverted*) into an autoregressive form (of infinite order). However, analogous to the stationarity condition described above, this can only be done if the moving average parameters follow certain conditions, that is, if the model is *invertible*. Otherwise, the series will not be *stationary*. Again, the *Time Series* program will automatically check whether the invertibility requirement is met.

The ARIMA Model

Autoregressive moving average model. The general model introduced by Box and Jenkins (1976) includes autoregressive as well as moving average parameters, and explicitly includes differencing in the formulation of the model. Specifically, the three types of parameters in the model are:

- The autoregressive parameters (*p*),
- The number of differencing passes (*d*), and
- Moving average parameters (*q*).

In the notation introduced by Box and Jenkins, models are summarized as ARIMA (*p, d, q*); so, for example, a model described as (*0, 1, 2*) means that it contains 0 (zero) autoregressive (*p*) parameters and 2 moving average (*q*) parameters which were computed for the series after it was differenced once.

Identification. As mentioned earlier, the input series for ARIMA needs to be *stationary*, that is, it should have a constant mean, variance, and autocorrelation through time. Therefore, usually the series first needs to be differenced until it is *stationary* (this also often requires log transforming the data to stabilize the variance). The number of times the series needs to be differenced to achieve stationarity is reflected in the *d* parameter (see the previous paragraph). In order to determine the necessary level of differencing, one should examine the plot of the data and autocorrelogram. Significant changes in level (strong upward or downward changes) usually require first order non-seasonal (lag=*1*) differencing; strong changes of slope usually require second order non-seasonal differencing. Seasonal patterns require respective seasonal differencing (see below). If the estimated autocorrelation coefficients decline slowly at longer lags, first order differencing is usually needed. However, one should keep in mind that some time series may require little or no differencing, and that *over differenced* series produce less stable coefficient estimates.

At this stage (which is usually called *Identification* phase, see below) you also need to decide how many autoregressive (*p*) and moving average (*q*) parameters are necessary to yield an effective but still *parsimonious* model of the process (*parsimonious* means that it has the fewest parameters and greatest number of degrees of freedom among all models that fit the data). In practice, the numbers of the *p* or *q* parameters very rarely need to be greater than 2 (see below for more specific recommendations).

Estimation and forecasting. At the next step (*Estimation*), the parameters are estimated (using function minimization procedures, see below; for more information on minimization procedures see also *Nonlinear Estimation*, Chapter 1), so that the sum of squared residuals is minimized. The estimates of the parameters are used in the last stage (*Forecasting*) to calculate new values of the series (beyond those included in the input data set) and confidence intervals for those predicted values. The estimation process is performed on transformed (differenced) data; before the forecasts are generated, the series needs to be *integrated* (integration is the inverse of differencing) so that the forecasts are expressed in values compatible with the input data. This automatic integration feature is represented by the letter *I* in the name of the methodology (ARIMA = Auto-Regressive Integrated Moving Average).

The constant in ARIMA models. In addition to the standard autoregressive and moving average parameters, ARIMA models may also include a constant, as described above. The interpretation of a (statistically significant) constant depends on the model that is fit. Specifically,

(1) If there are no autoregressive parameters in the model, then the expected value of the constant is μ, the mean of the series;

(2) If there are autoregressive parameters in the series, then the constant represents the intercept.

If the series is differenced, then the constant represents the mean or intercept of the differenced series. For example, if the series is differenced once, and there are no autoregressive parameters in the model, then the constant represents the mean of the differenced series and therefore the *linear trend slope* of the un-differenced series.

Model Identification

Number of parameters to be estimated. Before the estimation can begin, you need to decide on (identify) the specific number and type of ARIMA parameters to be estimated. The major tools used in the identification phase are plots of the series, correlograms of autocorrelation (ACF), and partial autocorrelation (PACF). The decision is not straightforward and in less typical cases requires not only experience but also a good deal of experimentation with alternative models (as well as the technical parameters of ARIMA). However, a majority of empirical time series patterns can be sufficiently approximated using one of the 5 basic models (see below) that can be identified based on the shape of the autocorrelogram (ACF) and partial autocorrelogram (PACF).

The following brief summary of these models is based on practical recommendations of Pankratz (1983); for additional practical advice, see also Hoff (1983), McCleary and Hay (1980), McDowall, McCleary, Meidinger, and Hay (1980), and Vandaele (1983). Also, note that since the number of parameters (to be estimated) of each kind is almost never greater than 2, it is often practical to try alternative models on the same data.

(1) *One autoregressive (p) parameter:* ACF - exponential decay; PACF - spike at lag *1*, no correlation for other lags.

(2) *Two autoregressive (p) parameters:* ACF - a sine-wave shape pattern or a set of exponential decays; PACF - spikes at lags *1* and *2*, no correlation for other lags.

(3) *One moving average (q) parameter:* ACF - spike at lag *1*, no correlation for other lags; PACF - damps out exponentially.

(4) *Two moving average (q) parameters:* ACF - spikes at lags *1* and *2*, no correlation for other lags; PACF - a sine-wave shape pattern or a set of exponential decays.

(5) *One autoregressive (p) and one moving average (q) parameter:* ACF - exponential decay starting at lag *1*; PACF - exponential decay starting at lag *1*.

Seasonal models. Multiplicative seasonal ARIMA is a generalization and extension of the method introduced in the previous paragraphs to series in which a pattern repeats seasonally over time. In addition to the non-seasonal parameters, seasonal parameters for a specified lag (established

in the identification phase) need to be estimated. Analogous to the simple ARIMA parameters, these are: seasonal autocorrelations (*ps*), seasonal differencing (*ds*), and seasonal moving average parameters (*qs*).

The full seasonal ARIMA model can thus be summarized as ARIMA (*p,d,q*)(*ps,ds,qs*). For example, the model *(0,1,2)(0,1,1)* describes a model that includes no autoregressive parameters, 2 regular moving average parameters and 1 seasonal moving average parameter, and these parameters are computed for the series after it is differenced once with lag *1*, and once seasonally differenced. The seasonal lag used for the seasonal parameters is usually determined during the identification phase and must be explicitly specified.

The general recommendations concerning the selection of parameters to be estimated (based on ACF and PACF) also apply to seasonal models (for an illustration of seasonal ARIMA, see the *Example* in the ARIMA section of this chapter, page 3341). The main difference is that in seasonal series, ACF and PACF will show sizable coefficients at multiples of the seasonal lag (in addition to their overall patterns reflecting the non-seasonal components of the series).

Parameter Estimation

The *Time Series* module of *STATISTICA* includes different methods for estimating the parameters. Each method should produce very similar estimates, but may be more or less efficient for any given model. In general, during the parameter estimation phase, a function minimization algorithm is used (the so-called *quasi-Newton* method; refer to the description in the *Nonlinear Estimation* chapter, see Chapter 1) to maximize the likelihood (probability) of the observed series, given the parameter values. In practice, this requires the calculation of the (conditional) sums of squares (*SS*) of the residuals, given the respective parameters. Different methods

have been proposed to compute the *SS* for the residuals; in *STATISTICA* you can choose between:

(1) The approximate maximum likelihood method according to McLeod and Sales (1983),

(2) The approximate maximum likelihood method with backcasting, and

(3) The exact maximum likelihood method according to Melard (1984).

Comparison of methods. In general, all methods should yield very similar parameter estimates. Also, all methods are about equally efficient in most real-world time series applications. However, method *1* above (approximate maximum likelihood, no backcasts) is the fastest and should be used in particular for very long time series (e.g., with more than 30,000 observations; note that the *Time Series* module will take full advantage of your computer's memory, and it will not impose fixed limitations on the lengths of time series that can be analyzed).

Melard's exact maximum likelihood method (number *3* above) may also become inefficient when used to estimate parameters for seasonal models with long seasonal lags (e.g., with yearly lags of 365 days). On the other hand, the *Time Series* module will always use the approximate maximum likelihood method first, in order to establish initial parameter estimates that are very close to the actual final values; thus, usually only a few iterations with the exact maximum likelihood method (*3*, above) are necessary to finalize the parameter estimates.

Parameter standard errors. For all parameter estimates, the *Time Series* module will compute so-called *asymptotic standard errors*. These are computed from the matrix of second-order partial derivatives that is approximated via finite differencing (see also the respective discussion in *Nonlinear Estimation*, Chapter 1).

Penalty value. As mentioned above, the estimation procedure requires that the (conditional) sums of squares of the ARIMA residuals be minimized. If the model is inappropriate, it may happen during the iterative estimation process that the parameter estimates become very large, and, in fact, invalid. In that case, the program will assign a very large value (a so-called *penalty value*) to the SS. This usually "entices" the iteration process to move the parameters away from invalid ranges. However, in some cases even this strategy fails, and you may see on the screen (during the estimation procedure) very large values for the SS in consecutive iterations. In that case, carefully evaluate the appropriateness of your model. If your model contains many parameters, and perhaps an intervention component (see below), you may try again with different parameter start values.

Evaluation of the Model

Parameter estimates. The *Time Series* module will report approximate *t* values, computed from the parameter standard errors (see above). If not significant, the respective parameter can in most cases be dropped from the model without substantially affecting the overall fit of the model.

Other quality criteria. Another straightforward and common measure of the reliability of the model is to compare the forecasts of "known data" which were generated based on a part of the series to the known (original) observations (this test is illustrated in the ARIMA *Example* in this chapter, page 3341). However, a good model should not only provide sufficiently accurate forecasts, it should also be parsimonious and produce statistically independent residuals that contain only noise and no systematic components (e.g., the correlogram of residuals should not reveal any serial dependencies).

A good test of the model is (a) to plot the residuals and inspect them for any systematic trends, and (b) to examine the autocorrelogram of residuals (there should be no serial dependency between residuals).

Analysis of residuals. The major concern here is that the residuals are systematically distributed across the series (e.g., they could be negative in the first part of the series and approach zero in the second part) or that they contain some serial dependency which may suggest that the ARIMA model is inadequate. The analysis of ARIMA residuals constitutes an important test of the model. The estimation procedure assumes that the residuals are not (auto-) correlated and that they are normally distributed. The *Time Series* module will automatically compute the residuals and make them available for further analyses with all other methods available in this module. The residuals can also be saved to a data file, together with the original series or its transformations.

Limitations. To reiterate, the ARIMA method is appropriate only for a time series that is stationary (i.e., its mean, variance, and autocorrelation should be approximately constant through time), and it is recommended that there be at least 50 observations in the input data. It is also assumed that the values of the estimated parameters are constant throughout the series.

Interrupted Time Series ARIMA

A common research question in time series analysis is whether an outside event affected subsequent observations. For example, did the implementation of a new economic policy improve economic performance; did the a new anti-crime law affect subsequent crime rates; and so on. In general, the evaluation of the impact of one or more discrete events on the values in the time series is desired. This type of *interrupted time series analysis* is described in detail in McDowall, McCleary, Meidinger, and Hay (1980). McDowall, et al., distinguishes between three major types of impacts that are possible:

(1) Permanent abrupt,

(2) Permanent gradual, and

(3) Abrupt temporary.

These different impact patterns are described below.

Abrupt Permanent Impact

A permanent abrupt impact pattern simply implies that the overall mean of the time series shifted after the intervention; the overall shift is denoted by ω (*omega*). This impact pattern is illustrated below:

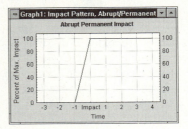

Gradual Permanent Impact

The gradual permanent impact pattern implies that the increase or decrease due to the intervention is gradual, and that the final permanent impact becomes evident only after some time (illustrated below).

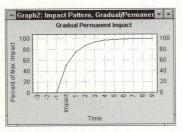

This type of intervention can be summarized by the expression:

```
Impact_t = δ * Impact_(t-1) + ω
(for all t ≥ time of impact, else = 0)
```

Note that this impact pattern is defined by the two parameters δ *(delta)* and ω *(omega)*. If δ is near *0* (zero), then the final permanent amount of impact will be evident after only a few more observations; if δ is close to *1*, then the final permanent amount of impact will only be evident after many more observations.

As long as the δ parameter is greater than *0* and less than *1* (the bounds of system stability), the impact will be gradual and result in an asymptotic change (shift) in the overall mean by the quantity:

```
Asymptotic change in level = ω/(1-δ)
```

The *Time Series* module will automatically compute the asymptotic change for gradual permanent impacts.

Note that, when evaluating a fitted model, it is important that both parameters are statistically significant; otherwise one could reach paradoxical conclusions. For example, suppose the ω parameter is not statistically significant from *0* (zero) but the δ parameter is. This would mean that an intervention caused a significant gradual change, the final result of which was not significantly different from zero.

Abrupt Temporary Impact

The abrupt temporary impact pattern implies an initial abrupt increase or decrease due to the intervention which then slowly decays, without permanently changing the mean of the series.

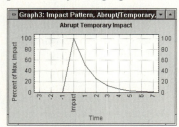

This type of intervention can be summarized by the expressions:

- Prior to intervention: $Impact_t = 0$
- At time of intervention: $Impact_t = \omega$
- After intervention: $Impact_t = \delta * Impact_{t-1}$

Note that this impact pattern is again defined by the two parameters δ *(delta)* and ω *(omega)*. As long as the δ parameter is greater than *0* and less than *1* (the bounds of system stability), the initial abrupt impact will gradually decay. If δ is near *0* (zero), then the decay will be very quick, and the impact will have entirely disappeared after only a few observations. If δ is close to *1*, then the decay will be slow, and the intervention will affect the series over many observations. Note that, when evaluating a fitted model, it is again important that both parameters are statistically significant; otherwise one could reach paradoxical conclusions. For example, suppose the ω parameter is not significantly different from *0* (zero) but the δ parameter is significant; this would mean that an intervention did not cause an initial abrupt change, which then showed significant decay.

Exponential Smoothing

Exponential smoothing has become very popular as a forecasting method for a wide variety of time series data. Historically, the method was independently developed by Brown and Holt. Brown worked for the US Navy during World War II, where his assignment was to design a tracking system for fire-control information to compute the location of submarines. Later, he applied this technique to the forecasting of demand for spare parts (an inventory control problem). He described those ideas in his 1959 book on inventory control. Holt's research was sponsored by the Office of Naval Research. Independently, he developed exponential smoothing models for constant processes, processes with linear trends, and for seasonal data.

The implementation of exponential smoothing methods in *STATISTICA* closely follows the survey of techniques presented by Gardner (1985), who proposed a "unified" classification of exponential smoothing methods. Excellent introductions can also be found in Makridakis, Wheelwright, and McGee (1983), Makridakis and Wheelwright (1989), Montgomery, Johnson, and Gardiner (1990).

Simple Exponential Smoothing

A simple and pragmatic model for a time series would be to consider each observation as consisting of a constant (*b*) and an error component ε (*epsilon*), that is: $X_t = b + \varepsilon_t$. The constant *b* is relatively stable in each segment of the series, but may change slowly over time. If appropriate, then one way to isolate the true value of *b*, and thus the systematic or predictable part of the series, is to compute a kind of moving average, where the current and immediately preceding ("younger") observations are assigned greater weight than the respective older observations. Simple exponential smoothing accomplishes exactly such weighting, where exponentially smaller weights are assigned to older observations. The specific formula for simple exponential smoothing is:

$$S_t = \alpha * X_t + (1-\alpha) * S_{t-1}$$

When applied recursively to each successive observation in the series, each new smoothed value (forecast) is computed as the weighted average of the current observation and the previous smoothed observation. The previous smoothed observation was computed in turn from the previous observed value and the smoothed value before the previous observation, and so on. Thus, in effect, each smoothed value is the weighted average of the previous observations, where the weights decrease exponentially depending on the value of parameter α (*alpha*).

If α is equal to *1* (one), then the previous observations are ignored entirely; if α is equal to *0* (zero), then the current observation is ignored entirely, and the smoothed value consists entirely of the previous smoothed value (which in turn is computed from the smoothed observation before it, and so on; thus all smoothed values will be equal to the initial smoothed value S_0). Values of α in-between will produce intermediate results.

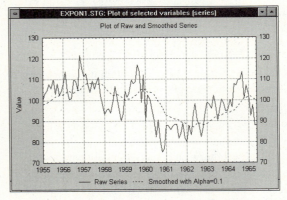

Even though significant work has been done to study the theoretical properties of (simple and complex) exponential smoothing (e.g., see Gardner, 1985; Muth, 1960; see also McKenzie, 1984, 1985), the method has gained popularity mostly because of its usefulness as a forecasting tool.

For example, empirical research by Makridakis et al. (1982; Makridakis, 1983) has shown simple exponential smoothing to be the best choice for one-period-ahead forecasting, from among 24 other time series methods and using a variety of accuracy measures (see also Gross and Craig, 1974, for additional empirical evidence). Thus, regardless of the theoretical model for the process underlying the observed time series, simple exponential smoothing will often produce quite accurate forecasts.

Choosing the Best Value for Parameter α (*alpha*)

Gardner (1985) discusses various theoretical and empirical arguments for selecting an appropriate smoothing parameter. Obviously, looking at the formula presented above, α should fall into the interval between *0* (zero) and *1* (although, see Brenner et al., 1968, for an ARIMA perspective, implying $0<\alpha<2$). Gardner (1985) reports that among practitioners, an α smaller than *.30* is usually recommended. However, in the study by Makridakis et al., (1982), α values above *.30* frequently yielded the best forecasts. After reviewing the literature on this topic, Gardner (1985) concludes that it is best to estimate an optimum α from the data (see below), rather than to "guess" and set an artificially low value.

Estimating the best α value from the data. In practice, the smoothing parameter is often chosen by a *grid search* of the parameter space; that is, different solutions for α are tried starting, for example, with $\alpha = 0.1$ to $\alpha = 0.9$, with increments of *0.1*. Then α is chosen so as to produce the smallest sums of squares (or mean squares) for the residuals (i.e., observed values minus one-step-ahead forecasts; this mean squared error is also referred to as *ex post* mean squared error, *ex post* MSE for short). The *Time Series* module provides an option to perform a grid search, and also allows the user to automatically search for the best α parameter via a general function minimization procedure (see below).

Indices of Lack of Fit (Error)

The most straightforward way of evaluating the accuracy of the forecasts based on a particular α value is to simply plot the observed values and the one-step-ahead forecasts. In the *Time Series* module, this plot will also include the residuals (scaled against the right-*Y* axis), so that regions of better or worst fit can also easily be identified. Shown below is an example of this plot.

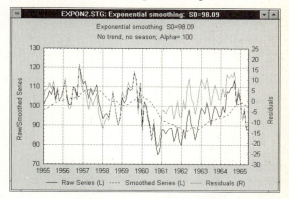

This visual check of the accuracy of forecasts is often the most powerful method for determining whether or not the current exponential smoothing model fits the data. In addition, besides the *ex post* MSE criterion (see previous paragraph), there are other statistical measures of error that can be used to determine the optimum α parameter (see Makridakis, Wheelwright, and McGee, 1983; all measures will automatically be computed by the *Time Series* module):

Mean error. The mean error (ME) value is simply computed as the average error value (average of observed minus one-step-ahead forecast). Obviously, a drawback of this measure is that positive and negative error values can cancel each other out, so this measure is not a very good indicator of overall fit.

Mean absolute error. The mean absolute error (MAE) value is computed as the average *absolute* error value. If this value is *0* (zero), the fit (forecast) is perfect. As compared to the mean *squared* error value, this measure of fit will "de-emphasize" outliers, that is, unique or rare large error values will affect the MAE less than the MSE value.

Sum of squared error (SSE), Mean squared error (MSE). These values are computed as the sum (or average) of the squared error values. This is the most commonly used lack of fit indicator in statistical fitting procedures.

Percentage error (PE). All of the above measures rely on the actual error value. It may seem reasonable to rather express the lack of fit in terms of the *relative* deviation of the one-step-ahead forecasts from the observed values, that is, relative to the magnitude of the observed values.

For example, when trying to predict monthly sales that may fluctuate widely (e.g., seasonally) from month to month, you may be satisfied if your prediction "hits the target" with about ±10% accuracy. In other words, the absolute errors may not be of as much interest as the relative errors in the forecasts. To assess the relative error, various indices have been proposed (see Makridakis, Wheelwright, and McGee, 1983). The first one, the percentage error value, is computed as:

$$PE_t = 100*(X_t - F_t)/X_t$$

where X_t is the observed value at time t, and F_t is the forecasts (smoothed values).

Mean percentage error (MPE). This value is computed as the average of the PE values.

Mean absolute percentage error (MAPE). As is the case with the mean error value (ME, see above), a mean percentage error near *0* (zero) can be produced by large positive and negative percentage errors that cancel each other out. Thus, a better measure of relative overall fit is the mean *absolute* percentage error. Also, this measure is usually more meaningful than the mean squared error. For example, knowing that the average forecast is "off" by ±5% is a useful result in and of itself, whereas a mean squared error of 30.8 is not immediately interpretable.

Automatic search for best parameter. The *Time Series* module effectively takes the guessing out of the parameter search process. A quasi-Newton function minimization procedure (the same as in ARIMA) is used to minimize either the mean squared error, mean absolute error, or mean absolute percentage error. In most cases, this procedure is more efficient than the grid search (particularly when more than one parameter must be determined), and the optimum α parameter can quickly be identified.

The first smoothed value S_0. A final issue that has been neglected up to this point is the problem of the initial value, or how to start the smoothing process. If you look back at the formula above, it is evident that one needs an S_0 value in order to compute the smoothed value (forecast) for the first observation in the series. Depending on the choice of the α parameter (i.e., when α is close to zero), the initial value for the smoothing process can affect the quality of the forecasts for many observations. As with most other aspects of exponential smoothing, it is recommended to choose the initial value that produces the best forecasts. On the other hand, in practice, when there are many leading observations prior to a crucial actual forecast, the initial value will not affect that forecast by much, since its effect will have long "faded" from the smoothed series (due to the exponentially decreasing weights, the older an observation the less it will influence the forecast). The *Time Series* module allows for user-defined initial values, but will also automatically compute initial values (see the description of actual smoothing methods below).

Seasonal and Non-seasonal Models with or without Trend

The discussion above in the context of simple exponential smoothing introduced the basic procedure for identifying a smoothing parameter, and for evaluating the goodness of fit of a model. In addition to simple exponential smoothing, more complex models have been developed to accommodate time series with seasonal and trend components. The general idea here is that forecasts are not only computed from consecutive previous observations (as in simple exponential smoothing), but an independent (smoothed) trend and seasonal component can be added. The *STATISTICA Time Series* module follows the two-way classification system proposed by Gardner (1985), who discusses the different models in terms of seasonality (none, additive, or multiplicative) and trend (none, linear, exponential, or damped).

Additive and multiplicative seasonality.

Many time series data follow recurring seasonal patterns (see also the *Classical Seasonal Decomposition* section, page 3284). For example, annual sales of toys will probably peak in the months of November and December, and perhaps during the summer (with a much smaller peak) when children are on their summer break. This pattern will likely repeat every year; however, the relative amount of increase in sales during December may slowly change from year to year. Thus, it may be useful to smooth the seasonal component independently with an extra parameter, usually denoted as δ (*delta*). Seasonal components can be additive in nature or multiplicative.

For example, during the month of December the sales for a particular toy may increase by 1 million dollars every year. Thus, you could add to your forecasts for every December, the amount of 1 million dollars (over the respective annual average) to account for this seasonal fluctuation. In this case, the seasonality is additive. Alternatively, during the month of December the sales for a particular toy may increase by 40%, that is, increase by a factor of 1.4. Thus, when the sales for the toy are generally weak, then the absolute (dollar) increase in sales during December will be relatively weak (but the percentage will be constant); if the sales of the toy are strong, then the absolute (dollar) increase in sales will be proportionately greater. Again, in this case the sales increase by a certain factor, and the seasonal component is thus multiplicative in nature

(i.e., the multiplicative seasonal component in this case would be 1.4).

In plots of the series, the distinguishing characteristic between these two types of seasonal components is that in the additive case, the series shows steady seasonal fluctuations, regardless of the overall level of the series; in the multiplicative case, the size of the seasonal fluctuations vary, depending on the overall level of the series.

The seasonal smoothing parameter δ.

In general the one-step-ahead forecasts are computed as (for no trend models, for linear and exponential trend models a trend component is added to the model; see below):

Additive model:

$$\text{Forecast}_t = S_t + I_{t-p}$$

Multiplicative model:

$$\text{Forecast}_t = S_t * I_{t-p}$$

In this formula, S_t stands for the (simple) exponentially smoothed value of the series at time t, and I_{t-p} stands for the smoothed seasonal factor at time t minus p (the length of the season). Thus, compared to simple exponential smoothing, the forecast is "enhanced" by adding or multiplying the simple smoothed value by the predicted seasonal component. This seasonal component is derived analogously to the S_t value from simple exponential smoothing as:

Additive model:

$$I_t = I_{t-p} + \delta * (1-\alpha) * e_t$$

Multiplicative model:

$$I_t = I_{t-p} + \delta * (1-\alpha) * e_t/S$$

Put into words, the predicted seasonal component at time t is computed as the respective seasonal component in the last seasonal cycle plus a portion of the error (e_t; the observed minus the forecast value at time t). Considering the formulas above, it is clear that parameter δ can assume values between *0* and *1*. If it is zero, then the seasonal component for a particular point in time is predicted to be identical to the predicted seasonal component for the respective time during the previous seasonal cycle, which in turn is predicted to be identical to that from the previous cycle, and so on. Thus, if δ is zero, a constant seasonal component is used to generate the one-step-ahead forecasts. If the δ parameter is equal to *1*, then the seasonal component is modified "maximally" at every step by the respective forecast error (times *1-α*, which will be ignored for the purpose of this brief introduction). In most cases, when seasonality is present in the time series, the optimum δ parameter will fall somewhere between *0* (zero) and *1* (one).

Linear, exponential, and damped trend. To return to the toy example above, the sales for a toy can show a linear upward trend (e.g., each year, sales increase by 1 million dollars), exponential growth (e.g., each year, sales increase by a factor of 1.3), or a damped trend (during the first year sales increase by 1 million dollars; during the second year the increase is only 80% over the previous year, i.e., $800,000; during the next year it is again 80% less than the previous year, i.e., $800,000 * .8 = $640,000; etc.). Each type of trend leaves a clear "signature" that can usually be identified in the series. In general, the trend factor may change slowly over time, and, again, it may make sense to smooth the trend component with a separate parameter [denoted as γ (*gamma*) for linear and exponential trend models, and φ (*phi*) for damped trend models].

The trend smoothing parameters γ (linear and exponential trend) and φ (damped trend). Analogous to the seasonal component, when a trend component is included in the exponential smoothing process, an independent trend component is computed for each time and modified as a function of the forecast error and the respective parameter. If the γ parameter is *0* (zero),

then the trend component is constant across all values of the time series (and for all forecasts). If the parameter is *1*, then the trend component is modified "maximally" from observation to observation by the respective forecast error. Parameter values that fall in-between represent mixtures of those two extremes. Parameter φ is a trend modification parameter and affects how strongly changes in the trend will affect estimates of the trend for subsequent forecasts, that is, how quickly the trend will be "damped" or increased.

Classical Seasonal Decomposition (Census Method I)

Suppose you recorded the monthly passenger load on international flights for a period of 12 years (such data are included in the example data file *Series_g.sta*, see also Box and Jenkins, 1976). If you plot those data, it is apparent that (1) there appears to be a linear upwards trend in the passenger loads over the years, and (2) there is a recurring pattern or *seasonality* within each year (i.e., most travel occurs during the summer months, and a minor peak occurs during the December holidays). The purpose of the seasonal decomposition method is to isolate those components, that is, to decompose the series into the trend effect, seasonal effects, and remaining variability. The "classic" technique designed to accomplish this decomposition is known as the *Census I* method. This technique is described and discussed in detail in Makridakis, Wheelwright, and McGee (1983), and Makridakis and Wheelwright (1989).

General Model

The general idea of seasonal decomposition is straightforward. In general, a time series like the one described above can be thought of as consisting of four different components:

(1) A seasonal component (denoted as S_t, where t stands for the particular point in time),

(2) A trend component (T_t),

(3) A cyclical component (C_t), and

(4) A random, error, or irregular component (I_t).

The difference between a cyclical and a seasonal component is that the latter occurs at regular (seasonal) intervals, while cyclical factors usually have a longer duration that varies from cycle to cycle. In the Census I method, the trend and cyclical components are customarily combined into a *trend-cycle component* (TC_t). The specific functional relationship between these components can assume different forms. However, two straightforward possibilities are that they combine in an *additive* or a *multiplicative* fashion:

Additive model:

$$X_t = TC_t + S_t + I_t$$

Multiplicative model:

$$X_t = T_t * C_t * S_t * I_t$$

Here X_t stands for the observed value of the time series at time t. Given some *a priori* knowledge about the cyclical factors affecting the series (e.g., business cycles), the estimates for the different components can be used to compute forecasts for future observations. (However, the *Exponential smoothing* method, which can also incorporate seasonality and trend components, is the preferred technique for forecasting purposes.)

Additive and Multiplicative Seasonality

Consider the difference between an additive and multiplicative seasonal component in the following illustration: The annual sales of toys will probably peak in the months of November and December, and perhaps during the summer (with a much smaller peak) when children are on their summer break. This seasonal pattern will likely repeat every year. Seasonal components can be additive or multiplicative in nature.

For example, during the month of December the sales for a particular toy may increase by 3 million dollars every year. Thus, you could *add* to your forecasts for every December the amount of 3 million to account for this seasonal fluctuation. In this case, the seasonality is *additive*. Alternatively, during the month of December the sales for a particular toy may increase by 40%, that is, increase by a *factor* of 1.4. Thus, when the sales for the toy are generally weak, then the absolute (dollar) increase in sales during December will be relatively weak (but the percentage will be constant); if the sales of the toy are strong, then the absolute (dollar) increase in sales will be proportionately greater. Again, in this case the sales increase by a certain *factor*, and the seasonal component is thus *multiplicative* in nature (i.e., the multiplicative seasonal component in this case would be 1.4).

In plots of series, the distinguishing characteristic between these two types of seasonal components is that in the additive case, the series shows steady seasonal fluctuations, regardless of the overall level of the series; in the multiplicative case, the size of the seasonal fluctuations vary, depending on the overall level of the series.

Additive and Multiplicative Trend-cycle

The previous example can be extended to illustrate the additive and multiplicative trend-cycle components. In terms of the toy example, a "fashion" *trend* may produce a steady increase in sales (e.g., a trend towards more educational toys in general); as with the seasonal component, this trend may be additive (sales increase by 3 million dollars per year) or multiplicative (sales increase by 30%, or by a factor of 1.3, annually) in nature. In addition, cyclical components may impact sales; to reiterate, a cyclical component is different from a seasonal component in that it usually is of longer duration and that it occurs at irregular intervals. For example, a particular toy may be particularly "hot" during a summer season (e.g., a particular doll which is tied to the release of a major children's movie, and is promoted with extensive advertising). Again such a cyclical component can effect sales in an additive manner or multiplicative manner.

Computations

Computationally, the *Seasonal Decomposition (Census I)* procedure of the *Time Series* module closely follows the standard formulas, as shown in Makridakis, Wheelwright, and McGee (1983), and Makridakis and Wheelwright (1989).

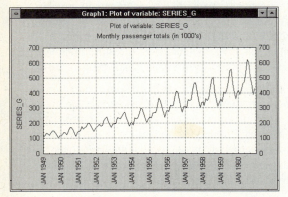

Shown above is the "classic" *Series G* data set of international airline passenger loads (Box and Jenkins, 1976), which will be used to illustrate the different components that can be extracted. The seasonal decomposition of a series proceeds as follows.

Moving average. First a moving average is computed for the series, with the moving average window width equal to the length of one season. If the length of the season is even, then the user can choose to use either equal weights for the moving average or unequal weights can be used, where the first and last observation in the moving average window are averaged [select the *Centered* option in the *Seasonal Decomposition (Census I)* dialog, see page 3395].

Ratios or differences. In the moving average series, all seasonal (within-season) variability will be eliminated, thus, the differences (in additive models) or ratios (in multiplicative models) of the observed and smoothed series will isolate the seasonal component (plus irregular component). Specifically, the moving average is subtracted from the observed series (for additive models) or the observed series is divided by the moving average values (for multiplicative models).

Seasonal components. The seasonal component is then computed as the average (for additive models) or medial average (for multiplicative models) for each point in the season.

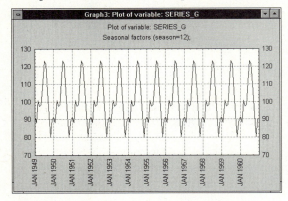

The medial average of a set of values is the mean after the smallest and largest values are excluded. The resulting values represent the (average) seasonal component of the series.

Seasonally adjusted series. The original series can be adjusted by subtracting from it (additive models) or dividing it by (multiplicative models) the seasonal component. The resulting series is the seasonally adjusted series (i.e., the seasonal component will be removed).

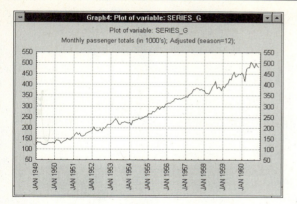

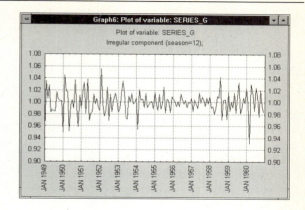

Trend-cycle component. Remember that the cyclical component is different from the seasonal component in that it is usually longer than one season, and different cycles can be of different lengths. The combined trend and cyclical component can be approximated by applying to the seasonally adjusted series a 5 point (centered) weighted moving average smoothing transformation with the weights of 1, 2, 3, 2, 1.

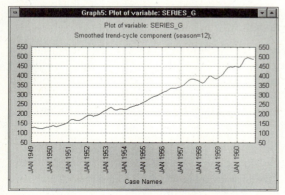

Random or irregular component. Finally, the random or irregular (error) component can be isolated by subtracting from the seasonally adjusted series (additive models) or dividing the adjusted series by (multiplicative models) the trend-cycle component.

X-11 Census Method II Seasonal Adjustment

The general ideas of seasonal decomposition and adjustment are discussed in the context of the Census I seasonal adjustment method above. The Census method II (2) is an extension and refinement of the simple adjustment method. Over the years, different versions of the Census method II evolved at the US Census Bureau; the method that has become most popular and is used most widely in government and business today is the so-called X-11 variant of the Census method II (see Shiskin, Young, and Musgrave, 1967). Subsequently, the term X-11 has become synonymous with this refined version of the Census method II.

In addition to the documentation that can be obtained from the Census Bureau, a detailed summary of this method is also provided in Makridakis, Wheelwright, and McGee (1983) and Makridakis and Wheelwright (1989).

Seasonal Adjustment: Basic Ideas and Terms

Suppose you recorded the monthly passenger load on international flights for a period of 12 years (such data are included in the example data file *Series_g.sta*, see also Box and Jenkins, 1976). If you plot those data, it is apparent that (1) there appears to be an upwards linear trend in the passenger loads over the years, and (2) there is a recurring pattern or *seasonality* within each year (i.e., most travel occurs during the summer months, and a minor peak occurs during the December holidays). The purpose of seasonal decomposition and adjustment is to isolate those components, that is, to de-compose the series into the trend effect, seasonal effects, and remaining variability. The "classic" technique designed to accomplish this decomposition was developed in the 1920's and is also known as the *Census I* method (see the previous section). This technique is also described and discussed in detail in Makridakis, Wheelwright, and McGee (1983), and Makridakis and Wheelwright (1989).

General model. The general idea of seasonal decomposition is straightforward. In general, a time series like the one described above can be thought of as consisting of four different components:

(1) A seasonal component (denoted as S_t, where t stands for the particular point in time),

(2) A trend component (T_t),

(3) A cyclical component (C_t), and

(4) A random, error, or irregular component (I_t).

The difference between a cyclical and a seasonal component is that the latter occurs at regular (seasonal) intervals, while cyclical factors usually have a longer duration that varies from cycle to cycle. The trend and cyclical components are customarily combined into a *trend-cycle component* (TC_t). The specific functional relationship between these components can assume different forms. However, two straightforward possibilities are that they combine in an *additive* or a *multiplicative* fashion:

Additive model:

$$x_t = TC_t + S_t + I_t$$

Multiplicative model:

$$x_t = T_t * C_t * S_t * I_t$$

where

x_t represents the observed value of the time series at time t

Given some *a priori* knowledge about the cyclical factors affecting the series (e.g., business cycles), the estimates for the different components can be used to compute forecasts for future observations. (However, the *Exponential smoothing* method, which can also incorporate seasonality and trend components, is the preferred technique for forecasting purposes.)

Additive and multiplicative seasonality.

Consider the difference between an additive and multiplicative seasonal component in the following example: The annual sales of toys will probably peak in the months of November and December, and perhaps during the summer (with a much smaller peak) when children are on their summer break. This seasonal pattern will likely repeat every year. Seasonal components can be additive or multiplicative in nature.

For example, during the month of December the sales for a particular toy may increase by 3 million dollars every year. Thus, you could *add* to your forecasts for every December the amount of 3 million to account for this seasonal fluctuation. In this case, the seasonality is *additive*.

Alternatively, during the month of December the sales for a particular toy may increase by 40%, that is, increase by a *factor* of 1.4. Thus, when the sales for the toy are generally weak, then the absolute (dollar) increase in sales during December will be relatively weak (but the percentage will be constant); if the sales of the toy are strong, then the absolute (dollar) increase in sales will be proportionately greater. Again, in this case the sales increase by a certain *factor*, and the seasonal component is thus *multiplicative* in nature (i.e., the multiplicative seasonal component in this case would be 1.4).

In plots of series, the distinguishing characteristic between these two types of seasonal components is that in the additive case, the series shows steady seasonal fluctuations, regardless of the overall level of the series; in the multiplicative case, the size of the seasonal fluctuations vary, depending on the overall level of the series.

Additive and multiplicative trend-cycle. The previous example can be extended to illustrate the additive and multiplicative trend-cycle components. In terms of the toy example, a "fashion" *trend* may produce a steady increase in sales (e.g., a trend towards more educational toys in general); as with the seasonal component, this trend may be additive (sales increase by 3 million dollars per year) or multiplicative (sales increase by 30%, or by a factor of 1.3, annually) in nature. In addition, cyclical components may impact sales. To reiterate, a cyclical component is different from a seasonal component in that it usually is of longer duration, and that it occurs at irregular intervals.

For example, a particular toy may be particularly "hot" during a summer season (e.g., a particular doll which is tied to the release of a major children's movie, and is promoted with extensive advertising). Again such a cyclical component can effect sales in an additive manner or multiplicative manner.

The Census II Method

The basic method for seasonal decomposition and adjustment outlined above can be refined in several ways. In fact, unlike many other time-series modeling techniques (e.g., *ARIMA*) which are grounded in some theoretical model of an underlying process, the X-11 variant of the Census II method simply contains many *ad hoc* features and refinements, that over the years have proven to provide excellent estimates for many real-world applications (see Burman, 1979, Kendall and Ord, 1990, Makridakis and Wheelwright, 1989; Wallis, 1974). Some of the major refinements are listed below.

Trading-day adjustment. Different months have different numbers of days, and different numbers of trading-days (i.e., Mondays, Tuesdays, etc.). When analyzing, for example, monthly revenue figures for an amusement park, the fluctuation in the different numbers of Saturdays and Sundays (peak days) in the different months will surely contribute significantly to the variability in monthly revenues. The X-11 variant of the Census II method allows the user to test whether such trading-day variability exists in the series, and, if so, to adjust the series accordingly.

Extreme values. Most real-world time series contain outliers, that is, extreme fluctuations due to rare events. For example, a strike may affect production in a particular month of one year. Such extreme outliers may bias the estimates of the seasonal and trend components.

The X-11 procedure includes provisions to deal with extreme values through the use of "statistical control principles," that is, values that are above or below a certain range (expressed in terms of multiples of *sigma*, the standard deviation) can be modified or dropped before final estimates for the seasonality are computed.

Multiple refinements. The refinement for outliers, extreme values, and different numbers of trading-days can be applied more than once, in order to obtain successively improved estimates of the components. The X-11 method applies a series of successive refinements of the estimates to arrive at the final trend-cycle, seasonal, and irregular components, and the seasonally adjusted series.

Tests and summary statistics. In addition to estimating the major components of the series, various summary statistics can be computed. For example, analysis of variance tables can be prepared to test the significance of seasonal variability and trading-day variability (see above) in the series; the

8. TIME SERIES ANALYSIS - INTRODUCTORY OVERVIEW

X-11 procedure will also compute the percentage change from month to month in the random and trend-cycle components.

As the duration or span in terms of months (or quarters for quarterly X-11) increases, the change in the trend-cycle component will likely also increase, while the change in the random component should remain about the same. The width of the average span at which the changes in the random component are about equal to the changes in the trend-cycle component is called the *month (quarter) for cyclical dominance*, or MCD (QCD) for short.

For example, if the MCD is equal to 2, then one can infer that over a 2 month span the trend-cycle will dominate the fluctuations of the irregular (random) component. These and various other results are discussed in greater detail below.

Result Tables Computed by the X-11 Method

The computations performed by the X-11 procedure are best discussed in the context of the results tables that are reported. The adjustment process is divided into seven major steps, which are customarily labeled with consecutive letters *A* through *G*.

A Prior adjustment (monthly seasonal adjustment only). Before any seasonal adjustment is performed on the monthly time series, various prior user-defined adjustments can be incorporated. The user can specify a second series that contains prior adjustment factors; the values in that series will either be subtracted (additive model) from the original series, or the original series will be divided by these values (multiplicative model). For multiplicative models, user-specified trading-day adjustment weights can also be specified. These weights will be used to adjust the monthly observations depending on the number of respective trading-days represented by the observation.

B Preliminary estimation of trading-day variation (monthly X-11) and weights. Next, preliminary trading-day adjustment factors (monthly X-11 only) and weights for reducing the effect of extreme observations are computed.

C Final estimation of trading-day variation and irregular weights (monthly X-11). The adjustments and weights computed in *B* above are then used to derive improved trend-cycle and seasonal estimates. These improved estimates are used to compute the final trading-day factors (monthly X-11 only) and weights.

D Final estimation of seasonal factors, trend-cycle, irregular, and seasonally adjusted series. The final trading-day factors and weights computed in *C* above are used to compute the final estimates of the components.

E Modified original, seasonally adjusted, and irregular series. The original and final seasonally adjusted series and the irregular component are modified for extremes. The resulting modified series allow the user to examine the stability of the seasonal adjustment.

F Month (quarter) for cyclical dominance (MCD, QCD), moving average, and summary measures. In this part of the computations, various summary measures (see below) are computed to allow the user to examine the relative importance of the different components, the average fluctuation from month-to-month (quarter-to-quarter), the average number of consecutive changes in the same direction (average number of runs), etc.

G Charts. Finally, the program will compute various charts (graphs) to summarize the results. For example, the final seasonally adjusted series

will be plotted, in chronological order (see below).

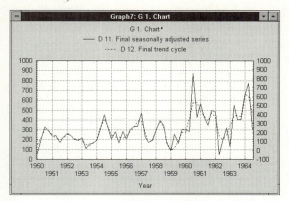

This series can also be plotted by month or quarter (see below).

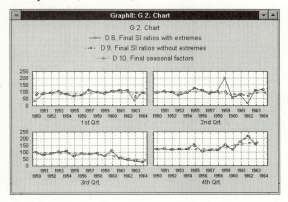

Specific Description of all Result Tables Computed by the X-11 Method

In each part *A* through *G* of the analysis (see above), different result tables are computed. Customarily, these tables are numbered and also identified by a letter to indicate the respective part of the analysis. For example, table *B 11* shows the initial seasonally adjusted series; *C 11* is the refined seasonally adjusted series, and *D 11* is the final seasonally adjusted series. Shown below is a list of all available tables. Those tables identified by an asterisk (*) are not available (applicable) when analyzing quarterly series. Also, for quarterly adjustment, some of the computations outlined below are slightly different. For example instead of a 12-term (monthly) moving average, a 4-term (quarterly) moving average is applied to compute the seasonal factors; the initial trend-cycle estimate is computed via a centered 4-term moving average, and the final trend-cycle estimate in each part is computed by a 5-term Henderson average.

Following the convention of the US Bureau of the Census version of the X-11 method, the *STATISTICA* implementation offers three levels of printout detail: *Standard* (17 to 27 tables), *Long* (27 to 39 tables), and *Full* (44 to 59 tables). An option to only produce selected tables (Scrollsheets) is also provided. In the description of each table below, the letters *S*, *L*, and *F* are used next to each title to indicate which tables will be displayed and/or printed at the respective setting of the output option. (For the charts, two levels of detail are available: *Standard* and *All*.)

*A 1. **Original series.** (S) This table will show the original series, prior to any initial user-defined or trading-day adjustment. Note that for quarterly series, no prior adjustment factors can be specified, and the original series will be shown as table *B 1* (see below).

*A 2 **Prior monthly adjustment factors.** (S) For monthly series, the user may specify a second series that contains prior monthly adjustment factors, for example, in order to adjust for an unusual holiday etc. The factors specified here will be subtracted from the original series for additive models or will be used to divide the original series if multiplicative seasonal adjustment was requested (thus, the values in this series must be unequal to zero in that case).

*A 3. **Original series adjusted by prior monthly adjustment factors.** (S) The factors specified in *A 2* will be subtracted from the original series (additive adjustment) or they will be used to divide the values in the original series (multiplicative adjustment). The resulting adjusted series is shown in this table.

*A 4. Prior trading day adjustment factors. *(S)* This table is only available (applicable) when prior trading-day adjustment factors were specified (this option is only available for the multiplicative model). The user may specify a weight for each day (Monday through Friday); those weights are then proportionately adjusted so that they add to 7. The series (*A1* or *A3*) is then divided by monthly calendar factors that are computed based on the number of the respective days in the respective month. Note that by default, the calendar factors are also adjusted for different lengths of different months; however, the length of month variability can also be included in the calendar factors (in which case a constant length of month of 30.4375 is used).

B 1. Prior adjusted series or original series. *(S)* This table shows the original series, or the initial adjusted series, depending on whether or not prior monthly adjustment factors and/or trading day adjustment factors were specified (for quarterly X-11, *B 1* is always the original series).

B 2. Trend cycle. *(L)* The initial trend-cycle estimate is computed as a centered 12-term moving average of *B 1*.

B 3. Unmodified S-I differences or ratios. *(F)* An initial estimate of the combined irregular and seasonal component is obtained by subtracting *B 2* from *B 1* (additive model) or dividing *B 1* by *B 2* (multiplicative model).

B 4. Replacement values for extreme S-I differences (ratios). *(F)* First a preliminary estimate of the seasonal component is computed by applying a weighted 5-term moving average separately to the *B 3* values for each month. Then a centered 12-term moving average of the preliminary factors for the entire series is computed, and the resulting values are adjusted to sum to zero (additive model) or 12.0 (multiplicative model) within each year. Next an initial estimate of the irregular component is obtained by subtracting from the S-I differences (additive model) or dividing the S-I ratios by the initial estimate for the seasonal component. For the resulting initial estimate of the irregular component, a 5-year sliding standard deviation (σ, sigma) is computed, and extreme values in the central year that are beyond $2.5*\sigma$ are removed. The 5-year sliding σ is then recomputed and the process repeated; however, this time a zero weight is assigned to irregular values beyond $2.5*\sigma$, a full weight is assigned to values within $1.5*\sigma$, and linearly graduated weights between zero and one are assigned to values between 1.5 and $2.5*\sigma$. Values receiving less than full weights are then recomputed as the average of the respective value times its weight and the nearest two full-weight values preceding and following the respective value in that month. Table *B 4* shows the final replaced (recomputed) values, and the sliding 5-year σ's.

B 5. Seasonal factors. *(F)* The extreme values in the *B 3* series are replaced by the values shown in *B 4*. From this series, preliminary seasonal factors are derived by applying a 5-term moving average to each month separately; then a 12-term moving average is computed for the entire series, and the resulting values adjusted to sum to zero (additive model) or 12.0 (multiplicative model) within each year.

B 6. Seasonally adjusted series. *(F)* The preliminary seasonally adjusted series is obtained by subtracting from *B 1* (additive model) or dividing *B 1* (multiplicative model) by the seasonal factors in *B 5*.

B 7. Trend cycle. *(L)* The seasonally adjusted series (*B 6*) is then smoothed via a variable moving average procedure (see Shiskin, Young, and Musgrave, 1967, for details). Optionally, extremes can be removed from the smoothed series by a process analogous to that described under *B 4* above. In general, the so-called Henderson curve moving average is applied, which is a weighted moving average with the magnitudes of the weights following a bell-shaped curve (see, for example, Makridakis and Wheelwright, 1978, or Shiskin, Young, and Musgrave, 1967). The choice of the appropriate length of the moving average is an important issue in the seasonal decomposition (i.e., the computation of the trend-cycle component). The general idea is to choose a longer moving average when there is a lot of random fluctuation in the data relative to the trend-cycle component, and to choose a shorter moving average when there is only relative little random fluctuation. By default the program will select a moving average transformation automatically. Specifically, first a preliminary 13-term Henderson (weighted) moving average of the seasonally adjusted series is computed (without extending to the ends of the series). A preliminary estimate of the irregular component is then computed by subtracting this series from (additive model) or dividing it into (multiplicative model) the seasonally adjusted series. Next, the average month-to-month difference (percent change) without regard to sign is computed for both the estimated irregular and trend-cycle components. The ratio of the average month-to-month differences (percent changes) in the two series reflects the relative importance of the irregular variations relative to the movements in the trend-cycle component. Depending on the value of this ratio, either a 9-term Henderson moving average is selected (if the ratio is between 0.0 and .99), a 13-term Henderson moving average is selected (if the ratio is between 1.0 and 3.49) or a 23-term Henderson moving average is selected (if the ratio is greater than 3.5).

B 8. Unmodified S-I differences (ratios). *(F)* This table is the same as *B 3* except that it is based on the trend-cycle values computed in *B 7*.

B 9. Replacement values for extreme S-I differences (ratios). *(F)* This table is the same as *B 4* except that the differences (ratios) in *B 8* are used to which a 7 term moving average is applied (to estimate the seasonal factors).

B 10. Seasonal factors. *(L)* After replacing extreme values by the corresponding *B 9* values, a 7-term weighted moving average is applied to the S-I differences (ratios) in *B 8*. The resulting estimate of the seasonal factors is then adjusted so that the sum for each year is equal to zero (additive model) or 12.0 (multiplicative model).

B 11. Seasonally adjusted series. *(F)* This table is the same as *B 6*, except that the seasonal factors in *B 10* are used.

B 12. (not used).

B 13. Irregular series. *(L)* The trend-cycle estimates in *B 7* are subtracted from the seasonally adjusted series in *B 11* (additive model), or the *B 7* values are used to divide the series in *B 11*. The resulting series is an improved estimate of the irregular series.

Tables B 14 through B 16, B 18, and B 19: Adjustment for trading-day variation. These tables are only available when analyzing monthly series. Different months contain different numbers of days of the week (i.e., Mondays, Tuesdays, etc.). In some series, the variation in the different numbers of trading days may contribute significantly to monthly fluctuations (e.g., the monthly revenues of an amusement park will be greatly influenced by the number of Saturdays/Sundays in each month). The user can specify initial weights for each trading day (see *A 4* above), and/or these weights can be estimated from the data (the user can also choose to apply those weights conditionally, i.e., only if they explain a significant proportion of variance).

***B 14. Extreme irregular values excluded from trading-day regression.** *(L)* The months in the series are sorted into different groups, depending on the particular day when the month begins (30-day, 31-day months, and Februarys are treated separately). Then, extreme values (beyond 2.5*σ; different σ values can also be specified) are identified within each type of month in a two-step procedure. The final extreme values that will be excluded are shown in this table.

***B 15. Preliminary trading-day regression.** *(L)* After removing the *B 14* extreme values from *B 13*, least squares estimates for the seven daily weights are computed.

***B 16. Trading-day adjustment factors derived from regression coefficients.** *(F)* From the trading-day regression weights, monthly adjustment factors are computed based on the number of particular trading days (i.e., Mondays, Tuesdays, etc.) in the respective months. These factors are printed in this table, and are then used to adjust (i.e., subtracted from or divided into) the *B 13* irregular series for trading-day variation.

B 17. Preliminary weights for irregular component. *(L)* The estimates of the irregular component (in *B 13* or adjusted by *B 16*, depending on whether or not a trading-day adjustment was performed) are further refined by computing graduated weights for extreme values, depending on their relative (in terms of a sliding 5-year σ) distance from 0. Specifically, a process analogous to that described in *B 4* above is used. This table (*B 17*) contains the resulting adjustment factors.

***B 18. Trading-day factors derived from combined daily weights.** *(F)* This table contains the final trading day adjustment factors, computed from the least squares trading-day weights in *B 15* and/or the prior trading-day weights in *A 4*.

***B 19. Original series adjusted for trading-day and prior variation.** *(F)* The values in *B 18* are used to adjust the original (adjusted) series (in *A 1, A 3*, or *B 1*, depending on whether or not prior adjustment factors were specified). Specifically, the values in *B 18* are subtracted from (additive model) or divided into the original series.

C 1. Original series modified by preliminary weights and adjusted for trading-day and prior variation. *(L)* The series in *B 19* (or *B 1* if no trading-day adjustment was requested) is adjusted for extreme values by the weights computed in *B 17*. The resulting modified series is shown in this table.

C 2. Trend-cycle. *(F)* An estimate of the combined trend-cycle component is computed from *C 1* by applying a centered 12-term moving average.

C 3. (not used).

C 4. Modified S-I differences (ratios). *(F)* To obtain the refined S-I differences (ratios), the values in *C 2* are subtracted from (additive model) or divided into (multiplicative model) the modified series in *C 1*.

C 5. Seasonal factors. *(F)* These values are the same as those in *B 5*, except that the *C 4* differences (ratios) are used.

C 6. Seasonally adjusted series. *(F)* The preliminary seasonally adjusted series is computed by subtracting *C 5* from (or dividing *C 5* into) *C 1*.

C 7. Trend cycle. *(L)* The seasonally adjusted series (*C 6*) is smoothed via a variable moving average procedure (the same procedure used for *B 7*, see also Shiskin, Young, and Musgrave, 1967, for details) to derive the preliminary estimate of the trend-cycle component.

C 8. (not used).

C 9. Modified S-I differences (ratios). *(F)* The modified S-I differences (ratios) are computed by subtracting *C 7* from (additive model) or dividing *C 7* into (multiplicative model) the *C 1* series.

C 10. Seasonal factors. *(L)* The seasonal factors are computed analogously to *B 10*, but based on the *C 9* S-I differences (ratios).

C 11. Seasonally adjusted series. *(F)* The refined seasonally adjusted series is computed by subtracting from *B 1* (additive model) or dividing *B 1* by (multiplicative model) the values in *C 10*.

C 12. (not used).

C 13. Irregular series. *(S)* The refined estimate of the irregular (random) component is computed by subtracting from *C 11* (additive model) or dividing *C 11* by (multiplicative model) the values in *C 7*.

Tables C 14 through C 16, C 18, and C 19: Adjustment for trading-day variation. These tables are only available when analyzing monthly series, and when adjustment for trading day variation is requested. In that case, the trading day adjustment factors are computed from the refined adjusted series, analogously to the adjustment performed in part *B* (*B 14* through *B 16*, *B 18* and *B 19*, see above).

*C 14. **Extreme irregular values excluded from trading-day regression.** *(S)* This table is analogous to table *B 14*, and it shows the extreme irregular values (usually beyond $2.5*\sigma$) after reapplying the trading-day routine (based on the monthly trading-day factors shown in *B 16*).

*C 15. **Final trading-day regression.** *(S)* This table is the same as *B 15*, except that the computations are based on the values from table *C 13*.

*C 16. **Final trading-day adjustment factors derived from regression coefficients.** *(S)* This table is analogous to *B 16*, except that the factors are subtracted from (additive case) or divided into (multiplicative case) the values from table *C 13*.

C 17. Final weights for irregular component. *(S)* This table is analogous to table *B 17*, except that it is computed based on the values in *C 16* (or *C 13* if no trading-day adjustment is requested).

*C 18. **Final trading-day factors derived from combined daily weights.** *(S)* This table is analogous to *B 18*, except that the final weights shown in *C 15* are used in the computations.

*C 19. **Original series adjusted for trading-day and prior variation.** *(S)* The values in *C 18* are used to adjust the original (adjusted) series (*in A 3* or *B 1*).

Specifically, the values in *C 18* are subtracted from (additive model) or divided into (multiplicative model) the original series.

D 1. **Original series modified by final weights and adjusted for trading day and prior variation.** *(L)* This table is analogous to *C 1*, except that the *C 17* weights and *C 19* adjusted series are used in the computations.

D 2. **Trend Cycle.** *(F)* A 12-term moving average of *D 1* is computed to estimate the trend-cycle component.

D 3. (not used).

D 4. **Modified S-I differences (ratios).** *(F)* The modified S-I differences (ratios) are computed by subtracting *D 2* from (additive model) or dividing *D 2* into (multiplicative model) the values in *D 1*.

D 5. **Seasonal factors.** *(F)* This table is computed analogously to *B 5*, except that the computations are based on the values in *D 4*.

D 6. **Seasonally adjusted series.** *(F)* The values in this table are computed by subtracting *D 5* from *D 1* (additive model) or dividing *D 1* by *D 5* (multiplicative model).

D 7. **Trend cycle.** *(L)* The values in this table are computed analogously to those in *B 7*, except that the computations are based on the values in *D 6*.

D 8. **Final unmodified S-I differences (ratios).** *(S)* The values in the *D 7* series are subtracted from (additive model) or divided into (multiplicative model) the values in *C 19* (or *B 1* if no adjustment for trading day variation is applied). Then an analysis of variance by month (or quarter) is performed on this series, in order to test for the presence of stable significant seasonality.

D 9. **Final replacement values for extreme S-I differences (ratios).** *(S)* The values in *D 7* are subtracted from (additive model) or divided into (multiplicative model) *D 1*. Values that are not identical to the corresponding entries in *D 8* are then reported. Also, for each month, the year-to-year difference (additive model) or percent change (multiplicative model) in the estimates of the irregular and the seasonal components and their ratio (called *MSR*, moving seasonality ratio) are computed. The MSR may be useful in order to determine the amount of moving seasonality present in each month.

D 10. **Final seasonal factors.** *(S)* This table is computed analogously to the values in *B 10*, except that it is computed based on the values reported in *D 8* and *D 9*.

D 11. **Final seasonally adjusted series.** *(S)* The final seasonally adjusted series is computed by subtracting *D 10* from *C 19* (additive model) or dividing *C 19* by *D 10*.

D 12. **Final trend-cycle.** *(S)* These values are computed by subtracting *D 10* from *D 1* (additive model) or by dividing *D 1* by *D 10* (multiplicative model).

D 13. **Final irregular.** *(S)* These values are computed by subtracting *D 12* from *D 11* (additive model) or by dividing *D 11* by *D 12* (multiplicative model).

E 1. **Modified original series.** *(S)* The values in this table are computed by replacing in the original series, extreme values (identified by a zero weight in *C 17*) by the values predicted from the final trend-cycle, seasonal, trading-day (if applicable), and prior adjustment (if applicable) components.

E 2. **Modified seasonally adjusted series.** *(S)* These values are computed by replacing

8. TIME SERIES ANALYSIS - INTRODUCTORY OVERVIEW

in the final seasonally adjusted series (*D 11*) extreme values (identified by a zero weight in *C 17*) with the *D 12* final trend-cycle values.

E 3. **Modified irregular series.** *(S)* The values in this table are computed by replacing the values in *D 13* with zero (additive model) or 1.0 (multiplicative model) if they were identified as extremes (i.e., assigned zero weight) in *C 17*.

E 4. **Differences (ratios) of annual totals.** *(S)* These values are computed as the differences (additive model) or ratios (multiplicative model) of the annual totals of (a) the original series *B 1* and the final seasonally adjusted series *D 11*, and (b) the modified original series *E 1* and the modified seasonally adjusted series *E 2*.

E 5. **Differences (percent changes) in original series.** *(S)* The values in this table are computed as the month-to-month (quarter-to-quarter) differences (additive model) or percent changes (multiplicative model) in *B 1*.

E 6. **Differences (percent changes) in final seasonally adjusted series.** *(S)* These values are the month-to-month (quarter-to-quarter) differences (additive model) or percent changes (multiplicative model) in *D 11*.

F 1. **MCD (QCD) moving average.** *(S)* The values in this series are computed by applying an unweighted moving average to the final seasonally adjusted series (*D 11*). The width of the smoothing window is determined by the *month (quarter) for cyclical dominance*, or MCD (QCD) for short. The MCD (QCD) is computed as the average span at which the changes in the random component are equal to the changes in the trend-cycle component.

F 2. **Summary measures.** *(S)* Several final summary tables are computed:

(1) The average differences (additive model) or percent changes (multiplicative model) are computed without regard to sign across spans 1, 2, 3 ..., 12 months (or four quarters) for the following series: Original series *A 1* (*B 1*), final seasonally adjusted series (*D 11*), final irregular series (*D 13*), final trend-cycle (*D 12*), final seasonal factors (*D 10*), final prior monthly adjustment factors (*A 2,* monthly X-11 only), final trading-day adjustment factors (*C 18,* monthly X-11 only), modified original series (*E 1*), modified seasonally adjusted series (*E 2*), modified irregular series (*E 3*).

(2) Next a table of relative contributions of the different components to the differences (additive model) or percent changes (multiplicative model) in the original series is computed.

(3) The next table reports the average duration of run (the average number of consecutive monthly changes in the same direction; "no change" is counted as a change in the same direction) for the following series: Final seasonally adjusted series (*D 11*), final irregular series (*D 13*), final trend-cycle (*D 12*), and the MCD (QCD) moving average (*F 1*).

(4) Finally, the means and standard deviations of differences (additive model) or percent changes (multiplicative model) are computed across different spans for each of the series mentioned above.

G1. **Chart.** *(S)* This line graph will show the final seasonally adjusted series and final trend-cycle components (*D 11* and *D 12,* respectively).

8. TIME SERIES ANALYSIS - INTRODUCTORY OVERVIEW

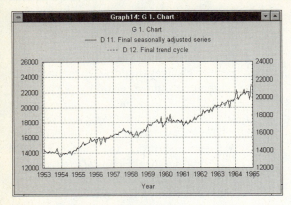

G2. Chart. *(S)* This line graph will show the final S-I differences (additive model) or ratios (multiplicative model) *with* the extremes, the final S-I differences (ratios) *without* extremes, and the final seasonal factors (i.e., *D 8*, *D 9*, and *D 10*, respectively), categorized by month (X-11 monthly) or quarter (X-11 quarterly).

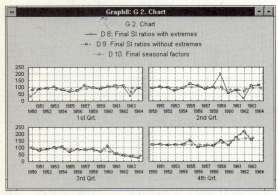

G3. Chart. *(A)* This plot shows the same values as *G 2*; however, this line plot shows those values in chronological order.

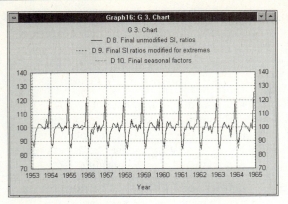

G 4. Chart. *(A)* This is a line graph of the final irregular and final modified irregular series (*D 13* and *E 3*, respectively).

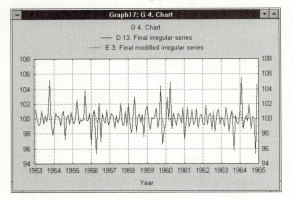

Distributed Lags Analysis

General Purpose

Distributed lags analysis is a specialized technique for examining the relationships between variables that involve some delay. For example, suppose that you are a manufacturer of computer software, and you want to determine the relationship between the number of inquiries that are received and the number of orders that are placed by your customers. You could record those numbers monthly for a one year period and then correlate the two variables.

However, obviously inquiries will precede actual orders, and one can expect that the number of orders will follow the number of inquiries with some delay. Put another way, there will be a (time) *lagged* correlation between the number of inquiries and the number of orders that are received.

Time-lagged correlations are particularly common in econometrics. For example, the benefits of investments in new machinery usually only become evident after some time. Higher income will change people's choice of rental apartments; however, this relationship will be lagged because it will take some time for people to terminate their current leases, find new apartments, and move. In general, the relationship between capital appropriations and capital expenditures will be lagged, because it will require some time before investment decisions are actually acted upon. In all of these cases, there is an independent or *explanatory* variable that affects the *dependent* variables with some lag. The distributed lags method allows you to investigate those lags.

Detailed discussions of distributed lags correlation can be found in most econometrics textbooks, for example, in Judge, Griffith, Hill, Luetkepohl, and Lee (1985), Maddala (1977), and Fomby, Hill, and Johnson (1984). In the following paragraphs, a brief description of these methods will be presented. It is assumed that you are familiar with the concept of correlation (see *Basic Statistics and Tables*, Volume I) and the basic ideas of multiple regression (see *Multiple Regression*, Volume I).

General Model

Suppose you have a dependent variable *y* and an independent or explanatory variable *x* which are both measured repeatedly over time. In some textbooks, the dependent variable is also referred to as the *endogenous* variable, and the independent or explanatory variable the *exogenous* variable. The simplest way to describe the relationship between the two would be in a simple linear relationship:

$$y_t = \Sigma \beta_i * x_{t-i}$$

In this equation, the value of the dependent variable at time *t* is expressed as a linear function of *x* measured at times *t, t-1, t-2*, etc. Thus, the dependent variable is a linear function of *x*, and *x* is *lagged* by *1, 2*, etc. time periods. The *beta* coefficients (β_i) can be considered slope parameters in this equation. You may recognize this equation as a special case of the general linear regression equation (see *Multiple Regression*, Volume I). If the coefficients for the lagged time periods are statistically significant, then you can conclude that the *y* variable is predicted (or explained) with the respective lag.

Almon Distributed Lag

A common problem that often arises when computing the coefficients for the multiple linear regression model shown above is that the values of adjacent (in time) values in the *x* variable are highly correlated. In extreme cases, their independent contributions to the prediction of *y* may become so redundant that the correlation matrix of measures can no longer be inverted, and thus, the *beta* coefficients cannot be computed. In less extreme cases, the computation of the *beta* coefficients and their standard errors can become very imprecise, due to round-off error. In the context of multiple regression, this general computational problem is discussed as the *multicollinearity* or *matrix ill-conditioning* issue.

Almon (1965) proposed a procedure that will reduce the multicollinearity in this case. Specifically, suppose you express each coefficient in the linear regression equation in the following manner:

$$\beta_i = \alpha_0 + \alpha_1 * i + \ldots + \alpha_q * i^q$$

Almon could show that in many cases it is easier (i.e., it avoids the multicollinearity problem) to estimate the *alpha* values than the *beta* coefficients directly. Note that with this method, the precision of

8. TIME SERIES ANALYSIS - INTRODUCTORY OVERVIEW

the *beta* coefficient estimates is dependent on the degree or order of the *polynomial approximation*.

Misspecifications. A general problem with this technique is that, of course, the lag length and correct polynomial degree are not known *a priori*. The effects of misspecifications of these parameters are potentially serious (in terms of biased estimation). This issue is discussed in greater detail in Frost (1975), Schmidt and Waud (1973), Schmidt and Sickles (1975), and Trivedi and Pagan (1979).

Spectrum and Cross-Spectrum (Fourier) Analysis

Spectrum analysis is concerned with the exploration of cyclical patterns of data. The purpose of the analysis is to decompose a complex time series with cyclical components into a few underlying sinusoidal (sine and cosine) functions of particular wavelengths. The term "spectrum" provides an appropriate metaphor for the nature of this analysis: Suppose you study a beam of white sun light, which at first looks like a random (white noise) accumulation of light of different wavelengths. However, when put through a prism, you can separate the different wave lengths or cyclical components that make up white sun light. In fact, via this technique you can now identify and distinguish between different sources of light. Thus, by identifying the important underlying cyclical components, you have learned something about the phenomenon of interest. In essence, performing spectrum analysis on a time series is like putting the series through a prism in order to identify the wave lengths and importance of underlying cyclical components. As a result of a successful analysis, one might uncover just a few recurring cycles of different lengths in the time series of interest, which at first looked more or less like random noise.

A much cited example for spectrum analysis is the cyclical nature of sun spot activity (e.g., see Bloomfield, 1976, or Shumway, 1988). It turns out that sun spot activity varies over 11 year cycles. Other examples of celestial phenomena, weather patterns, fluctuations in commodity prices, economic activity, etc. are also often used in the literature to demonstrate this technique.

To contrast this technique with ARIMA or exponential smoothing, the purpose of spectrum analysis is to identify the seasonal fluctuations of different lengths, while in the former types of analysis, the length of the seasonal component is usually known (or guessed) *a priori* and then included in some theoretical model of moving averages or autocorrelations.

On the following pages, the basic notation and principles of spectrum analysis and cross-spectrum analysis will be reviewed. The classic text on spectrum analysis is Bloomfield (1976); however, other detailed discussions can be found in Jenkins and Watts (1968), Brillinger (1975), Brigham (1974), Elliott and Rao (1982), Priestley (1981), Shumway (1988), or Wei (1989).

Frequency and Period

The "wave length" of a sine or cosine function is typically expressed in terms of the number of cycles per unit time (*Frequency*), often denoted by the Greek letter *nu* (ν; some text books also use *f*). For example, the number of letters handled in a post office may show 12 cycles per year: On the first of every month a large amount of mail is sent (many bills come due on the first of the month); then the amount of mail decreases in the middle of the month; and then it increases again towards the end of the month. Therefore, every month the fluctuation in the amount of mail handled by the post office will go through a full cycle. Thus, if the unit of analysis is one year, then ν would be equal to 12, as there would be 12 cycles per year. Of course, there will likely be other cycles with different frequencies. For example, there might be annual cycles ($\nu=1$), and perhaps weekly cycles ($\nu=52$ weeks per year).

The *period T* of a sine or cosine function is defined as the length of time required for one full cycle. Thus, it is the reciprocal of the frequency, or: $T = 1/\nu$. To return to the mail example in the previous paragraph, the monthly cycle, expressed in yearly terms, would be equal to $1/12 = 0.0833$. Put into words, there is a period in the series of length 0.0833 years. In the *Time Series* module the frequency is computed in terms of cycles per observations, since each observation represents one unit of time.

The General Structural Model

As mentioned before, the purpose of spectrum analysis is to decompose the original series into underlying sine and cosine functions of different frequencies, in order to determine those that appear particularly strong or important. One way to do so would be to cast the issue as a linear *Multiple Regression* problem (see Volume I), where the dependent variable is the observed time series, and the independent variables are the sine functions of all possible (discrete) frequencies. Such a linear multiple regression model may be written as:

```
For k = 1 to q:
    x_t = a_0 + Σ[a_k*cos(λ*t)+b_k*sin(λ*t)]
```

Following the common notation from classical harmonic analysis, in this equation λ (*lambda*) is the frequency expressed in terms of *radians* per unit time, that is: $\lambda = 2*\pi*\nu$, where π is the constant $pi = 3.1416$. What is important here is to recognize that the computational problem of fitting sine and cosine functions of different lengths to the data can be cast in terms of multiple linear regression.

Note that the cosine parameters a_k and sine parameters b_k are regression coefficients that tell you the degree to which the respective functions are correlated with the data. Overall there are q different sine and cosine functions; intuitively, it should be clear that you cannot have more sine and cosine functions than there are data points in the series. Without going into detail, if there are n data points, then there will be $n/2+1$ cosine functions and $n/2-1$ sine functions. In other words, there will be as many different sinusoidal waves as there are data points, and you will be able to completely reproduce the series from the underlying functions.

Note that if the number of cases in the series is odd, then the last data point will usually be ignored; in order for a sinusoidal function to be identified, you need at least two points: the high peak and the low peak.

To summarize, spectrum analysis will identify the correlation of sine and cosine functions of different frequency with the observed data. If a large correlation (sine or cosine coefficient) is identified, one can conclude that there is a strong periodicity of the respective frequency (or period) in the data.

Complex numbers (real and imaginary numbers). In many text books on spectrum analysis, the structural model shown above is presented in terms of complex numbers; that is, the parameter estimation process is described in terms of the Fourier transform of a series into real and imaginary parts. Complex numbers are the superset that includes all real and imaginary numbers.

Imaginary numbers, by definition, are numbers that are multiplied by the constant i, where i is defined as the square root of -1. Obviously, the square root of -1 does not exist, hence the term *imaginary* number; however, meaningful arithmetic operations on imaginary numbers can still be performed [e.g., $(i*2)**2 = -4$]. It is useful to think of real and imaginary numbers as forming a two dimensional plane, where the horizontal or *X*-axis represents all real numbers, and the vertical or *Y*-axis represents all imaginary numbers. Complex numbers can then be represented as points in the two-dimensional plane. For example, the complex number $3+i*2$ can be represented by a point with coordinates {3,2} in this plane. One can also think of complex numbers as angles; for example, one can connect the point representing a complex number in the plane with the origin (complex number 0+i*0), and measure the angle of that vector to the horizontal line. Thus, intuitively one can see how the spectrum decomposition formula shown above, consisting of sine and cosine functions, can be rewritten in terms of operations on complex numbers. In fact, in this manner the mathematical discussion and required computations are often more elegant and easier to perform, which is why many text books prefer the presentation of spectrum analysis in terms of complex numbers.

A Simple Example

Shumway (1988) presents a simple example to clarify the underlying "mechanics" of spectrum analysis. For this example, create a series with 16 cases following the equation shown below, and then see how you may "extract" the information that was put in it. First, create a variable and define it as:

```
x = 1 * cos[2*pi*.0625*(v0-1)] +
    .75 * sin[2*pi*.2*(v0-1)]
```

You can produce this variable by typing in the formula as the long variable label (in the *Spreadsheet Edit (Current) Specs* dialog; see *General Conventions*, Volume I, Chapter 1). This variable is made up of two underlying periodicities: The first at the frequency of v=.0625 (or period 1/v=16; one observation completes 1/16'th of a full cycle, and a full cycle is completed every 16 observations) and the second at the frequency of v=.2 (or period of 5). The cosine coefficient (1.0) is larger than the sine coefficient (.75). The spectrum analysis summary results Scrollsheet computed by the *Time Series* module is shown below.

Now review the columns of this Scrollsheet. Clearly, the largest cosine coefficient can be found for the .0625 frequency.

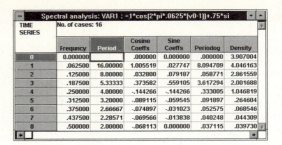

A smaller sine coefficient can be found at frequency = .1875. Thus, the two sine/cosine frequencies which were "inserted" into the example data file are reflected in the Scrollsheet.

Periodogram

The sine and cosine functions are mutually independent (or orthogonal); thus you may sum the squared coefficients for each frequency to obtain the *periodogram*. Specifically, the periodogram values in the above Scrollsheet are computed as:

$$P_k = \text{sine coeff.}_k^2 * \text{cosine coeff.}_k^2 * n/2$$

where P_k is the periodogram value at frequency v_k and n is the overall length of the series. The periodogram values can be interpreted in terms of variance (sums of squares) of the data at the respective frequency or period. Customarily, the periodogram values are plotted against the frequencies or periods, and many such plots are available in the *Time Series* module.

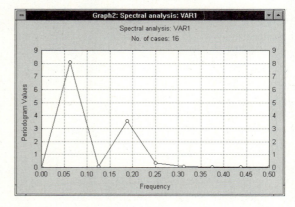

The Problem of Leakage

In the example above, a sine function with a frequency of 0.2 was "inserted" into the series. However, because of the length of the series (16), none of the frequencies reported in the Scrollsheet exactly "hits" on that frequency. In practice, what often happens in those cases is that the respective frequency will "leak" into adjacent frequencies.

For example, one may find large periodogram values for two adjacent frequencies, when, in fact, there is only one strong underlying sine or cosine function at a frequency that falls in-between those implied by the length of the series.

There are three ways in which one can approach the problem of leakage:

(1) By padding the series one may apply a finer frequency "roster" to the data,

(2) By *tapering* the series prior to the analysis, one may reduce leakage, or

(3) By smoothing the periodogram one, may identify the general frequency "regions" or (*spectral densities*) that significantly contribute to the cyclical behavior of the series.

Padding the Time Series

Because the frequency values are computed as *n/t* (the number of units of times), one may simply *pad* the series with a constant (e.g., zeros) and thereby introduce smaller increments in the frequency values. In a sense, padding allows one to apply a finer roster to the data. In fact, if you padded the data file described in the example above with ten zeros, the results would not change; that is, the largest periodogram peaks would still occur at the frequency values closest to .0625 and .2. (Padding is also often desirable for computational efficiency reasons; see below.)

Tapering

The so-called process of *split-cosine-bell tapering* is a recommended transformation of the series prior to the spectrum analysis. It usually leads to a reduction of leakage in the periodogram. The rationale for this transformation is explained in detail in Bloomfield (1976, page 80-94). In essence, a proportion (*p*) of the data at the beginning and at the end of the series is transformed via multiplication by the weights:

$w_t = 0.5*\{1-\cos[\pi*(t-0.5)/m]\}$
(for $t = 0$ to $m-1$)

$w_t = 0.5*\{1-\cos[\pi*(n-t+0.5)/m]\}$
(for $t = n-m$ to $n-1$)

where *m* is chosen so that 2*m/n* is equal to the proportion of data to be tapered (*p*).

Data Windows and Spectral Density Estimates

In practice, when analyzing actual data, it is usually not of crucial importance to exactly identify the frequencies for particular underlying sine or cosine functions. Rather, because the periodogram values are subject to substantial random fluctuation, one is faced with the problem of many "chaotic" periodogram spikes. In that case, one would like to find the frequencies with the greatest *spectral densities*, that is, the frequency regions, consisting of many adjacent frequencies, that contribute most to the overall periodic behavior of the series. This can be accomplished by smoothing the periodogram values via a weighted moving average transformation. Suppose the moving average window is of width *m* (which must be an odd number); the following are the most commonly used smoothers (note: *p = (m-1)/2*).

Daniell (or equal weight) window. The Daniell window (Daniell 1946) amounts to a simple (equal weight) moving average transformation of the periodogram values; that is, each spectral density estimate is computed as the mean of the $m/2$ preceding and subsequent periodogram values.

Tukey window. In the Tukey (Blackman and Tukey, 1958) or Tukey-Hanning window (named after Julius Von Hann), for each frequency, the weights for the weighted moving average of the periodogram values are computed as:

$$w_j = 0.5 + 0.5 * \cos(\pi * j/p) \quad \text{(for } j=0 \text{ to } p\text{)}$$
$$w_{-j} = w_j \quad \text{(for } j \neq 0\text{)}$$

Hamming window. In the Hamming (named after R. W. Hamming) window or Tukey-Hamming window (Blackman and Tukey, 1958), for each frequency, the weights for the weighted moving average of the periodogram values are computed as:

$$w_j = 0.54 + 0.46 * \cos(\pi * j/p) \quad \text{(for } j=0 \text{ to } p\text{)}$$
$$w_{-j} = w_j \quad \text{(for } j \neq 0\text{)}$$

Parzen window. In the Parzen window (Parzen, 1961), for each frequency, the weights for the weighted moving average of the periodogram values are computed as:

$$w_j = 1 - 6*(j/p)^2 + 6*(j/p)^3 \quad \text{(for } j=0 \text{ to } p/2\text{)}$$
$$w_j = 2*(1-j/p)^3 \quad \text{(for } j=p/2+1 \text{ to } p\text{)}$$
$$w_{-j} = w_j \quad \text{(for } j \neq 0\text{)}$$

Bartlett window. In the Bartlett window (Bartlett, 1950) the weights are computed as:

$$w_j = 1 - (j/p) \quad \text{(for } j=0 \text{ to } p\text{)}$$
$$w_{-j} = w_j \quad \text{(for } j \neq 0\text{)}$$

With the exception of the Daniell window, all weight functions will assign the greatest weight to the observation being smoothed in the center of the window and increasingly smaller weights to values that are further away from the center. (Note also that the *Time Series* module will standardize the weights so that they sum to 1.) In many cases, all of these data windows will produce very similar results; in the *Time Series* module it is very easy to experiment with different windows and different widths and to see the effect on the interpretability of results.

Preparing the Data for Analysis

Now consider a few other practical points in spectrum analysis. Usually, one wants to subtract the mean from the series and detrend the series (so that it is stationary) prior to the analysis. Otherwise the periodogram and density spectrum will mostly be "overwhelmed" by a very large value for the first cosine coefficient (for frequency 0.0). In a sense, the mean is a cycle of frequency 0 (zero) per unit time; that is, it is a constant.

Similarly, a trend is also of little interest when one wants to uncover the periodicities in the series. In fact, both of those potentially strong effects may mask the more interesting periodicities in the data, and thus both the mean and the trend (linear) should be removed from the series prior to the analysis (by default, they will be removed in the *Time Series* module). Sometimes, it is also useful to smooth the data prior to the analysis, in order to "tame" the random noise that may obscure meaningful periodic cycles in the periodogram.

Results when no Periodicity in the Series Exists

Finally, what if there are no recurring cycles in the data, that is, if each observation is completely independent of all other observations? If the distribution of the observations follows the normal distribution, such a time series is also referred to as a *white noise* series (like the white noise one hears on the radio when tuned in-between stations). A white noise input series will result in periodogram values that follow an exponential distribution (i.e, the *Chi-square* distribution with two degrees of freedom, see Bloomfield, 1976). Thus, by testing the distribution of periodogram values against the exponential distribution, one may test whether the input series is different from a white noise series.

The *Time Series* module will produce a histogram of the periodogram values and fit the exponential distribution to the histogram. The user can also request to compute the Bartlett Kolmogorov-Smirnov one-sample *d* statistic (Fuller, 1976; see also *Nonparametrics and Distributions*, Volume I, for more details). Specifically, the Kolmogorov-Smirnov *d* value is computed for the normalized cumulative periodogram, which is tested against the uniform distribution. In addition, Fisher's *Kappa* will be computed, which is the ratio of the largest periodogram value over the average periodogram value (note that the first and last periodogram values are excluded from the computations). Refer to Fuller (1976) for details concerning the interpretation of Fisher's *Kappa* statistic.

Testing for white noise in certain frequency bands. Note that you can also plot the periodogram values for a particular frequency range only. Again, if the input is a white noise series with respect to those frequencies (i.e., if there are no significant periodic cycles of those frequencies), then the distribution of the periodogram values should again follow an exponential distribution.

Fast Fourier Transforms (FFT)

It has not yet been discussed how the spectrum analysis is done computationally. Up until the mid-1960s the standard way of performing the spectrum decomposition was to use explicit formulas to solve for the sine and cosine parameters. The computations involved required at least N^2 (complex) multiplications. Thus, even with today's high-speed computers, it would be very time consuming to analyze even small time series (e.g., 8,000 observations would result in at least 64 million multiplications).

The time requirements changed drastically with the development of the so-called *fast Fourier transform* algorithm, or *FFT* for short. In the mid-1960s, J.W. Cooley and J.W. Tukey (1965) popularized this algorithm which, in retrospect, had in fact been discovered independently by various individuals. Various refinements and improvements of this algorithm can be found in Monro (1975) and Monro and Branch (1976). Readers interested in the computational details of this algorithm may refer to any of the texts cited at the beginning of this section. Suffice it to say that via the FFT algorithm, the time to perform a spectral analysis is proportional to $n*\log_2(n)$ -- a huge improvement.

However, a draw-back of the standard FFT algorithm is that the number of cases in the series must be equal to a power of 2 (i.e., 16, 64, 128, 256, ...). Usually, this necessitated padding of the series, which, as described above, will in most cases not change the characteristic peaks of the periodogram or the spectral density estimates. In cases, however, where the time units are meaningful, such padding may make the interpretation of results more cumbersome. As described below, the implementation of spectral analysis in the *Time Series* module does not impose any restrictions on the length of the input series.

Computation of FFT in *Time Series*

The implementation of the FFT algorithm in the *Time Series* module allows the user to take full advantage of the savings afforded by this algorithm. On most standard computers, series with over 100,000 cases can easily be analyzed. However, there are a few things to remember when analyzing series of that size.

As mentioned above, the standard (and most efficient) FFT algorithm requires that the length of the input series is equal to a power of 2. If this is not the case, additional computations have to be performed. The *Time Series* module will use the simple explicit computational formulas as long as the input series is relatively small and the number of computations can be performed in a relatively short amount of time. For long time series, in order to still

8. TIME SERIES ANALYSIS - INTRODUCTORY OVERVIEW

utilize the FFT algorithm, an implementation of the general approach described by Monro and Branch (1976) is used. This method requires significantly more storage space; however, series of considerable length can still be analyzed very quickly, even if the number of observations is not equal to a power of 2.

For time series of lengths not equal to a power of 2, consider the following recommendations: If the input series is small to moderately sized (e.g., only a few thousand cases), then do not worry. The analysis will typically only take a few seconds anyway. In order to analyze moderately large and large series (e.g., over 100,000 cases), pad the series to a power of 2 and then taper the series during the exploratory part of your data analysis. If necessary, you may later analyze the exact-length series to derive the final estimates.

Cross-Spectrum Analysis

Cross-spectrum analysis is an extension of *Single Spectrum (Fourier) Analysis* to the simultaneous analysis of two series. It is assumed that you have already read the introduction to single spectrum analysis above. The following paragraphs will provide an introduction to cross-spectrum analysis. Detailed discussions of this technique can be found in Bloomfield (1976), Jenkins and Watts (1968), Brillinger (1975), Brigham (1974), Elliott and Rao (1982), Priestley (1981), Shumway (1988), or Wei (1989).

The purpose of cross-spectrum analysis is to uncover the correlations between two series at different frequencies. For example, sun spot activity may be related to weather phenomena here on earth. If so, then if you were to record those phenomena (e.g., yearly average temperature) and submit the resulting series to a cross-spectrum analysis together with the sun spot data, you may find that the weather indeed correlates with the sunspot activity at the 11 year cycle. That is, you may find a periodicity in the weather data that is "in-sync" with the sun spot cycles. One can easily think of other areas of research where such knowledge could be very useful; for example, various economic indicators may show similar (correlated) cyclical behavior; various physiological measures likely will also display "coordinated" (i.e., correlated) cyclical behavior, and so on.

A simple example. Consider the following two series with 16 cases:

	VAR1	VAR2
1	1.000	-.058
2	1.637	-.713
3	1.148	-.383
4	-.058	.006
5	-.713	-.483
6	-.383	-1.441
7	.006	-1.637
8	-.483	-.707
9	-1.441	.331
10	-1.637	.441
11	-.707	-.058
12	.331	-.006
13	.441	.924
14	-.058	1.713
15	-.006	1.365
16	.924	.266

At first sight it is not easy to see the relationship between the two series. However, as shown below, the series were created so that they would contain two strongly correlated periodicities. Shown below are parts of the summary Scrollsheet from the cross-spectrum analysis (the spectral estimates were smoothed with a Parzen window of width 3).

	Frequncy	Period	X Density	Y Density	Cross Density	Cross Quad	Cross Amplit.
0	0.000000		.000000	0.000000	0.00000	0.00000	0.000000
1	.062500	16.00000	8.094709	7.798284	2.35583	-7.58781	7.945114
2	.125000	8.00000	.058771	.100936	-.04755	.06059	.077020
3	.187500	5.33333	3.617294	3.845154	-2.92645	2.31191	3.729484
4	.250000	4.00000	.333005	.278685	-.26941	.14221	.304637
5	.312500	3.20000	.091897	.067630	-.07435	.02622	.078835
6	.375000	2.66667	.052575	.036056	-.04253	.00930	.043539
7	.437500	2.28571	.040248	.026633	-.03256	.00342	.032740
8	.500000	2.00000	.037115	0.000000	0.00000	0.00000	0.000000

The complete summary Scrollsheet contains all spectrum statistics computed for each variable, as described in the context of single spectrum (Fourier) analysis. Looking at the results shown above, it is

TIM - 3306

Copyright © StatSoft, 1995

clear that both variables show strong periodicities at the frequencies *.0625* and *.1875*.

The Cross-periodogram, Cross-density, Quadrature-density, and Cross-amplitude

Analogous to the results for the single variables, the complete summary Scrollsheet will also display periodogram values for the cross periodogram. However, the cross-spectrum consists of complex numbers that can be divided into a real and an imaginary part. These can be smoothed to obtain the cross-density and quadrature density (quad density for short) estimates, respectively. The square root of the sum of the squared cross-density and quad-density values is called the *cross-amplitude*. The cross-amplitude can be interpreted as a measure of covariance between the respective frequency components in the two series. Thus you can conclude from the results shown in the table above that the .0625 and .1875 frequency components in the two series covary.

Squared Coherency, Gain, and Phase Shift

There are additional statistics that will be displayed in the complete summary Scrollsheet.

Squared coherency. One can standardize the cross-amplitude values by squaring them and dividing by the product of the spectrum density estimates for each series. The result is called the *squared coherency*, which can be interpreted similarly to the squared correlation coefficient (see *Basic Statistics and Tables*, Volume I); that is, the coherency value is the squared correlation between the cyclical components in the two series at the respective frequency. However, the coherency values should not be interpreted by themselves; for example, when the spectral density estimates in both series are very small, large coherency values may result (the divisor in the computation of the coherency values will be very small), even though there are no strong cyclical components in either series at the respective frequencies.

Gain. The gain value is computed by dividing the cross-amplitude value by the spectrum density estimates for one of the two series in the analysis. Consequently, two gain values are computed, which can be interpreted as the standard least squares regression coefficients for the respective frequencies

Phase shift. Finally, the phase shift estimates are computed as tan^{-1} of the ratio of the quad density estimates over the cross-density estimate. The phase shift estimates (usually denoted by the Greek letter ψ) are measures of the extent to which each frequency component of one series leads the other.

How the Example Data were Created

Now, return to the example data set presented above. The large spectral density estimates for both series, and the cross-amplitude values at frequencies $\nu = 0.0625$ and $\nu = .1875$ suggest two strong synchronized periodicities in both series at those frequencies. In fact, the two series were created as:

```
v1 = cos[2*pi*.0625*(v0-1)] +
     .75*sin[2*pi*.2*(v0-1)]

v2 = cos[2*pi*.0625*(v0+2)] +
     .75*sin[2*pi*.2*(v0+2)]
```

(where *v0* is the case number). Indeed, the analysis presented in this overview reproduced the periodicity "inserted" into the data very well.

8. TIME SERIES ANALYSIS - INTRODUCTORY OVERVIEW

PROGRAM OVERVIEW

The *STATISTICA Time Series* module contains a wide range of descriptive, modeling, decomposition, and forecasting methods for both time and frequency domain models. The user can perform a wide variety of specialized (time series) transformations, examine autocorrelation, partial autocorrelation, and crosscorrelation functions, fit seasonal and interrupted seasonal ARIMA models (intervention analysis), perform seasonal and non-seasonal exponential smoothing with or without (linear, nonlinear) trend, perform classical (Census method I) seasonal decomposition (also known as *ratio-to-moving-averages* method) as well as X-11 (Census method II) seasonal decomposition (of monthly and quarterly data), perform distributed lags analysis, and perform spectral (Fourier) analysis of one or two series (cross-spectrum analysis).

These procedures are fully integrated; that is, the results of one analysis (e.g., residuals from ARIMA) can directly be used in subsequent analysis (e.g., to compute the autocorrelation function of ARIMA residuals).

Also, numerous flexible options are provided to plot single or multiple time series; for example, different series can be plotted against differentially scaled left-*Y* and right-*Y* axes (to compare the patterns of time series measured on different scales, e.g., to compare the exponentially smoothed series against the residuals).

Analysis of Multiple Series, Capacity

Analyses can be performed on even very long series (e.g., with over 100,000 observations). Multiple series can be maintained in the "*active work area*" of the program (e.g., multiple raw input data series or series resulting from different stages of the analysis); the series can be reviewed and compared. The program will automatically keep track of successive analyses and maintain a log of transformations and other results (e.g., ARIMA residuals, seasonal component, etc.). Thus, the user can always return to prior transformations or compare (plot) the original series together with its transformations. The number of such backups that can be kept simultaneously in the *active work area* can be controlled be the user. All series and their backups can be saved in a new data file for later analyses with the *Time Series* or other *STATISTICA* modules.

Transformations, Modeling, Plots

The available time series transformations allow the user to fully explore patterns in the input series, and to perform all common time series transformations, including: trend subtraction (de-trending), removal of autocorrelation, moving average smoothing (unweighted and weighted, with user-defined or Daniell, Tukey, Hamming, Parzen, or Bartlett weights), moving median smoothing, simple exponential smoothing (see also description of all exponential smoothing options below), differencing, integrating, residualizing, shifting, 4253H smoothing, tapering, Fourier transformation (real and imaginary parts), inverse Fourier transformation, and others.

Autocorrelation Analysis

The *Time Series* module will compute autocorrelation, partial autocorrelation, and cross-correlation functions. These can be displayed in Scrollsheets, where significant autocorrelations will be highlighted, or they can be plotted via customized autocorrelogram plots. The autocorrelation options are available on most analysis specification dialogs; for example, they are available on the ARIMA (see below) specification dialog to aid in model

identification and the ARIMA results dialog to test for residual autocorrelation.

ARIMA

The ARIMA procedure in the *Time Series* module is a full implementation of the Autoregressive Integrated Moving Average Model (Box and Jenkins, 1976). Models may include a constant, and the series can be transformed prior to the analysis; these transformation will automatically be "undone" when ARIMA forecasts are computed, so that the forecasts and their standard errors are expressed in terms of the values of the original input series.

Two different methods for computing the conditional sums of squares are provided (approximate and exact maximum likelihood), and the ARIMA implementation in the *Time Series* module is uniquely suited to models with long seasonal periods (e.g., monthly periods of 30 days).

Results. Standard results include the parameter estimates and their standard errors and the parameter correlations. Forecasts and their standard errors can be computed and plotted, and appended to the input series. In addition, numerous options for examining the ARIMA residuals (for model adequacy) are available, including autocorrelograms and partial autocorrelograms, histograms, normal, half-normal, and detrended normal probability plots of residuals.

Interrupted time series analysis. The implementation of ARIMA in the *Time Series* module allows the user to perform interrupted time series analysis (intervention analysis). Several simultaneous interventions may be modeled, which can either be single-parameter abrupt-permanent interventions, or two-parameter gradual or temporary interventions (graphs of different impact patterns can be reviewed). Forecasts can be computed for all intervention models, which can be plotted (together with the input series) as well as appended to the original series.

Seasonal and Non-Seasonal Exponential Smoothing

The *Time Series* module contains a complete implementation of all common exponential smoothing models. Models can be specified to contain an additive or multiplicative seasonal component and/or linear, exponential, or damped trend; thus, available models include the popular Holt-Winter linear trend models. The user may specify the initial value for the smoothing transformation, initial trend value, and seasonal factors (if appropriate). Separate smoothing parameters can be specified for the trend and seasonal components.

The user can also perform a grid search of the parameter space in order to identify the best parameters; the respective results Scrollsheet will report for all combinations of parameter values the mean error, mean absolute error, sum of squares error, mean square error, mean percentage error, and mean absolute percentage error. The smallest value for these fit indices will be highlighted in the Scrollsheet.

In addition, the user can also request an automatic search for the best parameters with regard to the mean square error, mean absolute error, or mean absolute percentage error (a general function minimization procedure is used for this purpose).

Results. The results of the respective exponential smoothing transformation, the residuals, as well as the requested number of forecasts, are available for further analyses and plots. A summary plot is also available to assess the adequacy of the respective exponential smoothing model; that plot will show the original series together with the smoothed values and forecasts, as well as the smoothing residuals plotted separately against the right-Y axis.

8. TIME SERIES ANALYSIS - PROGRAM OVERVIEW

Classical Seasonal Decomposition (Census Method I)

This method is also known as the *ratio-to-moving averages* method. The user may specify the length of the seasonal period and choose either the additive or multiplicative seasonal model. The program will compute the moving averages, ratios or differences, seasonal factors, the seasonally adjusted series, the smoothed trend-cycle component, and the irregular component. Those components are available for further analysis; for example, the user may compute histograms, normal probability plots, etc. for any or all of these components (e.g., to test model adequacy). These series may also be saved in a data file for further analysis with other modules.

X-11 Monthly and Quarterly Seasonal Decomposition and Adjustment (Census Method II)

The *Time Series* module is a full featured implementation of the US Bureau of the Census X-11 variant of the Census method II seasonal adjustment procedure. The arrangement of options and dialogs follows closely the definitions and conventions described in the Bureau of the Census documentation. Additive and multiplicative seasonal models may be specified. The user may also specify prior trading day factors and seasonal adjustment factors. Trading day variation can be estimated via regression (controlling for extreme observations), and used to adjust the series (conditionally if requested). The standard options are provided for graduating extreme observations, for computing the seasonal factors, and for computing the trend-cycle component (the user can choose between various types of weighted moving averages; optimal lengths and types of moving averages can also automatically be chosen by the program).

Results. The final components (seasonal, trend-cycle, irregular) and the seasonally adjusted series are automatically available for further analyses and plots, and those components can also be saved for further analyses with other programs. Following the US Bureau of Census conventions, the user can choose between standard output (reporting up to 27 result Scrollsheets), long output (reporting up to 39 Scrollsheets), and full output (reporting up to 59 Scrollsheets); in addition, specific result tables (Scrollsheets) can also be requested. The program will produce the standard plots of the different components, in addition, categorized plots by months (or quarters) are also available.

Polynomial Distributed Lag Time Series Models

The implementation of the polynomial distributed lag methods in the *Time Series* module allow the user to fit both unconstrained polynomial lags models and Almon (constrained) distributed lags models. A variety of graphs are available to examine the distribution of the variables in the model.

Spectrum (Fourier) and Cross-Spectrum Analysis

The *Time Series* module includes a full implementation of spectrum (Fourier decomposition) analysis and cross spectrum analysis techniques. The program is particularly suited for the analysis of unusually long time series (e.g., with over 100,000 observations), and it will not impose any constraints on the length of the series (i.e., the length of the input series does not have to be a power of 2). However, the user may also choose to pad or truncate the series prior to the analysis. Standard pre-analysis transformations include tapering, subtraction of the mean, and detrending.

Results. For single spectrum analysis, the standard results include the frequency, period, sine and cosine

TIM - 3311

Copyright © StatSoft, 1995

8. TIME SERIES ANALYSIS - PROGRAM OVERVIEW

coefficients, periodogram values, and spectral density estimates. The density estimates can be computed using Daniell, Hamming, Bartlett, Tukey, Parzen, or user-defined weights and user-defined window width. An option that is particularly useful for long input series is to display only a user-defined number of the largest periodogram or density values in descending order; thus, the most salient periodogram or density peaks can easily be identified in long series.

The user can compute the Bartlett Kolmogorov-Smirnov *d* test and Fisher's *Kappa* for the periodogram values to test whether the input is a white-noise series. Numerous plots are available to summarize the results; the user can plot the sine and cosine coefficients, periodogram values, log-periodogram values, spectral density values, and log-density values against the frequencies, period, or log-period. For long input series, the user can choose the segment (period) for which to plot the respective periodogram or density values, thus enhancing the "resolution" of the periodogram or density plot.

For cross-spectrum analysis, in addition to the single spectrum results for each series, the program computes the cross-periodogram (real and imaginary part), co-spectral density, quadrature spectrum, cross-amplitude, coherency values, gain values, and the phase spectrum. All of these can also be plotted against the frequency, period, or log-period, either for all periods (frequencies) or only for a user-defined segment.

A user-defined number of the largest cross-periodogram values (real or imaginary) can also be displayed in a Scrollsheet in descending order of magnitude to facilitate the identification of salient peaks when analyzing long input series. As with all other *Time Series* procedures, all of these result series can be appended to the *active work area* and will be available for further analyses with other *Time Series* methods or *STATISTICA* modules.

Related Procedures

Linear regression-based time series and forecasting. *STATISTICA* also contains a comprehensive selection of descriptive, modeling, and forecasting options based on linear regression models using lagged or non-lagged variables (including fixed nonlinear models, regression through the origin, and ridge regression, see the *Multiple Regression* module, Volume I). The input and output data can easily be shared between these two complementary modules by saving the respective series to *STATISTICA* data files. For example, the predicted and residual scores produced by *Multiple Regression* can be saved to a data file and then accessed by the *Time Series* module for further smoothing.

Nonlinear regression-based time series and forecasting. A wide selection of nonlinear model fitting techniques are available in the *Nonlinear Estimation* module (Chapter 1), where you can fit to time series practically any function including piecewise models that are different for different periods or segments of the data set.

General purpose modeling, curve fitting, smoothing. A selection of general purpose modeling, curve fitting and data smoothing techniques are available as part of the graphics procedures (which are accessible from each *STATISTICA* module; see Volume II). Those techniques offer powerful tools for time series modeling. They include such procedures as distance weighted least squares smoothing, negative exponentially weighted smoothing, bicubic spline smoothing, polynomial smoothing of various orders, and others. Also, multiple custom designed functions can be overlaid on the data, multiple functions can be fitted to a data set in one graph, multiple graphs (panels) can be displayed on one screen or printout, and many other techniques are available that can facilitate the identification of patterns in time series.

TIM - 3312

StatSoft

Copyright © StatSoft, 1995

8. TIME SERIES ANALYSIS - GENERAL CONVENTIONS

GENERAL CONVENTIONS AND OPENING DIALOG

Active Work Area

All variables (series) and their transformations that are currently available for analysis are stored in the *active work area* and are listed in the scrollable edit fields at the top of the specification dialogs for the respective analyses.

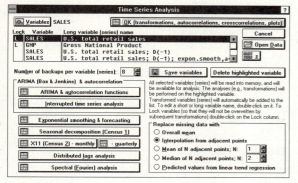

When you select new variables, the *active work area* will first be cleared, and then the selected variables will be read into the *active work area* (after all "holes" with missing data have been "patched;" see below for details).

Highlighted Variable

All subsequent analyses will be performed on the highlighted variable.

For example, when you perform a transformation, then the currently highlighted variable will be transformed, and a new (transformed) variable will be appended to the *active work area*. To highlight a variable, simply click on it in any of the scrollable edit fields.

Naming Conventions

When a new (e.g., transformed) variable is appended to the *active work area*, it is assigned (1) the same short variable name as the original variable (that was transformed), and (2) a new long variable name that consists of the old long variable name (as much of it as will fit) and a brief description of the respective transformation that was performed. In this manner, as you perform successive transformations or analyses (e.g., successively difference a series), an automatic log of transformations will be maintained in the long variable names. For example, see the variable and long variable names below:

SERIES_G   [The original series]

SERIES_G x+3;   [After adding a constant]

SERIES_G x+3;ln(x);   [After the logarithmic transformation was performed]

SERIES_G x+3;ln(x);D(-12);   [After the series was differenced with the lag of 12]

Editing Variable Names

Double-click in the column labeled *Variable* or *Long variable (series) name* to edit the short or long variable name for the series in the *active work area*. Note that the short and long names will only be changed in the *active work area*, not in the file (use the respective data spreadsheet operations to permanently change those names).

Number of Backups per Variable (Series)

All dialogs that contain the scrollable edit fields for highlighting a variable (series) for analysis also contain a field for specifying the desired *Number of backups per variable (series)*. As described above, after a transformation (or other analysis) was performed on a series, the resulting transformed series (or residuals, forecasts, etc.; e.g. in ARIMA)

TIM - 3313

Copyright © StatSoft, 1995

will be appended to the *active work area*, and the values of the series prior to the transformation will be maintained as a backup. The number of such backups that will be maintained in the *active work area* is controlled by this parameter. Thus, for example, if this parameter is set to *3* (default), and you have just performed the fourth transformation of an original variable, then the series with the data after the first transformation will be dropped from the *active work area* and replaced by the new (fourth) transformation. Thus, successive transformations will be appended to the *active work area* until there are as many backups as specified in this parameter; at that point the respective "oldest" transformation will be replaced by the new one. Up to 99 backups can be kept of a single original variable.

Locking Variables (Series)

The first column of the scrollable edit fields carries the header *Lock*. When you double-click in that column for a transformed variable, that variable will be locked in the *active work area* (or unlocked, if it was previously locked). An *L* will appear in that column to indicate that the variable is now locked. Locked variables will not be replaced as successive transformations exceed the current maximum number of backups (as described in the paragraph above). Note that original (untransformed) variables are always locked, and they cannot be unlocked.

Deleting Variables (Series) from the Current Work Area

To delete a transformed variable from the *active work area*, use the *Delete highlighted variable* button. Original (untransformed) variables cannot be deleted.

Saving Variables (Series) in the Current Work Area

Use the *Save variables* button to save the variables (series) in the *active work area* into a standard *STATISTICA* data file. You can save all variables or only selected variables.

Missing Data

Practically all time series analyses require that all data are observed, and that there are no "holes" with missing data in the time series. As long as the missing data are at the end of the series (trailing missing data) or the beginning of the series (leading missing data), the missing data will simply be ignored. Missing data embedded in the series have to be replaced in some way. *Time Series* offers a range of different methods for dealing with missing data in that case:

Overall Mean

In this method, all missing data will simply be replaced by the overall mean of the series. Very often, when the series is not stationary (see *Introductory Overview*, page 3274), or when there are large systematic fluctuations in the values of the series, this method may not be appropriate. On the other hand, the overall mean is often the best *a priori* (unbiased) guess for the missing data.

Interpolation from Adjacent Points

In this method, the missing data are computed by interpolation from the adjacent non-missing points. Graphically, this method amounts to replacing missing data by connecting with a straight line the point just prior to the missing data with the point just following the missing data. Thus, this method in a sense assumes that there is some serial correlation in the data, that is, that each observation is to some extent related to and therefore most similar to the previous observation.

Mean of *n* Adjacent Points

In this method the missing data are computed from the mean of the *n* adjacent points on both sides of the "hole" of missing data. For example, when *n* is left at its default value of *1,* then missing data will be replaced by the average of the value just prior to the missing data and the value immediately following the missing data. In general, this method implies that the data in the region or window specified by the *n* parameter are more similar to each other than points that are further away.

Median of *n* Adjacent Points

This method is essentially the same as that described above, except that missing data are replaced by the *median* of the *n* non-missing adjacent points.

Predicted Values from Linear Trend Regression

In this method, the program will fit a least-squares regression line to the time series. The missing data will then be replaced by the values predicted by this regression line. This method implies that the most salient (or strongest) feature of the series is its linear trend across time.

Display/Plot Options

All dialogs containing options to display or plot one or more time series also provide access to the display/plot options.

Those options will affect (1) how the horizontal *X*-axes in line plots are scaled and labeled, and (2) how the rows of Scrollsheets displaying the values of the series will be labeled.

Label Data Points with...

The choice of radio button in this box determines the labeling of the data points in displays or plots. The setting of these radio buttons will be retained across other analytical procedures in the *Time Series* module, for example, when displaying residuals from an ARIMA analysis.

Case numbers. By default (radio button *Case numbers*), each point will be identified by the respective case number, and case numbers will be used on the *X*-axis in plots of available series.

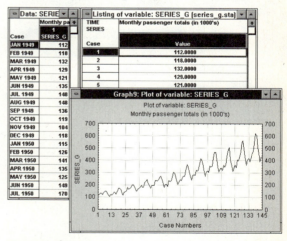

Case names. If there are *Case names* available in the data file, you may use them to label the data points (rows in Scrollsheets and the *X*-axis in plots).

8. TIME SERIES ANALYSIS - GENERAL CONVENTIONS

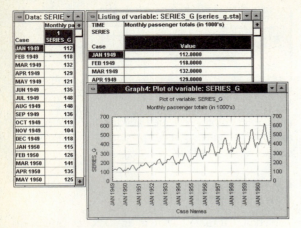

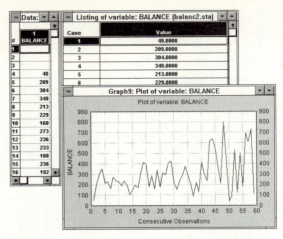

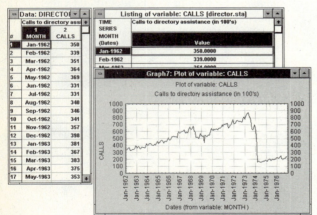

Dates. If you choose to use *Dates* (for a description of date variables see *General Conventions*, Volume I), a single variable selection dialog will come up where you can select the variable containing dates (i.e., values formatted as dates). Those dates will then be used to label points in Scrollsheets and the *X*-axis in plots.

Consecutive integers. Finally, you may use *Consecutive integers*, starting from 1, to label the points and *X*-axis in plots.

Note that in the *Consecutive integers* setting what will be displayed is the offset (plus one) of the respective data point from the first data point of any variable in the *active work area* (see above). For example, suppose you have in the *active work area* two series called *S1* and *S2*. *S1* has its first valid observation at case number 5, and *S2* has its first valid observation at case number 8. If you were to use the option *Review multiple variables* to display both variables and choose the *Consecutive integers* option, then the first valid observation for variable *S1* will be labeled *1*, and the first valid observation for variable *S2* will be labeled *4*.

Scale X-Axis in Plots Manually (Min, Step)

The setting of this check box will affect the scaling of the *X*-axis in line plots of single or multiple variables (series). The choice of scaling will be retained across other analytical procedures in this module, for example, when plotting forecasts from an ARIMA analysis. By default, if this check box is not set, the *X*-axis in all plots will always be scaled so that (1) all valid data points are displayed in the plot, and (2) the tickmarks (steps) on the *X*-axis are "neat" (i.e., the significant digit of the step size is either 1, 2, or 5). However, if there is a known periodicity in the data that is of interest, such as 12 months per year or 10 observations per day, etc., then it may be useful to set this check box and to scale the *X*-axis for plots accordingly.

TRANSFORMATIONS OF VARIABLES

EXAMPLE

General Conventions and Options

In this section, the general conventions used in the *Time Series* module to maintain the *active work area* (which functions as a queue of successive transformations of the input series; refer to the *General Conventions* section, page 3313) will be reviewed. Thus, transformations or the results of other analyses can be undone, saved, etc.

Transformations and the Active Work Area

File *Stocks.sta* contains the closing prices for two stocks over a 200-day period. Each trading week consists of exactly five trading days, and closing quotes for holidays (when the stock market was closed) were estimated. In this example, the two time series will be read into memory, some smoothing operations will be performed, several useful time series graphs will be produced, and an autocorrelation analysis of the stock prices will be performed. The first few cases in the data file are shown below.

	STOCK1	STOCK2	DATE
2/1/91	69.750	55.625	1-Feb-91
2/5/91	72.000	55.500	5-Feb-91
2/6/91	70.000	55.500	6-Feb-91
2/7/91	69.000	54.875	7-Feb-91
2/8/91	70.500	55.125	8-Feb-91
2/11/91	68.500	54.500	11-Feb-91
2/12/91	70.250	54.500	12-Feb-91
2/13/91	68.000	54.375	13-Feb-91
2/14/91	70.000	54.625	14-Feb-91
2/15/91	67.750	54.500	15-Feb-91
2/18/91	69.750	54.500	18-Feb-91

Note that this file contains dates in two places, in the variable *Date* (variable 3) and as case names (in the first column of the spreadsheet). The dates were included in those two places to show how they can be used in plots and Scrollsheets. To start the analysis, first start the *Time Series* module and open the data file *Stocks.sta*. Then, click on the *Variables* button and select variables *Stock1* and *Stock2*. The opening dialog of the *Time Series* module will now look like this.

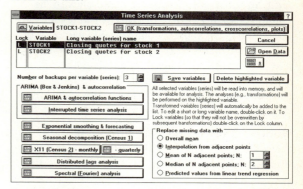

All variables (series) and their transformations that are currently available for analysis are stored in the *active work area* and are listed in the scrollable edit fields at the top of the dialog. When you select new variables, the *active work area* will first be cleared, and then the selected variables will be read into the *active work area* (after all "holes" with missing data have been "patched;" see below and page 3314).

Highlighted variable. All subsequent analyses will be performed on the highlighted variable. For example, when you perform a transformation, then the currently highlighted variable will be transformed, and a new (transformed) variable will be appended to the *active work area*. To highlight a variable, simply click on it in any of the scrollable edit fields. For this example, highlight the variable *Stock1*.

Naming conventions. When a new (e.g., transformed) variable is appended to the *active work area*, it is assigned (1) the same short variable name as the original variable (that was transformed), and (2) a new long variable name that consists of the old long variable name (as much of it as will fit) and a brief description of the respective transformation that was performed. In this manner, as you perform successive transformations or analyses (e.g., successively difference a series), an automatic log of transformations will be maintained in the long variable names.

Editing variable names. Double-click in the column labeled *Variable* or *Long variable (series) name* to edit the short or long variable name for the series in the *active work area*. Note that the short and long names will only be changed in the *active work area*, not in the file (use the respective data spreadsheet operations to permanently change those names; see *General Conventions*, Volume I).

Number of backups per variable (series). All dialogs that contain the scrollable edit fields for highlighting a variable (series) for an analysis also contain a field for specifying the desired *Number of backups per variable (series)*. As described above, after a transformation (or other analysis) is performed on a series, the resulting transformed series (or residuals, forecasts, etc. in ARIMA) will be appended to the *active work area*, and the values of the series prior to the transformation will be maintained as a backup. The number of such backups that will be maintained in the *active work area* is controlled by this parameter. Thus, for example, if this parameter is set to *3*, and you have just performed the fourth transformation of an original variable, then the series with the data after the first transformation will be dropped from the *active work area* and replaced by the new (fourth) transformed series. Thus, series created by successive transformations will be appended to the *active work area* until there are as many backups as specified in this parameter; at that point the respective "oldest" transformed series will be replaced by the new one. Up to 99 backups can be kept of a single original variable. For this example, accept the default value of *3* backups.

Locking variables (series). The first column of the scrollable edit fields carries the header *Lock*. When you double-click in that column for a transformed variable, that variable will be locked in the *active work area* (or unlocked, if it was previously locked). An *L* will appear in that column to indicate that the variable is now locked. Locked variables will not be replaced as successive transformations exceed the current maximum number of backups (as described in the paragraph above). Note that original (untransformed) variables are always locked, and they cannot be unlocked.

Deleting variables (series) from the current work area. To delete a transformed variable from the *active work area*, use the *Delete highlighted variable* button. Original (untransformed) variables cannot be deleted.

Saving variables (series) in the current work area. Use the *Save variables* button to save the variables (series) in the *active work area*. You can save all variables or only selected variables.

Missing Data

Practically all time series analyses require that all data are observed, and that there are no "holes" with missing data in the time series. As long as the missing data are at the end of the series (trailing missing data) or the beginning of the series (leading missing data), the missing data will simply be ignored. Missing data embedded in the series have to be replaced in some way. The *Time Series* module offers a range of different methods for dealing with missing data in this case, which are described on page 3314. For this example, select the missing data option *Interpolation from adjacent points*.

Note that the chosen missing data replacement method will be used not only when reading selected variables from the data file into the *active work area*, but it will also be used when time series transformations result in embedded missing data.

For example, suppose an input series contains a few 0's (zeros), and you request a *log* transformation. Since the *log* of 0 is undefined, the program will replace those observations with missing data; then, in a second pass through the series, those missing data will be replaced according to the method chosen on the opening dialog.

Overall mean. In this method, all missing data will simply be replaced by the overall mean of the series. Very often, when the series is not stationary (see *Introductory Overview*, page 3274), or when there are large systematic fluctuations in the values of the series, this method may not be appropriate. On the other hand, the overall mean is often the best *a priori* (unbiased) guess for the missing data.

Interpolation from adjacent points. In this method, the missing data are computed by interpolation from the adjacent non-missing points. Graphically, this method amounts to replacing missing data by connecting with a straight line the point just prior to the missing data with the point just following the missing data. Thus, this method in a sense assumes that there is some serial correlation in the data, that is, that each observation is to some extent related to and therefore most similar to the previous observation.

Mean of *n* adjacent points. In this method the missing data are computed from the mean of the *n* adjacent points on both sides of the "hole" of missing data. For example, when *n* is left at its default value of *1,* then missing data will be replaced by the average of the value just prior to the missing data and the value immediately following the missing data. In general, this method implies that the data in the region or window specified by the *n* parameter are more similar to each other than points that are further away.

Median of *n* adjacent points. This method is essentially the same as that described above, except that missing data are replaced by the *median* of the *n* non-missing adjacent points.

Predicted values from linear trend regression. In this method, the program will fit a least-squares regression line to the time series. The missing data will then be replaced by the values predicted by this regression line. This method implies that the most salient (or strongest) feature of the series is its linear trend across time.

Reviewing the Time Series

Now proceed with the analysis and review the closing quotes for the two series. Click on the *OK (Transformations)* button to bring up the *Transformations of Variables* dialog.

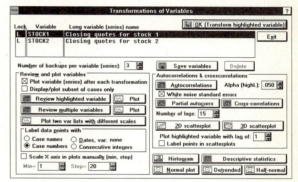

Several options for reviewing time series in the *active work area* are shown in *Review and plot variables* box. The *Label data points* box contains options that will determine how the horizontal (time) axis is scaled and labeled. Note that those options will affect all plots in the *Time Series* module (there are numerous options for plotting time series available with all procedures in the *Time Series* module). The example data file *Stocks.sta* contains dates in variable *Date* and in the case names and either of these can be used to label the horizontal axes in plots (set the appropriate radio button).

Note that case names may also contain information other than dates; for example, significant discrete events affecting the series (e.g., release of news affecting the stock prices) could also be noted in the case names and be used as labels in plots. For this example, label the horizontal axis in the plots with the dates in the variable *Date*. Click on the *Dates* radio button, and select that variable from the subsequent variable selection dialog. Because in this example series, each trading week consists of five days (Monday through Friday), set the *Scale the X-axis in plots manually* option and enter as the minimum (*Min=*) *1* (start with the first day), and the step size (*Step=*) *5*. Then, click on the *Review highlighted variable* button to produce the following Scrollsheet.

TIME SERIES DATE (Dates)	Closing quotes for stock 1 Value
1-Feb-91	69.75000
5-Feb-91	72.00000
6-Feb-91	70.00000
7-Feb-91	69.00000
8-Feb-91	70.50000
11-Feb-91	68.50000
12-Feb-91	70.25000
13-Feb-91	68.00000
14-Feb-91	70.00000
15-Feb-91	67.75000
18-Feb-91	69.75000

You can plot this series by either selecting the default *Quick Stats Graph* for this Scrollsheet (click with the right-mouse-button on the Scrollsheet and select the *Quick Stats Graphs - Line Graph* option from the flying menu) or by clicking on the *Plot* button next to the *Review highlighted variable* button on the *Transformations of Variables* dialog.

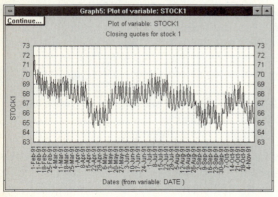

To plot both stocks simultaneously, click on the *Review (Plot) multiple variables* button, and then select the variables (series) to be displayed or plotted.

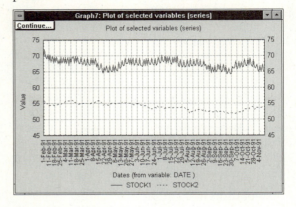

Plotting Two Series with Different Scales

As you can see, the closing quotes for *Stock2* are generally lower than those for *Stock1*. You can independently scale the vertical axes for those two series to obtain the best vertical resolution possible for each series. Use option *Plot two var lists with different scales* and select to plot *Stock1* against the left-*Y* axis, and *Stock2* against the right-*Y* axis.

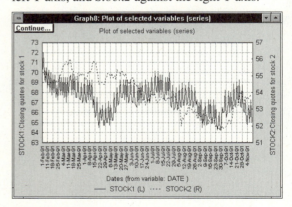

This plot allows you to compare the pattern or movement of the two series across time more clearly.

8. TIME SERIES ANALYSIS - TRANSFORMATIONS

Transforming Time Series

Now perform a few transformations. Bring up the *Transformations of Variables* dialog again (click on the *Continue* button in the line plot) and highlight the first series (*Stock1*). Click *OK (Transform highlighted variable)* to open the *Time Series Transformations* dialog.

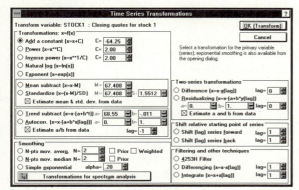

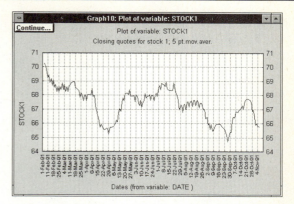

Now click *Continue* to return to the *Transformations of Variables* dialog.

The Updated Active Work Area

As described above, the transformed (smoothed) series has been appended to the *active work area*.

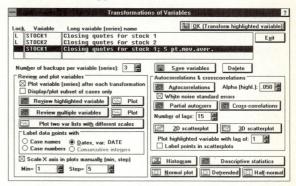

This dialog shows all common transformations for time series data. Some of those transformations require that you select a second variable, for example, for *Residualizing* a time series. For this example, select a simple (unweighted) 5-point moving average transformation for series *Stock1*. Click on the radio button labeled *N pts. mov. averg.*, and specify *5* as the window width in the *N=* edit field.

Then, click *OK*, and the moving average transformation will be performed. When all cases have been transformed, then, by default (i.e., if the *Plot variable (series) after each transformation* check box is set in the *Transformation of Variables* dialog), the transformed series will be plotted.

As you can see below, compared to the plot of the raw (untransformed) series (see above), the transformed series is much less "jagged," and the general trend over the trading days reflected by the data is much clearer.

Following the naming conventions described earlier, the transformed variable has the same short name (*Stock1*) and will have the same long name, except that a brief description of the transformation (*5 pt. mov. aver.*) was added to the existing title. Note that if the original title had been much longer, so that the description of the transformation couldn't fit, then the original long variable name would have been deleted.

TIM - 3323

Copyright © StatSoft, 1995

8. TIME SERIES ANALYSIS - TRANSFORMATIONS

Further Processing of the Transformed Series

The transformed series in the work area has the same "status" as those series that were originally selected and read into the *active work area* from the file. For example, they can be plotted, saved, or used as input into further analyses. Now, compare the smoothed series with the original input series. Click on the *Plot* button next to the *Review multiple variables* button. By default, the original series and the smoothed series will be highlighted in the subsequent variable selection dialog.

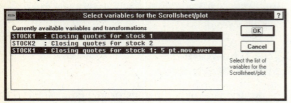

Simply accept this default selection and click *OK*. Shown below is the joint plot of the raw input series and the smoothed (transformed) series.

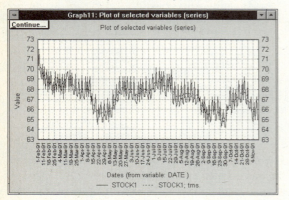

Multiple Successive Transformations

Click *Continue* to return to the *Transformations of Variables* dialog. Now, continue to transform the transformed variable and observe how successive series are appended to the *active work area*. Highlight the transformed series (the one at the bottom of the list) by clicking on it. For now, turn off the automatic graphing option, that is, de-select the *Plot variable (series) after each transformation* check box. Click *OK* to bring up the list of available transformations again, and select *Simple exponential smoothing* with the default parameter $\alpha = .2$ and click *OK*. Click *OK* once again and select the *4253H Filter* (this is a powerful smoothing/filtering technique that applies several moving average and moving median transformations in succession; refer to page 3336 for details).

There are now 3 transformed series that were appended to the *active work area*. Assuming that you have not changed the *Number of backups per variable* parameter from the default value of *3*, the *active work area* is now "full." The next transformation of any of the variables derived from *Stock1* will replace the "oldest" transformation for that variable. Thus, if you now transform variable *Stock2*, another (the first) backup of that variable will be added. Try this by applying the *4253H Filter* to *Stock2*. After the transformation is complete, the *active work area* will look like this.

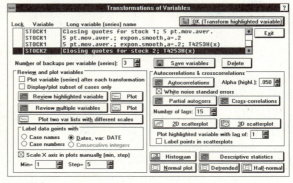

As you can see, all transformations of variable *Stock1* are still in place. However, now highlight the original variable *Stock1* again (scroll the edit window until *Stock1* is visible), and apply to it, for example, a *5-pt. mov. median* transformation. Now the "oldest" or first transformation that you performed on variable *Stock1* (the 5-point moving average transformation) will be replaced by the 5-

TIM - 3324

point moving median transformation of series *Stock1*.

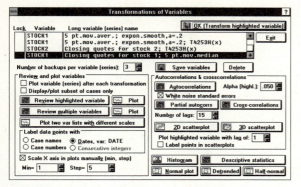

Perhaps an appropriate "mental model" of the way in which series are managed in the *active work area* is that of a *carousel*: Successive transformations of a series are placed in successive positions of the carousel. The number of places on the carousel is determined by the *Number of backups* parameter. Once all places have been taken, then the next transformation will replace the first transformation, as the carousel starts the next go-around.

Locking. Suppose you would like to keep a transformation in the *active work area*; that is, you would like to prevent it from being replaced by another transformation.

To accomplish this, the respective transformation should be *Locked*: Double-click on the respective series in the *Lock* column, and an *L* will appear in the respective row of that column. The respective series is now *locked*; that is, it will not be overwritten by successive transformations of the same variable, or, put another way, it will stay in the same place on the carousel.

For example, the next "oldest" transformation of *Stock1* that will be replaced is the *Exponential smoothing* transformation. Now, lock that series, then highlight the original *Stock1* series again and apply to it a *3-pts. moving average* transformation. As you can see, the locked transformation was not overwritten.

Saving the Series in the Active Work Area

Now save the transformations in the *active work area*. Suppose you would like to keep only the combined moving average, exponential smoothing, and *4253H* transformations of *Stock1* for further analysis. First, delete the series that you don't want to save from the *active work area* by highlighting them and then clicking on the *Delete* button.

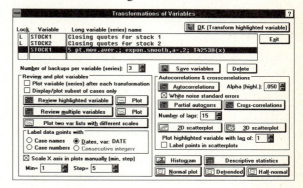

Then, save the variables in the *active work area* using the *Save* option. You will be prompted for a file name under which the program will save all data currently in the *active work area*.

Autocorrelation Analysis

Thus far, the interpretation of the transformations has not been discussed. In general, the analysis of time series data requires a good deal of experience not only with the available techniques, but also with the nature of the data. For example, stock prices often follow what is called a *random walk* model. Simply stated, each observation is equal to the previous observation plus some random component. In a sense, the process behaves like "a drunken man whose position at time *t* is his position at time *t-1* plus a step in a random direction at time *t*" (Wei, 1990, page 71). If so, you may expect that the simple autocorrelation is highest for a lag of *1*, next highest for lag *2*, etc., that is, that the autocorrelation function will show a slow decay. Put another way, the "drunken man" will be closest to where he was immediately before, a bit farther away from where he was before that, and so on. Technically, this process can be expressed as an autoregressive process, with the autoregressive parameter (ϕ in ARIMA terminology) approaching *1.0*.

Plotting the Autocorrelation Function

Now examine whether the closing quotes stored in *Stock2* follow this simple model. First, plot *Stock2* (highlight *Stock2* and click on the *Plot* button next to the *Review highlighted variable* button).

It appears that *Stock2* shows a downward trend. Such trends will bias the autocorrelation function; that is, if the stock is generally going down, then obviously, each quote will be more similar to the adjacent quotes as compared to those that are farther away.

Therefore, you can detrend the series by bringing up the list of *Time Series Transformations* (click *OK (Transform)* in the *Transformation of Variables* dialog) and choosing the *Trend subtract* option. If you plot the transformed variable again, you can see that the trend was removed.

Now click on the *Autocorrelation* button to display a Scrollsheet and plot of the autocorrelation function.

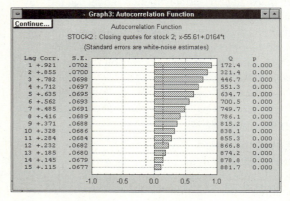

The correlation for lag *1* is large, and decays slowly thereafter; the plot of the partial autocorrelation function also supports the random walk model:

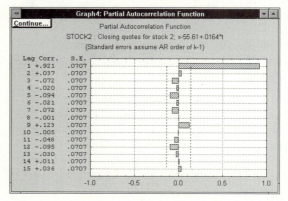

Above and beyond the very strong autocorrelation at lag *1*, none of the partial autocorrelations are significant. Put into words, each observations is mostly similar to the previous observation, plus some random shock -- which represents the random walk model. You can "remove" the strong single autocorrelation by differencing the series. To do this, bring up the list of *Time Series Transformations* again and select simple *Differencing (x = x-x(lag))*; then click on the *Autocorrelation* option in the *Transformation of Variables* dialog to bring up the plot for the differenced series.

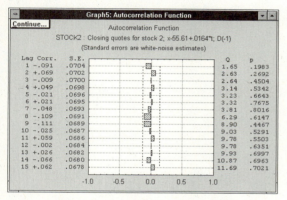

As you can see, none of the autocorrelations is significant. Thus, your initial guess from prior experience with the "behavior" of stock prices has been confirmed; *Stock2* indeed follows the random walk model, and, unfortunately, given a particular quote at a particular time, there is no way to predict whether the stock will go up or down.

DIALOGS, OPTIONS, STATISTICS

Click on the *Variable* or *Long variable (series) name* column to highlight the variable (series) in the *active work area* (see *General Conventions and Opening Dialog*, page 3313) that is to be transformed or analyzed. Click *OK* to select a transformation. Transformed variables will be appended to the *active work area*. The number of backups that will be kept for any one variable is determined by the *Number of backups per variable (series)* parameter (see page 3313, see also below).

Transformation of Variables Dialog

This dialog contains options to explore the time series currently in the *active work area* and allows the user to transform or smooth selected series.

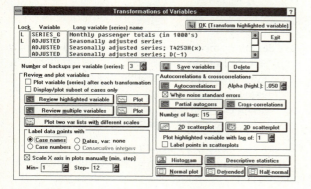

Number of Backups per Variable

This parameter determines the maximum number of backups that is kept in the *active work area* for each variable. When a new transformation is about to be appended, and the maximum number of backups for the respective original variable has already been exceeded, the "oldest" previous backup will be replaced. You can lock a variable (prevent it from being overwritten) by double-clicking on the respective variable in the *Lock* column. For more details, refer to the *General Conventions and Opening Dialog* section (page 3313).

Delete

This option allows you to delete the currently highlighted variable. To highlight a variable, click on it either in the *Lock*, *Variable*, or *Long variable (series) name* column.

Save Variables

This option allows you to save the variables (series) that are currently available in the *active work area* into a standard STATISTICA data file.

Review and Plot Variables

These options allow you to review or plot the currently highlighted variable (together with other variables). The labeling of the individual observations (that is, of the respective points in time represented by the rows in Scrollsheets or the horizontal *X*-axis in plots) is determined by the setting of the radio buttons in the *Label data points* box. For the plots, you can choose to *Scale the X-Axis in the plots manually*. This is often useful when the series consists of several regularly recurring time intervals, for example, if it is made up of 10 years with 12 (monthly) observations each, then it may be useful to scale the *X*-axis in the plot so that each major tickmark indicates exactly 1 year.

Plot Variables (Series) After Each Transformation

If this check box is set, then after each transformation, the transformed variable will automatically be plotted.

Display/Plot
Subset of Cases Only

If this check box is set, then, before plotting or displaying one or more series, you will be prompted to select a range of observations to include in the plot or display.

Review/Plot
Highlighted Variable

This button will bring up a Scrollsheet (or a plot) with the values for the currently highlighted variable.

Review/Plot
Multiple Variables

After clicking on this button, you will first be prompted to select a list of variables to be displayed or plotted. When plotting multiple variables the vertical (*Y*) axis will have a common scale; thus, if the series that are plotted contain values of different magnitudes, then use the option *Plot two var lists with different scales* (see below). In that option, two variable lists are plotted against the left-*Y* and right-*Y* axes using different scales.

Plot Two
Var. Lists with
Different Scales

This option allows you to plot two lists of variables using different scales for the vertical (*Y*) axes in the plot. The first list will be plotted against the left-*Y* axis, and the second list will be plotted against the right-*Y* axis.

For example, suppose you want to plot two variables (series) called *S1* and *S2*, and that these variables contain values of different magnitudes: The values of *S1* range from 1-100 and the values of *S2* range from 1000 to 10000. If you plot *S1* against the left-*Y* axis (scaled from 1-100) and *S2* against the right-*Y* axis (scaled from 1000-10000), then both variables will be plotted with optimal resolution, and thus

relationships between patterns across time (the *X*-axis) may become more clearly identifiable. Shown below is an example plot.

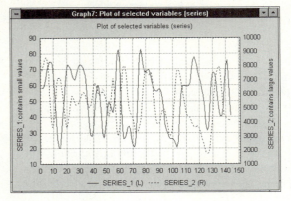

Label Data
Points with ...

The choice of radio button in this box determines the labeling of the data points in displays or plots. The setting of these radio buttons will be retained across other analytical procedures in this module, for example, when displaying residuals from an ARIMA analysis.

By default (radio button *Case numbers*), each point will be identified by the respective case number, and case numbers will be used on the *X*-axis in plots of available series. If there are *Case names* available in the data file, you may use them to label the data points (and the *X*-axis in plots).

If you choose to use *Dates,* a single variable selection window will open in which you can select the variable containing dates (i.e., values formatted as dates). Those dates will then be used to label the rows in Scrollsheets and the *X*-axis in plots. Finally, you may use *Consecutive integers,* starting from 1, to label the rows in Scrollsheets and *X*-axis in plots.

Note that in the *Consecutive integers* setting, what will be displayed is the offset (plus one) of the respective data point from the first data point of any variable in the *active work area* (see *General*

Conventions and Opening Dialog, page 3313). For example, suppose you have in the *active work area*, two series called *S1* and *S2*. *S1* has its first valid observation at case number 5, and *S2* has its first valid observation at case number 8. If you were to use the option *Review multiple variables* to display both variables, and choose the *Consecutive integers* option, then the first valid observation for variable *S1* will be labeled *1*, and the first valid observation for variable *S2* will be labeled *4*.

Scale X-Axis in Plots Manually (Min, Step)

The setting of this check box will affect the scaling of the *X*-axis in line plots of single or multiple variables (series). The choice of scaling will be retained across other analytical procedures in this module, for example, when plotting forecasts from an ARIMA analysis.

By default, if this check box is not set, the *X*-axis in all plots will always be scaled so that:

(1) All valid data points are displayed in the plot, and

(2) The tickmarks (steps) on the *X*-axis are "neat" (i.e., the significant digit of the step size is either 1, 2, or 5).

However, if there is a known periodicity in the data that is of interest, such as 12 months per year or 10 observations per day, etc., then it may be useful to set this check box and to scale the *X*-axis for plots accordingly.

Autocorrelations

This option will bring up a Scrollsheet and plot of the autocorrelations, for a lag of *1* through the number specified in the *Number of lags* edit field. The Scrollsheet will report the autocorrelations, their standard errors, the so-called *Box-Ljung statistic*, and the significance level of that statistic.

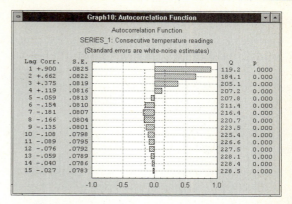

An autocorrelation is the correlation of a series with itself, shifted by a particular lag of *k* observations. The plot of autocorrelations for various lags is a crucial tool for determining an appropriate model for ARIMA analysis (see the *ARIMA Introductory Overview,* page 3273). The computations of the autocorrelation coefficients r_k follow the standard formulas, as described in most time series references (e.g., Box and Jenkins, 1976).

Standard error of r_k. Under the assumption that the true moving average process in the series is of the order *k-1*, then the approximate standard error of r_k is defined as:

```
StdErr(r_k) = √{(1/n) * [1+2*Σ(r_i^2)]}
(for i = 1 to k-1)
```

Here *n* is the number of observations in the series. However, under the assumption that the series is a white noise process, that is, that all autocorrelations are equal to zero, the standard error of r_k is defined as:

```
StdErr(r_k) = √{(1/n) * [(n-k)/(n+2)]}
```

Check the *White noise standard errors* option to compute the standard errors in this manner.

Box-Ljung Q. At a given lag *k* the Box-Ljung *Q* statistic is defined by:

$$Q_k = n*(n+2)*\Sigma[r_i^2(n-1)]$$

When the number of observations is large, then the Q statistic has a *Chi-square* distribution with k-p-q degrees of freedom, where p and q are the number of autoregressive and moving average parameters, respectively.

Alpha for Highlighting

Significant autocorrelations (significant *Box-Ljung* statistics) will be highlighted in the Scrollsheet. This edit field allows you to specify the significance level that is used for highlighting.

White Noise Standard Errors

If this check box is set, then the standard errors for the autocorrelations will be computed based on the assumption that the input series is a white noise series (see above).

Partial Autocorrelations

This option will bring up a Scrollsheet and plot of the partial autocorrelations, for a lag of *1* through the number specified in the *Number of lags* edit field.

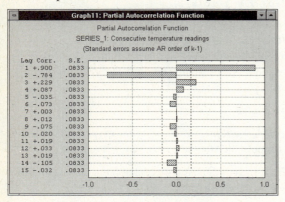

The Scrollsheet will report the partial autocorrelations and their standard errors. In general, the partial autocorrelation is the partial correlation of a series with itself, lagged by a particular number of observations, and controlling for all correlations for lags of lower order. For example, the partial autocorrelation for a lag of *2* represents the unique correlation of the series with itself at that lag, after controlling for the correlation at lag *1*. The computations of the partial autocorrelation coefficients ϕ_k follow the standard formulas, as described in most time series references (e.g., Box and Jenkins, 1976).

Standard error of ϕ_k. Under the assumption that the true autoregressive process in the series is of order $p \leq k-1$, then the approximate standard error of ϕ_k is defined as:

$$\texttt{StdErr}(\phi_k) = \sqrt{(1/n)}$$

Here n is the number of observations in the series.

Crosscorrelations

This option will bring up a Scrollsheet and plot of crosscorrelations, for a lag of -k through +k, where k is the number specified in the *Number of lags* edit field (see below). The Scrollsheet will report the crosscorrelations and their standard errors. In general, the crosscorrelation is the correlation of a series with another series, shifted by a particular number of observations. Note that the crosscorrelation function is not symmetrical about the lag of *0*; that is, different correlations will emerge depending on whether the *X*-axis is shifted forward or backwards. The crosscorrelation coefficient is computed following the standard formulas, as described in most time series references (e.g., Box and Jenkins, 1976):

```
r_xy = (1/n)*Σ[(X_t-X-bar)*(Y_t+1-Y-bar)]
(for t=1 to n-1; l=0 to k)

r_xy = (1/n)*Σ[(Y_t-Y-bar)*(X_t+1-X-bar)]
(for t=1 to n+1; l=-1 to -k)
```

Where *X-bar* and *Y-bar* are the means of the respective two series.

Standard error of $r_{xy(k)}$. Under the assumption that X and Y are independent, and that one of the series consists only of white noise, then the standard error of the crosscorrelation is approximately equal to:

$$\text{StdErr}(r_{xy(k)}) = \sqrt{[1/(n-k)]}$$

Where *n* is the number of observations in the series.

Number of Lags

This parameter determines the maximum number of lags for which the auto-, partial auto-, and crosscorrelations will be computed.

2D Scatterplot

This option will bring up a 2D scatterplot of the highlighted variable with another variable in the *active work area*; in this plot the values of the highlighted variable will be lagged by *k* observations, as specified in the field *Plot highlighted variable with lag of...* After clicking on this button, you will be prompted to select the second variable for the scatterplot. To review specific autocorrelations, choose the same variable for both axes of the plot.

3D Scatterplot

This option will bring up a 3D scatterplot of the highlighted variable with two other variables in the *active work area*; in this plot the values of the highlighted variable will be lagged by *k* observations, as specified in the field *Plot highlighted variable with lag of...* After clicking on this button, you will be prompted to select two additional variables for the scatterplot.

Label Points in Scatterplots

If this check box is set, the points in the scatterplot will be labeled with either case names, case numbers, dates, or consecutive integers, depending on the setting of the radio buttons in the *Label data points* box.

Histogram

This option will bring up a histogram of the currently highlighted variable.

Descriptive Statistics

This option will bring up a Scrollsheet with descriptive statistics for all variables currently available in the *active work area*.

Normal, Detrended, and Half-Normal Probability Plots

These options will bring up the respective normal probability plot for the highlighted variable. The way the *standard normal probability* plot is constructed is as follows. First the values are rank ordered. From these ranks, *z* values (i.e., standardized values of the normal distribution) can be computed *based on the assumption* that the data come from a normal distribution. These *z* values are plotted on the *Y*-axis in the plot. If the data values (plotted on the *X*-axis) are normally distributed, then all points should fall onto a straight line in the plot. If the values are not normally distributed, they will deviate from the line. Outliers may also become evident in this plot.

The *detrended normal probability plot* is constructed in the same way as the standard normal probability plot, except that the linear trend is removed. This often "spreads out" the plot, thereby allowing the user to detect patterns of deviations more easily.

8. TIME SERIES ANALYSIS - TRANSFORMATIONS

The *half-normal probability plot* is constructed in the same way as the standard normal probability plot, except that only the positive half of the normal curve is considered. Consequently, only positive normal values will be plotted on the *Y*-axis. This plot is basically used when one wants to ignore the sign of the residual, that is, when one is mostly interested in the distribution of absolute values, regardless of sign.

General Transformations

Select the desired type of transformation in this dialog. The new transformed variable will be appended to the *active work area* (see the *General Conventions*, page 3313).

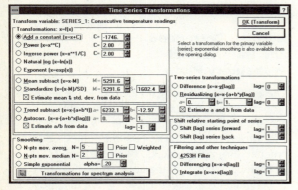

If a transformation produces an invalid value (e.g., square root of a negative value, division by zero, etc.), the program will assign a missing data value; then, in a second pass, those missing data will be replaced according to the choice in the *Replace missing data* box on the startup panel (see the *General Conventions* section, page 3314).

Simple Transformations: x=f(x)

Add a constant. A constant *C* will be added to all values in the *Transform variable*. By default, this constant will be equal to minus the smallest value in the series, so that after the transformation the minimum value of the series will be equal to *0* (zero).

Power. Each value in the series will be transformed as: $X=X^C$.

Inverse power. Each value in the series will be transformed as: $X=X^{(1/C)}$.

Natural log. Each value in the series will be transformed as: $X=natural\ log(X)$.

Exponent. Each value in the series will be transformed as: $X=e^X$ (where *e* is Euler's constant).

Mean subtract. Each value in the series is transformed as $X=X-M$; where *M* is either (1) the overall mean for the untransformed series, or (2) the value indicated in the *M*= edit field, depending on the setting of the *Estimate mean & std. dev. from data* check box (see below).

Standardize. Each value in the series is standardized: $X=(X-M)/SD$; where *M* and *SD* are either (1) the overall mean and standard deviation for the untransformed series, or (2) the values indicated in the respective edit fields to the right, depending on the setting of the *Estimate mean & std. dev. from data* check box (see below).

Estimate mean & std. dev. from data. This check box determines whether, for the two transformations described in the previous two paragraphs, the mean and/or standard deviation is to be computed from the data, or whether user-defined values are to be used.

Trend subtract. The values in the series will be transformed to remove the trend over time: $X=X-(a+b*t)$. In this formula, *t* refers to the case number and *a* and *b* are constants which are either user-defined or computed from the data, depending on the setting of the *Estimate a/b from data* check box (see below).

Autocorr. The values in the series will be transformed to remove an autocorrelation of a particular lag: $X=X-(a+b*X(lag))$. In this formula, a and b are constants which are either user-defined or computed from the data (via least squares regression), depending on the setting of the *Estimate a/b from data* check box (see below). The lag in the computations is determined by the value in the *Lag* edit field below.

Estimate a/b from data. This check box determines whether, for the two transformations described in the previous two paragraphs, the intercept (a) and slope (b) are to be computed from the data (via least squares regression), or whether user-defined values are to be used.

Smoothing

The general purpose of smoothing techniques is to "bring out" the major patterns or trends in a time series, while de-emphasizing minor fluctuations (random noise). Visually, as a result of smoothing, a jagged line pattern should be transformed into a smooth curve.

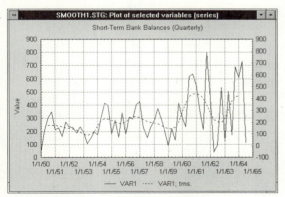

A number of techniques are available:

N points moving average. Each point in the transformed series is computed as the mean of n adjacent points (the so-called moving average *window*). If the n parameter is odd, then the moving average is naturally centered in the middle of the moving average window. If n is even, then the moving average is centered by averaging each pair of uncentered means. If the *Prior* check box is set, then the moving average will be computed from the n preceding values. If the *Weighted* check box is set, then a weighted moving average will be computed. In that case, after clicking *OK*, a dialog will come up for specifying the respective weights.

N points moving median. Each point in the transformed series is computed as the median of n adjacent points. If the n parameter is odd, then the moving median is naturally centered in the middle of the moving median window. If n is even, the moving median is centered by averaging each pair of uncentered medians. If the *Prior* check box is set, then the moving median will be computed from the n preceding values.

Simple exponential smoothing. In this transformation, each point is computed as a weighted average of all preceding observations, where greater weight is assigned to more recent observations (specifically, the weights decrease geometrically with the "age" of prior observations). Algorithmically, this is accomplished by:

$$S_t = \alpha * X_t + (1-\alpha) * S_{t-1}$$

where S_t is the value of the transformed series at time t, S_{t-1} is the value of the transformed series at time $t-1$, X_t is the value of the untransformed series at time t, and α (*alpha*) is a constant ($0<\alpha<1$). If α is close to *0* (zero), then more emphasis is placed on (greater weight is assigned to) observations prior to a particular time t, resulting in a smoother curve; if α is close to *1*, then more emphasis is placed on the respective untransformed observation at time t, resulting in a curve that is less smooth, but more closely following the actual (untransformed) data. Note that a complete implementation of exponential smoothing and forecasting is also available from the startup panel. See also the *Exponential Smoothing* overview section (page 3279).

Transformations for spectrum analysis.

When you click on this button, the *Transformations for Spectrum Analysis* dialog will open (see page 3337), in which you can select from among several transformations commonly used in the context of spectrum analysis (see also the *Spectrum Analysis* overview, page 3300).

Two-Series Transformations

These transformations require two variables (series). Thus, after selecting any of these transformations and clicking *OK*, you will be prompted to select the second (*Y*) variable for the respective transformation (the first or *X* variable will be the currently highlighted variable in the *active work area*).

Difference. The highlighted series (*X*) is transformed by: $X=X-Y(lag)$. The *Lag* can be specified in the edit field to the right of this option.

Residualizing. The highlighted series (*X*) is transformed by: $X=X-(a+b*y(lag))$. The *Lag* can be specified in the edit field to the right of this option. The parameters *a* and *b* are the intercept and slope, respectively, of the (lagged) crosscorrelation. These parameters can either be estimated via least squares regression from the data (set the *Estimate a and b from data* check box), or they can be entered manually.

Shift Relative Starting Point of Series

These transformations allow you to shift the series forward or backward, relative to all other series currently in the *active work area*.

Shift (lag) series forward. The series will be shifted forward by the number of observations specified in the *Lag* edit field.

Shift (lag) series backward. The series will be shifted backward by the number of observations specified in the *Lag* edit field.

Filtering and Other Techniques

4253H filter. This transformation consists of several passes of moving average/median smoothing and is a powerful filter for smoothing a series. The following transformations will be performed:

(1) A 4 points moving median centered by a moving median of 2,

(2) A 5 point moving median,

(3) A 3 point moving median, and

(4) A 3-point weighted moving average using Hanning weights (.25, .5, .25),

(5) Residuals are computed by subtracting the transformed series from the original series,

(6) Steps 1 through 4 are then repeated for the residuals,

(7) The transformed residuals are added to the transformed series.

In practice, this filtering method often produces a smooth series while maintaining the salient characteristics of the original series; here is an example:

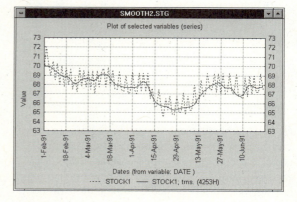

Differencing. The series will be transformed as: $X=X-X(lag)$. After differencing, the resulting series will be of length $N-lag$ (where n is the length of the original series).

Integrate. This transformation is the reverse of differencing. The series will be transformed as: $X=X+X(lag)$. After clicking *OK* to begin the transformations you will be prompted to select a variable with the *leading observations*. If none is selected, then all leading observations will be equal to *0* (zero).

For example, suppose the series consists of the values: 3, 4, 2, etc. If you integrate the series with a lag of *1*, and choose a variable without leading observations, then the first value of the new series will thus be *0* (zero); the next value will be 0+3, the next 0+3+4, and so on. The resulting series will be of length $n+1$, where n is the number of cases in the original series.

Now, suppose you selected a variable with leading observations, and that the value found in that series for case *1-lag* (that is, in this example, for the case just preceding the first case in the series to be transformed) is equal to 5. Then the first value of the new series will be 5, the next value will be 5+3, the next 5+3+2, and so on. In this manner, one can difference a series and fully restore it to its original values by integrating it, provided one uses the values of the original (untransformed) series for leading observations.

Transformations for Spectrum Analysis

You can select the desired type of transformation in this dialog. The new transformed variable will be appended to the *active work area* (see *General Conventions*, page 3313, for details). If a transformation produces an invalid value (e.g., square root of a negative value, division by zero, etc.), the program will assign a missing data value; then, in a second pass, those missing data will be replaced according to the choice in the *Replace missing data* box on the startup panel (see *General Conventions*, page 3314).

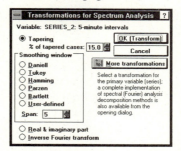

The transformations on this dialog are customarily performed on series (results) produced by a spectrum (Fourier) analysis (see *Introductory Overview*, page 3300, for a discussion of this technique). Forward Fourier analysis and inverse Fourier analysis is also used for constructing filters at particular frequencies (see below).

Tapering

The process of so-called *split-cosine-bell tapering* is a recommended transformation of a series prior to spectrum analysis. It usually leads to a reduction of leakage in the periodogram. The rationale for this transformation is explained in detail in Bloomfield (1976, page 80-94), see also the respective part of the *Introductory Overview*, page 3303.

Smoothing Window

As described in greater detail in the *Single Spectrum (Fourier) Analysis* overview section, the periodogram values computed in a spectrum analysis are subject to substantial random fluctuation. A clearer picture of underlying periodicities often only emerges when examining the spectral densities, that is, the frequency regions, consisting of many adjacent frequencies, that contribute most to the overall periodic behavior of the series. The spectral density estimates can be computed by smoothing the periodogram values with a weighted moving average. Suppose the moving average window is of width *m* as specified in the *Span* edit field (which must be an odd number, see below); the following smoothers listed below are those most commonly cited in the literature.

Daniell (or equal weight) window. This transformation amounts to a simple (equal weight) moving average transformation of the periodogram values, that is, each spectral density estimate is computed as the mean of the *m*/2 preceding and subsequent periodogram values. Note also that the *Time Series* module will standardize the weights so that they sum to 1.

Tukey window. In the Tukey (Blackman and Tukey, 1958) or Tukey-Hanning window (named after Julius Von Hann), for each frequency, the weights for the weighted moving average of the periodogram values are computed as:

```
w_j  = 0.5+0.5*cos(π*j/p)  (for j=0 to p)
w_-j = w_j                  (for j≠0)
```

Hamming window. In the Hamming (named after R. W. Hamming) window or Tukey-Hamming window (Blackman and Tukey, 1958), for each frequency, the weights for the weighted moving average of the periodogram values are computed as:

```
w_j  = 0.54+0.46*cos(π*j/p) (for j=0 to p)
w_-j = w_j                   (for j≠0)
```

Parzen window. In the Parzen window (Parzen, 1961), for each frequency, the weights for the weighted moving average of the periodogram values are computed as:

```
w_j  = 1-6*(j/p)^2 +6*(j/p)^3 (for j=0 to p/2)
w_j  = 2*(1-j/p)^3            (for j=p/2+1 to p)
w_-j = w_j                     (for j≠0)
```

Bartlett window. In the Bartlett window (Bartlett, 1950) the weights are computed as:

```
w_j = 1-(j/p)  (for j=0 to p)
```

```
w_-j = w_j    (for j≠0)
```

User-defined window. If this radio button is selected, then you are prompted to directly enter the weights used for smoothing (a dialog will open prompting you to specify the weights).

Span. This edit field determines the width of the smoothing window. The number entered here must be odd and greater than or equal to 3.

Real & Imaginary Part

This option will perform a Fourier transformation on the input series, resulting in a real and imaginary part (i.e., two new series are created, representing complex numbers; refer to the *Spectrum (Fourier) Analysis* introductory overview, page 3300, for details). The transformation is accomplished via the so-called *fast Fourier transform*, or *FFT* for short (first popularized by J.W. Cooley and J.W. Tukey, 1965). The implementation of the FFT algorithm in the *Time Series* module allows the user to take full advantage of the savings afforded by this algorithm.

On most standard computers, series with over 100,000 cases can easily be analyzed. However, there are a few things to remember when analyzing series of that size. The standard (and most efficient) FFT algorithm requires that the length of the input series is a power of 2. If this is not the case, then additional computations have to be performed. In order to still utilize the FFT algorithm, an implementation of the general approach described by Monro and Branch (1976) is used in the *Time Series* module. This method requires significantly more storage space; however, series of considerable length can still be analyzed very quickly, even if the number of observations is not equal to a power of 2.

The Fourier transform and its inverse transformation (see below) are often used to construct frequency filters (see, for example, Bloomfield, 1976). For example, if a single spectrum (Fourier) analysis of a series indicates no significant periodicities at high

8. TIME SERIES ANALYSIS - TRANSFORMATIONS

frequencies, then one may want to filter the series so as to completely eliminate all high frequency fluctuations. This can be accomplished by first performing the Fourier transformation, and then modifying the real and imaginary parts of the series at the respective frequencies, and finally recomputing the series via the inverse Fourier transform (see below).

Inverse Fourier Transform

This is the inverse of the Fourier transform described above (using the same computational algorithms). Note that the currently highlighted variable will be interpreted as the real part of the series, and you will be prompted to select the imaginary part after clicking *OK*.

ARIMA

EXAMPLES

Example 1: Single Series ARIMA

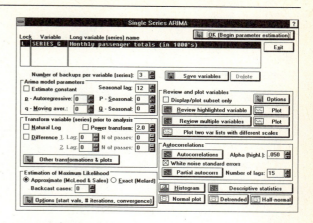

The following example is based on Chapter 9 from the classic book on ARIMA by Box and Jenkins (1976). The data are the monthly passenger totals (measured in thousands) in international air travel, for twelve consecutive years: 1949-1960 (see Box and Jenkins, 1976, page 531, "Series G").

The data are partially listed below; they are also included with your program in example data file *Series_g.sta* (variable *Series_g*).

Data: SERIES	Monthly pas SERIES_G
JAN 1949	112
FEB 1949	118
MAR 1949	132
APR 1949	129
MAY 1949	121
JUN 1949	135
JUL 1949	148
AUG 1949	148
SEP 1949	136
OCT 1949	119
NOV 1949	104
DEC 1949	118
JAN 1950	115

The series shows a clear growing trend over the years but at the same time there is strong seasonality present in the data (e.g., March figures are usually higher than February and April).

After starting the *Time Series* module, first open the data file *Series_g.sta*, and then click on the *Variables* button to select the first variable. Now, in the *Time Series Analysis* dialog, select the *ARIMA & autocorrelation functions* option.

After all 144 data points are read, the *Single Series ARIMA* dialog will be displayed.

Identification Phase

Before you can specify the parameters to be estimated in ARIMA, the model first has to be identified (see also the *Introductory Overview*, page 3275). Although most of the necessary options such as *Autocorrelations* or *Partial autocorrs.* are available on this dialog, the transformation options will be used to illustrate how the nature of the ARIMA process can be identified. Therefore, click on the *Other transformations & plots* button to bring up the *Transformations of Variables* dialog.

On this dialog, first select a more appropriate scaling for the horizontal *X*-axis in subsequent line plots. Set the *Scale X-axis in plots manually* check box, and then enter *1* as the *Min*, and *12* as the *Step* (12 months per year). Also, the data file *Series_g.sta* contains case names with the respective dates for each observation. You can use those labels to identify the different years in subsequent line plots; thus, set the *Case names* radio button in the *Label data points* box. These choices for the labeling and scaling of the horizontal *X*-axis in line plots are "module-wide;" that is, they will be used throughout the *Time Series* module whenever line plots of series are requested.

Refer to the description of the *Transformations of Variables* dialog (page 3329) for a detailed discussion of how line plots can be scaled and

8. TIME SERIES ANALYSIS - ARIMA

labeled in the *Time Series* module. The *Transformations of Variables* dialog will now look like this:

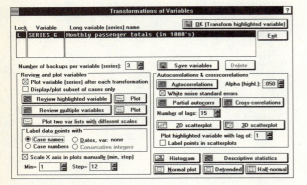

To produce a plot of the time series, now click on the *Plot* button next to the *Review highlighted variable* button.

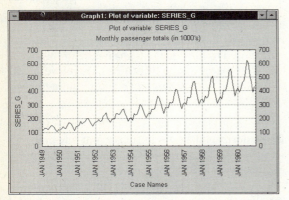

Both the trend and seasonality of the series are very apparent. In order to identify the specific lags for ARIMA differencing, autocorrelations and partial autocorrelations will be used (refer also to the *Introductory Overview*, page 3273, for a general discussion of these statistics).

Multiplicative seasonality. It is also clear from the plot of the series that the amplitude of the seasonal changes increases over time (i.e., there is evidence of *multiplicative seasonality*, see the *Introductory Overview*, page 3270), which may bias the values of autocorrelations. A natural-log transformation of the data will be performed to stabilize this variability.

Logarithmic transformation. Click *Continue* on the time series graph to bring the *Transformations of Variables* dialog back up; then click *OK (Transform highlighted variable)* to list the available *Time Series Transformations*.

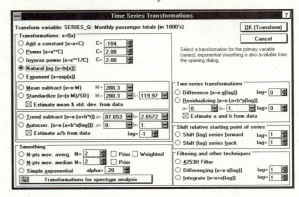

In this dialog, select the *Natural log (x=ln(x))* transformation and click *OK*. After all observations have been transformed, the transformed series will automatically be plotted (by default, i.e., if you did not de-select the option *Plot variable (series) after each transformation* in the *Transformation of Variables* dialog).

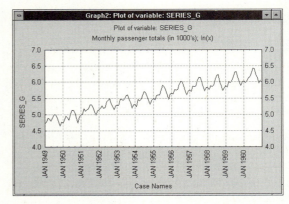

Note that the desired effect has been achieved, as the amplitude of changes is now fairly stable and the series is ready for further analyses with autocorrelations.

Autocorrelations. Click *Continue* to once again bring up the *Transformations of Variables* dialog. Change the default value for the *Number of lags* parameter in the *Autocorrelations & cross-correlations* box from *15* to *25*. Then, click on the *Autocorrelations* button to bring up a Scrollsheet with the autocorrelations, and the autocorrelation plot.

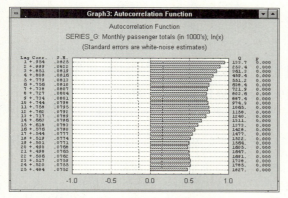

The plot indicates strong serial dependencies for lags of *1* to *12* with the highest value of autocorrelation for a lag of one.

Differencing. In order to remove the serial dependency, a nonseasonal differencing transformation will first be performed on the series, that is, difference it with lag of *1*. Click *Continue* to bring up the *Transformations of Variables* dialog again.

Note that the transformed (*log*-ed) series is automatically selected (highlighted) in the *active work area* (refer to the *General Conventions*, page 3313, for a description of the memory management in the *active work area*). Thus, simply click *OK (Transform highlighted variable)*, select transformation *Differencing (x=x-x(lag))* (do not change the default *lag* value of *1*) in the *Filtering and other techniques* box, and click *OK (Transform)*. After all cases have been transformed, the transformed (differenced) series will again (by default) be plotted.

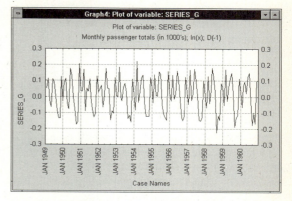

Now, each element of the transformed series represents the difference between its original (i.e., previous) value and the original value of its adjacent element. Note that the series is now shorter (by the number of elements equal to the lag, i.e., *1*) since the first element of the series could not be differenced.

Click *Continue* to return to the *Transformations of Variables* dialog and again select the *Autocorrelations* option.

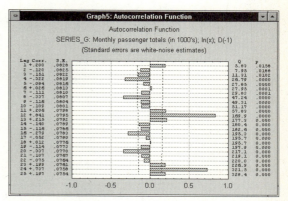

Not only the first order (i.e., for lag of *1*) but also most other serial dependencies have disappeared (as explained before, autocorrelations for consecutive lags are interdependent).

8. TIME SERIES ANALYSIS - ARIMA

Seasonal serial dependency. However, as it often happens, removal of the lower order serial dependencies exposed a higher order seasonality (for lag of *12*). There is also a clear (seasonal) dependency for a lag of *24* (and other multiples of 12, such as 36, 48, etc.). This indicates a strong seasonal pattern. This dependency reflects the seasonality of airline traffic (there are months when people travel more and months when they travel less).

Seasonal differencing. Seasonal differencing with a lag of *12* will take care of this dependency. Return to the *Transformations of Variables* dialog and click *OK*. Select again *Differencing (x=x-x(lag))*, but change the *lag* value to *12*. Then, click *OK* to perform the differencing transformation. Again, by default the transformed series will be plotted (de-select the *Plot variable (series) after each transformation* check box if you do not want to plot the series after each transformation). As before, bring the *Transformations of Variables* dialog back up and select the *Autocorrelations* options.

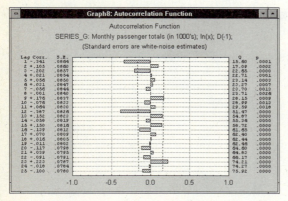

Most of the strong autocorrelations seemed to have been removed now. Even though there are still some autocorrelations which are larger than 2 times their standard errors (as indicated by the dotted line in the autocorrelation plot), one should be careful not to over-difference because this may cancel out the effects of moving average parameters.

Click *Continue* and then select the *Partial autocorrelations* option (i.e., autocorrelations controlling for all "within-lag" correlations).

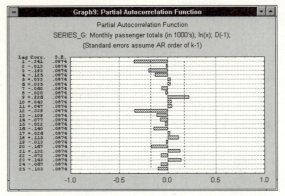

Parameters to be estimated. The correlogram looks well and the series is now ready for ARIMA. Based on the exploration of the nature of the series (i.e., the *identification* phase of ARIMA), a seasonal ARIMA (lag=*12*) will be run on data that will be differenced nonseasonally (lag=*1*) and seasonally (lag=*12*), and transformed (with natural logarithms) within ARIMA. Two moving average ARIMA parameters will be estimated: one seasonal (Q_s) and one nonseasonal (q). No autoregressive parameters will be estimated. A review of the general rules of the identification phase of ARIMA is provided in the *Introductory Overview* section at the beginning of this chapter (page 3273). Refer to the book by Box and Jenkins (1976) for a more comprehensive discussion of this example (Box and Jenkins, 1976, Chapter 9). Hoff (1983), McCleary and Hay (1980), McDowall, McCleary, Meidinger, and Hay (1980), Pankratz (1983), and Vandaele (1983) discuss numerous examples of how to identify ARIMA models based on data plots and plots of autocorrelation and partial autocorrelation functions.

ARIMA Integrated Transformations

Previously, it was found that the data required logarithmic transformation and two types of

TIM - 3344

Copyright © StatSoft, 1995

8. TIME SERIES ANALYSIS - ARIMA

differencing (nonseasonal and seasonal). All these modifications of the data were already performed (via *Transformations*) and their results were reviewed using the line plots and plots of autocorrelations. The modified series (currently residing in the *active work area*) can now be submitted directly to ARIMA. However, in cases like this, it is recommended to rather analyze the original series and select the necessary modifications of the data from within ARIMA (as part of the ARIMA specifications). This way, ARIMA will "know" about the modifications. If you wish to calculate forecasts (after the ARIMA parameters are estimated), these forecasts will be calculated from integrated (i.e., "re-differenced") and "re-modified" data, and they will be compatible with the original raw data (i.e., much easier to interpret).

Note that only logarithmic/power transformations and seasonal/nonseasonal differencing are available from within ARIMA. In some cases, some other data transformations are recommended before ARIMA. In those cases, complete the transformations before entering ARIMA. These transformations (e.g., smoothing) usually do not change the range of data and they do not have to be re-modified.

ARIMA Specifications Dialog

Now return to the *Single Series ARIMA* dialog by clicking *Exit* on the *Transformation of Variables* dialog. In this dialog, highlight the original (untransformed) variable *Series_G*. The *Single Series ARIMA* dialog allows you to specify autoregressive and moving average parameters to be estimated (seasonal or nonseasonal). You cannot proceed to the next step until at least one of the autoregressive or moving average parameters that are to be estimated (*P*, *p*, *Q*, or *q*) is selected. However, first you will have to specify the transformations and differencing.

In the *Transform variable (series) prior to analysis* box, select the *Natural Log* and *Difference* options. Then, select a *Lag* of *1* and set the *No. of passes* to *1*. You have now specified the *log*-transformation and the non-seasonal simple differencing transformation. To specify the seasonal differencing transformation, specify in the *2. Lag* edit field a lag of *12* and again set the *No. of passes* parameter to *1*.

ARIMA parameters. You still need to specify the ARIMA model parameters. You cannot start ARIMA until at least one of the autoregressive or moving average parameters to be estimated (*p*, *P*, *q*, or *Q*) has been requested. In the identification phase of ARIMA, it was decided to estimate two moving average parameters, one regular (*q*) and one seasonal (*Q*), and no autoregressive parameters. Shown below is the *Single Series ARIMA* dialog with all necessary settings in place.

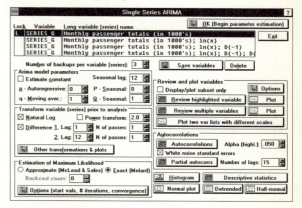

Parameter estimation. As described in the *Introductory Overview* section (page 3276), parameter estimation in ARIMA models is accomplished by maximizing the likelihood (probability) of the data, given particular values of the parameters. The *Time Series* module offers two methods for computing the *Maximum Likelihood* value for a particular ARIMA model: *Approximate (McLeod & Sales)* (with or without backcasting) and *Exact (Melard)*. The pro's and con's of each method are discussed in the *Introductory Overview*

8. TIME SERIES ANALYSIS - ARIMA

section. For this example, select the *Exact* method. [Note that Box and Jenkins, 1976, used a recursive method with backcasting; to reproduce their results (page 319) choose the *Approximate* method and set *Backcast cases* to *13*.] Next, click *OK (Begin parameter estimation)*, and the iterative parameter estimation procedure will begin (refer to the *Introductory Overview*, page 3276, for computational details).

Results

After the estimation process converges, click *OK* to bring up the *Single Series ARIMA Results* dialog.

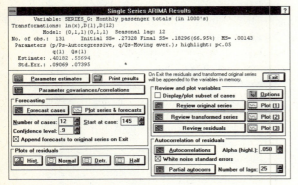

ARIMA output. The info box at the top of the *Single Series ARIMA Results* dialog summarizes the ARIMA specifications (ARIMA model, transformations), the parameter estimates, and their standard errors. Click on the *Parameter estimates* button to review the parameter estimates in a Scrollsheet.

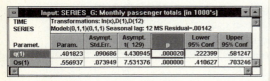

Both the non-seasonal and seasonal moving average parameters are highly statistically significant.

Forecast options. The *Forecasting* box contains options for the calculation of forecasts. By default, the program will compute forecasts for one full seasonal cycle following the last observed value, that is, following case 144 (starting at case 145). First look at the forecasts in a Scrollsheet; click on the *Forecast cases* button. The Scrollsheet will contain the forecasts and their confidence intervals; note that, had you requested forecasts for cases that were also observed, the Scrollsheet would also display the observed and residual values.

CaseNo.	Forecast	Lower 90.0000%	Upper 90.0000%
145	450.3939	423.1511	479.3906
146	425.6906	395.8395	457.7928
147	478.9716	441.3850	519.7590
148	492.3758	450.0683	538.6603
149	509.0141	461.8264	561.0232
150	583.2924	525.5728	647.3510
151	669.9440	599.7494	748.3542
152	667.0088	593.4766	749.6517
153	558.1345	493.7240	630.9478
154	497.1610	437.3513	565.1500
155	429.8310	376.1138	491.2202
156	477.1891	415.4215	548.1407

Plot of forecasts. A much better "picture" of how well the forecasts extend the observed series can be gained by plotting the forecasts, together with the observed values. Click *Continue* to bring up the *Single Series ARIMA Results* dialog again, and click on the *Plot series & forecasts*.

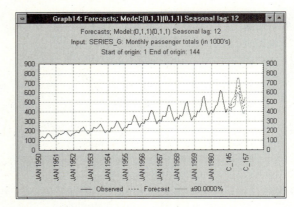

The plot displays the original series, forecasts, and their confidence limits (for the probability set in the *Single Series ARIMA Results* dialog); remember that earlier, you requested to label the horizontal *X*-axis

TIM - 3346

Copyright © StatSoft, 1995

of line plots with case names, and to use a step increment of 12 to accurately reflect the successive years. Additionally in the plot above, the *Value layout* list box on the *Graphics Scale Options: X* dialog was set to *Normal*, in order to make the *X*-axis labels more legible (refer to the *Graphics* options, Volume II, for details). Reviewing the graph, the forecasts generated by your ARIMA model seem to extend the observed series in a reasonable manner.

Exit the graph now (click *Continue*) and return to the *Single Series ARIMA Results* dialog in order to see how well the current ARIMA model will predict the last 12 observed cases of the series. Set the *Start at case* edit field to *133* (i.e., 144-12+1) and then click again on the *Plot series & forecasts* button.

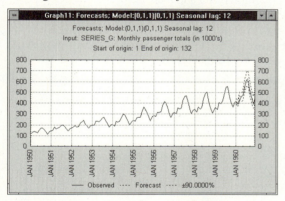

Again, the two-parameter ARIMA model seems to fit the series very well, and the observed values are well within the confidence band of predicted values.

Analysis of Residuals

So far, it looks like the model produces an adequate fit to the data. However, there are other important diagnostics that one should review. Two important assumptions of the ARIMA model are that (1) the residuals (observed minus predicted values) are normally distributed, and (2) that they are independent of each other, that is, that there is no residual serial correlation "left in the data." If the latter condition is not met, then you probably have overlooked an additional parameter that is "driving" the series (the underlying ARIMA process that produces the time series values).

Normal probability plots. The first assumption -- normal distribution of residuals -- can be tested by examining the normal probability plots of residuals. Shown below are the *Normal* and *Detrended* normal probability plots.

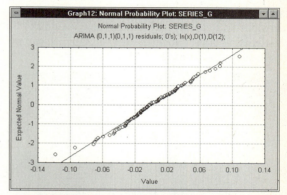

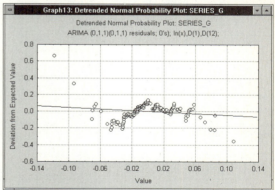

The way the standard *Normal plot* is constructed is as follows. First the residuals are rank ordered. From these ranks you can compute z values (i.e., standardized values of the normal distribution) *based on the assumption* that the residuals come from a normal distribution. These z values are plotted on the *Y*-axis in the plot. If the residuals

8. TIME SERIES ANALYSIS - ARIMA

(plotted on the *X*-axis) are normally distributed, then all points should fall onto a straight line in the plot, as is the case in the plot shown above.

The *Detrended* normal probability plot is constructed in the same way as the standard normal probability plot, except that, before the plot is produced, the linear trend is removed. This often "spreads out" the plot, thereby allowing the user to detect patterns of deviations more easily. In the plot above, most residuals cluster closely around the horizontal line, and again it looks like the residuals are indeed normally distributed. The *Histogram* of the residuals shows how well the normal distribution fits the actual distribution of residuals.

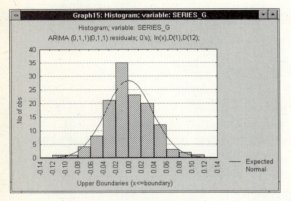

Autocorrelation of residuals. Now, turn your attention to the second assumption of ARIMA -- that the residuals are independent of each other.

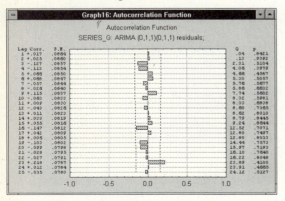

This can be tested by plotting the autocorrelation function (click on the *Autocorrelations* button on the *Single Series ARIMA Results* dialog). It is apparent in the plot above, that there is practically no residual autocorrelation left after you fit the current ARIMA model to the data. Therefore, you can be satisfied that the condition of independent residuals is also met.

Further Analyses

When you exit the *Single Series ARIMA Results* dialog, the ARIMA residuals will automatically be appended to the *active work area*. Also, if the *Append forecasts to original series on Exit* check box is set (which it is by default), another series with the original data and the forecasts will be appended to the *active work area*. Now, exit the *Single Series ARIMA Results* dialog by clicking on the *Exit* button; the *Single Series ARIMA* specification dialog will again be displayed.

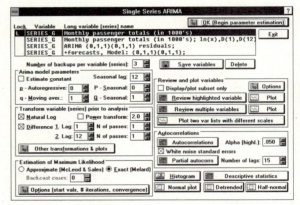

As you can see, both the residuals and forecasts were added to the *active work area*. To "finish up" your analysis, look at another informative summary plot. Plot the original series, forecasts, and residuals all in the same plot. Such a plot may reveal other "problems" with your ARIMA model, for example, if the residuals are particularly large, and the fit particularly poor in one region of the series (e.g., there may be a three-year period where the ARIMA model consistently predicts more international airline passengers than were observed). Because the values of the residuals and observed series (and forecasts) are not compatible (remember that the residuals pertain to the *log*-ed and twice differenced series, while the forecasts are generated in terms of the unmodified series), you need to *Plot two var lists with different scales*. Select as the first variable to plot, the original series with the forecasts added and as the second variable (in the second window) select the ARIMA residuals.

Now click *OK* to see the graph.

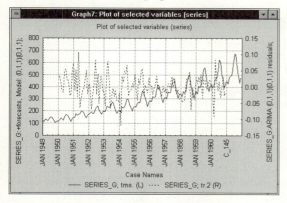

Again, the fit of the ARIMA model seems very good; that is, the residuals show about equal variability across the entire length of the series, and there is no evidence of any trend or drift.

Example 2: Interrupted ARIMA

The following example is based on data presented in McDowall, McCleary, Meidinger, and Hay (1980), who provide an excellent applications-oriented introduction to interrupted time series analysis (see also the *Introductory Overview*, page 3278, for a discussion).

The data in the example file *Director.sta* will be used. That file contains the number of monthly calls (in 100's) to Cincinnati Directory Assistance over the period from January, 1962, through December, 1976. In March 1974 (the 147'th month in the series), Cincinnati Bell initiated a new 20 cent charge for calls to Directory Assistance. This caused a marked drop-off in the number of requests for assistance, and the purpose of this analysis is to fit a model to the series that takes this abrupt change into account. Shown below is a listing of the first few cases in the data file *Director.sta*.

	MONTH	CALLS
1	Jan-1962	350
2	Feb-1962	339
3	Mar-1962	351
4	Apr-1962	364
5	May-1962	369
6	Jun-1962	331
7	Jul-1962	331
8	Aug-1962	340
9	Sep-1962	346
10	Oct-1962	341
11	Nov-1962	357
12	Dec-1962	398

Model Identification

As in *Example 1* above, the first step is to identify an appropriate model for the series. In this case, a regular ARIMA model will first be fit to the observations prior to the introduction of the service charge; then, an intervention component will be added to analyze the complete series.

After starting the *Time Series* module, open the data file *Director.sta*. There are 146 months of observations in the series prior to the introduction of the service charge. To open only those observations into the *active work area*, click on the *Selection Conditions* button (below the *Open Data* button on

8. TIME SERIES ANALYSIS - ARIMA

the opening panel) and specify the condition *Include if v0<147*.

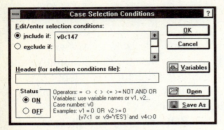

Chapter 1 of Volume I of the manual describes all general *STATISTICA* conventions, including the case selection conditions. *V0* in the selection condition shown above refers to the case numbers, so only cases numbers less than 147 will be included in the analysis. Now click *OK*, then click on the *Variables* button and select variable *Calls*. Then select the *ARIMA & autocorrelation functions* option to bring up the *Single Series ARIMA* specification dialog.

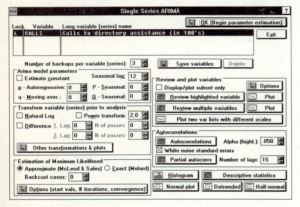

First plot the series; file *Director.sta* contains a variable with dates that can be used to label the horizontal *X*-axis of line plots. Click on the *Options* button in the *Review and plot variables* box to bring up the *Display Options* dialog, then select *Dates*, and select variable *Month* as the date variable. Since each year consists of 12 observations, also set the scaling to start at *1* with increments of *12*.

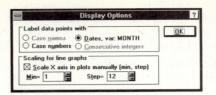

Now click *OK* to accept these settings and then click on the *Plot* button next to the *Review highlighted variable* button.

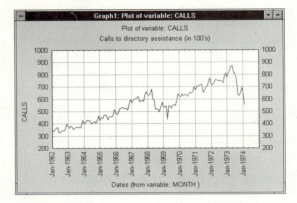

The series shows a linear upward trend, so some differencing will probably be necessary. Click *Continue* to bring up the *Single Series ARIMA* dialog again, and then click on the *Other transformations & plots* option to bring up the *Transformations of Variables* dialog. To perform simple (non-seasonal) differencing, click on the *OK (Transform highlighted variable)* button and select *Differencing (x=x-x(lag))*. Be sure that the *lag* parameter is set to *1*, and then click *OK (Transform)*. After all cases have been transformed, the differenced series will be plotted (unless you previously de-selected the *Plot variable (series) after each transformation* option on the *Transformations of Variables* dialog).

8. TIME SERIES ANALYSIS - ARIMA

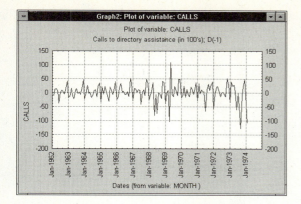

Autocorrelation. As described in the previous example and the *Introductory Overview* section (page 3275), the autocorrelation and partial autocorrelation functions are the most important tools for identifying an appropriate ARIMA model. Click *Continue* and bring up the *Transformations of Variables* dialog again. Because the series consists of monthly observations, you may suspect that any seasonality will occur with a seasonal lag of *12*. If seasonal differencing is necessary you would expect substantial autocorrelations at multiples of the seasonal lag (by contrast, a stationary series would be characterized by an autocorrelation function that dies out, the longer the lag); therefore, set the *Number of lags* parameter in the *Autocorrelations & crosscorrelations* box to *25* so that you may detect such a pattern. Then click on the *Autocorrelations* button.

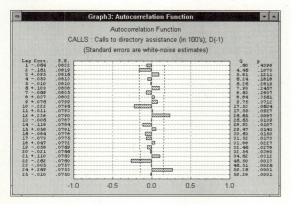

Indeed, the autocorrelation function shows no sign of decaying but suggests that further (seasonal) differencing is necessary. Bring up the *Time Series Transformations* dialog again and perform *Differencing (x-x-x(lag))* with *lag=12*. Shown below are the autocorrelation and partial autocorrelation functions for the resulting series.

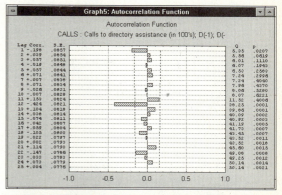

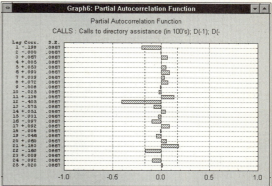

Both plots show a small spike at lag *1* and another stronger spike at *12*. The two spikes show no sign of gradual decay in the autocorrelation function, and so McDowall et al. (1980, page 69) suggest a moving-average process rather than an autoregressive process. Specifically, they initially fit a model to the first 146 observations of the series that includes non-seasonal and seasonal differencing, one seasonal autoregressive parameter, and a constant.

8. TIME SERIES ANALYSIS - ARIMA

In standard terminology (see *Introductory Overview*, page 3274), the model *(0, 1, 0) (0, 1, 1)* will be fit with seasonal lag *12* and a constant.

Specifying the ARIMA Model

Now return to the *Single Series ARIMA* dialog (*Exit* the *Transformations of Variables* dialog). *Delete* the differenced series from the *active work area,* and select (highlight) the original untransformed series *Calls*. Next, specify one *Seasonal* moving average parameter and a *Seasonal lag* of *12*, and click on the *Estimate constant* check box. Then, click on the *Difference* check box in the *Transform variable (series) prior to analysis* box and specify differencing with *1. Lag: 1* (*1* pass) and *2. Lag: 12* (*1* pass). (Make sure that the *Approximate (McLeod & Sales)* option is selected in the *Estimation of Maximum Likelihood* box.)

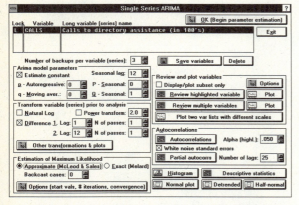

You are now ready to proceed; click *OK (Begin parameter estimation)*. After the parameter estimation finishes, bring up the *Single Series ARIMA Results* dialog.

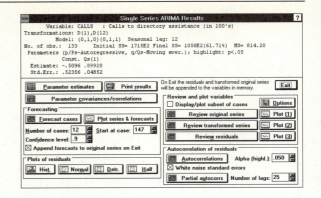

Preliminary Results

Click on the *Parameter estimates* button to bring up a Scrollsheet with the parameter estimates.

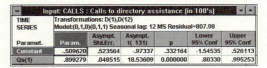

The constant is not statistically significant; therefore, McDowall et al. (1980) suggest to drop it from further analyses. Now, quickly look at the *Autocorrelations* and *Partial autocorrelations* of residuals.

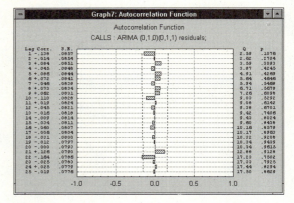

8. TIME SERIES ANALYSIS - ARIMA

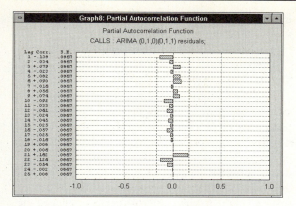

There is no evidence of any significant residual serial dependency in the data. Therefore, you can be satisfied that you have identified a reasonable model for the series.

Specifying Interrupted ARIMA

Now, exit the *Single Series ARIMA Results* dialog to specify an ARIMA model for the entire series, accounting for the introduction of the 20 cents user-fee for calls to Directory Assistance. Return to the *Time Series Analysis* opening dialog and click on the *Case selection conditions* button. Note that, after exiting from the *Single Series Results* dialog, the differenced series as well as the ARIMA residuals were automatically appended to the *active work area*. When you change the selection conditions, the *active work area* will be cleared, and a respective message box will appear to warn you that this is about to happen; when you subsequently select another type of time series analysis, the selected original variables will be read again, using the new selection conditions. Simply click *Yes* in response to the warning, and then turn off the case selection conditions. Next click on the *Interrupted time series analysis* button to bring up the *Interrupted Time Series ARIMA* dialog.

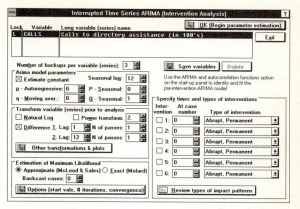

As you can see, all prior specifications are still in place. However, before proceeding, look at the entire series, that is, including the observations following the introduction of the service charge. Click on the *Other transformations & plots* button and then plot the series (use the *Plot* option as before).

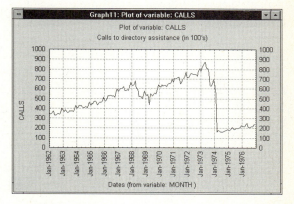

Evidently, the service charge caused a sharp abrupt decline in the number of calls to Directory Assistance. Now *Exit* the *Transformations of Variables* dialog back to the *Interrupted Time Series ARIMA* dialog.

Specifying the intervention. The *Time Series* module allows you to model three types of interventions: *Abrupt-Permanent, Gradual-Permanent,* and *Abrupt-Temporary.* The *Introductory Overview* (page 3278) discusses these

TIM - 3353

types of interventions in detail. In this case, looking back at the plot of the entire series, it certainly looks like the introduction of the service charge caused an abrupt and permanent change in the number of calls to Directory Assistance. However, in other cases the result of an intervention may not be as apparent; McDowall et al., (1980, pages 83-85) discuss how one might proceed when the researcher has no *a priori* hypotheses concerning the nature of the impact. Now, specify the intervention. Click on the *1* check box in the *Specify times and types of interventions* box. The service charge was introduced at the 147'th month in the series; thus specify *147* in the *At case number* edit field. Finally, remember that you (and McDowall et al.) concluded that a constant is not necessary for these data (i.e., after differencing); therefore de-select the *Estimate constant* check box. You are now ready to proceed with the parameter estimation. Click *OK*, and when the parameter estimation is done, bring up the *Interrupted Time Series ARIMA Results* dialog.

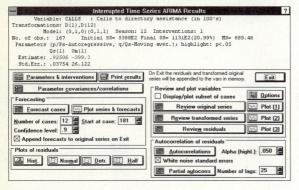

Reviewing the Final Results

Click on the *Parameters & interventions* button to bring up a Scrollsheet with the final parameter estimates.

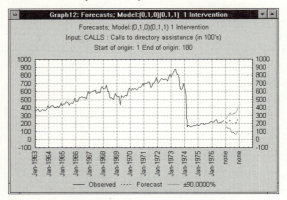

The seasonal moving average parameter is highly significant. Parameter *Omega* is the intervention parameter, and it is also highly significant. For *Abrupt, Permanent* interventions, *Omega* can simply be interpreted as the amount of permanent change that occurred at the point of intervention. Thus, you may conclude that the introduction of the 20 cents service charge for Directory Assistance reduced the number of such calls by about 399 * 100 = 39,000 calls.

Forecasts. As you can see by reviewing the options available on the results dialog, you may compute forecasts, taking into account the interventions in the series. Click on the *Plot series & forecasts* button to plot the series together with one seasonal cycle (one year) of forecasts.

Residual analysis. Before concluding the analysis, perform some final model checks. Shown below are the autocorrelation and partial autocorrelation functions for the residuals.

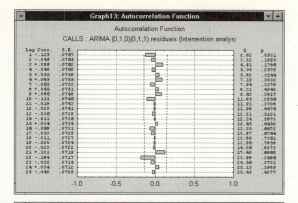

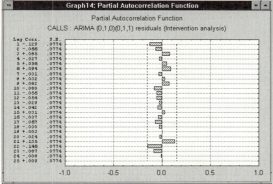

There is no evidence of any residual serial dependency in the data. Click on the *Histogram* button to review the distribution of residuals.

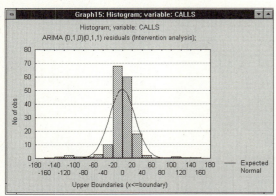

Even though the distribution of the residuals seems leptokurtic (i.e., more "peaked" than the standard normal distribution), they approximate the normal distribution reasonably well. Now click *Exit* to bring up the *Interrupted Time Series ARIMA* dialog again; the ARIMA residuals as well as the differenced series and the ARIMA forecasts will be appended to the *active work area*. As in *Example 1*, look at a final summary graph that includes the original series as well as the residuals. Click on the *Other transformations & plots* button, and then select to *Plot two var lists with different scales;* specifically, plot the original series *Calls* and the *ARIMA residuals*.

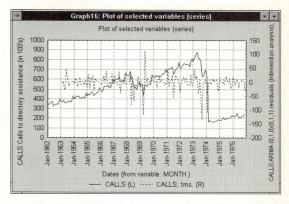

It looks like the fit of the ARIMA model is as good or better following the intervention as it is prior to the intervention. In other words, there is no evidence that after the intervention the ARIMA residuals show some kind of non-random (white noise) pattern. If that were the case, one might suspect that the ARIMA model you identified from the series prior to the intervention (i.e., based on the first 146 observations only) does not fit the series after the intervention; however, in your case there is no evidence of this.

DIALOGS, OPTIONS, STATISTICS

Single Series ARIMA

When you click on the *ARIMA & autocorrelation functions* button on the opening dialog, the following *Single Series ARIMA* dialog will open.

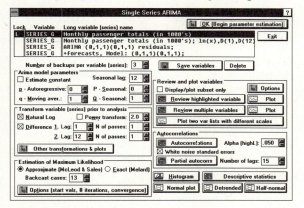

Click on the *Variable* or *Long variable (series) name* column to highlight the variable (series) in the *active work area* (see *General Conventions*, page 3313) that is to be analyzed. Select at least one seasonal or non-seasonal moving average or autoregressive parameter. If seasonal parameters are specified, a seasonal lag must also be specified. Click *OK* to begin parameter estimation.

Note that when returning from the *ARIMA Results* dialog, up to three new series will be created and appended to the *active work area*:

(1) The transformed original series (e.g., differenced, logged, etc., depending on the selections in the *Transform variable....* box, see below),

(2) The ARIMA residuals (from the series after transformations and differencing), and

(3) The original series plus forecasts.

Thus, make sure that there is sufficient "space" in the *active work area* for the respective variable, that is, that there are at least 3 (not locked) backups available. Increase the *Number of backups* parameter if necessary. Also, be sure to *Lock* all important backups of the original variable (the ones you want to retain) if there are less than 3 unused backups available; otherwise, those backups will be overwritten (by the 3 series created by the ARIMA procedure). To lock a backup, double-click in the *Lock* column for the respective variable. For a general discussion of the ARIMA model, refer to the *Overview* section (page 3273).

ARIMA Model Parameters

Estimate constant. In addition to the autoregressive and moving average parameters, ARIMA models may also include a constant. The interpretation of a (statistically significant) constant depends on the model that is fit. Specifically, (1) if there are no autoregressive parameters in the model, then the expected value of the constant is μ, the mean of the series; (2) if there are autoregressive parameters in the series, then the constant represents the intercept. If the series is differenced, then the constant represents the mean or intercept of the differenced series. For example, if the series is differenced once, and there are no autoregressive parameters in the model, then the constant represents the mean of the differenced series, and therefore the *linear trend slope* of the un-differenced series.

Seasonal lag. This parameter determines the seasonal lag that is applied to the seasonal autoregressive and/or moving average parameters (option *P-Seasonal* and *Q-Seasonal*, respectively). Note that for very long seasonal lags (e.g., 365 days per year), it is recommended to use the *Approximate* rather than *Exact* maximum likelihood method, which is less efficient in those cases. Refer to the *Overview* section (page 3277) for a discussion of the different methods for computing the likelihood for ARIMA models.

Autoregressive parameters. Select the number of autoregressive parameters in the model.

Seasonal autoregressive parameters. Select the number of seasonal autoregressive parameters in the model. If any seasonal parameters are selected, a *Seasonal lag* must also be specified.

Moving average parameters. Select the number of moving average parameters in the model.

Seasonal moving average parameters. Select the number of seasonal moving average parameters in the model. If any seasonal parameters are selected, a *Seasonal lag* must also be specified.

Transform Variable (Series) Prior to Analysis

The transformations that can be selected via check boxes will be performed prior to the analysis, and the ARIMA parameters will be estimated for the transformed series. Before forecasts are computed, those transformations will be "undone" and, therefore, those forecasts can be interpreted in terms of the metric of the untransformed series. The *Other transformations & plots* button will bring up the *General transformations* dialog (see page 3334), and those transformations will not automatically be undone when forecasts are computed.

Natural log. The natural log for each value will be computed.

Power transform. Each value in the series will be raised to the power of *C*, where *C* is the value specified in the edit field on the right.

Difference. If this check box is set, then the series will be differenced. In keeping with the standard notation introduced by Box and Jenkins (1976), non-seasonal (*1*) and seasonal (*2*) differencing can be requested. Specify the respective *Lag* and the number of difference passes that are to be performed.

Other transformations & plots. As described above, this option will bring up the *General transformations* dialog which allows you to perform a wide variety of transformations on the data. The transformed series will be appended to the *active work area*.

Estimation of Maximum Likelihood

The ARIMA *Overview* section (page 3273) discusses the procedure for estimating the ARIMA model parameters. In general, the estimation procedure will maximize the likelihood of the data, given the respective model (i.e., number and type of ARIMA parameters). Three different approaches are implemented in the *Time Series* module for computing the likelihood for an ARIMA model:

(1) The approximate maximum likelihood method according to McLeod and Sales (1983),

(2) The approximate maximum likelihood method with backcasting, and

(3) The exact maximum likelihood method according to Melard (1984). All methods will usually yield similar parameter estimates.

Also, all methods are about equally efficient in most real-world time series applications. However, method *1* (approximate maximum likelihood, no backcasts) is the fastest and should be used in particular for very long time series (e.g., with more than 30,000 observations; note that the *Time Series* module is unique in that it will take full advantage of your computer's memory, and it will not impose fixed limitations on the lengths of time series that can be analyzed). Melard's exact maximum likelihood method (number *3* above) may also become inefficient when used to estimate parameters for seasonal models with long seasonal lags (e.g., with yearly lags of 365 days). On the other hand, the *Time Series* module will always use the approximate maximum likelihood method first in order to establish initial parameter estimates that are very close to the actual final values; thus, usually only a few iterations with the exact maximum likelihood method (*3* above) are necessary to finalize the parameter estimates.

Backcast cases. This edit field is only active if the *Approximate* method is chosen. The backcast method is somewhat slower than the *Approximate* maximum likelihood method without backcasts; however, usually when the *n* of the series is relatively small and/or some parameter values are close to 1.0 (or -1.0), the estimates derived in this manner tend to be closer to the equivalent *Exact* maximum likelihood parameters. The optimum setting of this parameter depends on the type of model and the actual magnitude of the parameter values; in practice, McLeod and Sales (1983) recommend to choose the number of backcasts so that:

```
No.of backcasts = q + s*qs + 20*(p + s*ps)
```

where *p*, *ps*, *q*, and *qs* are the non-seasonal and seasonal autoregressive and moving average parameters, respectively, and *s* is the seasonal lag.

Options. This option will bring up the *ARIMA parameter estimation options* dialog (see page 3361). The estimation of ARIMA parameters is an iterative procedure. On that dialog you can specify the maximum number of iterations, the convergence criterion, and request to specify start values for the parameter estimation. If you request to manually specify start values, then, after clicking *OK*, you will be prompted to specify start values for all parameters in the model.

Review and Plot Variables

These options allow you to review or plot the currently highlighted variable (together with other variables). The labeling of the individual observations (that is, of the respective points in time represented by the rows in Scrollsheets or the horizontal *X*-axis in plots) is determined by the selections in the *Display/plot options* dialog (click on the *Options* button below; these options are also available on the *Transformations* dialog, and they are described in the *General Conventions* section of this chapter, see page 3315). For the plots, you can choose to *Scale the X-Axis in the plots manually*. This is often useful when the series consists of several regularly recurring time intervals, for example, if it is made up of 10 years with 12 (monthly) observations each. Then it may be useful to scale the *X*-axis in the plot so that each major tickmark indicates one year.

Display/plot subset of cases only. If this check box is set, then, before plotting or displaying one or more series, you will be prompted to select a range of observations to include in the plot or display.

Options (scaling, labels, etc.). This option will bring up the *Display/plot options* dialog (see the *General Conventions* section, page 3315). That dialog allows you to change the labeling of the observations (i.e., the rows in Scrollsheets and the *X*-axis in line plots can be labeled with case names, case numbers, dates from another variable, etc.) and the scaling of the *X*-axis in line plots (minimum, step size).

Review/Plot highlighted variable. This button will bring up a Scrollsheet (or a plot) with the values for the currently highlighted variable.

Review/Plot multiple variables. After clicking on this button, you will first be prompted to select a list of variables to be displayed or plotted. When plotting multiple variables the vertical (*Y*) axis will have a common scale; thus, if the series that are plotted contain values of completely different magnitudes use option *Plot two var lists with different scales* (see below), instead.

Plot two var. lists with different scales. This option allows you to plot two lists of variables using different scales for the vertical (*Y*) axes in the plot. The first list will be plotted against the left-*Y* axis and the second list will be plotted against the right-*Y* axis. For example, suppose you want to plot two variables (series) called *S1* and *S2*, and that these variables contain values of different magnitudes: the values of *S1* range from 1-100 and the values of *S2* range from 1000 to 10000. If you plot *S1* against the left-*Y* axis (scaled from 1-100) and *S2* against the right-*Y* axis (scaled from 1000-10000), then both variables will be plotted with optimal resolution, and thus relationships between patterns across time (the *X*-axis) may become more clearly identifiable.

Autocorrelations

These options will compute the *autocorrelations* and *partial autocorrelations* for the currently highlighted variable; these statistics will be displayed in a Scrollsheet and drawn in a correlogram graph.

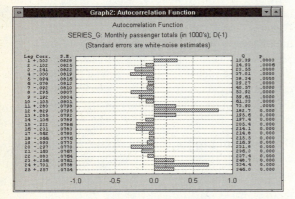

Refer to the *Overview* section (page 3275) for a brief description of how the pattern of (partial) autocorrelations aids in the determination of an appropriate ARIMA model.

Autocorrelations button. This option will bring up a Scrollsheet and plot of the autocorrelations for a lag of *1* through the number specified in the *Number of lags* edit field. The Scrollsheet will report the autocorrelations, their standard errors, the so-called *Box-Ljung statistic* (see the description of the *Transformation of Variables* dialog, page 3331, for computational details), and the significance level of that statistic. In general, the autocorrelation is the correlation of the series with itself, lagged by a particular number of observations. For details regarding the computations of those statistics refer also to the *autocorrelation* section (page 3271).

Alpha for highlighting. Significant autocorrelations (significant *Box-Ljung statistics*) will be highlighted in the Scrollsheet. This edit field allows you to specify the significance level that is used for highlighting.

White noise standard errors. Under the assumption that the true moving average process in the series is of the order *k-1*, the approximate standard error of the autocorrelation r_k is computed as:

$$\text{StdErr}(r_k) = \sqrt{(1/n) * [1+2*\Sigma(r_i^2)]}$$
(for i = 1 to k-1)

Here *n* is the number of observations in the series. Under the assumption that the series is a white noise process, that is, that all autocorrelations are equal to zero, the standard error of r_k is computed as:

$$\text{StdErr}(r_k) = \sqrt{(1/n) * [(n-k)/(n+2)]}$$

Check the *White noise standard errors* option to compute the standard errors in this manner.

Partial autocorrelations. This option will bring up a Scrollsheet and plot of the partial autocorrelations, for a lag of *1* through the number specified in the *Number of lags* edit field. The Scrollsheet will report the partial autocorrelations and their standard errors. In general, the partial autocorrelation is the partial correlation of a series with itself, lagged by a particular number of observations, and controlling for all correlations for lags of lower order. For example, the partial autocorrelation for a lag of *2* represents the unique correlation of the series with itself at that lag, after controlling for the correlation at lag *1*. For details regarding the computations of the partial autocorrelation and its standard error refer to the description of the *Transformation of Variables* dialog (page 3332). Partial autocorrelations that are larger than two times their respective standard errors will be highlighted in the Scrollsheet.

Number of lags. This parameter determines the maximum number of lags for which the auto-, and partial autocorrelations will be computed.

Histogram

This option will bring up a histogram of the currently highlighted variable.

Descriptive Statistics

This option will bring up a Scrollsheet with descriptive statistics for all variables currently available in the *active work area*.

Normal, Detrended, and Half-Normal Probability Plots

These options will bring up the respective normal probability plot for the highlighted variable. The way the *standard normal probability* plot is constructed is as follows. First the values are rank ordered. From these ranks you can compute z values (i.e., standardized values of the normal distribution) *based on the assumption* that the data come from a normal distribution. These z values are plotted on the *Y*-axis in the plot. If the data values (plotted on the *X*-axis) are normally distributed, then all points should fall onto a straight line in the plot. If the values are not normally distributed, they will deviate from the line. Outliers may also become evident in this plot. The *detrended normal probability plot* is constructed in the same way as the standard normal probability plot, except that, before the plot is produced, the linear trend is removed. This often

"spreads out" the plot, thereby allowing the user to detect patterns of deviations more easily.

The *half-normal probability plot* is constructed in the same way as the standard normal probability plot, except that only the positive half of the normal curve is considered. Consequently, only positive normal values will be plotted on the *Y*-axis. This plot is basically used when one wants to ignore the sign of the residual, that is, when one is mostly interested in the distribution of absolute values, regardless of sign.

Parameter Estimation Options

The choices on this dialog pertain to some technical aspect of the parameter estimation procedure.

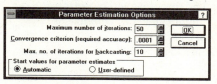

Usually, the default settings do not need to be changed, unless the model that you are trying to fit is "unusual" in some way, and those settings do not produce valid parameter estimates. In general, parameter estimation of ARIMA models is an iterative procedure. In each iteration the program will compute the conditional sums of squares for the current parameter values (refer to the *Overview* section, page 3276, for a description of the approximate and exact maximum likelihood methods for estimating the conditional sums of squares). The goal of the estimation procedure is to find a set of parameters that will minimize the sums of squares.

The *quasi-Newton* method (see Fletcher and Powell, 1963; Fletcher, 1969) is used to minimize the Sums of Squares. This method does not require explicit derivatives of the function to be minimized but will approximate them via differencing. Thus, this is a very efficient general method. Refer to *Nonlinear*

Estimation, Chapter 1, for additional details concerning the estimation of non-linear models. Iterations will continue until:

(1) The *Maximum number of iterations* have been exceeded (note that you can then request additional iterations and the procedure will continue where it left off; thus there is no point in requesting very many iterations in this dialog), or

(2) The parameter accuracy is less than the *Convergence criterion* (see below for details).

Maximum Number of Iterations

This parameter determines the maximum number of *quasi-Newton* iterations that will be performed. When the maximum number of iterations is exceeded during estimation, you will be prompted and asked whether to perform additional iterations.

Convergence Criterion

This parameter determines the accuracy or precision with which the parameters will be estimated, or more specifically, the accuracy of the parameters with respect to the minimum of the conditional SS as computed by the respective approximate or exact maximum likelihood method (refer to the *Overview* section, page 3276). The estimation procedure will terminate when the changes in the ARIMA parameters over consecutive iterations are less than this value. For the ω (*omega*) parameters in an interrupted time series analysis, the convergence parameter is interpreted so that sufficient accuracy is defined as: $.05*(convergence\ criterion)*\Delta\omega$; where $\Delta\omega$ stands for the change in the *omega* parameters from iteration to iteration.

Maximum Number of Iterations for Backcasting

This parameter determines the maximum number of iterations that is used in the process of backcasting, when *Approximate (McLeod & Sales)* maximum likelihood estimation with backcasting is chosen on the *Single Series ARIMA* dialog. The default value of *10* will be sufficient for most analyses. More iterations may only be required when the model contains a moving average parameter with roots near the unit circle (see McLeod and Sales, 1983), which often indicates model-inadequacy.

Start Values for Parameter Estimates

The choice of radio buttons in this box determines how the parameter values will be computed prior to the estimation procedure, that is, with which values to start the iterations.

Automatic. If *Automatic* is chosen, then, for simple ARIMA models, all parameters are started with *0.1*. For *interrupted time series* analysis (see the *Overview* section, page 3278), (1) the ARIMA parameters are started at their previously estimated value if an identical ARIMA model (with respect to the number and types of parameters) was estimated just prior to the intervention analysis; otherwise they are started at *0.1*; (2) the ω (*omega*) parameters for *abrupt* intervention types will be started as the difference between the values of the series prior to the intervention and after the intervention.

User-defined. If *User-defined* is chosen, then before beginning the parameter estimation [after clicking *OK (Begin parameter estimation)*] the *Specify start values* dialog will be displayed,

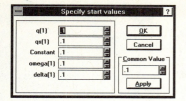

allowing you to manually specify start values for the parameter estimation.

Parameter Estimation Procedure

The ARIMA parameter estimation procedure is iterative in nature; that is, parameter estimates are refined in successive iterations (see also the *Overview* section, page 3276). A *quasi-Newton* algorithm is used to minimize the conditional residual sums of squares, given the respective ARIMA parameter estimates.

The sums of squares value is listed in the *Loss* column in the window; the parameter values are listed to the right.

If the Parameter Estimation Procedure Fails to Converge

There are a number of reasons why the parameter estimation procedure does not converge. One major reason can be that the model parameters are highly redundant (and intercorrelated); thus when one parameter is moved in one direction, then another parameter becomes less than optimal and must also be moved; that move in turn may cause the first parameter to be less than optimal, and so on. Often,

this type of problem occurs when the model is misspecified. Therefore, one should further study the patterns of (partial) autocorrelations and try to fit a simpler model with less parameters. Also, this problem may occur when there is a local minimum in the sums of squares function, "holding" some parameter values in a place where they are very redundant with other parameters. Therefore, you may also want to try different start values (see the *Parameter estimation options* panel, page 3361).

Another reason for the failure to converge could be that the estimation procedure repeatedly attempts to move the parameters outside their legal bounds. In that case, a very large value ("penalty function") is assigned to the SS to "entice" the estimation procedure to move the parameters away from the illegal values. When that fails, you may see very large values for the sums of squares, indicating that the model is inappropriate. This can happen, for example, when the series is clearly non-stationary, and a single autoregressive parameter is requested (see the *Overview* section, page 3273, for a discussion of the ARIMA model).

Interrupted Time Series

This dialog allows you to specify an interrupted time series ARIMA analysis.

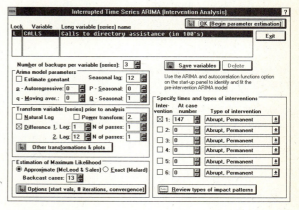

8. TIME SERIES ANALYSIS - ARIMA

This technique has also been called *impact analysis* or *intervention analysis*. In general, additional parameters (in addition to the autoregressive and moving average parameters) are estimated to account for one or more discrete events that either temporarily or permanently change the overall "level" or mean of the series. The general logic of this type of analysis and the different impact patterns are also discussed in the *Overview* section (page 3278). An excellent, brief and applications-oriented introduction to this method is provided by McDowall, McCleary, Meidinger, and Hay (1980).

The options for specifying the ARIMA model in this dialog are identical to those for the *Single Series Analysis* (page 3357, and those options are described there). Described below are the options specific for interrupted time series analysis.

Specify Times and Types of Interventions

Up to 6 interventions can be specified. To specify an intervention, first click on the respective check box and then enter the case number where the intervention occurred. Note that the case number should reference the actual case number in the file, not the offset (plus 1) from the first valid case. So for example, if the intervention occurs at case number 100 *in the data file*, but the first 20 cases of the series are missing, then you would still specify *100* as the time of intervention. Then specify the desired *Type of intervention*.

Review Type of Impact Patterns

This option will bring up the *Impact patterns* dialog, allowing you to review the different types of impact patterns in graphs, and to try different values for the δ parameter.

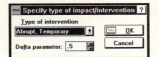

McDowall et al., (1980) distinguish between three major types of impacts that are possible:

(1) Permanent abrupt,

(2) Permanent gradual, and

(3) Abrupt temporary.

Abrupt permanent impact. A permanent abrupt impact pattern simply implies that the overall mean of the time series shifted after the intervention; the overall shift is denoted by ω (*omega*).

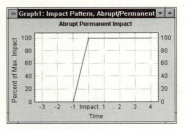

Gradual permanent impact. The gradual permanent impact pattern implies that the increase or decrease due to the intervention is gradual, and that the final permanent impact becomes evident only after some time.

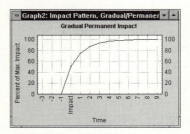

This type of intervention can be summarized by the expression:

```
Impact_t = δ * Impact_(t-1) + ω
(for all t ≥ time of impact, else = 0)
```

TIM - 3364

Copyright © StatSoft, 1995

Note that this impact pattern is defined by the two parameters δ *(delta)* and ω *(omega)*. If δ is near *0* (zero), then the final permanent amount of impact will be evident after only a few more observations; if δ is close to *1*, then the final permanent amount of impact will only be evident after many more observations. As long as the δ parameter is greater than *0* and less than *1* (the bounds of system stability), the impact will be gradual and result in an asymptotic change (shift) in the overall mean by the quantity:

`Asymptotic change in level = ω/(1-δ)`

The *Time Series* module will automatically compute the asymptotic change for gradual permanent impacts. Note that, when evaluating a fitted model, it is important that both parameters are statistically significant; otherwise one could reach paradoxical conclusions. For example, suppose the ω parameter is not statistically significant from *0* (zero) but the δ parameter is; this would mean that an intervention caused a significant gradual change, the final result of which was not significantly different from zero.

Abrupt temporary impact.
The abrupt temporary impact pattern implies an initial abrupt increase or decrease due to the intervention which then slowly decays, without permanently changing the mean of the series.

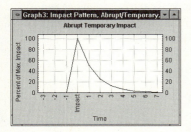

This type of intervention can be summarized by the expressions:

- Prior to intervention: $Impact_t = 0$
- At time of intervention: $Impact_t = ω$
- After intervention: $Impact_t = δ*Impact_{t-1}$

Note that this impact pattern is again defined by the two parameters δ *(delta)* and ω *(omega)*. As long as the δ parameter is greater than *0* and less than *1* (the bounds of system stability), the initial abrupt impact will gradually decay. If δ is near *0* (zero), then the decay will be very quick, and the impact will have entirely disappeared after only a few observations. If δ is close to *1* then the decay will be slow, and the intervention will affect the series over many observations. Note that, when evaluating a fitted model, it is again important that both parameters are statistically significant; otherwise one could reach paradoxical conclusions. For example, suppose the ω parameter is not statistically significant from *0* (zero) but the δ parameter is; this would mean that an intervention did not cause an initial abrupt change, which then showed significant decay.

ARIMA Results

This dialog provides access to the ARIMA results.

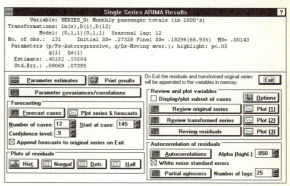

When exiting the dialog via *Cancel*, the ARIMA residuals will be appended to the *active work area* (see *General Conventions*, page 3313). Note that those residuals pertain to the differenced and transformed series. If during the review of the results ARIMA forecasts were computed, those forecasts will also be appended to the *active work area*.

Parameter Estimates

This option will bring up a Scrollsheet with the parameter estimates, their standard errors, the *t* values (parameter estimates divided by standard errors), and the respective *p*-levels. Note that the standard errors are *approximations*, computed from the inverse of the matrix of partial derivatives approximated via finite differencing (refer to the *Overview* section).

Parameters for interrupted time series.

The general logic of interrupted time series ARIMA is described in the *Overview* section (page 3278, refer also to the description of the *Interrupted time series* dialog, page 3363). Following the standard notation discussed in McDowall, McCleary, Meidinger, and Hay (1980), three major types of impacts are possible:

(1) A *permanent abrupt* impact pattern simply implies that the overall mean of the time series shifted after the intervention; the overall shift is denoted by ω (parameter name *omega(i)* in the Scrollsheet).

(2) The *gradual permanent* impact pattern implies that the increase or decrease due to the intervention is gradual, and that the final permanent impact becomes evident only after some time. This type of intervention can be summarized by the expression: $Impact_t = \delta*Impact_{t-1} + \omega$ (for all $t \geq$ time of impact, else = 0). The respective parameter names used in the Scrollsheet are *delta(i)* and *omega(i)*.

If δ is greater than *0* and less than *1* (the bounds of system stability), the impact will be gradual and result in an asymptotic change (shift) in the overall mean by the quantity: *Asymptotic change in level*=ω/(1-δ). The asymptotic change for gradual permanent impacts will also be displayed in the Scrollsheet. Note that, when evaluating a fitted model, it is important that both parameters are statistically significant; otherwise one could reach paradoxical conclusions.

For example, suppose the ω parameter is not statistically significant from *0* (zero) but the δ parameter is; this would mean that an intervention caused a significant gradual change, the final result of which was not significantly different from zero.

(3) The *abrupt temporary* impact pattern implies an initial abrupt increase or decrease due to the intervention which then slowly decays, without permanently changing the mean of the series. This type of intervention can be summarized by the expressions:

Prior to intervention: $Impact_t = 0$

At time of intervention: $Impact_t = \omega$

After intervention: $Impact_t = \delta*Impact_{t-1}$

Note that this impact pattern is again defined by the two parameters δ *(*parameter name *delta(i)* in the Scrollsheet) and ω (parameter name *omega(i)* in the Scrollsheet). Note that, when evaluating a fitted model, it is again important that both parameters are statistically significant; otherwise one could reach paradoxical conclusions.

For example, suppose the ω parameter is not statistically significant from *0* (zero) but the δ parameter is; this would mean that an intervention did not cause an initial abrupt change, which then showed significant decay.

Print Results

This button will print a summary of the results to the current printer, disk file, and/or Text/output Window, depending on your choice in the *Page/Output Setup* dialog.

Parameter Covariances and Correlations

The variance/covariance matrix of parameter estimates is computed from a finite difference approximation of the Hessian matrix of partial derivatives. The resulting estimates are asymptotic in nature. The higher the correlation between two parameters, the greater is their redundancy of their contribution to the fit of the model. The parameter covariance matrix cannot be computed if the Hessian matrix cannot be inverted, which usually indicates that one or more parameters are completely redundant. In terms of the ARIMA model, this usually means that the model is inappropriate (misspecified). In that case, carefully review the patterns of (partial) autocorrelation and try to fit a simpler model.

Forecasting

ARIMA forecasts are computed after all transformations and differencing (selected on the *ARIMA specification dialog*) have been "undone;" that is, the computed forecasts will be expressed in the units of the original time series. The computations of the forecasts and their standard errors closely follows the algorithm described in Box and Jenkins (1976; Program 4).

Forecast cases. This button will bring up a Scrollsheet with:

(1) The ARIMA forecasts,

(2) Their standard errors (if no *log* or *power* transformation was requested),

(3) Their confidence limits according to the value in the *Confidence level* edit field, and

(4) The observed values and residuals (observed minus predicted), if applicable (i.e., if forecasts are requested for cases that were also observed; see below).

Plot series & forecasts. This button will bring up a line graph showing:

(1) The values of the observed series,

(2) The values of the forecasts, and

(3) The confidence limits around the forecasts, according to the value in the *Confidence level* edit field.

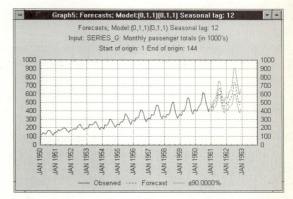

The scaling and labeling of the horizontal (*X*) axis in this plot depends on the current settings in the *Display/plot options* dialog (click on the *Options* button in the *Review and plot variables* box, see page 3315).

Number of cases...; Start at case. These parameters determine (1) how many forecasts will be computed, and (2) at which point in the series the forecasts begin. Ordinarily, one wants to predict the future, so by default the program will compute the forecasts starting with the first observation in the future (after the last valid observed case) through at least one seasonal cycle. However, a very powerful visual way of evaluating the quality of the fit of the current ARIMA model to the data is to evaluate how well the predicted values follow the actual observed values. Therefore, it is sometimes useful to compute forecasts for two or three seasonal cycles in the middle of the series.

Note that the *Start at* parameter should always be set in terms of the original case numbers in the file, *not* in terms of valid case numbers for the respective series. For example, if the variable being analyzed has, in the data file, missing data for the first 20 cases, and then 100 valid cases, then the first future case will be 20+100+1=121 (even though there are only 80 valid cases in the series).

Confidence level. This parameter determines the confidence band around the forecasts in the Scrollsheets and plots.

Append forecasts to original series on Cancel. If this check box is set (it is by default), then when exiting this dialog (via *Cancel*), the forecasts (the ones computed the last time prior to *Cancel*ing) will be combined with the observed values (up to the forecasts), and this new series will be appended to the *active work area*. So, for example, if you requested 20 forecasts starting with case number 101, then the first 101 observations for the new series will be copied from the original series, and then 20 forecasts (case 101 to 120) will be added.

Plots of Residuals

These plots allow you to review the distribution of the ARIMA residuals. Note that the residuals are computed for the transformed and differenced series, and thus they may be different than those that can be computed via the *Forecasting* options above. The ARIMA model assumes that the residuals (random shock component of the model: ε) are normally distributed. Quick visual checks of the validity of this assumption can be performed via these plots.

Histogram. This option will bring up a histogram for the residuals with the normal distribution curve superimposed.

Normal, detrended normal, and half-normal probability plots. These options will bring up the respective normal probability plot of the residuals. The way the *standard normal probability* plot is constructed is as follows. First the values are rank ordered. From these ranks you can compute z values (i.e., standardized values of the normal distribution) *based on the assumption* that the data come from a normal distribution. These z values are plotted on the Y-axis in the plot. If the residuals (plotted on the X-axis) are normally distributed, then all points should fall onto a straight line in the plot. If the residuals are not normally distributed, they will deviate from the line. Outliers may also become evident in this plot. The *detrended normal probability plot* is constructed in the same way as the standard normal probability plot, except that, before the plot is produced, the linear trend is removed. This often "spreads out" the plot, thereby allowing the user to detect patterns of deviations more easily. The *half-normal probability plot* is constructed in the same way as the standard normal probability plot, except that only the positive half of the normal curve is considered. Consequently, only positive normal values will be plotted on the Y-axis. This plot is basically used when one wants to ignore the sign of the residual, that is, when one is mostly interested in the distribution of absolute values, regardless of the sign.

Review and Plot Variables

These options allow you to review or plot the original series, the transformed and differenced series, and the residuals (computed for the transformed and differenced series; thus those residuals may be different than those computed by the *Forecasting* options above). The labeling of the individual observations (that is, of the respective points in time represented by the rows in Scrollsheets or the horizontal X-axis in plots) is determined by the selections in the *Display/plot options* dialog (click on the *Options* button below; see also the *General Conventions* section, page 3315). For the plots, you can choose to *Scale the X-Axis in the plots manually*. This is often useful when the series consists of several regularly recurring time intervals, for example, if it is made up of 10 years with 12 (monthly) observations each.

Then it may be useful to scale the X-axis in the plot so that each major tickmark indicates one year.

Display/plot subset of cases only. If this check box is set, then, before plotting or displaying a series, you will be prompted to select a range of observations to include in the plot or display.

Options. This option will bring up the *Display/plot options* dialog (see *General Conventions*, page 3315). That dialog allows you to change the labeling of the observations (e.g., the X-axis in line plots can be labeled with case names, case numbers, dates from another variable, etc.) and the scaling of the X-axis in line plots (minimum, step size).

Review/Plot original series. This button will bring up a Scrollsheet (or a plot) with the values for the original series (prior to transformations and differencing).

Review/Plot transformed series. This button will bring up a Scrollsheet (or a plot) with the values for the transformed and differenced series.

Review/Plot residuals. This button will bring up a Scrollsheet (or a plot) with the values for the residuals, computed from the transformed and differenced series.

Autocorrelations of Residuals

These options will compute the *autocorrelation* and *partial autocorrelation* for the residuals (computed for the transformed and differenced series); these statistics will be displayed in a Scrollsheet and drawn in a correlogram graph. Refer to the *Introductory Overview* section (page 3275) for a brief description of how the pattern of (partial) autocorrelations aids in the determination of an appropriate ARIMA model. In general, if the ARIMA model sufficiently reproduces the observed values in the series, then no (partial) autocorrelations should remain, that is, be present in the residuals.

Autocorrelations. This option will bring up a Scrollsheet and plot of the autocorrelations for the residuals, for a lag of *1* through the number specified in the *Number of lags* edit field. The Scrollsheet will report the autocorrelations, their standard errors, the so-called *Box-Ljung* statistic, and the significance level of that statistic. In general, the autocorrelation is the correlation of the series with itself, lagged by a particular number of observations. For details regarding the computations of those statistics, refer to the *Transformation of Variables* dialog (page 3331).

Alpha for highlighting. Significant autocorrelations (significant *Box-Ljung statistics*) will be highlighted in the Scrollsheet. This edit field allows you to specify the significance level that is used for highlighting.

White noise standard errors. Under the assumption that the true moving average process in the series is of the order $k-1$, the approximate standard error of the autocorrelation r_k is computed as:

```
StdErr(r_k) = √{(1/n) * [1+2*Σ(r_i^2)]}
(for i = 1 to k-1)
```

Here n is the number of observations in the series. Under the assumption that the series is a white noise process, that is, that all autocorrelations are equal to zero, the standard error of r_k is computed as:

```
StdErr(r_k) = √{(1/n) * [(n-k)/(n+2)]}
```

Check the *White noise standard errors* option to compute the standard errors in this manner.

Partial autocorrelations. This option will bring up a Scrollsheet and plot of the partial autocorrelations of the residuals, for a lag of *1* through the number specified in the *Number of lags* edit field. The Scrollsheet will report the partial autocorrelations and their standard errors. In general, the partial autocorrelation is the partial correlation of a series with itself, lagged by a particular number of observations, and controlling for all correlations for lags of lower order. For example, the partial autocorrelation for a lag of *2* represents the unique correlation of the series with itself at that lag, after controlling for the correlation at lag *1*. For details regarding the computations of the partial autocorrelation and its standard error, refer to the *partial autocorrelation* section. Partial autocorrelations that are larger than two times their respective standard errors will be highlighted in the Scrollsheet.

Number of lags. This parameter determines the maximum number of lags for which the auto-, and partial autocorrelations will be computed.

SEASONAL AND NON-SEASONAL EXPONENTIAL SMOOTHING

EXAMPLE

Example 1 of the ARIMA section (page 3341) discusses the analysis of a data set from the classic book on ARIMA by Box and Jenkins (1976). The data are monthly passenger totals (measured in thousands) in international air travel, for twelve consecutive years: 1949-1960 (see Box and Jenkins, 1976, page 531, "Series G"). They are partially listed below; they are also included with your program in example data file *Series_g.sta* (variable *Series_g*).

Data: SERIES	Monthly pas SERIES_G
JAN 1949	112
FEB 1949	118
MAR 1949	132
APR 1949	129
MAY 1949	121
JUN 1949	135
JUL 1949	148
AUG 1949	148
SEP 1949	136
OCT 1949	119
NOV 1949	104
DEC 1949	118
JAN 1950	115

The ARIMA analysis required a good deal of preparatory work during the identification stage. In fact, it usually requires a lot of experience and familiarity not only with ARIMA but also with the nature of the data, in order to identify satisfactory models. Often, the purpose of ARIMA is mostly to derive forecasts, and the interpretation of the nature of the model (i.e., the number and types of parameters) is only of secondary interest. In those cases, exponential smoothing provides a much easier alternative, one that usually produces forecasts of equal or better quality (see the *Introductory Overview*, page 3279, for a discussion of this point).

In this example, exponential smoothing will be performed on the same series used in ARIMA *Example 1* and the forecasts derived by the two methods will be compared.

Choosing a Model

Even though exponential smoothing is, in a way, a simpler method than ARIMA, some choices still have to be made. After starting the *Time Series* module, open the data file *Series_g.sta*. Click on the *Variables* button and select the variable *Series_G* (note that if the data file *Series_g.sta* is the currently open data file, and since *Series_G* is the only variable in that data file, then when the *Time Series Analysis* dialog opens, *Series_G* will automatically be selected). Now, choose the *Exponential smoothing & forecasting* option and proceed to the *Seasonal and Non-Seasonal Exponential Smoothing* dialog.

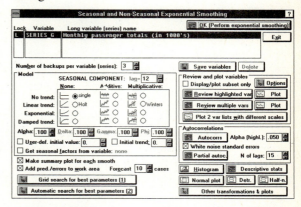

As described in the *Introductory Overview* section (page 3279), there are different exponential smoothing models available. In general, in all models the smoothed or forecasted values are computed as a weighted average of the preceding values. The difference between the models listed in the *Model* box is whether or not a trend and/or seasonal component are smoothed with extra smoothing parameters. Examine the differences between models by looking at the results of smoothing with the different techniques and parameters. However, first plot the series. Because *Series_g.sta* contains dates in the case names, use those to label the horizontal *X*-axis in line plots. Click on the *Options* button in the *Review and plot variables* box and click on the *Case names* radio button; then select to *Scale X-axis in plots manually* and specify *Min = 1* and *Step = 12* (as there are 12 months in a year). Click *OK* to close the *Display Options* dialog and click on the *Plot* button next to the *Review highlighted var* button.

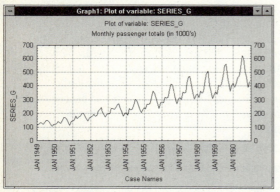

The data in this series are easily matched up with the general "model shapes" shown on the icons in the *Model* box of the *Seasonal and Non-Seasonal Exponential Smoothing* dialog. Clearly, there is a trend, which is more or less linear. Second, there is seasonal fluctuation; that is, every year the number of airline passengers follows an almost identical pattern (e.g., most travel occurs during the summer vacation months). This seasonality is multiplicative rather than additive in nature: The higher the overall level of the series the greater is the seasonal fluctuation. Put another way, the increase in airline passenger loads during the summer months each year can best be expressed by a *factor*; for example, each summer the passenger load increases by a

factor of 1.1, or 10%. Thus, the *Winters* model (*Linear trend, Multiplicative*) is probably the best exponential smoothing model to use for this series. However, first look at some other models.

Simple Exponential Smoothing

The *Forecast* edit field will show *10 cases* by default; change this to *12* and forecast one full year. Then, accept all other defaults and click *OK*. Shown below is the plot of the original and smoothed series, and the residuals.

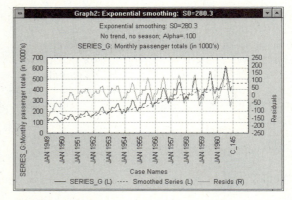

Two things are immediately apparent. First, the smoothed series traces the general linear trend but fails to follow the seasonal cycles. Second, all forecasts are the same. In fact, if you look back over the description of simple exponential smoothing in the *Introductory Overview* section (page 3280), this could be expected: Each smoothed value or forecast S_t is computed as $S_t = S_{t-1} + \alpha * e$, where e is the *error* or observed minus predicted (smoothed or forecasted) value. When there are no observed values available (e.g., when computing forecasts), then e is assumed to be *0* (zero). Thus, all forecasts are the same from then on.

The α (*alpha*) parameter. Now look at the effect of the smoothing parameter α (*alpha*). Looking at the formula above, it is clear that as α approaches *0* (zero), all smoothed values will become very similar; when α approaches *1*, then the smoothed values should very closely follow the actual observed data. Click *Continue* and set the *Alpha* parameter to *0.9*. Then click *OK (Perform exponential smoothing)* again.

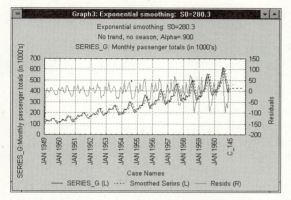

Now the smoothed values follow the observed values very closely; it almost looks like the original series is simply moved by one observation to the right. Indeed, if you were to set the α parameter to *1*, then each smoothed value would be equal to the previous observation. Therefore, in a sense, the α parameter can be considered a *stiffness* parameter. The smaller the α the "stiffer" the smoothed line; that is, the smoothed line will not be affected as much by the random observation-to-observation variability. The larger the α the more flexible the smoothed line; that is, the more closely will it follow the fluctuations in the observed values. This is generally true for all exponential smoothing models, and this principle applies equally to the seasonality and trend smoothing parameters of the more complex models (see below).

Exponential Smoothing with Linear Trend

Now select *Holt Linear trend* smoothing (without seasonality). In this model, a trend component is independently smoothed with parameter γ (*gamma*). If γ is set to *0*, then a constant slope will be included in the computation of smoothed values and forecasts. If γ is set to *1*, then the slope is recomputed at each observation from the respective immediately preceding smoothed value; thus, the slope is allowed to change as much as necessary from observation to observation, in order to approximate the observed values. Shown below is the summary plot for two smoothing trials, the first with α = *0.1* and γ = *0.1*, the second with α = *0.9* and γ = *0.9*.

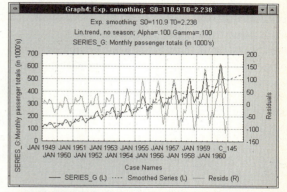

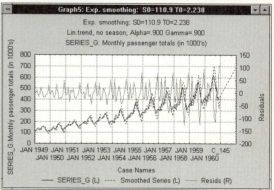

As expected, the smoothed values follow the observed values more closely in the second graph shown above. However, looking at the forecasts it is evident that in this model (without seasonality), the predicted values simply consist of a straight line. Thus, using the *Holt* two-parameter (*Linear trend*)

model, you would "miss" the significant seasonal increase of airline passengers during the summer months.

Now look at the model that seems most appropriate here, that is, the *Winters* three-parameter model with *Linear trend* and *Multiplicative* seasonality.

Triple Exponential Smoothing: Winters' Method

In this method, a third parameter δ (*delta*) is added to the model to smooth the multiplicative seasonal component. Again, if δ is *0* (zero), then a constant stable seasonal component is included in the computation of the smoothed values and forecasts; if δ is set to *1*, then the seasonal component is recomputed from observation to observation. Shown below are the summary plots for α = *0.1*, δ = *0.1*, and γ = *0.1*, and for α = *0.9*, δ = *0.9*, and γ = *0.9*.

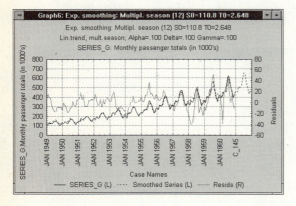

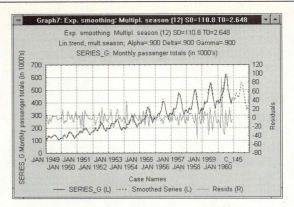

In this case, there is hardly any difference between the two summary plots. The reason for this is that the series indeed consists of a stable linear trend, strong stable seasonal fluctuation, and only little random fluctuation. Therefore, even though by setting δ and γ to *0.9*, you "allow" the seasonal and trend components to be modified substantially from observation to observation, no such modification is required. In fact, the automatic search for the best parameters discussed below will arrive at the same conclusion.

Parameter Grid Search

As discussed in the *Introductory Overview* section (page 3279), in practice, when one wants to compute forecasts, one is best advised to estimate optimum smoothing parameters from the data (e.g., see Gardner, 1985). This can be done in two ways. One common method is to perform a *grid search* of the parameter space. Click on the *Grid search for best parameters* button to bring up the *Parameter Grid Search* dialog.

The program will increment each parameter from the minimum (*Start parameter at*) by the value specified in the *Increment by* column, up to the value specified in the *Stop at* column.

For each combination of parameter values, the program will compute the *Sums of Squares* (*SS*) for the residuals (observed values minus smoothed values). By default, when the *Display parameters for 10 smallest mean squares* check box is set, then the "best" 10 solutions; that is, the combinations of parameters that yield the smallest residual variability will be displayed in a Scrollsheet. Accept all defaults and examine that Scrollsheet (see below). As suspected, the best-fitting models are those with parameter values for δ and γ near *0* (zero), that is, models with constant stable linear trend and seasonality.

TIME SERIES	Model: Linear trend, mult.season(12); S0=110.8 T0=2.648 SERIES_G: Monthly passenger totals (in 1000's)					
Model Number	Alpha	Delta	Gamma	Mean Error	Mean Abs Error	Sums of Squares
568	.800000	.100000	.100000	-.002499	7.943659	17686.50
649	.900000	.100000	.100000	-.004916	7.870651	17796.40
658	.900000	.200000	.100000	-.018611	7.928614	18006.46
487	.700000	.100000	.100000	.017182	8.115193	18130.23
667	.900000	.300000	.100000	-.031932	7.992037	18248.06
577	.800000	.200000	.100000	-.028284	8.083341	18272.95
676	.900000	.400000	.100000	-.044706	8.062034	18515.45
235	.300000	.900000	.100000	.171303	8.411577	18544.03
226	.300000	.800000	.100000	.249966	8.447845	18734.08
685	.900000	.500000	.100000	-.056970	8.135697	18804.76

Note that in addition to the *Sums of Squares* and *Mean Square*, several other indices of goodness of fit are listed in this Scrollsheet. All of these are discussed in the *Introductory Overview* (page 3281); of particular interest is often the *Mean absolute percentage error* (*MAPE*). This value expresses the average (absolute) difference between the observed and smoothed (predicted) values relative to the observed values. For example, for the first model with α = *0.8*, δ = *0.1*, and γ = *0.1*, the *MAPE* value is *2.97*. This means that on average the smoothed (predicted) values computed by this model only deviated 2.97% from the actual observed values.

Automatic Parameter Search

The second way to determine the optimum parameters for smoothing is to minimize the *Sums of Squares* of the residuals or some similar index of goodness of fit. This can be done in the *Time Series* module by using a general nonlinear function minimization algorithm (the so-called *quasi-Newton* method; the same method used to estimate the ARIMA parameters). Now, click on the *Automatic search for best parameters* button on the *Seasonal and Non-Seasonal Exponential Smoothing* dialog. This dialog (see below) contains several technical parameters that pertain to the quasi-Newton method; those parameters are described in detail in the *Dialogs, Options, Statistics* section (page 3387). The *Lack of fit indicator* box shows the different quantities that can be minimized.

As *Parameter start values*, you specified *0.1* for all three parameters (it is always a good idea to start the minimization procedure with small parameter values). By default, the *Unconstrained parameter estimation* check box is set. This means that in the course of the function minimization, you may see parameter values that become smaller than *0* (which is not permissible for the smoothing parameters). However, before the final results are reported, the parameters will automatically be set to the closest valid value, so you do not have to change this setting (again, refer to the description of this option in the *Dialogs, Options, Statistics* section, page 3387, for a more detailed discussion). Now click *OK* and after the iterative parameter search procedure finishes,

8. TIME SERIES ANALYSIS - EXPONENTIAL SMOOTHING

click *OK* again to look at the summary graph for the best parameter values.

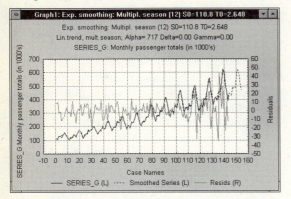

Once again, as in the grid search, the best model is one that contains a stable constant linear trend and seasonality. The remaining random variability is best smoothed with a very "flexible"(large) α parameter value (*0.72*) that allows the smoothed values to follow closely the observed values.

Final Results

Now return to the original goal, namely to compare the forecasts produced by exponential smoothing with those from ARIMA. You could now click *Continue, Lock* the exponentially smoothed values and then compute the ARIMA analysis as described in *Example 1* of the *ARIMA* section. After you completed the ARIMA analysis, go to the *Transformations of Variables* dialog and plot the exponentially smoothed series together with the original series plus ARIMA forecasts.

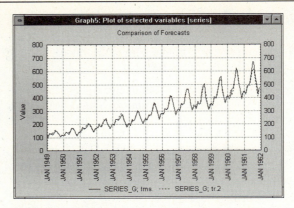

If you set the *Display/plot subset of cases only* check box on the *Transformations of Variables* dialog, you can specify to plot the 12 forecasts only.

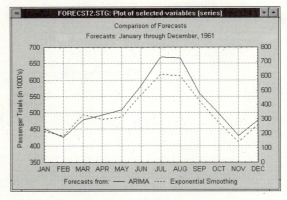

Note that the default *X*-axis labels in this plot have been changed to reflect the future (forecast) dates. Even though the two lines diverge slightly, looking back at the full plot of all cases, that divergence is relatively minor.

DIALOGS, OPTIONS, STATISTICS

Exponential Smoothing Specifications

The options on this dialog allow you to perform exponential smoothing for time series with or without a seasonal component and trend.

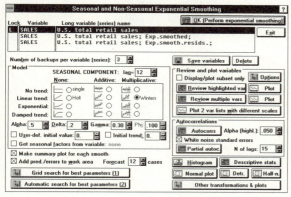

The implementation of these techniques in the *STATISTICA Time Series* module closely follows the comprehensive survey by Gardner (1985; see the *Introductory Overview*, page 3279, for details). Click in the *Variable* or *Long variable (series) name* column to highlight the variable (series) in the *active work area* (see *General Conventions*, 3313) that is to be analyzed. Then, select the appropriate exponential smoothing model from among the options in the *Model* box (see below). For seasonal models, specify the seasonal lag (length of seasonal period) or accept the default lag of *12*. Click *OK* to begin the exponential smoothing process. By default, 10 forecast values will be computed (past the last observed value in the series). You can change the number of forecasts in the *Forecast ... cases* edit field of this dialog. If the *Add pred./errors to work area* check box is set, then (on *OK*) the smoothed (forecast) values as well as residuals (observed minus smoothed values) will be appended to the *active work area*. If you choose to append those series, make sure that there is sufficient "space" in the *active work area* for the respective variable, that is, that there are at least two (not locked) backups available. You can increase the *Number of backups* parameter if necessary. Also, be sure to *Lock* all important backups of the original variable (those that you want to retain); otherwise, those backups may be overwritten (refer to the *General Conventions* section describing the memory management in the *active work area*, page 3313). To lock a backup, double-click in the *Lock* column on the respective variable.

Exponential Smoothing Models

The general ideas of exponential smoothing and forecasting are explained in the *Exponential Smoothing* overview section (*Introductory Overview*, page 3279). The exponential smoothing models available in this box are arranged in a two-way classification, as proposed by Gardner (1985): The time series may either contain no seasonal component, an additive component, or a multiplicative seasonal component. In addition, there may be no upward or downward trend in the series, a linear trend, an exponential trend, or a damped trend. The combination of these characteristics results in 12 different exponential smoothing models (which include the classical models developed by Winters, Holt, and Brown).

Nonseasonal, no trend. This model is equivalent to the simple exponential smoothing model (see *Introductory Overview*, page 3280). Note that by default, the first smoothed value will be computed based on an initial S_0 value equal to the overall mean of the series.

Additive season, no trend. This model is partially equivalent to the simple exponential smoothing model; however, in addition, each forecast is "enhanced" by an additive seasonal component that is smoothed independently (see the *Introductory Overview*, page 3279). This model would, for example, be adequate when computing forecasts for monthly expected amount of rain. The amount of rain will be stable from year to year, or change only slowly. At the same time, there will be seasonal changes ("rainy seasons"), which again may change slowly from year to year. To compute the smoothed values for the first season, initial values for the seasonal components are necessary. By default, the *Time Series* module will estimate those values (for all models including a seasonal component) from the data via *classical seasonal decomposition* (or the *Census I* method; see *Introductory Overview*, page 3284). The initial smoothed value S_0 will by default be computed as the mean for all values included in complete seasonal cycles.

Multiplicative season, no trend. This model is partially equivalent to the simple exponential smoothing model; however, in addition, each forecast is "enhanced" by a multiplicative component that is smoothed independently (see *Introductory Overview*, page 3279). This model would, for example, be adequate when computing forecasts for monthly expected sales for a particular toy. The level of sales may be stable from year to year, or change only slowly; at the same time, there will be seasonal changes (e.g., greater sales during the December holidays), which again may change slowly from year to year. The seasonal changes may affect the sales in a multiplicative fashion, for example, depending on the respective overall level of sales, December sales may always be greater by a *factor* of 1.4. To compute the smoothed values for the first season, initial values for the seasonal components are necessary.

By default, the *Time Series* module will estimate those values (for all models including a seasonal component) from the data via *classical seasonal decomposition* (or the *Census I* method; see *Introductory Overview*, page 3284). The initial

smoothed value S_0 will by default be computed as the mean for all values included in complete seasonal cycles.

Nonseasonal, linear trend (Holt's two-parameter method). In this model, the simple exponential smoothing forecasts are "enhanced" by a linear trend component that is smoothed independently via the γ (*gamma*) parameter (see also the *Introductory Overview* section, 3279). This model is also referred to as *Holt's two parameter* method. This model would, for example, be adequate when producing forecasts for spare parts inventories. The need for particular spare parts may slowly increase or decrease over time (the trend component), and the trend may slowly change as different machines etc. age or become obsolete, thus affecting the trend in the demand for spare parts for the respective machines.

Computations. In order to compute the smoothed value (forecast) for the first observation in the series, both estimates of S_0 and T_0 (initial trend) are necessary. By default, these values are computed as:

$$T_0 = (X_n - X_1)/(n-1)$$

where

n   is the length of the series, and

$$S_0 = X_1 - T_0/2$$

Additive season, linear trend. In this model, the simple exponential smoothing forecasts are "enhanced" both by a linear trend component [independently smoothed with parameter γ (*gamma*)] and an *additive* seasonal component [smoothed with parameter δ (*delta*)]. For example, suppose you were to predict the monthly budget for snow-removal in a community. There may be a trend component (as the community grows, there is a steady upward trend for the cost of snow removal from year to year). At the same time, there is obviously a seasonal component, reflecting the differential likelihood of snow during different months of the year. This seasonal component could be additive, meaning that a particular fixed additional amount of money is necessary during the winter months, or (see below) multiplicative, that is, given the respective budget figure, it may increase by a *factor* of, for example, 1.4 during particular winter months.

Computations. To compute the smoothed values for the first season, initial values for the seasonal components are necessary. By default, the *Time Series* module will estimate those values (for all models including a seasonal component) from the data via *Classical seasonal decomposition* (or the *Census I* method; see *Introductory Overview*, page 3284). Also, to compute the smoothed value (forecast) for the first observation in the series, both estimates of S_0 and T_0 (initial trend) are necessary. By default, these values are computed as:

$$T_0 = (M_k - M_1) / [(k-1) * p]$$

where

- k is the number of complete seasonal cycles
- M_k is the mean for the last seasonal cycle
- M_1 is the mean for the first seasonal cycle
- p is the length of the seasonal cycle, and

$$S_0 = M_1 - T_0 / 2$$

Multiplicative season, linear trend (triple exponential smoothing or Winters' three-parameter trend and seasonality method).

In this model, the simple exponential smoothing forecasts are "enhanced" both by a linear trend component [independently smoothed with parameter γ (*gamma*)] and a *multiplicative* seasonal component [smoothed with parameter δ (*delta*)].

For example, suppose you were to predict the monthly budget for snow-removal in a community. There may be a trend component (as the community grows, there is an upward trend for the cost of snow removal from year to year). At the same time, there is obviously a seasonal component, reflecting the differential likelihood of snow during different months of the year. This seasonal component could be multiplicative, meaning that given a respective budget figure, it may increase by a *factor* of, for example, *1.4* during particular winter months; or it may be additive (see above), that is, a particular fixed additional amount of money is necessary during the winter months.

Computations. To compute the smoothed values for the first season, initial values for the seasonal components are necessary. By default, the *Time Series* module will estimate those values (for all models including a seasonal component) from the data via *classical seasonal decomposition* (or the *Census I* method; see *Introductory Overview*, page 3284). Also, to compute the smoothed value (forecast) for the first observation in the series, both estimates of S_0 and T_0 (initial trend) are necessary. By default, these values are computed as:

$$T_0 = (M_k - M_1) / [(k-1) * p]$$

where

- k is the number of complete seasonal cycles
- M_k is the mean for the last seasonal cycle
- M_1 is the mean for the first seasonal cycle
- p is the length of the seasonal cycle, and

$$S_0 = M_1 - T_0 / 2$$

Nonseasonal, exponential trend.

In this model, the simple exponential smoothing forecasts are "enhanced" by an exponential trend component [smoothed with parameter γ (*gamma*)]. For example, suppose you wanted to predict the overall monthly costs of repairs to a production facility. There could be an exponential trend in the cost; that is, from year to year the costs of repairs may increase by a certain percentage or *factor,* resulting in a gradual exponential increase in the absolute dollar costs of repairs.

Computations. To compute the smoothed value (forecast) for the first observation in the series, both estimates of S_0 and T_0 (initial trend) are necessary. By default, these values are computed as:

`T₀ = (X₂/X₁)`

and

`S₀ = X₁/√T₀`

Additive season, exponential trend. In this model, the simple exponential smoothing forecasts are "enhanced" both by an exponential trend component [independently smoothed with parameter γ (*gamma*)] and an additive seasonal component [smoothed with parameter δ (*delta*)]. For example, suppose you wanted to forecast the monthly revenue for a resort area. Every year, revenue may increase by a certain percentage or *factor*, resulting in an exponential trend in overall revenue. In addition, there could be an additive seasonal component, for example a particular fixed (and slowly changing) amount of added revenue during the December holidays.

Computations. To compute the smoothed values for the first season, initial values for the seasonal components are necessary. By default, the *Time Series* module will estimate those values (for all models including a seasonal component) from the data via *classical seasonal decomposition* (or the *Census I* method; see *Introductory Overview*, page 3284). Also, to compute the smoothed value (forecast) for the first observation in the series, both estimates of S_0 and T_0 (initial trend) are necessary. By default, these values are computed as:

`T₀ = exp{[log(Mₖ)-log(M₁)]/p}`

where

`k`   is the number of complete seasonal cycles

`Mₖ`   is the mean for the last seasonal cycle

`M₁`   is the mean for the first seasonal cycle

`p`   is the length of the seasonal cycle, and

`S₀ = exp{[log(M₁)-p*log(T₀)]/2}`

Multiplicative season, exponential trend. In this model, the simple exponential smoothing forecasts are "enhanced" both by an exponential trend component [independently smoothed with parameter γ (*gamma*)] and a multiplicative seasonal component [smoothed with parameter δ (*delta*)]. For example, suppose you wanted to forecast the monthly revenue for a resort area. Every year, revenue may increase by a certain percentage or *factor*, resulting in an exponential trend in overall revenue. In addition, there could be a multiplicative seasonal component; that is, given the respective annual revenue, each year 20% of the revenue is produced during the month of December; that is, during Decembers the revenue grows by a particular (multiplicative) *factor*.

Computations. To compute the smoothed values for the first season, initial values for the seasonal components are necessary. By default, the *Time Series* module will estimate those values (for all models including a seasonal component) from the data via *classical seasonal decomposition* (or the *Census I* method; see *Introductory Overview*, page 3284). Also, to compute the smoothed value (forecast) for the first observation in the series, both estimates of S_0 and T_0 (initial trend) are necessary. By default, these values are computed as:

`T₀ = exp{[log(Mₖ)-log(M₁)]/p}`

where

`k`   is the number of complete seasonal cycles

`Mₖ`   is the mean for the last seasonal cycle

`M₁`   is the mean for the first seasonal cycle

`p`   is the length of the seasonal cycle, and

`S₀ = exp{[log(M₁)-p*log(T₀)]/2}`

Nonseasonal, damped trend. In this model, the simple exponential smoothing forecasts are "enhanced" by a damped trend component [independently smoothed with parameters γ

8. TIME SERIES ANALYSIS - EXPONENTIAL SMOOTHING

(*gamma*) for the trend, and φ (*phi*) for the damping effect]. For example, suppose you wanted to forecast from month to month the percentage of households that own a particular consumer electronics device (e.g., a VCR). Every year, the proportion of households owning a VCR will increase; however, this trend will be damped (i.e., the upward trend will slowly disappear) over time as the market becomes saturated.

Computations. To compute the smoothed value (forecast) for the first observation in the series, both estimates of S_0 and T_0 (initial trend) are necessary. By default, these values are computed as:

$$T_0 = (1/\phi)*(X_n - X_1)/(n-1)$$

where

- n   is the number of cases in the series,
- φ   is the smoothing parameter for the damped trend, and

$$S_0 = X_1 - T_0/2$$

Additive season, damped trend. In this model, the simple exponential smoothing forecasts are "enhanced" both by a damped trend component (independently smoothed with the single parameter φ, this model is an extension of Brown's one-parameter linear model, see Gardner, 1985, page 12-13) and an additive seasonal component (smoothed with parameter δ).

For example, suppose you wanted to forecast from month to month the number of households that purchase a particular consumer electronics device (e.g., VCR). Every year, the number of households that purchase a VCR will increase, however, this trend will be damped (i.e., the upward trend will slowly disappear) over time as the market becomes saturated. In addition, there will be a seasonal component, reflecting the seasonal changes in consumer demand for VCR's from month to month (demand will likely be smaller in the summer and greater during the December holidays). This seasonal component may be additive; for example, a relatively stable number of additional households may purchase VCR's during the December holiday season.

Computations. To compute the smoothed values for the first season, initial values for the seasonal components are necessary. By default, the *Time Series* module will estimate those values (for all models including a seasonal component) from the data via *classical seasonal decomposition* (or the *Census I* method; see *Introductory Overview*, page 3284). Also, to compute the smoothed value (forecast) for the first observation in the series, both estimates of S_0 and T_0 (initial trend) are necessary. By default, these values are computed as:

$$T_0 = (1/\phi)*(M_k - M_1)/[(k-1)*p]$$

where

- φ   is the smoothing parameter
- k   is the number of complete seasonal cycles
- M_k is the mean for the last seasonal cycle
- M_1 is the mean for the first seasonal cycle
- p   is the length of the seasonal cycle, and

$$S_0 = M_1 - p*T_0/2$$

Multiplicative season, damped trend. In this model, the simple exponential smoothing forecasts are "enhanced" both by a damped trend component (independently smoothed with the single parameter φ; this model is an extension of Brown's one-parameter linear model, see Gardner, 1985, page 12-13) and a multiplicative seasonal component (smoothed with parameter δ).

For example, suppose you wanted to forecast from month to month the number of households that purchase a particular consumer electronics device (e.g., VCR). Every year, the number of households that purchase a VCR will increase; however, this trend will be damped (i.e., the upward trend will slowly disappear) over time as the market becomes saturated. In addition, there will be a seasonal component, reflecting the seasonal changes in consumer demand for VCR's from month to month (demand will likely be smaller in the summer and greater during the December holidays). This seasonal component may be multiplicative; for example, sales during the December holidays may increase by factor of 1.4 (or 40%) over the average annual sales.

Computations. To compute the smoothed values for the first season, initial values for the seasonal components are necessary. By default, the *Time Series* module will estimate those values (for all models including a seasonal component) from the data via *classical seasonal decomposition* (or the *Census I* method; see *Introductory Overview*, page 3284). Also, to compute the smoothed value (forecast) for the first observation in the series, both estimates of S_0 and T_0 (initial trend) are necessary. By default, these values are computed as:

```
T₀ = (1/φ)*(Mₖ-M₁)/[(k-1)*p]
```

where

- φ is the smoothing parameter
- k is the number of complete seasonal cycles
- M_k is the mean for the last seasonal cycle
- M_1 is the mean for the first seasonal cycle
- p is the length of the seasonal cycle, and

```
S₀ = M₁-p*T₀/2
```

Model

Seasonal lag. This edit field is only available (active) when a seasonal model is chosen (see below). This value will determine the assumed length of one seasonal cycle. The default value is *12* (e.g., there are 12 months in each year). When changed, this parameter will be retained (remembered) for other time series analyses involving a seasonal component [e.g., in *ARIMA* models or in *Seasonal decomposition (Census I)*].

Alpha (α). This is the constant process smoothing parameter. The α (*alpha*) parameter is necessary for all models. If α is *0* (zero), then all smoothed values will be equal to the initial value (S_0, see also *User-defined initial value* below). If this parameter is *1*, then each smoothed value (forecast) will be equal to the respective previous observation. Values greater than *0* (zero) and less than *1* will produce a forecast that is a weighted average of the previous observations, with the weights decreasing exponentially the "older" the previous observation. The closer the α parameter is to *0*, the more slowly will the weights decrease (that is, the more slowly will the effect of prior observations disappear), the closer it is to *1*, the faster will the weights decrease (and the greater will be the effect of immediately preceding or "younger" observations). In plots of the series, small values of α will produce smooth fitted lines (forecasts) that only follow major trends or fluctuations spanning many observations; larger values of α (closer to *1*) will produce more jagged lines that are greatly influenced by even minor disturbances in the series.

Delta (δ). Parameter δ (*delta*) is the seasonal smoothing parameter and only needs to be specified for seasonal models. In general, the one-step-ahead forecasts are computed as (for no trend models, for linear and exponential trend models a trend component is added to the model; see below):

Additive model:

```
Forecastₜ = Sₜ+Iₜ₋ₚ
```

Multiplicative model:

```
Forecastₜ = Sₜ*Iₜ₋ₚ
```

In this formula, S_t stands for the (simple) exponentially smoothed value of the series at time *t*, and I_{t-p} stands for the smoothed seasonal factor at time *t* minus *p* (the length of the season). Thus, compared to simple exponential smoothing, the forecast is "enhanced" by adding or multiplying the simple smoothed value by the predicted seasonal

component. This seasonal component is derived analogously to the S_t value from simple exponential smoothing as:

Additive model:

$$I_t = I_{t-p} + \delta*(1-\alpha)*e_t$$

Multiplicative model:

$$I_t = I_{t-p} + \delta*(1-\alpha)*e_t/S_t$$

Put into words, the predicted seasonal component at time t is computed as the respective seasonal component in the last seasonal cycle plus a portion of the error (e_t; the observed minus the forecast value at time t). Parameter δ can assume values between 0 and 1. If it is zero, then the seasonal component for a particular point in time is predicted to be identical to the predicted seasonal component for the respective time during the previous seasonal cycle, which in turn is predicted to be identical to that from the previous cycle, and so on. Thus, if δ is zero, a constant unchanging seasonal component is used to generate the one-step-ahead forecasts. If the δ parameter is equal to 1, then the seasonal component is modified "maximally" at every step by the respective forecast error. In most cases, when seasonality is present in the time series, the optimum δ parameter will fall somewhere between 0 (zero) and 1 (one).

Gamma (γ) *and phi* (ϕ). These are the trend smoothing parameters. Parameter γ needs to be specified for linear and exponential trend models, and for damped trend models without seasonality. Parameter ϕ must be specified for damped trend models. Analogous to the seasonal component, when a trend component is included in the exponential smoothing process, an independent trend component is computed for each time and modified as a function of the forecast error and the respective parameter. If the γ parameter is 0 (zero), then the trend component is constant across all values of the time series (and for all forecasts). If the parameter is 1, then the trend component is modified "maximally" at every step by the respective forecast error. Parameter values that fall in-between represent mixtures of those two extremes. Parameter ϕ is a trend modification parameter and affects how strongly changes in the trend will affect estimates of the trend for subsequent forecasts, that is, how quickly the trend will be "damped" or increased.

User-defined initial value. If this check box is set, then the initial value of the smoothed series (S_0, necessary in order to compute the forecast for the first observation) is taken from the edit field; otherwise S_0 is computed from the data, depending on the model chosen. Refer to the description of *Exponential Smoothing Models* above for details of how S_0 is computed in that case.

User-defined initial trend value. This option is only available if a model with trend is chosen. In that case an initial estimate of the trend (T_0) is necessary in order to compute the forecast for the first observation. If this check box is set, then the initial value for the trend (T_0) is taken from the edit field; otherwise it is computed from the data, depending on the model chosen. Refer to the description of *Exponential Smoothing Models* above for details of how T_0 is computed in that case.

Get seasonal factors from variable. This option is only available if a model with a seasonal component is chosen. In that case, in order to compute the forecasts for the values in the first seasonal cycle of the series, an initial estimate of the seasonal component is necessary. If this check box is not set (default), then the initial seasonal factors will be computed from the data (following the procedures described in the *Seasonal Decomposition (Census I)* overview section, page 3284). If this check box is set, then you will be prompted to select a variable that contains the initial seasonal factors in the first p values in the file (where p is the length of the seasonal cycle).

8. TIME SERIES ANALYSIS - EXPONENTIAL SMOOTHING

Note that the program will always take the first p cases in the specified variable as the initial seasonal component, regardless of selection conditions that are in effect. For additive seasonality, the initial values for the seasonal component must add to 0 (zero); for multiplicative seasonality, the initial values for the seasonal component must add to the length of the seasonal cycle (p).

Make Summary Plot for Each Smooth

If this check box is set, then after each analysis (on *OK*) a summary plot is produced showing the values of the observed series, the smoothed series (forecasts), and the residuals (scaled against the right-Y axis).

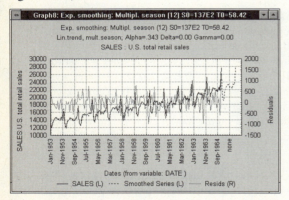

This plot provides an immediate visual check of the adequacy of the forecasts, and the distribution of the errors (residuals) across the series.

Add Pred./Errors to Work Area

If this check box is set, then (on *OK*) the smoothed (forecast) values as well as residuals (observed minus smoothed values) will be appended to the *active work area*. If you choose to append those series, make sure that there is sufficient "space" in the *active work area* for the respective variable, that is, that there are at least two (not locked) backups

available (see the *General Conventions* for details concerning the memory management in the *active work area*, 3313). Increase the *Number of backups* parameter if necessary. Also, be sure to *Lock* all important backups of the original variable (those that you want to retain); otherwise, those backups may be overwritten. To lock a backup, double-click in the *Lock* column on the respective variable.

Grid Search

This option will bring up a dialog for specifying a grid search of the parameter space (see the description of the *Parameter Grid Search* dialog below).

Strategies for identifying the optimum smoothing parameters are described in the *Introductory Overview* section of this chapter (page 3281). In general, this option allows you to systematically evaluate the goodness of fit for sets of parameters by starting each parameter at a particular value and then increasing it by a small amount in consecutive steps. You can view the results either for all steps or only the 10 most successful combinations of parameter values (lack of fit indices are also described in the *Introductory Overview* section, page 3281).

Automatic Search for Best Parameters

As recommend by Gardner (1985), in order to obtain the best (most accurate) forecasts one should determine the best smoothing parameters from the data. This option will allow you to perform an automatic search for the best set of parameters (given a user-defined lack of fit index; these are also described in the *Introductory Overview* section, page 3281).

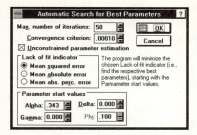

A quasi-Newton function minimization procedure (the same as in *ARIMA*, see also the description of the *Automatic parameter search*) is used to minimize either the mean squared error, mean absolute error, or mean absolute percentage error. In most cases, this procedure is more efficient than the grid search, in particularly, when more than one parameter must be determined.

Note that, by default, the parameter search is unconstrained; that is, parameter values outside the 0/1 boundaries may be estimated. When clicking *OK* after the parameter search converged, invalid parameters will automatically be set to their respective minimum or maximum (e.g., $\alpha = -0.2$ is set to $\alpha = 0$; $\alpha = 1.2$ is set to $\alpha = 1$); and the full set of results (like those produced by clicking *OK* on this dialog) will be displayed for the best parameters.

Review and Plot Variables

These options allow you to review or plot the currently highlighted variable (together with other variables). The labeling of the individual observations (that is, of the respective points in time represented by the rows in Scrollsheets or the horizontal *X*-axis in plots) is determined by the selections in the *Display/plot options* dialog (click on the *Options* button below; these options are also available on the *Transformations* dialog, and are described in the *General Conventions* section, page 3315).

For the plots, you can choose to *Scale the X-Axis in the plots manually* (select this option in the *Display/plot options* dialog). This is often useful when the series consists of several regularly recurring time intervals, for example, if it is made up of 10 years with 12 (monthly) observations each. Then it may be useful to scale the *X*-axis in the plot so that each major tickmark indicates one year.

Display/plot subset of cases only. If this check box is set, then, before plotting or displaying one or more series, you will be prompted to select a range of observations to include in the plot or display.

Options (scaling, labels, etc.). This option will bring up the *Display/plot options* dialog (see the *General Conventions* section, page 3315). That dialog allows you to change the labeling of the observations (i.e., the rows in Scrollsheets and the *X*-axis in line plots can be labeled with case names, case numbers, dates from another variable, etc.) and the scaling of the *X*-axis in line plots (minimum, step size).

Review/Plot highlighted variable. This button will bring up a Scrollsheet (or a plot) with the values for the currently highlighted variable.

Review/Plot multiple variables. After clicking on this button, you will first be prompted to select a list of variables to be displayed or plotted. When plotting multiple variables the vertical (*Y*) axis will have a common scale; thus, if the series that are plotted contain values of completely different magnitudes, use option *Plot two var lists with different scales* (see below), instead.

Plot two var. lists with different scales. This option allows you to plot two lists of variables using different scales for the vertical (*Y*) axes in the plot. The first list will be plotted against the left-*Y* axis, and the second list will be plotted against the right-*Y* axis. For example, suppose you want to plot two variables (series) called *S1* and *S2,* and that these variables contain values of different magnitudes: the

8. TIME SERIES ANALYSIS - EXPONENTIAL SMOOTHING

values of *S1* range from 1-100 and the values of *S2* range from 1000 to 10000. If you plot *S1* against the left-*Y* axis (scaled from 1-100) and *S2* against the right-*Y* axis (scaled from 1000-10000), then both variables will be plotted with optimal resolution, and thus relationships between patterns across time (the *X*-axis) may become more clearly identifiable.

Autocorrelations

These options will compute the *autocorrelations* and *partial autocorrelations* for the currently highlighted variable; these statistics will be displayed in a Scrollsheet and drawn in a correlogram graph. Refer to the *Introductory Overview* section (page 3275) for a brief description of how the pattern of (partial) autocorrelations aids in the determination of an appropriate ARIMA model.

Autocorrelations. This option will bring up a Scrollsheet and plot of the autocorrelations, for a lag of *1* through the number specified in the *Number of lags* edit field. The Scrollsheet will report the autocorrelations, their standard errors, the so-called *Box-Ljung statistic* (see the description of the *Transformation of Variables* dialog, page 3331, for computational details), and the significance level of that statistic. In general, the autocorrelation is the correlation of the series with itself, lagged by a particular number of observations.

***Alpha* for highlighting.** Significant autocorrelations (significant *Box-Ljung statistics*) will be highlighted in the Scrollsheet. This edit field allows you to specify the significance level that is used for highlighting.

White noise standard errors. Under the assumption that the true moving average process in the series is of the order *k-1*, the approximate standard error of the autocorrelation r_k is computed as:

$$\text{StdErr}(r_k) = \sqrt{\{(1/n) * [1+2*\Sigma(r_i^2)]\}}$$
$$(\text{for } i = 1 \text{ to } k-1)$$

Here *n* is the number of observations in the series. Under the assumption that the series is a white noise process, that is, that all autocorrelations are equal to zero, the standard error of r_k is computed as:

$$\text{StdErr}(r_k) = \sqrt{\{(1/n) * [(n-k)/(n+2)]\}}$$

Check the *White noise standard errors* option to compute the standard errors in this manner.

Partial autocorrelations. This option will bring up a Scrollsheet and plot of the partial autocorrelations, for a lag of *1* through the number specified in the *Number of lags* edit field. The Scrollsheet will report the partial autocorrelations and their standard errors. In general, the partial autocorrelation is the partial correlation of a series with itself, lagged by a particular number of observations, and controlling for all correlations for lags of lower order. For example, the partial autocorrelation for a lag of *2* represents the unique correlation of the series with itself at that lag, after controlling for the correlation at lag *1*. For details regarding the computations of the partial autocorrelation and its standard error, refer to the description of the *Transformation of Variables* dialog (page 3329). Partial autocorrelations that are larger than two times their respective standard errors will be highlighted in the Scrollsheet.

Number of lags. This parameter determines the maximum number of lags for which the auto-, and partial autocorrelations will be computed.

Histogram

This option will bring up a histogram of the currently highlighted variable.

Descriptive Statistics

This option will bring up a Scrollsheet with descriptive statistics for all variables currently available in the *active work area*.

Normal, Detrended, and Half-Normal Probability Plots

These options will bring up the respective normal probability plot for the highlighted variable. The way the *standard normal probability* plot is constructed is as follows. First the values are rank ordered. From these ranks you can compute *z* values (i.e., standardized values of the normal distribution) *based on the assumption* that the data come from a normal distribution. These *z* values are plotted on the *Y*-axis in the plot. If the data values (plotted on the *X*-axis) are normally distributed, then all points should fall onto a straight line in the plot. If the values are not normally distributed, they will deviate from the line. Outliers may also become evident in this plot.

The *detrended normal probability plot* is constructed in the same way as the standard normal probability plot, except that, before the plot is produced, the linear trend is removed. This often "spreads out" the plot, thereby allowing the user to detect patterns of deviations more easily. The *half-normal probability plot* is constructed in the same way as the standard normal probability plot, except that only the positive half of the normal curve is considered. Consequently, only positive normal values will be plotted on the *Y*-axis. This plot is basically used when one wants to ignore the sign of the residual, that is, when one is mostly interested in the distribution of absolute values, regardless of sign.

Other Transformations and Plots

This option will bring up the *Transformations of Variables* dialog (see page 3329) which allows you to perform a wide variety of transformations on the data. The transformed series will be appended to the *active work area* (see the *General Conventions* section, page 3313).

Parameter Grid Search

These options allow you to specify a grid search of the parameter space.

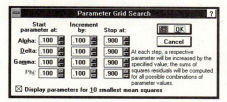

Strategies for identifying the optimum smoothing parameters are described in the *Introductory Overview* section (page 3281). In general, these options allow you to systematically evaluate the goodness of fit for sets of parameters by starting each parameter at a particular value (as specified in the *Start parameter at* column) and then increasing it by a small amount in consecutive steps (as specified in the *Increment by* column), all the way up to the value specified in the *Stop at* column.

You can view the results either for all steps (i.e., for all combinations of parameter values), or only the 10 combinations of parameter values that resulted in the smallest mean squared error for the residuals (other lack of fit indices are also described in the *Introductory Overview* section, page 3281).

Automatic Parameter Search

As recommend by Gardner (1985), in order to obtain the best (most accurate) forecasts, one should determine the best smoothing parameters from the data. A quasi-Newton function minimization procedure (the same as in *ARIMA*) is used to minimize either the mean squared error, mean absolute error, or mean absolute percentage error. In most cases, this procedure is more efficient than the grid search, in particularly when more than one parameter must be determined.

8. TIME SERIES ANALYSIS - EXPONENTIAL SMOOTHING

Note that, by default, the parameter search is unconstrained; that is, parameter values outside the 0/1 boundaries may be estimated. When clicking *OK* after the parameter search converged, invalid parameters will automatically be set to their respective minimum or maximum (e.g., $\alpha = -0.2$ is set to $\alpha = 0$; $\alpha = 1.2$ is set to $\alpha = 1$), and the full set of results will be displayed for the best parameters.

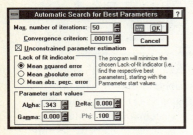

Max. Number of Iterations

The value in this edit field determines the maximum number of quasi-Newton iterations that will be performed. Note that if the maximum number of iterations is exceeded during the automatic parameter search, additional iterations can be requested at that point. Thus, there is no need to specify an excessively large number in this field.

Convergence Criterion

This parameter determines the accuracy or precision with which the parameters will be estimated, or more specifically, the accuracy of the parameters with respect to the *Lack of fit indicator* that is selected (see below, see also the *Introductory Overview*, page 3281). The estimation procedure will terminate when the changes in the parameters over consecutive iterations are less than this value. Usually, the default value (.*0001*) is adequate to produce estimates for the best parameters with sufficient accuracy.

Unconstrained Parameter Estimation

By default (when this check box is set), the search algorithm will be unconstrained; that is, parameter estimates outside the 0/1 boundaries my be produced (if they minimize the respective *Lack of fit* indicator). When clicking *OK* after the parameter search converges, invalid parameters will automatically be set to the respective minimum or maximum (e.g., $\alpha = -0.2$ is set to $\alpha = 0$; $\alpha = 1.2$ is set to $\alpha = 1$).

If this check box is not set, then a large (penalty) value will be assigned to the *Lack of fit* value during the iterative parameter estimation whenever an invalid parameter estimate results. This will usually "entice" the search algorithm to move away from the respective minima and maxima of the parameters. However, if, in fact, the best parameter values are near their minima or maxima (e.g., a constant slope, resulting in δ close to zero), then the parameter search may prematurely converge with less-than-optimal values for the other parameters. Therefore, the unconstrained minimization procedure is the recommended default.

Lack of Fit Indicator

These options determine which lack of fit indicator will be minimized during the parameter estimation process. Various such indices have been proposed (see Makridakis, Wheelwright, and McGee, 1983), and they are also described in the *Introductory Overview* section (page 3281).

Mean squared error. These values (MSE) are computed as the average of the squared error values. This is the most commonly used lack of fit indicator in most statistical fitting procedures.

Mean absolute error. The mean absolute error (MAE) value is computed as the average *absolute* error value. If this value is *0* (zero), the fit (forecast) is perfect. As compared to the mean *squared* error value, this measure of fit will "de-emphasize" outliers; that is, unique or rare large error values will affect the MAE less than the MSE value.

Mean absolute percentage error. Another measure of *relative* overall fit is the mean absolute percentage error (MAPE). Both the above measures rely on the actual error value. It may seem reasonable to rather express the lack of fit in terms of the *relative* deviation of the one-step-ahead forecasts from the observed values, that is, relative to the magnitude of the observed values. For example, when trying to predict monthly sales that may fluctuate widely (e.g., seasonally) from month to month, you may be satisfied if your prediction "hits the target" with about ±10% accuracy. In other words, the absolute errors may be not so much of interest as are the relative errors in the forecasts. To assess the relative error the so-called percentage error (PE) can be computed as:

$$PE_t = 100*(X_t - F_t)/X_t$$

where X_t is the observed value at time t, and F_t is the forecast (smoothed value). The mean percentage error is the average of the PE values for the entire series. This measure is often more meaningful than the mean squared error. For example, knowing that the average forecast is "off" by ±5% is a useful and easily interpretable result, whereas a mean squared error of 30.8 is not immediately interpretable.

Parameter Start Values

These values will determine the initial value for the parameter search process. The meaning of the parameters is explained in the context of the *Exponential Smoothing* dialog and in the *Introductory Overview* section (page 3279). In most cases, if the selected model is adequate for the data, the parameter values will converge at the same values, regardless of the parameter start values chosen.

8. TIME SERIES ANALYSIS - EXPONENTIAL SMOOTHING

8. TIME SERIES ANALYSIS - SEASONAL DECOMPOSITION (CENSUS I)

SEASONAL DECOMPOSITION (CENSUS METHOD I)

EXAMPLE

This example is based on a series reporting the monthly US total retail sales from 1953 to 1964. The data set is reported in Shiskin, Young, and Musgrave (1967) to illustrate the results of the X-11 (Census method II) seasonal adjustment procedure. In this example, the older Census method I seasonal decomposition method will be used to analyze this series; refer to the *X-11 Seasonal Decomposition Example* (page 3401) to see the results of the Census method II decomposition of this series.

Data File

Shown below is a partial listing of the file *Retail.sta* that contains this series (note that, as reported in Shiskin, Young, and Musgrave, 1967, these numbers are not directly comparable to the official published retail sales figures).

After starting the *Time Series* module, open the example data file *Retail.sta*. Click on the *Variables* button and select variable *Sales*; then click on the *Seasonal decomposition (Census I)* button to bring up the *Ratio-To-Moving Averages Classical Seasonal Decomposition (Census Method I)* dialog.

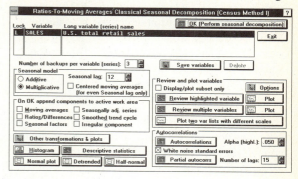

Determining the Seasonal Model

First plot this series. To use the dates recorded in variable *Date* in the plots, click on the *Options* button in the *Review and plot variables* box, click on the *Dates* radio button and select variable *Date*. Next click on the check box *Scale X-axis in plots manually* and specify *Min = 1* and *Step = 12*.

Now, click *OK* and then click on the *Plot* button next to the *Review highlighted variable* button.

TIM - 3391

Copyright © StatSoft, 1995

The retail sales show an upward trend and clear seasonal fluctuation. As described in the *Introductory Overview* section (page 3284), the goal of seasonal decomposition in general is to derive estimates of the seasonal, trend/cycle, and irregular components that make up the series. The seasonal component can be used to compute a seasonally adjusted series, that is, an estimate of the series after removing seasonal fluctuations.

Additive and multiplicative seasonality. The difference between the additive and multiplicative model is discussed in several places in the *Introductory Overview* section (page 3269). With regard to retail sales figures, it is conceivable that during the December holiday season retail sales increase by a relatively fixed amount every year; that is, each year you could *add* a certain amount to the average sales that year to arrive at an estimate for the December figures. Thus, the seasonal fluctuation would be *additive* in nature. Alternatively, each year the sales during December may increase by a *factor*; for example, they may increase by a factor of 1.3 or 30% over the respective yearly average. In that case, the seasonal fluctuations are *multiplicative* in nature. The two types of seasonal fluctuation leave distinctive "signatures" in the series. If the seasonality is additive in nature, the fluctuations are constant around the series, regardless of the overall level of the series. If the seasonality is multiplicative in nature, then the fluctuations are greater when the level of the series is higher. In this example, it rather looks like the seasonality is multiplicative in nature; this would also make more sense intuitively: Most likely, retail sales will increase by a certain percentage each December rather than by a fixed amount. Therefore, accept the default *Multiplicative* setting in the *Seasonal model* box.

Seasonal Decomposition

If you would like to plot later all components that are extracted, you will need to append them to the *active work area* (see *General Conventions*, page 3313, for details concerning the memory management in the *active work area*). Currently (by default), the *Number of backups per variable* parameter is set to *3*. There are six different series that can be appended; thus, set the *Number of backups* parameter to 6. Then select all components in the box entitled *On OK append components to active work area*.

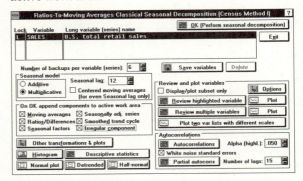

You are now ready to perform the analysis. Click *OK (Perform seasonal decomposition)*, and the result Scrollsheet with all series will be displayed.

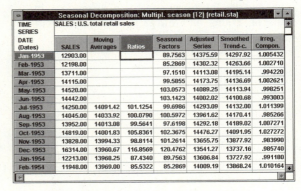

Now, click *Continue*, and the series will be appended to the *active work area*.

8. TIME SERIES ANALYSIS - SEASONAL DECOMPOSITION (CENSUS I)

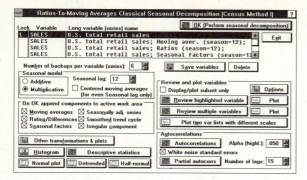

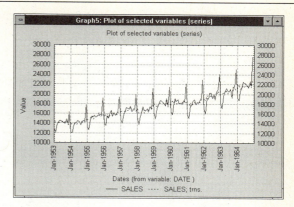

Plot the components that were extracted. Highlight (click on) the *Seasonal factors* series in the *active work area* and then click on the *Plot* button next to the *Review highlighted variable* button.

As you can see, the adjusted series shows no seasonal fluctuation any more. One can further smooth that series in order to remove the remaining random (irregular) fluctuations.

The smoothed seasonally adjusted series represents the *Trend/Cycle* component, because it will show the overall trend and cycles in the series (the *cycle* component is different from the seasonal component in that cycles usually last longer than a single seasonal period, and they occur at irregular intervals). Shown below is the plot of the trend/cycle and seasonally adjusted series.

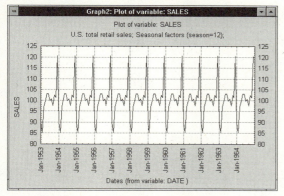

This is the stable seasonal variability that was extracted from the series. This seasonal component can be used to adjust the original series (i.e., it is either subtracted from the original series if the model is additive, or it is divided into the original series if the model is multiplicative).

Click on the *Plot* button next to the *Review multiple variables* button and then select the original series *Sales* and the *Adjusted* series.

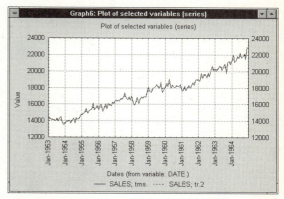

Finally, look at the irregular series with the residual fluctuations in the series. The residuals along with the original and seasonally adjusted series can be shown.

TIM - 3393

Copyright © StatSoft, 1995

8. TIME SERIES ANALYSIS - SEASONAL DECOMPOSITION (CENSUS I)

Use the option *Plot two var lists with different scales* and select the original and seasonally adjusted series for the first list (to be plotted against the left-*Y* axis), and select the *Irregular* series for the second list. Shown below is this summary plot.

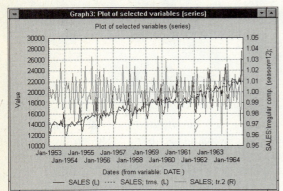

DIALOGS, OPTIONS, STATISTICS

Seasonal Decomposition (Census I)

The options on this dialog allow you to perform classical seasonal decomposition, a technique also known as the *Census I* method (refer to the *Introductory Overview* section, page 3284, for details).

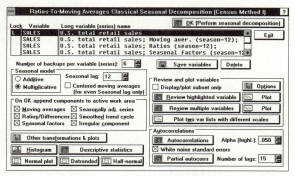

Click in the *Variable* or *Long variable (series) name* column to highlight the variable (series) in the *active work area* that is to be analyzed (see *General Conventions*, page 3313). Then, specify the seasonal lag (length of seasonal period) or accept the default value of *12* and click *OK* to begin the seasonal decomposition. To append any or all of the components that are computed to the *active work area*, set the respective check boxes in the *Append components to active work area* box (see below). If you choose to append components, make sure that there is sufficient "space" in the *active work area* for the respective variable, that is, that there are at least as many (not locked) backups available as there are components to be saved. Increase the *Number of backups* parameter if necessary. Also, be sure to *Lock* all important backups of the original variable (the ones you want to retain); otherwise, those backups may be overwritten. To lock a backup, double-click in the *Lock* column on the respective variable.

In order to compute forecasts, use the *Exponential smoothing* facilities of the *Time Series* module (see page 3279), which provide various options for computing forecasts based on additive or multiplicative seasonal models with or without trend.

Seasonal Model

The settings in this box determine the seasonal model. In general, the seasonality in the series can be *Additive* or *Multiplicative*. For example, when tracking the sales for a toy, it can be expected that during the December holiday season sales will increase; another seasonal increase could possibly occur during the summer vacation months.

Additive/multiplicative. This seasonal fluctuation (e.g., the increase in sales during December) can be *additive* (e.g., on average, sales increase in December by about 1 million dollars over the year's average), or it can be *multiplicative* (on average, sales increase in December by 30%, that is, by a *factor* of 1.3). In plots of a series, the distinguishing characteristic between these two types of seasonal components is that in the additive case, the series shows steady seasonal fluctuations, regardless of the overall level of the series; in the multiplicative case, the size of the seasonal fluctuations vary, depending on the overall level of the series. Refer to the *Introductory Overview* section (page 3285) for additional details.

Seasonal lag. This value will determine the assumed length of one seasonal cycle. The default value is 12 (e.g., there are 12 months in each year). When changed, this parameter will be retained (remembered) for other time series analyses involving a seasonal component (e.g., in *ARIMA* models or in *Exponential smoothing*).

Centered moving averages. As the first step in the decomposition, a moving average is computed for the series, with the moving average window width equal to the length of one season. If the length of the season is even, then the user can choose to use either equal weights for the moving average, or unequal weights can be used, where the first and last observation in the moving average window are averaged; this latter method is used when the *Centered moving averages* check box is set. If the length of the season is odd, the setting of this check box will not affect the computations.

On *OK* Append Components to Active Work Area

The setting of these check boxes determine which components will be added to the *active work area* (see also *General Conventions*, page 3313). If you choose to append components, make sure that there is sufficient "space" in the *active work area* for the respective variable, that is, that there are at least as many (not locked) backups available as there are components to be saved. Increase the *Number of backups* parameter if necessary. Also, be sure to *Lock* all important backups of the original variable (those that you want to retain); otherwise, those backups may be overwritten. To lock a backup, double-click in the *Lock* column on the respective variable.

Moving average. First a moving average is computed for the series, with the moving average window width equal to the length of one season (as specified in the *Seasonal lag* edit field). If the length of the season is even, then the user can choose to use either equal weights for the moving average, or unequal weights can be used, where the first and last observation in the moving average window are averaged (set the *Centered moving averages* check box; see above).

Ratios or differences. In the moving average series (see above), all seasonal (within-season) variability will be eliminated; thus, the differences (in the additive model) or ratios (in the multiplicative model) of the observed and smoothed series will isolate the seasonal component (plus irregular component). Specifically, the moving average is subtracted from the observed series (for the additive model), or the observed series is divided by the moving average values (for the multiplicative model).

Seasonal components. The seasonal component is then computed as the average (for additive models) or medial average (for multiplicative models) for each point in the season. (The medial average of a set of values is the mean after the smallest and largest values are excluded). The resulting values represent the (average) seasonal component of the series.

Seasonally adjusted series. The original series can be adjusted by subtracting from it (additive models) or dividing it by (multiplicative models) the seasonal component. The resulting series is the seasonally adjusted series (i.e., the seasonal component will be removed).

Trend-cycle component. The cyclical component is different from the seasonal component in that it is usually longer than one season, and different cycles can be of different lengths. The combined trend and cyclical component can be approximated by applying to the seasonally adjusted series a 5 point (centered) weighed moving average smoothing transformation with the weights of 1, 2, 3, 2, 1.

Random or irregular component. Finally, the random or irregular (error) component can be isolated by subtracting from the seasonally adjusted series (additive models) or dividing the adjusted series by (multiplicative models) the trend-cycle component.

Other Transformations and Plots

This option will bring up the *Transformations of Variables* dialog (see page 3329) which allows you to perform a wide variety of transformations on the data. The transformed series will be appended to the *active work area* (see *General Conventions*, page 3313).

Histogram

This option will bring up a histogram of the currently highlighted variable.

Descriptive Statistics

This option will bring up a Scrollsheet with descriptive statistics for all variables currently available in the *active work area*.

Normal, Detrended, and Half-Normal Probability Plots

These options will bring up the respective normal probability plot for the highlighted variable. The way the *standard normal probability* plot is constructed is as follows. First the values are rank ordered. From these ranks you can compute z values (i.e., standardized values of the normal distribution) *based on the assumption* that the data come from a normal distribution. These z values are plotted on the Y-axis in the plot. If the data values (plotted on the X-axis) are normally distributed, then all points should fall onto a straight line in the plot. If the values are not normally distributed, they will deviate from the line. Outliers may also become evident in this plot. The *detrended normal probability plot* is constructed in the same way as the standard normal probability plot, except that, before the plot is produced, the linear trend is removed. This often "spreads out" the plot, thereby allowing the user to detect patterns of deviations more easily.

The *half-normal probability plot* is constructed in the same way as the standard normal probability plot, except that only the positive half of the normal curve is considered. Consequently, only positive normal values will be plotted on the Y-axis. This plot is basically used when one wants to ignore the sign of the residual, that is, when one is mostly interested in the distribution of absolute values, regardless of sign.

Review and Plot Variables

These options allow you to review or plot the currently highlighted variable (together with other variables). The labeling of the individual observations (that is, of the respective points in time represented by the rows in Scrollsheets or the horizontal X-axis in plots) is determined by the selections in the *Display/plot options* dialog (click on the *Options* button below; these options are also available on the *Transformations* dialog and are described in the *General Conventions* section, page 3315). For the plots, you can choose to *Scale the X-Axis in the plots manually* (select this option in the *Display/plot options* dialog). This is often useful when the series consists of several regularly recurring time intervals, for example, if it is made up of 10 years with 12 (monthly) observations each. Then it may be useful to scale the X-axis in the plot so that each major tickmark indicates one year.

Display/plot subset of cases only. If this check box is set, then, before plotting or displaying one or more series, you will be prompted to select a range of observations to include in the plot or display.

Options (scaling, labels, etc.). This option will bring up the *Display/plot options* dialog (see *General Conventions*, page 3315). That dialog allows you to change the labeling of the observations (i.e., the rows in Scrollsheets and the X-axis in line plots can be labeled with case names, case numbers, dates from another variable, etc.) and the scaling of the X-axis in line plots (minimum, step size).

8. TIME SERIES ANALYSIS - SEASONAL DECOMPOSITION (CENSUS I)

Review/Plot highlighted variable. This button will bring up a Scrollsheet (or a plot) with the values for the currently highlighted variable.

Review/Plot multiple variables. After clicking on this button, you will first be prompted to select a list of variables to be displayed or plotted. When plotting multiple variables the vertical (Y) axis will have a common scale; thus, if the series that are plotted contain values of completely different magnitudes use option *Plot two var lists with different scales* (see below), instead.

Plot two var. lists with different scales. This option allows you to plot two lists of variables using different scales for the vertical (Y) axes in the plot. The first list will be plotted against the left-Y axis and the second list will be plotted against the right-Y axis. For example, suppose you want to plot two variables (series) called *S1* and *S2*, and that these variables contain values of different magnitudes: the values of *S1* range from 1-100 and the values of *S2* range from 1000 to 10000. If you plot *S1* against the left-Y axis (scaled from 1-100) and *S2* against the right-Y axis (scaled from 1000-10000), then both variables will be plotted with optimal resolution, and thus relationships between patterns across time (the X-axis) may become more clearly identifiable.

Autocorrelations

These options will compute the *autocorrelations* and *partial autocorrelations* for the currently highlighted variable; these statistics will be displayed in a Scrollsheet and drawn in a correlogram graph. Refer to the *Introductory Overview* section (page 3271, 3275) for a brief description of how the pattern of (partial) autocorrelations aids in the determination of an appropriate ARIMA model.

Autocorrelations. This option will bring up a Scrollsheet and plot of the autocorrelations, for a lag of *1* through the number specified in the *Number of lags* edit field. The Scrollsheet will report the autocorrelations, their standard errors, the so-called *Box-Ljung statistic* (see the description of the *Transformation of Variables* dialog, page 3331, for computational details), and the significance level of that statistic. In general, the autocorrelation is the correlation of the series with itself, lagged by a particular number of observations.

***Alpha* for highlighting.** Significant autocorrelations (significant *Box-Ljung statistics*) will be highlighted in the Scrollsheet. This edit field allows you to specify the significance level that is used for highlighting.

White noise standard errors. Under the assumption that the true moving average process in the series is of the order $k-1$, the approximate standard error of the autocorrelation r_k is computed as:

```
StdErr(r_k) = √{(1/n) * [1+2*Σ(r_i^2)]}
(for i = 1 to k-1)
```

Here n is the number of observations in the series. Under the assumption that the series is a white noise process, that is, that all autocorrelations are equal to zero, the standard error of r_k is computed as:

```
StdErr(r_k) = √{(1/n) * [(n-k)/(n+2)]}
```

Check the *White noise standard errors* option to compute the standard errors in this manner.

Partial autocorrelations. This option will bring up a Scrollsheet and plot of the partial autocorrelations, for a lag of *1* through the number specified in the *Number of lags* edit field. The Scrollsheet will report the partial autocorrelations and their standard errors. In general, the partial autocorrelation is the partial correlation of a series with itself, lagged by a particular number of observations, and controlling for all correlations for lags of lower order. For example, the partial autocorrelation for a lag of *2* represents the unique correlation of the series with itself at that lag, after controlling for the correlation at lag *1*. For details regarding the computations of the partial autocorrelation and its standard error, refer to the

description of the *Transformation of Variables* dialog (page 3332). Partial autocorrelations that are larger than two times their respective standard errors will be highlighted in the Scrollsheet.

Number of lags. This parameter determines the maximum number of lags for which the auto-, and partial autocorrelations will be computed.

8. TIME SERIES ANALYSIS - SEASONAL DECOMPOSITION (CENSUS I)

X-11 SEASONAL DECOMPOSITION (CENSUS METHOD II)

EXAMPLE

This example is based on a series reporting the monthly US total retail sales from 1953 to 1964. The data set is reported in Shiskin, Young, and Musgrave (1967) to illustrate the results of the X-11 (Census method II) seasonal adjustment procedure. Note that the same example series is used in the *Seasonal Decomposition Example* (page 3391); there, the older Census method I seasonal decomposition method was used to analyze this series.

Data File

Shown below is a partial listing of the file *Retail.sta* that contains this series (note that, as reported in Shiskin, Young, and Musgrave, 1967, these numbers are not directly comparable to the official published retail sales figures).

	SALES	DATE
1	12903	Jan-1953
2	12198	Feb-1953
3	13711	Mar-1953
4	14115	Apr-1953
5	14520	May-1953
6	14442	Jun-1953
7	14250	Jul-1953
8	14045	Aug-1953
9	13952	Sep-1953
10	14819	Oct-1953
11	13828	Nov-1953

After starting the *Time Series* module, open the example data file *Retail.sta*. Click on the *Variables* button and select variable *Sales*, and then click on the *X-11 (Census) monthly* button to bring up the dialog entitled *X-11 Monthly Seasonal Adjustment (Census Method II)*.

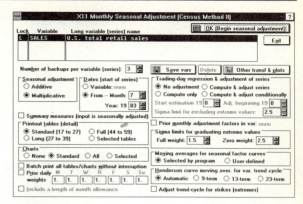

The options available on this dialog closely follow the options for the US Census Bureau X-11 procedure described by Shiskin, Musgrave, and Young (1967), and they are described in detail in the *Dialogs, Options, Statistics* section (page 3405, 3410). In general, the X-11 variant of the Census method II is the result of numerous refinements to the simple seasonal adjustment process (e.g., as described in the *Introductory Overview* for *Seasonal Decomposition (Census I)*, page 3284; most of these refinements pertain to the treatment of outliers, months with different numbers of trading days, and choices of moving average transformations for isolating the different components. The *Introductory Overview* section, page 3287, discusses those refinements in greater detail.

Determining the Seasonal Model

First plot this series. Click on the *Other transf & plots* button to bring up the *Transformation of Variables* dialog. Use the dates recorded in variable *Date* in the plots, so click on the *Dates* radio button in the *Label data points* box and select the variable *Date*. Next, click on the check box *Scale X-axis in plots manually* and specify *Min = 1* and *Step = 12*. Then, click on the *Plot* button next to the *Review highlighted variable* button to produce the following plot.

8. TIME SERIES ANALYSIS - X-11 SEASONAL DECOMPOSITION

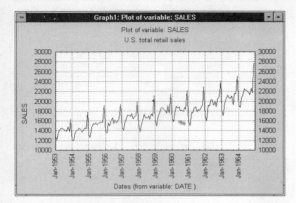

The retail sales show an upward trend and clear seasonal fluctuation. As described in the *Introductory Overview* section (page 3287), the goal of seasonal decomposition in general is to derive estimates of the seasonal, trend/cycle, and irregular components that make up the series. The seasonal component can be used to compute a seasonally adjusted series, that is, an estimate of the series after removing seasonal fluctuations.

Additive and multiplicative seasonality. The difference between the additive and multiplicative model is discussed in several places in the *Introductory Overview* section (page 3269). With regard to retail sales figures, it is conceivable that during the December holiday season, retail sales increase by a relatively fixed amount every year; that is, each year you could *add* a certain amount to the average sales that year to arrive at an estimate for the December figures. Thus, the seasonal fluctuation would be *additive* in nature.

Alternatively, each year the sales during December may increase by a *factor*; for example, they may increase by a factor of 1.3 or 30% over the respective yearly average. In that case, the seasonal fluctuations are *multiplicative* in nature.

The two types of seasonal fluctuations leave distinctive "signatures" in the series. If the seasonality is additive in nature, the fluctuations are constant around the series, regardless of the overall level of the series. If the seasonality is multiplicative in nature, then the fluctuations are greater when the level of the series is higher. In this example, it rather looks like the seasonality is multiplicative in nature; this would also make more sense intuitively: Most likely, retail sales will increase by a certain percentage each December rather than by a fixed amount. Therefore, accept the default *Multiplicative* setting in the *Seasonal adjustment* box of the X-11 dialog. Now return to that dialog; *Exit* from the *Transformations of Variables* dialog.

Seasonal Decomposition

It is desired to plot later all components that are extracted. The program will automatically append the seasonal factors, seasonally adjusted series, trend/cycle, and irregular components to the *active work area* (refer to the *General Conventions* for details concerning the memory management in the *active work area*, page 3313). Currently (by default), the *Number of backups per variable* parameter is set to *3*. There are four different series that can be appended; thus set the *Number of backups* parameter to 4.

Start date. The X-11 technique can adjust for months with different numbers of trading days (i.e., Mondays, Tuesdays, etc.). For example, when analyzing the revenues for an amusement park, the number of Saturdays and Sundays in different months will influence greatly the monthly totals. Therefore, the X-11 method requires that you specify a start date for the series (so that the program can then determine the different number of trading days in each month). You may either enter the start month and year into the respective edit fields in the *Dates (start of series)* box, or you can simply take the date from the first observation in the file. For this example, do the latter, that is, click on the *Variable* button in the *Dates (start of series)* box and then select variable *Date*.

Trading day regression. The variability due to trading-day variation (i.e., different numbers of trading days in different months) can be estimated (via least squares regression) from the data. For this example, select to compute the trading-day regression, but only use it to adjust the series if it is statistically significant (as was done in the illustration in Shiskin, Young, and Musgrave, 1967). Thus, click on the *Compute & adjust conditionally* radio button. You may leave all other entries in this box at their defaults (e.g., the *Start estimation* and *Adjust beginning* ... edit fields will automatically be set to the first observation in the series if no other numbers are entered).

Reviewing result tables. You are now ready to begin the analysis; click *OK (Begin seasonal adjustment)*. The implementation of the X-11 seasonal adjustment method in the *Time Series* module follows closely the US Bureau of Census procedure described by Shiskin, Musgrave, and Young (1967). The results consist of a series of tables and graphs. Shown below are a few selected tables.

Table *C 15 (*final trading-day regression results) above, shows that there is highly significant trading-day variation in the series. As expected, the test for stable seasonality is also highly significant; the respective ANOVA table is reported as table *D 8*.

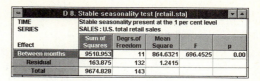

Here are the final estimates of the major components and the seasonally adjusted series. (Note: only the results for the first 6 months of each year are shown).

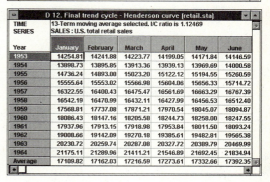

8. TIME SERIES ANALYSIS - X-11 SEASONAL DECOMPOSITION

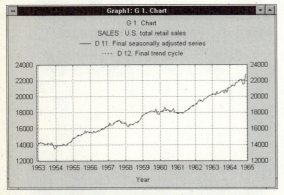

Continue through the list of output tables by clicking on the *Continue* button on the Scrollsheet. By default, two graphs will also be displayed. The first shows the final trend/cycle component and the seasonally adjusted series. Clearly, there is only relatively little variability of the adjusted series around the trend/cycle estimate.

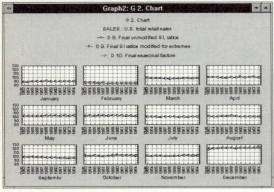

The second graph shows different components plotted (categorized) by month.

Click on the *Cont* icon to bring up the X-11 dialog again. The final estimates of the components as well as the final seasonally adjusted series have been added to the *active work area*.

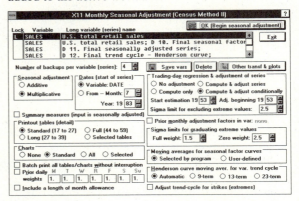

Finally, look at a summary plot of the original series, the seasonally adjusted series (*D 11*), and the irregular component (*D 13*).

Click on the *Other transf & plots* button to bring up the *Transformations of Variables* dialog. Then use option *Plot two var lists with different scales* and select the original and seasonally adjusted series for the first list (to be plotted against the left-*Y* axis), and select the *Irregular* series for the second list. Shown below is this summary plot.

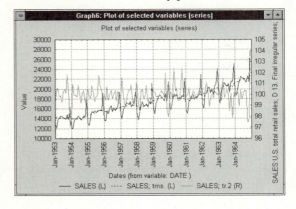

TIM - 3404

8. TIME SERIES ANALYSIS - X-11 SEASONAL DECOMPOSITION

DIALOGS, OPTIONS, STATISTICS

X-11 Monthly Seasonal Decomposition

The options on this dialog allow you to perform X-11 monthly seasonal decomposition, a technique also know as the *Census II* method (see the *Introductory Overview* section, page 3287, for details). The arrangement of options is specifically designed to closely follow the terminology and options of the X-11 procedure of the Bureau of the Census (Shiskin, Young, and Musgrave, 1967).

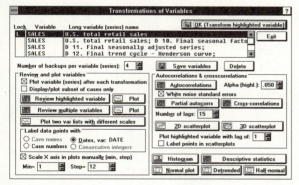

To select the variable (series) to be analyzed, click on the *Variable* or *Long variable (series) name* column to highlight the variable (series) in the *active work area* (see *General Conventions*, page 3313). The successive observations in that series should contain data (e.g., sales, revenue figures, etc.) for consecutive months.

Next, specify the starting month of the series (either specify a date variable, or type the respective month and year into the *From -- month/year* edit fields), additional options if appropriate (or accept the defaults), and click *OK* to begin the seasonal decomposition.

After the last output table has been displayed (see the listing of X-11 result tables in the *Introductory Overview*, page 3290, for a description of the tables produced by this method), the program will automatically append to the *active work area* the following series/components:

- The *final seasonal factors* (table *D 10*),
- The *final seasonally adjusted series* (*D 11*),
- The *final trend-cycle component* (*D 12*),
- The *final irregular (random) component* (*D 13*).

Be sure to *Lock* all important backups of the original variable (those which you want to retain); otherwise, those backups may be overwritten (refer to the section describing the memory management in the *active work area* in the *General Conventions* section, page 3313). To lock a backup, double-click in the *Lock* column on the respective variable. Note that if the display of consecutive results tables is interrupted, then none of these components will be appended to the *active work area*.

Other Transf. & Plots

This option will bring up the *Transformations of Variables* dialog, which allows you to perform a wide variety of transformations on the data. The transformed series will be appended to the *active work area*.

Seasonal Adjustment

The setting of radio buttons in this box determines whether an *Additive* or *Multiplicative* seasonal adjustment will be performed (refer to the *Introductory Overview* section, page 3287, for details). In short, in the additive model, the seasonal, trend-cycle, and irregular (random) components are assumed to combine in an additive fashion (e.g., the seasonal fluctuation adds

TIM - 3405

Copyright © StatSoft, 1995

$1,000,000 to the sales in the holiday month of December). In the multiplicative model, they are thought to combine in a multiplicative fashion (e.g., due to the seasonal component, in December sales increase by 30%, i.e., 1.3 times the baseline).

Dates (Start of Series)

The setting in this box determines the start date of the series. The start date will not only affect the labeling of the output tables but can also substantially affect the computations. The X-11 monthly adjustment procedure allows the user to adjust the series for fluctuations in the lengths of different months as well as the numbers of different days (i.e., Mondays, Tuesdays, etc.) in each month. When you select the *Variable* option in this box, a variable selection dialog will open in which you can select the variable containing the starting date (i.e., the date stored in this variable in the first case of the respective series will be used as the start date of the series). Selecting the *From* option will allow you to specify the specific month and year in which the series started.

Summary Measures (Input is Seasonally Adjusted)

If this check box is set, then the program will assume that the input is already a seasonally adjusted series. In that case, the X-11 procedure can be used to test for residual seasonality and trading-day variation by completing the standard steps in parts *B* and *C* of the computations (see the listing of X-11 result tables in the *Introductory Overview* section, page 3290, for details).

Note that part *A* and tables *E 2*, *E 4*, and *E 6* are not applicable in that case; also, table *D 11* (final seasonally adjusted series) will be identical to *B 1* (the input series), and the final trend-cycle moving average is applied to table *D 1*.

Printout Tables (Detail)

The setting of radio buttons in this box will determine the detail of results that will be reported after clicking *OK*. The number of specific X-11 results tables that will be computed depends on the choices in this box (see the listing of X-11 result tables in the *Introductory Overview*, page 3290, for a description of the tables produced by this method). Note that, if the *Batch print all tables/charts without interruption* check box is set (see below), then all selected tables will automatically be printed without interruption. Refer to the *X-11 Result Tables* dialog for a description of the specific tables (Scrollsheets) that are reported for the *Standard*, *Long*, or *Full* printout detail.

Standard. When you select this radio button, 17 to 27 tables will be displayed.

Long. When you select this radio button, 27 to 39 tables will be displayed.

Full. When you select this radio button, 44 to 59 tables will be displayed.

Selected tables. If you click on this radio button, the *Select the X-11 tables and charts to be printed/displayed* dialog will open in which the identifiers for all available tables are listed. In this dialog, you can select the specific tables (Scrollsheets) that you want displayed and/or printed. If no tables are selected (i.e., do not select any tables and click *OK* in this dialog), then only the final estimates of the seasonally adjusted series, the seasonal component, irregular component, and trend-cycle component will be appended to the *active work area*.

Charts

The setting of radio buttons in this box will determine the detail of graphics results that will be reported after clicking *OK* in this dialog. Note that, if the *Batch print all tables/charts without*

interruption check box is set and *Printer* output is selected in the *Page/Output Setup* dialog, then the respective graphs will automatically be printed. Refer also to the *Introductory Overview* section (page 3290, 3297) for a description of the charts (line plots) computed by the procedure.

None. If this radio button is set, then no plots will be produced.

Standard. If this radio button is set, then charts *G 1* (final seasonally adjusted series and final trend-cycle components; *D 11* and *D 12*, respectively) and *G 2* (final S-I differences or ratios *with* the extremes, the final S-I differences (ratios) *without* extremes, and the final seasonal factors; *D 8*, *D 9*, and *D 10*) will be computed.

All. If this radio button is set, then all charts *G 1* through *G 4* will be displayed (see the *Introductory Overview* section, page 3297, for details).

Selected. If this radio button is set, then the *Select the X-11 tables and charts to be printed/displayed* dialog will open in which all available output charts (and tables) are listed. The user can then select the desired charts (line plots) to be computed.

Batch Print All Tables/Charts without Interruption

If this check box is set, then all tables that are displayed will automatically be printed (to the printer, file, and/or Text/output Window, whichever is selected in the *Page/Output Setup* dialog) without requiring any additional user input. If *Printer* output is selected, all graphs will also automatically be printed.

Prior Daily Weights

This option is only available if *Multiplicative* adjustment was selected (see above). Different months consist of different numbers of trading-days (i.e., Mondays, Tuesdays, etc.), which may greatly affect the monthly values (see the *Introductory Overview*, page 3287, for details). By specifying prior daily weights, the monthly observations can be adjusted so as to reflect the differences in the number of different days in each month (this adjustment can also be performed via least squares regression; see *Trading-day regression & adjustment of series* below).

For example, suppose the input series contains monthly revenues for an amusement park. In this case, one may want to assign greater *a priori* weights to the Saturdays and Sundays (since most revenue is generated on those days) in order to adjust the monthly series prior to the seasonal decomposition.

When you set this check box, you will be able to enter the seven weights into the edit fields to the right. Note that the program will automatically adjust the weights entered here so that they will sum to 7.

Include a Length of Month Allowance

This option is only available (meaningful) if *Prior daily weights* (see above) and/or *Trading-day regression & adjustment* (see below) are requested and when *Multiplicative* adjustment is requested (see above). When this check box is set, then the differences in the lengths of different months (30-day months, 31-day months, and Februarys) are not used to adjust the daily weights. Consequently, the variability arising from differences in the lengths of months will be included in the trading-day factors.

Trading-Day Regression & Adjustment of Series

Different months consist of different numbers of trading-days (i.e., Mondays, Tuesdays, etc.), which may greatly affect the monthly values (see the

8. TIME SERIES ANALYSIS - X-11 SEASONAL DECOMPOSITION

Introductory Overview, page 3287, for details). The options in this box will determine whether and how a least-squares trading-day adjustment will be included in the analysis.

No adjustment. If this radio button is set, then no automatic trading-day adjustment is performed.

Compute only. If this radio button is set, then the trading-day weights are computed and reported; however, the series will not be adjusted by these weights.

Compute & adjust series. If this radio button is set, then the trading-day weights are computed (starting with the year indicated, see below) and applied to adjust the input series (starting with the year indicated, see below).

Compute & adjust conditionally. If this radio button is set, then the trading-day weights are computed (starting with year indicated, see below) and applied to adjust the input series (starting with the year indicated, see below). However, in part *C* of the analysis (see the description of X-11 result tables in the *Introductory Overview* section, page 3290), the trading-day weights are only used if they explain statistically significant variation.

Start estimation, Adjust beginning. These edit fields determine which years of the input series are used to compute the trading-day weights (by default all years) and to which years they are applied (by default to all years).

***Sigma* limit for excluding extreme values.** If trading-day regression is requested, then the value specified in this edit field determines how outliers in the irregular component are to be treated. Specifically, all irregular values will be excluded from the computations if they deviate from *0* (additive model) or *1* (multiplicative model) by more than $x * \sigma$ (where x is the value in the edit field and *sigma* is the estimate of the trading-day standard deviation).

Prior Monthly Adjustment Factors

The user can specify a second series with prior monthly adjustment factors. Prior to the analysis, this series will be subtracted from the input series (additive model) or divided into the input series (multiplicative model). When setting this check box, a dialog will open listing all variables (series) in the *active work area* where the user can select the second series with the prior adjustment factors.

Sigma Limits for Graduating Extreme Values

One of the strengths of the X-11 method is in the treatment of outliers. When estimating the seasonal and trend-cycle components, irregular values are assigned different weights depending on their deviation from *0* (zero; additive model) or *1* (multiplicative model). See also the *Introductory Overview* section (i.e., the description of table *B 4*, page 3292) for details concerning this procedure. The *Full weight* value determines the multiple of σ (*sigma*; the estimate of the trend-cycle standard deviation) below which a value will be assigned a full weight. The *Zero weight* value determines the multiple of σ above which a zero weight will be assigned. In-between these two cut-off values, linearly graduated weights are assigned.

Moving Averages for Seasonal Factor Curves

The radio buttons in this box determine the lengths of moving averages that are used to compute the seasonal factors (seasonal factor curve moving average).

Selected by program. By default (*Selected by program*) the program will select a 3x3 moving average for the first estimate in each part (see tables *B 5*, *C 5*, and *D 5* in the *Introductory Overview*,

TIM - 3408

Copyright © StatSoft, 1995

page 3287) and a 3x5 average for the second estimate (see tables *B 10*, *C 10*, and *D 10* in the *Introductory Overview*, page 3290*).* A 3x3 moving average is a 3-term moving average applied to a series previously smoothed by a 3-term moving average, which is equivalent to a 5-term weighted moving average; a 3x5 moving average is 3-term moving average applied to a series previously smoothed by a 5-term moving average, which is equivalent to a 7-term weighted moving average; for a discussion of different types of moving averages see Makridakis and Wheelwright (1983).

User-defined. Clicking on the *User-defined* radio button will open the *Moving Averages for Seasonal Factor Curves* dialog in which you can select the length of moving average for each month (the same length will be used for both parts).

Henderson Curve Moving Average for Variable Trend-Cycle

The radio buttons in this box determine the choice of the moving average for the variable trend-cycle routine, which is used in the computation of the trend-cycle component from the seasonally adjusted series (e.g., tables *B 7, C 7,* and *D 7;* see *Introductory Overview*, page 3290). In general, the Henderson curve moving average is a weighted moving average with the magnitudes of the weights following a bell-shaped curve (see, for example, Makridakis and Wheelwright, 1978, or Shiskin, Young, and Musgrave, 1967).

The choice of the appropriate length of the moving average is an important issue in the seasonal decomposition (i.e., the computation of the trend-cycle component). The general idea is to choose a longer moving average when there is a lot of random fluctuation in the data relative to the trend-cycle component, and to choose a shorter moving average when there is only relative little random fluctuation. The *Automatic* option, below, implements this reasoning to choose the optimum length for the Henderson moving average.

Automatic. If this radio button is set, then the program will select a moving average transformation automatically. Specifically, first a preliminary 13-term Henderson (weighted) moving average of the seasonally adjusted series is computed (without extending to the ends of the series). A preliminary estimate of the irregular component is then computed by subtracting this series from (additive model) or dividing it into (multiplicative model) the seasonally adjusted series.

Next, the average month-to-month difference (percent change) without regard to sign is computed for both the estimated irregular and trend-cycle components. The ratio of the average month-to-month differences (percent changes) in the two series reflects the relative importance of the irregular variations relative to the movements in the trend-cycle component.

Depending on the value of this ratio, either a 9-term Henderson moving average is selected (if the ratio is between 0.0 and .99), a 13-term Henderson moving average is selected (if the ratio is between 1.0 and 3.49) or a 23-term Henderson moving average is selected (if the ratio is greater than 3.5). Alternatively, the user can also specifically select one of those moving average transformations (see below).

9-term. If this radio button is set, then a 9-term Henderson moving average will be selected to compute the trend-cycle component.

13-term. If this radio button is set, then a 13-term Henderson moving average will be selected to compute the trend-cycle component.

23-term. If this radio button is set, then a 23-term Henderson moving average will be selected to compute the trend-cycle component.

8. TIME SERIES ANALYSIS - X-11 SEASONAL DECOMPOSITION

Adjust Trend-Cycle for Strikes

An additional adjustment for strikes or other extreme outliers can be incorporated in the computation of the trend-cycle component (table *B 7*; see *Introductory Overview*, page 3293). This adjustment for extremes may substantially reduce the effect of major prolonged strikes or similar irregular occurrences on the *B 7* and subsequent trend-cycle estimates. However, Shiskin, Young, and Musgrave (1967) caution that estimates near sharp business cycle peaks or troughs will also be affected by this adjustment.

X-11 Quarterly Seasonal Adjustment

The options on this dialog allow you to perform X-11 quarterly seasonal decomposition, a technique also know as the *Census II* method (see the *Introductory Overview* section, page 3287, for details). The arrangement of options is specifically designed to closely follow the terminology and options of the X-11 procedure of the Bureau of the Census (Shiskin, Young, and Musgrave, 1967).

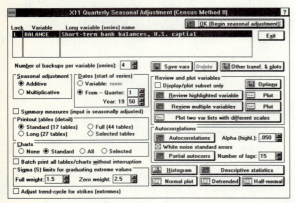

To select the variable (series) to be analyzed, click on the *Variable* or *Long variable (series) name* column to highlight the variable (series) in the *active work area* (see *General Conventions*, page 3313). The consecutive observations in that series should contain data (e.g., sales, revenue figures, etc.) for consecutive quarters. Next, specify the starting quarter of the series (either specify a date variable, or type the respective quarter and year into the *From -- quarter/year* edit fields), and additional options if appropriate (or accept the defaults), and then click *OK* to begin the seasonal decomposition.

After the last output table has been displayed (see the listing of X-11 result tables in the *Introductory Overview*, page 3290, for a description of the tables produced by this method), the program will automatically append to the *active work area* the following series/components:

- The *final seasonal factors* (table *D 10*)
- The *final seasonally adjusted series* (*D 11*)
- The *final trend-cycle component* (*D 12*)
- The *final irregular (random) component* (*D 13*).

Be sure to *Lock* all important backups of the original variable (those which you want to retain); otherwise, those backups may be overwritten (refer to the section describing the memory management in the *active work area* in the *General Conventions* section, page 3313).

To lock a backup, double-click in the *Lock* column on the respective variable. Note that if the display of consecutive results tables is interrupted, then none of these components will be appended to the *active work area*.

Other Transf. & Plots

This option will bring up the *Transformations of Variables* dialog (see page 3329) which allows you to perform a wide variety of transformations on the data. The transformed series will be appended to the *active work area*.

Seasonal Adjustment

The setting of radio buttons in this box determines whether an *Additive* or *Multiplicative* seasonal adjustment will be performed (refer to the *Introductory Overview* section, page 3287, for details).

In short, in the additive model the seasonal, trend-cycle, and irregular (random) components are assumed to combine in an additive fashion (e.g., the seasonal fluctuation adds $1,000,000 to the sales in the holiday month of December). In the multiplicative model, they are thought to combine in a multiplicative fashion (e.g., due to the seasonal component, in December sales increase by 30%, i.e., 1.3 times the baseline).

Dates (Start of Series)

The setting in this box determines the start date of the series. When you select the *Variable* option in this box, a variable selection dialog will open in which you can select the variable containing the starting dates. Selecting the *From* option will allow you to specify the specific quarter and year in which the series started.

Summary Measures (Input is Seasonally Adjusted)

If this check box is set, then the program will assume that the input is already a seasonally adjusted series. In that case, the X-11 procedure can be used to test for residual seasonal variation by completing the standard steps in parts *B* and *C* of the computations (see the listing of X-11 result tables in the *Introductory Overview* section, page 3290, for details).

Printout Tables (Detail)

The setting of radio buttons in this box will determine the detail of results that will be reported after clicking *OK*. The number of specific X-11 result tables that will be computed depends on the choices in this box.

Note that, if the *Batch print all tables/charts without interruption* check box is set (see below), then all tables will automatically be printed without interruption. Refer to the *X-11 Result Tables* dialog (page 3415) or the *Introductory Overview* (page 3290) for a description of the specific tables (Scrollsheets) that are reported for the *Standard*, *Long*, or *Full* printout detail.

Standard. When you select this radio button, 17 to 27 tables will be displayed.

Long. When you select this radio button, 27 to 39 tables will be displayed.

Full. When you select this radio button, 44 to 59 tables will be displayed.

Selected tables. If you click on this radio button, the *Select the X-11 tables and charts to be printed/displayed* dialog will open in which the identifiers for all available tables are listed. In this dialog, you can select the specific tables (Scrollsheets) that you want displayed and/or printed. If no tables are selected (i.e., do not select any tables and click *OK* in this dialog), then only the final estimates of the seasonally adjusted series, the seasonal component, irregular component, and trend-cycle component will be appended to the *active work area*.

Charts

The setting of radio buttons in this box will determine the detail of graphics results that will be reported after clicking *OK* in this dialog. Note that, if the *Batch print all tables/charts without interruption* check box is set and *Printer* output is selected in the *Page/Output Setup* dialog, then the respective graphs will automatically be printed. Refer also to the *Introductory Overview* section (page 3290, 3297) for a description of the charts (line plots) computed by the procedure.

None. If this radio button is set, then no plots will be produced.

Standard. If this radio button is set, then charts *G 1* (final seasonally adjusted series and final trend-cycle components; *D 11* and *D 12*, respectively) and *G 2* (final S-I differences or ratios *with* the extremes, the final S-I differences (ratios) *without* extremes, and the final seasonal factors; *D 8*, *D 9*, and *D 10*) will be computed.

All. If this radio button is set, then all charts *G 1* through *G 4* will be displayed (see the *Introductory Overview* section, page 3297, for details)

Selected. If this radio button is set, then the *Select the X-11 tables and charts to be printed/displayed* dialog will open in which all available output charts (and tables) are listed. The user can then select the desired charts (line plots) to be computed.

Batch Print All Tables/Charts without Interruption

If this check box is set, then all tables that are displayed will automatically be printed (to the printer, file, and/or Text/output Window, whichever is selected in the *Page/Output Setup* dialog) without requiring any additional user input. If *Printer* output is selected, all graphs will also automatically be printed.

Sigma Limits for Graduating Extreme Values

One of the strengths of the X-11 method is in the treatment of outliers. When estimating the seasonal and trend-cycle components, irregular values are assigned different weights depending on their deviation from *0* (zero; additive model) or *1* (multiplicative model). See also the *Introductory Overview* section (i.e., the description of table *B 4*, page 3292) for details concerning this procedure. The *Full weight* value determines the multiple of σ (*sigma*; the estimate of the trend-cycle standard deviation) below which a value will be assigned a full weight. The *Zero weight* value determines the multiple of σ above which a zero weight will be assigned. In-between these two cut-off values, linearly graduated weights are assigned.

Adjust Trend-Cycle for Strikes

An additional adjustment for strikes or other extreme outliers can be incorporated in the computation of the trend-cycle component (table *B 7*; see *Introductory Overview*, page 3293). This adjustment for extremes may substantially reduce the effect of major prolonged strikes or similar irregular occurrences on the *B 7* and subsequent trend-cycle estimates. However, Shiskin, Young, and Musgrave (1967) caution that estimates near sharp business cycle peaks or troughs will also be affected by this adjustment.

Review and Plot Variables

These options allow you to review or plot the original series, the transformed and differenced series, and the residuals (computed for the transformed and differenced series; thus those residuals may be different than those computed by the *Forecasting* options above). The labeling of the individual observations (that is, of the respective points in time represented by the rows in Scrollsheets or the horizontal *X*-axis in plots) is determined by the selections in the *Display/plot options* dialog (click on the *Options* button below;

see also the *General Conventions* section, page 3315).

For the plots, you can choose to *Scale the X-Axis in the plots manually*. This is often useful when the series consists of several regularly recurring time intervals, for example, if it is made up of 10 years with 12 (monthly) observations each. Then it may be useful to scale the *X*-axis in the plot so that each major tickmark indicates one year.

Display/plot subset of cases only. If this check box is set, then, before plotting or displaying one or more series, you will be prompted to select a range of observations to include in the plot or display.

Options (scaling, labels, etc.). This option will bring up the *Display/plot options* dialog (see the *General Conventions* section, page 3315). That dialog allows you to change the labeling of the observations (i.e., the rows in Scrollsheets and the *X*-axis in line plots can be labeled with case names, case numbers, dates from another variable, etc.) and the scaling of the *X*-axis in line plots (minimum, step size).

Review/plot highlighted variable. This button will bring up a Scrollsheet (or a plot) with the values for the currently highlighted variable.

Review/plot multiple variables. After clicking on this button, you will first be prompted to select a list of variables to be displayed or plotted. When plotting multiple variables the vertical (*Y*) axis will have a common scale; thus, if the series that are plotted contain values of completely different magnitudes use option *Plot two var lists with different scales* (see below), instead.

Plot two var. lists with different scales. This option allows you to plot two lists of variables using different scales for the vertical (*Y*) axes in the plot. The first list will be plotted against the left-*Y* axis and the second list will be plotted against the right-*Y* axis.

For example, suppose you want to plot two variables (series) called *S1* and *S2*, and that these variables contain values of different magnitudes: the values of *S1* range from 1-100 and the values of *S2* range from 1000 to 10000. If you plot *S1* against the left-*Y* axis (scaled from 1-100) and *S2* against the right-*Y* axis (scaled from 1000-10000), then both variables will be plotted with optimal resolution, and thus relationships between patterns across time (the *X*-axis) may become more clearly identifiable.

Autocorrelations

These options will compute the *autocorrelations* and *partial autocorrelations* for the currently highlighted variable; these statistics will be displayed in a Scrollsheet and drawn in a correlogram graph. Refer to the *Introductory Overview* section (page 3271, 3275) for a brief description of how the pattern of (partial) autocorrelations aids in the determination of an appropriate ARIMA model.

Autocorrelations. This option will bring up a Scrollsheet and plot of the autocorrelations, for a lag of *1* through the number specified in the *Number of lags* edit field. The Scrollsheet will report the autocorrelations, their standard errors, the so-called *Box-Ljung statistic* (see the description of the *Transformation of Variables* dialog, page 3331, for computational details), and the significance level of that statistic. In general, the autocorrelation is the correlation of the series with itself, lagged by a particular number of observations.

Alpha for highlighting. Significant autocorrelations (significant *Box-Ljung statistics*) will be highlighted in the Scrollsheet. This edit field allows you to specify the significance level that is used for highlighting.

White noise standard errors. Under the assumption that the true moving average process in the series is of the order *k-1*, the approximate standard error of the autocorrelation r_k is computed as:

8. TIME SERIES ANALYSIS - X-11 SEASONAL DECOMPOSITION

$$\text{StdErr}(r_k) = \sqrt{(1/n) * [1+2*\Sigma(r_i^2)]}$$
(for i = 1 to k-1)

Here *n* is the number of observations in the series. Under the assumption that the series is a white noise process, that is, that all autocorrelations are equal to zero, the standard error of r_k is computed as:

$$\text{StdErr}(r_k) = \sqrt{(1/n) * [(n-k)/(n+2)]}$$

Check the *White noise standard errors* option to compute the standard errors in this manner.

Partial autocorrelations. This option will bring up a Scrollsheet and plot of the partial autocorrelations, for a lag of *1* through the number specified in the *Number of lags* edit field. The Scrollsheet will report the partial autocorrelations and their standard errors. In general, the partial autocorrelation is the partial correlation of a series with itself, lagged by a particular number of observations, and controlling for all correlations for lags of lower order.

For example, the partial autocorrelation for a lag of *2* represents the unique correlation of the series with itself at that lag, after controlling for the correlation at lag *1*. For details regarding the computations of the partial autocorrelation and its standard error refer to the description of the *Transformation of Variables* dialog (page 3332). Partial autocorrelations that are larger than two times their respective standard errors will be highlighted in the Scrollsheet.

Number of lags. This parameter determines the maximum number of lags for which the auto-, and partial autocorrelations will be computed.

Histogram

This option will bring up a histogram of the currently highlighted variable.

Descriptive Statistics

This option will bring up a Scrollsheet with descriptive statistics for all variables currently available in the *active work area*.

Normal, Detrended, and Half-Normal Probability Plots

These options will bring up the respective normal probability plot for the highlighted variable. The way the *standard normal probability* plot is constructed is as follows. First the values are rank ordered. From these ranks you can compute *z* values (i.e., standardized values of the normal distribution) *based on the assumption* that the data come from a normal distribution. These *z* values are plotted on the *Y*-axis in the plot. If the data values (plotted on the *X*-axis) are normally distributed, then all points should fall onto a straight line in the plot. If the values are not normally distributed, they will deviate from the line. Outliers may also become evident in this plot.

The *detrended normal probability plot* is constructed in the same way as the standard normal probability plot, except that, before the plot is produced, the linear trend is removed. This often "spreads out" the plot, thereby allowing the user to detect patterns of deviations more easily.

The *half-normal probability plot* is constructed in the same way as the standard normal probability plot, except that only the positive half of the normal curve is considered. Consequently, only positive normal values will be plotted on the *Y*-axis. This plot is basically used when one wants to ignore the sign of the residual, that is, when one is mostly interested in the distribution of absolute values, regardless of sign.

8. TIME SERIES ANALYSIS - X-11 SEASONAL DECOMPOSITION

Moving Averages for Seasonal Factor Curves

The radio buttons in this dialog determine the lengths of moving averages that are used to compute the preliminary and final seasonal factors for each month in each part of the X-11 monthly analysis (*B*, *C*, and *D*; see the *Introductory Overview*, page 3290, for a description of tables).

By default (*Automatic*) the program will select for each month a 3x3 moving average for the first estimate in each part (tables *B 5*, *C 5*, and *D 5*) and a 3x5 moving average for the second estimate (tables *B 10*, *C 10*, and *D 10*). A 3x3 moving average is a 3-term moving average applied to a series previously smoothed by a 3-term moving average, which is equivalent to a 5-term weighted moving average; a 3x5 moving average is 3-term moving average applied to a series previously smoothed by a 5-term moving average, which is equivalent to a 7-term weighted moving average; for a discussion of different types of moving averages see Makridakis and Wheelwright (1983). Alternatively, you can also choose a different type of moving average for each month. In that case, the same type of moving average will be applied for both the preliminary (e.g., *B 5*) and final estimates (e.g., *B 10*) of the seasonal component in each part. However, different moving averages can be applied to different months.

Selecting Specific Output Tables and/or Charts

Select the desired output tables and/or charts (line plots *G 1* through *G 4*, see below). If no tables or charts are selected, then no Scrollsheets or line plots will be displayed; however, the final components *D 10* through *D 13* (see below) will still be appended to the *active work area* (see the *General Conventions* section, page 3313, for a description of the memory management in the *Time Series* module).

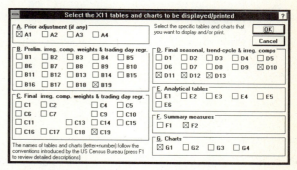

For details concerning the X-11 Census II method refer to the *Introductory Overview* section, page 1027. A complete listing and description of all available tables is also given in the *Introductory Overview*, page 1027. Those tables identified by an asterisk (*) are not available (applicable) when analyzing quarterly series. Following the convention of the US Bureau of the Census version of the X-11 method, the *STATISTICA* implementation offers three levels of printout detail: *Standard* (17 to 27 tables), *Long* (27 to 39 tables), and *Full* (44 to 59 tables). An option to only produce selected tables (Scrollsheets) is also provided. In the listing below, the tables listed under the level of printout headings indicate which tables will be displayed and/or printed at the respective setting of the output option. (For the charts, two levels of detail are available: *Standard* and *All*.) For more detailed information on each of the tables, see the respective pages.

Standard Printout Detail (*S*)

The following tables may be included as output when *Standard* printout detail is selected.

8. TIME SERIES ANALYSIS - X-11 SEASONAL DECOMPOSITION

*A 1.   Original series (see page 3291)

*A 2.   Prior monthly adjustment factors (see page 3291)

*A 3.   Original series adjusted by prior monthly adjustment factors (see page 3291)

*A 4.   Prior trading-day adjustment factors (see page 3292)

B 1.   Prior adjusted series or original series (see page 3292)

C 13.   Irregular series (see page 3295)

*C 14. Extreme irregular values excluded from trading-day regression (see page 3295)

*C 15. Final trading-day regression (see page 3295)

*C 16. Final trading-day adjustment factors derived from regression coefficients (see page 3295)

C 17.   Final weights for irregular component (see page 3295)

*C 18. Final trading-day factors derived from combined daily weights (see page 3295)

*C 19. Original series adjusted for trading-day and prior variation (see page 3295)

D 8.   Final unmodified S-I differences (ratios; see page 3296)

D 9.   Final replacement values for extreme S-I differences (ratios; see page 3296)

D 10.   Final seasonal factors (see page 3296)

D 11.   Final seasonally adjusted series (see page 3296)

D 12.   Final trend-cycle (see page 3296)

D 13.   Final irregular (see page 3296)

E 1.   Modified original series (see page 3296)

E 2.   Modified seasonally adjusted series (see page 3296)

E 3.   Modified irregular series (see page 3297)

E 4.   Differences (ratios) of annual totals (see page 3297)

E 5.   Differences (percent changes) in original series (see page 3297)

E 6.   Differences (percent changes) in final seasonally adjusted series (see page 3297)

F 1.   MCD (QCD) moving average (see page 3297)

F 2.   Summary measures (see page 3297)

Standard Printout Detail for Charts (S)

The following charts may be included as output when *Standard* printout detail is selected.

G1.   Chart (see page 3297)

G2.   Chart (see page 3298)

Long Printout Detail (L)

In addition to the tables listed under the *Standard* printout detail, the following tables may be included as output when *Long* printout detail is selected.

B 2.   Trend-cycle (see page 3292)

B 7.   Trend-cycle (see page 3293)

B 10.   Seasonal factors (see page 3293)

B 13.   Irregular series (see page 3293)

*B 14. Extreme irregular values excluded from trading-day regression (see page 3294)

*B 15. Preliminary trading-day regression (see page 3294)

B 17.   Preliminary weights for irregular component (see page 3294)

C 1.   Original series modified by preliminary weights and adjusted for trading-day and prior variation (see page 3294)

8. TIME SERIES ANALYSIS - X-11 SEASONAL DECOMPOSITION

C 7.   Trend-cycle (see page 3295)

C 10. Seasonal factors (see page 3295)

D 1.   Original series modified by final weights and adjusted for trading-day and prior variation (see page 3296)

D 7.   Trend-cycle (see page 3296)

Full Printout Detail (*F*)

In addition to the tables listed under the *Standard* and *Long* printout details, the following tables may be included as output when *Full* printout detail is selected.

B 3.   Unmodified S-I differences or ratios (see page 3292)

B 4.   Replacement values for extreme S-I differences (ratios) (see page 3292)

B 5.   Seasonal factors (see page 3292)

B 6.   Seasonally adjusted series (see page 3293)

B 8.   Unmodified S-I differences (ratios) (see page 3293)

B 9.   Replacement values for extreme S-I differences (ratios) (see page 3293)

B 11. Seasonally adjusted series (see page 3293)

*B 16. Trading-day adjustment factors derived from regression coefficients (see page 3294)

*B 18. Trading-day factors derived from combined daily weights (see page 3294)

*B 19. Original series adjusted for trading-day and prior variation (see page 3294)

C 2.   Trend-cycle (see page 3294)

C 4.   Modified S-I differences (ratios; see page 3295)

C 5.   Seasonal factors (see page 3295)

C 6.   Seasonally adjusted series (see page 3295)

C 9.   Modified S-I differences (ratios; see page 3295)

C 11. Seasonally adjusted series (see page 3295)

D 2.   Trend-Cycle (see page 3296)

D 4.   Modified S-I differences (ratios; see page 3296)

D 5.   Seasonal factors (see page 3296)

D 6.   Seasonally adjusted series (see page 3296)

All Printout Detail for Charts (*A*)

The following charts may be included as output when *All* printout detail is selected.

G 1.   Chart (see page 3297)

G 2.   Chart (see page 3298)

G 3.   Chart (see page 3298)

G 4.   Chart (see page 3298)

DISTRIBUTED LAGS ANALYSIS

EXAMPLE

Overview and Data File

The current example is based on data published by the US Education Department. The file *Teachers.sta* contains data describing;

(1) The number of students enrolled in public schools (variable *Children*),

(2) The number of public school teachers (*Teachers*), and

(3) The average salary of public school teachers (*Salary*).

These data are presented for the period from 1900 through 1980, in ten-year intervals.

	CHILDREN	TEACHERS	SALARY
1900	15503110	423062	325
1910	17813852	523210	485
1920	21578316	679302	871
1930	25678015	854263	1420
1940	25433542	875477	1441
1950	25111427	913671	3010
1960	36086771	1355000	5174
1970	45909088	2061115	9570
1980	40984093	21835000	17600

It is reasonable to assume that the number of teachers employed in the public schools will be a function of the number of students that are in the schools. However, you may expect some lag in the relationship. When, due to demographic changes, there are many students, there will be more hiring of teachers; however, it will take some time to "produce" those teachers. Greater demand for teachers should also drive up the salaries, again, probably with some lag.

Specifying the Analysis

After starting the analysis, open the data file *Teachers.sta*. Click on the *Variables* button and select all variables, then select *Distributed lags analysis*.

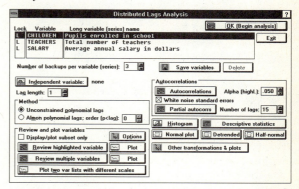

Begin this example by analyzing the number of teachers as the dependent variable; the independent or explanatory variable is the number of school children. To select the dependent variable, highlight *Teachers* in the *active work area*; then click on the *Independent variable* button and select *Children* as the independent variable.

Set the *Lag length* to *2* to look at a 10-year and 20-year lag.

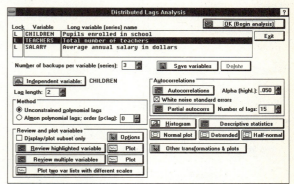

8. TIME SERIES ANALYSIS – DISTRIBUTED LAGS ANALYSIS

Reviewing Results

Click *OK (Begin analysis)* to begin the analysis. The results will be displayed in two Scrollsheets.

Polyn. Distr. Lags; Analysis of Variance (teachers.sta)
Indep: CHILDREN: Pupils enroll Dep: TEACHERS: Total number
Lag: 2 R=.8774 R-square=.7698 N:7

TIME SERIES Effect	Sums of Squares	df	Mean Square	F	p
Regress.	37386E10	3	12462E10	4.459354	.091353
Residual	11178E10	4	279459E8		
Total	48564E10				

Polyn. Distr. Lags; Regression Coefficients (teachers.sta)
Indep: CHILDREN: Pupils enroll Dep: TEACHERS: Total number
Lag: 2 R=.8774 R-square=.7698 N:7

TIME SERIES Lag	Regressn Coeff.	Standard Error	t(4)	p
0	-.74285629	.35593690	-2.0870449	.10518149
1	1.33554839	.61843500	2.1595615	.09693672
2	-.40262978	.63659171	-.6324773	.56142521

The results show that there is a strong but only marginally significant correlation between variables ($R = .88$). However, note that the regression computations in distributed lags analysis do not allow for an intercept in the equation (as is apparent in the equation presented in the *Introductory Overview* section, page 3298). As with many econometric models, the intercept of the regression line is assumed to be zero, since, if there are no students, there would not be any teachers either.

Regression coefficients. Judging from the results shown above, there is indication of a 10-year lagged effect; that is, the *t* value for *Lag = 1* (and for *Lag = 0*) is greater than *2*. Of course, because of the small number of cases in this file, this *t* value is not statistically significant.

Salary. Now, repeat these analyses, this time with *Salary* as the dependent variable. Click on the Scrollsheet *Continue* button to bring the *Distributed Lags Analysis* dialog back up. Then, highlight *Salary* in the *active work area* and click *OK*.

Polyn. Distr. Lags; Regression Coefficients (teachers.sta)
Indep: CHILDREN: Pupils enroll Dep: SALARY : Average annu
Lag: 2 R=.9174 R-square=.8416 N:7

TIME SERIES Lag	Regressn Coeff.	Standard Error	t(4)	p
0	-.00021727	.00028170	-.7712905	.48356828
1	.00089565	.00048945	1.8298942	.14124491
2	-.00051226	.00050382	-1.0167403	.36677597

As before, the largest *t* value is in the second row of the Scrollsheet, denoting the 10-year lag.

Conclusion. The results of these analyses lend some support to the hypothesis that the number of teachers and their salaries "respond" to the number of students with a lag.

Almon Distributed Lag

As described in the *Introductory Overview* section (page 3299), the standard multiple regression estimates for lags analysis sometimes suffer from multicollinearity problems. Now repeat the analyses using the *Almon distributed lags* method. On the *Distributed Lags Analysis* dialog, set the *Almon polynomial lags* radio button. Next, highlight variable *Teachers* in the *active work area.*

Specifying polynomial order. As described in the *Introductory Overview*, page 3299, this technique approximates the regression weights with a polynomial series of length smaller than the lag length. For this example, set the *order* edit field to *1*. You are now ready to proceed. Click *OK* to review the results.

Almon Polyn. Distr.Lags; Regression Coefficients
Indep: CHILDREN: Pupils enroll Dep: TEACHERS: Total number
Lag: 2 Polyn. order: 1 R=.7366 R-square=.5426 N:7

TIME SERIES Lag	Regressn Coeff.	Standard Error	t(4)	p
0	-.40174586	.39315202	-1.0218588	.36461872
1	.10905913	.04932277	2.2111314	.09151078
2	.61986411	.47260848	1.3115806	.25987650

In this case, the *t* value for the 10-year lag is much larger than that of the other lags, further supporting the hypothesis. If you repeat this analysis for the salary data, you will obtain the following results.

Almon Polyn. Distr.Lags; Regression Coefficients
Indep: CHILDREN: Pupils enroll Dep: SALARY : Average annu
Lag: 2 Polyn. order: 1 R=.8560 R-square=.7327 N:7

TIME SERIES Lag	Regressn Coeff.	Standard Error	t(4)	p
0	.00000803	.00028674	.02799390	.97900800
1	.00008556	.00003597	2.37847266	.07611563
2	.00016309	.00034469	.47316093	.66076346

Again, the regression weight for the 10-year lag is the most significant one of the three.

TIM - 3420

Copyright © StatSoft, 1995

DIALOGS, OPTIONS, STATISTICS

Distributed Lags Analysis

Select the dependent variable by highlighting the respective series in the *active work area* (see *General Conventions*, page 3313); select the independent variable with the *Independent variable* button. After clicking *OK*, Scrollsheets will be computed:

(1) For the *alpha* weights (if *Almon polynomial distributed lags* was selected, see below),

(2) With the analysis of variance table, and

(3) With the regression weights.

For an overview over this method refer to the *Introductory Overview* section, page 3298.

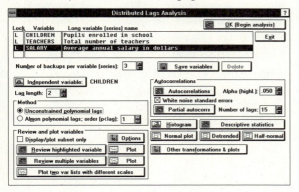

Independent Variable

This option will bring up a standard variable selection window for selecting the independent (explanatory or exogenous) variable.

Lag Length

This edit box specifies the maximum lag length to be used in the analysis; the default (and minimum value) is *1*; the maximum lag is *48*. However, note that the maximum lag is also constrained by the number of cases in your data file. Specifically, the valid number of cases in the data file must be larger than: *2 * lag length + 3*. Otherwise, reliable parameter estimates cannot be obtained.

Method

The *Time Series* module will estimate two types of distributed lags models: The *unconstrained lags model* and the *Almon polynomial distributed lag* model (Almon, 1965; refer to the *Introductory Overview* section, page 3299, for details).

Polynomial order. This edit box is only active if *Almon distributed lags* was selected. You may specify the polynomial order for the parameter estimation (refer to the *Introductory Overview* section, page 3299); this value must be smaller than the *Lag length*.

Review and Plot Variables

These options allow you to review or plot the currently highlighted variable (together with other variables). The labeling of the individual observations (that is, of the respective points in time represented by the rows in Scrollsheets or the horizontal *X*-axis in plots) is determined by the selections in the *Display/plot options* dialog (click on the *Options* button below; these options are also available on the *Transformations* dialog and are described in the *General Conventions* section, page 3315). For the plots, you can choose to *Scale the X-Axis in the plots manually* (select this option in the *Display/plot options* dialog). This is often useful when the series consists of several regularly recurring time intervals, for example, if it is made up of 10 years with 12 (monthly) observations each.

TIM - 3421

Then it may be useful to scale the *X*-axis in the plot so that each major tickmark indicates one year.

Display/plot subset of cases only. If this check box is set, then, before plotting or displaying one or more series, you will be prompted to select a range of observations to include in the plot or display.

Options (scaling, labels, etc.). This option will bring up the *Display/plot options* dialog (see *General Conventions*, 3315). That dialog allows you to change the labeling of the observations (i.e., the rows in Scrollsheets and the *X*-axis in line plots can be labeled with case names, case numbers, dates from another variable, etc.) and the scaling of the *X*-axis in line plots (minimum, step size).

Review/Plot highlighted variable. This button will bring up a Scrollsheet (or a plot) with the values for the currently highlighted variable.

Review/Plot multiple variables. After clicking on this button, you will first be prompted to select a list of variables to be displayed or plotted. When plotting multiple variables the vertical (*Y*) axis will have a common scale; thus, if the series that are plotted contain values of completely different magnitudes use option *Plot two var lists with different scales* (see below), instead.

Plot two var. lists with different scales. This option allows you to plot two lists of variables using different scales for the vertical (*Y*) axes in the plot. The first list will be plotted against the left-*Y* axis, and the second list will be plotted against the right-*Y* axis. For example, suppose you want to plot two variables (series) called *S1* and *S2*, and that these variables contain values of different magnitudes: The values of *S1* range from 1-100 and the values of *S2* range from 1000 to 10000. If you plot *S1* against the left-*Y* axis (scaled from 1-100) and *S2* against the right-*Y* axis (scaled from 1000-10000), then both variables will be plotted with optimal resolution, and thus relationships between patterns across time (the *X*-axis) may become more clearly identifiable.

Autocorrelations

These options will compute the *autocorrelations* and *partial autocorrelations* for the currently highlighted variable; these statistics will be displayed in a Scrollsheet and drawn in a correlogram graph. Refer to the *Introductory Overview* section (pages 3271, 3275) for a brief description of how the pattern of (partial) autocorrelations aids in the determination of an appropriate ARIMA model.

Autocorrelations. This option will bring up a Scrollsheet and plot of the autocorrelations, for a lag of *1* through the number specified in the *Number of lags* edit field. The Scrollsheet will report the autocorrelations, their standard errors, the so-called *Box-Ljung statistic* (see the description of the *Transformation of Variables* dialog, page 3331, for computational details), and the significance level of that statistic. In general, the autocorrelation is the correlation of the series with itself, lagged by a particular number of observations.

***Alpha* for highlighting.** Significant autocorrelations (significant *Box-Ljung statistics*) will be highlighted in the Scrollsheet. This edit field allows you to specify the significance level that is used for highlighting.

White noise standard errors. Under the assumption that the true moving average process in the series is of the order *k-1*, the approximate standard error of the autocorrelation r_k is computed as:

$$\text{StdErr}(r_k) = \sqrt{(1/n) * [1+2*\Sigma(r_i^2)]}$$
(for i = 1 to k-1)

Here *n* is the number of observations in the series. Under the assumption that the series is a white noise process, that is, that all autocorrelations are equal to zero, the standard error of r_k is computed as:

$$\text{StdErr}(r_k) = \sqrt{(1/n) * [(n-k)/(n+2)]}$$

Check the *White noise standard errors* option to compute the standard errors in this manner.

Partial autocorrelations. This option will bring up a Scrollsheet and plot of the partial autocorrelations, for a lag of *1* through the number specified in the *Number of lags* edit field. The Scrollsheet will report the partial autocorrelations and their standard errors. In general, the partial autocorrelation is the partial correlation of a series with itself, lagged by a particular number of observations, and controlling for all correlations for lags of lower order. For example, the partial autocorrelation for a lag of *2* represents the unique correlation of the series with itself at that lag, after controlling for the correlation at lag *1*. For details regarding the computations of the partial autocorrelation and its standard error refer to the description of the *Transformation of Variables* dialog (page 3332). Partial autocorrelations that are larger than two times their respective standard errors will be highlighted in the Scrollsheet.

Number of lags. This parameter determines the maximum number of lags for which the auto-, and partial autocorrelations will be computed.

Histogram

This option will bring up a histogram of the currently highlighted variable.

Descriptive Statistics

This option will bring up a Scrollsheet with descriptive statistics for all variables currently available in the *active work area*.

Normal, Detrended, and Half-Normal Probability Plots

These options will bring up the respective normal probability plot for the highlighted variable. The way the *standard normal probability* plot is constructed is as follows. First the values are rank ordered. From these ranks you can compute z values (i.e., standardized values of the normal distribution) *based on the assumption* that the data come from a normal distribution. These z values are plotted on the *Y*-axis in the plot. If the data values (plotted on the *X*-axis) are normally distributed, then all points should fall onto a straight line in the plot. If the values are not normally distributed, they will deviate from the line. Outliers may also become evident in this plot. The *detrended normal probability plot* is constructed in the same way as the standard normal probability plot, except that, before the plot is produced, the linear trend is removed. This often "spreads out" the plot, thereby allowing the user to detect patterns of deviations more easily.

The *half-normal probability plot* is constructed in the same way as the standard normal probability plot, except that only the positive half of the normal curve is considered. Consequently, only positive normal values will be plotted on the *Y*-axis. This plot is basically used when one wants to ignore the sign of the residual, that is, when one is mostly interested in the distribution of absolute values, regardless of sign.

Other Transformations and Plots

This option will bring up the *Transformations of Variables* dialog (see page 3329) which allows you to perform a wide variety of transformations on the data. The transformed series will be appended to the *active work area* (see *General Conventions*, 3313).

SPECTRUM (FOURIER) ANALYSIS

EXAMPLE

The *Introductory Overview* section (page 3300) discusses two very simple examples (based on Shumway, 1988) to illustrate the nature of spectrum analysis and the interpretation of results. If you are not familiar with this technique, it is recommended that you first review that section of this chapter.

Overview and Data File

File *Sunspot.sta* contains part of the famous Wolfer sunspot numbers for the years 1749 through 1924 (Anderson, 1971). Shown below is a listing of the first few cases in the example file.

	Wolfer's SPOTS
1749	809
1750	834
1751	477
1752	478
1753	307
1754	122
1755	96
1756	102
1757	324
1758	476

The number of sunspots are believed to affect the weather on earth, and thus human activities such as agriculture, telecommunications, etc. In this analysis, you will try to find out whether sunspot activity is cyclical in nature (which it is, this data set is widely discussed in the literature; see, for example, Bloomfield, 1976, or Shumway, 1988).

Specifying the Analysis

After starting the analysis, open the data file *Sunspot.sta*. Click on the *Variables* button and select the variable *Spots* (note that if the data file *Sunspot.sta* is the currently open data file, and since *Spots* is the only variable in that data file, then when the *Time Series Analysis* dialog opens, *Spots* will automatically be selected). Now, click on the *Spectral (Fourier) analysis* button to go to the *Fourier (Spectral) Analysis* dialog.

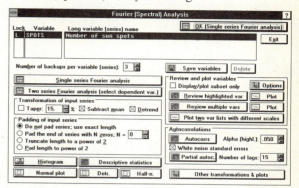

Before performing the spectrum analysis, first plot the sunspot numbers. Note that file *Sunspot.sta* contains the respective years in the case names. To use those case names in the line plots, click on the *Options* button in the *Review and plot variables* box. On the *Display Options* dialog, select *Case names* in the *Label data points* box. Also, select to *Scale X-axis in plots manually*, with *Min = 1* and *Step = 10*. Then click *OK*, and on the *Fourier (spectral) Analysis* dialog, click on the *Plot* button next to the *Review highlighted var* button.

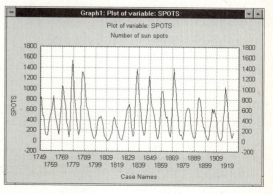

8. TIME SERIES ANALYSIS - SPECTRUM (FOURIER) ANALYSIS

The sunspot numbers clearly appear to follow a cyclical pattern. Also, there is no sign of trend; therefore, click on the *Continue* button and then de-select the *Detrend* check box in the *Transformation of input series* box.

Obviously, the mean of the series is greater than *0* (zero). Therefore, keep the *Subtract mean* option selected [otherwise the periodogram will be "overwhelmed" by a very large spike at frequency *0* (zero)].

You are now ready to begin the analysis. To learn more about the other options on this dialog, refer to the *Introductory Overview* (page 3300) and *Dialogs, Options, Statistics* sections (page 3429). Now, click *OK (Single series Fourier analysis)* to bring up the *Single Series Fourier (spectral) Analysis Results* dialog.

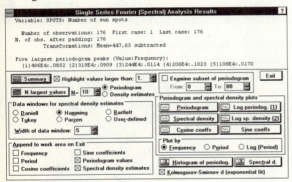

Results

The *Info* box at the top of the dialog shows some summary statistics for the series. It also shows the five largest periodogram peaks (by frequency). The largest three peaks are at .0852, .0909, and .0114. This information is often useful when analyzing very long series (e.g., with over 100,000 observations) that cannot readily be summarized in a single plot. In this case, however, the periodogram values can easily be reviewed; click on the *Periodogram* button in the *Periodogram and spectral density plots* box.

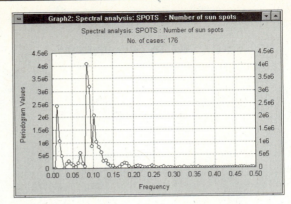

There are two clear "peaks" in the periodogram plot. The largest one is at approximately *0.9*. Click *Continue*, and then click on the *Summary* button to see all of the periodogram values (and other results) in a Scrollsheet. Shown below is the part of the Scrollsheet with the largest peak which was identified in the periodogram.

TIME SERIES	No. of cases: 176					
	Frequency	Period	Cosine Coeffs	Sine Coeffs	Periodog	Density
12	.068182	14.6667	3.701	-82.4991	600141.	371257.
13	.073864	13.5385	47.061	-.5986	194931.	403353.
14	.079545	12.5714	-29.058	6.9976	78612.	1196597.
15	.085227	11.7333	-213.104	27.3617	4062267.	2638459.
16	.090909	11.0000	154.572	110.9504	3185828.	2687325.
17	.096591	10.3529	65.701	-74.5249	868612.	1834630.
18	.102273	9.7778	111.164	105.1378	2060202.	1522726.
19	.107955	9.2632	10.687	108.1342	1039034.	1211949.
20	.113636	8.8000	.432	96.6095	821356.	851775.
21	.119318	8.3810	-84.414	2.2811	627526.	591681.

As discussed in the *Introductory Overview* section (page 3300), the *Frequency* is the number of cycles per unit time (where each observation is treated as one unit of time). Thus, the *Frequency* of *0.0909* corresponds to a *Period* (the number of units of time necessary to complete one full cycle) of *11*. Since the sunspot data in *Sunspot.sta* represent annual observations, you can conclude that there is a strong 11-year (perhaps a little longer than 11-year) cycle in sunspot activity. Now click *Continue* to return to the results dialog.

Spectral density. It is customary to smooth the periodogram -- so as to remove random fluctuations

-- in order to obtain *spectral density* estimates. The type of weighted moving average and the window width can be selected in the *Data windows for spectral density estimates* box. The *Introductory Overview* section (page 3300) discusses these options in detail. For this example, accept the default window (*5-point Hamming* window) and select the *Spectral density* plot.

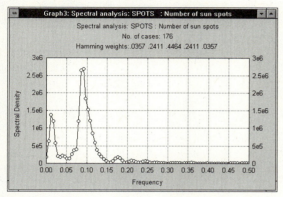

The two peaks are now even clearer. Look at the periodogram values by period. Click *Continue* and then click on the *Period* button in the *Plot by* box. Next, select the *Spectral density* plot.

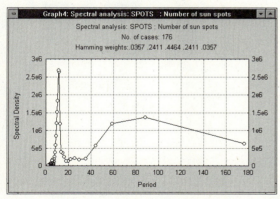

Again, it appears that there is a strong 11-year cycle in the sunspot activity; furthermore, there is some evidence of a longer, approximately 80- to 90-year cycle.

8. TIME SERIES ANALYSIS - SPECTRUM (FOURIER) ANALYSIS

DIALOGS, OPTIONS, STATISTICS

Spectrum (Fourier) Analysis

The options on this dialog allow you to perform spectrum (Fourier) analysis for single variables as well as cross-spectrum analysis (refer to the *Introductory Overview* section, page 3300, for a discussion of these methods). Click in the *Variable* or *Long variable (series) name* column to highlight the variable (series) in the *active work area* that is to be analyzed (see also the *General Conventions* section, page 3313, for a description of the memory management in the *Time Series* module). For cross-spectrum analysis, that variable will be treated as the *x* or independent variable. Then select the desired options (see below) and begin the analysis by clicking either on the *Single series Fourier analysis* or *Two series Fourier analysis* button.

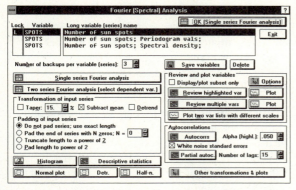

Note that the input series should contain an even number of valid observations. If this is not the case, then the last case of the series will automatically be ignored. The components computed as the result of the analyses can later be appended to the *active work area*. Thus, make sure that there is sufficient "space" in the *active work area* for the respective variable, that is, that there are at least as many (not locked) backups available as there are components that you want to save. Increase the *Number of backups* parameter if necessary. Also, be sure to *Lock* all important backups of the original variable (those which you want to retain); otherwise, those backups may be overwritten (refer to the *General Conventions* section, page 3313, describing the memory management in the *active work area*). To lock a backup, double-click in the *Lock* column on the respective variable.

Single Series Fourier Analysis

This option will perform a spectrum analysis for the currently highlighted variable (series), and then bring up the *Single Series Fourier (Spectral) Analysis Results* dialog. If selected, the transformations specified in the *Transformation of input series* box (see below) will be performed prior to the analysis.

Two Series Fourier Analysis

This option will perform a cross-spectrum analysis of the currently highlighted variable (series) and another variable. The currently highlighted variable will be treated as the independent or *x* variable in the analyses. After clicking on this button, a dialog will open, prompting the user to select the dependent variable for the analysis.

Transformation of Input Series

The transformations available in this box will be performed on the series prior to the analysis. The *Introductory Overview* section (page 3303) discusses the process of tapering and the reasons for de-meaning/trending the data.

Taper. The so-called process of *split-cosine-bell tapering* is a recommended transformation of the series prior to the spectrum analysis. It usually leads to a reduction of leakage in the spectral density plots. (*Leakage* describes the condition when a strong periodicity at a particular frequency results in large spectral density estimates in several adjacent frequencies). The rationale for this transformation is explained in detail in Bloomfield (1976, pages 80-94).

Subtract mean. If this check box is set, then the overall mean will be subtracted from the series prior to the analysis. Because the goal of spectrum analysis is to detect underlying periodicity, the overall mean is usually not of interest. If the mean is not removed, then it will "show up" as the cosine coefficient at frequency *0* (zero; the mean can be thought of as a periodic cycle with a frequency of *0* per unit time). Often, this will lead to an extremely large periodogram value at that frequency, which makes it hard to identify other spikes in the periodogram or spectral density plots.

Detrend. If this check box is set, then the linear trend will be removed from the series prior to the analysis. Like the mean, an overall trend is also not of interest when one wants to uncover periodicities in the series. Therefore, it should be removed prior to the analysis.

Padding of Input Series

You may choose to *pad* the series prior to the analysis, that is, to add zeros to the end of the series. Padding of the input series may be useful for two reasons: First, in a sense, it allows you to apply a finer "frequency roster" to the series, checking for successive frequencies at smaller increments. Second, for moderate to large size series (e.g., with more than 100,000 cases), choosing to pad (or truncate) the length of the series to a power of 2 may substantially reduce memory requirements and increase speed (for small series, you will not notice any difference). Both of these issues are also discussed in the *Introductory Overview* section

(page 3300). If extensive padding is used, it is recommended that you also taper the series.

Do not pad series; use exact length. If this radio button is set, then the series will not be padded, and the resulting periodogram will show values for *n/2+1* distinct frequencies (where *n* is the even number of cases in the series). As described in the overview section, the standard (and most efficient) FFT (*fast Fourier transform*) algorithm requires that the length of the input series is a power of 2. If this is not the case, then additional computations have to be performed. The *Time Series* module will use the simple explicit computational formulas as long as the input series is relatively small, and the number of computations can be performed in a relatively short amount of time. For long time series, in order to still utilize the FFT algorithm, an implementation of the general approach described by Monro and Branch (1976) is used. This method requires significantly more storage space; however, series of considerable length can still be analyzed very quickly, even if the number of observations is not equal to a power of 2.

Pad the end of the series with *n* zeros. If this radio button is set, then *n* zeros will be added to the end of the input series (where *n* is the number of observations in the input series). Note that the zeros will be added after the series has been de-meaned/trended (see above).

Truncate length to a power of 2. If this radio button is set, then the length of the input series will be truncated so that the number of observations will be equal to a power of 2 (e.g., 8, 16, 32, 64, 128, ...). When the length of the input series is a power of 2, then the so-called *fast Fourier transform* (*FFT*) algorithm (widely popularized by Cooley and Tukey, 1965; for various refinements and improvements see Monro, 1975) can be used for the analysis (see also the *Introductory Overview* section, page 3300, for computational details). This is the fastest method for computing the spectrum analysis (the processing time is proportional to $n * log_2(n)$).

Pad length to a power of 2. If this radio button is set, then the length of the input series will be padded with 0's (zeros) so that the number of observations will be equal to a power of 2 (e.g., 8, 16, 32, 64, 128, ...). When the length of the input series is a power of 2, then the so-called *fast Fourier transform* algorithm (widely popularized by Cooley and Tukey, 1965; for various refinements and improvements, see Monro, 1975) can be used for the analysis. This is the fastest method for computing the spectrum analysis (the processing time is proportional to $n * log_2(n)$).

Histogram

This option will bring up a histogram of the currently highlighted variable.

Descriptive Statistics

This option will bring up a Scrollsheet with descriptive statistics for all variables currently available in the *active work area*.

Normal, Detrended, and Half-Normal Probability Plots

These options will bring up the respective normal probability plot for the highlighted variable. The way the *standard normal probability* plot is constructed is as follows. First the values are rank ordered. From these ranks you can compute z values (i.e., standardized values of the normal distribution) *based on the assumption* that the data come from a normal distribution. These z values are plotted on the Y-axis in the plot. If the data values (plotted on the X-axis) are normally distributed, then all points should fall onto a straight line in the plot. If the values are not normally distributed, they will deviate from the line. Outliers may also become evident in this plot. The *detrended normal probability plot* is constructed in the same way as the standard normal probability plot, except that, before the plot is produced, the linear trend is removed. This often "spreads out" the plot, thereby allowing the user to detect patterns of deviations more easily.

The *half-normal probability plot* is constructed in the same way as the standard normal probability plot, except that only the positive half of the normal curve is considered. Consequently, only positive normal values will be plotted on the Y-axis. This plot is basically used when one wants to ignore the sign of the residual, that is, when one is mostly interested in the distribution of absolute values, regardless of sign.

Review and Plot Variables

These options allow you to review or plot the currently highlighted variable (together with other variables). The labeling of the individual observations (that is, of the respective points in time represented by the rows in Scrollsheets or the horizontal X-axis in plots) is determined by the selections in the *Display/plot options* dialog (click on the *Options* button below; these options are also available on the *Transformations* dialog, and are described in the *General Conventions* section, page 3315). For the plots, you can choose to *Scale the X-Axis in the plots manually* (select this option in the *Display/plot options* dialog). This is often useful when the series consists of several regularly recurring time intervals, for example, if it is made up of 10 years with 12 (monthly) observations each. Then it may be useful to scale the X-axis in the plot so that each major tickmark indicates one year.

Display/plot subset of cases only. If this check box is set, then, before plotting or displaying one or more series, you will be prompted to select a range of observations to include in the plot or display.

Options (scaling, labels, etc.). This option will bring up the *Display/plot options* dialog (see *General Conventions*, page 3315). That dialog allows you to change the labeling of the observations (i.e., the rows in Scrollsheets and the *X*-axis in line plots can be labeled with case names, case numbers, dates from another variable, etc.) and the scaling of the *X*-axis in line plots (minimum, step size).

Review/Plot highlighted variable. This button will bring up a Scrollsheet (or a plot) with the values for the currently highlighted variable.

Review/Plot multiple variables. After clicking on this button, you will first be prompted to select a list of variables to be displayed or plotted. When plotting multiple variables the vertical (*Y*) axis will have a common scale; thus, if the series that are plotted contain values of completely different magnitudes use option *Plot two var lists with different scales* (see below), instead.

Plot two var. lists with different scales. This option allows you to plot two lists of variables using different scales for the vertical (*Y*) axes in the plot. The first list will be plotted against the left-*Y* axis and the second list will be plotted against the right-*Y* axis. For example, suppose you want to plot two variables (series) called *S1* and *S2,* and that these variables contain values of different magnitudes: the values of *S1* range from 1-100 and the values of *S2* range from 1000 to 10000. If you plot *S1* against the left-*Y* axis (scaled from 1-100) and *S2* against the right-*Y* axis (scaled from 1000-10000), then both variables will be plotted with optimal resolution, and thus relationships between patterns across time (the *X*-axis) may become more clearly identifiable.

Autocorrelations

These options will compute the *autocorrelations* and *partial autocorrelations* for the currently highlighted variable; these statistics will be displayed in a Scrollsheet and drawn in a correlogram graph. Refer to the *Introductory Overview* section (page 3271, 3275) for a brief description of how the pattern of (partial) autocorrelations aids in the determination of an appropriate ARIMA model.

Autocorrelations. This option will bring up a Scrollsheet and plot of the autocorrelations, for a lag of *1* through the number specified in the *Number of lags* edit field. The Scrollsheet will report the autocorrelations, their standard errors, the so-called *Box-Ljung statistic* (see the description of the *Transformation of Variables* dialog, page 3331, for computational details), and the significance level of that statistic. In general, the autocorrelation is the correlation of the series with itself, lagged by a particular number of observations.

***Alpha* for highlighting.** Significant autocorrelations (significant *Box-Ljung statistics*) will be highlighted in the Scrollsheet. This edit field allows you to specify the significance level that is used for highlighting.

White noise standard errors. Under the assumption that the true moving average process in the series is of the order *k-1*, the approximate standard error of the autocorrelation r_k is computed as:

```
StdErr(r_k) = √{(1/n) * [1+2*Σ(r_i^2)]}
(for i = 1 to k-1)
```

Here *n* is the number of observations in the series. Under the assumption that the series is a white noise process, that is, that all autocorrelations are equal to zero, the standard error of r_k is computed as:

```
StdErr(r_k) = √{(1/n) * [(n-k)/(n+2)]}
```

Check the *White noise standard errors* option to compute the standard errors in this manner.

Partial autocorrelations. This option will bring up a Scrollsheet and plot of the partial autocorrelations, for a lag of *1* through the number specified in the *Number of lags* edit field. The Scrollsheet will report the partial autocorrelations and their standard errors.

In general, the partial autocorrelation is the partial correlation of a series with itself, lagged by a particular number of observations, and controlling for all correlations for lags of lower order.

For example, the partial autocorrelation for a lag of 2 represents the unique correlation of the series with itself at that lag, after controlling for the correlation at lag 1. For details regarding the computations of the partial autocorrelation and its standard error refer to the description of the *Transformation of Variables* dialog (page 3332). Partial autocorrelations that are larger than two times their respective standard errors will be highlighted in the Scrollsheet.

Number of lags. This parameter determines the maximum number of lags for which the auto-, and partial autocorrelations will be computed.

Other Transformations and Plots

This option will bring up the *Transformations of Variables* dialog which allows you to perform a wide variety of transformations on the data. The transformed series will be appended to the *active work area* (see the *General Conventions* section 3313).

Single Variable Spectrum Analysis Results

This dialog provides options for computing all spectrum analysis results statistics. The interpretation of the different statistics is also discussed in the *Introductory Overview* section (page 3300).

When reviewing the results from long series, you may use the *Examine subset of periodogram* check box. When set, the results computed by the *Summary* option and all graph options (*Periodogram, Spectral density*, etc.) will be computed only for the range of cases specified in the edit boxes below the check box.

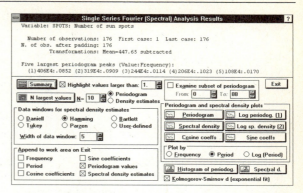

To append any of the computed statistics to the *active work area*, set the respective check boxes in the *Append to work area on Cancel* box. Note that if there are not enough unused or un-locked backups for the respective input variable available, then the program will automatically increase the *Number of backups per variable* parameter by as much as possible. Refer to the *General Conventions* section for a description of the memory management in the *active work area*.

Summary

This button will produce a summary Scrollsheet with the frequencies, periods, cosine and sine coefficients, periodogram values, spectral density estimates (computed according to the selection in the *Data windows* box; see below), and the weights used to produce the spectral density estimates. If the *Highlight values larger than...* check box is set, then all values in the periodogram and spectral density columns that are larger than the specified value will be highlighted.

Note that as the default graph for this Scrollsheet, you may select to plot the sine/cosine coefficients, the periodogram values (or log-periodogram), or the spectral density estimates (or the log-densities) against the frequencies or period.

STATISTICA uses the standard formulas to calculate the following statistics (see, for example, Shumway, 1988):

Frequency. The frequency is defined as the number of cycles per unit time. Since in the *Time Series* module, one observation is used as the time unit (i.e., frequency is expressed in terms of cycles per observation), the successive frequencies are computed as k/n (for $k=0$ to $n/2$) where n is the number observations in the series. Thus, for example, a frequency of .0833 would mean that each observation completes .0833 of the full cycle, or that 12 observations complete one full cycle (.0833*12=1). Thus, if the series contains monthly data collected over several years, the respective periodicity identifies an annual cycle.

Period. The period is computed as the inverse of the frequency. Thus, it can be interpreted as the number of observations that is necessary in order to complete one cycle at the respective frequency.

Cosine coefficients. The cosine coefficients can be interpreted as regression coefficients; that is, they tell the degree to which the respective cosine functions are correlated with the data at the respective frequencies.

Sine coefficients. The sine coefficients can be interpreted analogous to the cosine coefficients (see previous paragraph).

Periodogram. The periodogram values are computed as the sum of the squared sine and cosine coefficients at each frequency (times $n/2$). The periodogram values can be interpreted in terms of variance (sums of squares) of the data at the respective frequency or period.

Spectral density estimates. The spectral density estimates are computed by smoothing the periodogram values, using the specifications in the *Data windows for spectral density estimates* box. By smoothing the periodogram, one may identify the general frequency "regions" (or *spectral densities*) that significantly contribute to the cyclical behavior of the series.

Note that the weights used for the smoothing will always be standardized so that they add to 1.0. Also, at the beginning and end of the series, the smoothing is done via reflection.

Weights. This column will report the actual weights used in the smoothing window to produce the spectral density estimates (see above). The different smoothing windows are described below (*Data windows for spectral density estimates*). Note that the weights are standardized so that they will always sum to 1.

Examine Subset of Periodogram

When reviewing the results for long time series, it is often useful to examine particular frequency ranges. This can be accomplished by setting this check box and then specifying the range of observations to review in the edit boxes below (*From, To*).

Note that the numbers specified here pertain to the numbers of rows in the *Summary* Scrollsheet, so for example, if you specify 100 and 200, then rows 100 to 200 of the Scrollsheet will be displayed (or plotted). In addition to the *Summary* Scrollsheet option, the range selected here will also apply to all plots on this dialog.

n Largest Values

This option will bring up a Scrollsheet with the same statistics as those computed by the *Summary* option; however, the results will only be shown for the n (as specified in the edit box to the right) largest periodogram or spectral density values, depending on the setting of the radio buttons to the right.

Data Windows for Spectral Density Estimates

As described in greater detail in the *Introductory Overview* section (page 3300), the periodogram values themselves are subject to substantial random

fluctuation. A clearer picture of underlying periodicities often only emerges when examining the spectral densities, that is, the frequency regions, consisting of many adjacent frequencies, that contribute most to the overall periodic behavior of the series. The spectral density estimates are computed by smoothing the periodogram values with a weighted moving average. You can select the type of moving average smoothing via this option.

Daniell (or equal weight) window. The Daniell window (Daniell 1946) amounts to a simple (equal weight) moving average transformation of the periodogram values; that is, each spectral density estimate is computed as the mean of the $m/2$ preceding and subsequent periodogram values.

Tukey window. In the Tukey (Blackman and Tukey, 1958) or Tukey-Hanning window (named after Julius Von Hann), for each frequency, the weights for the weighted moving average of the periodogram values are computed as:

$w_j = 0.5+0.5*\cos(\pi*j/p)$ (for j=0 to p)
$w_{-j} = w_j$ (for j≠0)

Hamming window. In the Hamming (named after R. W. Hamming) window or Tukey-Hamming window (Blackman and Tukey, 1958), for each frequency, the weights for the weighted moving average of the periodogram values are computed as:

$w_j = 0.54+0.46*\cos(\pi*j/p)$ (for j=0 to p)
$w_{-j} = w_j$ (for j≠0)

Parzen window. In the Parzen window (Parzen, 1961), for each frequency, the weights for the weighted moving average of the periodogram values are computed as:

$w_j = 1-6*(j/p)^2 +6*(j/p)^3$ (for j=0 to p/2)
$w_j = 2*(1-j/p)^3$ (for j=p/2+1 to p)
$w_{-j} = w_j$ (for j≠0)

Bartlett window. In the Bartlett window (Bartlett, 1950) the weights are computed as:

$w_j = 1-(j/p)$ (for j=0 to p)
$w_{-j} = w_j$ (for j≠0)

User-defined window. If this radio button is selected, you will be prompted to enter the weights used for smoothing directly (a dialog box will come up prompting you to specify the weights). Note that in all cases the *Time Series* module will standardize the weights so that they always sum to 1. For example, if you entered *1 2 3 2 1* as the user-defined weights, they will be converted to: 1/9 2/9 3/9 2/9 1/9 (so that they add up to 1). In many cases, all of the standard data windows will produce very similar results; but it is often useful to experiment with different windows of different widths and to examine resulting density plots.

Width of data window. This edit field will determine the width of the moving average window used during smoothing.

Append to Work Area on Cancel

These check boxes will determine which of the results statistics will be appended to the *active work area* when exiting from this dialog. The different statistics are described under the *Summary* option above. If the currently available (unused and unlocked) number of backups in the *active work area* is less than the number of series that is to be appended, then the *Number of backups per variable* parameter will be increased by as much as possible (refer to the *General Conventions* section, page 3313, for details concerning the memory management in the *active work area*).

Periodogram and Spectral Density Plots

These options will produce line plots of the respective statistics; refer to the *Summary* option above for a description of the available results. Note that all plots are produced either for the entire range of frequencies or only for the range specified via the *Examine subset of periodogram* option (see above).

Periodogram. This option will bring up a line graph of the periodogram values, plotted either by frequency, period, or the log-period, depending on the setting of the radio buttons in the *Plot by* box below.

Log periodogram. This option will bring up a line graph of the log-periodogram values, plotted either by frequency, period, or the log-period, depending on the setting of the radio buttons in the *Plot by* box below.

Spectral density. This option will bring up a line graph of the spectral density values, plotted either by frequency, period, or the log-period, depending on the setting of the radio buttons in the *Plot by* box below.

Log spectral density. This option will bring up a line graph of the log-spectral density values, plotted either by frequency, period, or the log-period, depending on the setting of the radio buttons in the *Plot by* box below.

Cosine coefficients. This option will bring up a line graph of the cosine coefficient values, plotted either by frequency, period, or the log-period, depending on the setting of the radio buttons in the *Plot by* box below.

Sine coefficients. This option will bring up a line graph of the sine coefficient values, plotted either by frequency, period, or the log-period, depending on the setting of the radio buttons in the *Plot by* box below.

Plot by. The setting of the radio buttons in this box determines how the *X* (horizontal) axis in the line plots; the periodogram, spectral density, or sine/cosine coefficient values will be plotted (i.e., by *Frequency*, *Period*, or *Log (period)*).

Histogram of Periodogram Values

If the observations in the time series are independent of each other (i.e., there is no periodicity) and they follow the normal distribution, such a time series is also referred to as a *white noise* series (like the white noise one hears on the radio when tuned in-between stations). A white noise input series will result in periodogram values that follow an exponential distribution.

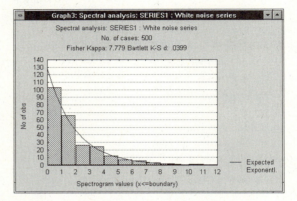

Thus, by testing the distribution of periodogram values against the exponential distribution, one may test whether the input series is different from a white noise series. This option will produce a histogram of the periodogram values and fit the exponential distribution to the histogram. In addition, the user can also request to compute the *Bartlett Kolmogorov-Smirnov d* statistic (see also *Nonparametrics and Distributions,* Volume I, for more details) and *Fisher's Kappa*, see Fuller, 1976, for details concerning these tests. Note that, as with the line graphs, the histogram will be produced either for all frequencies or for the range selected after setting the *Examine subset of periodogram* check box. In this manner one can test whether the series is a white noise series over selected frequency ranges.

Spectral Density

This option will bring up a histogram of the spectral density values.

Cross-Spectrum Analysis Results

This dialog provides options for computing all cross-spectrum analysis results statistics. The interpretation of the different statistics is also discussed in the *Introductory Overview* section (page 3300).

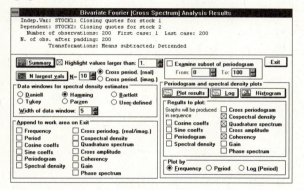

When reviewing the results from long series you may use the *Examine subset of periodogram* check box; when set, the results computed by the *Summary* option and all graphs will be computed only for the range of cases (frequencies) specified in the edit boxes below the check box. To append any of the computed statistics to the *active work area* set the respective check boxes in the *Append to work area on Cancel* box.

Note that if there are not enough unused or un-locked backups for the respective input variable available, then the program will increase the *Number of backups per variable* parameter by as much as possible. Refer to the *General Conventions* section (page 3313) for a description of the memory management in the *active work area*.

Summary

This button will produce a summary Scrollsheet with the frequencies, periods, cosine and sine coefficients, periodogram values, spectral density estimates (computed according to the selection in the *Data windows* box; see below), cross-periodogram values (real and imaginary), cospectral density and quadrature spectrum values (computed according to the selection in the *Data windows* box; see below), the cross amplitude values, coherency values, gain values, phase spectrum values, and the weights used to produce the spectral density estimates. If the *Highlight values larger than...* check box is set, then all values in the periodogram and spectral density columns that are larger than the specified value will be highlighted. The values in this Scrollsheet are computed as follows:

Frequency. The frequency is defined as the number of cycles per unit time. Since in the *Time Series* module, one observation is used as the time unit (i.e., frequency is expressed in terms of cycles per observation), the successive frequencies are computed as k/n (for $k=0$ to $n/2$) where n is the number observations in the series. Thus, for example, a frequency of .0833 would mean that each observation completes .0833 of the full cycle, or that 12 observations complete one full cycle (.0833*12=1). Thus, if the series contains monthly data collected over several years, the respective periodicity identifies an annual cycle.

Period. The period is computed as the inverse of the frequency. Thus it can be interpreted as the number of observations that is necessary in order to complete one cycle at the respective frequency.

Cosine coefficients. The cosine coefficients can be interpreted as regression coefficients; that is, they tell you the degree to which the respective cosine functions are correlated with the data at the respective frequencies. Cosine coefficients are computed for both the *X* and *Y* variable.

Sine coefficients. The sine coefficients can be interpreted analogous to the cosine coefficients (see previous paragraph). Sine coefficients are computed for both the *X* and *Y* variable.

Periodogram. The periodogram values are computed as the sum of the squared sine and cosine coefficients at each frequency (times *n/2*). The periodogram values can be interpreted in terms of variance (sums of squares) of the data at the respective frequency or period. Periodogram values are computed for both the *X* and *Y* variable.

Spectral density estimates. The spectral density estimates are computed by smoothing the periodogram values, using the specifications in the *Data windows for spectral density estimates* box. By smoothing the periodogram one may identify the general frequency "regions" or (*spectral densities*) that significantly contribute to the cyclical behavior of the series.

Note that the weights used for the smoothing will always be standardized so that they add to 1.0. Also, at the beginning and end of the series, the smoothing is done via reflection.

Cross-periodogram values. Analogous to the results for the single variables, the Scrollsheet will also display periodogram values for the cross periodogram. However, the cross-spectrum consists of complex numbers that can be divided into a real and an imaginary part.

Cospectral density. The cospectral density is computed by smoothing the real part of the cross-periodogram values (using the specifications in the *Data windows for spectral density estimates* box).

Quadrature spectrum. The quadrature spectrum values (*quad density* for short) are computed by smoothing the imaginary part of the cross-periodogram values (using the specifications in the *Data windows for spectral density estimates* box).

Cross amplitude. The cross amplitude values are computed as the square root of the sum of the squared cross-density and quad-density values. The cross-amplitude can be interpreted as a measure of covariance between the respective frequency components in the two series.

Coherency. One can standardize the cross-amplitude values by squaring them and dividing by the product of the spectrum density estimates for each series. The result is called the *squared coherency*, which can be interpreted similar to the squared correlation coefficient (see, for example, *Basic Statistics and Tables*, Volume I); that is, the coherency value is the squared correlation between the cyclical components in the two series at the respective frequency. However, the coherency values should not be interpreted by themselves.

For example, when the spectral density estimates in both series are very small, large coherency values may result (the divisor in the computation of the coherency values will be very small), even though there are no strong cyclical components in either series at the respective frequencies.

Gain. The gain value is computed by dividing the cross-amplitude value by the spectrum density estimates for one of the two series in the analysis. Consequently, two gain values are computed, which can be interpreted as the standard least squares regression coefficients for the respective frequencies

Phase spectrum. The phase spectrum estimates are computed as tan^{-1} of the ratio of the quad density estimates over the cross-density estimate. The phase shift estimates (usually denoted by Greek letter ψ) are measures of the extent to which each frequency component of one series leads the other.

Weights. This column will report the actual weights used in the smoothing window to produce the spectral density estimates (see above). The different smoothing windows are described below (*Data windows for spectral density estimates*). Note that the weights are standardized so that they will always sum to 1.

TIM - 3438

8. TIME SERIES ANALYSIS - SPECTRUM (FOURIER) ANALYSIS

Examine Subset of Periodogram

When reviewing the results for long time series, it is often useful to examine particular frequency ranges. This can be accomplished by setting this check box and then specifying the range of observations to review in the edit boxes below (*From, To*). Note that the numbers specified here pertain to the numbers of rows in the *Summary* Scrollsheet, so for example, if you specify 100 and 200, then rows 100 to 200 of the Scrollsheet will be displayed (or plotted). In addition to the *Summary* Scrollsheet option, the range selected here will also apply to all plots on this dialog.

n Largest Values

This option will bring up a Scrollsheet with the same statistics as those computed by the *Summary* option; however, the results will only be shown for the *n* (as specified in the edit box to the right) largest real cross-periodogram or imaginary cross-periodogram values, depending on the setting of the radio buttons to the right.

Data Windows for Spectral Density Estimates

As described in greater detail in the *Introductory Overview* section (page 3300), the periodogram values themselves are subject to substantial random fluctuation. A clearer picture of underlying periodicities often only emerges when examining the spectral densities, that is, the frequency regions, consisting of many adjacent frequencies, that contribute most to the overall periodic behavior of the series.

The spectral and co-spectral density estimates are computed by smoothing the periodogram values with a weighted moving average. You can select the type of moving average smoothing via this option.

Daniell (or equal weight) window. The Daniell window (Daniell 1946) amounts to a simple (equal weight) moving average transformation of the periodogram values; that is, each (co-) spectral density estimate is computed as the mean of the $m/2$ preceding and subsequent periodogram values.

Tukey window. In the Tukey (Blackman and Tukey, 1958) or Tukey-Hanning window (named after Julius Von Hann), for each frequency, the weights for the weighted moving average of the periodogram values are computed as:

$$w_j = 0.5 + 0.5 * \cos(\pi * j/p) \quad \text{(for } j=0 \text{ to } p\text{)}$$
$$w_{-j} = w_j \quad \text{(for } j \neq 0\text{)}$$

Hamming window. In the Hamming (named after R. W. Hamming) window or Tukey-Hamming window (Blackman and Tukey, 1958), for each frequency, the weights for the weighted moving average of the periodogram values are computed as:

$$w_j = 0.54 + 0.46 * \cos(\pi * j/p) \quad \text{(for } j=0 \text{ to } p\text{)}$$
$$w_{-j} = w_j \quad \text{(for } j \neq 0\text{)}$$

Parzen window. In the Parzen window (Parzen, 1961), for each frequency, the weights for the weighted moving average of the periodogram values are computed as:

$$w_j = 1 - 6*(j/p)^2 + 6*(j/p)^3 \quad \text{(for } j=0 \text{ to } p/2\text{)}$$
$$w_j = 2*(1-j/p)^3 \quad \text{(for } j=p/2+1 \text{ to } p\text{)}$$
$$w_{-j} = w_j \quad \text{(for } j \neq 0\text{)}$$

Bartlett window. In the Bartlett window (Bartlett, 1950) the weights are computed as:

$$w_j = 1 - (j/p) \quad \text{(for } j=0 \text{ to } p\text{)}$$
$$w_{-j} = w_j \quad \text{(for } j \neq 0\text{)}$$

User-defined window. If this radio button is selected, you are prompted to enter the weights used for smoothing directly (a dialog box will come up prompting you to specify the weights). Note that in all cases the *Time Series* module will standardize the weights so that they always sum to 1. For example, if you entered *1 2 3 2 1* as the user-defined weights, they will be converted to: 1/9 2/9 3/9 2/9 1/9 (so that

TIM - 3439

they add up to 1). In many cases, all of the standard data windows will produce very similar results; but it is often useful to experiment with different windows of different widths and to examine resulting density plots.

Width of data window. This edit field will determine the width of the moving average window used during smoothing.

Append to Work Area on *Cancel*

These check boxes will determine which of the results statistics will be appended to the *active work area* when exiting from this dialog. The different statistics are described under the *Summary* option above. If the currently available (unused and unlocked) number of backups in the *active work area* is less than the number of series that is to be appended, then the *Number of backups per variable* parameter will be increased by as much as possible (refer to the *General Conventions* section, page 3313, for details concerning the memory management in the *active work area*).

Periodogram and Spectral Density Plots

These options will produce line plots of the respective statistics; refer to the *Summary* option above for a description of the available results. Note that all plots are produced either for the entire range of frequencies or only for the range specified via the *Examine subset of periodogram* option (see above).

Plot results. This option will produce line plots of the selected statistics in sequence.

Log. This option will produce plots of the log-periodogram/densities; for other statistics the plots will be identical to those produced by the *Plot results* option.

Histogram. This option will produce histograms showing the distributions of the statistics selected in the *Results to plot* box. Note that the *Single Spectrum (Fourier) Analysis Results* dialog also contains options to fit an exponential distribution to the periodogram values, and to compute the Bartlett Kolmogorov-Smirnov *d* statistic to test whether the respective data represent a white-noise series (Fuller, 1976; see also *Nonparametric Statistics*, Volume I). If the respective series is a white noise series, then the distribution of the periodogram values will follow the exponential distribution.

Plot by. The setting of the radio buttons in this box determines the *X* (horizontal) axis in the line plots; the selected statistics can be plotted by *Frequency*, *Period*, or *Log (period)*.

INDEX

A

ACF (autocorrelation function), 3272, 3275, 3331
Active work area, in Time Series, 3313
Additive seasonality, 3285, 3288
Almon distributed lag, 3299, 3420
AR process, 3274
ARIMA, 3310
    abrupt-permanent impact, 3278, 3364
    abrupt-temporary impact, 3365
    ACF (autocorrelation function), 3275
    approximate max. likelihood method, 3277, 3358
    asymptotic standard errors, 3277
    autocorrelation, 3275, 3343, 3348, 3360
    autocorrelation of residuals, 3369
    autoregressive moving average model, 3274
    autoregressive parameters, 3358
    autoregressive process, 3274
    backcasting, 3359
    constant, 3274, 3275
    convergence criterion, 3362
    d parameter, 3274
    delta, 3279, 3364
    dialogs, options, statistics, 3357
    differencing, 3275, 3343, 3358
    ds parameter, 3276
    estimation, 3275, 3276, 3277, 3358, 3361, 3363
    exact max. likelihood method, 3277, 3358
    examples, 3341
    forecasting, 3275, 3277, 3346, 3354, 3367
    gradual-permanent impact, 3278, 3364
    identification, 3275, 3341, 3349
    impact analysis, 3363
    impact patterns, 3278, 3364
    interrupted time series, 3278, 3310, 3349, 3353, 3363
    intervention analysis, 3310, 3363
    intervention types, 3353, 3364
    invertibility, 3274

ARIMA (continued)
    iterations, 3362
    log transformation, 3342, 3358
    long seasonality, 3277
    MA process, 3274
    McLeod & Sales method, 3277, 3358
    Melard method, 3277, 3358
    missing data, 3314
    model, 3274
    model identification, 3275, 3341, 3349
    model parameters, 3357
    moving average parameters, 3358
    moving average process, 3274
    multiplicative seasonality, 3270, 3276, 3342
    normal probability plot, 3347
    omega, 3278, 3364
    overview, 3273
    p parameter, 3274, 3276
    PACF (partial autocorrelation function), 3275
    parameter covariances, 3367
    parameter estimates, 3366
    parameter estimation, 3276, 3345, 3361
    parameter standard errors, 3277
    parameter start values, 3362
    parameters, 3345
    partial autocorrelation, 3275, 3360
    partial autocorrelation of residuals, 3369
    penalty value, 3277
    permanent impact, 3278
    power transformation, 3358
    program overview, 3310
    ps parameter, 3276
    q parameter, 3274, 3276
    qs parameter, 3276
    quasi-Newton method, 3276, 3361
    random walk model, 3325
    residuals, 3278, 3347, 3354, 3368
    results, 3277, 3310, 3346, 3365
    seasonal autoregressive parameters, 3358
    seasonal differencing, 3344
    seasonal moving average parameters, 3358
    seasonality, 3276, 3342
    specifying the model, 3345, 3352

ARIMA (continued)
    standard error of autocorrelation, 3360
    standard errors, 3277
    start values, 3362
    stationary series, 3274, 3275
    temporary impact, 3279
    types of interventions, 3364
    white noise standard errors, 3360
Autocorrelation, 3271, 3275, 3309, 3325, 3326, 3331, 3343, 3360
Autoregressive moving average model, 3274
Autoregressive parameters, 3358
Autoregressive process, 3274

B

Backcasting, 3359
Bartlett window, 3304, 3338
Batch printing, in X-11 seasonal decomposition, 3406, 3407, 3412
Box-Jenkins ARIMA, 3273, 3341
Box-Ljung Q, 3331

C

Census method I, 3391, 3395
    additive seasonality, 3285, 3395
    additive trend-cycle component, 3285
    centered moving averages, 3395
    components, 3284, 3396
    computations, 3286
    dialogs, options, statistics, 3395
    example, 3391
    introductory overview, 3284
    irregular component, 3287, 3396
    missing data, 3314
    multiplicative seasonality, 3395
    multiplicative trend-cycle component, 3285
    program overview, 3311
    random component, 3287, 3396
    seasonality, 3391, 3395
    seasonally adjusted series, 3286, 3396
    trend-cycle component, 3284, 3396
Census method II (see also X-11), 3287, 3289, 3401
Coherency, 3307
Complex numbers, 3301
Constant, in ARIMA, 3274, 3275

8. TIME SERIES ANALYSIS - INDEX

Correlogram, 3272
Cross-amplitude, 3307
Cross-correlation function, 3332
Cross-density, 3307
Cross-spectrum analysis
    Bartlett window, 3439
    coherency, 3307, 3438
    cosine coefficients, 3437
    cospectral density, 3438
    cross-amplitude, 3307, 3438
    cross-density, 3307
    cross-periodogram values, 3438
    Daniell window, 3439
    detrending, 3430
    frequency, 3437
    gain, 3438
    Hamming window, 3439
    introductory overview, 3300, 3306
    n largest values, 3439
    padding, 3430
    Parzen window, 3439
    period, 3437
    periodogram, 3440
    phase spectrum, 3307, 3438
    plots, 3440
    program overview, 3311
    quadrature spectrum, 3307, 3438
    results, 3311, 3437
    sine coefficients, 3438
    subtract mean, 3430
    tapering, 3337
    transformations, 3337
    Tukey window, 3439

D

Damped trend, 3380
Daniell window, 3303, 3338
Data windows, in spectrum analysis, 3303
Delta, in ARIMA, 3279
Delta, in exponential smoothing, 3283
Differencing, 3273, 3275, 3337, 3343
Distributed lags analysis, 3298
    Almon distributed lag, 3299, 3420
    dialogs, options, statistics, 3421
    example, 3419
    independent variable, 3421
    introductory overview, 3298
    lag length, 3421
    method, 3421
    misspecification of lags, 3300

Distributed lags analysis (continued)
    model, 3299
    polynomial order, 3421
    program overview, 3311
    results, 3420
    specifying the analysis, 3419
Double exponential smoothing, 3378

E

Endogenous variable, 3299
Ex post MSE, 3281
Exogenous variable, 3299
Exponential smoothing, 3279, 3310, 3371
    additive seasonality, 3283, 3377
    alpha, 3280, 3372, 3382
    automatic parameter search, 3281, 3282, 3375, 3384, 3387
    damped trend, 3284, 3380
    delta, 3283, 3382
    dialogs, options, statistics, 3377
    double exponential smoothing, 3378
    error, 3281
    ex post MSE, 3281
    example, 3371
    exponential trend, 3379
    forecasting, 3279, 3280, 3371
    gamma, 3284, 3383
    grid search, 3374, 3384, 3387
    Holt's method, 3282, 3373, 3378
    initial value, 3282, 3383
    lack of fit indicators, 3281, 3388
    linear trend, 3373
    MAE (mean absolute error), 3281, 3388
    MAPE (mean absolute percentage error), 3282, 3389
    mean error, 3281
    mean percentage error, 3282
    mean square error, 3388
    missing data, 3314
    models, 3371, 3377
    MPE (mean percentage error), 3282
    MSE (mean square error), 3281, 3388
    multiplicative seasonality, 3283, 3378
    parameters, 3382
    percentage error, 3282
    phi, 3284, 3383
    program overview, 3310

Exponential smoothing (continued)
    results, 3310, 3376
    seasonality, 3282, 3374
    simple, 3280, 3372
    SSE (sum of squared error), 3281
    summary plot, 3384
    trend, 3282, 3284, 3373
    triple exponential smoothing, 3374, 3379
    unconstrained parameter estimation, 3388
    Winters' method, 3282, 3374, 3379
Extreme values, in X-11 decomposition, 3289

F

Fast Fourier transform algorithm, 3305, 3338
FFT (fast Fourier transforms), 3305
Forecasting, 3277, 3279, 3280, 3367, 3371
    in ARIMA, 3275, 3346
Forty two-fifty three (4253) H filter, 3324, 3336
Fourier analysis (see also spectrum analysis), 3300, 3338, 3425, 3429
Frequency, in spectrum analysis, 3300

G

Gain, 3307
Gamma, in exponential smoothing, 3284
Grid search, in exponential smoothing, 3374, 3384, 3387

H

Hamming window, 3304, 3338
Henderson curve moving average, 3409
Holt's exponential smoothing model, 3282, 3373, 3378

I

Imaginary numbers, 3301
Impact analysis, 3278, 3310
Initial value, in exponential smoothing, 3282
Integrating a time series, 3337
Interrupted time series, 3278, 3310, 3349

Intervention analysis, 3278, 3310, 3349
Inverse Fourier transform, 3339
Invertibility, 3274
Irregular component, 3287

L

Leakage, 3303, 3337, 3429

M

MA process, 3274
MAE (mean absolute error), 3281, 3388
MAPE (mean absolute percentage error), 3282, 3389
MCD (month for cyclical dominance), 3290
Mean absolute error (MAE), 3281
Mean absolute percentage error (MAPE), 3282
Mean error (ME), 3281
Median smoothing, 3271, 3335
Melard's exact maximum likelihood method, 3277
Missing data replacement
    interpolation from adjacent points, 3314
    mean of n adjacent points, 3315, 3321
    median of n adjacent points, 3315, 3321
    overall mean, 3314, 3321
    predicted values from linear trend regression, 3315, 3321
Model identification, ARIMA, 3275
Month for cyclical dominance, 3290
Moving average parameters, 3274, 3358
Moving average smoothing, 3271, 3335
MPE (mean percentage error), 3282
Multiplicative seasonality, 3270, 3276, 3285, 3288, 3342

N

Number of backups, in Time Series, 3313

O

Omega, in ARIMA, 3278

P

PACF (partial autocorrelation function), 3272, 3275, 3332
Partial autocorrelation, 3272, 3275, 3332, 3360
Parzen window, 3304, 3338
PE (percentage error), 3282
Period, in spectrum analysis, 3300
Periodogram, 3302, 3426
Phase spectrum, 3307
Phi, in exponential smoothing, 3284
Polynomial distributed lags, 3311
Prior daily weights, in X-11, 3407
Probability plots
    detrended normal probability plots, 3333, 3347
    half-normal probability plots, 3333
    normal probability plots, 3333, 3347

Q

QCD (quarter for cyclical dominance), 3290
Quadrature spectrum, 3307
Quarter for cyclical dominance, 3290
Quasi-Newton method, 3276, 3361

R

Random walk model, 3325
Ratio-to-moving-averages method, 3284, 3395
Real and imaginary numbers, 3301
Result tables, in X-11 decomposition, 3290

S

Seasonal decomposition, 3284, 3311, 3391, 3395, 3401
    additive trend-cycle component, 3285
    components, 3284
    irregular component, 3287
    multiplicative trend-cycle component, 3285
    random component, 3287
    seasonally adjusted series, 3286
    trend-cycle component, 3284, 3287
    X-11 method, 3287

Seasonal differencing, 3344
Seasonality, 3270, 3276, 3277, 3282, 3288, 3342, 3391
Serial correlation, 3271
Series G, 3270, 3341, 3371
Smoothing, 3335, 3337
    median smoothing, 3271
    moving average smoothing, 3271
Spectral density, 3426
Spectrum analysis, 3425
    Bartlett window, 3304, 3338, 3435
    coherency, 3307
    complex numbers, 3301
    computation, 3305
    cosine coefficients, 3434
    cross-amplitude, 3307
    cross-density, 3307
    cross-periodogram, 3307
    Daniell window, 3303, 3338, 3435
    data windows, 3303, 3337, 3434
    density estimates, 3426
    detrending, 3430
    dialogs, options, statistics, 3429
    example, 3302, 3425
    fast Fourier transformation algorithm, 3305, 3338
    frequency, 3300, 3434
    gain, 3307
    general model, 3301
    Hamming window, 3304, 3338, 3435
    imaginary numbers, 3301
    introductory overview, 3300
    leakage, 3303, 3337, 3429
    missing data, 3314
    model, 3301
    multiple regression estimates, 3301
    n largest values, 3434
    padding, 3303, 3430
    Parzen window, 3304, 3338, 3435
    period, 3300, 3434
    periodogram, 3302, 3426, 3434, 3435
    phase spectrum, 3307
    plots, 3435
    preparing data, 3304
    program overview, 3311
    quadrature-spectrum, 3307
    results, 3311, 3426, 3433
    sine coefficients, 3434

TIM - 3443

Copyright © StatSoft, 1995

8. TIME SERIES ANALYSIS - INDEX

Spectrum analysis (continued)
    single series Fourier analysis, 3429
    spectral density estimates, 3426, 3434, 3437
    split-cosine-bell tapering, 3303, 3429
    subtract mean, 3430
    sunspot data, 3425
    tapering, 3303, 3337, 3429
    transformations, 3337
    Tukey window, 3304, 3338, 3435
    two-series Fourier analysis, 3429
    white noise periodogram, 3304
Split-cosine-bell tapering, 3303, 3337, 3429
Stationary series, 3274, 3275
Summary measures, in X-11 seasonal decomposition, 3406

T

Tapering, 3303, 3337, 3429
Time series
    ACF (autocorrelation function), 3272
    active work area, 3313, 3319, 3323
    additive seasonality, 3285, 3288
    additive trend-cycle component, 3285
    adjustment for trading-day variation, 3294, 3295
    Almon distributed lag, 3299
    alpha, in exponential smoothing, 3280, 3372
    approximate max. likelihood method, 3277
    AR process, 3274
    ARIMA, 3273, 3274, 3275, 3310
    ARIMA estimation, 3276
    ARIMA examples, 3341
    ARIMA forecasting, 3277
    ARIMA parameters, 3345, 3357
    ARIMA residuals, 3278, 3368
    ARIMA results, 3277, 3310, 3346, 3365
    ARIMA standard errors, 3277
    autocorrelation, 3271, 3275, 3309, 3325, 3326, 3331, 3343, 3348
    autoregressive process, 3274
    backcasting, 3359
    backups, 3313, 3320, 3329

Time series (continued)
    Bartlett window, 3304, 3338
    Box-Jenkins ARIMA, 3273, 3341
    Box-Ljung Q, 3331
    Census method I, 3284
    coherency, 3307
    comparison of parameter estimation methods, 3277
    complex numbers, 3301
    components, 3284
    constant, in ARIMA, 3274, 3275
    conventions, 3319
    correlogram, 3272
    cross-amplitude, 3307
    cross-correlation, 3332
    cross-density, 3307
    cross-spectrum analysis, 3300, 3306
    cross-spectrum results, 3437
    d parameter, 3274
    damped trend, 3284, 3380
    Daniell window, 3303, 3338
    data windows, 3303, 3337
    deleting series, 3314, 3329
    delta, in ARIMA, 3279, 3364
    delta, in exponential smoothing, 3283
    dialog, options, statistics, 3329
    differencing, 3273, 3275, 3337, 3343
    display/plot options, 3315
    distributed lags analysis, 3298, 3311
    endogenous variable, 3299
    ex post MSE, 3281
    exact max. ARIMA likelihood, 3277
    exogenous variable, 3299
    exponential smoothing, 3279, 3310, 3335, 3371, 3377
    exponential smoothing parameters, 3382
    exponential trend, 3379
    extreme values, 3289
    fast Fourier transform algorithm, 3305, 3338
    fitting a function, 3271
    forecasting, 3280, 3346, 3354, 3367, 3371
    forty two-fifty three (4253) H filter, 3324, 3336
    Fourier analysis, 3311, 3338
    frequency, 3300
    gain, 3307

Time series (continued)
    general conventions, 3313, 3319
    general overview, 3269
    grid search, in exponential smoothing, 3374, 3387
    Hamming window, 3304, 3338
    Holt's exponential smoothing, 3282, 3373, 3378
    identification, in ARIMA, 3341
    identifying patterns, 3270
    imaginary numbers, 3301
    integrating a series, 3337
    interrupted time series, 3349, 3353, 3363
    intervention analysis, 3278, 3310, 3349
    inverse Fourier transform, 3339
    invertibility, 3274
    irregular component, 3287
    labeling plots, 3315, 3330
    lagging a series, 3336
    leakage, 3303
    locked series, 3314, 3320, 3325
    long seasonality, 3277
    MA process, 3274
    MAE (mean absolute error), 3281, 3388
    main goals, 3269
    MAPE (mean absolute percentage error), 3282, 3389
    MCD (month for cyclical dominance), 3290
    McLeod & Sales method, 3277
    mean absolute percentage error (MAPE), 3282
    mean error, 3281
    mean percentage error, 3282
    median smoothing, 3271, 3335
    missing data, 3314, 3320
    model identification, ARIMA, 3275
    moving average parameters, 3274, 3358
    moving average smoothing, 3271, 3335
    moving median smoothing, 3335
    MPE (mean percentage error), 3282
    MSE (mean square error), 3388
    multiplicative seasonality, 3270, 3276, 3288, 3342
    multiplicative trend-cycle component, 3285
    naming conventions, 3313, 3319

TIM - 3444

Copyright © StatSoft, 1995

8. TIME SERIES ANALYSIS - INDEX

Time series (continued)
- normal probability plot, 3333
- number of backups, 3313, 3320, 3329
- omega, 3278, 3364
- p parameter, 3274
- PACF (partial autocorrelation function), 3272
- parameter estimation, ARIMA, 3276
- partial autocorrelation, 3272, 3275, 3332, 3360
- Parzen window, 3304, 3338
- percentage error, 3282
- period, 3300
- periodogram, 3302
- phase spectrum, 3307
- plot options, 3315
- plotting time series, 3322, 3329
- prior daily weights, 3407
- program overview, 3309
- ps parameter, 3276
- q parameter, 3274
- QCD (quarter for cyclical dominance), 3290
- qs parameter, 3276
- quadrature spectrum, 3307
- quarterly seasonal adjustment, 3410
- quasi-Newton method, 3276, 3361
- random walk model, 3325
- ratio-to-moving-averages method, 3284
- related procedures, 3312
- removing serial dependency, 3273
- residuals, 3347
- reviewing the time series, 3321, 3322, 3330
- saving series in active work area, 3314, 3325, 3329
- scaling of plots, 3316, 3331
- seasonal autoregressive parameters, 3358
- seasonal decomposition, 3284, 3311, 3391
- seasonal differencing, 3344
- seasonal moving average parameters, 3358
- seasonality, 3270, 3276, 3282, 3288, 3342, 3391
- serial correlation, 3271
- Series G, 3270, 3341, 3371
- shifting a series, 3336

Time series (continued)
- simple exponential smoothing, 3372
- smoothing, 3271, 3335
- spectrum analysis, 3300, 3311, 3336, 3425, 3429
- split-cosine-bell tapering, 3303, 3337, 3429
- standard error of autocorrelation, 3331, 3360
- standard error of cross-correlation, 3333
- standard error of partial autocorrelation, 3332
- stationarity, 3275
- stationary series, 3274
- subset plots, 3330
- summary measures, 3406
- tapering, 3303, 3337, 3429
- trading-day regression, 3289, 3403
- transformations, 3309, 3319, 3323, 3329, 3334
- trend, 3270, 3284
- trend subtraction, 3334
- trend-cycle component, 3284, 3287, 3289
- triple exponential smoothing, 3374
- Tukey window, 3304, 3338
- white noise, 3270
- white noise standard errors, 3360
- Winters' exponential smoothing, 3282, 3379
- X-11 charts, 3297
- X-11 result tables, 3290, 3291, 3415
- X-11 seasonal decomposition, 3287, 3311, 3401

Trading-day adjustment, 3289, 3403
Transformations, in Time Series, 3319, 3323, 3329, 3334
Trend analysis, 3270
Trend component, 3270
Trend-cycle component, 3284, 3285, 3287, 3289
Triple exponential smoothing, 3374, 3379
Tukey window, 3304, 3338

W

White noise, 3270
White noise standard errors, 3360

Winters' exponential smoothing model, 3282, 3374

X

X-11 method
- additive seasonality, 3288, 3401, 3405
- batch printing, 3407, 3412
- Census method II, 3289
- charts, 3297, 3406, 3411
- dates (start of series), 3406, 3411
- dialogs, options, statistics, 3405
- example, 3401
- extreme values, 3289, 3408
- final estimation of components (Part D), 3290
- final trading-day variation (Part C), 3290
- full printout, 3406, 3411
- general model, 3288
- Henderson moving average, 3409
- introductory overview, 3287
- length of month allowance, 3407
- long printout, 3406, 3411
- MCD (month for cyclical dominance), 3290
- missing data, 3314
- modified series (Part E), 3290
- monthly adjustment, 3405
- moving averages for seasonal factors, 3408, 3415
- multiplicative seasonality, 3288, 3401, 3405
- outliers, 3289
- output detail, 3291
- preliminary estimate of trading day variation (part B), 3290
- printout detail, 3406, 3411
- prior adjustment (part A), 3290
- prior daily weights, 3407
- program overview, 3311
- QCD (quarter for cyclical dominance), 3290
- quarterly adjustment, 3410
- result tables, 3290, 3291, 3403, 3415
- results, 3311
- seasonality, 3288, 3401
- sigma limits for graduating extremes, 3408, 3412
- standard printout, 3406, 3411
- strikes, 3410, 3412
- summary measures, 3406, 3411

TIM - 3445

8. TIME SERIES ANALYSIS - INDEX

X-11 method (continued)
    summary statistics, 3289
    trading-day regression, 3289, 3403, 3407
    trend-cycle component, 3289, 3409
X-11 seasonal decomposition, 3287

Chapter 9:
LOG-LINEAR ANALYSIS OF FREQUENCY TABLES

Table of Contents

Introductory Overview	3449
Program Overview	3455
Example	3457
Dialogs, Options, Statistics	3465
Startup Panel	3465
Specifying Tables Directly (Frequency Table Input)	3466
Model Specification	3466
Reviewing Tables (Slicing Tables)	3469
Specifying and Testing Specific Models	3469
Automatic (Stepwise) Selection of Best Model	3470
Specifying Structural Zeros	3470
Results	3471
Technical Notes	3475
Index	3477

Detailed Table of Contents

INTRODUCTORY OVERVIEW 3449
General Purpose	3449
Two-Way Frequency Tables	3449
Multi-Way Frequency Tables	3450
The Log-Linear Model	3451
Goodness of Fit	3451
Automatic Model Fitting	3451
Interpreting the Final Model: Marginal Tables	3453
Structural Zeros	3454

PROGRAM OVERVIEW 3455
Methods	3455
Data Files	3455
Output	3455
Alternative Procedures	3455

9. LOG-LINEAR ANALYSIS - CONTENTS

EXAMPLE 3457
    Overview 3457
    Data Files 3457
    Specifying the Analysis 3459
    Results 3459
    Automatic Stepwise Model Selection 3463

DIALOGS, OPTIONS, STATISTICS 3465
    Startup Panel 3465
        Input File 3465
        Variables 3465
        Select Codes 3465
        Specify Table 3466
    Specifying Tables Directly (Frequency Table Input) 3466
    Model Specification 3466
        Review Complete Observed Table 3466
        Display Slices of Table 3467
        Specify Model to be Tested 3468
        Test All Marginal and Partial Association Models 3468
        Automatic Selection of Best Model 3468
        Save the Table 3468
        Define Structural Zeros 3468
        Parameter *Delta* 3468
        Maximum Number of Iterations 3468
        Convergence Criterion 3469
    Reviewing Tables (Slicing Tables) 3469
    Specifying and Testing Specific Models 3469
    Automatic (Stepwise) Selection of Best Model 3470
        $1 - p(1)$, $2 - p(2)$ 3470
        Starting Model 3470
    Specifying Structural Zeros 3470
    Results 3471
        Observed Table 3471
        Fitted Table 3471
        Residuals (Observed - Fitted) 3471
        Residuals Standardized 3472
        Components of Maximum Likelihood Ratio *Chi-square* 3472
        Freeman-Tukey Deviates 3472
        Marginal Tables 3472
        Plot Options 3472

TECHNICAL NOTES 3475
    Algorithms 3475

INDEX 3477

Chapter 9:
LOG-LINEAR ANALYSIS OF FREQUENCY TABLES

INTRODUCTORY OVERVIEW

General Purpose

One basic and straightforward method for analyzing data is via crosstabulation. For example, a medical researcher may tabulate the frequency of different symptoms by patients' age and gender; an educational researcher may tabulate the number of high school drop-outs by age, gender, and ethnic background; an economist may tabulate the number of business failures by industry, region, and initial capitalization; a market researcher may tabulate consumer preferences by product, age, and gender; etc. In all of these cases, the major results of interest can be summarized in a *multi-way frequency table*, that is, in a crosstabulation table with two or more factors. *STATISTICA* contains procedures within the *Basic Statistics and Tables* module for generating, analyzing, and reviewing different types of tables. The *Log-Linear* module provides a more "sophisticated" way of looking at crosstabulation tables. Specifically, this module allows the user to test the different factors that are used in the crosstabulation (e.g., gender, region, etc.) and their interactions for statistical significance (see *Elementary Concepts*, Volume I). The following text will present a brief introduction to these methods.

Two-Way Frequency Tables

The introduction will begin with the simplest possible crosstabulation, the 2 by 2 table. Suppose you were interested in the relationship between age and the graying of people's hair. You took a sample of 100 subjects, and determined who does and does not have gray hair. You also recorded the approximate age of the subjects. The results from this study may be summarized as follows:

```
              Age
Gray       ------------------------
Hair       Below 40    40 or older    Total
-----------------------------------------
No            40            5           45
Yes           20           35           55
-----------------------------------------
Total         60           40          100
```

While interpreting the results of this little study, terminology will be introduced that will allow you to generalize to complex tables more easily.

Design variables and response variables. In multiple regression (see *Multiple Regression*, Volume I) or analysis of variance (see *ANOVA/MANOVA*, Volume I), one customarily distinguishes between independent and dependent variables. Dependent variables are those that the user is trying to explain, that is, that the user hypothesizes to *depend* on the independent variables. The factors in the 2 by 2 table may be classified accordingly: think of hair color (gray, not gray) as the dependent variable, and age as the independent variable.

Alternative terms that are often used in the context of frequency tables are *response variables* and *design variables*, respectively. Response variables are those that vary in *response* to the design variables. Thus, in the example table above, hair color can be considered to be the response variable, and age the design variable.

Fitting marginal frequencies. Now, return to the analysis of your example table. You could ask what the frequencies would look like if there were no relationship between variables (the null hypothesis). Without going into detail, intuitively one could expect that the frequencies in each cell would proportionately reflect the *marginal* frequencies (*Total*s).

For example, consider the following table:

9. LOG-LINEAR ANALYSIS - INTRODUCTORY OVERVIEW

```
             Age
Gray     -------------------------
Hair     Below 40    40 or older    Total
         -------------------------  -----
No          27           18           45
Yes         33           22           55
         -------------------------  -----
Total       60           40          100
```

In this table, the proportions of the marginal frequencies are reflected in the individual cells. Thus, *27/33=18/22=45/55* and *27/18=33/22=60/40*. Given the marginal frequencies, these are the cell frequencies that one would expect if there were no relationship between age and graying. If you compare this table with the previous one you will see that the previous table *does* reflect a relationship between the two variables: There are more than expected (under the null hypothesis) cases below age 40 without gray hair, and more cases above age 40 with gray hair.

This example illustrates the general principle on which the log-linear analysis is based: Given the marginal totals for two (or more) factors, you can compute the cell frequencies that would be expected if the two (or more) factors are unrelated. Significant deviations of the observed frequencies from those expected frequencies reflect a relationship between the two (or more) variables.

Model fitting approach. The discussion of the 2 by 2 table so far can be rephrased as follows. It can be said that *fitting the model* of two variables that are not related (age and hair color) amounts to computing the cell frequencies in the table based on the respective marginal frequencies (totals). Significant deviations of the observed table from those fitted frequencies reflect the lack of fit of the independence (between two variables) model. In that case you would reject that model for your data, and instead accept the model that allows for a relationship or *association* between age and hair color.

Multi-Way Frequency Tables

The reasoning presented for the analysis of the 2 by 2 table can be generalized to more complex tables. For example, suppose there was a third variable in the study, namely whether or not the individuals in the sample experience stress at work. Because you are interested in the effect of stress on graying, *Stress* will be considered as another design variable. (Note that if your study was concerned with the effect of gray hair on subsequent stress, then stress would be the *response* variable and hair color would be the design variable.) The resultant table is a three-way frequency table.

Fitting models. The previous reasoning can be applied to analyze this table. Specifically, you could fit different models that reflect different hypotheses about the data. For example, you could begin with a model that hypothesizes independence between all factors. As before, the expected frequencies in that case would reflect the respective marginal frequencies. If any significant deviations occur, you would reject this model.

Interaction effects. Another conceivable model would be that age is related to hair color, and stress is related to hair color, but the two (age and stress) factors do not interact in their effect. In that case, you would need to simultaneously fit the marginal totals for the two-way table of age by hair color collapsed across levels of stress, and the two-way table of stress by hair color collapsed across the levels of age. If this model does not fit the data, you would conclude that age, stress, and hair color all are interrelated. Put another way, you would conclude that age and stress *interact* in their effect on graying.

The concept of interaction here is analogous to that used in analysis of variance (see *ANOVA/MANOVA*, Volume I). For example, the age by stress interaction could be interpreted such that the relationship of age to hair color is modified by stress. While age brings about only little graying in the absence of stress, age is highly related when stress is present. Put another way, the effects of age and stress on graying are *not additive*, but interactive.

If you are not familiar with the concept of interaction, then it is recommended that you read the *Introductory*

Overview to *ANOVA/MANOVA* (see Volume I). Many aspects of the interpretation of results from a log-linear analysis of a multi-way frequency table are very similar to ANOVA.

Iterative proportional fitting. The computation of expected frequencies becomes increasingly complex when there are more than two factors in the table. However, they *can* be computed, and, therefore, the reasoning developed for the 2 by 2 table can easily be applied to complex tables. The commonly used method for computing the expected frequencies is the *iterative proportional fitting* procedure.

The Log-Linear Model

The term *log-linear* derives from the fact that one can, through logarithmic transformations, restate the problem of analyzing multi-way frequency tables in terms that are very similar to ANOVA. Specifically, one may think of the multi-way frequency table as reflecting various main effects and interaction effects that add together in a *linear* fashion to bring about the observed table of frequencies. Bishop, Fienberg, and Holland (1974) provide details on how to derive log-linear equations to express the relationship between factors in a multi-way frequency table.

Goodness of Fit

In the previous discussion, reference to the "significance" of deviations of the observed frequencies from the expected frequencies was made. One can evaluate the *statistical* significance of the goodness of fit of a particular model via a *Chi-square* test. The *Log-Linear* module will compute two types of *Chi-squares*, the traditional Pearson *Chi-square* statistic and the maximum likelihood ratio *Chi-square* statistic.

In practice, the interpretation and magnitude of those two *Chi-square* statistics are essentially identical. Both tests evaluate whether the expected cell frequencies under the respective model are significantly different from the observed cell frequencies. If so, then the respective model for the table is rejected.

Reviewing and plotting residual frequencies. After one has chosen a model for the observed table, it is always a good idea to inspect the residual frequencies, that is, the observed minus the expected frequencies. If the model is appropriate for the table, then all residual frequencies should be "random noise," that is, consist of positive and negative values of approximately equal magnitudes that are distributed evenly across the cells of the table. The *Log-Linear* module also allows the user to produce various plots of residual frequencies and related statistics.

Statistical significance of effects. The *Chi-squares* of models that are hierarchically related to each other can be directly compared. In general, two models are hierarchically related to each other if one can be produced from the other by either adding terms (variables or interactions) or deleting terms (*but not both at the same time*).

For example, if you first fit a model with the age by hair color interaction and the stress by hair color interaction, and then fit a model with the age by stress by hair color (three-way) interaction, then the second model is a superset of the previous model. You could evaluate the difference between the *Chi-square* statistics for the two models, based on the difference in the degrees of freedom; if the differential *Chi-square* statistic is significant, then you would conclude that the three-way interaction model provides a significantly better fit to the observed table than the model without this interaction. Therefore, the three-way interaction is statistically significant.

Automatic Model Fitting

When analyzing four-way or higher-way tables, finding the best-fitting model can become increasingly difficult. The *Log-Linear* module contains automatic model fitting options to facilitate the search for a "good model" that fits the data. The general logic of this algorithm is as follows. First, the program will fit a model with no relationships between factors; if that model does not fit (i.e., the respective *Chi-square* statistic is significant), then it will fit a model with all two-way interactions. If that model does not fit either, then the program will fit all three-way interactions, and so on. Now, assume that this process found that the model with all two-way interactions fits the data. The program will then proceed to eliminate all two-way interactions that are not statistically significant. The resulting model will be the one that includes the least number of interactions necessary to fit the observed table.

Examining all k-factor interactions. One can also follow the logic of the automatic model fitting algorithm "manually." For example, one could examine the table of models that include no two-way interactions, no three-way interactions, etc. The *Log-Linear* module will automatically compute that table. Shown below are the results for the data set used in the *Example* section (page 3457).

LOG-LIN. ANALYSIS K-Factor	Degrs.of Freedom	Max.Lik. Chi-squ.	Probab. p	Pearson Chi-squ	Probab. p
1	8	632.1563	0.000000	881.2506	0.000000
2	23	134.4250	.000000	141.2284	.000000
3	28	30.9093	.321227	31.2334	.306927
4	12	9.0115	.701936	8.9275	.709100

Results of Fitting all K-Factor Interactions (center2.sta). These are simultaneous tests that all K-Factor Interactions are simultaneously Zero.

This Scrollsheet shows that the improvement in fit when including all 2-way interactions in the model (*k-factors = 2*) is highly significant. The improvement in fit when adding all 3-way interactions to the model (*k-factors = 3*) is not significant. Therefore, you would conclude that the best model is located somewhere "in-between," that is, that it will contain some two-way interactions and probably no three-way interactions.

Marginal and partial associations. The next step in the "manual" procedure would be to determine which two-way interactions are necessary in the model to fit the data. There are two ways in which one could proceed; both alternatives consider the two-way interactions one by one. First, one could fit all two-way interactions except the one under consideration. If the difference between the model with and without the respective two-way interaction is significant, then that interaction would be retained in the model. This type of test is also referred to as the test for *partial association* (akin to *partial correlation*, see *Multiple Regression*, Volume I). The second test that could be performed would be to compare a model without any interactions with a model including the two-way interaction under consideration. This test is referred to as the test for *marginal association* (conceptually, in *Multiple Regression* this test would be the equivalent of looking at the simple correlations, not the partial correlations). Shown below is the table reporting both of these tests for the data set analyzed in the *Example* section.

LOG-LIN. ANALYSIS	Degrs.of Freedom	Prt.Ass. Chi-sqr.	Prt.Ass. p	Mrg.Ass. Chi-sqr.	Mrg.Ass. p
1	3	370.1494	0.000000	370.1494	0.000000
2	1	152.8534	.000000	152.8534	.000000
3	2	100.2144	.000000	100.2144	.000000
4	2	8.9397	.011456	8.9397	.011456
12	3	10.1762	.017138	9.4863	.023491
13	6	1.4726	.961292	3.0924	.797161
14	2	34.2330	.000000	35.4078	.000004
23	2	4.1726	.124162	7.7174	.021105
24	2	7.7881	.020373	10.8884	.004325
34	4	66.8077	.000000	72.2174	.000000
123	6	4.8305	.565739	7.7128	.259942
124	6	4.1269	.659540	5.1878	.519973
134	12	11.4065	.494461	12.3497	.418057
234	4	7.4794	.112648	8.7272	.068316

Tests of Marginal and Partial Association (center2.sta)

From the above Scrollsheet, you would infer that the interaction between factors 3 and 4 (*34*) is highly significant for both the partial association and marginal association. Other significant two-way interactions are *12* (factors 1 and 2), *14* (factors 1 and 4), and *24* (factors 2 and 4).

The *23* interaction is only significant when the marginal association between the two factors is tested; the partial association is not significant. You can think of this mixed finding in terms analogous to *Multiple Regression* (see the *Introductory Overview* section to that chapter). When no other two-way interactions are in the model, the marginal association between factors 2 and 3 is significant. This is like looking at the simple correlation between those two variables. When other two-way interactions are in the model, the partial association ("partial correlation") between factors 2 and 3 is not significant; thus, whatever "non-additivity" in the observed table led to the significant marginal association, after all other two-way interactions were included in the model, this "non-additivity" was largely explained or accounted for by the other interactions.

Given the results shown above, the next step would probably be to fit a model with the *12*, *14*, *24*, and *34* interactions. (This, incidentally, is the same model that will be fit by the automatic model selection procedure, see page 3463).

Associations between design variables. To reiterate, design variables are those on which you want to base your explanation of the response variable. To return to the "gray hair" example introduced earlier, gray hair is the "response" to stress and age; therefore, hair color would be the response variable, and stress and age would be the design variables.

Usually one is not interested in any associations between design variables that do *not* involve the response variable. In the "gray hair" example, it would be of little interest if variables *stress* and *age* are associated (but not with gray hair). In other words, if there is generally more stress reported among older subjects than younger subjects, then this is not of interest as long as this "unevenness" is not associated with differences in the number of subjects with or without gray hair; if this association *is* associated with hair color, you would have a three-way interaction or association.

In practice, one usually includes in one's model all interactions between design variables. In this way, you avoid obtaining an overall lack of fit which may be entirely due to interactions between design variables when those interactions are of little interest to the researcher. The analysis presented in the *Example* section (page 3457) will further illustrate this point.

Interpreting the Final Model: Marginal Tables

In order to interpret the final model, you could look at the fitted (expected) frequencies in the table. The interpretation of factors and their interactions is analogous to that in analysis of variance (see *ANOVA/MANOVA*, Volume I). For example, suppose in your study of stress, age, and gray hair, you obtained a final model with two two-way interactions: one between age and hair color, and another between stress and hair color. Because the final model does not include the three-way interaction, you could look at the respective marginal frequency tables from which the expected frequencies were computed. These marginal frequency tables might, for example, look like this:

```
Marginal table: Age by Hair Color
                    Age
Gray    ----------------------------
Hair    Below 40    40 or older    Total
----------------------------------------
No      40          10             50
Yes     20          30             50
----------------------------------------
Total   60          40             100

Marginal table: Stress by Hair Color
                   Stress
Gray    ----------------------------
Hair    Low         High           Total
----------------------------------------
No      40          10             50
Yes     10          40             50
----------------------------------------
Total   50          50             100
```

Looking at the first table, you can clearly see that more subjects who are 40 or older have gray hair; the second table shows that there are more people under high stress with gray hair than under low stress. Therefore, you would conclude that both age and stress contribute to gray hair.

General case. In general, to interpret a model you would review the marginal tables that represent the factors and interactions in the model. The *Log-Linear* module will automatically compute the marginal tables for the respective user-specified model.

Structural Zeros

Some tables, by design, have empty cells. For example, suppose you tabulated the number of patients in a hospital by gender and diagnosis. If the hospital has a maternity ward, then the table will contain empty cells by design, namely, there will not be any males who are in the hospital to give birth. Such empty cells in the table that are empty by design are called *structural zeros*. When computing the expected frequencies, it would, of course, be silly to put expected frequencies that are larger than 0 in those cells, since you do *not* expect any cases in those cells. The *Log-Linear* module will automatically handle tables with structural zeros, after the user has specified which cells in the table are empty by design.

PROGRAM OVERVIEW

Methods

The *Log Linear* module offers a complete implementation of log-linear modeling procedures for multi-way frequency tables. The user can analyze up to 7-way tables. Both complete and incomplete tables (with structural zeros) can be analyzed. Because of the various ways in which the user can specify tables, and review the complete as well as marginal tables, the *Log Linear* module can also be considered to be a very flexible crosstabulation program, that is an alternative to the *Tables and Banners* procedure (see *Basic Statistics and Tables*, Volume I).

Model selection. To aid in the selection of models, the program includes an intelligent stepwise selection option that will automatically select the model that best fits the respective observed frequency table. In addition, the program will compute the table of tests of marginal and partial association, and a table of tests of all *k*-factor interactions.

Data Files

The user may specify raw data, in which case the program will first compute the crosstabulation table before beginning the analysis. The user may also analyze data files containing previously tabulated data, or files containing the actual frequency table. Like all other output in *STATISTICA*, frequency tables can be saved in standard *STATISTICA* data files.

Output

The standard output includes the maximum likelihood ratio *Chi-square* and standard Pearson *Chi-square* with the appropriate degrees of freedom and significance levels, the observed, expected (fitted), and residual frequency tables, the table of standardized residuals, components of the maximum likelihood ratio *Chi-square*, Freeman-Tukey deviates, etc. In all tables, the user has control over the arrangement of factors. The marginal tables for fitted models can also be displayed.

Plots. A wide selection of specialized graphics options are available to aid in the interpretation of results. The user can review scatterplots of the observed versus fitted frequencies, and of the fitted frequencies versus the residual frequencies, standardized residual frequencies, components of the maximum likelihood ratio *Chi-square*, and Freeman-Tukey deviates. Also, a large selection of descriptive graphs are available to aid in the interpretation of input data and the output. For example, all multi-way tables can be reviewed in categorized histograms and 3D bivariate histograms.

Alternative Procedures

The *Tables and Banners* option in the *Basic Statistics and Tables* module (see Volume I) is a general crosstabulation routine that will compute a wide range of statistics for two-way tables. That module also includes various facilities for producing publication-quality tables and banners. For 2 by 2 tables, *Nonparametrics and Distributions* will compute various specialized statistics. That module also contains a wide range of nonparametric correlation measures. Note that categorical variables in coded form can also be analyzed by many other modules such as *Multiple Regression*, *Nonlinear Estimation*, and *Discriminant Analysis*, to name only a few.

Mantel and Haenszel test. Your *STATISTICA* program contains an example *STATISTICA BASIC* program (file *Manthaen.stb* in the *STBASIC* subdirectory) for computing the Mantel-Haenszel test for comparing two groups (see Mantel and Haenszel, 1959; see also Lee, 1992). This test has been used in many clinical and epidemiological studies, in order to control for the effects of confounding variables. The test is based on the analysis of 2 by 2 tables (e.g., *Group 1/2* vs. *Survival*), stratified by another categorical variable (the confounding variable; e.g., *Location*). It allows you to test whether the two variables in the 2 by 2 tables (e.g., *Group* and *Survival*) are associated with each other, after controlling for the stratification variable (e.g., *Location*).

9. LOG-LINEAR ANALYSIS - EXAMPLE

EXAMPLE

Overview

The following example is based on a "classic" data set reported by Morrison, et al. (1973), and discussed by Bishop, Fienberg, and Holland (1975). The data are contained in the data file *Center.sta* (and *Center2.sta*; see below) that is included with your *STATISTICA* program. The data file contains a frequency table of the number of breast cancer patients who survived three years or longer after the diagnosis (obviously, those data are not representative of the chances of surviving breast cancer in the nineties).

The frequencies are reported separately for four different types of inflammation and appearance (*MIN_MAL*, *GRT_MAL*, *MIN_BEN*, *GRT_BEN*), three age groups (*under 50*, *50-69*, *over 69*), and separately for three diagnostic centers (*Tokyo*, *Boston*, and *Glamorgan*). The complete table was entered into the spreadsheet as follows:

			1 MIN_MAL	2 MIN_BEN	3 GRT_MAL	4 GRT_BEN
Tokyo	under 50	no	9	7	4	3
		yes	26	68	25	9
	50-69	no	9	9	11	2
		yes	20	46	18	5
	over 69	no	2	3	1	0
		yes	1	6	5	1
Boston	under 50	no	6	7	6	0
		yes	11	24	4	0
	50-69	no	8	20	3	2
		yes	18	58	10	3
	over 69	no	9	18	3	0
		yes	15	26	1	1
Glmrgn	under 50	no	16	7	3	0
		yes	16	20	8	1
	50-69	no	14	12	3	0
		yes	27	39	10	4
	over 69	no	3	7	3	0
		yes	12	11	4	1

Note that the case name column was used to denote the levels of three factors, that is, the *Location* of the diagnostic center (furthest to the left), the *Age* (in the middle of the case name column), and *Survival* (to the right in the case name column).

Goal of the analysis. In general, the goal of log-linear analysis of a frequency table is to uncover relationships between the categorical variables (factors) that make up the table. The *Introductory Overview* section of this chapter introduced the distinction between *design* variables and *response* variables (page 3457), a distinction that basically corresponds to that between independent and dependent variables, respectively. The major response variable of interest in this table is, of course, *Survival*. All other factors are treated as design factors. Thus, you will not be concerned with any interactions between, for example, the location of the diagnostic center and the age of the patients or the appearance of the cancer.

Data Files

Before the actual analysis of the table begins, the different ways in which data files can be specified in the *Log-Linear* module will first be demonstrated. Note that the data file *Center.sta* (shown in the above spreadsheet) contains only frequencies as values for the variables; there are no coding variables with text codes that identify the levels of the factors.

After opening the *Log-Linear* module, the *Log-Linear Analysis* startup panel will be displayed.

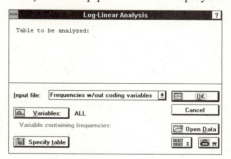

Select the *Frequencies w/out coding variables* option in the *Input file* combo box from this dialog. Then click on the *Variables* button and select all four variables.

9. LOG-LINEAR ANALYSIS - EXAMPLE

Specifying the table. In order to make sure that the program will understand how to interpret the data, that is, how to organize the numbers into the table, click on the *Specify table* button to open the *Specify dimensions of the table* dialog (see below). Internally, the program will simply read the frequencies in the four variables as one long string of numbers, reading row by row, starting from the left-most variable. The information supplied by the user via the *Specify dimensions of the table* dialog allows the program to "understand" the structure of the table. Now, enter each factor's name and its respective number of levels in the *Specify dimensions of the table* dialog. There are four factors that need to be entered into this dialog: *Appearnc* (and inflammation) of the cancer, *Survival*, *Age*, and *Location* of the diagnostic center. The number of levels in each factor also needs to be specified.

The program will interpret the first factor entered in this dialog to be the one with the "fastest-changing subscript," the second factor entered as the one with the second-fastest changing subscript, and so on. Because STATISTICA reads the frequencies across rows, the factor with the fastest-changing subscript in this example is the factor whose four levels are listed in the column headings of the spreadsheet -- *Appearnc*. Therefore, enter *Appearnc* as the *Factor Name* and *4* as the *No. of levels* for that factor in the first line. The next-fastest changing factor is *Survival*; the levels of this factor change from line to line in the spreadsheet. Therefore, enter the factor name *Survival* and *2* levels in line 2. Specify the remaining two factors as follows: *3* levels for factor *Age*, and *3* levels for factor *Location*.

The startup panel will now look like this:

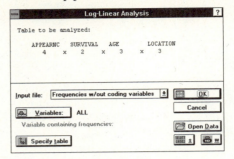

Now the table is ready to be analyzed.

Saving a table in an alternative format. In order to make the results more readable with meaningful text values denoting the levels of the factors, a different data file setup is preferred. When you save tables from within the *Log-Linear* module, they will automatically be saved in this preferred setup. You could now click *OK* as if you were to begin the analysis, and on the subsequent *Log-Linear Model Specification* dialog, click on the *Save the table* button to save the table. The resulting file would contain one line for each cell of the table. In addition to a variable containing the respective cell frequencies, that file will also include one variable for each factor in the table, with integer codes to denote the respective levels. Such a file was previously created and then appropriate text values (see *General Conventions*, Volume I) were added to it. This alternative way of representing the table was used in the data file *Center2.sta*. Below is a partial listing of that file.

This file will be used in subsequent analyses which will make the output more readable.

Specifying the Analysis

After opening the *Log-Linear* module, make sure that the data file *Center2.sta* is opened. Next, select the *Frequencies with coding variables* option (which is the way that the table is represented in the data file) in the *Input file* combo box from the startup panel. Now, click on the *Variables* button and select variable *Frequency* as the variable with the frequency counts and variables *1-Appearnc* through *4-Location* as the variables with the codes.

Finally, click on the *Select codes* button and specify the respective codes that were used to denote the levels of the factors. To select all codes, you may use the asterisk (*) convention in each of the codes selection windows, click on each of the *All* buttons, or click on the *Select All* button to select all of the codes for each of the variables.

The startup panel will now look like this:

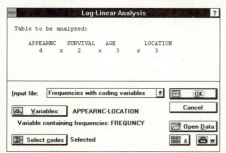

Results

You are now ready to begin the analysis; press *OK* and the *Log-Linear Model Specification* dialog will appear.

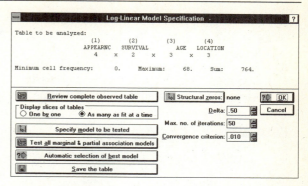

Observed table. You can review the observed table via the *Review complete observed table* button on this dialog. When you click on this button, the *Specify how Table is to be Reviewed* dialog will open in which you can flexibly specify the way in which to review the table.

For example, you may request to review the table with *Age* (factor 3) as the column variable and *Location* (factor 4) as the row variable (within each level of the other factors), etc.

Finding a model. Finding an appropriate model for a multi-way (more than two-way) frequency table is often not an easy task. The *Log-Linear* module provides a number of options to facilitate the search. In particular, the *Automatic selection of best model* option is useful because it will automatically find the least complex model that will fit the data. This option will be used later to see whether it will arrive at the same or similar conclusions (model) as you will when "left to your own devices."

Simultaneous test of all k-factor interactions. A first step towards understanding the degree of complexity of the table is to review the table of simultaneous tests for all *k*-factor interactions, and the tests of all marginal and partial association models. These tests will be computed when you click on the *Test all marginal & partial association models* button. The first time that you use this option, the computations involved in testing those models will require some time; thereafter, these results can be called up instantaneously.

LOG-LIN. ANALYSIS K-Factor	Degrs.of Freedom	Max.Lik. Chi-squ.	Probab. p	Pearson Chi-sqr.	Probab. p
1	8	632.1563	0.000000	881.2506	0.000000
2	23	134.4250	.000000	141.2284	.000000
3	28	30.9093	.321227	31.2334	.306927
4	12	9.0115	.701936	8.9275	.709100

Results of Fitting all K-Factor Interactions (center2.sta). These are simultaneous tests that all K-Factor Interactions are simultaneously Zero.

The interpretation of the tests of all *k*-factor interactions was discussed in the *Introductory Overview* section of this chapter (page 3452). In short, the Scrollsheet above shows that the improvement in fit when including all 2-way interactions in the model (*k*-factors = 2) is highly significant (i.e., the model provides a very poor fit). The improvement in fit when adding all 3-way interactions to the model (*k*-factors = 3) is not significant (i.e., the model provides an adequate fit). Therefore, one can conclude that the least complex model that will fit the observed table need not contain any three-way associations, but will contain one or more two-way associations.

Tests of all marginal and partial associations.

To see which of the two-way associations seem to be significant, the table of all partial and marginal association models will be reviewed.

LOG-LIN. ANALYSIS	Degrs.of Freedom	Prt.Ass. Chi-sqr.	Prt.Ass. p	Mrg.Ass. Chi-sqr.	Mrg.Ass. p
1	3	370.1494	0.000000	370.1494	0.000000
2	1	152.8534	.000000	152.8534	.000000
3	2	100.2144	.000000	100.2144	.000000
4	2	8.9397	.011456	8.9397	.011456
12	3	10.1762	.017138	9.4863	.023491
13	6	1.4726	.961292	3.0924	.797161
14	6	34.2330	.000006	35.4078	.000004
23	2	4.1726	.124162	7.7174	.021105
24	2	7.7881	.020373	10.8884	.004325
34	4	66.8077	.000000	72.2174	.000000
123	6	4.8305	.565739	7.7128	.259942
124	6	4.1269	.659504	5.1878	.519973
134	12	11.4065	.494361	12.3497	.418057
234	4	7.4794	.112648	8.7272	.068316

Tests of Marginal and Partial Association (center2.sta)

The interpretation of this table is also discussed in the *Introductory Overview* section of this chapter (page 3452). In short, the *partial association Chi-square* test evaluates the significance of the respective effect (indicated by the digits in the *Effect* column) by comparing the model that includes all interactions of the same order with the model without the respective effect.

For example, look at effect *12*. This effect represents the association or interaction between factors *1-Appearnc* and *2-Survival*. When dropping this effect from the model with all other two-way associations, the difference in the (maximum likelihood) *Chi-square* values is equal to *10.18*, with *3* degrees of freedom. This value is significant at the *p<.02* level. Therefore, the model fit becomes significantly worse when excluding this two-way interaction from the model; thus you would include it.

To make an analogy to multiple regression (see *Multiple Regression*, Volume I), the partial association test gauges the *unique* contribution of the respective effect (association or interaction) to the fit of the model. The logic of the test is analogous to that of partial correlations.

The *marginal association* test of effect *12* denotes the difference between the model without any two-way interactions and the model that includes the *12* interaction (and no other two-way interactions). As you can see, the model fit improves significantly when adding the association between factors *1-Appearnc* and *2-Survival* (*Chi-square = 9.49, df = 3, p<.03*). To continue the analogy to multiple regression, this test is the equivalent to the regular (zero-order) correlation coefficient.

Choosing the effects for the model.

If the table of marginal and partial associations is now reviewed for the significance of other two-way interactions, you will find that other associations that should be included in the model are:

(1) The association between factors *1-Appearnc* and *2-Survival* (effect *12*),

(2) The association between factors *1-Appearnc* and *4-Location* (effect *14*),

(3) The association between factors *2-Survival* and *4-Location* (effect *24*),

(4) The association between factors *3-Age* and *4-Location* (effect *34*).

The association between *2-Survival* and *3-Age* is not significant when evaluated with all other two-way associations (see partial association for effect *23*). Therefore, it will not be included in the model for now.

Rules for specifying a model. You are now ready to specify and test a particular model. However, before proceeding, please review a few important points. First, interaction or association effects automatically include lower-order effects. So, for example, if you specify the *12* association, you request to fit the marginal table of factors *1* by *2*. This table will obviously contain or reflect the marginal tables of *1* and *2* alone. Therefore, it is not necessary to explicitly include the lower-order effects when specifying a model. Second, make sure that all effects are reflected in the model. For example, if you specify only *12* and make no other reference to other factors, then you basically hypothesize that the marginal frequencies for all other factors are *equal*. In this example, it would be silly to hypothesize that there are an equal number of patients diagnosed in the three diagnostic centers. Therefore, fitting a table that forces those numbers to be equal unnecessarily worsens the fit of the model.

Specifying the model. To specify the model that was previously derived on page 3460, click on the *Specify model to be tested* button.

You can now type in the desired model in the *Specify Model to be Tested* dialog. In this case, you know that you need to include the two-way associations *12*, *14*, *24*, and *34*. However, as discussed in the *Introductory Overview* (page 3453), you generally include all interactions between design variables in the model, so that they will not contribute to the overall lack of model fit. It was assumed in this study that you were not interested in any interactions between the inflammation/appearance of the cancer, the age, and the diagnostic center. It may well be that the distribution of age is different in different diagnostic centers, or that the appearance of the cancer is different in different age groups. However, you are mostly interested in the factors that are associated with survival. Since you are not interested in any associations between design variables in this study, fit the three-way association (*134*) between all design factors, in addition to the *12* and *24* effects. To specify this model, type *12,24,134* into the edit box of the dialog.

Evaluating the goodness of fit. Now press *OK* and the *Results* dialog will appear. As you can see in the results summary below, the overall model fits the observed table (the *Chi-square* tests are not significant). Therefore, you can conclude that the specified model is sufficient to explain the frequencies in the table.

```
Results

Table to be analyzed:
              (1)        (2)        (3)        (4)
           APPEARNC   SURVIVAL      AGE     LOCATION
              4    x     2    x     3    x     3

Minimum cell frequency:    0.   Maximum:    68.   Sum:    764.

Model to be tested: 21,42,431

Delta: .5000 ;  Maximum iterations:  50 ;  Conv. criterion: .0100
Convergence reached after   5 iterations
                                        df         p
   Maximum Likelihood Chi-square:  31.74380   30    .3796245
              Pearson Chi-square:  32.07727   30    .3640367
```

You may click on the *Graph of observed vs. fitted* button to see whether there are any major discrepancies between the observed and the fitted frequencies in the table.

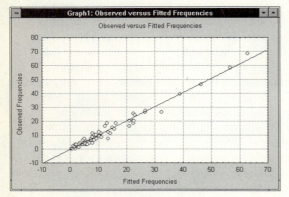

Most points in the graph above fall onto a straight line. Thus it appears that there are no major outliers ("misfitted" cells) in the table.

Hierarchical tests of alternative models.

Before interpreting the results, test the statistical significance of the *24* and *12* associations and the significance of the association between *Age* and *Survival* (*23*), which is not included in this model. As described in the *Introductory Overview* (page 3451), you can evaluate the statistical significance of effects by comparing the *Chi-square* of the model that includes the effect with the *Chi-square* of the model that excludes the effect. For example, to test the *24* association, fit the model *12,134* and compare this *Chi-square* with the previous model *Chi-square* (this model is the same as the current model, except that the *24* interaction was dropped).

If you fit the model *12,134*, the resulting maximum likelihood ratio *Chi-square* value will be *43.37* with *32* degrees of freedom (see the summary section of the *Results* dialog above). This *Chi-square* is significantly worse (i.e., larger) than the previous model (which included the *24* interaction): The *Chi-square* difference is equal to *43.37-31.74=11.63*, and the degrees of freedom difference is *32-30=2*; the resulting significance level is *p<.005*. Therefore, you can conclude that the *24* interaction is significant (i.e., there is a significant association between the survival rate and the diagnostic center). Following the same logic, you will find that the *Appearnc* (factor 1) by *Survival* (factor 2) association is also highly significant (*Chi-square* difference = *10.23*, *df* difference = *3*, *p<.025*). To assess the significance of the *32* (*Age* and *Survival*) association that is not in the current model, add it to the model and assess the significance of the improvement in model fit. As you will see, the *32* association does not significantly improve the fit of the model to the observed table.

Interpreting the results.

The analysis so far has yielded two significant effects, that is, associations between design variables and the response variable: (1) a relationship between *Appearnc* (factor 1) and *Survival* (factor 2), and (2) an association between *Location* (factor 4) and *Survival* (factor 2). Now, examine the nature of these effects. Remember that fitting a model involves the computations of the expected values, so that they reflect the relative frequencies of the respective marginal tables. Therefore, to interpret an effect, you would examine the marginal tables. Click on the *Marginal tables* button on the *Results* dialog for the *12,24,134* model in order to display those tables in individual Scrollsheets (including the three-way table of the *134* effect). First, look at the association *Appearnc by Survival* Scrollsheet:

LOG-LIN. ANALYSIS	APPEARNC MIN_MAL	APPEARNC MIN_BEGN	APPEARNC GRT_MAL	APPEARNC GRT_BEGN	Total
NO	80.5000	94.5000	41.5000	11.50000	228.0000
YES	150.5000	302.5000	89.5000	29.50000	572.0000
Total	231.0000	397.0000	131.0000	41.00000	800.0000

Careful examination of this table reveals that the survival rate of patients whose cancer was diagnosed as malignant (column headers *Min_Mal* and *Grt_Mal*) is roughly 2 to 1 (survival *YES* to *NO*); for benign cancer that rate is about 3 to 1.

Note that in order to simplify this example, the original appearance and inflammation factors were combined into the single 4-level factor *Appearnc*. In order to treat inflammation and appearance as separate factors in the table, you could split *Appearnc* into two variables (see *Data Management*, Volume I) and re-analyze this table.

The *Survival* by *Location* table looks like this:

LOG-LIN. ANALYSIS	SURVIVAL NO	SURVIVAL YES	Total
TOKYO	66.0000	236.0000	302.0000
BOSTON	88.0000	177.0000	265.0000
GLAMORGN	74.0000	159.0000	233.0000
Total	228.0000	572.0000	800.0000

It seems that the survival rate is highest for cancer patients diagnosed in *Tokyo*, about 3 to 1 (survival *YES* to *NO*). In *Boston* and *Glamorgan*, that rate stands at about 2 to 1. Of course, you cannot infer any specific cause for this effect. Obviously there are any number of differences (not measured in this study) between the patients in Tokyo and Boston or Glamorgan. However, the apparently differential survival rates would certainly warrant further study.

Note that the frequencies in the marginal table will include the *delta* constant as specified on the *Model Specification* dialog. By default, the program will add *0.5* (the *delta*) to each cell frequency before fitting any models (for details, see page 3468). Therefore, in order to obtain accurate marginal counts when using the *Log-Linear* module, be sure to set the *delta* constant to *0*.

Automatic Stepwise Model Selection

The more complex the table, the more difficult it will be to find a model that fits and at the same time includes all effects of importance (significance). In fact, the final conclusion was arrived at "the hard way;" you could have also used the *Automatic selection of best model* option on the *Log-Linear Model Specification* dialog. After clicking on this button, the following dialog appears:

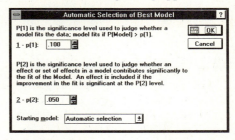

The algorithm. The algorithm used in the *Log-Linear* module to find a sufficient model for the observed table basically implements the same logic that you followed when you examined the *k*-level interactions table and the table of marginal and partial associations.

First, the program will determine the complexity or order of the interactions that need to be included in the model in order to make it fit to the observed table. Option *1 - p(1)* controls the *p*-level that is used at this stage of the search to decide whether or not a model fits.

Next, the program will remove associations (of the order found in step one) from the model, step by step. At this stage, if an effect is found to be more significant than is specified in *2 - p(2)*, then it is retained in the model.

The default settings for *p(1)* and *p(2)* are reasonable, so simply click *OK* to see which model will be chosen by the program.

Results. In the following illustration, you can see that the initial model consists of all two-way associations; this was also your starting point. The final model is basically the same one that you arrived at; namely, it includes the two major association effects of interest: *12* (*Appearnc* and *Survival*) and *24* (*Location* and *Survival*).

9. LOG-LINEAR ANALYSIS - EXAMPLE

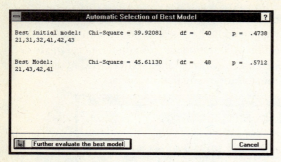

Note that this model is automatically "transferred" into the model specification dialog (by default, the "best" model selected by the program will be entered into the edit box in the *Specify Model to be Tested* dialog); thus, simply click on the *Further evaluate the best model* button in this dialog and then click *OK* in the resulting *Specify Model to be Tested* dialog and the *Results* dialog for the final model will be displayed.

Conclusions and Final Remarks

It can be concluded from the analysis of this table that the major factors associated with the 3-year survival of patients were the diagnosed malignancy of the cancer and the location of the center where it was diagnosed. Interestingly, age did not seem to be related to long-term survival.

As mentioned earlier, there exist any number of possible explanations for why the survival rate in Tokyo was higher than in the other diagnostic centers (time of diagnosis, dietary differences, cultural differences in "healthy behaviors," differences in the environment, etc.). However, the apparent differences revealed in this study would certainly be worthy of further investigation.

DIALOGS, OPTIONS, STATISTICS

Startup Panel

The options available on this dialog will differ slightly, depending on the setting of the *Input File* combo box (see below) in the *Log-Linear Analysis* startup panel.

Input File

The *Log-Linear* module can process three different types of data files: *Raw Data* with coding variables, *Frequencies with coding variables*, and *Frequencies without coding variables*.

Raw data. If the *Raw data* option is selected, then *STATISTICA* will compute the frequency table from the data. The program expects that the data file contains coding variables with integer values or text values that uniquely identify to which cell in the table each case belongs. This way of specifying the design of a table is very much like specifying the design of an ANOVA in the *ANOVA/MANOVA* module (see Volume I).

Frequencies with coding variables. If the file contains frequencies, then *STATISTICA* can read those directly. If *Frequencies with coding variables* is specified, then *STATISTICA* expects coding variables with integer codes or text values that

uniquely specify the cell in the table to which each frequency belongs and a variable containing the frequency. The example data file *Center2.sta* provided with *STATISTICA* and analyzed in the *Example* section (page 3458) is organized in this manner.

Frequencies without coding variables. If *Frequencies without coding variables* is specified, then the program will read the data variable by variable, and case by case, and interpret consecutive data points as frequencies of the table as specified by the *Specify Table* option (see below). Thus, the user can type tables directly into a new spreadsheet (without having to add any codes) and process them with the *Log-Linear* module.

Variables

Clicking on this button will bring up a standard variable selection dialog. If the *Input File* combo box is set to *Raw data* or *Frequencies with coding variables*, then use this option to select the variables denoting the factors; those variables should contain the codes or text values that identify the cells of the table. If the *Input File* combo box is set to *Frequencies with coding variables*, then clicking on the *Variables* button will bring up the two-variable list selection dialog. Select the variable containing the frequencies and the variables (factors) with the codes that specify the table. The different ways in which tables can be specified in data files, and read by the *Log-Linear* module, are demonstrated in the *Example* section of this chapter (page 3457).

Select Codes

This option is only available if the *Input File* combo box is set to either *Raw data* or *Frequencies with coding variables* (see above), and after variables (i.e., factors or "independent" variables) have been selected via the *Variables* option (see above). Clicking on this button will bring up the standard code selection dialog containing an edit box for each selected variable (factor). Note that you can use the

asterisk (*) option in each edit box to automatically select all codes or text values in the respective variable or factor, or you can click on the *Select All* button in this dialog.

Specify Table

This option is only available if the input data file was specified to contain *Frequencies without coding variables*. In this case, the program will interpret all consecutive data points that are read (variable by variable, case by case) as frequencies of the table. In order to instruct the program as to which cell of the table each frequency belongs, the user must specify the design (or "shape") of the table. The logic of specifying the design is as follows: The first factor that is specified should be the fastest-changing factor, the second factor the second-fastest changing factor, etc. (see below).

Specifying Tables Directly (Frequency Table Input)

It is assumed that you have typed the frequencies for the table directly into the data file. Internally, STATISTICA will simply read the frequencies in the selected variables as one long string of numbers, reading row by row, starting from the left-most variable. The information supplied here allows the program to "understand" the structure of the table.

For each factor in the table, specify the number of levels and a name in the *Specify the dimensions of the table* dialog.

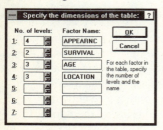

When reading the data (frequencies) from the data file, the program will interpret the first factor listed in this dialog to be the one with the "fastest-changing subscript," the second factor listed in this dialog as the one with the second-fastest changing subscript, and so on. Refer to the *Example* section (page 3458) for an example of how to specify tables in this manner.

Model Specification

The *Log-Linear Model Specification* dialog will allow you to review the observed tables, specify specific models, and explore other options.

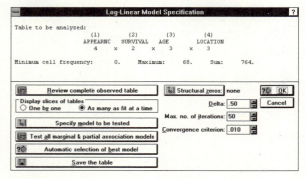

Review Complete Observed Table

Via the *Review complete observed table* button you can review the multi-way frequency table in the most flexible manner, slice by slice, as specified by the user. If there are more than two factors in the frequency table, the program allows the user to view any two-way table within each of the levels of the remaining factors. After clicking on this button (if the table contains more than 2 factors), the user is prompted to select the two factors by which to view the table (see page 3469). Scrollsheets will then be displayed showing the frequencies in two-way tables, within the levels of all remaining factors.

9. LOG-LINEAR ANALYSIS - DIALOGS AND OPTIONS

The first *Quick Stats Graph* for all tables (Scrollsheets) is the 3D histogram. This graph is very helpful in order to visualize the pattern of frequencies in the table.

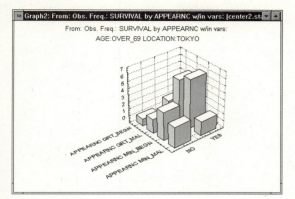

Display Slices of Table

The setting of the radio buttons in this group will determine how tables are displayed in this module, and it will affect the display of tables selected from the model specification dialog, as well as the *Results* dialog window.

As many as fit at a time. By default, the program will automatically display as many slices of the table (Scrollsheets) as can be simultaneously viewed. Note that you can easily increase the default number of Scrollsheets in the queue (3 Scrollsheets at a time) via the *Scrollsheet Manager* option on the *Window* pull-down menu.

One by one. Alternatively, the user may set the *One by one* radio button. In this case, only one slice at a time will be displayed, and the user will be prompted to *Continue* after each slice is displayed.

Note: If *printer*, disk *file,* or *Text/output Window* output is selected in the *Page/Output Setup* dialog, then before the first table is displayed, the user has the option to automatically print (to a disk or the printer) the entire table (without interruption).

Graphs from Scrollsheets. Several types of graphs are available from the results Scrollsheets right-mouse-button flying menu that will enable you to better interpret the output. For example, you may want to visualize the results, categorized by a grouping variable (click on the Scrollsheet with the right-mouse-button and select the *Quick Stats Graphs - 2D Histogram by* option).

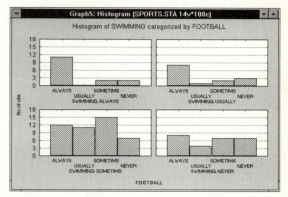

The default graph from the results Scrollsheets is the *3D Histogram* (click on the Scrollsheet with the right-mouse-button and select the *Quick Stats Graphs - 3D Histogram* option).

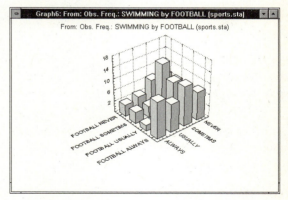

LOG - 3467

Copyright © StatSoft, 1995

Specify Model to be Tested

Clicking on this button will bring up a dialog in which the model to be tested can be specified (see page 3469).

Test All Marginal & Partial Models

The utility and interpretation of these tests is described in the *Introductory Overview* section (page 3452) and the *Example* section (page 3460). The partial association between, for example, two factors 1 and 2 (denoted by *12*) is computed by comparing the fit (i.e., evaluating the *Chi-square* difference) of the model that includes all two-way interactions with the model that excludes the interaction between factors 1 and 2. The marginal association between those factors is computed by comparing the fit of the model including all main effects (i.e., all effects of lower order than the one of interest) with the model including the interaction between factors 1 and 2.

Note that these tests are actually computed only the first time after this option is chosen. Thereafter (unless the user returns to the startup panel), these tests are stored in memory and are available instantaneously.

Automatic Selection of Best Model

This option will bring up the *Automatic Selection of Best Model* dialog (see page 3470). The logic of the automatic selection algorithm is described in the *Introductory Overview* section (page 3451).

Save the Table

This option allows the user to save the observed table in a *STATISTICA* data file. The file will be saved with coding variables (indicating the levels for each factor) and a variable named *Frequncy* that contains the frequencies for each cell in the table.

Define Structural Zeros

With the *Log-Linear* module, one can analyze incomplete tables, that is, tables containing structural zeros. Structural (as opposed to *sampling*) zeros are empty cells in the table that are empty by design (see the *Introductory Overview*, page 3454, and also page 3470). After pressing this button, the program will first "find" all zero frequencies in the table, after which you can select the structural zeros from that list.

Note that the *Log-Linear* module will correctly compute the degrees of freedom for models involving structural zeros as long as those zero frequencies do not result in zeros in the marginal tables of the respective model.

Delta

The constant *delta* (*0.5* by default) is added to all frequencies in the observed table before it is submitted to the actual analysis. This is a recommended procedure when the table contains several cells with low frequencies (e.g., less than 10); it is a correction analogous to the Yates' correction for two-way tables, and this correction does not affect the results unless the table contains low frequencies.

For further discussion of this correction, see Conover, 1974; Everitt, 1977; Hays, 1988; Kendall and Stuart, 1979; and Mantel, 1974.

Note: Be sure to set *delta* to zero if you want to use this module as a tabulation program. Otherwise, you will not obtain accurate (raw) cell frequencies and marginal counts when requesting those tables later from the *Results* dialog.

Maximum Number of Iterations

Fitting a model (i.e., marginal tables) to a multi-way frequency table is accomplished via the *iterative*

9. LOG-LINEAR ANALYSIS - DIALOGS AND OPTIONS

proportional fitting procedure. The default setting of 50 iterations will almost always ensure that this iterative procedure will converge, that is, that the respective model will be fitted to the table with sufficient accuracy.

Convergence Criterion

Fitting a model (i.e., marginal tables) to a multi-way frequency table is an iterative procedure (*iterative proportional fitting*; see above). At each iteration, the fitted marginal tables (i.e., the fitted model) will approximate more closely the observed marginal tables.

The iterative procedure will terminate when the difference between fitted and observed margins is no greater than the convergence criterion specified here. The default value (*0.01*) ensures an accuracy that is usually more than sufficient.

Reviewing Tables (Slicing Tables)

Tables can be reviewed slice by slice, as specified in the *Log-Linear Model Specification* dialog (page 3467). In the *Specify how Table is to be Reviewed* dialog, you can select the two factors (row and column variables) by which to view the table. Scrollsheets will be displayed showing the two-way tables within the levels of all remaining factors.

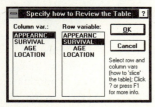

Reviewing all slices at once. By default, the program will automatically display as many slices of the table (Scrollsheets) as can be simultaneously viewed. Note that you can easily increase the default number of Scrollsheets in the queue (3 Scrollsheets at a time) via the *Scrollsheet Manager* option on the *Window* pull-down menu.

If you want to review the slices of the table one by one, set the *One by one* radio button on the *Model Specification* dialog (page 3467). The program will then prompt the user to *Continue* after each slice is displayed.

The default custom graph for all tables is the 3D histogram (see page 3467). This type of graph is very helpful in order to visualize the pattern of frequencies in the table.

Specifying and Testing Specific Models

When you click on the *Specify Model to be Tested* button, you will be able to directly enter the factors (including interactions) that you want to be included in the model.

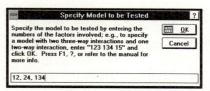

The logic of specifying models in *STATISTICA* is as follows:

- Refer to factors in the table by their respective numbers.

- Use those numbers in order to specify main effects and interactions. For example, to specify a model with no interactions but only main effects for each of three factors, enter *1 2 3* (separated by spaces or commas).

- In order to specify a model with a 2-way interaction between factors 1 and 3, and the main effect for factor 2, enter *13 2* (or *2 31*; the sequence does not matter).

- Higher-order interactions are specified analogously (e.g., *124*, *134*, *15*). After exiting

LOG - 3469

Automatic (Stepwise) Selection of Best Model

When you click on the *Automatic selection of best model* button in the *Log-Linear Model Specification* dialog, the following dialog will open.

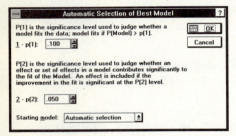

1 - p(1), 2 - p(2)

The logic of the automatic selection algorithm is described in the *Introductory Overview* section (page 3452). Briefly, the program will first fit a model with no interactions, then with all two-way interactions, three-way interactions, etc. until a model that fits at the *p*-level greater than *p(1)* has been found. You can change this *p*-level via option *1 - p(1)*. Note that this first part is skipped if an initial *user-defined* "starting model" is selected (see below). Next, the program will eliminate from this initial model in a stepwise manner all effects that do not contribute significantly to the overall fit at the *p(2)* level of significance.

Starting Model

This option allows you to skip the first step of the automatic selection procedure (see above). If you set this combo box to *User-defined*, then after clicking OK, the regular *Specify Model to be Tested* dialog will appear (page 3469). The model that you then specify will serve as the initial model from which non-significant effects [as specified by *p(2)*] will be eliminated.

This option is also useful if you want to perform statistical significance tests for all effects in a previously determined model. As described above, during the backwards elimination stage of the automatic model selection procedure, the program will drop out each effect, one by one, and test its statistical significance. You can print the results at each step by requesting *Long* output in the *Page/Output Setup* dialog.

Specifying Structural Zeros

With the *Log-Linear* module, one can analyze incomplete tables, that is, tables containing structural zeros. Structural (as opposed to *sampling*) zeros are empty cells in the table that are empty by design.

For example, suppose you tabulated the number of patients in a hospital by gender and diagnosis. If the hospital has a maternity ward, then the table will contain empty cells by design; namely, there will not be any males who are in the hospital to give birth.

Shown in this dialog are all cells in the table with zero frequencies. Select the structural zeros from that list.

Note that the correct computation of the degrees of freedom for models fit to tables with structural zeros can be extremely complex. The *Log-Linear* module will correctly compute the degrees of freedom for models involving structural zeros as long as those zero frequencies do not result in zeros in the marginal tables of the respective model.

Results

The *Results* dialog contains options to interactively view the analysis results via Scrollsheets or graphs.

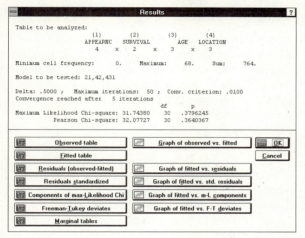

Note that the observed and marginal frequency tables that can be displayed from this dialog will include the *delta* constant, as specified in the *Model Specification* dialog. Therefore, when using the *Log-Linear* module as a crosstabulation program, make sure that *delta* is set to *0 (zero)*; otherwise you will not obtain accurate frequency counts.

The default custom graph for all results tables is the 3D histogram (see page 3467). This type of graph helps to visualize the pattern of frequencies, residuals, etc. in the table.

Observed Table

Via this option the user can review the multi-way frequency table in the most flexible manner, slice by slice, as specified by the user. If there are more than two factors in the frequency table, the program allows the user to view any two-way table within each of the levels of the remaining factors.

After clicking on this button (if the table contains more than 2 factors) the user is prompted to select the two factors by which to view the table. Scrollsheets will be displayed showing the frequencies in two-way tables, within the levels of all remaining factors.

Reviewing all slices at once. By default, *STATISTICA* will automatically display as many slices of the table as can be simultaneously viewed. Note that you can easily increase the default number of Scrollsheets in the queue (3 Scrollsheets at a time) via the *Scrollsheet Manager* option on the *Window* pull-down menu. If you want to review the slices of the table one by one, set the *One by one* radio button on the *Model Specification* dialog (page 3467). The program will then prompt the user to *Continue* after each slice is displayed.

Printing the entire observed table. If *printer*, disk *file*, or *Text/output Window* output is selected in the *Page/Output Setup* dialog, then before the first table is displayed, the user has the option to automatically print (to a disk or the printer) the entire table (without interruption).

Fitted Table

This option allows the user to review the fitted table slice by slice, as described above (*Observed table*).

Residuals (Observed - Fitted)

This option allows review of the residualized table (i.e., the table of observed minus fitted frequencies) in the same manner (slice by slice) as described above. Raw residuals ($r_{ijk..}$) are computed as:

$$r_{ijk..} = f_{ijk..} - F_{ijk..}$$

where

$F_{ijk..}$ denotes the fitted or expected cell frequency for cell *ijk..*

$f_{ijk..}$ denotes the observed frequency

Residuals Standardized

This option allows the user to review the table of standardized residuals slice by slice. The standardized residuals ($s_{ijk..}$) are computed as:

$$s_{ijk..} = (f_{ijk..} - F_{ijk..})/(F_{ijk..})^{\frac{1}{2}}$$

where

$F_{ijk..}$ denotes the fitted or expected cell frequency for cell *ijk..*

$f_{ijk..}$ denotes the observed frequency

Components of Max-Likelihood Ratio *Chi-square*

This option allows the user to review the table of components of the maximum likelihood ratio *Chi-square* slice by slice. These are the contributions of each cell to the overall maximum likelihood ratio *Chi-square* goodness of fit statistic:

$$c_{ijk..} = 2 * f_{ijk..} * \ln(f_{ijk..}/F_{ijk..})$$

where

$\ln$ denotes the natural logarithm

$F_{ijk..}$ denotes the fitted or expected cell frequency for cell *ijk..*

$f_{ijk..}$ denotes the observed frequency

Freeman-Tukey Deviates

This option allows the user to review the table of Freeman-Tukey deviates slice by slice. Freeman-Tukey deviates ($fr_{ijk..}$) represent a normalizing transformation that is appropriate when the frequencies in the table come from a Poisson distribution:

$$fr_{ijk..} = f_{ijk..}^{\frac{1}{2}} + (f_{ijk..}+1)^{\frac{1}{2}} - (4*F_{ijk..}+1)^{\frac{1}{2}}$$

where

$F_{ijk..}$ denotes the fitted or expected cell frequency for cell *ijk..*

$f_{ijk..}$ denotes the observed frequency

Marginal Tables

This option allows you to review the marginal tables, as specified by the current model, that were fit to the observed table. These tables will be displayed one by one, and slice by slice if marginal tables involving more than two factors are displayed. The number of simultaneously displayed tables (Scrollsheets) depends on the number of Scrollsheets in the queue (by default, 3 Scrollsheets). You can change the queue via the *Scrollsheet Manager* option on the *Window* pull-down menu. Note that the marginal tables will include the *delta* constant, as specified on the *Model Specification* dialog. Thus, when using the *Log-Linear* module as a crosstabulation program, make sure that *delta* is set to *0*; otherwise accurate marginal counts will not be obtained.

Plot Options

After choosing any of the plot options below, a scatterplot of the respective values will be displayed:

(1) A plot of the observed versus fitted (expected) frequencies;

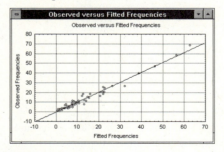

(2) A plot of the fitted frequencies versus residual frequencies (observed minus fitted);

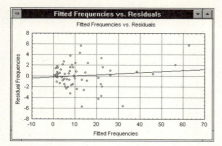

(3) A plot of fitted frequencies versus standardized residuals;

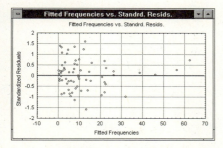

(4) A plot of fitted frequencies versus components of the maximum likelihood ratio *Chi-square*;

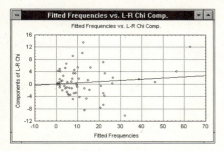

(5) A plot of the fitted frequencies versus Freeman-Tukey deviates.

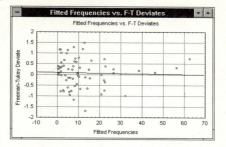

These graphs are useful for evaluating the overall fit of the model for different magnitudes of frequencies, and for detecting outliers.

9. LOG-LINEAR ANALYSIS - DIALOGS AND OPTIONS

TECHNICAL NOTES

Algorithms

The fitting of models (marginal tables) to observed frequency tables is accomplished via iterative proportional fitting (see Deming and Stephan, 1940; Brown, 1959; Ireland and Kullback, 1968; Haberman, 1972, 1974). The computations of the Pearson *Chi-square* and maximum likelihood ratio *Chi-square* statistics are described in detail by Bishop, Fienberg, and Holland (1975) and Fienberg (1978). The logic of the automatic selection algorithm is described in Goodman (1971).

INDEX

A

Associations between design variables, 3453
Automatic model fitting, 3451, 3468

C

Chi-square tests
    goodness of fit, 3451
    marginal association Chi-square, 3460
    maximum likelihood ratio Chi-square, 3451
    partial association Chi-square, 3460
    Pearson Chi-square, 3451
Crosstabulation of data, 3449

D

Dependent variables, 3449
Design variables, 3449, 3453

G

Goodness of fit tests, 3451, 3461

H

Hierarchical tests of alternative models, 3462
Histograms
    2D histograms, 3467
    3D histograms, 3467

I

Independent variables, 3449
Interaction effects, 3450
Iterative proportional fitting, 3451, 3475

K

k-factor interactions, 3452, 3459

L

Log-linear analysis
    automatic model fitting, 3451, 3468
    convergence criterion, 3469
    crosstabulation of data, 3449
    delta, 3468
    design variables, 3449, 3453
    example, 3457
    Freeman-Tukey deviates, 3472
    goal of the analysis, 3457
    goodness of fit, 3451, 3461
    hierarchical tests of alternative models, 3462
    interaction effects, 3450
    iterative proportional fitting, 3451, 3475
    k-factor interactions, 3452, 3459
    log-linear model, 3451
    Mantel-Haenszel test, 3455
    marginal associations, 3452, 3460, 3468
    marginal frequencies, 3449
    marginal tables, 3453, 3472
    maximum likelihood ratio Chi-square, 3472
    maximum number of iterations, 3468
    model fitting approach, 3450
    multi-way frequency tables, 3450
    partial associations, 3452, 3460, 3468
    residual frequencies, 3451
    response variables, 3449, 3453
    rules for specifying a model, 3461
    startup panel, 3465
    stepwise model selection, 3463, 3470
    structural zeros, 3454, 3468
    two-way frequency tables, 3449

M

Mantel-Haenszel test, 3455
Marginal associations, 3452, 3460, 3468
Marginal frequencies, 3449
Marginal tables, 3453, 3472
Multi-way frequency tables, 3450

P

Partial associations, 3452, 3460, 3468

R

Residual frequencies, 3451
Response variables, 3449, 3453

S

Scatterplots
    fitted frequencies vs. components of the maximum likelihood ratio Chi-square, 3473
    fitted frequencies vs. Freeman-Tukey deviates, 3473
    fitted frequencies vs. standardized residuals, 3473
    observed vs. expected frequencies, 3472
    residual frequencies, 3473
Stepwise model selection, 3463, 3470
Structural zeros, 3454, 3468

T

Two-way frequency tables, 3449

Chapter 10:

SURVIVAL ANALYSIS & CENSORED REGRESSION

Table of Contents

Introductory Overview 3485
Program Overview 3495
Examples 3499
    Example 1: Actuarial Life Table 3499
    Example 2: Kaplan-Meier Product-Limit Estimates 3503
    Example 3: Comparing Survival in Two or More Groups 3504
    Example 4: Regression Models 3506
    Example 5: Cox Model with Time-Dependent Covariates 3508
Dialogs, Options, Statistics 3511
    Startup Panel 3511
    Life Table Analysis (Analyzing Raw Data) 3512
    Life Tables (Analyzing Tables) 3513
    Life Tables & Distribution Fitting Results 3514
    Kaplan-Meier Product-Limit Analysis 3515
    Kaplan-Meier Product-Limit Analysis - Results 3516
    Comparing Two Groups 3517
    Comparing Survival in Two Groups - Results 3517
    Comparing Multiple Groups 3520
    Comparing Multiple Groups - Results 3520
    Regression Models 3522
    Cox Proportional Hazard Model with Time-Dependent Covariates 3523
    Setting Parameters for Estimation 3525
    Regression Models for Survival Data - Results 3526
Technical Notes 3531
Index 3537

Detailed
Table of Contents

INTRODUCTORY OVERVIEW **3485**
    General Purpose 3485
    Life Table Analysis 3485
    Distribution Fitting 3486

10. SURVIVAL ANALYSIS - CONTENTS

Kaplan-Meier Product-Limit Estimator	3487
Comparing Samples	3488
Regression Models	3489
Cox Proportional Hazard Model	3490
Cox Proportional Hazard Model with Time-Dependent Covariates	3490
Exponential Regression	3493
Normal and Log-Normal Regression	3493
Stratified Analyses	3493
References	3494

PROGRAM OVERVIEW 3495

Overview	3495
Dates and Survival Times	3495
Actuarial Life Tables	3495
Survival Distributions	3495
Kaplan-Meier Product-Limit Method	3495
Comparison of Survival in Two or More Samples	3496
Regression Models and Estimation	3496
Alternative Procedures	3496

EXAMPLES 3499

Overview and Data File	3499
Example 1: Actuarial Life Table	3499
Specifying the Analysis	3499
Reading an Aggregated Life Table	3500
Reviewing Results	3501
Fitting a Theoretical Survival Distribution	3501
Example 2: Kaplan-Meier Product-Limit Estimates	3503
Specifying the Analysis	3503
Reviewing Results	3503
Example 3: Comparing Survival in Two or More Groups	3504
Specifying the Analysis	3504
Comparing Survival in Two Groups	3505
Example 4: Regression Models	3506
Specifying the Analysis	3506
Estimating the Parameters	3506
Results	3507
Example 5: Cox Model with Time-Dependent Covariates	3508
Overview	3508
Specifying the Analysis	3508
Specifying Time-Dependent Covariates	3509
Reviewing Results	3509

DIALOGS, OPTIONS, STATISTICS 3511

Startup Panel	3511

10. SURVIVAL ANALYSIS - CONTENTS

Life Tables and Distributions 3511
Kaplan-Meier Product-Limit Method 3511
Comparing Two Samples 3511
Comparing Multiple Samples 3511
Regression Models 3512
Life Table Analysis (Analyzing Raw Data) 3512
Input Data 3512
Variables 3512
Codes for Complete and Censored Responses 3513
Compute Table Based on 3513
Correct Intervals Containing No Terminations/Deaths 3513
Life Tables (Analyzing Tables) 3513
Input Data 3513
Variables 3513
Correct Intervals Containing No Terminations/Deaths 3514
Life Tables & Distribution Fitting Results 3514
Life Table 3514
Model 3514
Parameter Estimates 3515
Estimates of Hazard, Survival, and Probability Density Function 3515
Graph of Hazard, Survival, and Probability Density Function 3515
Kaplan-Meier Product-Limit Analysis 3515
Variables 3515
Codes for Complete and Censored Responses 3516
Kaplan-Meier Product-Limit Analysis - Results 3516
Product-Limit Survival Analysis 3516
Percentiles of the Survival Function 3516
Graphs of Survival Functions 3516
Comparing Two Groups 3517
Variables 3517
Codes for Complete and Censored Responses 3517
Codes for First and Second Group 3517
Comparing Survival in Two Groups - Results 3517
Gehan's Wilcoxon Test 3518
Cox's *F*-test 3518
Cox-Mantel Test 3518
Log-Rank Test 3518
Peto and Peto's Wilcoxon Test 3519
Display Proportion Surviving for Each Group 3519
Graph of Proportion Surviving for Each Group 3519
Cumulative Proportion Surviving by Group (Kaplan-Meier) 3519
Comparing Multiple Groups 3520
Variables 3520
Codes for Complete and Censored Responses 3520

10. SURVIVAL ANALYSIS - CONTENTS

Codes (for Groups) 3520
Comparing Multiple Groups - Results 3520
    Survival Times and Scores 3520
    Scores for Each Group 3521
    Percent Surviving in Each Group 3521
    Descriptive Statistics 3521
    Plot Histogram of the Sum of Scores for Each Group 3521
    Plot of Proportion Surviving in Each Group 3521
    Cumulative Proportion Surviving by Group (Kaplan-Meier) 3522
Regression Models 3522
    Model 3522
    Variables 3522
    Codes for Complete and Censored Responses 3523
    Codes (for Groups) 3523
    Equal coefficients, different baseline h0 3523
Cox Proportional Hazard Model with Time-Dependent Covariates 3523
    Model 3523
    Variables (Survival Times, Censoring, [Optional] Grouping) 3524
    Codes for Complete and Censored Responses 3524
    Independent Variables (Covariates) 3524
    Codes (for Groups) 3525
Setting Parameters for Estimation 3525
    Maximum Numbers of Iterations 3525
    Convergence Criterion 3526
    Start Values 3526
    Missing Data Deletion 3526
Regression Models for Survival Data - Results 3526
    Parameter Estimates 3526
    Parameter Variance/Covariance Matrix 3527
    Stratified Analysis 3527
    Means and Standard Deviations 3527
    Graphs 3527
    Cox Proportional Hazard Model 3527
    Exponential Regression Model 3528
    Normal and Log-Normal Regression Models 3529

TECHNICAL NOTES 3531
    Life Table 3531
    Fitting Theoretical Survival Distributions 3531
    Two Sample Tests 3532
    Mantel and Haenszel Test 3532
    Cox Proportional Hazard Model 3532
    Estimating the Parameters for Time-Dependent Covariates 3533
    Syntax Rules for Specifying (Time- Dependent) Covariates 3533

Wald Statistic	3535
Exponential Regression	3535
Normal and Log-normal Regression	3535
INDEX	**3537**

10. SURVIVAL ANALYSIS - CONTENTS

Chapter 10:
SURVIVAL ANALYSIS & CENSORED REGRESSION

INTRODUCTORY OVERVIEW

General Purpose

The techniques available in this module were primarily developed in the medical and biological sciences, but they are also widely used in the social and economic sciences, as well as in engineering (reliability and failure time analysis).

Imagine that you are a researcher in a hospital who is studying the effectiveness of a new treatment for a generally terminal disease. The major variable of interest is the number of days that the respective patients survive. In principle, one could use the standard parametric and nonparametric statistics for describing the average survival, and for comparing the new treatment with traditional methods (see *Basic Statistics and Tables*, Volume I, and *Nonparametrics and Distributions*, Volume I). However, at the end of the study there will be patients who survived over the entire study period, in particular among those patients who entered the hospital (and the research project) late in the study; there will be other patients with whom contact has been lost. Surely, one would not want to exclude all of those patients from the study by declaring them to be missing data (since most of them are "survivors" and, therefore, they reflect on the success of the new treatment method). Those observations which contain only partial information are called *censored* observations (e.g., "patient *A* survived at least 4 months before he moved away and contact was lost").

Censored observations. In general, censored observations arise whenever the dependent variable of interest represents the time to a terminal event, and the duration of the study is limited in time. Censored observations may occur in a number of different areas of research. For example, in the social sciences you may study the "survival" of marriages, high school drop-out rates (time to drop-out), turnover in organizations, etc. In each case, by the end of the study period, some subjects will still be married, will not have dropped out, or are still working at the same company; thus, those subjects represent censored observations.

In economics you may study the "survival" of new businesses or the "survival" times of products such as automobiles. In quality control research, it is common practice to study the "survival" of parts under stress (failure time analysis).

Analytic techniques. Essentially, the methods offered in this module address the same research questions as many of the other *STATISTICA* procedures; however, all methods in the *Survival Analysis* module will handle censored data. The *life table* (see below), *survival distribution fitting* (page 3486), and *Kaplan-Meier* (page 3487) survival function estimation are all descriptive methods for estimating the distribution of survival times from a sample. Several techniques are included for comparing the survival in two or more groups (page 3489). Finally, the *Survival Analysis* module includes several regression models for estimating the relationship of (multiple) continuous variables to survival times (page 3489).

Life Table Analysis

The most straightforward way to describe the survival in a sample is to compute the *Life Table*. The life table technique is one of the oldest methods for analyzing survival (failure time) data (e.g., see Berkson and Gage, 1950; Cutler and Ederer, 1958;

Gehan, 1969). This table can be thought of as an "enhanced" frequency distribution table. Shown below is an example (of a partial table).

SURVIVAL ANALYSIS Interval	Number Entering	Number Withdrwn	Number Dying	Proportn Dead	Proportn Surviving	Cum.Prop Surviving	Probity Density
Intno.1	65	14	19	.327586	.672414	1.000000	.002030
Intno.2	32	4	4	.133333	.866667	.672414	.000556
Intno.3	24	4	0	.022727	.977273	.582759	.000082
Intno.4	20	4	1	.055556	.944444	.569514	.000196
Intno.5	15	1	1	.068966	.931035	.537874	.000230
Intno.6	13	3	1	.086957	.913043	.500780	.000270
Intno.7	9	1	2	.235294	.764706	.457234	.000667
Intno.8	6	1	0	.090909	.909091	.349649	.000197
Intno.9	5	1	1	.222222	.777778	.317863	.000438
Intno.10	3	2	0	.250000	.750000	.247227	.000383
Intno.11	1	0	0	.500000	.500000	.185420	.000575
Intno.12	1	1	0	1.000000	0.000000	.092710	

The distribution of survival times is divided into a certain number of intervals. For each interval, you can then compute the number and proportion of cases or objects that entered the respective interval "alive," the number and proportion of cases that failed in the respective interval (i.e., number of terminal events, or number of cases that "died"), and the number of cases that were lost or censored in the respective interval.

Based on those numbers and proportions, several additional statistics can be computed:

Number of cases at risk. This is the number of cases that entered the respective interval alive, minus half of the number of cases lost or censored in the respective interval.

Proportion failing. This proportion is computed as the ratio of the number of cases failing in the respective interval divided by the number of cases at risk in the interval.

Proportion surviving. This proportion is computed as 1 minus the proportion failing.

Cumulative proportion surviving (survival function). This is the cumulative proportion of cases surviving up to the respective interval. Since the probabilities of survival are assumed to be independent across the intervals, this probability is computed by multiplying out the probabilities of survival across all previous intervals. The resulting function is also called the *survivorship* or *survival function*.

Probability density. This is the estimated probability of failure in the respective interval, computed per unit of time, that is:

$$F_i = (S_i - q_i)/h_i$$

In this formula, F_i is the respective probability density in the i'th interval, S_i is the estimated cumulative proportion surviving at the end of the i_t interval, q_i is the proportion failing in the respective interval, and h_i is the width of the respective interval.

Hazard rate. The hazard rate is defined as the probability per time unit that a case that has survived to the beginning of the respective interval will fail in that interval. Specifically, it is computed as the number of failures per time units in the respective interval divided by the average number of surviving cases at the mid-point of the interval.

Median survival time. This is the survival time at which the cumulative survival function is equal to *0.5*. Other percentiles (25th and 75th percentile) of the cumulative survival function can be computed accordingly.

Required sample sizes. In order to arrive at reliable estimates of the three major functions (survival function, probability density, and hazard) and their standard errors at each time interval, the minimum recommended sample size is 30.

Distribution Fitting

In summary, the life table gives one a good indication of the distribution of failures over time. However, for predictive purposes it is often desirable to understand the shape of the underlying survival function in the population. The major distributions that have been proposed for modeling survival or failure times are the exponential (and linear exponential) distribution, the Weibull distribution of extreme events, and the Gompertz distribution. The *Survival Analysis* module will fit all of these theoretical distributions to the observed life table.

Estimation. The parameter estimation procedure (for estimating the parameters of the theoretical survival functions) used in *STATISTICA* is a least squares linear regression algorithm (see Gehan and Siddiqui, 1973). A linear regression algorithm can be used because all four theoretical distributions can be "made linear" by appropriate transformations (see the *Technical Notes* section, page 3531). Such transformations sometimes produce different variances for the residuals at different times, leading to biased estimates. Therefore, the fitting algorithm in the *Survival Analysis* module also computes two types of weighted least squares estimates.

Goodness of fit. Given the parameters for the different distribution functions and the respective model, the likelihood of the data can be computed. One can also compute the likelihood of the data under the null model, that is, a model that allows for different hazard rates in each interval. Without going into detail, these two likelihoods can be compared via an incremental *Chi-square* test statistic. If this *Chi-square* is statistically significant, then it can be concluded that the respective theoretical distribution fits the data significantly worse than the null model; that is, you reject the respective distribution as a model for your data.

Plots. *Survival Analysis* will produce plots of the survival function, hazard, and probability density for the observed data and the respective theoretical distributions. These plots provide a quick visual check of the goodness of fit of the theoretical distribution. The example plot below shows an observed survival function and the fitted Weibull distribution.

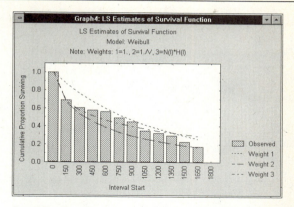

Specifically, the three lines in this plot denote the theoretical distributions that resulted from three different estimation procedures (least squares and two methods of weighted least squares).

Kaplan-Meier Product-Limit Estimator

Rather than classifying the observed survival times into a life table, you can estimate the survival function directly from the continuous survival or failure times. Intuitively, imagine that you create a life table so that each time interval contains exactly one case. Multiplying out the survival probabilities across the "intervals" (i.e., for each single observation) you would get for the survival function:

$$S(t) = \Pi[(n-j)/(n-j+1)^{\delta(j)}]$$

In this equation, $S(t)$ is the estimated survival function, n is the total number of cases, and Π denotes the multiplication (geometric sum) across all cases less than or equal to t; $\delta(j)$ is a constant that is either *1* if the *j*'th case is uncensored (complete), or *0* if it is censored. This estimate of the survival function is also called the *product-limit estimator*, and was first proposed by Kaplan and Meier (1958). Shown below is the Kaplan-Meier estimate of the survival function for the same data shown in the life table above (only a part of the Scrollsheet is shown below).

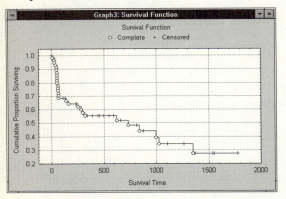

The plot of this function is shown below.

Note that censored and uncensored observations are identified by different point markers in this plot. The advantage of the Kaplan-Meier Product-Limit method over the life table method for analyzing survival and failure time data is that the resulting estimates do not depend on the grouping of the data (into a certain number of time intervals). Actually, the Product-Limit method and the life table method are identical if the intervals of the life table contain at most one observation.

Comparing Samples

One can compare the survival or failure times in two or more samples. In principle, because survival times are not normally distributed, nonparametric tests that are based on the *rank ordering* of survival times should be applied. The *Nonparametrics and Distributions* module (Volume I) offers a wide range of nonparametric tests that can be used in order to compare survival times; however, the tests in that module cannot "handle" censored observations.

Available tests. The *Survival Analysis* module contains five different (mostly nonparametric) tests for censored data: Gehan's generalized Wilcoxon test, the Cox-Mantel test, the Cox *F* test, the log-rank test, and Peto and Peto's generalized Wilcoxon test. A nonparametric test for the comparison of multiple groups is also available. Most of these tests are accompanied by appropriate z-values (values of the standard normal distribution); these z-values can be used to test for the statistical significance of any differences between groups. However, note that most of these tests will only yield reliable results with fairly large samples sizes; the small sample "behavior" is less well understood.

Plot of survival function. The *Survival Analysis* module allows you to plot the Kaplan-Meier estimates for the survival function, broken down by groups.

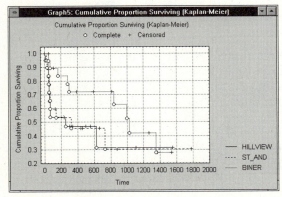

This plot facilitates the identification and interpretation of differences between groups.

Mantel and Haenszel test. Your *STATISTICA* program contains an example *STATISTICA BASIC* program (file *Manthaen.stb*) for computing the Mantel-Haenszel test for comparing two groups (see Mantel and Haenszel, 1959; see also Lee, 1992). This test has been used in many clinical and epidemiological studies, in order to control for the effects of confounding variables. The test is based on the analysis of 2 x 2 tables (e.g., *Group 1/2* vs. *Survival*), stratified by another categorical variable (the confounding variable; e.g., *Location*). It allows you to test whether the two variables in the 2 x 2 tables (e.g., *Group* and *Survival*) are associated with each other, after controlling for the stratification variable (e.g., *Location*).

Choosing a two-sample test. There are no widely accepted guidelines concerning which test to use in a particular situation. Cox's *F* test tends to be more powerful than Gehan's generalized Wilcoxon test when:

(1) Sample sizes are small (i.e., *n* per group less than 50);

(2) If samples are from an exponential or Weibull distribution;

(3) If there are no censored observations (see Gehan and Thomas, 1969).

Lee, Desu, and Gehan (1975) compared Gehan's test to several alternatives and showed that the Cox-Mantel test and the log-rank test are more powerful (regardless of censoring) when the samples are drawn from a population that follows an exponential or Weibull distribution; under those conditions there is little difference between the Cox-Mantel test and the log-rank test. Lee (1980) discusses the power of different tests in greater detail.

Multiple sample test. The multiple-sample test implemented in the *Survival Analysis* module is an extension (or generalization) of Gehan's generalized Wilcoxon test, Peto and Peto's generalized Wilcoxon test, and the log-rank test. First, a score is assigned to each survival time using Mantel's procedure (Mantel, 1967); next a *Chi-square* value is computed based on the sums (for each group) of this score. If only two groups are specified, then this test is equivalent to Gehan's generalized Wilcoxon test, and the computations will default to that test in this case.

Unequal proportions of censored data. When comparing two or more groups it is very important to examine the number of censored observations in each group. Particularly in medical research, censoring can be the result of, for example, the application of different treatments: Patients who get better faster or get worse as the result of a treatment may be more likely to drop out of the study, resulting in different numbers of censored observations in each group. Such systematic censoring may greatly bias the results of comparisons.

Regression Models

The *Survival Analysis* module contains several specialized multiple regression techniques for survival/failure time data.

Purpose. A common research question in medical, biological, or engineering (failure time) research is to determine whether or not certain continuous (independent) variables are correlated with the survival or failure times. There are two major reasons why this research issue cannot be addressed via straightforward multiple regression techniques (as available in the *Multiple Regression* module, Volume I). First, survival times are usually not a simple linear function of the respective predictor variables, and thus the analysis via *Multiple Regression* may produce very misleading results (e.g., not identify strong predictors which have a non-linear relationship to the survival times).

Second, there is the problem of censoring; that is, some observations will be incomplete (as described earlier, page 3485).

Methods. *Survival Analysis* includes the five common regression models for censored survival-/failure time data:

(1) Cox's proportional hazard model (Cox, 1972)

(2) Cox's proportional hazard model with time-dependent covariates;

(3) Exponential regression model (Prentice, 1973; Feigl and Zelen, 1965; Glasser, 1967; and Zippin and Armitage, 1966);

(4) Normal linear regression model (Wolynetz, 1979a, 1979b);

(5) Log-normal linear regression model [which is a modification of (4)].

For each model, the program will compute the maximum likelihood parameter estimates and evaluate the overall goodness of fit. Each of these models will be discussed below.

Cox Proportional Hazard Model

The proportional hazard model is the most general of the regression models because it is not based on any assumptions concerning the nature or shape of the underlying survival distribution. The model assumes that the underlying hazard *rate* (rather than survival time) is a function of the independent variables (covariates); no assumptions are made about the nature or shape of the hazard function. Thus, in a sense, Cox's regression model may be considered to be a nonparametric method. The model may be written as:

```
h[(t),(z_1, z_2, ..., z_m)] =
    h_0(t)*exp(b_1*z_1 +...+ b_m*z_m)
```

where $h(t,...)$ denotes the resultant hazard, given the values of the m covariates for the respective case (z_1, z_2, ..., z_m) and the respective survival time (t). The term $h_0(t)$ is called the *baseline hazard*; it is the hazard for the respective individual when all independent variable values are equal to zero. You can linearize this model by dividing both sides of the equation by $h_0(t)$ and then taking the natural logarithm of both sides:

```
log{h[(t),(z...)]/h_0(t)} = b_1*z_1 +...+ b_m*z_m
```

Now you have a fairly "simple" linear model that can be readily estimated.

Assumptions. While no assumptions are made about the shape of the underlying hazard function, the model equations shown above do imply two assumptions. First, they specify a multiplicative relationship between the underlying hazard function and the log-linear function of the covariates. This assumption is also called the *proportionality assumption*. In practical terms, it is assumed that, given two observations with different values for the independent variables, the ratio of the hazard functions for those two observations does not depend on time (to relax this assumption, use *time-dependent* covariates; see below). The second assumption of course is that there is a log-linear relationship between the independent variables and the underlying hazard function.

Cox Proportional Hazard Model with Time-Dependent Covariates

The validity of the proportionality assumption may often be questionable. For example, consider a hypothetical study where the covariate consists of a categorical (grouping) variable, namely, whether or not a patient underwent surgery for a particular ailment. Patient 1 in the study had the surgery while patient 2 did not. According to the proportionality assumption, the hazard for those two patients should be constant over time, and only depend on the covariate; that is, in this example, on whether or not the patient underwent surgery. In other words, depending on whether or not a patient underwent surgery, his or her probability of dying at any given time after surgery is higher or lower by a fixed proportion relative to a patient who did not undergo surgery.

Usually, a more realistic possibility is that shortly after surgery, the risk to the patient is higher; however, over time, due to the benefits of the surgery, the risk of dying decreases and will eventually become lower as compared to that of a patient who did not undergo surgery. Thus, the proportionality assumption is most likely violated; rather, the effect of the covariate on the hazard is dependent on time.

There are many other relatively obvious examples that one might think of where the proportionality assumption is likely violated. For example, age is often included in studies of physical health. Suppose you studied survival after surgery. It is likely that age is a more important predictor of risk immediately after surgery, than at some time after the surgery (after initial recovery). In accelerated life testing one sometimes uses a stress covariate (e.g., amount of voltage) that is slowly increased over time until failure occurs (e.g., until the electrical insulation fails; see Lawless, 1982, page 393). Again, in this case the impact of the covariate is clearly dependent on time.

Stratification. In the case of categorical covariates, such as whether or not a patient has had surgery, Kalbfleish and Prentice (1980) recommend to perform a stratified analysis. The *Survival Analysis* module allows you to fit the proportional hazard model to the data, separately for each group in a stratified analysis. In that manner, one can explicitly allow the hazard function to be different in each group.

Testing the Proportionality Assumption. As indicated by the previous examples, there are many applications where it is likely that the proportionality assumption does not hold. In that case, one can explicitly define covariates as functions of time. The *Example* section (page 3508) will briefly discuss the analysis of a data set presented by Pike (1966) which consists of survival times for two groups of rats that had been exposed to a carcinogen (see also Lawless, 1982, page 393, for a similar example). Following the notation used earlier, suppose that *z* is a grouping variable with codes *1* and *0* to denote whether or not the respective rat was exposed. One could then fit the proportional hazard model:

$$h(t,z) = h_0(t)*\exp\{b_1*z + b_2*[z*\log(t) - 5.4]\}$$

Thus, in this model the conditional hazard at time *t* is a function of (1) the baseline hazard h_0, (2) the covariate *z*, and (3) of *z* times the logarithm of time. Note that the constant *5.4* is used here for scaling purposes only: the mean of the logarithm of the survival times in this data set is equal to *5.4*. In other words, the conditional hazard at each point in time is a function of the covariate and time; thus, the effect of the covariate on survival is dependent on time; hence the name *time-dependent covariate*. Shown below are the results of fitting this model computed by *STATISTICA*.

SURVIVAL ANALYSIS	Dependent Variable: SURVIVAL (pike.sta) Censoring var.: CENSORED Chi² = 2.89413 df = 2 p = .23527					
N=40 Exp.Name	Beta	Standard Error	t-value	exponent beta	Wald Statist.	p
GROUP	-.599723	.348375	-1.72149	.548964	2.963528	.085172
TIMDEP	-.229527	1.824892	-.12578	.794909	.015820	.899910

This model allows one to specifically test the proportionality assumption. If parameter b_2 (labeled *Timdep* in the Scrollsheet above) is statistically significant (e.g., if it is at least twice as large as its standard error), then one can conclude that, indeed, the effect of the covariate *z* on survival is dependent on time, and, therefore, that the proportionality assumption does not hold. In the Scrollsheet shown above, the parameter for the time-dependent covariate *Timdep* is estimated as *-0.23* with a standard error of *1.82*. Thus, in this case, there is no indication that the simpler model with the single fixed covariate is inappropriate.

Specifying Time-Dependent Covariates. The *Survival Analysis* module allows the user to type in arithmetic expressions to define the time-dependent covariates. These expressions may contain all standard arithmetic and logical operators and functions, and thus a wide variety of models can be specified. Refer to the description of the *Cox Proportional Hazard Model with Time Dependent Covariates* dialog (page 3523) for additional details.

Estimating Parameters for Time-Dependent Covariates. The *Technical Notes* section (page 3533) contains a description of the parameter estimation procedure for Cox proportional hazard regression models with and without time-dependent covariates. In general, the partial likelihood (see Breslow, 1974) for these types of models is slightly modified, to reflect the respective transformation of the covariates. As usual, the partial likelihood for a given set of parameters is the geometric sum of the likelihood across cases. For time-dependent covariates, to compute the partial likelihood for a particular case, the program must process all cases with survival times as long or longer. Thus, when the data set is large, these computations may require noticeably more time than those necessary to estimate models with fixed covariates only.

Categorical Variables and Coding. The arithmetic expressions that define the covariates do not have to include references to survival time. Instead, you may specify some functions of two or more other covariates. This may be, for example, a convenient method for evaluating models for data collected in multi-factor experiments. For each factor, one can create a variable in the data file to define the desired contrasts. The logic and selection of *a priori* contrast coefficients is explained in detail in Volume I, *General ANOVA/MANOVA*. When specifying the covariates for the proportional hazard regression model, one can then type in the respective multiplications to define the interaction terms.

For example, suppose Factor A has two levels. All individuals assigned to the first level of the factor were given a *-1* in the respective variable in the data file (variable *A*), all individuals assigned to the second level of that factor where given a *+1*. A second Factor B, also with two levels, was coded in the same manner (variable *B*). One could now specify as the covariates variables *A* and *B*, and the expression *A * B* as a third covariate to test for the interaction between the two factors in the experiment.

However, remember that the estimation of parameters for time-dependent covariates (i.e., using expressions) can be much slower (for large files) than using simple fixed covariates; thus, it is sometimes quicker to define the appropriately coded covariates using the *STATISTICA BASIC* language, and then to treat them as fixed covariates.

Segmented Time-Dependent Covariates. When specifying the arithmetic expressions for the time-dependent covariates, you may follow the same syntax as that used for entering spreadsheet formulas for transforming individual variables in the file (see page 3533 for an overview). For the most updated complete reference of all available operators and functions, use the *Electronic Manual* (i.e., press F1 or click on the button).

In some cases one may hypothesize that the effect of one or more covariates on the hazard is a non-continuous function of time. For example, the hazard for a patient after surgery may depend on age during the first two days after the operation, and thereafter on some other factors. In that case, you can use the same logical operators that are also supported in spreadsheet formulas. For example, you could specify a time dependent covariate as:

```
Age * (T_ <= 2)
```

Note that the logical expression $T\_ <= 2$ will evaluate to *0* (false) if the survival time for an individual is greater than *2*, and to *1* (true) otherwise. Thus, the parameter estimated for this time-dependent covariate pertains to the effect of age during the first two days only.

Exponential Regression

Basically, this model assumes that the survival time distribution is exponential and contingent on the values of a set of independent variables (z_i). The rate parameter of the exponential distribution can then be expressed as:

$$S(z) = \exp(a + b_1 * z_1 + b_2 * z_2 + \ldots + b_m * z_m)$$

$S(z)$ denotes the survival times, a is a constant, and the b_i's are the regression parameters.

Goodness of fit. The *Chi-square* goodness of fit value is computed as a function of the log-likelihood for the model with all parameter estimates (L_1) and the log-likelihood of the model in which the coefficients for all covariates are forced to 0 (zero; L_0). If this *Chi-square* value is significant, reject the null hypothesis and assume that the independent variables are significantly related to survival times.

Standard exponential order statistic. One way to check the exponentiality assumption of this model is to plot the residual survival times against the standard exponential order statistic *alpha*.

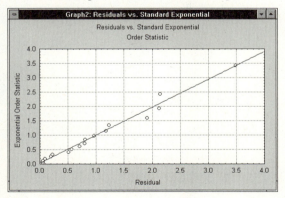

If the exponentiality assumption is met, then all points in this plot will be arranged roughly in a straight line.

Normal and Log-Normal Regression

In this model, it is assumed that the survival times (or log survival times) come from a normal distribution; the resulting model is basically identical to the ordinary multiple regression model, and may be stated as:

$$t = a + b_1 * z_1 + b_2 * z_2 + \ldots + b_m * z_m$$

where t denotes the survival times. If log-normal regression is requested, t is replaced by its natural logarithm. The normal regression model is particularly useful because many data sets can be transformed to yield approximations of the normal distribution. Thus, in a sense this is the most general fully parametric model (as opposed to Cox's proportional hazard model which is non-parametric), and estimates can be obtained for a variety of different underlying survival distributions.

Goodness of fit. The *Chi-square* value is computed as a function of the log-likelihood for the model with all independent variables (L_1), and the log-likelihood of the model in which the coefficients for all independent variables are forced to 0 (zero; L_0).

Stratified Analyses

The purpose of a stratified analysis is to test the hypothesis whether identical regression models are appropriate for different groups, that is, whether the relationships between the independent variables and survival are identical in different groups. If a stratified analysis is requested, the *Survival Analysis* module will first fit the respective regression model separately within each group. The sum of the log-likelihoods from these analyses represents the log-likelihood of the model with different regression coefficients (and intercepts where appropriate) in different groups. The program will then fit the requested regression model to all data in the usual manner (i.e., ignoring group membership) and compute the log-likelihood for the overall fit. The difference between the log-likelihoods is then tested for statistical significance (via the *Chi-square* statistic).

Separate baseline hazard, equal coefficients. For Cox proportional hazard models, the program provides the option to fit to the stratified data a model with common coefficients for the different groups (strata), but with different baseline hazards. In effect, observations in the same group (stratum) are assumed to have proportional hazard functions, but this is not necessarily the case for observations in different groups. For more information about stratification of proportional hazard models, see Lawless (1982, page 365).

References

Thorough introductions to survival analysis can be found in, for example, Bain (1978), Barlow and Proschan (1975), Cox and Oakes (1984), Elandt-Johnson and Johnson (1980), Gross and Clark (1975), Lawless (1982), Lee (1980, 1992), Miller (1981), and Nelson (1982). Engineering applications of these techniques are discussed in Hahn and Shaprio (1967), Haviland (1964), Henley and Kumamoto (1980), Kivenson (1971), Lipson and Sheth (1973), Lloyd and Lipow (1977), Sandler (1963), and Smith (1972).

PROGRAM OVERVIEW

Overview

This module is designed for the analysis of censored survival or failure time data. It contains procedures for describing survival times and for estimating the survival, hazard, and probability density functions, for fitting theoretical survival distributions to the data, and for comparing survival in two or more samples. The *Survival Analysis* module also contains regression procedures for fitting explanatory models to data sets with censored observations (Cox's proportional hazard model, Cox's proportional hazard model with time-dependent covariates, exponential regression, and normal and log-normal regression).

Dates and Survival Times

All analyses in the *Survival Analysis* module will automatically convert dates into numbers of days. Thus, rather than specifying actual survival times, the user can record the date when the observation began and the date when it was terminated, either due to failure or censoring.

Actuarial Life Tables

The *Survival analysis* module allows the user to tabulate the survival data into a user-specified number of intervals. In addition to various descriptive statistics, for each interval, the program will then compute

(1) The proportion of observations failing (dying),

(2) The proportion of observations surviving,

(3) The cumulative proportion surviving up to the respective interval and the corresponding standard error,

(4) The survival probability density and its standard error,

(5) The hazard rate and its standard error,

(6) The median life expectancy and its standard error.

Data files. Life tables can be computed from raw data files as well as aggregated (pre-tabulated) data.

Survival Distributions

The program will fit the four common theoretical survival functions to the data: exponential, linear exponential, Weibull, and Gompertz. Least squares and two types of weighted least squares parameter estimates are computed for each distribution.

The goodness of fit of each model can be evaluated via incremental *Chi-square* tests based on the log-likelihood for the fitted distribution and the null model (i.e., the log-likelihood of the data).

Plots. The observed and fitted distributions can be plotted; specifically, the user can plot the observed and fitted hazard function, survival function, and probability density function, based on least squares or weighted least squares parameter estimates.

Kaplan-Meier Product-Limit Method

The *Survival Analysis* module will also estimate the survival function based on the Kaplan-Meier product-limit method. The results will include the estimates of the survival function and their standard errors, and estimates of the quartiles (25th, median, 75th) of the survival function.

Plots. The user can plot the survival and log-survival function against the survival times as well as the log-survival times.

Comparison of Survival in Two or More Samples

The program includes five tests for comparing the survival in two groups: Gehan's generalized Wilcoxon test, the Cox-Mantel test, the Cox *F* test, the log-rank test, and Peto and Peto's generalized Wilcoxon test. A nonparametric test for the comparison of multiple groups is also available. Note that *STATISTICA* also includes a *STATISTICA BASIC* program (*Manthaen.stb*) for computing the Mantel-Haenszel test for comparing two groups, controlling for the effects of a confounding variable.

Plots. To summarize the comparisons between groups, the user can produce plots of the survival functions by group and other graphs.

Regression Models and Estimation

Models. Available models are the Cox proportional hazard model, Cox proportional hazard model with time-dependent covariates, the exponential regression model, and the normal and log-normal linear regression models. Models can be fit separately to different groups (*stratified analysis*); thus, differences between groups can be examined.

Estimation. All regression models will produce maximum likelihood parameter estimates. The proportional hazard model and the exponential regression model are estimated via unconstrained Newton-Raphson iterations; the maximum likelihood parameters for the normal and log-normal regression model are estimated via an expectation maximization (*EM*) algorithm. This algorithm was first proposed by Dempster, Laird, and Rubin (1977) and is discussed in Cox and Oakes (1984).

Output. The overall significance of the regression model will be evaluated via a *Chi-square* test computed from the log-likelihoods of the respective null model and the fitted model. The asymptotic standard errors for the parameter estimates are also computed. In addition, the user can review the covariance matrix of the parameter estimates.

Plots. Various plots are available for assessing the adequacy of the model fit. For proportional hazard models, the user can plot the survival functions for various values of the independent variables. For the exponential regression model, the user can plot the residuals against the exponential order statistic (Lawless, 1982), against the predicted survival times, and the log-transformed observed survival times. For the normal and log-normal linear regression model, the user can produce scatterplots of the observed versus predicted survival times, the predicted versus the residual survival times, and the normal probability plot of residuals.

Alternative Procedures

The *Survival Analysis* module contains four common regression models for estimating the relationship between continuous independent variables and survival in two or more groups. It is not an uncommon procedure to dummy-code group membership in a variable, and then to use that variable in a regression analysis. If the survival-/failure time data are not censored, most of the nonparametric correlation statistics in the *Nonparametrics and Distributions* module (Volume I) are applicable (to cases with single independent variables). For non-censored data, you can also use the *Nonlinear Estimation* module (Chapter 1) to fit any kind of regression model to the data (including probit, logit, and exponential models). If survival or failure is treated as a binary variable, the logit or probit regression models in the *Nonlinear Estimation* module may be applicable for investigating continuous explanatory variables. Note that any type of linear or nonlinear regression model can be estimated via that program. Another common method for comparing survival in different groups is via frequency tables. If the survival or failure times can be categorized into two or more intervals, the general log-linear model in the *Log-*

Linear module (Chapter 9) can be used to evaluate the relationship of different categorical independent (grouping) variables to that measure.

EXAMPLES

Overview and Data File

The following example is based on a data set reported by Crowley and Hu (1977) pertaining to the survival of heart transplant patients. The data are included on your *STATISTICA* disks in the example data file *Heart.sta*.

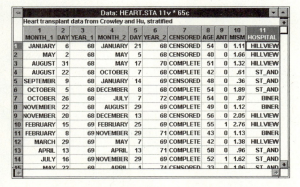

The first six variables in this data set are dates, that is, the date of the heart transplant (in the order month-day-year) and the date when the respective patient either died or dropped out of the study (could no longer be contacted). Variable *Censored* is the censoring indicator variable with the codes that identify whether a respective time represents an observation that is completely specified or a censored observation (*0-Complete*; *1-Censored*). The variable *Hospital* is a (fictitious) grouping variable which identifies to which one of three different hospitals a respective case belongs.

Instead of recording the dates in six variables (month, day, and year for the beginning of the respective observation and month, day, and year representing the termination of the observation), you can enter the survival time in one variable or enter the beginning and ending dates in two separate variables (see below; see also page 3512).

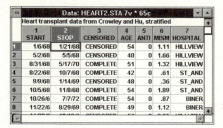

Please refer to Volume I (or the *Electronic Manual*) for information on date and time representation in *STATISTICA*.

Example 1: Actuarial Life Table

In this example, you will compute a life (survival) table for these data, estimate the survival, probability, and hazard functions for different time intervals, and see which theoretical distribution best fits the survival function.

Specifying the Analysis

After starting the *Survival Analysis* module, open the data file *Heart.sta*.

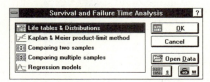

Next, click on the *Life Tables & Distributions* option in the *Survival and Failure Time Analysis Startup Panel* in order to open the *Life Table & Distribution of Survival Times* dialog.

10. SURVIVAL ANALYSIS - EXAMPLES

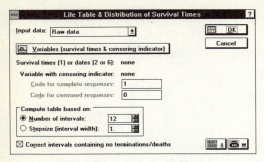

The *Survival Analysis* module will automatically "understand" dates as well as any other measurements of survival times. If you click on the *Variables* button and select 6 variables, then the program will interpret the first three variables as the month, day, and year, respectively, marking the beginning of the respective observation, and the subsequent three variables as the month, day, and year, marking the termination of the observation (due to failure or censoring). The program will automatically add 1900 to the years in the dates if those values are less than 100 (as they were entered in the example data file *Heart.sta* shown above).

Now, click on the *Variables* button and select the first 6 variables in the first list. As explained above, the program will interpret the first and fourth variable in the list as months, the second and fifth as days, and the third and sixth as years.

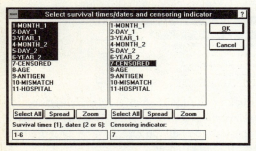

Next, you need to specify variable *Censored* as the censoring indicator variable in the second list of this variable selection dialog. The *Life Table & Distribution of Survival Times* dialog will now look like this:

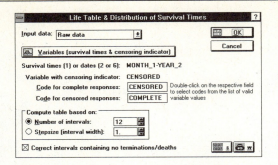

Because the default codes were used for the values of the censoring indicator variable (*0-Complete*, *1-Censored*), STATISTICA will automatically display the *Code for complete response* and *Code for censored response* in the dialog. In addition, you can specify the *Number of intervals* for the life table, or the *Interval width*. You can also specify whether the intervals in which there are no deaths-/terminations will be adjusted so that survival distributions can be fitted. Set the *Correct intervals containing no terminations/deaths* check box when fitting survival distributions, and de-select this check box when generating a life table for descriptive purposes only.

Reading an Aggregated Life Table

Note that instead of raw data, the *Survival Analysis* module will also accept already tabulated survival times as input (select *Table of survival times* in the *Input data* combo box).

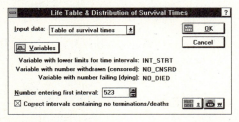

Specifically, a file with tabulated data should contain 3 variables with the following information:

10. SURVIVAL ANALYSIS - EXAMPLES

(1) The lower limits for each time interval,

(2) The number of individuals withdrawn alive from each interval, and

(3) The number of individuals dying in each interval.

After selecting the *Table of survival times*, a different *Life Tables & Distribution of Survival Times* dialog will open in which you can select these variables in exactly the order described above.

Reviewing Results

Returning to the previous *Life Table & Distributions of Survival Times* dialog (for raw data), you are now ready to begin the analysis. Accept all other default selections and click *OK*. After all cases have been processed the *Life Table & Distribution of Survival Times Results* dialog will appear.

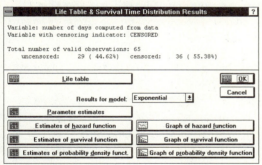

Click on the *Life Table* button in order to display a Scrollsheet of the complete life table.

Note that only a partial listing of the complete life table is shown in the Scrollsheet illustration above.

Fitting a Theoretical Survival Distribution

The *Survival Analysis* module will fit the major theoretical survival time distributions to the data, using ordinary and two methods of weighted least squares estimation. Now, to choose the best fitting distribution, look first at the exponential distribution (select *Exponential* in the *Results for model* combo box). Click on the *Parameter estimates* button to display the parameter estimates for that distribution as well as the goodness of fit *Chi-square* in a Scrollsheet.

Goodness of fit. The logic of this goodness of fit *Chi-square* test is described in the *Introductory Overview* section of this chapter (page 3487). In short, the test is based on the comparison of the likelihood of the respective model with the null model; that is, the model that allows for separate hazard estimates in each interval. If this test is significant, you can conclude that the fitted distribution is significantly different from the observed data, and therefore, you reject it as a model for the survival times. In the illustration above, none of the different parameter estimates for the exponential distribution seems to fit the observed survival distribution.

Plot of survival function. To see the lack of fit, click on the *Graph of survival function* button from the results dialog. As you can see below, none of the lines approximates the observed distribution very well. It seems that the observed survival times drop off faster than what would be expected under this distribution.

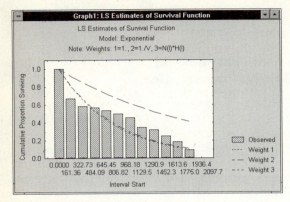

It appears that the third set of parameters (*Weight 3*) provides a reasonable fit to the data; the *Chi-square* test for that model is not significant *(p=.56)*. Therefore, you would conclude that the Weibull distribution with the third set of parameters provides a good theoretical model for the data.

Hazard and probability density function. The *Introductory Overview* section (page 3486) describes the computation of the hazard rate and probability density function. In short, the hazard rate is an estimate of the probability (per time unit) that an observation that has not failed prior to a particular interval will fail in that interval; the probability density function is an estimate of the probability density of failure per time unit in the respective interval.

In order to evaluate the goodness of fit of the chosen theoretical distribution, you can also review these functions in plots, together with the values for the observed distribution (click on the *Graph of hazard function* and *Graph of probability density function* buttons). Usually, the hazard rate will increase over time (see below), because the probability of failure generally increases as time progresses.

Choosing a distribution. You can review the parameter estimates for the different distributions by first selecting the distribution from the *Results for model* combo box and then clicking on the *Parameter estimates* button. If you review all of the distributions, you will find that the only one yielding a non-significant fit is the Weibull distribution with weighted least squares parameter estimates.

SURVIVAL ANALYSIS Estimatn Method	Lambda	Varianc Lambda	Std.Err Lambda	Gamma	Chi-Sqr	df	p
Weight 1	.00031	.00000	.00057	1.14217	31.3240	9	.00026
Weight 2	.01600	.00032	.01795	.64432	13.5076	9	.14101
Weight 3	.05110	.00522	.07223	.42768	7.7570	9	.55881

Parameter Estimates, Model: Weibull (heart.sta). Note: Weights: 1=1., 2=1./V, 3=N(I)*H(I)

Shown below is a plot of the *survival* function with the expected values under the Weibull distribution indicated as lines in the plot.

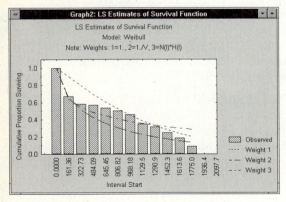

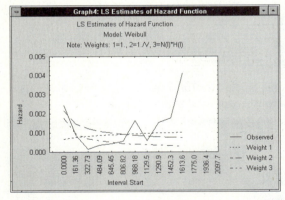

The probability density will usually decrease over time (see below), reflecting the fact that, overall, the probability (density) of failure is greater in the earlier time intervals.

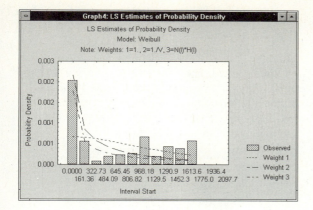

Click *OK* to begin the analysis.

Reviewing Results

From the *Product-Limit (Kaplan & Meier) Analysis Results* dialog, click on the *Product-limit survival analysis* button to review the estimates of the *survival* function. Note that censored observations in this table are marked by a plus (+) sign.

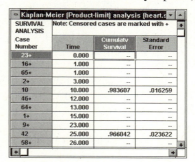

You can plot the estimated *survival* function via the *Graph of survival times vs. cum. proportion surviving* button.

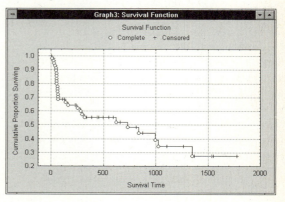

Example 2: Kaplan-Meier Product-Limit Estimates

The life table method is the oldest and most commonly used technique for estimating the *survival* function (and the hazard and probability density functions). However, the exact estimates from the life table will depend on the choice of the number and widths of survival time intervals. The *Kaplan-Meier product-limit* method estimates the survival function directly from the survival times, without tabulation. The *Introductory Overview* (page 3487) explains the general computational approach of this method. In this example these estimates will be computed for the example data file *Heart.sta*.

Specifying the Analysis

After returning to the *Survival Analysis Startup Panel* (click *Cancel* in each dialog until you return to the *Startup Panel*), select the *Kaplan & Meier product-limit method* option. The *Product-Limit (Kaplan & Meier) Analysis* dialog will open in which you can select the variables with the survival times, the censoring variable, and the censoring codes (once again, in this case the program automatically scanned the censoring variable and entered the codes for the censoring variable).

The characteristics of this particular function now seem much clearer than before: The survival function drops off sharply during the first 100 or so days after the heart transplant. Thereafter, the distribution declines much less sharply. Therefore, you can conclude that the first 100 days after the transplant are the most critical to survival.

Percentiles. Click on the *Percentiles of survival function* button to compute these percentiles and display them in a Scrollsheet.

SURVIVAL ANALYSIS	Percentiles of (heart.sta)
	the Survival Function
Percentiles	Survival Time
25'th percentile (lower quartile)	63.5140
50'th percentile (median)	679.1255
75'th percentile (upper quartile)	--

These percentages again reflect the nature of the distribution. Twenty-five percent of all patients die within the first 64 days after the transplant (note that the *Survival Analysis* module will interpolate values to estimate the percentiles). Fifty percent of all patients survive longer than 679 days. The 75th percentile cannot be computed from these data because only censored observations show very long survival times (see the survival table shown earlier).

Example 3: Comparing Survival in Two or More Groups

Variable *Hospital* in the *Heart.sta* data file is a (fictitious) grouping variable that identifies the hospital where each patient received the heart transplant. There are three hospitals represented in this study: *Hillview* (coded *1*), *St_Andreas* (coded *2*), and *Biner* (coded *3*). After comparing the survival functions at the three hospitals, some of the two-sample tests that are available in the *Survival Analysis* module will be reviewed.

Specifying the Analysis

From the *Startup Panel*, select the *Comparing multiple samples* option. In the *Comparing Survival in Multiple Groups* dialog, click on the *Variables* button and select the survival times, censoring indicator (as in *Example 1*, page 3499), and the grouping variable (variable *Hospital*).

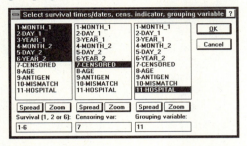

Now, you need to select the codes for the censoring indicator and grouping variable. If the default codes (*0=complete* and *1=censored*) are used in the censoring indicator variable, *STATISTICA* will automatically include those codes in the dialog. If this is not the case, double-click on the *Codes for complete response* and *Codes for censored response* edit fields and select the respective codes from the list of variable values. To enter the codes for the grouping variable, click on the *Codes (for groups)* button and then in the resulting dialog, first click on the *All* button (to select all of the codes) and then click *OK* (to accept these codes and exit the dialog). The *Comparing Survival in Multiple Groups* dialog will now look like this:

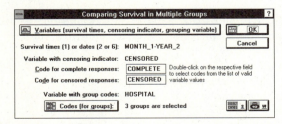

You are now ready to proceed, so just click *OK* and the *Results* dialog will open.

10. SURVIVAL ANALYSIS - EXAMPLES

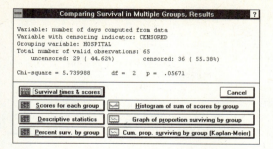

The test for multiple samples is explained in the *Introductory Overview* section (page 3489). The overall *Chi-square* test is almost significant in this example ($p<.06$). Therefore, you would tentatively conclude that there are some differences between hospitals.

To see those differences, you can plot the survival functions for the different groups, that is, click on the *Cumul. prop. surviving by group (Kaplan-Meier)* button.

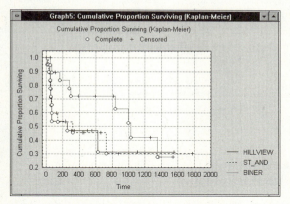

Clearly, the survival function in the *Biner* hospital shows a slower initial drop-off as compared to the other hospitals. Therefore, you would conclude that, somehow, heart transplant patients in the Biner hospital have a greater chance of survival, particularly during the critical first 100 days after the transplant.

You can return to the *Comparing Survival in Multiple Groups, Results* dialog to review the life tables for the three hospitals (click on the *Percent surviving by group* button to display a separate Scrollsheet for each hospital), which will confirm the above conclusion.

Comparing Survival in Two Groups

The *Survival Analysis* module includes various methods for comparing two samples. Select the *Comparing two samples* option on the *Startup Panel* and specify all of the variables as before (see *Example 1*, page 3499).

To compare groups, double-click on the *Code for first group* edit box and select the first group to compare (in this case, select *Hillview*); then, double-click on the *Code for second group* edit box and select this code (*Biner*) from the list.

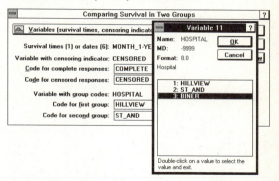

Now, click *OK* in the *Comparing Survival in Two Groups* dialog to go to the *Two Sample Tests* dialog.

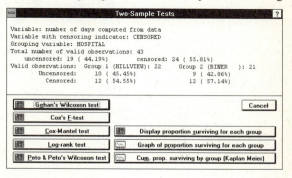

10. SURVIVAL ANALYSIS - EXAMPLES

The different two-sample tests are described and compared in the *Introductory Overview* section (page 3488). In this example, some of the tests yield significance levels at around $p=.05$, while others are not significant. Therefore, you may only tentatively conclude that, indeed, the survival of heart transplant patients from these two hospitals differ significantly, with the Biner hospital producing the better results (survival).

Example 4: Regression Models

The data file *Heart.sta* contains some additional variables: the age of the patient at the time of the transplant (variable *Age*), a measure of antigen mismatch (variable *Antigen*), and a tissue mismatch score (variable *Mismatch*).

It is of interest to determine the relationship between variables *Age*, *Antigen*, and *Mismatch*, and survival times. The most general regression model (that does not make any assumptions about the nature or shape of the underlying survival function) is Cox's proportional hazard model. You can estimate the regression coefficient for these three independent variables in the prediction of survival times using the proportional hazard model.

Specifying the Analysis

Click on the *Regression models* option in the *Startup Panel* to open the *Regression Models for Censored Data* dialog. Now, to select the variables for the analysis, click on the variables button and specify all of the survival time variables and censoring indicator variable as before (see *Example 1*, page 3499). You will also need to select independent variables (*Age*, *Antigen*, and *Mismatch*).

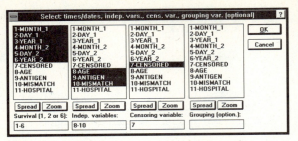

Now, select the codes for the censoring indicator and grouping variable. If the default codes ($0=complete$ and $1=censored$) are used in the censoring indicator variable, *STATISTICA* will automatically include those codes in the dialog. If this is not the case, double-click on the *Codes for complete response* and *Codes for censored response* edit fields and select the respective codes from the list of variable values. The *Regression Methods for Censored Data* dialog will appear as follows:

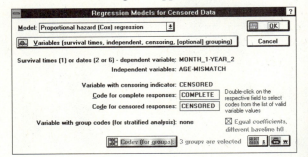

Estimating the Parameters

Because the *Model* combo box is set (by default) to *Proportional hazard (Cox) regression*, you are now ready to begin the analysis. Click *OK* and the *Regression Model Estimation* dialog will open.

10. SURVIVAL ANALYSIS - EXAMPLES

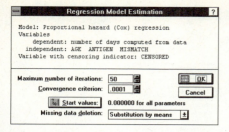

This dialog allows the user to control some of the technical parameters for the estimation procedure (see below). The defaults are usually appropriate, thus, simply press *OK* and the estimation procedure will begin.

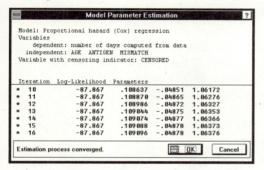

The estimation procedure maximizes the log-likelihood of the regression model via Newton-Raphson iterations. After the best parameters have been found by the program, the iterative procedure stops and the user is prompted to click *OK* to advance to the *Results* dialog.

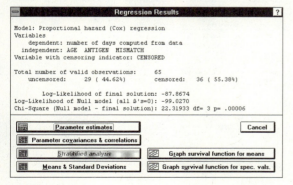

Results

This dialog gives the overall *Chi-square* value for the model; because the *Chi-square* shown above is highly significant, you can conclude that at least some of the independent variables are significantly related to survival. Click on the *Parameter estimates* button to review the parameter estimates and their standard errors.

SURVIVAL ANALYSIS	Dependent Variable: Survival times in days [heart.sta]					
	Censoring var.: CENSORED Chi² = 22.3193 df = 3 p = .00006					
N=65	Beta	Standard Error	t-value	exponent beta	Wald Statist.	p
AGE	.109096	.033293	3.276836	1.115269	10.73766	.001051
ANTIGEN	-.048782	.471644	-.103431	.952388	.01070	.917622
MISMATCH	1.063761	.394599	2.695804	2.897246	7.26736	.007026

The standard errors are computed as part of the estimation procedure, and they are *asymptotic* in nature. Specifically, they are computed from the second-order partial derivatives of the log-likelihood function. This means that the *t* values should also be considered to be approximations.

Usually, any parameter estimates that is at least two times larger than its standard error ($t>2.0$) can be considered to be statistically significant (at the $p<.05$ level); the Scrollsheet also reports the *Wald test* for each coefficient (see Rao, 1973; this test is based upon the asymptotic normality of maximum likelihood estimates; see the *Technical Notes* section, page 3535). Therefore, you would conclude from the Scrollsheet above that age and tissue mismatch are the most important (significant) predictors of hazard.

Plots. In addition to the parameter estimates, you may review graphs of survival as a function of the independent variables, that is, conditional on certain values of the independent variable.

Specifically, you may examine the survival function:

(1) When all independent variables are at their mean (click on the *Graph survival for means* button); or

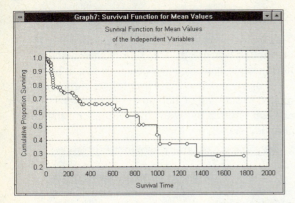

(2) When the covariates have user-specified values (click on the *Graph survival for specific values* button).

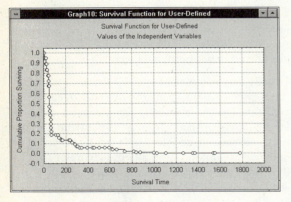

Example 5: Cox Model with Time-Dependent Covariates

Overview

The following example is based on a data set reported by Pike (1966) describing the survival times for two groups of rats that had been exposed to a carcinogen. Shown below is a partial listing of that file:

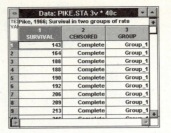

Suppose we suspect that the effect of the treatment (exposure to carcinogen, coded in variable *Group*) on the underlying hazard is not constant; that is, that the proportionality assumption may be violated. To test whether this assumption is tenable, a model will be fit to the data that includes both the fixed covariate *Group* as well as a time-dependent variable defined as *Group * (Log(Time)-5.4)*. Note that the value *5.4* is used for scaling purposes here, since the mean of the log-survival times is approximately equal to *5.4*.

Specifying the Analysis

After starting the *Survival Analysis* module, open the data file *Pike.sta*. Then click on the *Regression Models* option button to bring up the *Regression Models for Censored Data* dialog. On that dialog, in the *Model* combo box, select option *Prop. hazard (Cox) /w time dep. covariates*.

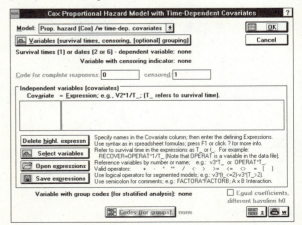

Click on the button labeled *Variables (survival times, censoring [optional] grouping)* and select as the dependent variable *Survival* (variable 1), and as the censoring indicator variable *Censored*.

Specifying Time-Dependent Covariates

Since the default code for censored observations is the same as the one used in the data file, all that is left to do is to specify the covariates. The dialog shown above has two editable (and scrollable) fields. The left column (*Covariate*) can be used to enter a label for the covariate that will be used later in the output. The right (wider) column (*Expression*) can be used to define the respective covariate via an arithmetic expression.

Rather than typing in the names of the covariates, begin by clicking on the *Select variables* button in the *Independent variables (covariates)* box and a standard single variable list selection dialog will come up. The variables that you select here will be automatically transferred into both edit fields, i.e., into the *Covariate* field and the *Expression* field. Thus, this is a quick way to specify fixed covariates, or to transfer variable names one wants to transform into the edit fields.

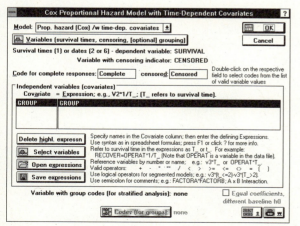

Select variable *Group* and click *OK*. Variable *Group* will be transferred into the first line of the edit fields. Repeat the selection of variable *Group*, and identical text will appear in the second line of the edit fields; then modify the second line as shown below.

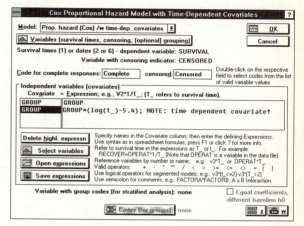

Note that the text in the second line past the semicolon will be interpreted as a comment, so you may annotate your formulas. Also, for repetitive analyses you can save and retrieve definitions via options *Save expressions* and *Open expressions*, respectively.

You are now ready to begin the analysis.

Reviewing Results

Click *OK* to bring up the intermediate *Regression Model Estimation* dialog.

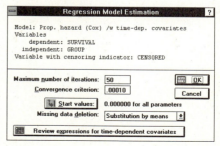

Accept all defaults in this dialog and simply click *OK* to begin the parameter estimation. Shown below is the dialog after the estimation process has converged.

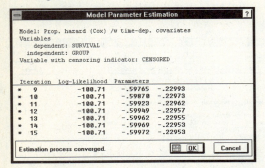

Click *OK*, and from the *Regression Results* dialog choose the option *Parameter estimates*.

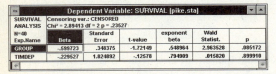

As you can see, the parameter estimate for the time-dependent covariate is much smaller than its standard errors, and not statistically significant. Thus, the model with the single fixed covariate does not appear to be inappropriate.

10. SURVIVAL ANALYSIS - DIALOGS AND OPTIONS

DIALOGS, OPTIONS, STATISTICS

Startup Panel

Essentially, the methods offered in this module address the same research questions as many of the other *STATISTICA* procedures; however, all methods in the *Survival Analysis* module will handle censored data. The *Startup Panel* will allow you to select the desired type of analysis.

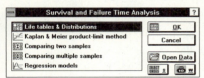

Life Tables and Distributions

This method is the most straightforward way to describe the survival in a sample (see page 3485). This table can be thought of as an "enhanced" frequency distribution table. The distribution of survival times is divided into a certain number of intervals. For each interval, you can compute the number and proportion of cases or objects that entered the respective interval "alive," the number and proportion of cases that failed in the respective interval (i.e., number of terminal events, or number of cases that "died"), and the number of cases that were lost or censored in the respective interval.

Also, use this option to fit four common theoretical survival functions to the life table: exponential, linear exponential, Weibull, and Gompertz. Least squares and two types of weighted least squares parameter estimates are computed for each distribution.

Kaplan-Meier Product-Limit Method

This method estimates the survival function directly from the continuous survival or failure times (see page 3487). The advantage of the *Kaplan-Meier Product-Limit* method over the *Life Table* method (see above) for analyzing survival and failure time data is that the resulting estimates do not depend on the grouping of the data (into a certain number of time intervals). Actually, the Product-Limit method and the life table method are identical if the intervals of the life table contain at most one observation.

Comparing Two Samples

You can compare the survival or failure times in two or more samples when you select this method (see page 3488).

Note that your *STATISTICA* program also contains an example *STATISTICA BASIC* program (file *Manthaen.stb*) for computing the Mantel-Haenszel test for comparing two groups (see Mantel and Haenszel, 1959; Lee, 1992; see also the *Technical Notes* section, page 3532).

Comparing Multiple Samples

The multiple-sample test implemented in the *Survival Analysis* module is an extension (or generalization) of Gehan's generalized Wilcoxon test, Peto and Peto's generalized Wilcoxon test, and the log-rank test (see page 3489). When you select this method, a score is first assigned to each survival time using Mantel's procedure (Mantel, 1967); next a *Chi-square* value is computed based on the sums (for each group) of this score. If only two groups are specified, then this test is equivalent to Gehan's generalized Wilcoxon test, and the computations will default to that test in this case.

SUR - 3511

10. SURVIVAL ANALYSIS - DIALOGS AND OPTIONS

Regression Models

Selecting this method will allow you to choose between five different regression models: Cox's proportional hazard regression, Cox's proportional hazard regression with time-dependent covariates, exponential regression, log-normal linear regression, and normal linear regression (see page 3489).

Life Table Analysis (Analyzing Raw Data)

When you select *Life Tables & Distributions* from the *Startup Panel*, the *Life Table and Distribution of Survival Times* dialog will open. You can choose between *Raw Data* as the input data or a *Table of Survival Times*. If you select *Raw Data*, then the following options will be available to you (when you select *Table of Survival Times*, a separate dialog will display those options, see *Life Table Analysis (Analyzing Tables)*, page 3513):

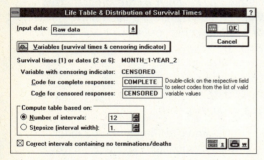

In this dialog you will be able to select one variable with survival times or two or six variables with dates (see below); select the censoring indicator variable and the codes (or text values) used in this variable to denote whether a respective case is complete or censored (defaults are *0* and *1*, respectively).

Input Data

The *Life Tables* routine in the *Survival Analysis* module will accept as input both *Raw Data* as well as *Tables of Survival Times* (see above).

Variables

This button will bring up a standard variable selection dialog for specifying (1) the variable(s) with the survival times and (2) the censoring indicator variable.

Specifying survival times. There are three ways for specifying survival times. You may

(1) Select one variable with survival times (e.g., number of weeks, days, years, etc., surviving),

(2) Select two variables with dates (see page 3499) that specify when the respective observation began and when it was terminated (due to death/failure or censoring), or

(3) Select six variables that identify dates; specifically, these variables should contain the month (1 to 12), day (1 to 31), and year when the particular observation began (e.g., when the patient was admitted to the hospital), and the month, day, and year when the observation was terminated (due to death/failure or censoring, for example, when a patient was dismissed from the hospital).

When specifying dates via the latter method, the six variables containing the dates must be specified in exactly this order (month, day, year at entry; month, day, year at termination). While processing the data, the *Survival Analysis* module will compute the number of days that elapsed between dates and perform the analysis on this measure. Note that if the value of the year is less than 100, the program will automatically assume that the year refers to the 20th century; for example, the year *88* would be converted into *1988*.

Censoring indicator variable. The censoring indicator is the variable containing integer codes or text values that uniquely identify complete and censored observations.

SUR - 3512

10. SURVIVAL ANALYSIS - DIALOGS AND OPTIONS

Codes for Complete and Censored Responses

Specify the codes or text values that were used in the censoring indicator variable to uniquely identify complete (uncensored) and incomplete (censored) observations. To review all codes in the respective variable, double-click on the respective edit box. The default codes for uncensored and censored data are *0* and *1*, respectively.

Compute Table Based on

You can compute the table based on a particular number of intervals or on a specified step size.

Number of intervals. This option allows you to determine the number of intervals for the life table.

Step size (Interval width). Via this option you can specify the step size that is to be used for tabulating the survival times.

Correct Intervals Containing No Terminations/Deaths

This check box determines whether intervals in which there are no deaths/terminations will be adjusted so that survival distributions can be fitted. Parameters cannot be computed for intervals where no deaths/terminations occur; by default (box is checked) the proportion surviving in those intervals is therefore adjusted by the program (the proportion surviving is set to .5/number exposed). Do not check this box (do not adjust) when generating a life table for descriptive purposes only.

Life Tables (Analyzing Tables)

When you select *Life Tables & Distributions* from the *Startup Panel*, the *Life Table and Distribution of Survival Times* dialog will open.

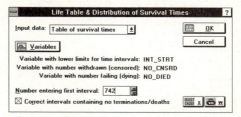

You can choose between *Raw Data* as the input data or a *Table of Survival Times*. If you select *Table of Survival Times*, then the above options will be available to you (when you select *Raw Data*, a separate dialog will display those options, see *Life Table Analysis (Raw Data)*, page 3512).

In this dialog you can specify the respective variables (see below) and the number of individuals (observations) entering the first interval.

Input Data

The *Life Tables & Distributions* routine in *Survival Analysis* will accept as input both *Raw Data* as well as a *Table of Survival Times* (see above).

Variables

This button will bring up a standard variable selection dialog for specifying the variables of the survival table. Specifically, select 3 variables with the following information:

(1) The lower limits for each time interval;

(2) The number of individuals withdrawn alive from each interval;

(3) The number of individuals dying in each interval.

10. SURVIVAL ANALYSIS - DIALOGS AND OPTIONS

The variables that you select here will be appropriately displayed in the statements below the *Variables* button.

Correct Intervals Containing No Terminations/Deaths

This check box determines whether intervals in which there are no deaths/terminations will be adjusted so that survival distributions can be fitted. Parameters cannot be computed for intervals where no deaths/terminations occur; by default (box is checked) the proportion surviving in those intervals is therefore adjusted by the program (the proportion surviving is set to .5/number exposed). Do not check this box (do not adjust) when generating a life table for descriptive purposes only.

Life Tables & Distribution Fitting Results

The *Life Table and Survival Time Results* dialog offers many options (described below) to analyze the survival data and visually display the results.

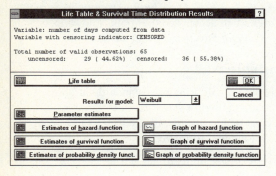

Life Table

Clicking on this button will bring up a Scrollsheet with the complete life table (shown below is an example, in three segments).

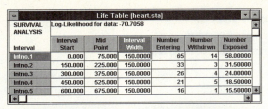

In addition to the standard descriptive statistics, this Scrollsheet will contain estimates of the cumulative survival, the hazard rate, survival probability density, and the median life expectancy at each interval (see the *Introductory Overview* section, page 3485, for a definition of these estimates). The standard errors of these estimates are also computed.

Model

The *Life Tables & Distributions* routine in *Survival Analysis* will automatically fit to the data the four common theoretical distributions of survival times: Exponential, Weibull, Gompertz, and Linear Hazard (see the *Introductory Overview* section, page 3486). This combo box allows the user to choose between these distributions, that is, the setting of this box will determine the output (parameters, graphs) displayed after selecting any of the following options.

SUR - 3514

Copyright © StatSoft, 1995

10. SURVIVAL ANALYSIS - DIALOGS AND OPTIONS

Parameter Estimates

This option will bring up a Scrollsheet with the parameter estimates for the respective theoretical survival distribution.

SURVIVAL ANALYSIS Estimatn Method	Parameter Estimates, Model: Exponential (heart.sta) Note: Weights: 1=1., 2=1./V, 3=N(I)*H(I)						
	Lambda	Varianc Lambda	Std.Err Lambda	Log-Likelhd	Chi-Sqr	df	p
Weight 1	.00107	.00000	.00034	-80.886	20.3614	10	.02605
Weight 2	.00058	.00000	.00015	-86.035	30.6583	10	.00067
Weight 3	.00126	.00000	.00022	-80.979	20.5466	10	.02452

The particular model (theoretical distribution) to which the parameter estimates pertain can be chosen via the *Model* combo box, above. The *Survival Analysis* module will fit the theoretical distributions via ordinary unweighted least squares (i.e., all weights equal to 1) and two methods of weighted least squares (see *Technical Notes*, page 3531).

Estimates of Hazard, Survival, and Probability Density Function

Choosing any of these output options will bring up a Scrollsheet with the respective estimates for each time interval, based on the respective theoretical distribution. The particular model (theoretical distribution) to which the parameter estimates pertain can be chosen via the option *Model* combo box, above. Estimates based on the unweighted and two weighted least squares methods will be reported in the Scrollsheet (see *Technical Notes*, page 3531).

Graph of Hazard, Survival, and Probability Density Function

Choosing any of the graphics output options will bring up a graph of the respective estimated function and the actual observed data (see below).

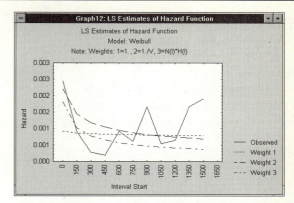

The particular model (theoretical distribution) to which the estimated functions pertain can be chosen via the option *Model* combo box above.

Kaplan-Meier Product-Limit Analysis

When you select *Kaplan & Meier Product-Limit Method* from the *Startup Panel*, the *Product-Limit (Kaplan & Meier) Analysis* dialog will open in which you can specify the respective variables (see below) for the analysis.

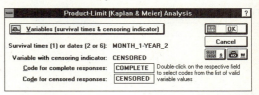

Variables

Clicking the *Variables* button will bring up a standard variable selection dialog for specifying:

(1) The variable(s) with the survival times, and

(2) The censoring indicator variable.

For more information on these two types of variables, see page 3512.

SUR - 3515

10. SURVIVAL ANALYSIS - DIALOGS AND OPTIONS

Codes for Complete and Censored Responses

Specify the codes or text values that were used in the censoring indicator variable to uniquely identify complete (uncensored) and incomplete (censored) observations. To review all codes in the respective variable, double-click on the respective edit box. The default codes for uncensored and censored observations are *0* and *1*, respectively.

Kaplan-Meier Product-Limit Analysis - Results

The *Product-Limit (Kaplan-Meier) Analysis Results* dialog offers many options (described below) to analyze the survival data and visually display the results.

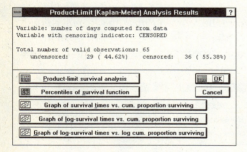

Product-Limit Survival Analysis

Clicking on this button will bring up a Scrollsheet with the Product-Limit (Kaplan-Meier) estimates of the cumulative survival curve. Note that censored observations are marked by a + (plus) sign next to the survival times displayed in the left-most column of the Scrollsheet.

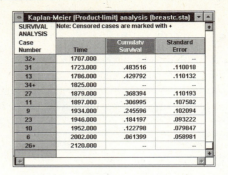

Percentiles of the Survival Function

This option will bring up a Scrollsheet with estimates of the 25th, 50th, and 75th percentile of the survival function.

Note that the program will interpolate values if the respective quartile boundary is not exactly defined by an uncensored case.

Graphs of Survival Functions

Choosing any of the graphics output options will bring up a graph of the survival function.

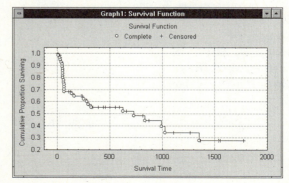

SUR - 3516

Copyright © StatSoft, 1995

Note that this plot will show both complete and censored observations, with different point markers.

You can select to plot the cumulative survival across time, cumulative survival versus the natural logarithm (*log*) of the survival times, or *log* cumulative survival versus the *log* of the survival times.

Comparing Two Groups

When you select *Comparing Two Samples* from the *Startup Panel*, the *Comparing Survival in Two Groups* dialog will open in which you can specify the respective variables (see below) for the analysis.

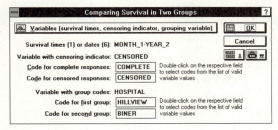

Variables

This option will bring up a standard variable selection dialog for specifying:

(1) The variable(s) with the survival times,

(2) The censoring indicator variable, and

(3) The grouping variable.

For information on the first two types of variables, see page 3512. The grouping variable should contain the integer codes or text values that uniquely identify to which group (sample) each observation belongs.

Codes for Complete and Censored Responses

Specify the codes or text values that were used in the censoring indicator variable to uniquely identify complete (uncensored) and incomplete (censored) observations. To review all codes in the respective variable, double-click on the respective edit box. The default codes for uncensored and censored observations are *0* and *1*, respectively.

Codes for First and Second Group

Specify the codes or text values that were used in the grouping variable to uniquely identify to which one of the two groups each observation belongs. To review all codes in the respective variable, double-click on the respective edit box.

Comparing Survival in Two Groups - Results

The *Survival Analysis* module will compute five different nonparametric tests for comparing the survival function in two groups.

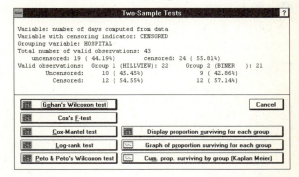

Brief descriptions of these methods are provided in the *Introductory Overview* section (page 3488).

Gehan's Wilcoxon Test

This option will bring up a Scrollsheet with the results of Gehan's Wilcoxon test. In addition to the Gehan-Wilcoxon test statistic, the Scrollsheet lists all observations [survival times; note that censored observations are marked by a + (plus) sign], the group to which each belongs, the number of observations that are smaller than the respective observation (under the column header $R1$), and the number of observations that are larger than the respective observation (under the column header $R2$).

Cox's *F*-test

This option will bring up a Scrollsheet with the results of the Cox *F*-test.

In addition to the *F*-test, the Scrollsheet will list all distinct failure times, the number of observations at risk at the respective failure time [under the column header $R(I)$], the multiplicity of each distinct failure time [under the header $M(I)$], the ratio of the two, and the Kaplan-Meier product-limit estimate of the survival function (in the right-most column of the Scrollsheet).

Cox-Mantel Test

This option will bring up a Scrollsheet with the results of the Cox-Mantel test.

In addition to the Cox-Mantel test statistic, the Scrollsheet will list the risk set at each failure time (the number of observations whose failure times are at least this long or longer), and the proportion of observations in group 2 that belong to the respective risk set.

Log-Rank Test

This option will bring up a Scrollsheet with the results of the log-rank test.

In addition to the log-rank test statistic, the Scrollsheet will list all observations (survival times; note that censored observations are marked by a + sign), the group to which each belongs, and the scores that are used in the computation of this test.

Peto and Peto's Wilcoxon Test

This option will bring up a Scrollsheet with the results of Peto and Peto's Wilcoxon test (Peto and Peto, 1972; see also Lee, 1980).

Peto & Peto Wilcoxon Test (heart.sta)	
SURVIVAL ANALYSIS	WW = 3.0281 Sum = 10.389 Var = 2.6577 Test statistic = 1.857437 p = .06325
Survival Time	Score
1.0000+	0.000000
3.0000+	0.000000
10.000	.975610
12.000+	-.024390
13.000	-.024390
15.000+	-.024390
23.000+	-.024390
39.000	.924119
48.000+	-.051491
50.000	.841224

In addition to the test statistic, the Scrollsheet will list all observations (survival times; note that censored observations are marked by a + sign), the group to which each belongs, and the scores that are used in the computation of this test.

Display Proportion Surviving for Each Group

This option will bring up a Scrollsheet with the comparative life tables for the two groups.

Life Table for Group 1 and Group 2 (heart.sta)						
SURVIVAL ANALYSIS Lower Limit	Group 1: HILLVIEW Group 2: BINER					
	Group 1: No.Enter	Group 2: No.Enter	Group 1: No.Cnsrd	Group 1: % Srvng	Group 2: % Srvng	Group 1: Cum.%.Sr	Group 2: Cum.%.Sr
1.0000	22	21	5	58.9744	84.2105	100.0000	100.0000
173.00	9	14	3	86.6667	85.7143	58.9744	84.2105
345.00	5	12	1	100.0000	100.0000	51.1111	72.1805
517.00	4	11	1	71.4286	100.0000	51.1111	72.1805
689.00	2	9	0	100.0000	87.5000	36.5079	72.1805
861.00	2	6	0	100.0000	66.6667	36.5079	63.1579
1033.0	2	4	1	100.0000	100.0000	36.5079	42.1053
1205.0	1	4	0	100.0000	66.6667	36.5079	42.1053
1377.0	1	1	0	100.0000	100.0000	36.5079	28.0702
1549.0	1	0	1	100.0000	0.0000	36.5079	28.0702

Graph of Proportion Surviving for Each Group

This option will bring up a graph of the survival function for each group.

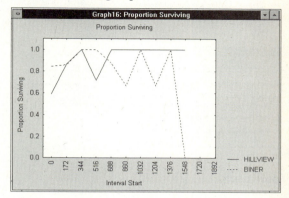

Cumulative Proportion Surviving by Group (Kaplan-Meier)

This option will bring up a graph of the cumulative survival function for each group.

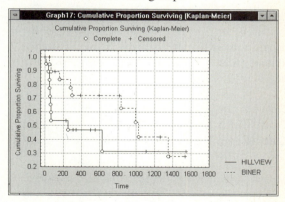

Note that complete and censored observations in this plot are indicated by different point markers.

10. SURVIVAL ANALYSIS - DIALOGS AND OPTIONS

Comparing Multiple Groups

When you select *Comparing Multiple Samples* from the *Startup Panel*, the *Comparing Survival in Multiple Groups* dialog will open in which you can specify the respective variables (see below) for the analysis.

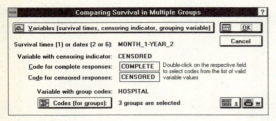

Variables

This option will bring up a standard variable selection dialog for specifying:

(1) The variable(s) with the survival times,

(2) The censoring indicator variable, and

(3) The grouping variable.

For more information on the first two types of variables, see page 3512. The grouping variable should contain the integer codes or text values that uniquely identify to which group (sample) each observation belongs.

Codes for Complete and Censored Responses

Specify the codes or text values that were used in the censoring indicator variable to uniquely identify complete (uncensored) and incomplete (censored) observations. To review all codes in the respective variable, double-click on the respective edit box. The default codes for uncensored and censored observations are *0* and *1*, respectively.

Codes (for Groups)

Specify the codes or text values that were used in the grouping variable to uniquely identify to which one of the two groups each observation belongs. To review all codes in the respective variable, double-click on the respective edit box.

Comparing Multiple Groups - Results

Several options (described below) are available in the *Comparing Survival in Multiple Groups* dialog.

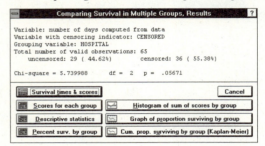

Survival Times and Scores

This option will bring up a Scrollsheet with all survival times and the scores (computed according to Mantel's procedure; see Mantel, 1967) used for the computation of the test statistic.

Survival Time	Group	Score
0.0000+	ST_AND	0.0000
1.0000+	ST_AND	0.0000
1.0000+	BINER	0.0000
3.0000+	HILLVIEW	0.0000
10.000	HILLVIEW	-60.0000
12.000+	HILLVIEW	1.0000
13.000+	HILLVIEW	1.0000
15.000+	HILLVIEW	1.0000
23.000+	HILLVIEW	1.0000
25.000	ST_AND	-54.0000
26.000+	ST_AND	2.0000
29.000	ST_AND	-51.0000
30.000+	ST_AND	3.0000
39.000	HILLVIEW	-48.0000

SUR - 3520

Copyright © StatSoft, 1995

Note that censored observations are marked by a + (plus) sign.

Scores for Each Group

This option will bring up a Scrollsheet with the sum of scores for each group (computed according to Mantel's procedure; see Mantel, 1967) used for the computation of the test statistic; in addition, the number and percent of complete and censored cases in each group are displayed.

SURVIVAL ANALYSIS	Code for Group	Sum of Scores	N.uncsrd	Percent uncensrd	N.censrd	Percent censored	Total N
HILLVIEW	1	-116.000	10	45.45454	12	54.54546	22
ST_AND	2	-115.000	10	45.45454	12	54.54546	22
BINER	3	231.000	9	42.85714	12	57.14286	21

Percent Surviving in Each Group

This option will bring up a life table for each group. Note that you can use the *Scrollsheet Manager* option from the *Window* pull-down menu to adjust the number of Scrollsheets that can simultaneously be displayed on the screen (see *General Conventions*, Volume I).

Descriptive Statistics

This option will bring up a Scrollsheet with descriptive statistics for each group and for all groups combined.

SURVIVAL ANALYSIS	Median	Mean	Std.Dv.	No.uncsd	N.censrd	Total N
HILLVIEW	63.0000	269.0000	397.6063	10	12	22
ST_AND	64.5000	281.1364	423.2123	10	12	22
BINER	589.0000	608.1429	503.6648	9	12	21
Total	161.0000	382.6769	463.2328	29	36	65

These statistics are computed according to the standard formulas (disregarding censoring) and include the mean and median survival times in each group.

Plot Histogram of the Sum of Scores for Each Group

This option is only available if there are more than two groups in the analysis.

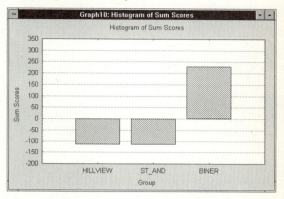

Clicking on this button will bring up a histogram (see above) of the sum of scores for each group (computed according to Mantel's procedure, Mantel, 1967) used for the computation of the test statistic.

Plot of Proportion Surviving in Each Group

This option will bring up a graph of the survival function for each group.

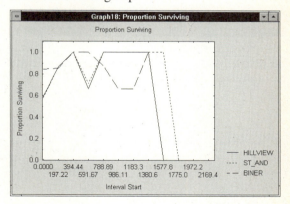

Cumulative Proportion Surviving by Group (Kaplan-Meier)

This option will bring up a graph of the cumulative survival function for each group (see below).

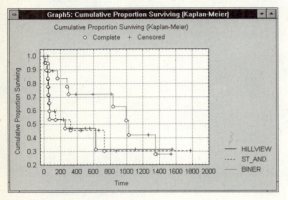

Note that complete and censored observations in this plot are indicated by different point markers.

Regression Models

When you select *Regression Models* from the *Startup Panel*, the *Regression Models for Censored Data* dialog will open in which you can specify the respective variables (see below) for the analysis.

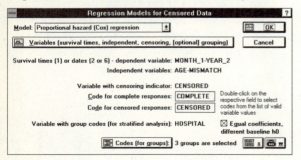

Model

This combo box allows you to choose between the five different regression models available in *Survival Analysis*: Cox's proportional hazard regression, Cox's proportional hazard regression with time-dependent covariates, exponential regression, log-normal linear regression, and normal linear regression (see *Regression Models*, page 3489).

Variables

Clicking on this button will bring up a standard variable selection dialog for specifying:

(1) The variable(s) with the survival times,

(2) A list of independent variables (covariates),

(3) The censoring indicator variable, and

(4) An optional grouping variable (for a stratified, by-group analysis).

For information on specifying survival times and a censoring indicator variable, see page 3512.

Variable with group codes. This optional variable should contain the integer codes or text values that uniquely identify to which group (sample) each observation belongs. If no grouping variable is specified, a regular analysis will be performed on all data. If a variable is specified here, then a stratified analysis will be performed.

In general, in a stratified analysis, separate regression models are first fit to each group and the log-likelihoods for those models are summed up. This log-likelihood is then compared to that of the overall model (collapsed across groups). However, if the current regression model is *Proportional hazard (Cox) regression*, and the *Equal coefficients, different baseline h0* check box is set (see below), then the program will estimate a common (for all groups) set of parameters for the Cox proportional hazard model, but allow for different baseline hazards in each group.

Codes for Complete and Censored Responses

Specify the codes or text values that were used in the censoring indicator variable to uniquely identify complete (uncensored) and incomplete (censored) observations. To review all codes in the respective variable, double-click on the respective edit box. The default codes for uncensored and censored observations are *0* and *1*, respectively.

Codes (for Groups)

These codes can only be specified if a stratified (by-group) analysis was requested, that is, if a grouping variable was selected. Specify the codes or text values that were used in the grouping variable to uniquely identify to which one of the groups each observation belongs. To review all codes in the respective variable, double-click on the respective edit box.

Equal coefficients, different baseline h0

This option is only available if (1) the current *Model* is *Proportional hazard (Cox) regression*, and (2) after you selected a stratification variable. If this checkbox is set then the program will fit a stratified model with constant coefficient vector for each group, but allowing for different baseline hazard (h_0) in each group. In effect, observations in the same group (stratum) are assumed to have proportional hazard functions, but this is not necessarily the case for observations in different groups. If this checkbox is deselected, then the program will allow for different baseline hazards as well as coefficient vectors in each group. For more information about stratification of proportional hazard models, see Lawless (1982, p. 365).

Cox Proportional Hazard Model with Time-Dependent Covariates

This dialog will come up when you select *Regression Models* from the startup panel, and then set the *Model* combo box to *Prop. hazard (Cox) /w time-dep. covariates*.

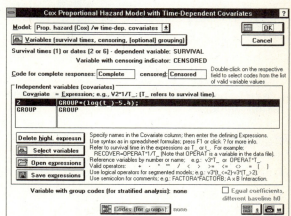

In this dialog you can specify the dependent variable for the analysis, the censoring indicator, an optional stratification (grouping) variable, and the time-dependent covariates. The syntax rules for the arithmetic expressions defining the time-dependent covariates are identical to those for the spreadsheet formulas. A brief overview is presented in the *Technical Notes* section (page 3533). For the complete listing of all supported operators and functions, refer to the *Electronic Manual* accessible by pressing F1 or by clicking on the ? button.

Model

This combo box allows you to choose between the five different regression models available in *Survival Analysis*. When this combo box is set to *Prop. hazard (Cox) /w time-dep. covariates*, then the dialog shown earlier will be displayed.

Variables (Survival Times, Censoring, [Optional] Grouping)

Clicking on this button will bring up a standard variable selection window for specifying:

(1) The variable(s) with the survival times,

(2) The censoring indicator variable, and

(3) An optional grouping variable (for a stratified by-group analysis).

Use the edit fields in the *Independent variables (covariates)* box to specify the independent variables for the analysis.

Codes for Complete and Censored Responses

Specify the codes or text values that were used in the censoring indicator variable to uniquely identify complete (uncensored) and incomplete (censored) observations. To review all codes in the respective variable, double-click on the respective edit box. The default codes for uncensored and censored observations are *0* and *1*, respectively.

Independent Variables (Covariates)

Use the two edit lists to specify the covariates in the model. In the left edit field, under the header *Covariate*, enter a label for the covariate; that label will be used in the results Scrollsheets. Use the *Expression* edit field to define the respective covariate by entering an arithmetic expression.

Syntax. The syntax rules for the expressions are identical to those for spreadsheet formulas. A brief summary of the rules is given in the *Technical Notes* section (see page 3533). The complete set of available operators and functions (including probability functions) can be reviewed in the *Electronic Manual*, available via F1 or by clicking on the ? button in this dialog.

In addition to the available arithmetic operators and functions, the expression interpreter will also recognize the symbol $t\_$ or $T\_$ (i.e., letter *t* and the underscore) to denote the survival times. By using this symbol, you can define time-dependent covariates (see also the *Introductory Overview*, page 3490, or the *Example* section, page 3508). Shown below are a few examples:

```
Covariate = Expression        ; opt. comment
---------   ----------------------------------
timdep      v2*(log(t_)-5.4); time-dependent
Covar2      Age*(Group=1) - Age*(Group=2)
FactorA     FactorA
FactorB     FactorB
AXB         FactorA*FactorB
```

As illustrated in the first example (*timdep* = ...), you may include comments in the expressions by placing them after the expressions, separated by a semicolon. Variables may be referenced by name, or via the standard *Vxxx* convention where *xxx* is the respective variable number in the data file.

The second expression illustrates the use of logical operators: If a case belongs to *Group* 1 (where *Group* is a variable name), then *Covar2* is defined as +*Age*; if a case belongs to *Group 2*, then *Covar2* is defined as -*Age*.

The third and fourth expressions define the fixed covariates *FactorA* and *FactorB*; the fifth expression defines their interaction.

Note that in the examples above, only the first expression actually defines a time-dependent covariate; while the others make no reference to the survival times (i.e., they do not include the symbol $t\_$ or $T\_$).

Editing expressions. To edit expressions, you can highlight them, and then either double-click on them or press F2 to place the cursor in the respective field. The standard Windows conventions for copying (CTRL+C) and pasting (CTRL+V) are also supported. You can use the arrow keys to move up and down through the list of expressions. To add an expression, use the INS (*Insert*) key on your keyboard; to delete an expression, highlight it and then click on the *Delete highl. expression* button, or use the DEL key on the keyboard. Note that you can also save the expressions, and later use them in future analyses (see below).

Delete highl. expression. Use this button to delete the highlighted expression.

Select variables. When you click on this button, a standard variable selection dialog will come up. The variables (i.e., their names) that you select in this dialog will be moved to both the left edit box (as the *Covariate* labels) and the right edit box (as the *Expression*). Thus, by default the selected variables will be defined as fixed covariates.

Open expression. When you click on this button, a file selection dialog will come up where you can select a previously saved file with expressions. By default, those files are saved with the file name extension *.ini* (see below). Note that the program will automatically append the expressions in the file to the ones that were previously entered.

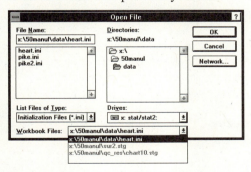

Workbook options. Note that if you do not remember the name (of the file) under which you saved the previous expressions for the current data set, then you can retrieve the name from the Workbook list available in every *Open/Save* dialog (refer to Volume I for details on Workbooks).

Save expressions. Use this option to save the current expressions in a file; by default the program will suggest the file name extension *.ini*.

Codes (for Groups)

These codes can only be specified if a stratified (by-group) analysis was requested, that is, if a grouping variable was selected. Specify the codes or text values that were used in the grouping variable to uniquely identify to which of the groups each observation belongs.

Setting Parameters for Estimation

You can set the parameters for the regression model estimation in the *Regression Model Estimation* dialog.

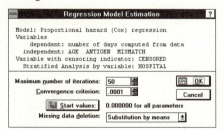

Maximum Numbers of Iterations

The estimation of the maximum likelihood parameters is an iterative procedure; this option allows the user to specify the maximum number of iterations (default is *50*).

Convergence Criterion

The iterative estimation procedure will terminate when either the maximum number of iterations has been exceeded or when the change in the parameter estimates from iteration to iteration does not exceed the convergence criterion. The smaller the value of the convergence criterion, the more iterations will be performed by the program (within the limits of the maximum number of iterations).

Start Values

This option is only available if the *Proportional hazard (Cox) regression* model (with or without time-dependent covariates) was requested on the previous dialog. When you click on this button, the *Specify Start Values* dialog will open.

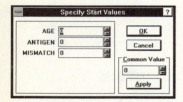

An iterative procedure is used to estimate the parameters for the proportional hazard regression model. In this dialog you can specify the parameter values that are to be used in the first iteration of the computations.

Missing Data Deletion

There are two ways to handle missing data: Missing data may either be casewise deleted, or substituted by means.

Casewise deletion. When *Casewise deletion* of missing data is selected, then only cases that do not contain any missing data for any of the variables selected for the analysis will be included in the analysis. In the case of correlations, all correlations are calculated by excluding cases that have missing data for any of the selected variables (all correlations are based on the same set of data).

Mean Substitution. When you select *Mean substitution*, the missing data will be replaced by the means for the respective variables during an analysis.

Regression Models for Survival Data - Results

After you have set the parameters for the regression model estimation in the *Regression Model Estimation* dialog and the iterative procedure is complete, the *Regression Results* dialog will open.

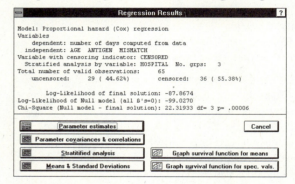

This dialog displays a summary of the design and options to review the regression results. From this dialog, the user can also review plots of the survival function when all independent variable (covariate) values are zero, at their respective means, and at user-specified values.

Parameter Estimates

This option will bring up a Scrollsheet with parameter estimates, their standard errors, and the ratio of the parameter estimates divided by their standard errors (under the heading *t-value*); for Cox proportional hazard models, the Scrollsheet will also show the Wald statistic (see also the *Technical Notes* section, page 3535), and the statistical significance of that statistic.

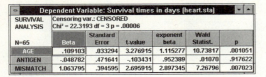

Note that the standard errors (and *t*-values) are computed from the second-order partial derivatives of the log-likelihood function, resulting in asymptotic standard errors (and *t*-values) for the parameters.

Parameter Variance-/Covariance Matrix

This option will bring up a Scrollsheet with the variance/covariance matrix of the parameter estimates,

and another Scrollsheet with the correlation matrix of the parameter estimates.

Stratified Analysis

This option is only available if a stratified analysis (analysis by groups) was requested (and the *Equal coefficients, different baseline h0* option was deselected for Cox proportional hazard models). Clicking on this button will bring up a Scrollsheet with the results of the stratified analysis; specifically, the Scrollsheet will show the sum of the log-likelihoods of the model, fitted within each group, and the log-likelihood of the model fitted to all observations combined.

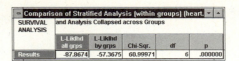

In order to examine the parameter estimates for a particular group, specify the appropriate case selection conditions (i.e., select only cases belonging to the group of interest), and then fit the respective model in the usual manner.

Means and Standard Deviations

This option will bring up a Scrollsheet with the means and standard deviations for the survival times and the independent variables (covariates).

Graphs

The graphs that are available from the *Regression Results* dialog depend on the currently selected regression model. The following paragraphs describe the graphs for each of the models.

Cox Proportional Hazard Model

The following options are available with this model.

Graph of survival function for mean values of covariates. This option will bring up a graph of the survival function when the values for the independent variables (covariates) are at their respective means.

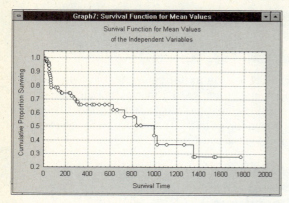

Graph of survival function for user-defined values of the independent variables. This option allows the user to specify values for the independent variables (covariates); the survival function will then be drawn based on those values. When you click on the *Graph survival for specific values* button, the *Independent Variable Values* dialog will open.

Specify in this dialog the desired values for the independent variables. The program will then compute and draw the respective survival function based on those values.

Exponential Regression Model

The following options are available with this model.

Graph of residuals vs. exponential order statistic (*alpha*). This option will bring up a scatterplot of the residual survival times and the standard exponential order statistic.

If all points are arranged roughly on a straight line, then the data are consistent with the assumption of exponentiality.

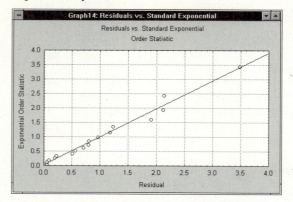

Graph of residual vs. predicted survival times. This option will bring up a scatterplot of the residual versus predicted survival times.

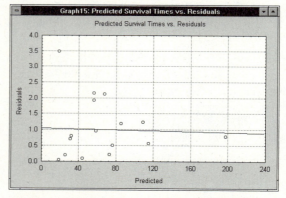

Graph of residuals vs. log-survival times. This option will bring up a scatterplot of the residual survival times versus the observed log-survival times.

10. SURVIVAL ANALYSIS - DIALOGS AND OPTIONS

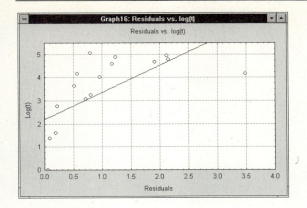

Normal and Log-Normal Regression Models

The following options are available with this model.

Graph of observed vs. predicted survival times. This option will bring up a scatterplot of the observed survival times and the estimated survival times.

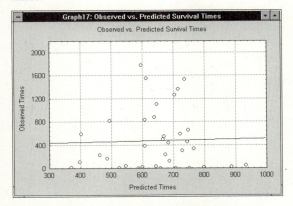

Graph of predicted vs. residual survival times. Pressing this button will bring up a scatterplot of the predicted versus residual survival times.

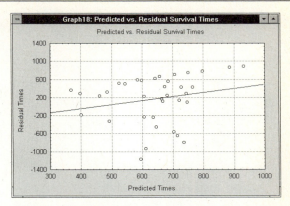

Normal probability plot of residuals. This option will bring up a normal probability plot of the residual survival times.

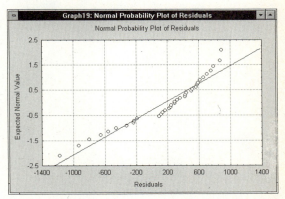

If the residual survival times (or log-survival times) are normally distributed, all plotted points in this graph should be arranged roughly in a straight line.

SUR - 3529

TECHNICAL NOTES

Life Table

The life table technique is one of the oldest methods for analyzing survival (failure time) data (e.g., see Berkson and Gage, 1950; Cutler and Ederer, 1958; Gehan, 1969). In order to arrive at reliable estimates of the three major functions (survival, probability density, and hazard), the minimum recommended sample size is 30 (see Lee, 1980). The life table estimates are computed according to standard formulas as described in Lawless (1982), Lee (1980), Nelson (1982), and others (refer also to the *Introductory Overview*, page 3485).

Specifying tabulated survival data as input.

The life table routine in *Survival Analysis* will accept already tabulated data as input. For example, suppose you had 750 observations in a study, and have already tabulated the survival times in the following manner:

```
Var_1:      Var_2:         Var_3:
--------    -----------    -----------
Interval    No. of Obs.    No. of Obs.
Start       Censored       Failed
--------    -----------    -----------
0           10             185
1           10             88
2           10             55
3           10             43
4           14             32
5           52             31
6           38             20
7           24             7
8           25             6
9           24             6
```

In the *Life Table & Distribution of Survival Time* dialog, you would first set the *Input data* option to *Table of survival times*. Then, specify the three variables in exactly this order (*1-3*) in the variable selection dialog (click on the *Variables* button). Finally, you would enter the number of observations at the beginning of the study (that entered the first interval) via the edit box in this dialog.

Fitting Theoretical Survival Distributions

The regression procedure for fitting the four theoretical distributions to the life table is based on algorithms proposed by Kennedy and Gehan (1971), and discussed in detail in Lee (1980). Basically, the hazard functions (specifically, the logarithmic transforms of the hazard functions) of all four theoretical distributions are linear functions of the survival times (or log-survival times). Thus, the hazard functions [$h(t)$] may be expressed in terms of linear regression functions as:

```
h(t) = L              Exponential
h(t) = L'*g*t^(g-1)   Weibull  (where L'=L^g)
h(t) = exp(L+g*t)     Gompertz
h(t) = L+g*t          Linear exponential
```

If you set $y=h(t)$ or $y=log\ h(t)$, then all four models can be stated in the general form:

```
y = a + b*x
```

Weighted least squares.
Gehan and Siddiqui (1973) suggest weighted least squares methods to fit the parameters of the respective models to the data. Specifically, the program will minimize the quantity:

```
WSS = Σ[w_i(y_i-a-b*x_i)^2]
```

Three different weights are used in the estimation:

```
w_i = 1           (unweighted least squares)
w_i = 1/v_i
w_i = n_i*h_i
```

where v_i is the variance of the hazard estimate, and h_i and n_i are the interval width and number of observations exposed to risk in the i'th interval, respectively.

Correcting intervals without terminations (deaths).
The *Correct intervals containing no terminations/deaths* check box on the *Life Table & Distribution of Survival Times* dialog determines whether intervals in which there are no deaths/terminations will be adjusted so that survival

distributions can be fitted. Parameters cannot be computed for intervals where no deaths/terminations occur; by default (box is checked) the proportion surviving in those intervals is therefore adjusted by the program (the proportion surviving is set to .5/number exposed). De-select this option when generating a life table for descriptive purposes only.

Two Sample Tests

There are five two-sample tests that can be computed with this module, namely, Gehan's generalized Wilcoxon test (Gehan, 1965a, 1965b, computed via Mantel's procedure; Mantel, 1967), Cox's *F*-test (Cox, 1964), Cox-Mantel's test (Cox 1959, 1972; Mantel, 1966), the log-rank statistic (Peto and Peto, 1972), and Peto and Peto's generalized Wilcoxon test (Peto and Peto, 1972). All of these procedures test the statistical significance of survival in two groups.

Analyzing large samples. Because of the nature of the algorithms used in the two-sample comparison routines, this program imposes stricter limitations on the maximum group size than other routines available in *Survival Analysis*. Use the multiple sample comparison option to analyze large data sets; if only two groups are specified for the multiple sample procedure, the computations and resulting statistics will be identical to those of Gehan's generalized Wilcoxon test.

Mantel and Haenszel Test

Your *STATISTICA* program contains an example *STATISTICA BASIC* program (file *Manthaen.stb* in the *STBASIC* subdirectory) for computing the Mantel-Haenszel test for comparing two groups (see Mantel and Haenszel, 1959; see also Lee, 1992). This test has been used in many clinical and epidemiological studies, in order to control for the effects of confounding variables. The test is based on the analysis of 2 x 2 tables (e.g., *Group 1/2* vs. *Survival*), stratified by another categorical variable (the confounding variable; e.g., *Location*). It allows you to test whether the two variables in the 2 x 2 tables (e.g., *Group* and *Survival*) are associated with each other, after controlling for the stratification variable (e.g., *Location*).

Cox Proportional Hazard Model

The proportional hazard model (Cox, 1972) is the most general of the regression models because it does not make any assumptions about the nature or shape of the underlying survival distribution. The model assumes that the underlying hazard *rate* (rather than survival time) is a function of the independent variables (covariates); thus, in a sense, Cox's regression model may be considered a nonparametric method. The model may be expressed as:

$$h(t,z) = h_0(t)*exp(b'z)$$

In the above expression $h(t,z)$ is the hazard rate, contingent on a particular covariate vector z; $h_0(t)$ is referred to as the *baseline hazard*; that is, it is the hazard rate when the values for all independent variables (i.e., in z) are equal to zero; b is the vector of regression coefficients.

The program uses the Newton-Raphson method to maximize the simplified partial likelihood (proposed by Breslow, 1974):

$$L = \Pi\{exp(b's_i)/[\Sigma exp(b'z_j)]^{d(i)}\}$$

In this formula, $d(i)$ stands for the number of cases observed to fail at time t_i, s_i stands for the sum (over the $d(i)$ cases observed to fail at time t_i) of the covariates, z_j stands for the covariate vector for a case j in the risk set at time t_i. (The geometric sum Π is over all k distinct failure times, the regular sum Σ is over all cases $j \in R_i$ in the respective *risk set* R_i, that is, cases that are observed to fail at or after the respective failure time t_i.)

To estimate the survival function S contingent on a particular covariate vector z, the algorithm uses the relationship:

$$S(t,z) = S_0(t)\exp(b'z)$$

In this formula, S_0 is the *baseline survivorship* which is independent of the covariates. Breslow (1974) proposed the following estimator of the baseline survival function:

$$S_0(t_i) = \Pi\{1-[d/\sum\exp(b'z)]\}$$

The *Chi-square* value is computed as a function of the log-likelihood for the model with all covariates (l_1), and the log-likelihood of the model in which the coefficients for all covariates are forced to 0 (L_0); specifically:

$$\text{Chi-square} = 2*(L_0 - L_1)$$

Assumptions. While no assumptions are made about the shape of the underlying hazard function, the model equations shown above do imply two assumptions. First, they specify a multiplicative relationship between the underlying hazard function and the log-linear function of the covariates. This assumption is also called the *proportionality assumption*. In practical terms, it is assumed that, given two observations with different values for the independent variables, the ratio of the hazard functions for those two observations does not depend on time. The second assumption of course is that there is a log-linear relationship between the independent variables and the underlying hazard function.

Estimating the Parameters for Time-Dependent Covariates

To accommodate time-dependent covariates, the following modification to the simplified partial likelihood (Breslow, 1974) is introduced (see Lawless, 1982, page 393):

$$L = \Pi\{\exp(b's_i(t_i))/[\sum\exp(b'z_j(t_i))]^{d(i)}\}$$

In this formula, $s_i(t_i)$ stands for the sum of the transformed (time-dependent) covariates at a particular time t_i, $z_j(t_i)$ stands for the transformed (time-dependent) covariate vector for a case j in the risk set at time t_i, and $d(i)$ stands for the number of cases observed to fail at time t_i. (The geometric sum Π is over all k distinct failure times, the regular sum Σ is over all cases $j \in R_i$ in the respective *risk set R_i*, that is, cases that are observed to fail at or after the respective failure time t_i.)

This partial likelihood function is minimized using the Newton-Raphson method. Note that in order to compute the likelihood for a given set of parameters, for each case i, all cases with survival times greater than or equal to that of case i have to be processed. Thus, fitting models with time-dependent covariates may require extensive computations, particularly when there are many cases in the data file.

Syntax Rules for Specifying (Time-Dependent) Covariates

The following is a summary of syntax conventions for the arithmetic expressions for the time-dependent covariates in *Survival Analysis*. For the most updated complete reference of all available operators and functions, use the *Electronic Manual* (i.e., press F1 or click on the [?] button in the respective dialog).

The general conventions are similar to those used in spreadsheet formulas for transforming individual variables in the data file.

- Refer to variables either by their numbers (e.g., *v1*, *v2*, etc.) or name (e.g., *Hospital*, *Survival*, etc.)

- Denote survival times by *t_* or *T_* in order to define time-dependent covariates (e.g., *v2*(log(t_)-5.4)*)

SUR - 3533

10. SURVIVAL ANALYSIS - TECHNICAL NOTES

- *v0* is the case number
- A semicolon starts a comment: e.g.,
  Int = Age*Gender ; interaction between factors
- Operators: +, -, *, /, ** or ^ (exponentiation), <, >, >=, <=, <>, ( )

Constants. The following constants are supported in these expressions:

`Pi`   - π (3.14159265358979)

`Euler` - *e* (2.71828182845905)

Functions. Functions supported in these expressions are:

`abs(x)`	- absolute value of x
`sign(x)`	- sign of x: if x>0 then +1, if x<0 then -1, if x=0 remains 0
`sin(x)`	- sine of x
`cos(x)`	- cosine of x
`tan(x)`	- tangent of x
`arcsin(x)`	- arc sine of x
`sinh(x)`	- hyperbolic sine of x
`exp(x)`	- e to the power of x
`log(x)`	- natural logarithm of x
`log2(x)`	- binary logarithm of x
`log10(x)`	- common logarithm of x
`sqrt(x)`	- square root of x
`trunc(x)`	- truncate x to the nearest integer
`rnd(x)`	- random number in the range of 0 to x

If the value of any variable used in the formula is missing (in the current case), then the expression evaluates to missing data (for the current case).

Logical operators can be used to define conditional transformation expressions (e.g., *Age*(Hospital=1) - Age*(Hospital=2)*).

For more extensive data transformations, use *STATISTICA BASIC* (a general programming language which supports a variety of special facilities for statistical data management, graphics, and transformations).

Distributions. Distributions (and their integrals) supported in these expressions are listed below. (Note that *x* represents the point at which the distribution will be computed; *parameter1* and *parameter2* depend on the distribution and can represent the mean, standard deviation, etc., or other distribution specific parameters):

Distribution	Density or Probability Function	Distribution Function	Inverse Distribution Function
Beta	beta(x,ν,ω)	ibeta(x,ν,ω)	vbeta(x,ν,ω)
Binomial	binom(x,p,n)	ibinom(x,p,n)	
Cauchy	cauchy(x,η,θ)	icauchy(x,η,θ)	vcauchy(x,η,θ)
Chi-square	chi2(x,ν)	ichi2(x,ν)	vchi2(x,ν)
Exponential	expon(x,λ)	iexpon(x,λ)	vexpon(x,λ)
Extreme	extreme(x,a,b)	iextreme(x,a,b)	vextreme(x,a,b)
F	F(x,ν,ω)	iF(x,ν,ω)	vF(x,ν,ω)
Gamma	gamma(x,c)	igamma(x,c)	vgamma(x,c)
Geometric	geom(x,p)	igeom(x,p)	
Laplace	laplace(x,a,b)	ilaplace(x,a,b)	vlaplace(x,a,b)
Logistic	logis(x,a,b)	ilogis(x,a,b)	vlogis(x,a,b)
Lognormal	lognorm(x,μ,σ)	ilognorm(x,μ,σ)	vlognorm(x,μ,σ)
Normal	normal(x,μ,σ)	inormal(x,μ,σ)	vnormal(x,μ,σ)
Pareto	pareto(x,c)	ipareto(x,c)	vpareto(x,c)
Poisson	poisson(x,λ)	ipoisson(x,λ)	
Rayleigh	rayleigh(x,b)	irayleigh(x,b)	vrayleigh(x,b)
Student's	student(x,df)	istudent(x,df)	vstudent(x,df)
Weibull	weibull(x,b,c,θ)	iweibull(x,b,c,θ)	vweibull(x,b,c,θ)

For more information on these distributions, see the *Electronic Manual* or the *Appendix* in Volume I.

Wald Statistic

The results Scrollsheet with the parameter estimates for the Cox proportional hazard regression model includes the so-called *Wald* statistic (see Rao, 1973), and the *p* level for that statistic. This statistic is based on the asymptotic normality property of maximum likelihood estimates, and is computed as:

```
W = β * 1/Var(β) * β
```

In this formula, β stands for the parameter estimates, and *Var(β)* stands for the asymptotic variance of the parameter estimates. The Wald statistic is tested against the *Chi-square* distribution.

Exponential Regression

The regression model for censored exponential survival data has been described by Feigl and Zelen (1965), Glasser (1967), Prentice (1973), and Zippin and Armitage (1966). The algorithm for obtaining the maximum likelihood estimates uses the Newton-Raphson method and is based on Lagakos and Kuhns (1978). The model assumes that the survival time distribution, contingent on the values of a covariate vector *z*, is exponential, and that the rate parameter can be expressed as:

```
L(z) = exp(a + b*z)
```

The *Chi-square* value is computed as usual, that is, as a function of the log-likelihood for the model with all covariates (L_1), and the log-likelihood of the model in which the coefficients for all covariates are forced to 0 (zero; L_0).

Computation of the standard exponential order statistic.

One way to check the exponentially assumption of this model is to plot the residual survival times against the standard exponential order statistic α (see Lawless, 1982). Specifically, α is computed as:

```
αᵢ = S(n-j+1) - 1
```

If the exponentiality assumption is met, all points in this plot should be arranged roughly in a straight line.

Normal and Log-normal Regression

In this model, it is assumed that the survival time (or log survival times) come from a normal distribution, and that the model may be stated as:

```
tᵢ = a + zᵢ*b
```

where z_i is the covariate vector for subject *i*, and *a* is a constant. If log-normal regression is requested, t_i is replaced by its natural logarithm.

Estimation. The program uses a very efficient method - the so-called expectation maximization (*EM*) algorithm - for obtaining the maximum likelihood estimates for this model. This algorithm was first proposed by Dempster, Laird, and Rubin (1977) and is discussed in Cox and Oakes (1984). Specifically, *Survival analysis* uses routines that follow the logic developed by Wolynetz (1979a, 1979b).

Goodness of fit. The *Chi-square* value is computed as usual, that is, as a function of the log-likelihood for the model with all covariates (L_1), and the log-likelihood of the model in which the coefficients for all covariates are forced to 0 (zero; L_0). Note that the null model contains two parameters, namely maximum likelihood estimates of the variance and the mean (also estimated via the EM algorithm), and the full model contains parameters for all covariates, a constant (the intercept), and the variance. The normal regression model is particularly useful because many data sets can be transformed to yield approximations of the normal distribution. Thus, in a sense this is the most general fully parametric model (as opposed to Cox's proportional hazard model which is non-parametric), and estimates can be obtained for a variety of different underlying survival distributions (see

Schneider, 1986, for a discussion of the normal distribution and its transforms).

INDEX

A

Actuarial life table, 3499
Aggregated life table, 3500
Assumptions of the Cox proportional hazard model, 3490, 3533

C

Casewise deletion of missing data, 3526
Categorical variables, in Cox regression models, 3492
Censored observations, 3485
Chi-square tests
    exponential regression model, 3535
    goodness of fit, 3487, 3493, 3495, 3501
    goodness of fit, regression models, 3507
    multiple sample test, 3489, 3511
    normal regression model, 3535
    proportional hazard model, 3533
    stratified analysis, regression models, 3494
Comparing survival in two or more groups, 3504
Cox F test, 3488, 3518
Cox proportional hazard model, 3527, 3532
    assumptions, 3490, 3533
    categorical variables (factors), 3492
    editing expressions for time-dependent covariates, 3524
    equal coefficients, different baseline h0, 3523
    experimental designs, 3492
    likelihood function, 3533
    overview, 3490
    parameter estimation, 3492, 3533
    retrieving expressions, 3525
    reviewing results, 3509
    saving expressions, 3525
    segmented time-dependent covariates, 3492

Cox proportional hazard model (cont.)
    separate baseline hazards in different groups, 3494, 3523
    specifying time-dependent covariates, 3509, 3523
    stratification, 3491
    syntax for specifying time-dependent covariates, 3524, 3533
    technical notes, 3533
    testing the proportionality assumption, 3491
    time-dependent covariates, 3490
    time-dependent covariates example, 3508, 3524
    Wald statistic, 3526, 3535
Cox-Mantel test, 3489, 3518
Cumulative proportion surviving, 3486

D

Distributions
    exponential distribution, 3531
    fitting, 3486, 3499, 3501, 3514, 3531
    Gompertz distribution, 3531
    linear exponential distribution, 3531
    Weibull distribution, 3502, 3531

E

Editing expressions for time-dependent covariates, 3524
Exponential distribution, 3531
Exponential order statistic, 3493, 3528, 3535
Exponential regression, 3493, 3528, 3535

G

Gehan's Wilcoxon test, 3518
Gompertz distribution, 3531
Goodness of fit, 3487, 3493, 3495, 3501

H

Hazard rate, 3486, 3502, 3515

K

Kaplan-Meier product-limit estimator, 3487, 3503, 3511, 3515, 3516

L

Life table analysis, 3485, 3495, 3499, 3500, 3511, 3512, 3513, 3514, 3531
Likelihood function, 3533
Linear exponential distribution, 3531
Log-normal regression, 3493, 3529, 3535
Log-rank test, 3518

M

Mantel-Haenszel test, 3489, 3496, 3532
Mean substitution of missing data, 3526
Median survival time, 3486
Missing data deletion
    casewise deletion of missing data, 3526
    mean substitution of missing data, 3526

N

Normal regression, 3493, 3529, 3535
Number of cases at risk, 3486

P

Parameter estimation, time-dependent covariates, 3492, 3533
Percentiles of survival function, 3504, 3516
Peto and Peto's Wilcoxon test, 3519
Probability density function, 3502, 3515
Product-limit estimator, 3487, 3495, 3503, 3511, 3515, 3516
Proportion failing, 3486
Proportion surviving, 3486

R

Retrieving expressions for time-dependent covariates, 3525

10. SURVIVAL ANALYSIS - INDEX

S

Saving expressions for time-dependent covariates, 3525
Segmented time-dependent covariates, 3492
Specifying time-dependent covariates, 3523
Standard exponential order statistic, 3493, 3535
Stratified analyses, 3491, 3493
Survival analysis
    aggregated life table, 3500
    censored observations, 3485
    censoring indicator variable, 3512
    codes for complete and censored responses, 3513, 3516, 3517, 3520, 3523
    comparing multiple groups, 3488, 3504, 3511, 3520
    comparing two groups, 3488, 3504, 3505, 3511, 3517
    computation of the standard exponential order statistic, 3535
    correcting intervals without terminations (deaths), 3531
    Cox proportional hazard model, 3490, 3527, 3532
    Cox-Mantel test, 3489, 3518
    cumulative proportion surviving, 3486
    distribution fitting, 3486, 3501, 3514, 3531
    equal coefficients, different baseline h0, 3523
    exponential distribution, 3531
    exponential order statistic (alpha), 3528, 3535
    exponential regression, 3493, 3528, 3535
    F test, 3518
    Gehan's Wilcoxon test, 3518
    Gompertz distribution, 3531
    goodness of fit, 3487, 3493, 3495, 3501
    hazard rate, 3486, 3502, 3515
    Kaplan-Meier product-limit estimator, 3487, 3503, 3511, 3515, 3516
    life table analysis, 3485, 3511, 3512, 3513, 3514, 3531
    linear exponential distribution, 3531

Survival analysis (continued)
    log-normal regression, 3493, 3529, 3535
    log-rank test, 3518
    Mantel-Haenszel test, 3489, 3496, 3532
    median survival time, 3486
    multiple sample test, 3489
    normal regression, 3493, 3529, 3535
    number of cases at risk, 3486
    percentiles of the survival function, 3516
    Peto and Peto's Wilcoxon test, 3519
    plot of survival function, 3501
    probability density function, 3486, 3502, 3515
    proportion failing, 3486
    proportion surviving, 3486
    regression models, 3489, 3506, 3512, 3522, 3526
    saving Workbook files, 3525
    segmented time-dependent covariates, 3492
    separate baseline hazards in different groups, 3494, 3523
    setting parameters for estimation, 3525
    specifying survival times, 3512
    specifying tabulated survival data as input, 3531
    specifying time-dependent covariates, 3509
    standard exponential order statistic, 3493, 3535
    startup panel, 3511
    stratified analysis, 3493, 3527
    survival for user-defined values of the independent variable, 3528
    survival function, 3486, 3515
    survival function for mean values of covariates, 3527
    survival times and scores, 3520
    syntax for specifying time-dependent covariates, 3533
    time-dependent covariates, 3490, 3508, 3523
    two-sample test, 3489
    unequal proportions of censored data, 3489
    Weibull distribution, 3502, 3531
    weighted least squares, 3531

Survival distributions, 3501
Survival function, 3486, 3515
Syntax for specifying time-dependent covariates, 3524, 3533

T

Testing the proportionality assumption, 3491
Time-dependent covariates
    categorical variables (factors), 3492
    editing expressions, 3524
    example, 3508, 3524
    likelihood function, 3533
    overview, 3490
    parameter estimation, 3492, 3533
    retrieving expressions, 3525
    reviewing results, 3509
    saving expressions, 3525
    segmented time-dependent covariates, 3492
    specifying, 3509, 3523
    stratification, 3491
    syntax for specifying, 3524, 3533
    technical notes, 3533
    testing the proportionality assumption, 3491

W

Wald statistic, 3526, 3535
Weibull distribution, 3502, 3531
Weighted least squares, for fitting survival distributions, 3531
Workbook files, 3525

Chapter 11:
STRUCTURAL EQUATION MODELING

James H. Steiger
University of British Columbia

Table of Contents

Introductory Overview	3545
Program Overview	3549
Introductory Examples	3551
Dialogs and Options	3571
Examples	3599
Monte Carlo Methods	3625
Evaluating Multivariate Normality	3643
Solving Iteration Problems	3645
Technical Aspects of *SEPATH*	3651
Index	3685

Detailed Table of Contents

INTRODUCTORY OVERVIEW 3545
Conceptual Overview 3545
The Basic Idea Behind Structural Modeling 3546
Program Operation: A Basic Overview 3547

PROGRAM OVERVIEW 3549
Efficient Interface 3549
Fully Integrated Monte Carlo Module 3549
Analysis of Correlations 3549
New Standardization Technique 3549
Multiple Sample Analysis 3550
Structured Means Analysis 3550
Robust Estimation 3550

INTRODUCTORY EXAMPLES 3551
Overview 3551
A Confirmatory Factor Analysis Example 3551

11. STRUCTURAL MODELING - CONTENTS

Structural Equation Modeling and the Path Diagram 3557
Rules for *SEPATH* Path Diagrams 3559
Resolving Ambiguities in Path Diagrams 3560
Inputting Path Diagrams with the *PATH1* Language 3561
Dual Representations for Latent Variable Variances 3563
A Structural Equation Model Example 3563
    Specifying the Measurement Model for Exogenous Factors 3565
    Specifying the Measurement Model for Endogenous Factors 3566
    Specifying Structural Paths 3566
    Simple Modifications to a Structural Model 3568

DIALOGS AND OPTIONS 3571
Startup Panel 3571
    Basic Options 3571
    Path Model Group 3571
    Options Group 3572
    Monte Carlo Group 3572
Specify Groups 3572
    Multiple Group Matrix Format 3573
Analysis Parameters 3573
    Data to Analyze 3573
    Discrepancy Function 3574
    Standardization 3575
    Manifest Exogenous 3575
    Initial Values 3576
    Convergence Criteria 3576
    Global Iteration Parameters 3576
    Line Search Method 3577
    Line Search Parameters 3577
    Output Options 3578
Confirmatory Factor Model Wizard 3578
    Overview 3578
    Step 1 Dialog 3578
    Correlate Factors 3579
    Correlate Residuals 3580
    Closing Dialog 3580
Structural Modeling Wizard 3580
    Define Exogenous Variables 3581
    Correlate Factors 3581
    Define Endogenous Variables 3582
    Define Structural Equation Paths 3582
    Correlated Disturbances 3582
    Correlated Residuals 3583
Path Construction Tool 3583

Editing the Latent Variables List	3584
Adding Arrows and Wires	3584
Multiple Path Commands	3584
Adding Paths to the Model	3585
Editing the Paths Already Created	3585
Removing All Paths Involving a Latent Variable	3585
Controlling Parameter Numbers	3585
Controlling Start Values - The Start Value Group	3586
Adding Comments to the Model File	3586
Adding Group and Endgroup Commands	3586
Adjusting Analysis Parameters	3587
Monte Carlo Setup	3587
Monte Carlo Seeds	3587
Monte Carlo Replications	3587
Specifying Information to be Stored	3587
Specifying the Monte Carlo Population	3588
Creating Special Data Types	3589
Changing Group Sample Sizes and Outlier Characteristics	3589
Changing Group Distributional Characteristics	3590
Iteration Dialog	3590
Analysis Results	3591
Results Text Window	3592
Model Summary	3593
Alpha level for highlighting	3593
Basic Summary Statistics	3594
Goodness of Fit Statistics	3594
Iteration History	3595
Tests of Assumptions	3595
Measures of Univariate Skewness	3595
Measures of Univariate Kurtosis	3595
Measures of Multivariate Kurtosis	3595
Reflector Matrix	3596
Analysis of Residuals	3596
Monte Carlo Results	3596
Display Overall Results	3597
Generate Data	3598

EXAMPLES 3599

Example 1: Stability of Alienation	3599
Example 2: A Confirmatory Factor Analysis	3599
Example 3: Confirmatory Factor Analysis with Identifying Constraints	3601
Example 4: Effect of Peer Influence on Ambition	3602
Example 5: Standardized Solutions for the Effect of Peer Influence on Ambition	3604
Example 6: Factor Analysis with an Intercept Variable	3607

11. STRUCTURAL MODELING - CONTENTS

Example 7: Comparing Factor Structure in Two Groups 3608
Example 8: Testing for Circumplex Structure 3609
Example 9: Testing for Stability of a Correlation Matrix over Time.................................... 3611
Example 10: A Multiple Regression Model for Home Environment and Math Achievement..... 3612
Example 11: Structural Models for Home Environment and Mathematics Achievement..... 3613
Example 12: Test Theory Models for Sets of Congeneric Tests 3614
Example 13: Comparing Dependent Variances 3616
Example 14: A Multi-Trait, Multi-Method Model 3617
Example 15: A Longitudinal Factor Model 3619
Example 16: A Structural Model for 10 Personality and Drug Use Variables 3620
Example 17: A Test for Compound Symmetry.. 3621
Example 18: Testing the Equality of Correlation Matrices from Different Populations...... 3622

MONTE CARLO METHODS 3625
Introduction 3625
Adequacy of Sample Size and Heywood Cases in Factor Analysis 3626
Performance of GLS and ML Estimation in the Comparison of Correlation Matrices 3633
Bootstrapping 3638

EVALUATING MULTIVARIATE NORMALITY 3643
Introduction 3643
Constructing Histograms 3643
Assessing Departures from Normality 3644

SOLVING ITERATION PROBLEMS 3645
Introduction 3645
How Iteration Procedures Work 3645
When Iteration Procedure "Hangs Up" 3645
An Example 3647

TECHNICAL ASPECTS OF *SEPATH* 3651
Models and Methods 3651
    The LISREL Model 3651
    The COSAN Model 3653
    McArdle's RAM Model 3653
    The Bentler-Weeks Model 3655
    The *SEPATH* Model 3656
Statistical Estimation 3657
    General Properties of Discrepancy Functions 3658
    OLS Estimation 3659
    GLS Estimation 3659
    Maximum Wishart Likelihood Estimation 3659
    Iteratively Reweighted GLS Estimation 3659
    Chi-square Test Statistics 3660
    ADF Estimation 3660

11. STRUCTURAL MODELING - CONTENTS

- Model Identification 3662
- Unconstrained Minimization Techniques 3664
    - Steepest Descent Iterations 3665
    - Line Search Options 3665
    - Boundary Constraints 3666
    - Convergence Criteria 3666
- Noncentrality-Based Indices of Fit 3666
    - General Theoretical Orientation 3667
    - Noncentrality-Based Parameter Estimates and Confidence Intervals 3667
    - Steiger-Lind RMSEA Index 3669
    - Population *Gamma* Index 3670
    - Adjusted Population *Gamma* Index 3672
    - McDonald's Index of Noncentrality 3672
    - Extensions to Multiple Group Analysis 3672
- Other Indices of Fit 3674
    - Jöreskog-Sörbom GFI 3674
    - Jöreskog-Sörbom Adjusted GFI 3674
    - Rescaled Akaike Information Criterion 3674
    - Schwarz's Bayesian Criterion 3674
    - Browne-Cudeck Single Sample Cross-Validation Index 3674
    - Independence Model *Chi-square* and *df* 3675
    - Bentler-Bonett Normed Fit Index 3675
    - Bentler-Bonett Non-Normed Fit Index 3675
    - Bentler Comparative Fit Index 3675
    - James-Mulaik-Brett Parsimonious Fit Index 3675
    - Bollen's *Rho* 3676
    - Bollen's *Delta* 3676
- New Method for Standardizing Endogenous Latent Variables 3676
- Analyzing Correlation Matrices 3677
- Fully Standardized Path Models 3679
- Analyzing Invariance Properties 3679
    - Introduction 3679
    - Types of Invariance 3680
    - Analyzing Invariance of Fitted Covariance Structures 3680
    - Reflector Matrices 3680
    - Using Reflector Matrices 3680
- Multiple Sample Models 3681
- Models with Structured Means or Intercept Variables 3681
- Indices of Skewness and Kurtosis 3682
    - Indices of Univariate Skewness 3682
    - Indices of Univariate Kurtosis 3683
    - Indices of Multivariate Kurtosis 3683
- Monte Carlo Data Generation Techniques 3684

11. STRUCTURAL MODELING - CONTENTS

INDEX **3685**

Chapter 11:
STRUCTURAL EQUATION MODELING

James H. Steiger[1]
University of British Columbia

INTRODUCTORY OVERVIEW

Conceptual Overview

Structural Equation Modeling is a very general, very powerful multivariate analysis technique that includes specialized versions of a number of other analysis methods as special cases. In this chapter, it is assumed that you are familiar with the basic logic of statistical reasoning as described in *Elementary Concepts* (Volume I) and the concepts of variance, covariance, and correlation; if not, it is advised that you read the *Basic Statistics and Tables* chapter (Volume I) at this point. Although it is not absolutely necessary, it is highly desirable that you have some background in *Factor Analysis* (Chapter 6) before attempting to use structural modeling.

Major applications of structural equation modeling include:

(1) *Causal modeling*, or *path analysis*, which hypothesizes causal relationships among variables and tests the causal models with a linear equation system;

(2) *Confirmatory factor analysis*, an extension of factor analysis in which specific hypotheses about the structure of the factor loadings and intercorrelations are tested;

(3) *Second order factor analysis*, a variation of *Factor Analysis* in which the correlation matrix of the common factors is itself factor analyzed to provide *second order factors*;

(4) *Regression models*, an extension of *Multiple Linear Regression* analysis (Volume I) in which regression weights may be constrained to be equal to each other, or to specified numerical values;

(5) *Covariance structure models*, which hypothesize that a covariance matrix has a particular form. For example, you can test the hypothesis that a set of variables all have equal variances with this procedure;

(6) *Correlation structure models*, which hypothesize that a correlation matrix has a particular form. A classic example is the hypothesis that the correlation matrix has the structure of a circumplex (Guttman, 1954; Wiggins, Steiger, and Gaelick, 1981);

(7) *Structured means models,* which model the means of the variables as well as their variances and covariances.

Many different kinds of models fall into each of the above categories, so it is difficult to delineate the precise boundaries of what constitutes "structural modeling."

Structural equation models involving only linear relationships among manifest and latent variables can be expressed as path diagrams. This program uses a command language (*PATH1*) that looks very much like a path diagram. Consequently even beginners to structural modeling can perform complicated analyses with a minimum of training.

---

[1] J. H. Steiger is currently at the Department of Psychology, University of British Columbia, Vancouver, B.C., Canada V6T 1Z4. SEPATH was programmed by J. H. Steiger with assistance from StatSoft's Research and Development Department.

11. STRUCTURAL MODELING - INTRODUCTORY OVERVIEW

The Basic Idea Behind Structural Modeling

One of the fundamental ideas taught in intermediate applied statistics courses is the effect of additive and multiplicative transformations on a list of numbers. Students are taught that, if you multiply every number in a list by some constant K, you multiply the mean of the numbers by K. Similarly, you multiply the standard deviation by the absolute value of K.

For example, suppose you have the list of numbers *1, 2, 3*. These numbers have a *mean* of *2* and a *standard deviation* of *1*. Now, suppose you were to take these *3* numbers and multiply them by *4*. Then the mean would become *8*, and the standard deviation would become *4*, the variance thus *16*.

The point is, if you have a set of numbers X related to another set of numbers Y by the equation $Y = 4X$, then the variance of Y must be *16* times that of X, so you can test the hypothesis that Y and X are related by the equation $Y = 4X$ *indirectly* by comparing the variances of the Y and X variables.

This idea generalizes, in various ways, to several variables inter-related by a group of linear equations. The rules become more complex, the calculations more difficult, but the basic message remains the same — *you can test whether variables are interrelated through a set of linear relationships by examining the variances and covariances of the variables.*

Statisticians have developed procedures for testing whether a covariance matrix fits a specified structure. The way structural modeling works is as follows:

(1) You state (often with the aid of a *path diagram*) a model that is consistent with the way that you believe the variables are inter-related;

(2) The program works out, via some complex internal rules, what the implications of this model are for the variances and covariances of the variables;

(3) The program tests whether the variances and covariances fit this model;

(4) The program reports the results of the statistical testing, and also returns parameter estimates and standard errors for the numerical coefficients in the linear equations, as well as a variety of other diagnostic indices;

(5) On the basis of this information, you decide whether the model seems like a good fit to your data.

There are some important and very basic logical points to remember about this process, which is diagrammed below.

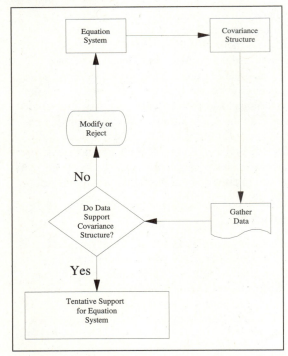

First, although the mathematical machinery required to perform structural equations modeling is extremely complicated, the basic logic is embodied in the above 5 steps.

Second, you must remember that it is unreasonable to expect a structural model to fit perfectly — for a number of reasons. A structural model with linear relations is only an approximation. The world is unlikely to be linear. Indeed, the true relations between variables are probably nonlinear. Moreover, many of the statistical assumptions are somewhat questionable as well. The real question is not so much, "Does the model fit perfectly?" but rather, "Does it fit well enough to be a useful approximation to reality, and a reasonable explanation of the patterns we have observed in our data?"

Third, keep in mind that an excellent fit of model to data does not necessarily mean that the model is correct. One cannot prove that a model is true — to assert the truth of the model is the fallacy of affirming the consequent. For example, you could say "If Joe is a cat, Joe has hair." However, "Joe has hair" does not imply Joe is a cat. Similarly, you might say that "If a certain causal model is true, it will fit the data." However, the model fitting the data does not necessarily imply the model is the correct one. There may be another model that fits the data equally well.

Program Operation: A Basic Overview

The *Structural Equation Modeling* module (*SEPATH*) offers a large number of options. Although program operation is simplified in many ways, relative to similar programs, there are many options to master. The basic operation of fitting a structural model to data can be broken down into the following basic steps:

(1) Open the appropriate data file. This task can be done from the startup panel (see page 3571).

(2) Decide on the appropriate structural model. This decision requires some familiarity with the basic concepts of structural equation modeling (see page 3557).

(3) Express your model as a *path diagram* (see page 3559).

(4) Translate the *path diagram* into the *PATH1* language (see page 3561).

(5) Enter the *PATH1* commands into a program instruction file (usually with filename extension **.cmd*). You can do this task by hand, using a *Text/output Window* (see Volume I) in the *SEPATH* module (or any ASCII text editor). However, you will find that model entry can be completed in a few minutes, with virtually no typing, using the program's user interface, which has been specifically designed to make this task as effortless as possible. For simple structural models or confirmatory factor models, use the *Path Wizards* (see page 3578). For more complex or unusual models, use the *Path Construction Tool* (see page 3583). The basic operation of these user-interface tools will be examined in detail in the *Introductory Examples* section (see page 3551).

(6) Set the program *Analysis Parameters* to the analysis options required for your particular model (see page 3573).

(7) Start program execution.

(8) Examine program results. To enhance the interactive nature of the process, the program will automatically open a *Text/output Window* and display your path model in the *PATH1* language with numerical estimates for all parameters included. If you click *OK* after the convergence of a series of iterations, the program will open a *Results* dialog (see page 3591), in which you may examine a wide variety of indices reflecting the fit of your model.

11. STRUCTURAL MODELING - INTRODUCTORY OVERVIEW

PROGRAM OVERVIEW

The *Structural Equation Modeling* module (*SEPATH*) is a full-featured program for analyzing structural equation models, causal models, and other models that can be conceptualized as special cases of path analysis. The program includes special capabilities as described below.

Efficient Interface

SEPATH features a specially designed, user-friendly interface that is highly intuitive and easy-to-use, yet both powerful and fast, even for extremely complex models. *SEPATH* uses several techniques:

The *PATH1* language. This language, (introduced by Steiger [1989] in the program EzPATH™) allows specification of path models in a highly portable, yet readable format.

Path construction tool. Typing of the *PATH1* commands is virtually eliminated through the use of the *Path Construction Tool*, a powerful editor designed especially to create commands in *PATH1*.

Path Wizards. The process of model specification is streamlined still further through the use of *Path Wizards*, which guide you step-by-step through the model specification process for the most common types of models.

Efficient display of results. Structural modeling generates a wide variety of analysis results, often spanning several pages for the fit of a single model. Other structural modeling programs scroll all this information out in a bewildering array for the user to sort through. *SEPATH* uses *STATISTICA*'s Scrollsheet technology to provide hierarchical random access to program results. At the click of a mouse button, you can quickly assess model fit, continue with a new model, or examine a large number of complex analytical results. You see only the results you want to see, immediately, when you want to see them. This technology makes the process of model selection and evaluation much more efficient and convenient.

Fully Integrated Monte Carlo Module

SEPATH includes facilities for Monte Carlo analyses, allowing you to perform complex statistical simulations of your models, to enable you to analyze, before the fact, the impact of violations of assumptions, limited sample size, model misspecification, etc. A number of integrated graphics procedures can be used to visualize the Monte Carlo results.

Analysis of Correlations

A number of authors have pointed out that some structural modeling software gives incorrect standard error estimates when a correlation matrix is analyzed as if it were a covariance matrix. This fault can be tested for by analyzing the classic Lawley example (see page 3599) with such programs. *SEPATH* can directly, and automatically analyze a correlation matrix *and* provide correct standard errors. We provide, for evaluative purposes, both the correct standard errors and the incorrect values given by some software.

New Standardization Technique

Using constrained estimation, *SEPATH* can automatically compute a completely standardized solution *with standard errors*. *SEPATH* can compute a standardized solution with all variables (latent and observed) standardized, *and* provide standard errors for such a solution.

Multiple Sample Analysis

With its GROUP command, *SEPATH* handles multiple samples routinely. *SEPATH*'s *Path Construction Tool* allows even the most complex multiple sample model to be specified fully in a few minutes.

Structured Means Analysis

SEPATH can analyze models with structured means or "intercept variables" effortlessly (see the example on page (3607), in one or several samples. One simply uses the reserved variable name *CONSTANT* to specify means and/or intercepts, and selects *Moments* as the *Data to Analyze*. No other special manipulations of data or model are required.

Robust Estimation

SEPATH can compute Asymptotically Distribution-Free (ADF) estimates and test statistics, using the standard weighted least squares approach of Browne (1984). The program can also perform bootstrap analysis.

INTRODUCTORY EXAMPLES

Overview

The *Structural Equation Modeling* module (*SEPATH*) can appear very intimidating at first glance. In this introductory example, you will be taken step-by-step through the model specification and fitting process, to demonstrate how *SEPATH* is designed to make the structural modeling process as easy and efficient as possible.

Two examples will be presented here, one a *confirmatory factor analysis* model, the other a classic structural equation model. These are the types of models processed most frequently as structural equation models. The purpose of the example is to give you a feeling for the general way the program operates. A discussion of technical specifics will be presented later in the chapter (see page 3578).

A Confirmatory Factor Analysis Example

Confirmatory Factor Analysis is an extension of *factor analysis* in which specific hypotheses about the structure of the factor loadings and intercorrelations are tested. In confirmatory factor models the factor loadings, factor correlations, and/or residual variances and covariances can be specified to be equal to each other, or to specified numerical values. Confirmatory factor models are sometimes tested as a follow-up to the standard factor analysis procedures (sometimes referred to as *exploratory factor analysis*) performed by the *Factor Analysis* module of *STATISTICA*.

As preparation for this example, turn to the *Introductory Overview* section of the *Factor Analysis* module (Chapter 6 of the manual). Work through the example, using the *Factor.sta* data file, until you finish the section titled *Interpreting the Factor Structure*.

At that stage, you will have completed a typical *exploratory factor analysis*, and arrived at a factor pattern manifesting *simple structure*. That is, some of the factor loadings are high, some are low, and each column of the factor pattern is characterized by several high and several low loadings. Moreover, the variables with high loadings in the first column have low loadings in the second column, and vice-versa. The basic form of the pattern is shown in the following table.

In the table, X stands for a "meaningful" loading, 0 for a "meaningless" loading, where "meaningless" is defined roughly as follows. A factor loading is said to be meaningless if setting it to zero would have no important effect on the ability of the factor model to reproduce the correlations among the observed variables.

Case	1 FACTOR_1	2 FACTOR_2
WORK_1	X	0
WORK_2	X	0
WORK_3	X	0
HOME_1	0	X
HOME_2	0	X
HOME_3	0	X

Suppose that, entering the factor analysis, you already had a good understanding of the underlying structure of the 6 variables, and that you hypothesized, in advance, that the structure of the factor pattern would be as shown in the above table. That is, the common factor *Factor 1* loads only on the variables WORK_1, WORK_2, and WORK_3, while the common factor *Factor 2* loads only on HOME_1, HOME_2, and HOME_3. Furthermore, you wish to allow the factors to be correlated, in order to increase the chances of obtaining a factor pattern with this structure by removing the zero correlation constraint. You could test this model as a confirmatory factor model using the *Confirmatory Factor Analysis Wizard* in *SEPATH*.

11. STRUCTURAL MODELING - INTRODUCTORY EXAMPLES

To do this, start the *Structural Equation Modeling* module either by clicking on its icon in the *STATISTICA Program Group* (if you have installed individual icons for your modules) or by selecting it from the *STATISTICA* module switcher list of modules.

When the program has opened, you will see the *Startup Panel* shown below. The *Startup Panel* is the "command center" for all activities in *SEPATH*. Most activities in *SEPATH* return control automatically to the *Startup Panel* after completion. If the *Startup Panel* is not active, you can use the CTRL+T key sequence to open it.

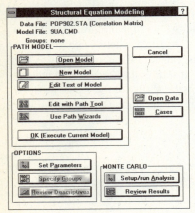

At startup, you will see two other important windows in the background. They are the data file (**.sta*) and the model file (**.cmd*) that were open when the program was last used. To perform this example, you will need to open the data file *Factor.sta* from the example files. You should also start with a new (clean) model file.

If the current data file is not *Factor.sta*, click the *Open Data* button in the *Startup Panel* and open *Factor.sta* from the *STATISTICA Examples* directory.

If the model file is not *New.cmd*, close the *Text/output Window* by double-clicking the upper-left corner. Then, click the *New Model* button on the

Startup Panel. At this point, the screen should look like this:

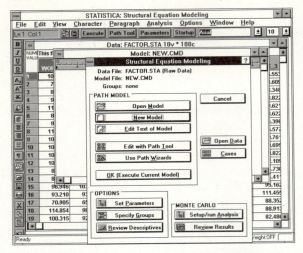

Now you are ready to set up the confirmatory factor model, using *SEPATH*'s *Confirmatory Factor Wizard*. However, before you begin, open the *Page/Output Setup* dialog (double-click on the *Output* area of the *Status Bar* or select the *Print* option from the *Options* pull-down menu) and check the *Suppl. Info. - Medium* option. Selecting this option will control the amount of supplementary information that is printed, saved to a file, or displayed in the *Text/output Window* along with the results of the analysis.

For these introductory examples, we suggest that you direct your output to a *Text/output Window*, rather than to the printer or to a file. In that case, the *Print* options dialog should look like this when you are about to exit.

11. STRUCTURAL MODELING - INTRODUCTORY EXAMPLES

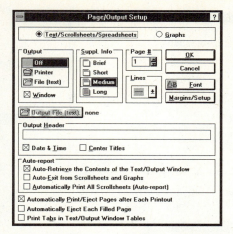

Restore the *Startup Panel* by double-clicking its icon or using the CTRL+T key sequence. Click the *Use Path Wizards* button on the *Startup Panel*. You will see the following dialog box:

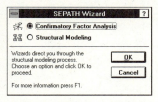

Select *Confirmatory Factor Analysis* and click *OK* to bring up the Wizard. The first dialog box you will see is designed to allow you to specify the factor names and the factor pattern quickly and efficiently.

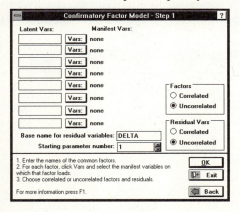

You can specify the names for up to 8 factors by simply typing them in to the appropriate edit fields. In this case, there will be two factors. Since one factor is common to the WORK variables, and one is common to the HOME variables, enter *Work* and *Home,* respectively, into the first two edit fields.

Once you have named your factors, the next step is to choose which variables will load on each factor. To select the variables that will load on the *Work* factor, click the *Vars* button next to the edit field containing its name. A variable selection dialog will open to allow you to select the variables that load on *Work*. Use the mouse to select variables *WORK_1*, *WORK_2*, and *WORK_3*. Then click *OK*. In similar fashion, select variables *HOME_1*, *HOME_2*, and *HOME_3* to load on the factor *Home*. Correlated factors (but uncorrelated residuals) are desired in this model, so click the *Correlated* radio button in the *Factors* box and the *Uncorrelated* radio button in the *Residual Vars* box of this dialog. At this point, the dialog box should appear as follows:

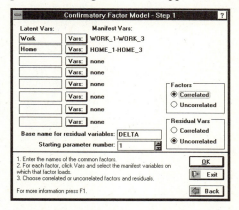

If it does, you are ready to proceed to the next step, specifying the factor intercorrelations.

Click *OK*, and the following dialog box will appear.

SEPATH - 3553

Copyright © StatSoft, 1995

11. STRUCTURAL MODELING - INTRODUCTORY EXAMPLES

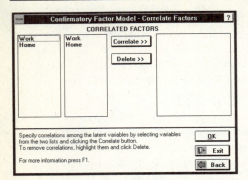

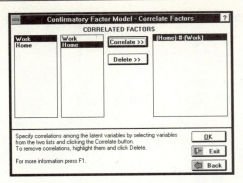

In this dialog box, you specify which inter-factor correlations you want to allow to be non-zero. Any possible correlation *not* specified at this stage will be constrained to zero during the parameter estimation process. In the dialog box, you see two lists on the left. To specify one or more correlations between factors, highlight factor names on the two lists, then click *Correlate*. All non-redundant correlations between selected factors on the left and those on the right will be added to the list on the far right. By non-redundant, we mean that only correlations of the form r_{ij} for $i>j$ will be added to the list. So, in the current example, you would obtain only the correlation between *Home* and *Work*, whether you highlighted *Home* in the left list and *Work* in the right list, *Work* in the left list and *Home* in the right list, or both *Home* and *Work* in both lists.

To specify *all possible correlations* among the factors, simply highlight all factors in both lists and click *Correlate*.

To complete the example, specify a correlation between *Home* and *Work*. Highlight *Work* on the left list, *Home* on the right, then click *Correlate>>*. The correlation path should appear in the list on the right side of the dialog, as follows:

When you are done, click *OK*. You will then see the final Wizard dialog box.

This dialog allows you to either (1) append the current model to one already in the current model file, or (2) replace the model already in the model file with the current one. This example began with an empty model file, so either choice will be fine. Click *OK* and you will return to the *Startup Panel*. Now, bring up the *Text/output Window* containing *New.cmd*, and examine its contents. The window contains the commands for specifying the model you have just created, in a special command language called *PATH1*. It should look like this:

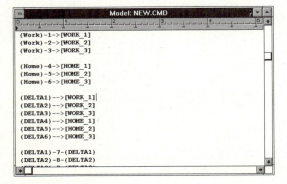

SEPATH - 3554

Later in this chapter, the *PATH1* language and how it is used to specify structural equation models will be described in detail. For now, simply note that each line of text stands for a path, and the integers in certain paths are placeholders for *free parameters*, numerical coefficients that *SEPATH* will estimate using an iterative procedure. You may want to use this file again, so use the *File Save As* menu option to save the file to disk as *Demo1.cmd*.

Before starting the statistical estimation process, you will need to adjust the analysis parameters. Covariance structural modeling procedures were originally designed to operate directly on a covariance matrix. However, unlike most other programs of its type, *SEPATH* has the ability to analyze either covariances or correlations correctly and routinely. In this case, it is far more convenient to analyze correlations in the confirmatory factor model. Traditional exploratory factor analysis procedures are generally applied to a correlation matrix, and adopting the same approach here makes the results much easier to compare.

In order to configure *SEPATH* to analyze correlations, open the *Analysis Parameters* dialog by clicking the *Parameters* button on the toolbar at the top of the screen. (If the *Startup Panel* is active, you can open the *Analysis Parameters* dialog directly by clicking on the button labeled *Set Parameters*.)

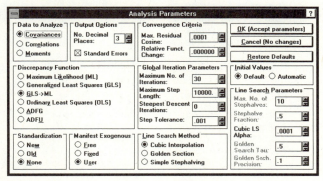

Click the *Correlations* radio button in the *Data to Analyze* group near the upper-left corner of the dialog, then click *OK* to return to the *Startup Panel*.

You are now ready to start the estimation process. You can do this in either of two ways. You can click on the *Startup Panel* icon to restore the dialog, then click on the button labeled *OK (Execute Current Model)* or you can click on the button on the *Execute* toolbar button.

When execution starts, an *Iteration* dialog will open and display the progress of the iterative estimation process.

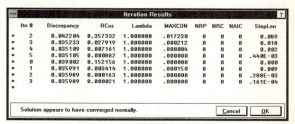

Once iteration converges, you can elect to *Cancel* and not examine program results, or proceed to the *Results* dialog. In this case, click *OK* to proceed to the *Results* dialog to examine program output.

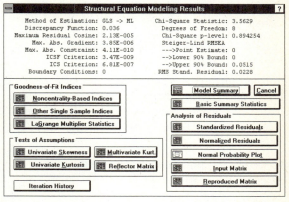

At this point in the analysis, results are actually available in two places. You will notice that an output window called *New.rtf* has opened automatically, and contains a fair amount of program output resulting from the iteration just completed. Click on this window to review the output displayed here (the *Results* dialog will minimize automatically).

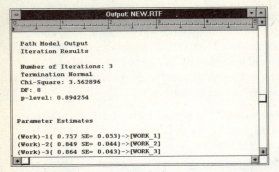

Scroll past the beginning of the output, where basic information about the analysis variables is described. There is a section indicating the major options for the analysis, i.e., the name of the data file, model file, type of data analyzed (remember, you chose to analyze the correlation matrix in this case), type of discrepancy function employed, and type of initial values used (these last two options will be discussed later in the chapter).

```
Path Model Output
Analysis Parameters

Data File: FACTOR.STA
Model File: DEMO1.CMD
Data to Analyze: Correlations
Discrepancy Function: GLS->ML
Initial Values: Default
```

This information is useful as an identification marker for the analysis, in case you perform several analyses in the same section, testing several different models (possibly with different analysis options) on the same data.

Following this information are several lines describing the basic outcome of the iteration.

```
Path Model Output
Iteration Results

Number of Iterations: 3
Termination Normal
Chi-Square: 3.562896
DF: 8
p-level: 0.894254
```

You can see from the low *Chi-square* value, and the high probability level, that the hypothesis of perfect fit for this model could not be rejected. The evidence so far suggests that this model fits these data quite well.

Next comes the output from the model estimation process. Notice that this output is in the same language as the input model, except that numerical coefficients are now reported for each model path. For example, the first 6 lines represent the factor loadings from common factor *Work* to the three WORK variables, and from factor *Home* to the three HOME variables. Beside each coefficient value, a standard error is also reported within the braces (see below). You can control whether standard errors are reported in the *Output Options* box in the *Analysis Parameters* dialog (see page 3578).

In this case, we see that all factor loadings are quite high, and greatly exceed their standard errors.

```
Parameter Estimates
(Work)-1{ 0.757 SE= 0.053}->[WORK_1]
(Work)-2{ 0.849 SE= 0.044}->[WORK_2]
(Work)-3{ 0.864 SE= 0.043}->[WORK_3]

(Home)-4{ 0.729 SE= 0.057}->[HOME_1]
(Home)-5{ 0.897 SE= 0.043}->[HOME_2]
(Home)-6{ 0.815 SE= 0.049}->[HOME_3]
```

Next come 6 paths for the residual variables, or "unique factors."

```
(DELTA1)-->[WORK_1]
(DELTA2)-->[WORK_2]
(DELTA3)-->[WORK_3]
(DELTA4)-->[HOME_1]
(DELTA5)-->[HOME_2]
(DELTA6)-->[HOME_3]
```

Finally, there are paths representing the variances of the unique factors, and the intercorrelation between the factors *Home* and *Work*.

```
(DELTA1)-7{ 0.428 SE= 0.080}-(DELTA1)
(DELTA2)-8{ 0.280 SE= 0.075}-(DELTA2)
(DELTA3)-9{ 0.253 SE= 0.075}-(DELTA3)
(DELTA4)-10{ 0.468 SE= 0.083}-(DELTA4)
(DELTA5)-11{ 0.195 SE= 0.078}-(DELTA5)
(DELTA6)-12{ 0.336 SE= 0.080}-(DELTA6)

(Home)-13{ 0.278 SE= 0.107}-(Work)
```

Now double-click on the *Results* icon to restore the *Results* dialog.

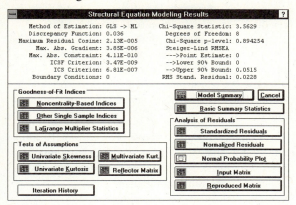

In the *Summary Box* of the dialog, you will see a text display of a number of indices designed to allow you to assess the quality of model fit quickly. Press F1 (or click on the ? button) to open the *Electronic Manual* to see a description of these indices and how they are interpreted (see also page 3666). In particular, numerical indices on the left side of the screen are, with the exception of the discrepancy function, generally close to zero if the model has been specified properly, and proper convergence of the iterative sequence has occurred.

With the *Results* dialog active, you have the option of examining a great deal of additional information. For example, click on the *Model Summary* button to see an overall summary of the model just fitted.

	Parameter Estimate	Standard Error	T Statistic	Prob. Level
(Work)-1->[WORK_1]	.757	.053	14.293	.000
(Work)-2->[WORK_2]	.849	.044	19.160	.000
(Work)-3->[WORK_3]	.864	.043	20.049	.000
(Home)-4->[HOME_1]	.729	.057	12.819	.000
(Home)-5->[HOME_2]	.897	.043	20.730	.000
(Home)-6->[HOME_3]	.815	.049	16.711	.000
(DELTA1)-->[WORK_1]	....	....	....	....
(DELTA2)-->[WORK_2]	....	....	....	....
(DELTA3)-->[WORK_3]	....	....	....	....
(DELTA4)-->[HOME_1]	....	....	....	....
(DELTA5)-->[HOME_2]	....	....	....	....
(DELTA6)-->[HOME_3]	....	....	....	....
(DELTA1)-7-(DELTA1)	.428	.080	5.338	.000
(DELTA2)-8-(DELTA2)	.280	.075	3.717	.000
(DELTA3)-9-(DELTA3)	.253	.075	3.390	.001
(DELTA4)-10-(DELTA4)	.468	.083	5.647	.000
(DELTA5)-11-(DELTA5)	.195	.078	2.508	.012
(DELTA6)-12-(DELTA6)	.336	.080	4.220	.000
(Home)-13-(Work)	.278	.107	2.600	.009

This summary shows, for each path, the estimate for the free parameter, the standard error, a *T-statistic* and the probability level. Paths with a probability level below *.05* are highlighted to indicate they are "significant." (In this example, all paths meet this criterion, and are highlighted.)

The noncentrality based index of fit is one class of statistic for evaluating the overall fit of a model to the data that is now gaining considerable favor with structural modeling experts. Some (but not all) fit indices based on noncentrality lend themselves naturally to a confidence interval approach to fit assessment. Rather than testing the overall hypothesis that fit is perfect (which often seems to work against you when sample size is high), these indices assess, with a confidence interval, how good fit is, and how accurately fit has been determined. If you click on the *Noncentrality Based Indices* button, you can obtain several of these indices, which are described in detail later in the manual. The results, in this case, show that fit of this model is excellent.

Feel free, at this stage, to explore the *Results* dialog for a while. When you are through examining numerical results from the analysis, click *Cancel* to return control to the *Startup Panel*.

The next two sections present a detailed discussion of the command language used by *SEPATH* to specify structural models, and how this language relates to a path diagram.

Structural Equation Modeling and the Path Diagram

Path Diagrams play a fundamental role in structural modeling. *Path diagrams* are like flowcharts. They show variables interconnected with lines that are used to indicate causal flow. Each *path* involves two variables (in either boxes or ovals) connected by either *arrows* (lines, usually straight, with an

11. STRUCTURAL MODELING - INTRODUCTORY EXAMPLES

arrowhead on the end) or *wires* (lines, usually curved, with no arrowhead).

One can think of a path diagram as a device for showing which variables cause changes in other variables. However, path diagrams need not be thought of strictly in this way. They may also be given a narrower, more specific interpretation.

Consider the classic linear regression equation

$Y = aX + E$

and its path representation shown below. Such diagrams establish a simple isomorphism. All variables in the equation system are placed in the diagram, either in boxes or ovals. Each equation is represented on the diagram as follows: All independent variables (the variables on the right side of an equation) have *arrows* pointing to the dependent variable. The weighting coefficient is placed in clear proximity to the arrow. The diagram below shows a simple linear equation system and its path diagram representation.

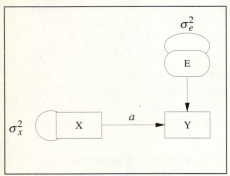

Notice that, besides representing the linear equation relationships with arrows, the diagrams also contain some additional aspects. First, the variances of the independent variables, which must be specified in order to test the structural relations model, are shown on the diagrams using curved lines without arrowheads attached. Such lines are referred to as *wires*. Second, some variables are represented in ovals, others in rectangular boxes. *Manifest variables* (i.e., those that can be measured directly) are placed in boxes in the path diagram. *Latent variables* (i.e., those that cannot be measured directly, like factors in factor analysis, or regression residuals) are placed in an oval or circle. For example, the variable E in the above diagram can be thought of as a linear regression residual when Y is predicted from X. Such a residual is not observed directly, but is in principle calculable from Y and X (if a is known), so it is treated as a *latent* variable and placed in an oval.

The example discussed above is an extremely simple one. Generally, one is interested in testing much more complicated models. As the equation systems under examination become increasingly complicated, so do the *covariance structures* they imply. Ultimately, the complexity can become so bewildering that one loses sight of some very basic principles. For one thing, the train of reasoning which supports testing causal models with linear structural equations testing has several weak links. The variables may be non-linear. They may be linearly related for reasons unrelated to what we commonly view as causality. The ancient adage, "correlation is not causation" remains true, even if the correlation is complex and multivariate. What causal modeling *does* allow you to do is examine the extent to which data fail to agree with one consequence (viz., the implied covariance structure) of a model of causality. If the linear equations system isomorphic to the path diagram does fit the data well, it encourages continued belief in the model, but does not prove its correctness.

Although path diagrams can be used to represent causal flow in a system of variables, they need not imply such a causal flow. Path diagrams may be viewed as simply an isomorphic representation of a linear equations system. As such, they can convey linear relationships whether or not causal relations are assumed. Hence, although one *might* interpret the diagram in the above figure to mean that "X causes Y," the diagram can also be interpreted as a visual representation of the linear regression relationship between X and Y.

Rules for *SEPATH* Path Diagrams

In this section, rules for path diagrams are established that will guarantee that the diagram will represent accurately any model which fully accounts for all variances and covariances of all variables, both manifest and latent. These rules are based on the following considerations.

Path diagrams consist of variables connected by *wires* and *arrows*, representing, respectively, *undirected* and *directed* relationships between variables. These variables must be either *endogenous* or *exogenous*. (An *endogenous variable* is one that is a dependent variable in at least one linear equation in the equation system under consideration; an *exogenous variable* is one that is never a dependent variable. In a path diagram, endogenous variables have at least one arrow pointing to them, exogenous variables have no arrows pointing to them.) The variables must also be either *manifest* or *latent*. Hence any variable can be classified into 4 categories: (a) manifest endogenous, (b) manifest exogenous, (c) latent endogenous, and (d) latent exogenous.

If random variables are related by linear equations, then variables which are endogenous have variances and covariances which are determinate functions of the variables on which they regress. For example, if X and Y are orthogonal and

$$W = aX + bY,$$

then

$$\sigma_W^2 = a^2\sigma_X^2 + b^2\sigma_Y^2.$$

Hence, one way of guaranteeing that a diagram can account for variances and covariances among all its variables is to require:

(1) representation of all variances and covariances among exogenous variables,

(2) no variances or covariances to be directly represented in the diagram for endogenous variables, and

(3) all variables in the diagram be involved in at least one relationship.

There is a significant practical problem with many path diagrams — lack of space. In many cases, there are so many exogenous variables that there is simply not enough room to represent, adequately, the variances and covariances among them. Diagrams which try often end up looking like piles of spaghetti. For a beautiful example of a spaghetti diagram, see page 147 of James, Mulaik, and Brett (1982).

One way of compensating for this problem is to include rules for default variances and covariances which allow a considerable number of them to be represented implicitly in the diagram.

These considerations lead to the following rules:

(1) *Manifest variables* are always represented in boxes (squares or rectangles) while *latent variables* are always in ovals or circles.

(2) Each directed relationship is represented explicitly by an *arrow* between two variables.

(3) Undirected relationships need not be represented explicitly. (See rule 9 below regarding implicit representation of undirected relationships.)

(4) Undirected relationships, when represented explicitly, are shown by a *wire* from a variable to itself, or from one variable to another.

(5) *Endogenous variables* may never have wires connected to them.

(6) *Free parameter* numbers for a *wire* or *arrow* are always represented with integers placed on, near, or slightly above the middle of the wire or arrowline. A *free parameter* is a number whose value is estimated by the program. Two free parameters having the same parameter number are required to have the same value.

11. STRUCTURAL MODELING - INTRODUCTORY EXAMPLES

(7) *Fixed values* for a *wire* or *arrow* are always represented with a floating point number containing a decimal point. The number is generally placed on, near, or slightly above the middle of the *wire* or *arrow* line. A fixed value is assigned by the user. (There are default values that are applicable in various situations.)

(8) Different statistical populations are represented by a line of demarcation and the words *Group 1* (for the first population or group), *Group 2*, etc., in each diagram section.

(9) All *exogenous* variables must have their variances and covariances represented either explicitly or implicitly by either free parameters or fixed values. If variances and covariances are not represented explicitly, then the following rules hold:

(9a) Amongst latent exogenous variables, variances not explicitly represented in the diagram are assumed to be fixed values of *1.0*, and covariances not explicitly represented are assumed to be fixed values of *0*.

(9b) Amongst manifest exogenous variables, variances and covariances not explicitly represented are assumed to be *free parameters* each having a different parameter number. These parameter numbers are not equal to any number appearing explicitly in the diagram.

By adopting a consistent standard for path diagrams, we can facilitate clear communication of path models, regardless of what system is used to analyze them. Besides standing on their own as a coherent standard for path diagrams, the above rules for *SEPATH* path diagrams are designed to match the *PATH1* language, and allow quick translation from the diagram to the language, and vice-versa.

These simplifying conventions will be adhered to within this manual. However, the typical *SEPATH* user will attempt to use the program to reproduce results from published papers employing a wide variety of standards for their path diagrams. In some cases this approach will create no problems, and the user will be able to translate directly to and from the published path diagram to a *PATH1* representation of the model. However, experience indicates that it is often useful to translate published diagrams into an *SEPATH* diagram, i.e., one which obeys rules 1-9 above, before coding the diagram in the *PATH1* language. Frequently the translation process will draw attention to errors or ambiguities in the published diagram. This issue will be discussed in the following section.

Resolving Ambiguities in Path Diagrams

The figure below shows a portion of a path diagram which is quite typical of what is found in the literature. This is not a complete diagram and it does not conform to *SEPATH* diagramming rules in the preceding section.

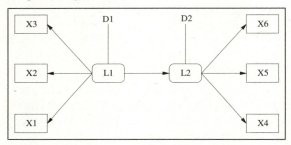

Some of the diagram is clear and routine, but what do we make of the symbols *D1* and *D2*? Variable *L1* is a latent exogenous variable. It has arrows pointing away from it and no arrows pointing to it. Since, by rule 9 for *SEPATH* diagrams (see above), all exogenous variables must have their variances and covariances explained, the most reasonable assumption is that *D1* stands for the variance of latent variable *L1*. Hence, the diagram is modified to make *D1* a parameter attached to a wire from *L1* to itself.

But what is the status of *D2*? In the diagram it looks just like *D1*, but closer inspection reveals it must mean something different. *D2* is connected to *L2*, and *L2* is an endogenous latent variable. Consequently, the most reasonable interpretation is that *D2* represents an error variance for latent variable *L2*. It is represented with an error latent variable *E2* with variance *D2*.

The revised path diagram, accurately reflecting the author's model, is shown in the figure below.

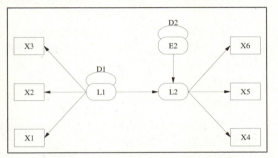

In some cases you will have to be creative, tenacious, and lucky to figure out what the author intended. Even the most accomplished and generally careful authors will leave out paths, forget to mention that some values were fixed rather than free parameters, or simply misrepresent the model actually tested. Sometimes the only way to figure out what the author actually did is to try several models with *SEPATH*, until you find coefficients which agree with the published values. These difficulties are compounded by the occasional typographical errors that appear in published covariance and correlation matrices.

It seems reasonable to conclude that if authors were to adopt *SEPATH* diagramming rules and/or report their models in the *PATH1* language, these problems would be reduced.

Some path diagrams do not represent the error variance attached to endogenous latent variables at all — they leave this to the reader to figure out for him/herself. Whenever an endogenous latent variable has no error term, you should suspect that an error latent variable has been left out, especially

if your degrees of freedom don't agree with those of the published paper.

Inputting Path Diagrams with the *PATH1* Language

PATH1 is a computer language designed to mimic the appearance of a path diagram as best as possible within a completely portable, character-based environment. *PATH1* programs are written in ordinary text files, yet they are easy to relate to a typical path diagram.

PATH1 was designed to allow quick conversion of *SEPATH* path diagrams to a form where they can be read by a computer. Here are the rules for the *PATH1* language.

(1) Each arrow and wire is represented on a separate line.

(2) Blanks never count. They are stripped from the line before parsing.

(3) Manifest variable names are represented as the full variable name enclosed within brackets. The name enclosed in brackets must follow the rules for a *STATISTICA* variable name. The name thus can be at most 8 alphanumeric characters in length. Characters must be upper case. The underscore character "_" is also allowed.

For example:

`[MATH1]`
`[HS_MATH]`
`[ANXIETY]`

(4) The manifest variable name `[CONSTANT]` is reserved for use with structured means models, or models with intercept variables. It stands for a variable with a mean of *1* and a variance of *0*.

(5) Latent variable names are represented by a variable name in parentheses. The name can be up to 20 characters in length. Upper and lower case characters are allowed, and are

distinguished. Underscores are allowed, but dashes are not allowed.

For example:

```
(Verbal_Intelligence)
(Explosive_Strength)
```

(6) Directed relationships (arrows) are represented in the following form in a *PATH1* input line:

`VNAME1-<#1>{<#2>}->VNAME2`

where *VNAME1* and *VNAME2* are valid manifest or latent variables adhering to rules 3 and 4 above, `<#1>` is an integer representing the coefficient number, and `<#2>` is a real value representing the start value.

If *VNAME1* is omitted, then the program will find the last preceeding line with two variable names and use the first variable name as *VNAME1*.

For example, the following two sets of 3 input lines are equivalent:

```
(IQ)-1->[X1]        (IQ)-1->[X1]
(IQ)-2->[X2]            -2->[X2]
(IQ)-3->[X3]            -3->[X3]
```

`<#1>` is the integer value for the parameter number. It must be between *1* and *30000* in value. `<#1>` is required if the path has a coefficient which is a free parameter. If the coefficient is a fixed value, `<#1>` is omitted.

If the coefficient for the path is a free parameter, then `<#2>` is the starting value used during iteration. If the coefficient is fixed, then `<#2>` represents the fixed value. If both `<#1>` and `<#2>` are omitted, then the path is assumed to have a fixed coefficient with a value of *1*.

For example:

```
(IQ_10)-1->[WECHSLER]
(X)-->[Y]
```

(7) Undirected relationships (wires) are presented in the following form

`VNAME1-<#1>{<#2>}-VNAME2`

where *VNAME1*, *VNAME2*, `<#1>`, and `<#2>` are the same as in the preceding section. If *VNAME1* is omitted, then the program will find the last preceeding line with two variable names and use the first variable name as *VNAME1*.

For example:

```
(L1)-1-(L2)
(Intelligence)-{.5}-(Success)
(L1)--(L1)
```

(8) Different statistical populations are denoted in the *PATH1* language by a **GROUP** statement of the form

GROUP `<#>`

where `<#>` is the number of the population.

All *PATH1* files begin (implicitly) with the statement **GROUP 1**. Groups must be stacked in order, starting with the first group. All statements are assumed to refer to the group referenced in the nearest preceding **GROUP** statement. Before beginning a new group, you must terminate a block of lines with the command

ENDGROUP

(9) Blank lines, and any lines beginning with an * (asterisk), are treated as comment lines, and are not analyzed as *PATH1* statements.

Each element in a *SEPATH* path diagram has an obvious corresponding element in *PATH1*. However there is one situation where the *SEPATH* diagram may represent information implicitly which must be expressed explicitly in the *PATH1* representation. This situation arises when the model has *manifest exogenous* variables. As noted in section 9b of the *Rules for SEPATH Path Diagrams* (see page 3560), the diagram allows you the option of not representing all the variances and covariances among exogenous manifest variables explicitly. However, for reasons of computational efficiency, the user must follow one of two courses of action. If either the *Fixed* or *Free* option has been checked in the *Manifest Exogenous* box in the *Analysis Parameters* dialog, then no variances and covariances among exogenous manifest variables are to be expressed explicitly in the diagram, as *SEPATH* will take them into account automatically. If the *User* option is selected, then all variances and covariances among exogenous manifest variables must be expressed explicitly in the *PATH1* representation. Covariances among latent and manifest exogenous variables are assumed to be zero, and must be represented explicitly if non-zero. This restriction in the program (rather than the *PATH1* language itself) presents few problems in practice.

Dual Representations for Latent Variable Variances

It is important to recognize that *PATH1* is a more flexible and precise input medium than other devices for conveying structural models. Consequently, there are often several ways to express the same model in *PATH1*. One situation arises frequently in practice and in some of the examples in this manual. There are two ways of representing (exogenous) latent variable variances. Consider the latent variable E in the portion of a path diagram shown below. E is a residual variance for the variable X.

In this version, E has a variance that, according to *PATH1* rules, is implicitly fixed at *1*.

The diagram below is equivalent.

In each case, the total path from the residual variable E has a variance a^2. In the first case, E itself has a variance of *1*, but the regression coefficient of a magnifies the variance by the square of a. In the

second case, the variance of E is a^2, and the regression coefficient of *1* leaves it unchanged. Notice that, in *PATH1*, it is actually more efficient to specify the variance using the first approach, since it requires only one *PATH1* statement, while the second approach requires two statements.

A Structural Equation Model Example

Now that you have been introduced to basic program operation, the *SEPATH* rules for path diagrams, and the command language used to input such diagrams, a more ambitious example will be presented. This example, from a paper on stability of alienation by Wheaton, Múthen, Alwin, and Summers (1977), is one of the "classic" examples of a full structural equation model discussed in many textbooks and program manuals. It is also the first example in the *SEPATH Electronic Manual*. The path diagram for this model is contained in the *Electronic Manual* (select the *SEPATH Examples* option from the *Help* pull-down menu and then select *Example 1 - Stability of Alienation*). The covariance matrix for the data, based on a sample size of *932* cases, is in a file called *Wheaton.sta*. Before proceeding, bring up the *Startup Panel* and open this data file.

Case	1 ANOMIA67	2 POWLES67	3 ANOMIA71	4 POWLES71	5 EDUCATN	6 SEINDEX
ANOMIA67	11.834	6.947	6.819	4.783	-3.839	-21.899
POWLES67	6.947	9.364	5.091	5.028	-3.889	-18.831
ANOMIA71	6.819	5.091	12.532	7.495	-3.841	-21.748
POWLES71	4.783	5.028	7.495	9.986	-3.625	-18.875
EDUCATN	-3.839	-3.889	-3.841	-3.625	9.610	35.522
SEINDEX	-21.899	-18.831	-21.748	-18.775	35.522	450.288
Means						
Std.Dev.						
No.Cases	932.000					
Matrix	4.000					

Jöreskog and Sörbom analyzed several models with these data. Below is the path diagram for the first model they analyzed. This diagram is similar to one in Jöreskog and Sörbom (1984). (To obtain a larger copy of this picture, click on *Help Examples*, go to

11. STRUCTURAL MODELING - INTRODUCTORY EXAMPLES

Example number 1 in the *Electronic Manual*, and print the topic from the Help facility.)

Remember, any arrow without any numerical index attached is assumed to have a fixed coefficient of *1*. Such arrows may be given in the *PATH1* language in a simplified form - for example, the arrow from *EPSILON1* to *ANOMIA67* is denoted as follows:

`(EPSILON1)-->[ANOMIA67]`

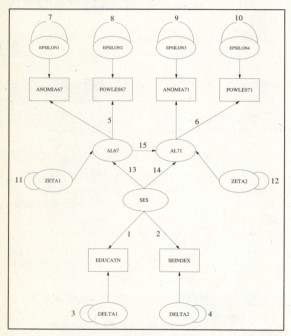

Before setting up this rather imposing looking model, try to "chop it down to size" conceptually. Like many structural equation models, this one can be conceptualized, according to the classic LISREL model of Karl Jöreskog, as being composed of three smaller models, two of them common factor models commonly referred to as "measurement models," and one a multiple regression model called the "structural model."

The basic goal of the study is to examine the regression relationships between socioeconomic status (SES) and personal alienation (AL), measured at two time points (1967) and (1971). As is frequently the case with social science data, the observed variables used to infer socioeconomic status and alienation have varying degrees of reliability of measurement. Hence correlations among the observed variables are attenuated by unreliability, and regression relationships among them may as a consequence be misleading.

To combat these problems, LISREL models postulate regression relationships among latent variables that, as common factors of the observed variables, have error of measurement "partialled out." There are two measurement models, one a factor model for the exogenous latent variables, one a factor model for the endogenous latent variables.

In this case, the two measurement models are at the top and bottom of the path diagram, the structural model is in the center. At the bottom of the diagram, you can see a factor model with two manifest variables, *EDUCATN* and *SEINDEX*, and common factor *SES*.

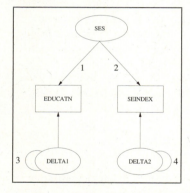

At the top of the diagram, there is a two factor model shown below, with common factors *AL67* and *AL71* loading on variables *ANOMIA67*, *POWLES67*, *ANOMIA71*, and *POWLES71*.

11. STRUCTURAL MODELING - INTRODUCTORY EXAMPLES

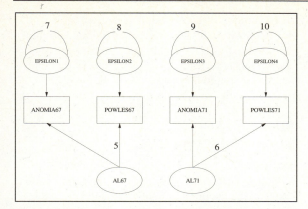

In the center of the diagram, there is a regression model shown below, relating *SES*, *AL67*, and *AL71*.

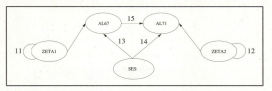

This is a good time to review a few aspects of the diagram.

First, the curved lines with the integers 3 and 4 attached to them are wires representing the variances for the residual variables *DELTA1* and *DELTA2*. Each integer in the diagram represents a free parameter to be estimated by the program. The integers 1 and 2 attached to the *arrows* from *SES* to *EDUCATN* and *SEINDEX* represent the factor loadings for those latter variables on the factor *SES*. Near the top of the diagram, you will notice that two factor loadings (the paths from *AL67* to *ANOMIA67* and *AL71* to *ANOMIA71*) do not have free parameters attached to them.

Recall that paths with neither an integer (representing a free parameter) nor a floating point number (representing a fixed numerical value) are assumed to have a fixed value of *1.0*. These paths are fixed at *1.0* in order to identify the variances of the endogenous latent factors in the diagram.

With most structural modeling programs, it would take a considerable amount of time to enter the specifications for this model. To demonstrate the power of *SEPATH*'s user interface, use the *Structural Modeling Wizard* to set up the model in the path diagram.

Make sure the data file *Wheaton.sta* is opened before you begin. Then bring up the *Startup Panel*, and click *New Model* to open a new model *Text/output Window*. Now, start the *Structural Modeling Wizard*. First click the *Use Path Wizards* button.

Then select *Structural Modeling* in the resulting *SEPATH Wizard* dialog, and click *OK*.

Specifying the Measurement Model for Exogenous Factors

The *Structural Modeling Wizard* prompts you for three submodels, in sequence. First, in the *Structural Modeling - Define Exogenous Variables* dialog, you specify the *measurement model* for the *exogenous* latent factors in your structural model. The exogenous latent factors are common factors that have no arrows pointing to them. In this case, there is only one exogenous factor in the center of the diagram, *SES*. The factor model for *SES* has the variable *SES* loading on two variables, *EDUCATN* and *SEINDEX*. It is this factor model that you specify in the first dialog.

The *Structural Modeling - Define Exogenous Variables* dialog that appears is very similar to the one in the *Confirmatory Factor Analysis* wizard (see page 3578), but is tailored specially for application in structural equation modeling. In this model, there

SEPATH - 3565

is only one exogenous variable to define. Type the name *SES* into the first edit field, then click *Vars*, and select the two variables *EDUCATN* and *SEINDEX* that load on *SES*, then click *OK*. This completes the specification of the measurement model for the exogenous factor.

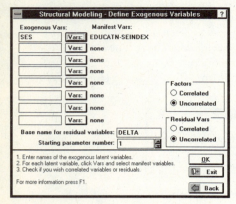

Specifying the Measurement Model for Endogenous Factors

After you click *OK*, the *Structural Modeling - Define Endogenous Variables* dialog will immediately come up for entering the latent *endogenous* factors.

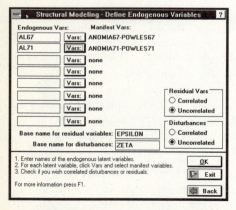

These are common factors in the diagram that have at least one arrow pointing to them. There are two such common factors in the diagram, *AL67* and *AL71*. For each factor, select all manifest variables loading on the factor. In this case, there are only two variables for each factor. When you are done, the dialog should appear as shown above. Click *OK* to move on to the next step.

Specifying Structural Paths

The next dialog that appears is the structural modeling *Path Tool*, which is used to specify paths between the latent factors in your model.

Note that the *Path Tool* has already parceled the latent variables into appropriate groups. The *From:* list on the left side of the *Variables* box in this dialog contains both endogenous and exogenous variables, because it is legal for paths to be drawn from one endogenous variable to another.

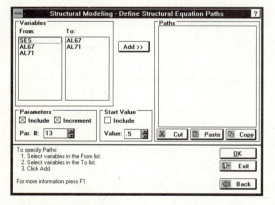

On the other hand, exogenous variables cannot receive arrows, so the exogenous latent variables are not included in the *To:* list on the right side in the *Variables* box in this dialog.

To add paths, simply highlight variables in the *From:* and *To:* lists and click on the *Add>>* button. All paths from the (highlighted) list on the left to the (highlighted) list on the right will be added to the *Paths* list on the far right of the dialog.

11. STRUCTURAL MODELING - INTRODUCTORY EXAMPLES

There are a number of options available in this dialog to allow you to control the way in which paths are numbered as they are added. Press F1 to open the *Electronic Manual* and learn more about these options (see also page 3585).

In this case, you simply wish to add 3 paths, each with a free parameter attached to it, so the procedure is simple. First, add the paths from *SES* to *AL67* and *AL71*. To do this, highlight *SES* in the *From:* list and both *AL67* and *AL71* in the *To:* list and click *Add>>*. Then add the path from *AL67* to *AL71*. As before, highlight *AL67* in the *From:* list and *AL71* in the *To:* list and again click *Add>>*.

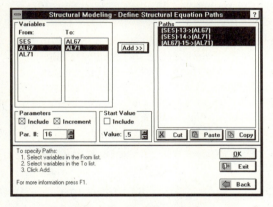

Now, click *OK* and you will, as before, see a final dialog asking how you want to save the completed model.

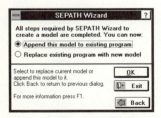

In this case, since you are operating on an empty *Text/output Window*, it doesn't really matter. Choose either option and click *OK*. At this point, you will be returned to the *Startup Panel*. Click on the *Edit Text of Model* button and you will enter the *New.cmd* window.

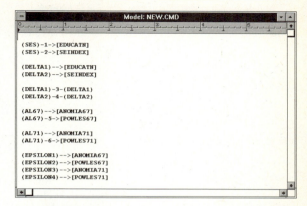

Scroll through the window, and carefully examine the paths, studying how they correspond to each path in the path diagram. Then click on the *Execute* toolbar button to begin executing the model-fitting process. The *Iteration Results* dialog will come up automatically, and the results of the iterative process will be displayed.

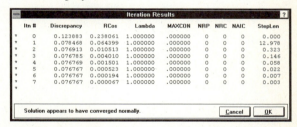

When iteration is finished, click *OK* to examine the analysis results.

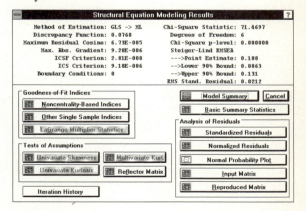

SEPATH - 3567

Copyright © StatSoft, 1995

11. STRUCTURAL MODELING - INTRODUCTORY EXAMPLES

You should see a *Chi-square* value of *71.4697* if the model has been specified according to the diagram. If you obtain some other value, you may be able to figure out what went wrong by comparing your current model with the model in the sample file *Wheatona.cmd*.

Simple Modifications to a Structural Model

Suppose you want to fit a slightly modified model with the same data. For example, suppose you want to allow the residuals attached to *ANOMIA67* and *ANOMIA71* to be correlated. You would do this by adding a single path, a wire from *EPSILON1* to *EPSILON3*. The path diagram for this model is shown below.

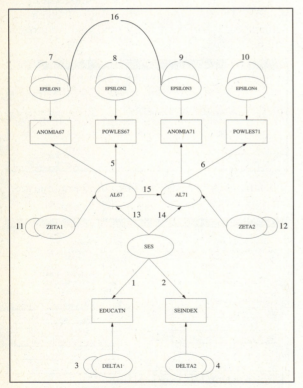

One approach to specifying such a model is to start from scratch, using the *Structural Modeling Wizard* again. However, you have several other options that may be better in this case, because the desired modification is so simple.

One option is to simply add the single path manually. You can do this by clicking *Edit Text of Model* in the *Startup Panel* and simply typing in the path:

```
(EPSILON1)-16-(EPSILON3)
```

If you do this, then click *Execute* on the toolbar and iteration will start again. Ultimately, you should obtain convergence, with a *Chi-square* value of *6.33071*, with this new model.

Another option is to use the *Path Construction Tool*, a powerful editor designed specifically to edit models specified in the *PATH1* language. Before trying the *Path Construction Tool*, go back to the *New.cmd Text/output Window*, and delete the path you just added. Then, click *Edit with Path Tool* in the *Startup Panel* to bring up the tool.

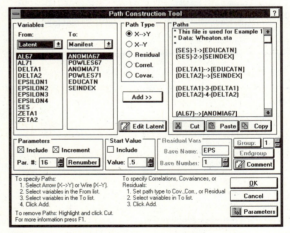

The *Path Construction Tool* is similar in general layout to the *Path Tool* in the *Structural Modeling Wizard*, but it is much more general and more powerful. Virtually any structural model can be

SEPATH - 3568

specified in just a few moments with this tool, with very little typing.

When the *Path Construction Tool* is opened, you will see all the current paths listed in the *Paths* area of the dialog. On the left are two user-selectable lists of variables (*From:*, *To:*). You can request *manifest* or *latent* variables in either list. There are a number of options for the types of paths you can construct. These are discussed in the description of the *Path Construction Tool* later in the manual (see page 3583).

For now, simply add the desired path. It is desired to add a path between 2 latent variables, so select *Latent* as the variable type in the *To:* list.

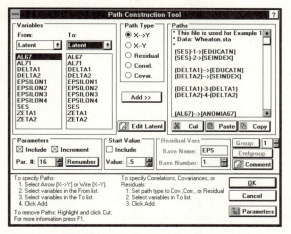

Now, highlight *EPSILON1* in the *From:* list of variables and *EPSILON3* in the *To:* list of variables. Next, select the wire path type by clicking on *X--Y* in the *Path Type* group. Then, scroll to the bottom of the list of current paths displayed in the *Paths* list on the far-right side of the dialog. Click on the area at the bottom of the list and a black bar will appear indicating where the next path will be inserted. Finally, click *Add>>*.

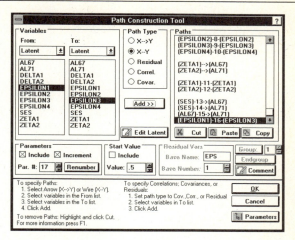

The new path will be added to the list immediately. To exit the dialog, click *OK*. You will return immediately to the *Startup Panel*. If you check the *PATH1* commands in *New.cmd*, you will see that the new path has been added to the window.

Clicking on the *Execute* button (or selecting *Execute Current Model* from the *Startup Panel* dialog), will result in this new model being fitted to the data.

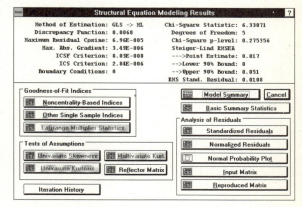

These introductory examples have given you some general notions about how *SEPATH* works, and how the program's user interface design makes structural modeling less difficult. In subsequent sections, you will learn about the many powerful features of the program, and how these features can be used to perform different kinds of analyses quickly and efficiently.

DIALOGS, OPTIONS, STATISTICS

Startup Panel

You can think of the *Startup Panel* dialog as a Master Control Panel that directs access to the many functions of *SEPATH*. Each structural modeling job requires a data (*.sta*) file, and a model (*.cmd*) file. The *Startup Panel* displays the names of these current files, and the number of groups currently selected for analysis.

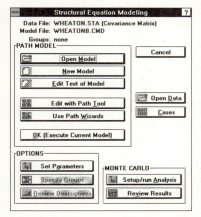

There is no variable selection dialog in this module. Whatever variables are included in the model file are selected automatically for analysis when the iterative estimation process begins.

Basic Options

Cancel. This button closes the *Startup Panel* dialog, and places the cursor in the top window underneath the panel.

Open data. This button opens a standard data file selection dialog. The data file is either *raw data* or a *covariance* or *correlation matrix* previously created via *Factor Analysis* or another *STATISTICA* module (e.g., *Basic Statistics*, *Canonical Analysis*, *Multiple Regression*, *Reliability and Item Analysis*, or directly via *Data Management*; see also, *Matrix File Format,* Volume I). Note that correlation matrix files (including multiple group correlation matrix files) can be created via the *Quick Basic Stats - Extended Options* dialog (select *Quick Basic Stats - More* from the *Analysis* pull-down menu; see Volume I for additional details).

Cases. This button opens a standard *Case Selection Conditions* dialog. For more information on the options available in this dialog, see Volume I.

Path Model Group

Open model. This button opens a standard *Open File* dialog, for opening a *.cmd* file for analysis and/or modification.

New model. This button opens a new model file, with the temporary name *New.cmd*, in a new *Text/output Window*.

Edit text of model. This button brings up the text edit window containing the current program (*.cmd*) file. You may edit the file directly in this window. This facility can be especially useful if you wish to format the text in the file to make it more readable, or add last minute comments. For more information about the editing commands available in this window, see page 1159.

Edit with path tool. This button opens the *Path Tool*, a user-friendly interface specially designed for fast, error-free entry of path analysis programs. The current program (*.cmd* file) is automatically opened by the *Path Tool* for editing. For more information, see page 3583.

Use path wizards. This button opens a dialog to allow you to select one of the special *Path Wizards*. These Wizards (see page 3578 for detailed discussions) lead you, step by step, through the construction of certain common types of path models.

11. STRUCTURAL MODELING - DIALOGS AND OPTIONS

OK (execute current model). This button causes the current path model to be tested on the current data (*.sta*) file, using the analysis options currently selected in the *Analysis Parameters* dialog (see page 3573).

Options Group

Specify groups. If your data file is a multiple group correlation or covariance matrix file, you may analyze one or several groups. This button opens a *Specify Groups* dialog in which you can define the groups to be analyzed. For more information, see below.

Set parameters. Before leaving the *Startup Panel*, you can use this option to open the *Analysis Parameters* dialog (see page 3573), in order to adjust the parameters for the analysis.

Review descriptives. This button calls up a dialog that allows you to review a variety of descriptive statistics if you have raw data to analyze.

Monte Carlo Group

SEPATH has extensive capabilities for performing *Monte Carlo* analyses.

Setup/Run analysis. This button starts the Monte Carlo experiment. First, you will be presented with the *Monte Carlo Setup* dialog (see page 3587). When you terminate the dialog by clicking *OK*, the Monte Carlo analysis will be executed, using the setup parameters you selected. Progress will be indicated in the status bar at the bottom of the screen.

Review results. This button opens a *Monte Carlo Results* dialog for reviewing the results from the most recently completed Monte Carlo experiment. For more information, see page 3596.

Specify Groups

If you are analyzing raw data, you may select a grouping variable, to be used to divide your sample into two or more groups. If your data are not in raw form, they must conform to the *Multiple Group Stacked Matrix Format* discussed below.

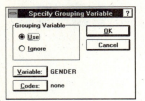

Use. If this radio button is checked, the current grouping variable and codes (see below) will be used for the analysis. The program will automatically compute the number of groups, and will expect the model fitted to the data to contain specifications for all the groups present.

Ignore. If this radio button is checked, the current grouping variable and codes will be ignored, and a single group analysis will be performed.

Grouping variable. Activating this option calls up a list of current variables. Select a categorical variable to be employed to divide your sample into two or more groups.

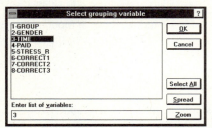

Clicking on the *Zoom* button will display summary information about a selected variable, including a list of numerical values, group mean and standard deviation, etc. Clicking on the *Spread* button will display detailed variable names. For more

SEPATH - 3572

Copyright © StatSoft, 1995

11. STRUCTURAL MODELING - DIALOGS AND OPTIONS

information about variable selection dialogs, refer to Volume I.

Codes. In this dialog you select values of the grouping variable to be used to divide your sample into groups.

One group will be created for each code value you select.

Multiple Group Matrix Format

Below is an example of a data file with correlation matrices for two groups.

NUMERIC VALUES	SEPATH Electronic Manual Example #7				
	1 X2	2 X1	3 Y2	4 Y1	5 Y3
X2	1.000				
X1	.292	1.000			
Y2	.282	.184	1.000		
Y1	.166	.383	.386	1.000	
Y3	.231	.277	.431	.537	1.000
Means	36.698	5.040	1.543	1.548	1.554
Std.Dev.	21.277	2.198	.640	.670	.627
No.Cases	432.000				
Matrix	1.000				
X2	1.000				
X1	.220	1.000			
Y2	.277	.268	1.000		
Y1	.183	.424	.550	1.000	
Y3	.142	.238	.482	.472	1.000
Means	23.467	4.041	1.288	1.129	1.300
Std.Dev.	16.224	2.097	.747	.814	.784
No.Cases	368.000				
Matrix	1.000				

If your data are not raw, and you wish to analyze multiple groups, you must construct your data file similar to the one above, in a format conforming to these rules:

(1) The variable names are the same for all groups.

(2) Data for each individual group is set up exactly the same as a matrix file for a single group. (See *Statistical Matrix File Format,* Volume I, for a detailed description of this format.)

(3) Data for the groups are stacked in a single column, with the first group data appearing first in the file, second group data appearing second, etc.

Note that correlation matrix files (including multiple group correlation matrix files) can be created via the *Quick Basic Stats - Extended Options* dialog (select *Quick Basic Stats - More* from the *Analysis* pull-down menu; see Volume I for additional details).

Analysis Parameters

This dialog allows you to specify the numerous parameters and options for the analysis.

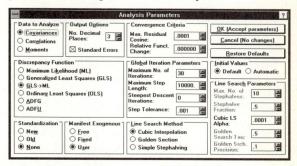

The following groups of options are available in this dialog.

Data to Analyze

Covariances. If you select this option, and the covariance matrix can be reconstructed from the data, the program will analyze the covariance matrix of the input variables, regardless of what kind of data are input. So, for example, if you input a correlation matrix file with standard deviations, *SEPATH* will calculate the covariance matrix and analyze it. If your input file is a covariance matrix file, *SEPATH* will analyze it directly. If your file contains raw data, *SEPATH* will analyze it and calculate the covariance matrix for you.

The statistical distribution of the elements of a covariance matrix is not the same as that of the elements of a correlation matrix. This fact is

SEPATH - 3573

Copyright © StatSoft, 1995

obvious if you consider the diagonal elements of a covariance matrix, which are the variances of the variables. These are random variables—they vary from sample to sample. On the other hand, the diagonal elements of a correlation matrix are not random variables—they are always *1*. The methods employed by most structural modeling programs are based on the assumption that a covariance matrix is being analyzed. The sampling distribution theory they employ is not applicable to a correlation matrix, except in special circumstances.

Recent research has emphasized that it is possible (indeed likely) that you will get some wrong results if you analyze a correlation matrix as if it were a covariance matrix. However, a number of currently distributed structural modeling programs will analyze a correlation matrix as though it were a covariance matrix. The fact that such programs yield incorrect results has been described in the literature (see, for example, Cudeck, 1989). In order to provide compatibility with these other programs, *SEPATH* will analyze a correlation matrix as if it were a covariance matrix, but it can, unlike most other programs, directly and automatically provide correct analysis of a correlation matrix as well.

Correlations. If you select this option, *SEPATH* will calculate the correlation matrix from the input data and analyze it. *SEPATH* analyzes the correlation matrix correctly, using constrained estimation theory developed by Michael Browne (see Browne, 1982; Mels, 1989; Browne and Mels, 1992). As a result, *SEPATH* gives the correct standard errors, estimates, and test statistics when a correlation matrix is analyzed directly. When combined with the *Standardization -- New* option (see *Standardization*, page 3575, for a discussion of this option) *SEPATH* can estimate a *completely standardized path model*, where all variables, both manifest and latent, are standardized to have unit variance, and standard errors of the path coefficients can be estimated as well.

Moments. If you select this option, *SEPATH* will analyze the *augmented product-moment matrix* instead of the covariance matrix. This option is selected if you are analyzing models involving intercepts or structured means. (See page 3607 for an example.) *SEPATH* will also add, to the input variable list, an additional variable called *CONSTANT* that always takes on the value *1*. *CONSTANT* is a dummy variable used to represent means in structured means models, i.e., those with an intercept variable. Because *CONSTANT* is always an exogenous variable, *SEPATH* automatically selects the *Manifest Exogenous Free* option (page 3576) when you choose to analyze moments.

Discrepancy Function

In this section of the *Analysis Parameters* dialog, the user selects which discrepancy function or functions will be minimized to yield parameter estimates. Consult the section on *Statistical Estimation* (page 3657) for an in-depth technical discussion of these methods.

Maximum likelihood (ML). When you select this option, *SEPATH* performs Maximum Wishart Likelihood estimation if *Correlations* or *Covariances* are analyzed, and Maximum Normal Likelihood estimation if *Moments* are analyzed.

Generalized least squares (GLS). If this option is selected, *SEPATH* performs Generalized Least Squares estimation.

GLS --> ML (GLS followed by ML). By default, this option is selected. *SEPATH* performs 5 iterations using the Generalized Least Squares estimation procedure, regardless of the current setting in the *Maximum No. of Iterations* field of the *Global Iteration Parameters* group in the *Analysis Parameters* dialog (see page 3576). At that point it shifts to Maximum Likelihood Estimation (see above).

Ordinary least squares (OLS). If this option is selected, *SEPATH* performs Ordinary Least Squares estimation.

Asymptotically distribution free Gramian (ADFG). *SEPATH* performs Asymptotically Distribution Free estimation, which does not require the assumption of multivariate normality. In this variant, the weight matrix is guaranteed to be Gramian (symmetric, positive semi-definite). Generally this matrix will be invertible as well. As a preliminary to ADFG estimation, *SEPATH* performs GLS estimation.

Asymptotically distribution free unbiased (ADFU). *SEPATH* performs Asymptotically Distribution Free estimation, which does not require the assumption of multivariate normality. In this variant, the weight matrix is an unbiased estimate of the true weight matrix. However, this estimated matrix may not be Gramian (i.e., symmetric positive semi-definite) in all cases. If it is not, the matrix will not be invertible, and so ADF estimation cannot be performed. In this case *SEPATH* will give an error message. At that point, the user should restart the estimation using the ADFG option. As a preliminary to ADFU estimation, *SEPATH* performs GLS estimation.

Standardization

In this section of the *Analysis Parameters* dialog, the user chooses to generate either a standardized solution (where latent variables all have unit variance) by one of two methods, or an unstandardized solution.

New. If this option is selected, *SEPATH* estimates a standardized solution via constrained estimation. This approach produces a solution where all latent variables, both independent and dependent, have variances of *1*. Unlike the old method, however, standard errors are available with this option. Combining this option with the *Data to Analyze - Correlations* option available in this dialog (see above) allows one to estimate a *completely standardized path model*, where all variables, manifest and latent, have unit variances, and standard errors can be estimated for all parameters.

Old. Some programs generate a standardized solution *after* iteration is complete, then perform a calculation *after the fact* to transform the solution to a standardized form. This method gets a solution faster than the *New* option described above, because it does not need to use constrained estimation. However, standard errors cannot be computed.

None. An unstandardized solution is calculated.

Manifest Exogenous

Most models do not have *manifest* variables that are *exogenous*. If a model does have such variables, their variances and covariances must be accounted for in the *PATH1* model statements. In most cases (but *not all*), the variances and covariances estimated under one of the standard estimation techniques will be identical to the observed variances and covariances, and consequently these parameters are not of much interpretive interest. *SEPATH* provides two approaches to calculating these parameters automatically and keeping them out of view (the *Fixed* and *Free* options below) and also allows them to be declared explicitly.

Fixed. In this case, the variances and covariances for *manifest exogenous* variables are fixed (at the value of the observed variances and covariances) during iteration, then treated as if they were free parameters at the end of iteration. This approach eliminates, in the case of several manifest exogenous variables, the addition of a number of extra free parameters, which can slow down iteration and make it less reliable. The user should take the output obtained by this approach, and resubmit it using the *Free* option described below to guarantee correct estimates. If you select this option and attempt (in the *PATH1* input) to specify the variance or covariance of a *manifest exogenous* variable, you will generate an error message.

Free. If this option is selected, the variances and covariances among the manifest exogenous variables are treated as free parameters and added to the model, although their values are not shown. Start values are the observed variances and covariances. If you select this option and attempt (in the *PATH1* input) to specify the variance or covariance of a *manifest exogenous* variable, you will generate an error message.

User. If this option is selected, you *must* account for the variances and covariances of all *manifest exogenous* variables, using standard *PATH1* syntax. This can be done easily in a few seconds, using the *Covar.* option in the *Path Type* group in the *Path Construction Tool* (see page 3585).

Initial Values

In this section of the *Analysis Parameters* dialog, you select the method employed to find initial values for free parameters.

Default. The default method uses .5 for all free parameters, except variances and covariances (or correlations) of manifest exogenous variables. These parameters are initialized at the values obtained from the sample data. If this option is selected, you may also specify starting values for any free parameter by including the desired value in braces immediately following the parameter number. So, for example, the following line would result in a starting value of -.47 for parameter number 5.

```
(F1)-5{-.47}->[X2]
```

Automatic. If this option is selected, the initial values are obtained by a method that is a minor adaptation of the technique described by McDonald and Hartmann (1992).

Convergence Criteria

In this box of the *Analysis Parameters* dialog you can adjust constants that can directly affect the point during iteration at which the program decides convergence has occurred. The default values produce desirable results for a wide variety of problems, and you will seldom need to adjust these criteria.

Maximum residual cosine criterion. This criterion becomes small when parameter values have stabilized. You may alter this criterion in the input field. The default value of .0001 works well across a wide variety of situations. For a technical definition of this criterion and a discussion of its merits, see the technical discussion of minimization methods on page 3664.

Relative function change criterion. This criterion (defined on page 3666) becomes small when the discrepancy function being minimized is no longer changing. On occasion, especially when *boundary cases* are encountered, it will not be possible for the program to iterate to a point where a local or global minimum is obtained. In that case, the *Maximum Residual Cosine* criterion may not fall below its tolerance value, but the discrepancy function will not change. At that point, this criterion will stop iteration. In general, it should be kept very low, or it may cause iteration to terminate prematurely.

Global Iteration Parameters

In this box of the *Analysis Parameters* dialog, you enter parameters that control the basic iterative process.

Maximum number of iterations. In this field, you enter the maximum number of iterations allowed. The default number is *30*. When this number of iterations is reached, the iterative process will automatically terminate, and a message will be issued to indicate that the maximum number of iterations was exceeded. Minimum is zero (in which case the discrepancy function and estimated covariance matrix will be computed, and control will be returned to the user), maximum is *1000*.

Maximum step length. In this field, you enter the maximum length of the step vector that will be allowed. See the section on *Unconstrained Minimization Techniques* (page 3664) for a discussion of this parameter. See the discussion of *Solving Iteration Problems* (page 3645) for suggestions on how to use this parameter to solve some problems encountered during iteration.

No. of steepest descent iterations. In this field, you enter a number of *Steepest Descent Iterations* to proceed the standard iterations. See the discussion of *Solving Iteration Problems* (page 3645) for suggestions on how to use this parameter to solve some problems encountered during iteration.

Step tolerance. In this field, you enter a tolerance value at which a parameter is temporarily eliminated from the iterative process. The tolerance value is basically one minus the squared multiple correlation of a parameter with the other parameters. If a parameter becomes highly redundant with other parameters during iteration, the *approximate Hessian* employed during Gauss-Newton iteration becomes unstable. Very low step tolerance values mean that a parameter will never be removed. Hence iteration may "blow up" if the approximate Hessian becomes nearly singular. High step tolerance values mean that parameters will not be varied while they are moderately correlated with other parameters. In these circumstances interation may "stall out" if parameters are fairly highly correlated at the solution point. Change this parameter seldom, and in small increments. The default value seems to work well on a wide range of problems.

Line Search Method

In this box of the *Analysis Parameters* dialog, you choose a basic line search method. Once the *step direction* has been chosen, the minimization problem is reduced from a problem in n unknowns to a problem in 1 unknown, i.e., the length of the step. There are 3 methods for choosing the length of the step. Their technical aspects are discussed on page 3665.

Cubic interpolation. This method is reasonably fast and rather robust. It works well in the vast majority of circumstances.

Golden section. This method tries to solve the one dimensional minimization problem *exactly* on each iteration. It often converges in slightly fewer iterations than cubic interpolation, but takes longer, because it requires more function evaluations on each iteration.

Simple stephalving. This method is the fastest, but will fail to converge for a fair number of problems on which the other two, more sophisticated methods succeed.

Line Search Parameters

In this box of the *Analysis Parameters* dialog, you choose numerical parameters that control the performance of the line search method you have chosen. Only a subset of the parameters are relevant to each line search method. Only relevant parameters will be enabled for the currently selected *Line Search Method*.

Max. no. of stephalves. This parameter sets the maximum number of stephalves allowed on a single iteration if the *Simple Stephalving* line search method is used.

Stephalve fraction. This parameter sets the fraction by which the current step is multiplied when *Simple Stephalving* is used as the line search method.

11. STRUCTURAL MODELING - DIALOGS AND OPTIONS

Cubic LS alpha. This parameter controls how large a reduction in the discrepancy function has to be made before a step is considered acceptable when the *Cubic Interpolation* line search method is used. The default value, *.0001*, allows virtually any improvement to be considered acceptable.

Golden section *tau*. This parameter controls the width of the range to which the *Golden Section* line search is limited.

Golden search precision. This parameter controls the precision of estimation in a *Golden Section* line search.

Output Options

The options in this box of the *Analysis Parameters* dialog allow you some preliminary control over the appearance of your output. You can preselect the number of decimal places to be printed in output Scrollsheets, and you can select whether or not to print estimated standard errors for model coefficients.

No. of decimal places. Here you can set the number of decimal places displayed by *default* in the text output, in the *Results Text Window* and output Scrollsheets. The number can be between *1* and *6*.

Standard errors. If this box is checked, the program will display estimated standard errors for all parameters in the *PATH1* text output, and *Model Summary* Scrollsheet. However, if OLS estimation is performed, or the "old" standardization method is employed, standard errors will *not* be available.

Confirmatory Factor Model Wizard

Overview

The *Path Wizards* are preprogrammed sequences of dialog boxes that take you, step by step, through the construction of the most common kinds of structural models. The Wizards do most of the work, and they structure the problem for you so that the chances of making a mistake in specifying a model are greatly reduced when a Wizard is guiding you.

Confirmatory factor analysis models are just like ordinary common factor models, except that some factor loadings and factor intercorrelations can be constrained to be equal to each other, or to zero. These constraints allow statistical testing of preconceived notions about the general form of the factor pattern.

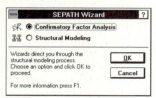

The confirmatory *Factor Analysis Wizard* allows you to construct any confirmatory factor model in just a few seconds, so long as the model has 8 common factors or less. Most models encountered in practice fall in this category. Larger models can be constructed, with a bit more effort, using the generalized *Path Construction Tool*.

Step 1 Dialog

Most of the information about the confirmatory factor model is entered in this step. First, you decide on factor names, and which *manifest* variables will load on each factor.

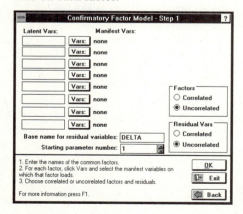

Latent vars. On the left side of the dialog, there is a column of edit fields. You enter the names of the common factors in these fields.

Vars. Click on this button to activate a list of currently available manifest variables. Use this list to select which variables will load on a particular factor.

Manifest vars. The variables you have already selected to load on a particular *latent* variable will appear in this field.

Next, you decide on the name for your residual variables, and the numbering of model parameters.

Base name for residual variables. Enter a *Base Name* for your residual variables in this field. The names of the residual variables will be constructed by adding an integer (starting at one) to this *Base Name*. So, for example, if you enter *EPS* as the base name, and you have 4 manifest variables, your residual variables will be named *EPS1*, *EPS2*, *EPS3*, and *EPS4*.

Starting parameter number. Choose a starting parameter number here. The free parameters in your model will be numbered consecutively, starting at this value.

Finally, you decide whether or not you want your common factors and residuals to be correlated.

Factors. Click on the appropriate radio button to indicate whether you want the common factors to be *Uncorrelated* or *Correlated*.

Residual vars. Click on the appropriate radio button to indicate whether you want the residuals to be *Uncorrelated* or *Correlated*.

If you decide to have correlations for either your factors or residual variables, additional dialog boxes will appear and prompt you for these correlations.

Correlate Factors

In this dialog, which comes up automatically if you checked the *Correlated* radio button in the *Factors* group in the *Confirmatory Factor Model - Step 1* dialog, you enter paths for correlations between factors. Any possible correlation *not* specified at this stage will be constrained to zero during the parameter estimation process.

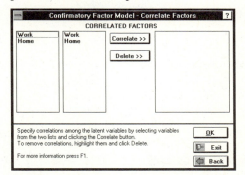

List boxes. In the dialog box, you see two lists on the left. To specify one or more correlations between factors, highlight factor names on the two lists, then click *Correlate*. All non-redundant correlations between selected factors on the left and those on the right will be added to the list on the far right. By non-redundant, we mean that only correlations of the form r_{ij} for $i>j$ will be added to the list. So, in the current example, you would obtain only the correlation between *Home* and *Work*, whether you highlighted *Home* in the left list and *Work* in the right list, *Work* in the left list and *Home* in the right list, or both *Home* and *Work* in both lists.

You may repeat the process described above until all desired correlations are specified. The far right list box will show all correlations currently queued to be added to the model.

To specify *all possible correlations* among the factors, simply highlight all variables in both lists and click *Correlate*.

Correlate. Click on this button to add correlations for the selected factors.

Delete. Click on this button to delete selected correlations from the box on the far right.

Correlate Residuals

In this dialog, which comes up automatically if you checked the *Correlated* radio button in *the Residual Vars.* group in the *Confirmatory Factor Model - Step 1* dialog, you enter paths for correlations between residual variables.

List boxes. Select the variables you want to correlate in the two boxes on the left. All *non-redundant* correlations between the variables selected in the first box and those selected in the second box will be added to the far right list box when you click on the *Correlate* button. The far right list box will show all correlations currently queued to be added to the model.

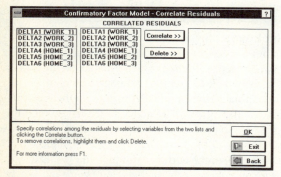

Correlate. Click on this button to add correlations for the selected variables.

Delete. Click on this button to delete selected correlations from the box on the far right.

Closing Dialog

In the final Wizard dialog, which comes up automatically after the model has been specified completely, you are given several options for terminating the Wizard or returning to a previous step in the model construction process.

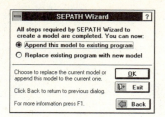

Append this model to existing program. If this radio button is selected, the model you have just specified with the Wizard will be appended to the *PATH1* model statements in the **.cmd* file that was active when you opened the Wizard.

Replace existing program with new model. If this radio button is selected, the model you have just specified with the Wizard will replace the *PATH1* model statements in the **.cmd* file that was active when you opened the Wizard.

OK. Click on this button to exit the Wizard and perform the action specified by the selected radio button.

Exit. Click on this button to exit the Wizard without saving the model statements you have just created.

Back. Click on this button to go back to a preceding step in the model construction process.

Structural Modeling Wizard

Many structural equation models can be thought of as two factor models linked by a multiple regression model. Indeed, the classic LISREL model formulation for structural equations is based on this idea.

If we envision "causal flow" in the latent variabels in the path diagram as going from left to right, we have a confirmatory factor, or "measurement" model for one set of *exogenous latent* variables on the left, a second confirmatory factor, or "measurement" model for a set of *endogenous latent* variables on the right. In the center is a "structural model," a multiple regression with arrows pointing to the *endogenous latent* variables from *exogenous latent* variables (or possibly from other endogenous latent variables). (For a discussion of these ideas in connection with an example, see the introductory example on page 3563.)

A general structural model fitting the above description actually consists of 3 models — two factor models and a regression model. In keeping with this idea, the *Structural Model Wizard* takes you, step by step, through the process of generating these 3 models.

Define Exogenous Variables

This dialog box is similar to *Step 1* in *Confirmatory Factor Analysis* (see page 3578). Here you define the Measurement Model for the independent, or *exogenous latent* variables in the path diagram. These are the variables with no *arrows* pointing to them.

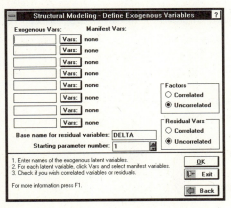

Correlate Factors

In this dialog, which comes up automatically if you checked the *Correlated* radio button in the *Factors* group in the *Structural Modeling - Define Exogenous Factors* dialog (see above), you enter paths for correlations between factors.

In this dialog box, you specify which inter-factor correlations you want to allow to be non-zero. Any possible correlation *not* specified at this stage will be constrained to zero during the parameter estimation process. In the dialog box, you see two lists on the left. To specify one or more correlations between variables, highlight variable names on the two lists, then click *Correlate*. All non-redundant correlations between selected variables on the left and those on the right will be added to the list on the far right. By non-redundant, we mean that only correlations of the form r_{ij} for $i>j$ will be added to the list. So, in the current example, you would obtain only the correlation between *Home* and *Work*, whether you highlighted *Home* in the left list and *Work* in the right list, *Work* in the left list and *Home* in the right list, or both *Home* and *Work* in both lists.

To specify *all possible correlations* among the factors, simply highlight all variables in both lists and click *Correlate*.

List boxes. Select the variables you want to correlate in these two boxes. All *non-redundant* correlations between the variables selected in the left box and those selected in the right box will be added to the model when you click on the *Correlate* button. The far right list box will show all correlations currently queued to be added to the model.

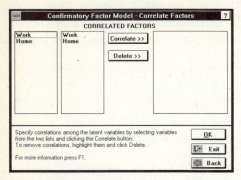

Correlate. Click on this button to add correlations for the selected variables.

Delete. Click on this button to delete selected correlations from the box on the far right.

Define Endogenous Variables

This dialog box is for definition of the Measurement Model for the dependent, or *endogenous latent* variables in your path diagram. These are the latent variables with at least one arrow pointing to them. This factor model is defined in essentially the same way as in the *Confirmatory Factor Analysis Wizard*.

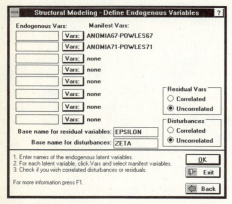

There is an important difference between the two dialog boxes, however. Each endogenous latent variable normally has a *disturbance* term, i.e., a residual attached to it. These are added automatically, with a name you have the option of changing in the dialog. You may specify whether these disturbances are correlated. If you select correlated disturbances in this dialog, you will be prompted automatically to define the correlations in a separate dialog (see *Correlated Disturbances*, below).

Define Structural Equation Paths

This dialog, which comes up automatically after the Endogenous and Exogenous variables have been selected in their respective dialogs, is used to define structural equation paths by selecting from a *From* list and a *To* list. The *From* list will automatically contain names of all the latent variables. The *To* list will automatically contain the *endogenous latent* variables, and only those variables.

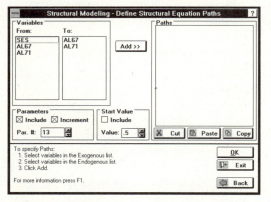

When you click the *Add* button, all paths from the currently selected *From* variables to the currently selected *To* variables will be generated.

Correlated Disturbances

This dialog is used to specify correlations among disturbance variables. This dialog will come up automatically if you have selected the *Correlated* radio button in the *Disturbances* group in the *Structural Modeling - Define Endogenous Variables* dialog (see page 3582).

11. STRUCTURAL MODELING - DIALOGS AND OPTIONS

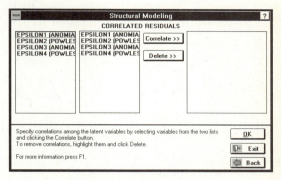

dialog (see page 3581) or the *Structural Modeling - Define Endogenous Variables* dialog (see page 3582).

The "disturbance" variables are residual latent variables with a single path to one *endogenous* latent variable. They may be correlated.

List boxes. Select the disturbances you want to correlate in these two boxes. All *non-redundant* correlations between the disturbance variables selected in the left box and those selected in the right box will be added to the model when you click on the *Correlate* button. The far right list box will show all correlations currently queued to be added to the model.

Correlate. Click on this button to add correlations for the selected disturbance variables.

Delete. Click on this button to delete selected correlations from the box on the far right.

Correlated Residuals

This dialog is used to enter paths for correlations between residual latent variables with arrows pointing to manifest variables in the two "measurement" factor models. A "residual latent variable" is an exogenous latent variable having a single arrow pointing from it to a manifest variable. Residual latent variables are used to denote residuals in factor analysis models and multiple regression models for manifest variables. The dialog will come up automatically if you have selected the *Correlated* radio button in the *Residual Vars* group in either the *Structural Modeling - Define Exogenous Variables*

List boxes. Select the residuals you want to correlate in these two boxes. All *non-redundant* correlations between the residual variables selected in the left box and those selected in the right box will be added to the model when you click on the *Correlate* button. The far right list box will show all correlations currently queued to be added to the model.

Correlate. Click on this button to add correlations for the selected residual variables.

Delete. Click on this button to delete selected correlations from the box on the far right.

Path Construction Tool

The *Path Construction Tool* allows you to construct and edit path model files in the *PATH1* language, with virtually no typing.

Because you type latent variable names only once, never have to type manifest variable names, and never have to enter paths manually, there is virtually no chance of a syntax error. Even extremely complicated models can be entered in a few minutes with this tool.

11. STRUCTURAL MODELING - DIALOGS AND OPTIONS

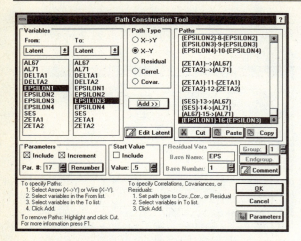

The *Path Construction Tool* also allows you to use the same facilities to alter models you have already constructed.

Editing the Latent Variables List

When the *Path Construction Tool* is first activated, the *Latent Variables* list will be presented in the *From* variables column. The variable names listed will be those *already in* the *.cmd file being edited. If the file is a new one, or if you wish to add latent variable names, you will have to edit the latent variable names list.

Edit Latent. This button activates the dialog for editing latent variables.

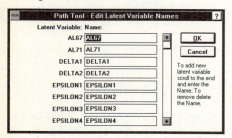

Within this dialog, you can add, delete, or change latent variable names.

Adding Arrows and Wires

There are two basic types of paths, the *arrow* and the *wire*. Basic operation is as follows:

(1) Choose a type of path by checking either the *X-->Y* or *X--Y* option in the *Path Type* group.

(2) Select the type of variable in the *From* column of the *Variables* group.

(3) Highlight the variable(s) from which to draw paths in the *From* list box.

(4) Select type of variable in the *To* column of the *Variables* group.

(5) Highlight the variables to which paths will be drawn in the *To* list box.

(6) Press the *Add>>* button.

At this point, the requested paths will appear in the *Paths List Box* on the far right side of the screen. Note that by using the *X--Y* option, one can specify correlations between latent exogenous and manifest exogenous variables, although this construction is unlikely to be used in practice.

Multiple Path Commands

Besides the basic arrow and wire path types, the *Path Type* group includes options for 3 additional *multiple path* commands. These commands automate operations which are (a) highly repetitive, and (b) common to most path diagrams. These commands can save substantial amounts of tedious typing. They are described below.

Residual. This option creates residual (or "disturbance") variables automatically for all currently highlighted variables in the *To* column. Suppose, for example, you wish to create residuals ("unique factors") for all the manifest variables (*X1, X2, X3, X4, X5, X6*) in the *Child.sta* file. You would simply:

11. STRUCTURAL MODELING - DIALOGS AND OPTIONS

(1) Select *Manifest* in the *To* variables column.

(2) Highlight all the variables.

(3) Select *Residual* as the *Path Type* option.

(4) Enter a *Base Name* (say, *DELTA*) and a *Base Number* (say, *1*) in the *Residual Vars* dialog.

(5) Click the *Add>>* button. At this point, paths for residual variables *DELTA1, DELTA2, DELTA3, DELTA4, DELTA5,* and *DELTA6* and their variances will be added automatically, and these residual variable names will be added to the latent variables list.

Base name. This name will be used for all residual variable names created during the current residual variables operation.

Base number. As residual variables are created, this number will be added to the *Base Name*, then incremented. So, if the *Base Name* is *DELTA*, and the *Base Number* is *5*, the first residual variable created will have the name *DELTA5*, the second *DELTA6*, etc.

Correl. Frequently in a structural model, you need to insert paths for the correlation coefficients among a set of latent or manifest variables. The program does this automatically for all variables highlighted in the *To* column. For example, if you highlight variables *X1, X2, X3* and click the *Correl.* button, you get the following paths:

```
[X2]-1-[X1]
[X3]-2-[X1]
[X3]-3-[X2]
[X1]-{1.0}-[X1]
[X2]-{1.0}-[X2]
[X3]-{1.0}-[X3]
```

Covar. This option creates paths for all the variances and covariances for all variables highlighted in the *To* variables column. It operates in virtually the same way as the *Correl.* option, but does not specify the variances to be *1.0* as above.

Adding Paths to the Model

Add>>. This button causes paths to be created, according to the currently created options. The new paths will appear immediately in the *Paths* box. If you do not like them, simply delete them with the *Cut* command (see below).

Editing the Paths Already Created

The program offers editing capabilities, to change and/or reorganize the paths already created. These paths are listed in the *Paths* list box on the right side of the screen. Available functions include:

Cut. Highlighted paths are deleted, and placed on the Clipboard.

Paste. Material currently on the Clipboard is pasted into the *Paths* window after the current cursor position.

Copy. A copy of the currently highlighted material is placed on the Clipboard.

Removing All Paths Involving a Latent Variable

The program offers an especially easy way to completely eliminate a *latent variable* from an analysis. Simply activate the *Edit Latent Variables* dialog, and delete the name of the *latent variable* that you wish to remove from the model. After you click *OK*, all paths involving this variable, and the variable itself, will be deleted from the model.

Controlling Parameter Numbers

The *Parameters* group contains functions that allow you to control if, and how, free parameter numbers are inserted in the paths you create.

Include. If this box is checked, a free parameter will be included in the path. The number will be the current value in the *Par #* field (see below).

Increment. If this box is checked, the number of the free parameter in the *Par #* field (see below) will be increased by 1 after each parameter number is added to a path.

Par #. This field specifies the parameter number to be added to the next path (if the *Include* box is checked). When the *Path Construction Tool* is first activated, or after any renumbering option, this value is set equal to one more than the highest parameter number currently in the active *.cmd* file.

Renumber. This button causes all *highlighted* paths to be renumbered, starting at the value in the *Par #* field. The renumbering preserves all order relationships, and same-difference relationships in the parameters of the highlighted paths. *It is important to remember that the action of the Renumber command is affected by the current settings of the Include and Increment boxes!* What this means is that, if you *Renumber* paths with the *Include* box *not checked*, all parameter numbers will be removed from the paths.

The *Renumber* command is a powerful and convenient one for making your programs easier to read. It is best, when finishing a program, to renumber the paths, starting at *1*. In this way, it is easy to verify the number of free parameters in a model.

The *Renumber* command can be especially useful when you are creating a multiple groups model with similar models in each group. You can simply *Copy* the model for the first group, then use the *Renumber* command to renumber the parameters.

Controlling Start Values - The Start Value Group

The fields in the *Start Value* group control whether a start value is added to a path, and what value is added.

Include. If this box is checked, a start value is added to each path.

Value. The value in this field is added as the start value.

Adding Comments to the Model File

You can add comments to the *Model File* using the *Comment* button.

Comment. This button calls up a dialog for adding a comment.

Type the text of the comment into the edit field. *Do not add the asterisk* to denote the comment field. The program will do this automatically.

Adding Group and Endgroup Commands

You can add *Group* and *Endgroup* commands, using the appropriate buttons.

Group. This button has an integer field to its right. If you click this button, the command

`Group #`

(where # is the integer in the field) will be added to the *Paths* file after the current line (see page 3562).

Endgroup. This button causes an *Endgroup* command to be added to the *Paths* file after the current line (see page 3562).

Adjusting Analysis Parameters

You can call up the *Analysis Parameters* (see page 3573), using the *Set Parameters* button.

Monte Carlo Setup

A Monte Carlo experiment involves executing a specified number of *replications* of a statistical procedure.

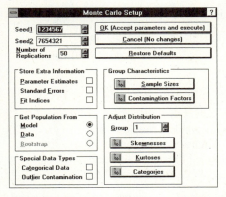

Each replication involves a complete analysis of a random sample created with certain characteristics, so Monte Carlo experiments can be very time consuming.

For technical details on Monte Carlo methods, and some examples of the valuable information you can obtain from Monte Carlo experimentation, see the section on *Monte Carlo Methods* (page 3625). You may terminate any Monte Carlo experiment prematurely with the ESC key or by clicking on the *Cancel* button. Replications completed at that point will be saved, and can be reviewed in the *Monte Carlo Results* dialog discussed on page 3596.

In the *Monte Carlo Setup* dialog, you can enter the parameters and conditions for a Monte Carlo experiment. Begin by entering the *Seed Values* and *Number of Replications*.

Monte Carlo Seeds

Monte Carlo methods generate numbers that act, in virtually all respects, like random numbers. However, they are in fact totally determinate. Once a starting value, or *seed* is specified, all subsequent numbers are completely determined. In *SEPATH*, either one or two seeds are used, depending on the type of data simulated by the random number generation procedure. The user is allowed to specify the seed values, so that any Monte Carlo experiment may be replicated, and interesting results examined in further detail.

Seed1. This is the main seed used in all Monte Carlo experiments. It must be an integer between *1* and *2,147,483,647*.

Seed2. This seed is used only in experiments involving generation of *contaminated normal* distributions. It must be an integer between *1* and *2,147,483,647*.

Monte Carlo Replications

In *SEPATH*, you may call for as many as *1000* Monte Carlo replications. Each replication involves generation of a complete simulated data set, fitting of a specified model to the data, and storage of the analysis results for later review. Since Monte Carlo results may be saved as Scrollsheets or *STATISTICA* data files, you may conduct experiments with more than *1000* replications by merging files with the *Data Management* module.

Number of replications. In this field you enter the number of Monte Carlo replications. This number is an integer between *1* and *1000*.

Specifying Information to be Stored

As a Monte Carlo experiment is performed, information about the analysis results for each Monte Carlo replication is stored in memory, and is displayed in an *Overall Results* Scrollsheet (see page 3597) at the end of the experiment. Results for each replication are stored in a new row of the Scrollsheet. The basic results, stored for *all* Monte Carlo experiments, include the Monte Carlo seeds used for each replication, the discrepancy function value, the number of iterations required, and a variety of indicators used to determine whether iteration converged satisfactorily. These are described in detail in the *Monte Carlo Results* section on page 3596.

In the *Store Extra Information* dialog, the user may also opt to store additional information about the results of each analysis. Checking the following boxes in this group causes this information to be added to the *Overall Results* stored during the experiment.

Parameter estimates. Checking this box causes the values for *free parameters* to be stored. They are given names corresponding to the numbers actually assigned in the *PATH1* program. The names are *PAR_#*, where # is the parameter number. So, if a program had free parameters *1, 2, 3, 6*, the variables *PAR_1, PAR_2, PAR_3,* and *PAR_6* would be added to the *Overall Results* Scrollsheet. To avoid a problem with an overly long variable name, avoid specifying parameter numbers greater than *9999* in your Monte Carlo analyses.

Standard errors. Checking this box causes the *estimated standard errors* to be stored for each free parameter, using the following naming scheme. Each standard error is named *SE_#*, where # is the free parameter number.

Fit indices. Checking this box causes the available fit indices to be stored for each analysis.

Specifying the Monte Carlo Population

In the *Get Population From* box, you are offered several options. In Monte Carlo analysis, the program generates data that simulate samples from a particular population structure. The option you check in this group determines how that population is obtained.

Model. If you select this option, the program generates a population structure corresponding to the current model. Whatever model is active in the current window, with the starting values specified for the parameters, is used to generate a population covariance matrix. You can specify the numerical value you want for various free parameters by specifying these values as *starting values* within braces after the free parameter number. So, for example, if you wanted your population to be based on a factor model where the first factor loads .5 on the first variable, you would have a line like this in your **.cmd* file:

`(F1)-1{.5}->[X1]`

Data. If you select this option, the program will take your data, calculate the appropriate matrix (correlation, covariance, augmented moment) and generate simulated data from it *based on the currently selected distributional characteristics*.

It is important that you understand the fine points of how this option works. If your input data file is a correlation or covariance matrix, the program will treat that as the population matrix, or convert it to the appropriate matrix type if necessary. If your input file contains raw data, the program will calculate the appropriate matrix type, and sample from that matrix.

This means that, if your file contains raw data, and you have selected the default distributional options, the program will calculate the sample means and covariance matrix corresponding to your data, and will generate simulated data from a *Multivariate Normal* distribution having the same means and covariance matrix.

If you have a raw data file, and you want a simulated population that exactly mirrors the characteristics of those raw data, you should use the *Bootstrap* option below.

Bootstrap. This option may only be selected if the data file contains raw data. In bootstrapping, a simulated sample of size *N* is created by treating the current data file as a discrete multivariate population, where each observation vector occurs equally often. The bootstrapped sample is obtained by sampling, randomly *with replacement* vectors of observations from the current data file. So, suppose the current data file contains 100 observations on 10 variables. If you request the *Bootstrap* option with *Sample Size* of *50*, the program will, 50 times, sample randomly with replacement from the integers from 1 to 100. The resulting list of 50 integers is used to select which observations from the data file are included in the bootstrapped sample. Note, it is possible (indeed likely in some cases) that some of the observations in the resulting data file will be identical. This pattern will occur if, during the sampling of integers, the same integer is selected more than once.

Creating Special Data Types

By default, the *Monte Carlo* procedure creates data that simulate a multivariate normal population. However, the system can also generate many other types of data. The *Special Data Types* group provides options that allow you to create special types of non-normal data. Besides allowing you to adjust the skewness and kurtosis of individual variables (see the *Group Characteristics* group below), the program also allows you to create Monte Carlo data of two special types:

Categorical data. If you check this option, the program will, *as a last step*, change the data from continuous to categorical, using categorization rules that you have specified. You may select between 2 and 10 categories, and you select the boundaries for the categorization. (See the *Categories* option under the *Adjust Distribution* group on page 3590 for details.)

Outlier contamination. This option allows you to create a distribution *contaminated by outliers*, using a standard mixture distribution technique. For details, see the *Contamination Factors* section (see below) of the *Group Characteristic* group.

Changing Group Sample Sizes and Outlier Characteristics

In this group, you enter group-specific characteristics for your Monte Carlo groups. You can alter sample sizes independently for each group, and the *contamination factors* used for controlling the simulation of data containing *outliers*.

Sample sizes. This button activates a dialog to allow you to enter the sample sizes for each group in the analysis.

Sample sizes should be greater than the number of variables in the analysis, but can be different for each group. Within the *Set Monte Carlo Sample Sizes* dialog, you may reset all sample sizes to a default size of *100* by clicking on the *Restore Defaults* button.

Contamination factors. In this dialog, you enter, for each group, a proportion of outliers, and a multiplier factor. Outlier contamination is simulated in the Monte Carlo module by a standard mixture distribution technique. Suppose your data come from a population with mean μ and covariance matrix Σ. We simulate contamination by substituting, a certain proportion *p* of the time (in the long run), observations with a mean μ and a covariance matrix $k\Sigma$, where *k* is a moderately large multiplier (say, *10*).

11. STRUCTURAL MODELING - DIALOGS AND OPTIONS

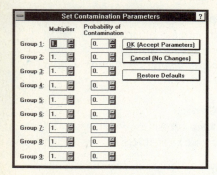

In effect, the outliers are obtained by taking "what they would have been before the mean was added on" (i.e., their deviation from the mean vector) and multiplying by the square root of k. For each group, you enter contamination factors k and p. For each corresponding statistical population, a proportion p of the time (in the long run) the observations will then have a mean μ and a covariance matrix $k\Sigma$.

Changing Group Distributional Characteristics

In this group, you adjust special distributional characteristics if you are simulating data that are not multivariate normal. Characteristics may be adjusted separately by group.

Group. In this field, you enter the number of the group whose characteristics you wish to alter.

Skewnesses. This button activates a dialog in which you enter the desired skewnesses for variables in each group, or a *Common Value* to be applied to all groups.

Kurtoses. This button activates a dialog in which you enter the desired kurtoses for variables in each group, or a *Common Value* to be applied to all groups.

Categories. This button activates an "interactive Scrollsheet" in which you enter the parameters for converting each continuous variable to categorical form.

Categorical Data Setup [wheaton.sta]					
Continue...	Number of Cuts	Cut1	Cut2	Cut3	Cut4	Cut5
GROUP 1					
ANOMIA67	4	-.842	-.253	.253	.842
POWLES67	4	-.842	-.253	.253	.842
ANOMIA71	4	-.842	-.253	.253	.842
POWLES71	4	-.842	-.253	.253	.842
EDUCATN	4	-.842	-.253	.253	.842
SEINDEX	4	-.842	-.253	.253	.842

For each variable, you enter the number of cuts. If you have k categories, you will have $k-1$ cutoff points. All continuous values below the first cutoff will be given the categorical score *1*. All continuous values exceeding the first cutoff but below the second cutoff will be given the categorical score *2*, etc. To set category cutoffs, simply enter the required values in the appropriate cells of the Scrollsheet. Click *Continue* when you have finished entering the values.

Iteration Dialog

During iteration, the *Iteration Dialog* opens to report the progress of iteration. The information reported here can be useful in diagnosing problems in convergence. When iteration is terminated, the user is presented with 2 options. These options give the user the opportunity to continue the analysis, or terminate without presenting the *Results* dialog.

SEPATH - 3590

Copyright © StatSoft, 1995

11. STRUCTURAL MODELING - DIALOGS AND OPTIONS

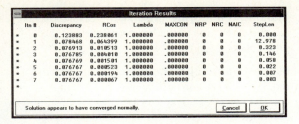

The following information is presented:

Itn #. The number of the iteration in the iterative sequence.

Discrepancy. The current value of the discrepancy function being minimized.

RCos. The current value of the Maximum Residual Cosine criterion.

Lambda. The value of the step multiplier used on that iteration. A value of *1.0* means that the first "full" step reduced the discrepancy function "sufficiently" to continue iteration. A value less than *1.0* means that the program had to use a *line search* along the chosen step direction in order to find a point where the discrepancy function was reduced. Very small values usually indicate that iteration is about to fail.

MAXCON. The maximum value of any constraint function. This value will only be non-zero during *constrained estimation*, used when the *New* option is chosen for *Standardization* or *Correlations* are chosen as the *Data to Analyze*. If iteration progresses satisfactorily, this value should decrease systematically to a value very close to zero.

NRP. Number of redundant parameters. If some parameters are redundant with others, the program may be able to detect the condition, and, if so, will set this value other than zero.

NRC. Number of redundant constraints. If some parameter constraints are redundant, the program may be able to detect the condition, and, if so, will set this value other than zero.

NAIC. Number of active inequality constraints, or "boundary conditions." During iteration, the program maintains certain inequality constraints on parameters to prevent some "impossible" values from arising.

For example, negative variances are not allowed to occur. If the program detects a variance will be set to a negative value on the next iteration, it constrains the value at the boundary point (i.e., zero) and only minimizes the discrepancy function relative to the other parameters. A *Heywood case* in common factor analysis occurs when the minimum of the *discrepancy function* is obtained with one or more *negative values* as estimates for the variance of the unique variables. Such values are of course impossible. Heywood cases occur frequently when too many factors are extracted, or the sample size is too small. If, for example, you have a *Heywood case* in a confirmatory factor model, you will have a non-zero value of this index at termination of iteration.

StepLen. The length of the current full iteration step. If an asterisk appears next to the value, it indicates that the maximum allowable step was taken. You can control the maximum step length allowed by setting a parameter in *Global Iteration Parameters* group in the *Analysis Parameters* dialog (see page 3577).

When iteration stops, the following commands are available.

Cancel. Clicking this button will terminate the current analysis and return to the *Startup Panel*.

OK. Click *OK* in order to calculate results, and present the *Results* dialog (see below).

Analysis Results

When iteration concludes, you can examine the results of the analysis in this dialog.

11. STRUCTURAL MODELING - DIALOGS AND OPTIONS

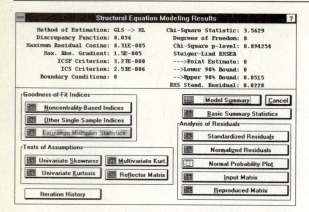

The top section of the dialog will present a number of results that can be scanned to determine, quickly and efficiently, (a) whether iteration concluded successfully, and (b) how well the model fit the data.

There are also several groups of buttons that allow the user to examine Scrollsheets containing various results.

Results Text Window

When the *Results* dialog opens, a considerable amount of information is displayed in a text window at the top of the dialog. The information on the left side of the text window is designed to enable you to evaluate, quickly and efficiently, whether iteration was successful. The information on the right side is useful for quickly assessing the quality of model fit. For technical definition and discussion of these indices, consult page 3666 in the section on *Technical Aspects of SEPATH*.

Method of estimation. At the top of the window is the *Method of Estimation* (i.e., the *discrepancy function* used).

Discrepancy function value. This number is the numerical value of the *discrepancy function*.

Maximum residual cosine. This numerical criterion, described by Browne (1982, eq. 1.9.5), will usually be close to zero if iteration was successful.

Maximum absolute gradient. This number is the absolute value of the largest element of the gradient.

Maximum absolute constraint. In options requiring constrained estimation, i.e., when correlations are analyzed, or when the *New Standardization* (see *Analysis Parameters,* page 3573, for details) method is employed, a number of constraint functions are evaluated, and must iterate to zero if the constraints are maintained. If this value is neither zero nor a very small value, it indicates iteration was unsuccessful.

A structural model is invariant under change of scale if model fit is not changed by rescaling the variables, i.e., by multiplying them by scale factors.

A structural model is invariant under a constant scale factor (ICSF) if model fit is not changed when all variables are multiplied by the same constant. Most, but not all, structural models that are of practical interest are ICSF. *SEPATH* reports two indices that allow the user to assess invariance properties of the model being fitted. These are discussed in detail on page 3680.

ICSF criterion. This criterion should be close to zero if the structural model is invariant under *a constant scaling factor* (ICSF). Most, but not all, structural models are invariant under a constant scaling factor.

ICS criterion. This criterion should be close to zero if the structural model is invariant under *changes of scale* (ICS). In general, when *correlations are analyzed*, this index should be close to zero.

Boundary conditions. This criterion indicates the number of inequality constraints that were operative at convergence. This number should be zero, unless there is a boundary case such as a *Heywood Case* in your model. If this number is *not zero*, then the *Chi-square* statistic will not necessarily have the proper distribution.

SEPATH - 3592

Copyright © StatSoft, 1995

The information on the right side of the window is basic statistical information about the fit of the model.

Chi-square statistic. Available for all discrepancy functions except OLS, this statistic has an asymptotic *Chi-square* distribution if the null hypothesis of perfect fit is true. In multiple group models, this is an overall statistic for all groups.

Degrees of freedom. This is the number of degrees of freedom for the *Chi-square* statistic.

Chi-square p level. This is the probability level for the *Chi-square* statistic.

Steiger-Lind RMSEA. Here the point estimate and 90% confidence interval for the *Steiger-Lind RMSEA* are given.

RMS standardized residual. This number is the root mean square standardized residual. Roughly, it must be less than *.05* for the fit to be "good" in a practical sense.

Model Summary

The model summary Scrollsheet presents model output in a Scrollsheet form convenient for analysis.

	Model Estimates (factor.sta)			
	Parameter Estimate	Standard Error	T Statistic	Prob. Level
(Work)-1->[WORK_1]	.757	.053	14.293	.000
(Work)-2->[WORK_2]	.849	.044	19.160	.000
(Work)-3->[WORK_3]	.864	.043	20.049	.000
(Home)-4->[HOME_1]	.729	.057	12.819	.000
(Home)-5->[HOME_2]	.897	.043	20.730	.000
(Home)-6->[HOME_3]	.815	.049	16.711	.000
(DELTA1)-->[WORK_1]	....	....	....	....
(DELTA2)-->[WORK_2]	....	....	....	....
(DELTA3)-->[WORK_3]	....	....	....	....
(DELTA4)-->[HOME_1]	....	....	....	....
(DELTA5)-->[HOME_2]	....	....	....	....
(DELTA6)-->[HOME_3]	....	....	....	....
(DELTA1)-7-(DELTA1)	.428	.080	5.338	.000
(DELTA2)-8-(DELTA2)	.280	.075	3.717	.000
(DELTA3)-9-(DELTA3)	.253	.075	3.390	.001
(DELTA4)-10-(DELTA4)	.468	.083	5.647	.000
(DELTA5)-11-(DELTA5)	.195	.078	2.508	.012
(DELTA6)-12-(DELTA6)	.336	.080	4.220	.000
(Home)-13-(Work)	.278	.107	2.600	.009

For each path in the model, the following information is given:

Path representation. Each row of the Scrollsheet begins with the *PATH1* representation of the model statement.

Parameter estimate. This number is the actual numerical value of the parameter estimate.

Standard error. This number is the estimated standard error of the parameter estimate.

T-statistic. Actually, this should more properly be called the "asymptotic normal statistic," as it does not actually have a Student's *t* distribution. It represents a test of the hypothesis that the parameter value is zero. Rejecting the hypothesis implies that the parameter value is not zero, and hence the parameter should be included in the model. Paths that are significant at the chosen significance level are highlighted. This significance level, which is *.05* by default, may be altered. (See *Alpha Level for Highlighting* below.)

Probability level. This is the standard normal probability level for the T-Statistic.

Alpha level for highlighting

The default value for *alpha* is *.05*. This value is used to determine the significance of a particular path. If the the probability level is less than or equal to the designated value for *alpha*, then the path is considered significant, and is highlighted in the *Model Summary* Scrollsheet (see above).

In this window (accessible from the *Options* button on the *Model Summary* Scrollsheet toolbar), set the desired *alpha* level ($0 < \alpha < 1$) that is to be used for highlighting significant results in the Scrollsheet.

11. STRUCTURAL MODELING - DIALOGS AND OPTIONS

Basic Summary Statistics

This button produces a Scrollsheet containing some of the basic summary statistics (also presented in the *Results Text Window*).

Basic Summary Statistics [wheaton.sta]	Value
Discrepancy Function	.077
Maximum Residual Cosine	.000
Maximum Absolute Gradient	.000
ICSF Criterion	.000
ICS Criterion	.000
ML Chi-Square	71.470
Degrees of Freedom	6.000
p-level	.000
RMS Standardized Residual	.021

Having the results in Scrollsheet form allows them to be printed, saved to a file, etc.

Goodness of Fit Statistics

The options in this box allow you to generate several of the best-known and most useful *indices of fit* for evaluating structural equation models. There are, literally, dozens of such indices, and we present only those we think are very widely used, especially valuable, or both. For a full technical discussion and definition of these indices, consult the section on *Technical Aspects of SEPATH,* page 3666).

There are three buttons for generating indices of fit.

Noncentrality-based indices. These indices are all based on the idea, first proposed by Steiger and Lind (1980), of basing goodness of fit assessment on an estimation of the *population noncentrality parameter*.

Noncentrality Fit Indices [wheaton.sta]	Lower 90% Conf. Bound	Point Estimate	Upper 90% Conf. Bound
Population Noncentrality Parameter	.045	.070	.103
Steiger-Lind RMSEA Index	.086	.108	.131
McDonald Noncentrality Index	.950	.966	.978
Population Gamma Index	.967	.977	.985
Adjusted Population Gamma Index	.884	.920	.949

Statistics presented include

(1) Population Noncentrality Parameter

(2) Steiger-Lind RMSEA Index

(3) McDonald's Index of Noncentrality

(4) Population *Gamma* Index

(5) Adjusted Population *Gamma* Index

Other single-sample indices. This button activates a Scrollsheet with a sampling of some of the better known single sample indices of fit, and some related measures.

Single Sample Fit Indices [wheaton.sta]	Value
Joreskog GFI	.975
Joreskog AGFI	.913
Akaike Information Criterion	.109
Schwarz's Bayesian Criterion	.187
Browne-Cudeck Cross Validation Index	.109
Independence Model Chi-Square	2131.433
Independence Model df	15.000
Bentler-Bonett Normed Fit Index	.966
Bentler-Bonett Non-Normed Fit Index	.923
Bentler Comparative Fit Index	.969
James-Mulaik-Brett Parsimonious Fit Index	.387
Bollen's Rho	.916
Bollen's Delta	.969

Statistics presented include

(1) Jöreskog GFI

(2) Jöreskog AGFI

(3) Akaike Information Criterion

(4) Schwarz's Bayesian Criterion

(5) Browne-Cudeck Cross Validation Index

(6) Independence Model *Chi-square* and *df*

(7) Bentler-Bonett (1980) Normed Fit Index

(8) Bentler-Bonett Non-Normed Fit Index

(9) Bentler Comparative Fit Index

(10) James-Mulaik-Brett Parsimonious Fit Index

(11) Bollen's *Rho*

(12) Bollen's *Delta*

LaGrange multiplier statistics. This button activates a Scrollsheet with LaGrange Multiplier statistics for each constraint required by the fitting procedure. If the *New* standardization method is employed, each *endogenous latent* variable will have its variance constrained to one, and there will be a Lagrange multiplier statistic for each variable. In the *Correlation* option of *Data to Analyze*, each *endogenous manifest* variable has a dummy *latent* variable (whose variance is constrained to one) attached to it.

The Lagrange multiplier statistics should be zero, if the constraints serve (as they are meant to) solely as identification constraints. If any are non-zero, or significantly exceed their estimated standard errors, then the model has probably been mis-specified, or estimation did not converge properly.

Iteration History

During iteration, the program saves the output that is displayed in the *Iteration Window* for the last executed sequence of iterations. By opening this Scrollsheet, you can review these results. You may of course also save them to a data file or analyze them graphically.

Tests of Assumptions

This group has options that allow you to analyze raw data to examine whether assumptions of multivariate normality have been violated, and to examine invariance properties of the model being tested. Formulas are given in the section on *Technical Aspects of SEPATH,* page 3682.

Measures of Univariate Skewness

Skewness. This number is the standard measure of univariate skewness. For normally distributed data, it should be zero in the population.

Corrected skewness. This number is a bias-corrected measure of univariate skewness.

Normalized skewness. This number is a sample based estimate which, under multivariate normality, has approximately a standard normal distribution

with large samples. This measure may be evaluated roughly like any other standard normal statistic.

Measures of Univariate Kurtosis

Kurtosis. This number is the standard measure of univariate kurtosis. For normally distributed data, it should be zero in the population.

Corrected kurtosis. This number is a bias-corrected measure of univariate kurtosis.

Normalized kurtosis. This number is a sample based estimate which, under multivariate normality, has approximately a standard normal distribution with large samples. This measure may be evaluated roughly like any other standard normal statistic.

Measures of Multivariate Kurtosis

Mardia's coefficient of multivariate kurtosis. If the sample comes from a multivariate normal distribution, this coefficient should be close to zero.

Normalized multivariate kurtosis. This normalized version of the Mardia coefficient has a distribution that is approximately standard normal for large samples. The measure may be evaluated roughly like any other standard normal statistic.

The following additional measures of kurtosis are provided for compatibility with other structural modeling programs.

Mardia-based *kappa*. The elliptical distribution family includes the multivariate normal distribution as a special case. In this distribution family, all variables have a common kurtosis parameter

$$\kappa = \frac{\sigma_{iiii}}{3\sigma_{ii}^2} - 1$$

This parameter can be used to rescale the *Chi-square* statistic if the assumption of an elliptical distribution is valid. The Mardia-based *kappa* is an estimate of *kappa* obtained by rescaling Mardia's coefficient of multivariate kurtosis. This parameter estimate should be close to zero if the population distribution is multivariate normal.

Mean scaled univariate kurtosis. This is an alternate estimate of *kappa*, obtained simply by averaging the rescaled univariate kurtoses.

Adjusted mean scaled univariate kurtosis. Distribution theory provides a lower bound for *kappa*. It must never be less than $-2/(p+2)$, where p is the number of variables. The preceding two estimates do not always obey this bound. This estimate averages the scaled univariate kurtosis, but adjusts each one that falls below this bound to be at the lower bound point. This coefficient should be close to zero if the distribution is multivariate normal.

Relative multivariate kurtosis. This measure is the Mardia-based *kappa*, rescaled to have a mean of *1*. It should be close to *1* in value if the distribution is multivariate normal.

Reflector Matrix

The reflector matrix is used for evaluating model invariance properties. Its use is discussed in the section on *Technical Aspects of SEPATH* (see page 3680).

Analysis of Residuals

This group allows you to examine model residuals, to help assess model fit, and, in particular, to find which observed variables are being reproduced poorly by the structural model.

Standardized residuals. Standardized residuals represent the difference between the input matrix and the reproduced matrix, standardized in the correlation metric. We express the residuals in standardized form to eliminate the effects of the scale of the variable on model residuals.

Normalized residuals. These residuals are divided by an estimate of their standard error, to provide a measure that, in large samples, is approximately normally distributed when the model fits perfectly in the population. These residuals may be interpreted (roughly) like any standardized normal statistic. Keep in mind that, when evaluating a large matrix of residuals, you need to correct for the fact that you are examining, *post hoc*, a large number of coefficients. For example, you expect approximately 5% of these to be "significant" by chance at the 5% level, so a matrix with 45 residuals would have, typically, 2 or 3 "significant" values by chance.

Normal probability plot. If the proposed structural model fits perfectly in the population, and if the population distribution is multivariate normal, the residuals will be approximately normally distributed. You can use the normal probability plot of the residuals to detect departures from normality, perfect fit, or both.

Input matrix. This matrix is the one actually analyzed by the structural modeling iterative routine, rather than the data matrix. So, for example, if you input *Raw Data*, but select *Moments* as your option under the *Data to Analyze* group of the *Analysis Parameters* dialog, you will see an augmented moment matrix here.

Reproduced matrix. This is the matrix calculated from the model, using the final parameter estimates as the numerical values.

Monte Carlo Results

This dialog gives you the opportunity to review the results of the most recently completed Monte Carlo analysis, to regenerate Monte Carlo data for individual replication, and to save those data for other types of analyses.

Display Overall Results

This button displays a *Monte Carlo Results* Scrollsheet containing an extensive summary of the outcome of the Monte Carlo experiment.

STRUCT. EQUAT.	SEED1	TERMCODE	DISCREP	RED_CON	BOUNDARY	CHI_SQR	DF
1	133358E4	0	.193075	0	0	19.11438	24
2	103296E4	0	.173316	0	0	17.15824	24
3	935706E3	0	.190657	0	0	18.87506	24
4	734541E3	1	.228669	0	1	22.63824	24
5	164416E4	0	.332603	0	0	32.92766	24
6	602855E3	0	.184011	0	0	18.21711	24
7	396193E3	3	.335800	0	1	33.24417	24
8	126333E4	0	.314676	0	0	31.15296	24
9	100909E3	0	.235267	0	0	23.29143	24
10	291326E3	3	.276922	0	1	27.41525	24
11	191291E3	0	.208214	0	0	20.61314	24
12	161212E4	0	.222509	0	0	22.02837	24
13	148928E4	0	.180271	0	0	17.84681	24
14	586818E3	1	.264694	0	1	26.20470	24
15	167424E4	0	.218813	0	0	21.66253	24
16	168375E4	1	.188615	0	1	18.67287	24
17	187897E4	1	.252848	0	1	25.03199	24
18	509420E3	0	.153930	0	0	15.23910	24

The Scrollsheet stores the results with the replications in rows, and the data for each replication in columns. The options that were selected in the *Monte Carlo Setup* dialog (see page 3587) determine which data are stored.

Here are the codes for the variable names:

SEED1 - The first of the two Monte Carlo seeds.

SEED2 - The second seed, used only in Contaminated Normal distribution generation.

TERMCODE - The Termination Code for the analysis. If this number is zero, the analysis apparently converged normally. If not, then the following codes apply:

1 - The *relative function change criterion* was below the criterion value. This signal can occur when the function has stabilized, but the gradient and relative cosine criteria do not go to zero, because one of the parameters is on a boundary value.

2 - The line search algorithm was unable to reduce the discrepancy function along the searched direction.

3 - The number of iterations reached the maximum permissible value. If necessary, this value may be altered in the *Analysis Parameters* dialog (see page 3573).

4 - Singular covariance matrix was encountered during iteration. On occasion, the parameters will be changed to values that yield a singular estimated covariance matrix. When this happens in *Maximum Likelihood* estimation, the discrepancy function cannot be evaluated, so iteration is stopped.

5 - (this value is currently not in use)

6 - The iteration was terminated by user request, i.e., the user stopped iteration with the ESC key or the *Cancel* button.

DISCREP - The value of the discrepancy function after iteration.

RCOS - The maximum residual cosine criterion.

GRADIENT - The maximum absolute value of the gradient elements after iteration.

NUM_ITER - The number of iterations required before termination.

ICSC - The ICSF invariance criterion.

ICS - The ICS invariance criterion.

RED_PAR - The number of redundant parameters.

RED_CON - The number of redundant constraints.

11. STRUCTURAL MODELING - DIALOGS AND OPTIONS

BOUNDARY - The number of active inequality constraints (NAIC), or "boundary cases," after iteration.

CHI_SQR - The *Chi-square* goodness-of-fit statistic.

DF - The number of degrees of freedom for the *Chi-square* statistic.

PLEVEL - The probability level for the *Chi-square* statistic.

PAR_# - The parameter values, numbered as they are in the *PATH1 *.cmd file*. So, for example, *PAR_23* is the value for the free parameter numbered 23 in the **.cmd* file.

SE_# - The standard errors, numbered in the same way as the parameter numbers.

RMS_LO - The lower endpoint of the 90% confidence interval for the Steiger-Lind (1980) RMS index.

RMS_PT - The point estimate for the Steiger-Lind (1980) RMS index.

RMS_HI - The upper endpoint of the 90% confidence interval for the Steiger-Lind (1980) RMS index.

NCP_LO - The lower endpoint of the 90% confidence interval for the population discrepancy function.

NCP_PT - The point estimate for the population discrepancy function.

NCP_HI - The upper endpoint of the 90% confidence interval for the population discrepancy function.

AIC - The rescaled Akaike information criterion.

BIC - The Schwarz Bayesian criterion.

BR_CUD - The Browne-Cudeck single sample cross-validation index.

GAMMA_LO - The upper endpoint of the 90% confidence interval for the population *gamma* index.

GAMMA_PT - The point estimate for the population *gamma* index.

GAMMA_HI - The upper endpoint of the 90% confidence interval for the population *gamma* index.

GAMAD_LO - The upper endpoint of the 90% confidence interval for the adjusted population *gamma* index.

GAMAD_PT - The point estimate for the adjusted population *gamma* index.

GAMAD_HI - The upper endpoint of the 90% confidence interval for the adjusted population *gamma* index.

IRGLS - The IRGLS discrepancy function, if maximum likelihood estimates were obtained.

Generate Data

This button generates a Scrollsheet containing the actual raw data employed in the Monte Carlo analysis, for the replication number currently entered in the *Replication* field. These data may, of course, be saved to a file for subsequent analyses.

SEPATH - 3598

Copyright © StatSoft, 1995

EXAMPLES

Example 1: Stability of Alienation

This example, from a paper on stability of alienation by Wheaton, Múthen, Alwin, and Summers (1977), is one of the "classic" examples of a full structural equation model discussed in many textbooks and program manuals. It is also the first example in the *SEPATH Electronic Manual*.

The covariance matrix for the data, based on a sample size of *932*, is in the data file *Wheaton.sta*.

Case	1 ANOMIA67	2 POWLES67	3 ANOMIA71	4 POWLES71	5 EDUCATN	6 SEINDEX
ANOMIA67	11.834	6.947	6.819	4.783	-3.839	-21.899
POWLES67	6.947	9.364	5.091	5.028	-3.889	-18.831
ANOMIA71	6.819	5.091	12.532	7.495	-3.841	-21.748
POWLES71	4.783	5.028	7.495	9.986	-3.625	-18.875
EDUCATN	-3.839	-3.889	-3.841	-3.625	9.610	35.522
SEINDEX	-21.899	-18.831	-21.748	-18.775	35.522	450.288
Means						
Std.Dev.						
No.Cases	932.000					
Matrix	4.000					

This covariance matrix file, and some others in the examples covered in this section, do not contain entries for the means and standard deviations of the variables. In many cases this will pose no problem. *SEPATH* can compute standard deviations and correlations from the variances in a covariance matrix **.sta* file, and variances and covariances from correlations and standard deviations in a correlation matrix file. *SEPATH* will examine the selection in *Data to Analyze* in the *Analysis Parameters* dialog (page 3573), and will evaluate whether the information in the current **.sta* file is sufficient to continue. If it is, *SEPATH* will automatically "fill in" any missing information it requires. If sufficient information is not available, an error message will be issued.

The reader should also notice that correlation and covariance matrix files may be entered in full matrix form, as in the *Wheaton.sta* file, or in *lower triangular* form, with missing values in the positions above the diagonal, as in the *Lawley.sta* file used in the next example. *SEPATH* can handle either format. Obviously, if you are entering a covariance matrix by hand from some other source, you can save substantial effort by entering only the lower triangular part of the matrix.

The model files for the example are *Wheatona.cmd* and *Wheatonb.cmd*.

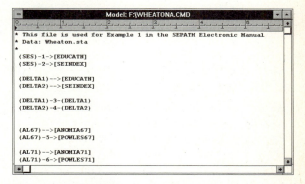

```
Model: F:\WHEATONA.CMD
* This file is used for Example 1 in the SEPATH Electronic Manual
* Data: Wheaton.sta
*
(SES)-1->[EDUCATN]
(SES)-2->[SEINDEX]

(DELTA1)-->[EDUCATN]
(DELTA2)-->[SEINDEX]

(DELTA1)-3-(DELTA1)
(DELTA2)-4-(DELTA2)

(AL67)-->[ANOMIA67]
(AL67)-5->[POWLES67]

(AL71)-->[ANOMIA71]
(AL71)-6->[POWLES71]
```

This example is discussed in detail in the *Introductory Examples* in the section, *A Structural Equation Model Example,* beginning on page 3563.

Example 2: A Confirmatory Factor Analysis

Lawley and Maxwell (1971, chapter 7) present techniques for confirmatory factor analysis. These authors are careful to point out that estimation of standard errors must be adjusted when a correlation matrix is analyzed instead of a covariance matrix. They give formulae for computing standard errors when a covariance matrix is analyzed, and provide an alternative method for computing standard errors when the sample correlation matrix is analyzed. The formulae are illustrated with a numerical example,

11. STRUCTURAL MODELING - EXAMPLES

the results of which are presented in their Tables 7.9 (page 99) and 7.10 (page 102).

The Lawley-Maxwell example was based on an analysis in a paper by Jöreskog and Lawley (1968), which used data from Holzinger and Swineford (1939). Nine psychological tests were administered to seventh and eighth grade students. The correlation matrix for these data are contained in the file *Lawley.sta*.

	Data: LAWLEY.STA 9v * 13c								
	SEPATH Electronic Manual Example #2								
Case	1 VIS_PERC	2 CUBES	3 LOZENGES	4 PAR_COMP	5 SEN_COMP	6 WRD_MNG	7 ADDITION	8 CNT_DOT	9 ST_CURVE
VIS_PERC	1.000								
CUBES	.245	1.000							
LOZENGES	.418	.362	1.000						
PAR_COMP	.282	.217	.425	1.000					
SEN_COMP	.257	.125	.304	.784	1.000				
WRD_MNG	.239	.131	.330	.743	.730	1.000			
ADDITION	.122	.149	.265	.185	.221	.118	1.000		
CNT_DOT	.253	.183	.329	.021	.139	-.027	.601	1.000	
ST_CURVE	.583	.147	.455	.381	.400	.235	.385	.462	1.000
Means									
Std.Dev.									
No.Cases	72.000								
Matrix	1.000								

The tests were (1) Visual Perception, (2) Cubes, (3) Lozenges, (4) Paragraph Comprehension, (5) Sentence Completion, (6) Word Meaning, (7) Addition, (8) Counting Dots, and (9) Straight-curved capitals. The Lawley-Maxwell example is a typical one, in that each factor is restricted to load on only 3 or 4 manifest variables, and the factors are allowed to correlate with each other. Moreover, by examining what the variables that load on each factor have in common, we can see why the factors are named *VISUAL*, *VERBAL*, and *SPEED*.

The model is portrayed in the diagram below. (To obtain a larger copy of this picture, click on *Help - SEPATH Examples*, go to example number 2 in the *Electronic Manual*, and print the topic from the *Help* facility.)

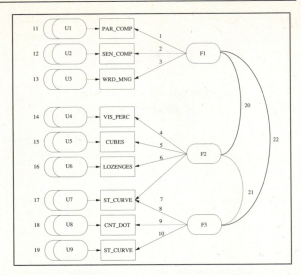

The model file corresponding to this diagram is *Lawley.cmd*. Open this file, using the *Open Model* command from the *Startup Panel*.

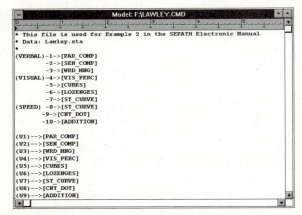

You can reproduce *both* sets of results in the Lawley and Maxwell tables. First, to simulate what happens with most covariance structure analysis programs, analyze the correlation matrix *incorrectly* as though it were a covariance matrix by using the *Covariances* option in the *Data to Analyze* section of the *Analysis Parameters* dialog. Set the number of decimal places to 2, to make the comparison with the Lawley-Maxwell tables easier.

11. STRUCTURAL MODELING - EXAMPLES

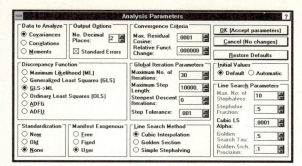

Before beginning iteration, the program will interrupt you with a warning message, that you are about to analyze a correlation matrix (incorrectly) as if it is a covariance matrix.

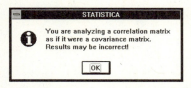

Click *OK* to proceed with the (incorrect) analysis. You will obtain the erroneous standard errors described by Lawley and Maxwell. (The curious reader might try running this example with other structural modeling programs and comparing their results to those shown immediately below.)

Model Estimates (lawley.sta)		
	Parameter Estimate	Standard Error	T Statistic
(VERBAL) -1-> [PAR_COMP]	.91	.10	9.51
-2-> [SEN_COMP]	.87	.10	8.87
-3-> [WRD_MNG]	.82	.10	8.23
(VISUAL) -4-> [VIS_PERC]	.68	.12	5.70
-5-> [CUBES]	.34	.13	2.63
-6-> [LOZENGES]	.66	.12	5.50
-7-> [ST_CURVE]	.67	.13	4.98
(SPEED) -8-> [ST_CURVE]	.19	.13	1.47
-9-> [CNT_DOT]	.92	.14	6.51
-10-> [ADDITION]	.65	.13	4.97
(U1) --> [PAR_COMP]	....	....	....
(U2) --> [SEN_COMP]	....	....	....

Now, return to the *Analysis Parameters* dialog, reset the *Data to Analyze* to *Correlations*.

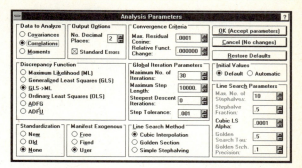

Then re-execute the iteration. You will see estimates and standard errors shown below.

Model Estimates (lawley.sta)		
	Parameter Estimate	Standard Error	T Statistic
(VERBAL) -1-> [PAR_COMP]	.91	.04	25.52
-2-> [SEN_COMP]	.87	.04	21.41
-3-> [WRD_MNG]	.82	.05	17.66
(VISUAL) -4-> [VIS_PERC]	.68	.09	7.87
-5-> [CUBES]	.34	.12	2.83
-6-> [LOZENGES]	.66	.09	7.44
-7-> [ST_CURVE]	.67	.11	5.96
(SPEED) -8-> [ST_CURVE]	.19	.13	1.48
-9-> [CNT_DOT]	.92	.11	8.30
-10-> [ADDITION]	.65	.10	6.29
(U1) --> [PAR_COMP]	....	....	....
(U2) --> [SEN_COMP]	....	....	....

These standard errors are identical to those reported by Lawley and Maxwell as the correct values, which they obtained by using corrected formulae.

Example 3: Confirmatory Factor Analysis with Identifying Constraints

Everitt (1984, pages 45-52) discusses, in considerable detail, a confirmatory factor analysis of a data set in Child (1970). This factor model is similar in some ways to the Lawley-Maxwell example. However, in this case one factor loads on only two manifest variables. This pattern, unfortunately, leads to a model which is not identified, unless some identifying constraints are applied. In this model, two factor loadings are constrained to be equal to each other, and two factor covariances are constrained to be zero. Data for this example are in the file *Child.sta*.

SEPATH - 3601

11. STRUCTURAL MODELING - EXAMPLES

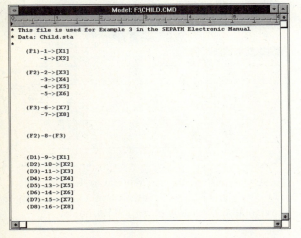

The *PATH1* specification for this model is contained in the file *Child.cmd*.

The path diagram corresponding to these commands shown below.

Everitt's analysis of this example is incorrect, because a correlation matrix was treated as if it was a covariance matrix. You can reproduce Everitt's incorrect results by specifying *Covariances* as the *Data to Analyze* in the *Analysis Parameters* dialog.

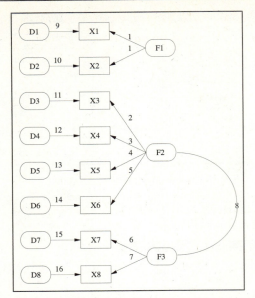

Correct results (shown below) may be obtained by specifying *Correlations* as the *Data to Analyze*.

	Parameter Estimate	Standard Error	T Statistic	Prob. Level
(F1)-1->[X1]	.735	.050	14.623	.000
-1->[X2]	.735	.050	14.623	.000
(F2)-2->[X3]	.712	.072	9.950	.000
-3->[X4]	.729	.070	10.386	.000
-4->[X5]	.721	.071	10.169	.000
-5->[X6]	.510	.092	5.544	.000
(F3)-6->[X7]	.818	.137	5.963	.000
-7->[X8]	.623	.120	5.212	.000
(F2)-8-(F3)	.490	.123	3.998	.000
(D1)-9->[X1]	.678	.054	12.457	.000
(D2)-10->[X2]	.678	.054	12.457	.000
(D3)-11->[X3]	.702	.073	9.651	.000
(D4)-12->[X4]	.685	.075	9.166	.000
(D5)-13->[X5]	.693	.074	9.407	.000
(D6)-14->[X6]	.860	.054	15.794	.000
(D7)-15->[X7]	.575	.195	2.945	.003
(D8)-16->[X8]	.782	.095	8.202	.000

Example 4: Effect of Peer Influence on Ambition

Duncan, Haller, and Portes (1968) analyzed the effect of peer influences on ambition. The correlation matrix from their study, based on 329 subjects, is contained in the file *Dhp.sta*.

11. STRUCTURAL MODELING - EXAMPLES

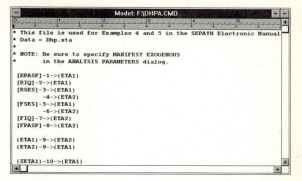

Their data have been analyzed in a number of publications, and the substantive content of the example will not be discussed here. Jöreskog and Sörbom (1984) present the results from several path models on these data. Here we analyze the model the results for which are given in Table III.12 of Jöreskog and Sörbom.

The path diagram is shown below.

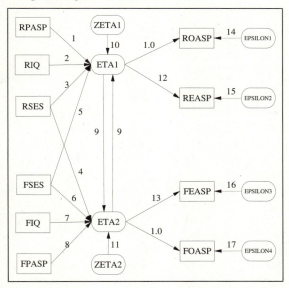

Corresponding *PATH1* statements for the model are in a file called *Dhpa.cmd*.

We begin by analyzing the model the way it was analyzed in Jöreskog and Sörbom (1984). Several comments are in order. First, note from the path diagram that the model has manifest exogenous variables. In the file *Dhpa.cmd*, you are reminded to use the *Manifest Exogenous Fixed* option in the *Analysis Parameters* dialog with this model. The command is used because of the existence in the model of a number of manifest exogenous variables (located on the left hand side of the diagram). We have taken the convenient option here and not specified *any* of the variance-covariance relationships for these variables. If you fail to activate the *Manifest Exogenous Fixed* command in the *Analysis Parameters* dialog, the program will give an error message and abort the analysis, unless you add paths for all of the variance-covariance relationships among these variables.

Second, note that some paths are constrained to have fixed coefficients of *1.0*. It is a relatively common technique in structural modeling to fix one path coefficient from each endogenous latent variable, in order to identify the variance of the latent variable at *some* fixed value. In this case, we see two path coefficients were fixed by Jöreskog and Sörbom (1984), ostensibly in order to identify the variances of the latent endogenous variables *ETA1* and *ETA2*.

Example 5: Standardized Solutions for the Effect of Peer Influence on Ambition

In a standardized solution, the latent variables all have unit variance. Setting the variances of exogenous latent variables to unity is trivial, because (a) these variances may be specified directly as part of a model, and (b) they are essentially arbitrary. Endogenous latent variable variances are a different matter. In some structural modeling programs, there is no convenient way to constrain the endogenous latent variables. For such programs to produce a standardized solution, a two stage procedure is employed. First, model parameters are estimated with the variances of the endogenous variables identified (at some unspecified value). After iteration is complete, the standardized coefficients are computed, using well known results from the algebra of multiple linear regression to transform the variances of endogenous latent variables to unity.

To produce such an old fashioned standardized solution, analyze the data in *Dhp.sta*.

Case	1 RIQ	2 RPASP	3 RSES	4 ROASP	5 REASP	6 FIQ	7 FPASP
RIQ	1.0000						
RPASP	.1839	1.0000					
RSES	.2220	.0489	1.0000				
ROASP	.4105	.2137	.3240	1.0000			
REASP	.4043	.2742	.4047	.6247	1.0000		
FIQ	.3355	.0782	.2302	.2995	.2863	1.0000	
FPASP	.1021	.1147	.0931	.0760	.0702	.2087	1.0000
FSES	.1861	.0186	.2707	.2930	.2407	.2950	−.0438
FOASP	.2598	.0839	.2786	.4216	.3275	.5007	.1988
FEASP	.2903	.1124	.3054	.3269	.3669	.5191	.2784
Means							
Std.Dev.							

Select the *Manifest Exogenous - Fixed* and *Standardization - Old* options in the *Analysis Parameters* dialog.

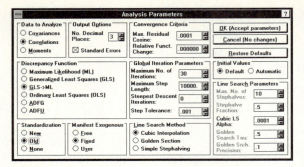

Use the model in *Dhpa.cmd*.

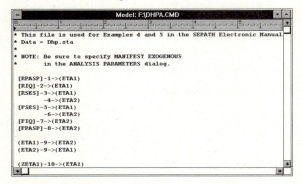

Execute the estimation, and compare the results with the unstandardized solution. One point is particularly noteworthy here. Some of the constraints which were specified for the unstandardized model are violated in the standardized model! For example, the path coefficients to *ROASP* and *FOASP* which were fixed at *1.0* in the unstandardized model now have values of *.767* and *.771*, respectively. This is because the *post hoc* adjustment which produces the standardized coefficients simply employs the (unstandardized) numerical values and some regression algebra. The procedure is not "aware" of any equality constraints on parameters, and cannot be guaranteed to maintain them.

This would appear to create no real problem, since the values were assigned an arbitrary value of *1.0* simply to identify the variances at *some* value. On the other hand, two paths were constrained to be equal (those with the coefficient label 9 in the path diagram) for reasons other than identification purposes. *These no longer have equal values in the standardized model estimates.*

There is a third issue which is probably already obvious to the reader after comparing standardized and unstandardized solutions. *Standard errors* are not provided in the "old" standardized solution. This is because the *post hoc* adjustment of the unstandardized coefficients produces values whose information matrix is not the same as that for the unstandardized values.

We have established three major points so far:

(1) The "old" approach to standardization produces the same discrepancy function as obtained in the unstandardized solution.

(2) Equality constraints on the parameters imposed on the unstandardized solution may not be maintained when the parameters are transformed to the standardized form.

(3) Standard errors are not available with the *Standardization - Old* option.

Now, we will explore the properties of the new standardization approach. This approach uses constrained estimation to force endogenous latent variables to have variances of one. Consequently, fixed parameter values of one which where used to identify the variances of endogenous latent variables should now be changed to free parameters before employing the *Standardization - New* option.

Call up *Dhpa.cmd* and find the following section of model.

```
(ETA1)-12->[REASP]
     -->[ROASP]
(ETA2)-->[FOASP]
     -13->[FEASP]
```

Change the fixed coefficients to free parameters, so the revised paths look like this:

```
(ETA1)-12->[REASP]
     -21->[ROASP]
```

```
(ETA2)-22->[FOASP]
     -13->[FEASP]
```

Use the options *Manifest Exogenous - Free,* and *Standardization - New* in the *Analysis Parameters* dialog. Also, set *No. Decimal Places* to 4.

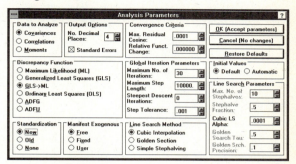

Execute the iteration, and observe the results. You will notice that the *Chi-square* statistic for the model is slightly lower than it was for the unstandardized solution. (*29.8964* to *29.8987*). This is because the *Standardization - New* option produces standardized coefficients by constraining the estimation process. All equality constraints you imposed in your model are maintained during estimation. Notice that the two coefficients constrained to be equal (those with the parameter number 9) are estimated to be the same number. In the *Standardization - Old* procedure, these are not actually equal once the *post hoc* standardization procedure is employed. This demonstrates an important point which is often lost in discussions of standardization. *Because constraints imposed on the unstandardized solution are usually not maintained in the standardized solution, the Chi-square statistic obtained in the old standardization approach may not, in fact, be correct.* Another way of putting it is that the "old" standardization approach may remove your model constraints without your consent. Some of these constraints may have been artificial ones designed to achieve identification. However, if the constraints have some important substantive purpose, you should be aware that the "old" standardization approach may remove them.

The Lagrange multiplier statistics for the variance constraints are an important diagnostic tool for assessing whether model constraints were serving merely to achieve identification, or were accomplishing some other purpose. In this case, we have removed two constraints on model coefficients. These constraints forced two parameters to be equal to a fixed value of one. We have now removed these constraints, and if forcing latent variable variances to be unity imposes no further constraints on the model fitting process, we should see two indications.

(1) The *Chi-square* statistic for the model in the *Standarize - New* option should be the same as in the *Standardization - Old* option.

(2) LaGrange multiplier statistics for all standardization constraints should all be zero.

We see that, in this case, *both* indications are *not* present. We have already noted that the *Chi-square* statistic is slightly different than in the *Standardization - Old* solution. The LaGrange multiplier statistics are slightly greater than zero, indicating that the standardization of latent endogenous variables actually imposes constraints on this model.

LaGrange Multiplier Statistics (dhp.sta)			
	Variance	LaGrange Multiplier	Standard Error
ETA1	1.0000	.0002	.0038
ETA2	1.0000	-.0002	.0038

If the unit variances were simply an identification condition (as they frequently are) these statistics would all be zero. But in this case, imposing the standardization restriction on the endogenous latent variables constrains the solution for the path coefficients. The value of the coefficients which minimizes the discrepancy function, *subject to the standardization constraints*, is not one of the solutions which is a global minimum when the latent variable variances are unconstrained.

The problem is that the constraints on the variances of the latent variables interact with the constraint on the two coefficients labeled 9. (Thorough analysis of this phenomenon would require an entire additional chapter in this manual.) If we allow the two paths labeled with coefficient 9 to vary freely, by renumbering one of them 99, we find a slightly lower *Chi-square* value, and zero Lagrange multiplier statistics (input for this example is in the file *Dhpa3.cmd*). For this solution, the coefficients will *not* agree with the old fashioned output produced by the commands in *Dhpa.cmd*.

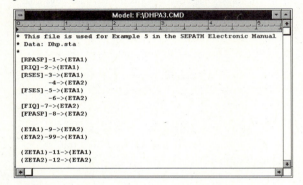

The old fashioned output was obtained by standardizing, *after the fact*, coefficients which were generated under the constraint that the two paths between *ETA1* and *ETA2* are equal. Standardization after the fact invalidates that constraint, but the solution will *not* be the same as it would be if you relaxed the constraint, computed a solution, and *then* standardized. If, however, you return to the original input file (*Dhpa.cmd*), relax the equality constraint for the two paths labeled with coefficient number 9, and recompute, you will find that the standardized coefficients and *Chi-square* statistic agree precisely with those computed by the new method with the commands in *Dhpa3.cmd*. Note that the new procedure allows standard errors to be calculated, while the old procedure does not.

Let us briefly review the points covered in the preceding examples.

(1) The old fashioned approach to standardization has one advantage -- computational simplicity.

However, confidence intervals for standardized parameters cannot be calculated with this method. Moreover, equality constraints and fixed value constraints which are satisfied by coefficients in the unstandardized solution may not be satisfied in the standardized solution.

(2) When converting an old model to the new standardized method, you should free up parameters which were fixed solely for the purpose of identifying the variances of endogenous latent variables. There should be one such parameter for each endogenous latent variable.

(3) If Lagrange multiplier statistics are not zero, relax all equality constraints on parameters, and double-check to make sure you have freed up the parameters which were fixed for identification purposes.

(4) If your solution has a significant number of manifest exogenous variables, you can speed up iteration using the *Manifest Exogenous Fixed* option, then restart the iteration with the *Manifest Exogenous Free* option after convergence has occurred in order to guarantee correct standard errors and *Chi-square* values.

(5) In the above examples, we analyzed a covariance matrix as though is was a correlation matrix in order to remain consistent with traditional analyses of the data. Consequently, standard errors are not generally correct. The program gave warning messages before computing its results.

By setting the *Data to Analyze* option in the *Analysis Parameters* dialog to *Correlations*, you can obtain a fully standardized path model with correct standard errors.

Example 6: Factor Analysis with an Intercept Variable

This example, discussed in the textbook by Bollen (1989, pages 308-311), shows, in a very simple context, the general technique for estimating factor means.

The data file for the example is *Boll310.sta*.

Case	1 X1	2 X2	3 X3	4 X4	5 X5
X1	14.185				
X2	4.383	1.459			
X3	7.505	2.344	4.209		
X4	4.546	1.423	2.447	1.502	
X5	4.006	1.244	2.173	1.307	1.200
Means	6.758	2.230	3.427	2.097	1.918
Std.Dev.					
No. Cases	60.000				
Matrix	4.000				

In this example, 5 judges were asked to estimate, to the nearest tenth of an inch, the length of lines drawn on 60 index cards, with one line per card.

Judge 1 gave his estimates to the nearest tenth of a centimeter, while the other four judges gave their ratings to the nearest tenth of an inch. The rating process is modeled with a single factor model. The ratings of each judge are modeled as arising through multiplicative (representing a rater-specific variance) and additive (representing rater-specific mean) transformations of the true length, represented by the factor F. So, for example, the ratings that we see for rater *X1* are the true length, multiplied by a factor loading, with a rater-specific constant added on.

The *PATH1* commands for this model are in the file *Boll310.cmd*.

11. STRUCTURAL MODELING - EXAMPLES

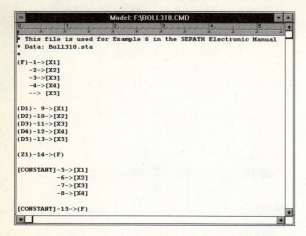

The primary purpose for using the factor model in this case is to gauge *differences* among the raters. Indeed, the scale for the underlying latent variable is arbitrary. We fix the scale by setting the factor loading for rater 5 to be fixed at *1.0*, and the additive constant for rater 5 to be fixed at zero.

The path diagram for the model is shown below.

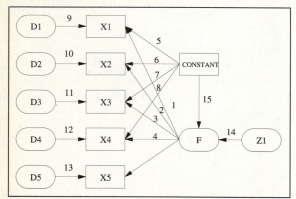

The manifest variable *CONSTANT* is actually an artificial manifest variable with a mean of *1* and a variance of zero. It contributes no variance, and adds a constant to variables that load on it.

To analyze this example using an intercept variable, you must select the *Moments* option for *Data to Analyze* in the *Analysis Parameters* dialog.

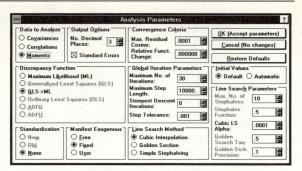

In that case, the augmented moment matrix is analyzed, and the artificial variable *CONSTANT* is added automatically to the list of available manifest variables.

Example 7: Comparing Factor Structure in Two Groups

This example, discussed in the textbook by Bollen (1989, pages 355-365), shows how to compare two groups for equivalence of factor structure. It provides a basic introduction to the techniques involved in fitting and comparing factor models in two or more populations. The textbook contains an extensive discussion of the example.

Data for this example are in the file *Boll362m.sta*.

Case	1 X2	2 X1	3 Y2	4 Y1	5 Y3
X2	1.000				
X1	.292	1.000			
Y2	.282	.184	1.000		
Y1	.166	.383	.386	1.000	
Y3	.231	.277	.431	.537	1.000
Means	36.698	5.040	1.543	1.548	1.554
Std.Dev.	21.277	2.198	.640	.670	.627
No.Cases	432.000				
Matrix	1.000				

The models are in files *Boll362a.cmd* and *Boll362b.cmd*.

11. STRUCTURAL MODELING - EXAMPLES

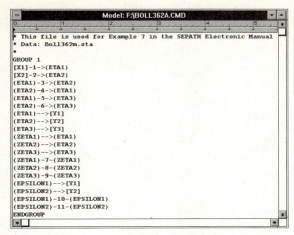

In order to process this example, you must select the *Fixed* option in the *Manifest Exogenous* box in the *Analysis Parameters* dialog (see page 3575).

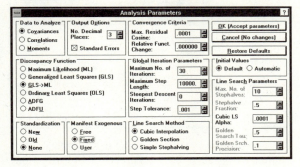

Example 8: Testing for Circumplex Structure

The data matrix, from Guttman (1954), is in the file *Guttman.sta*.

This data set, used frequently to demonstrate a Guttman *circumplex*, is a correlation matrix among 6 different kinds of abilities for 710 Chicago school children.

A perfect circumplex correlation matrix (Guttman, 1954) has equal correlations on sub-diagonal strips. For example, a 6x6 correlation matrix would be of the form:

$$
\begin{matrix}
1 & & & & & \\
\rho_1 & 1 & & & & \\
\rho_2 & \rho_1 & 1 & & & \\
\rho_3 & \rho_2 & \rho_1 & 1 & & \\
\rho_2 & \rho_3 & \rho_2 & \rho_1 & 1 & \\
\rho_1 & \rho_2 & \rho_3 & \rho_2 & \rho_1 & 1
\end{matrix}
$$

A circumplex hypothesis is an example of a correlational *pattern hypothesis.* (See Steiger, 1980a for a review of the literature on pattern hypothesis testing.) A *pattern hypothesis* is any hypothesis that specifies correlations are equal to each other, or to specified numerical values. *SEPATH* can perform any pattern hypothesis on 1 or more correlation matrices.

A circumplex hypothesis is a pattern hypothesis within a single population. Frequently, comparisons of correlations *across* populations are of interest as well as comparisons within populations. *SEPATH* can test *pattern hypotheses* across samples as well as within samples. (See *Testing the Equality of Correlation Matrices from Different Populations*, page 3622.)

The general approach to testing correlational pattern hypotheses with *SEPATH* is straightforward.

11. STRUCTURAL MODELING - EXAMPLES

(1) Use the *Correlations* option in the *Data to Analyze* section of the *Analysis Parameters* dialog (see page 3573).

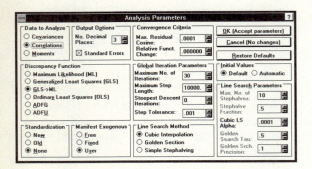

(2) Use the *Path Construction Tool* to create all the paths for a correlation matrix for the relevant variables. Click *Correl.* as the *Path Type,* then highlight all of the manifest variables you wish to test in the *To:* column.

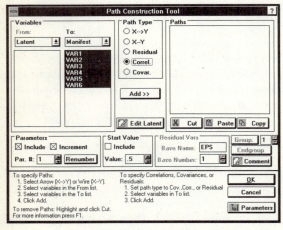

(3) Next, click *Add>>*. This creates all the paths for the full correlation matrix for the manifest variables.

(4) Click *OK* to save the model.

(5) Edit the model to apply the required constraints.

The commands for testing a hypothesis of circumplex structure are in the file *Circle.cmd*.

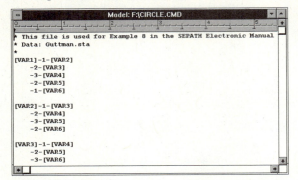

Test these commands, using the *Correlations* option of *Data to Analyze*.

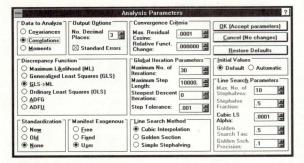

The sample size is very large in this example. Hence, we would expect the precision of estimation to be very high. At the same time, we would have to keep in mind that the "reject-support" approach of the *Chi-square* test would be of very limited usefulness in this situation. We recognize that a model with as many constraints as this one will almost certainly not fit perfectly in the population, and we have very high power to detect an imperfect fit.

The *Chi-square* statistic yields, in this case, a value of *27.05* with *12* degrees of freedom. The probability level is *.008*, indicating that the null hypothesis of perfect fit must be rejected. Jöreskog (1978), analyzing these data, remarked that they "do not fit a circumplex well."

This conclusion seems, in the 20-20 vision of hindsight, unjustified, and incorrect. A more reasonable conclusion is that it is highly probable that they do not fit a circumplex *perfectly*. Goodness-of-fit indices indicate how well these data actually do fit a circumplex. The 90% confidence interval for the Steiger-Lind (1980) RMS index is between *.021* and *.064*.

The corresponding confidence interval for the *adjusted* population *gamma* coefficient is between *.972* and *.997*.

A reasonable conclusion would seem to be that Guttman's data fit a circumplex *very* well, a fact which we have verified statistically.

Example 9: Testing for Stability of a Correlation Matrix over Time

Suppose you measured a set of variables twice, and wished to test the hypothesis that the correlation coefficient had not changed from time 1 to time 2. For example, suppose 120 individuals are measured twice on verbal, quantitative, and analytical ability. In this case, the covariance matrix would be 6x6.

A 6x6 correlation matrix is given in the file *Twocorr.sta*. Open this file to begin the example.

Case	1 VERBAL_1	2 QUANT_1	3 ANALY_1	4 VERBAL_2	5 QUANT_2	6 ANALY_2
VERBAL_1	1.000					
QUANT_1	.650	1.000				
ANALY_1	.540	.680	1.000			
VERBAL_2	.270	.300	.210	1.000		
QUANT_2	.320	.210	.270	.590	1.000	
ANALY_2	.180	.260	.220	.480	.550	1.000
Means						
Std.Dev.	2.100	3.000	2.400	1.600	3.300	2.200
No.Cases	120.000					
Matrix	1.000					

Here, we modify the general approach to setting up a *pattern hypothesis* described above in order to compare correlations measured at two times.

(1) Use the *Correlations* option in the *Data to Analyze* section of the *Analysis Parameters* dialog (see page 3573).

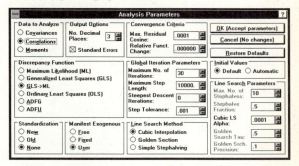

(2) Use the *Path Construction Tool* to create all the paths for a correlation matrix for the relevant variables. Click *Correl.* as the *Path Type*, then highlight all 6 of the manifest variables in the *To:* column.

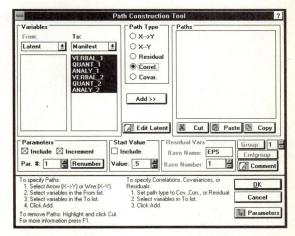

(3) Next, click *Add>>*. This action creates all the paths for the full correlation matrix for the 6 variables.

(4) You want to constrain the correlations between the variables at time two to be the same as those at time 1. Notice first that there are 3 correlations between the 3 variables at time 1, and they are assigned parameter numbers 1-3. Highlight the corresponding 3 correlations at

11. STRUCTURAL MODELING - EXAMPLES

time 2 by clicking on them while holding down the CTRL key.

(5) Set the *Renumber* command of the *Path Construction Tool* to renumber these 3 correlations starting at 1.

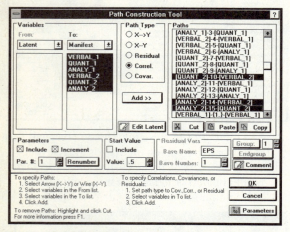

(6) Next, highlight all the paths and renumber them, starting at 1.

(7) Click *OK* to save the file to a *Text/output Window*.

Your final model should look like this.

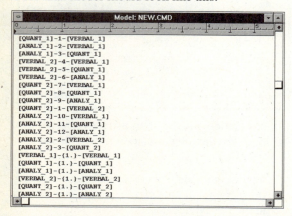

The commands (reorganized for easier reading) for performing the hypothesis test are in a file called *Twocorr.cmd*. The estimated correlations among the alias latent variables are maximum likelihood estimates of the correlations among the manifest variables, under the null hypothesis.

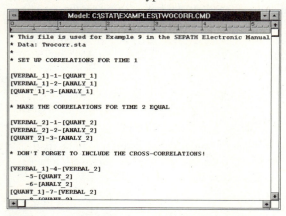

A note of caution should be added here. The tests performed by *SEPATH* on correlation matrices are valid asymptotic statistics. However, the very limited evidence available at this time suggests that (especially at smaller sample sizes) tests especially designed for correlational hypotheses (Steiger 1980a, 1980b; Steiger & Browne, 1984; Steiger & Hakstian, 1982; Wilson & Martin, 1983), which use the Fisher transform, *may* be somewhat more accurate and more powerful than tests performed with *SEPATH*.

Example 10: A Multiple Regression Model for Home Environment and Math Achievement

Jöreskog and Sörbom (1982) discuss several structural models which they fit to data from a study of home environment and school achievement by Keeves (1972). Keeves studied 215 sixth grade boys over a one year period, and measured initial mathematics achievement (Y_1), a structural dimension in the home (X_1), an attitudinal dimension of the home environment (X_2), a process dimension of the home environment (X_3), and final mathematics achievement (Y_2). The correlation matrix for the Keeves data is in a file called *Keeves.sta*.

	1 MATH68	2 STRUCT	3 ATTITUDE	4 PROCESS	5 MATH69
Case	SEPATH Examples #10 and #11				
MATH68	1.000				
STRUCT	.430	1.000			
ATTITUDE	.520	.670	1.000		
PROCESS	.540	.450	.630	1.000	
MATH69	.830	.450	.560	.520	1.000
Means	0.000	0.000	0.000	0.000	0.000
Std.Dev.	1.000	1.000	1.000	1.000	1.000
No.Cases	215.000				
Matrix	1.000				

According to Jöreskog and Sörbom,

> The three home environment dimensions were derived as the principal components of three sets of items which sampled each of the three home domains. Thus, the structural dimension was based on home interview items which determined the level of father's education, father's occupation, mother's occupation before marriage, religious affiliation, and number of children in the family. The attitudinal dimension was based on parents' attitudes toward the child's present education, future education, and occupation, and the parents' aspirations for themselves. The process dimension was based on relations between home and school, use of books and libraries, parents' help with formal school work, and arrangements for doing home assignments.

Cooley and Lohnes (1976) analyzed Keeves' data with a path model on the observed variables. This model, essentially a standardized multiple regression, is shown in the figure below.

PATH1 statements corresponding to this model can be found in a file *Keeves1.cmd*.

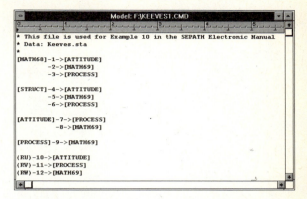

For other analyses of these data, see the next example on *Structural Models for Home Environment and Mathematics Achievement*.

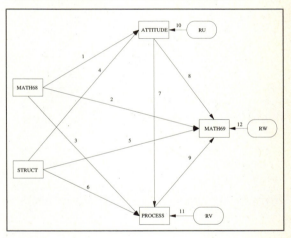

Example 11:
Structural Models for Home Environment and Mathematics Achievement

Jöreskog and Sörbom (1982) fit two structural models of their own to the Keeves (1972) data analyzed in a different manner by Cooley and Lohnes (1976). The Cooley and Lohnes approach is demonstrated in the example on *A Multiple Regression Model for Home Environment and*

Mathematics Achievement, above. The data are in the file *Keeves.sta*.

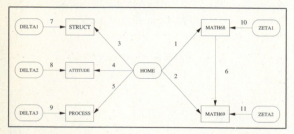

A path diagram for the first model fit by Jöreskog and Sörbom (1982) is shown below.

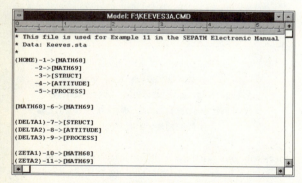

This model takes measurement errors in the home variables into account and treats them merely as fallible indicators of an aggregate construct variable "home." *PATH1* statements for this model are in a file called *Keeves3a.cmd*.

```
* This file is used for Example 11 in the SEPATH Electronic Manual
* Data: Keeves.sta

(HOME)-1->[MATH68]
      -2->[MATH69]
      -3->[STRUCT]
      -4->[ATTITUDE]
      -5->[PROCESS]

[MATH68]-6->[MATH69]

(DELTA1)-7->[STRUCT]
(DELTA2)-8->[ATTITUDE]
(DELTA3)-9->[PROCESS]

(ZETA1)-10->[MATH68]
(ZETA2)-11->[MATH69]
```

A somewhat different model is represented in the diagram below. This model incorporates errors of measurement in the measures of mathematics achievement. It assumes that reliabilities for the two measures are known to be *0.90*. Consequently the path coefficients representing true score and error variance coefficients are shown as fixed in the diagram.

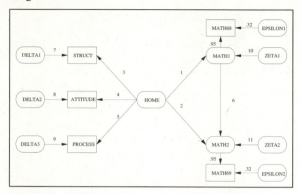

PATH1 statements for this model are in a file called *Keeves3b.cmd*.

```
* This file is used for Example 11 in the SEPATH Electronic Manual
* Data: Keeves.sta

(KSI)-1->[MATH68]
    -2->[MATH69]
    -3->[STRUCT]
    -4->[ATTITUDE]
    -5->[PROCESS]

[MATH68]-6->[MATH69]

(D1)-7->[STRUCT]
(D2)-8->[ATTITUDE]
(D3)-9->[PROCESS]

(ZETA1)-10->(MATH1)
(ZETA2)-11->(MATH2)
```

Example 12: Test Theory Models for Sets of Congeneric Tests

A variety of interesting test theory models can be tested and estimated using structural equations modeling. These models are all special cases of the common factor model, and are discussed in Jöreskog (1974) on pages 49-56.

The classical test theory model can be expressed as a common factor model. Suppose a group of tests are congeneric, i.e., have the same true scores underlying them. Then if the observed score is equal to the true score plus error, and error is uncorrelated with true scores, we have the path model shown in the diagram of Model A below for two sets of two congeneric tests.

Suppose you wished to assess whether two different tests measure the same trait. In classical test theory, we would say that the tests measure the same trait if their true scores are perfectly correlated, i.e., if the correlation coefficient between the tests, corrected for attenuation, is unity. Jöreskog (1978) refers to such tests as *tau-equivalent*.

Model A:

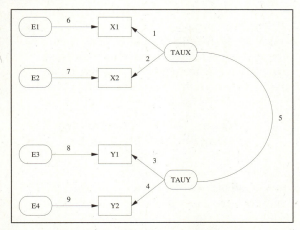

Lord (1957) proposed a test for tau-equivalence. His test required the assumption that the elements of each pair of tests have equal variance and equal reliability, i.e., that the tests are *parallel* within the pairs. The model corresponding to the hypothesis tested by Lord's method, that the tests are both parallel and tau-equivalent, is illustrated in the path diagram for Model B below. It is possible to test the *assumptions* underlying Lord's (1957) test directly, i.e., we can test the hypothesis that the elements of each pair of tests are parallel (i.e., have equal variance and equal reliability) *without testing the hypothesis that the disattenuated correlation coefficient is unity*.

Model B:

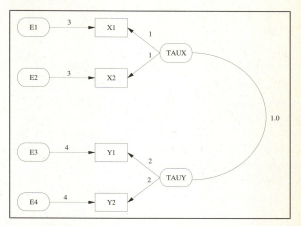

This hypothesis is shown in the path diagram for Model C below. Models B and C form a nested sequence, since B is a special case of C. If we compute the *Chi-square* statistics for Models B and C, and take their difference, we obtain a *Chi-square* difference statistic with one degree of freedom for testing the hypothesis that the disattenuated correlation is 1, given that the tests within pairs are parallel.

Model C:

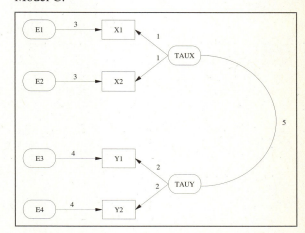

Lord's test is valid provided its assumptions are met. However, Model D shown below allows us to test whether the disattenuated correlation coefficient is unity *without requiring the restrictions of Lord's test.*

Model D hypothesizes that the tests are congeneric (but not *necessarily* parallel) and tau-equivalent.

Model D is a special case of model A, i.e., the two models form a nested sequence. Consequently, subtracting the *Chi-square* statistics for model A from that obtained for model D will produce a *Chi-square* statistic with one degree of freedom. This provides an alternative test that the disattenuated correlation is 1.

Model D:

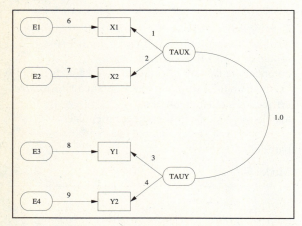

Jöreskog tested models A, B, C, and D on some data given in Lord's (1957) paper. The results of the testing are summarized both in Jöreskog (1974) and in Jöreskog (1978).

The covariance matrix for this example is in the file *Lord.sta*.

Case	SEPATH Example #12			
	1 X1	2 X2	3 Y1	4 Y2
X1	86.398			
X2	57.775	86.263		
Y1	56.865	59.318	97.285	
Y2	58.899	59.668	73.820	97.819
Means				
Std.Dev.				
No.Cases	649.000			
Matrix	4.000			

The *PATH1* statements for the 4 models are contained in the file *Lorda.cmd*, *Lordb.cmd*, *Lordc.cmd*, *Lordd.cmd*.

Both *Chi-square* difference tests yield the same conclusion: the hypothesis that the disattenuated correlation coefficient is 1 can be resoundingly rejected. For example, the *Chi-square* difference statistic (with one degree of freedom) for Model A versus Model D is *35.51*. On the other hand, the maximum likelihood estimate for the disattenuated correlation is quite high (.*899*). Moreover, the standard error for this value is only .*019*.

Example 13: Comparing Dependent Variances

SEPATH can be used to perform pattern hypotheses on a covariance matrix. As a simple example, consider some data gathered by William E. Coffman, and reported by Lord (1963). These data represent performance on the Stanford Achievement Test for 95 students measured in the seventh and eighth grades. The covariance matrix for the data is in a file called *Var2.sta*.

Case	SEPATH Electronic	
	1 X1	2 X2
X1	134.560	
X2	144.295	201.640
Means		
Std.Dev.		
No.Cases	95.000	
Matrix	4.000	

Suppose you wished to test the hypothesis that the population variance had not changed from seventh to eighth grade. The *PATH1* specification for such a hypothesis (contained in the file *Var2.cmd*) is as follows:

```
[X1]-1-[X1]
[X2]-1-[X2]
[X1]-2-[X2]
```

The *Chi-square* statistic indicates that the hypothesis of equal variances can be rejected, as the probability level is a very small value, i.e., around *.0001*.

Example 14: A Multi-Trait, Multi-Method Model

When personality traits or characteristics are measured, variation among people can occur for several reasons. Two obvious contributing factors are variation in the traits themselves, and variations in the way people react to a particular method.

When a trait is measured by only one method, there is a possibility that the variation observed is actually *method variance*, rather than trait variance. For example, if a particular questionnaire does not control for acquiescence response set, variation among people due to a problem in the method is confounded with actual trait variation.

One way around this problem is to try to measure both trait and method variation in the same experiment. The *multi-trait, multi-method* correlation matrix contains correlations between t traits or characteristics each measured by the same m methods. Campbell and Fiske (1959) suggested that the $mt \times mt$ multi-trait, multi-method correlation matrix should be examined to provide evidence of construct validity.

In their original work, Campbell and Fiske suggested that two kinds of validity, which they termed *convergent validity* and *discriminant validity*, could be evaluated by examining this matrix. There are 4 kinds of correlations in the matrix:

(1) same-trait, same-method

(2) same-trait, different-method

(3) different-trait, same-method

(4) different-trait, different-method.

Convergent validity is demonstrated if same-trait, different-method correlations are large. Discriminant validity is evidenced if same-trait different-method correlations are substantially higher than different-trait, different-method correlations.

Kenny (1979) analyzed data from Jaccard, Weber, and Lundmark (1975). Their study measured two traits, attitude toward cigarette smoking (C) and attitude toward capital punishment (P), with 4 different methods. The methods were:

(1) semantic differential

(2) Likert

(3) Thurstone

(4) Guilford

The correlation matrix, based on only 35 observations, is in a file called *Jaccard.sta*.

Case	1 C1	2 C2	3 C3	4 C4	5 P1	6 P2	7 P3	8 P4
C1	1.000							
C2	.780	1.000						
C3	.810	.770	1.000					
C4	.760	.710	.810	1.000				
P1	.290	.230	.190	.100	1.000			
P2	.280	.290	.180	.090	.840	1.000		
P3	.260	.310	.240	.080	.810	.890	1.000	
P4	.270	.240	.230	.150	.840	.910	.850	1.000
Means								
Std.Dev.								
No.Cases	35.000							
Matrix	1.000							

Kenny used a classic test theory approach to analyzing the data. The traits are factors whereas the disturbances or unique factors are allowed to be correlated across measures using the same method. Such a model is identified if there are at least two traits and three methods. Assuming the model fits the data, then convergent validation is assessed by high loadings on the trait factors, discriminant validation by low to moderate correlations between the trait factors, and method variance by highly correlated disturbances.

11. STRUCTURAL MODELING - EXAMPLES

The path diagram for the resulting model is shown below. The *PATH1* specification for the model is in the file *Jaccard.cmd*.

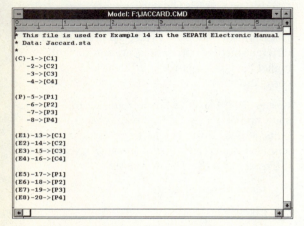

Kenny remarked that the data fit the model well in this case. With statistically-based fit indices, and the 20-20 vision of hindsight, we can see that the issue is very much in doubt. Run the example with *SEPATH* and see for yourself. First analyze the problem using the *Analyze Covariances* option.

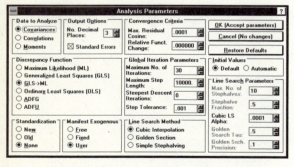

Examine the output, paying particular attention to the standard errors.

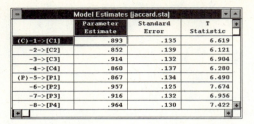

	Parameter Estimate	Standard Error	T Statistic
(C)-1->[C1]	.893	.135	6.619
-2->[C2]	.852	.139	6.121
-3->[C3]	.914	.132	6.904
-4->[C4]	.860	.137	6.280
(P)-5->[P1]	.867	.134	6.490
-6->[P2]	.957	.125	7.674
-7->[P3]	.916	.132	6.956
-8->[P4]	.964	.130	7.422

Then use the *Analyze Correlations* option to analyze the problem correctly.

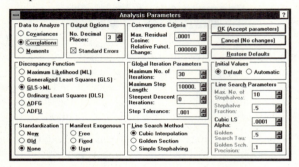

Again examine the standard errors.

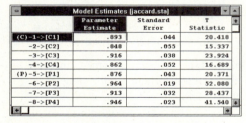

	Parameter Estimate	Standard Error	T Statistic
(C)-1->[C1]	.893	.044	20.418
-2->[C2]	.848	.055	15.337
-3->[C3]	.916	.038	23.924
-4->[C4]	.862	.052	16.689
(P)-5->[P1]	.876	.043	20.371
-6->[P2]	.964	.019	52.080
-7->[P3]	.913	.032	28.437
-8->[P4]	.946	.023	41.540

Notice how some of the standard errors differ dramatically when the sample correlation matrix is analyzed correctly.

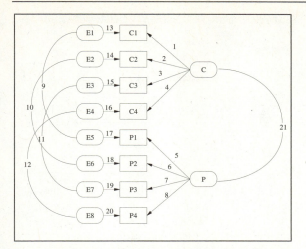

The output exhibits high loadings on the trait factors (coefficients 1 through 8), and a low correlation between the two trait factors. Disturbances are correlated, but not too highly.

The sample size is so small that the confidence intervals for the statistically-based fit indices are quite wide. For example, the 90% confidence interval for the Steiger-Lind RMS index ranges from *0* to *.1056*. In practice, we would prefer a substantially larger sample size.

Example 15: A Longitudinal Factor Model

Corballis and Traub (1970) presented a longitudinal factor analysis model, which stipulates that factorial structure underlying a set of tests remains constant over two or more administrations of the tests. An example of such a model is given by Everitt (1984, pages 52-55). The data were from a study by Meyer and Bendig (1961), who administered the 5 Thurstone Primary Mental Abilities tests to 49 boys and 61 girls in grades 8 and 11. The tests are Verbal Meaning (V), Space (S), Reasoning (R), Numerical (N), and Word Fluency (W). The correlation matrix for these data are in a file called *Meyer.sta*.

The model, analyzed by Everitt (1984) stipulates a single common factor underlying the scores on both occasions. The diagram for the model is in the figure below.

The *PATH1* translation of the diagram is contained in a file called *Meyer.cmd*.

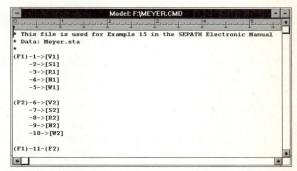

Everitt (1984) analyzed the correlation matrix with LISREL as though it were a covariance matrix. The standard errors reported in his Table 3.10 are not correct.

11. STRUCTURAL MODELING - EXAMPLES

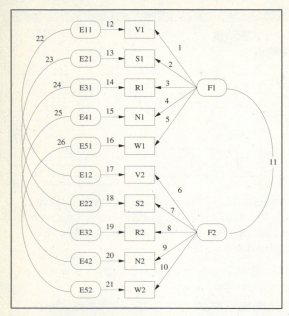

The *Chi-square* value of *51.46* allows us to reject the null hypothesis of *perfect* fit. On the other hand, the noncentrality-based fit indices indicate that the jury is, in a sense, still out regarding whether the fit of this model is acceptable. Consider, for example, the Steiger-Lind RMS index. The point estimate is *.076*, but the confidence interval ranges from *.03* to *.11*. Basically, this indicates that the sample size of 110 is insufficient to determine, with adequate precision, the quality of the population fit.

Example 16: A Structural Model for 10 Personality and Drug Use Variables

Huba and Harlow (1987) present a structural model relating personality characteristics to alcohol and marijuana consumption in adolescents. The correlation matrix (to the two-digit level of precision given in their printed article) for their data, based on 257 observations, is given in a file called *Hh.sta*.

Their first model corresponds to the path diagram below. (To obtain a larger copy of this diagram, click on *Help - SEPATH Examples*, go to example number 16 in the *Electronic Manual*, and print the topic from the *Help* facility.)

The file *Hh.cmd* contains a *PATH1* representation of the model. We urge the user to try to produce the *PATH1* statements corresponding to the figure *before* examining the contents of *Hh.cmd*. The file *Hh.cmd* is designed to replicate the analysis in the original article. Results obtained with *SEPATH* will correspond very closely, but not exactly, to those reported in the first column of Table 2 in Huba and Harlow (1987). We assume the discrepancy is due to the fact that we are using the correlations from their article, which they reported to only two digit accuracy.

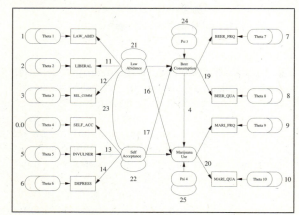

If you experiment with this example, you will discover a number of interesting things about it. For one thing, it is difficult to find the solution point for this problem. (See a discussion on this point in *Solving Interation Problems,* page 3645.) Use *Automatic* starting values and the *Correlation* option of *Data to Analyze.*

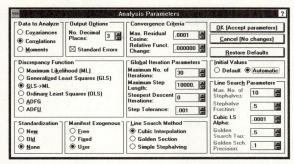

Try reversing some of the paths. If you experiment for a while, you will discover that there are other models, besides the one in the above figure, that fit the Huba-Harlow data as well as the one in their article. The theoretical reason for this can be found in an article by Lee and Hershberger (1990), who gave rules for visually inspecting path diagrams to detect paths that may be reversed without altering model fit.

Example 17: A Test for Compound Symmetry

A test for compound symmetry (i.e., equal variances and correlations for all pairs of variables) of the covariance matrix is sometimes performed in the context of repeated measures analysis of variance (see, e.g., Winer, 1971, pages 596-598). This hypothesis states that the covariance matrix has equal diagonal elements, and equal off-diagonal elements. The file *Winer.sta* contains the pooled covariance matrix analyzed by Winer.

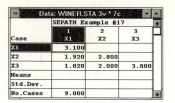

Because the matrix was obtained by pooling two samples of size 5, it would have the same distribution (assuming both populations have the same covariance matrix) as a covariance matrix based on a single sample of size 9. Hence, that is the sample size recorded in the data file. We test that the covariance matrix has the form

$$\begin{matrix} \sigma^2 & \rho\sigma^2 & \rho\sigma^2 \\ \rho\sigma^2 & \sigma^2 & \rho\sigma^2 \\ \rho\sigma^2 & \rho\sigma^2 & \sigma^2 \end{matrix}$$

Commands for testing the hypothesis are in the file *Winer.cmd.*

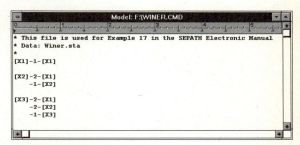

Note that the maximum likelihood estimates under the null hypothesis correspond to the results on page 598 of the Winer text.

	Parameter Estimate	Standard Error	T Statistic	Prob. Level
[X1]-1-[X1]	3.233	1.217	2.657	.008
[X2]-2-[X1]	1.913	1.187	1.612	.107
-1-[X2]	3.233	1.217	2.657	.008
[X3]-2-[X1]	1.913	1.187	1.612	.107
-2-[X2]	1.913	1.187	1.612	.107
-1-[X3]	3.233	1.217	2.657	.008

We find that the *Chi-square* statistic, with *4* degrees of freedom, is only *.694*. This value is not significant. In fact, the probability level is suspiciously high, as is sometimes seen with artificial data in ANOVA textbooks! This fact, coupled with the high values for the fit coefficients, suggests that the population matrix deviates only trivially from the hypothesized structure.

Example 18: Testing the Equality of Correlation Matrices from Different Populations

SEPATH can perform *correlational pattern hypothesis* tests either between or within samples (or both). In this example, using the file *Jennrich.sta*, we show you how to use *SEPATH* to compare correlations in two or more populations for equality.

Case	1 VAR1	2 VAR2	3 VAR3	4 VAR4	5 VAR5	6 VAR6	7 VAR7	VA
VAR1	1.000							
VAR2	.543	1.000						
VAR3	.560	.684	1.000					
VAR4	.532	.414	.484	1.000				
VAR5	.526	.485	.503	.870	1.000			
VAR6	.498	.602	.614	.755	.763	1.000		
VAR7	.444	.354	.521	.500	.572	.576	1.000	
VAR8	.336	.323	.264	.372	.423	.433	.611	
VAR9	.399	.427	.514	.493	.601	.612	.793	
VAR10	.398	.648	.491	.560	.662	.680	.466	
VAR11	.568	.527	.546	.619	.672	.650	.503	
VAR12	.654	.674	.712	.527	.616	.651	.570	
Means								
Std. Dev.								

The general procedure for comparing two or more correlation matrices for equality is extremely straightforward. In general, you specify all the off-diagonal elements of the correlation matrix as free parameters, then copy the identical model for as many groups as you want to compare.

Before setting up your model file, you will have to enter the appropriate group selection commands. (See the discussion of the *Specify Groups* dialog on page 3572). Once you have specified the groups you wish to compare, and opened the data file (in this case *Jennrich.sta*), set up your model file using the following approach.

(1) Using the *Path Construction Tool*, enter the **GROUP 1** command.

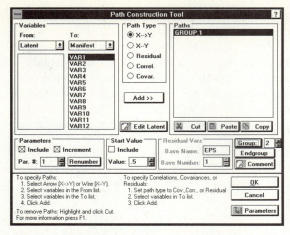

(2) Use the *Correl.* option in the *Path Construction Tool*, select *Manifest* in the right side variables list, and highlight all the variables for the correlation matrix to be tested. Then click *Add>>*.

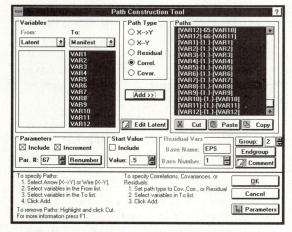

(3) Add an **ENDGROUP** command.

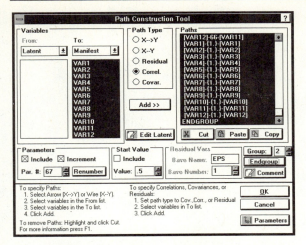

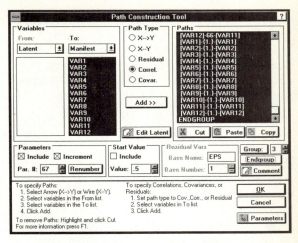

(4) Highlight the commands that specify the entire correlation matrix. Copy them to the Clipboard, using the *Copy* command.

(5) Then for each group, add a **GROUP** command. In this case, we will add only a second group.

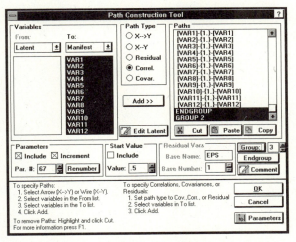

(6) Then paste in a copy of the correlation matrix commands, and add an **ENDGROUP** command.

(7) Finally click *OK* to save the model.

The example described here is taken from an article by Jennrich (1970). In the article, Jennrich describes a "dedicated" procedure for comparing two correlation matrices. His procedure, based on the method of generalized least squares, is somewhat different from the method employed by *SEPATH*. Jennrich gives an example in which two groups of subjects (40 mentally retarded, 89 normal), are both measured on 12 variables. The correlation matrices are then compared for equality.

The commands for comparing the two matrices for equality are in the file *Jennrich.cmd*.

Jennrich (1970) compares GLS and ML test statistics for this problem. Try running the example with the GLS discrepancy function, then with the ML discrepancy function. You will find that the GLS statistic with 66 degrees of freedom fails to reject at the .05 level of significance (*Chi-square = 56.27, p =.7977*), while the ML statistic does reject (*Chi-square = 102.25, p = .0028*). Depending on one's theoretical predisposition, one might be tempted to accept one result or the other as the "right" one in this case. However, such a judgement might be ill-advised and statistically naive. There are several possible reasons for these conflicting results. The ML test may be more powerful than the

GLS procedure. On the other hand, both procedures are *asymptotic* tests, and their exact sampling distributions are not known. It may be, therefore, that the ML procedure is more liberal at small to moderate sample sizes. That is, it may be that the ML procedure rejects too often at small sample sizes.

Hardly any information is available in the literature about the performance of correlational pattern hypothesis tests for comparisons across populations. (Steiger (1980b) presents some evidence in single sample tests that the ML test may be more liberal than the GLS test at small samples, and hence may tend to reject a true null hypothesis too often.) However, the interested experimenter may generate some evidence on his/her own by using *SEPATH*'s powerful *Monte Carlo* module (page 3625).

MONTE CARLO METHODS

In this section, we discuss the role of Monte Carlo methods in covariance structure analysis, and review some examples. For a discussion of the dialogs, menus, and options for the Monte Carlo module, see page 3587.

Introduction

The statistical testing procedures used in *SEPATH* are based on *asymptotic* distributional results. Asymptotic distribution theory describes how test statistics and parameter estimates are distributed as sample sizes become infinitely large. Such results are useful in practice because (1) exact sampling distributions cannot be derived for the whole family of covariance structure hypotheses, and (2) the asymptotic results may provide a reasonable approximation to the actual distribution of the test statistic.

Whenever asymptotic distributional results are employed, the natural question to ask is, "How fast does convergence of the sampling distribution to the asymptotic result occur?" Asymptotic distribution theory tells us the shape of the sampling distribution as N approaches infinity. But does this distributional form approximate the distribution when $N = 100$?

One of the best ways to answer such a question is to conduct a *Monte Carlo* study. Monte Carlo procedures use random number simulation techniques to simulate random samples in the situation you are interested in. By simulating *many* such samples, running the statistical procedure on each sample, and keeping track of the results, you can obtain an empirical estimate of the actual sampling distribution of the test statistic under study.

Ideally, Monte Carlo studies should investigate a broad range of conditions before reaching strong conclusions about a particular statistical procedure. Unfortunately, the more general and complex a statistical procedure is, the less likely one is to find an authoritative Monte Carlo investigation. For example, in the field of covariance structure analysis, only a handful of Monte Carlo studies have been published, and these have examined only a trivial range of situations, often with unrealistically small numbers of variables and unrealistically simple models. The primary reason for this state of affairs is practical. Covariance structure models take a fair amount of computation time to test. A Monte Carlo experiment must simulate hundreds, or even thousands of such tests. Until very recently, it was simply too expensive and time consuming to perform such experiments. A few years ago, one medium-sized Monte Carlo study might have exhausted an entire academic department's annual mainframe computing budget!

A second reason for the dearth of studies is the absence of appropriate tools for performing Monte Carlo analyses. Of the available covariance structure programs, only one offered an integrated Monte Carlo procedure. Consequently, those who wished to conduct a Monte Carlo analysis were forced to use extremely cumbersome methods to obtain their data. These methods involved writing computer routines to generate the simulated data (in a form readable by the covariance structure program), running the covariance structure program on each sample, then collating the results by hand. Huge amounts of time and tedious effort were required to run a Monte Carlo analysis in this manner.

In *SEPATH*, we provide you with very extensive, fully integrated tools for performing your own Monte Carlo analyses. You will find that, using these tools, you can generate studies of your own that are far more extensive and authoritative than any yet published. Moreover, you can also use the

Monte Carlo module to address questions that arise in the course of your own research.

The following examples represent only a tiny subset of the interesting questions you can address with Monte Carlo methods.

Adequacy of Sample Size and Heywood Cases in Factor Analysis

One of the questions that tends to arise in many contexts in statistics is "How big a sample do I need?" In standard classical testing situations, this is related frequently to the question of statistical power. If you need to reject the null hypothesis to prove a theoretical point, you certainly want to have adequate power to detect a false null hypothesis. Increasing sample size is the most straightforward way of manipulating power.

In covariance structure analysis, the experimenter is frequently in a somewhat different position, i.e., trying to show that a particular model fits the data well. Here, there can be a different reason for worrying about sample size. This reason is not emphasized sufficiently in most textbooks on structural modeling. Specifically, when sample size is insufficient, the iterative procedure may converge to a minimum that is simply impossible, i.e., represents estimates that are way out of line with reality.

For example, in factor analysis, one may encounter the *Heywood case*, in which one of the residual (or "unique") variances is estimated to be zero (or, in older structural modeling programs, a negative value).

A natural question to ask at the outset of a structural modeling study is, "If my model is a good approximation to reality, the basic statistical assumptions are met, and I gather a sample of size N, am I likely to obtain results from my analysis that agree with my model?" If the answer is "No," then you may as well not conduct the study! Yet, people seldom investigate such a question before conducting structural modeling studies, and no published textbook describes the use of Monte Carlo methods to investigate such a question.

Here, we examine how the use of Monte Carlo methods can reveal an insufficient sample size, i.e., an N that leaves a high prior probability of a misleading analysis.

Open the data file *Lawley.sta*, and imagine that you were Lawley, about to gather data to test a confirmatory factor model like the one in the file *Lawley.cmd*. In this example, you will use the Monte Carlo module of *SEPATH* to examine the long run performance of the testing procedure in a situation where the confirmatory factor model and underlying statistical assumptions are correct.

From the *Startup Panel,* click on the *New Model* button, then select *Edit with Path Tool* to bring up the *Path Construction Tool*.

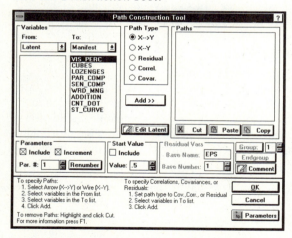

In the *Path Construction Tool* dialog, click *Edit Latent Vars.* and type in the names *Visual, Verbal,* and *Speed,* then click *OK*.

11. STRUCTURAL MODELING - MONTE CARLO METHODS

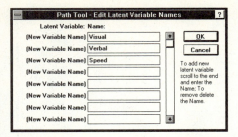

We are going to simulate a model in which all factor loadings are .6, all unique variances are .64, and all the factor intercorrelations are zero. We will do this by creating a model with these numbers inserted in braces as *initial values*. The Monte Carlo procedure will read these numbers, treat them as population parameters, and generate samples from the covariance matrix corresponding to these values.

First, create the factor loadings for the *Visual* factor. Highlight *Visual* on the left list. Highlight *VIS_PERC, CUBES,* and *LOZENGES* in the right side list. Click the *X-->Y* radio button in the *Path Type* group. Make sure the *Include* box is checked in the *Start Value* group, and type in the value .6. Now click *Add>>*.

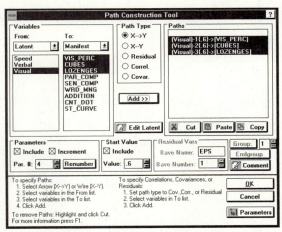

You should see the following 3 paths appear in the *Paths* window on the far right.

```
(Visual)-1{.6}->[VIS_PERC]
(Visual)-2{.6}->[CUBES]
(Visual)-3{.6}->[LOZENGES]
```

Next, create the factor loadings for the *Verbal* factor. Highlight *Verbal* on the left list, and *PAR_COMP, SEN_COMP,* and *WRD_MNG* in the right side list. Now click *Add>>*.

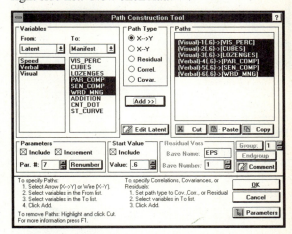

You should see the following 3 paths added to the previous paths in the *Paths* window.

```
(Verbal)-4{.6}->[PAR_COMP]
(Verbal)-5{.6}->[SEN_COMP]
(Verbal)-6{.6}->[WRD_MNG]
```

Next, in similar fashion, create loadings from the *Speed* factor to the variables *ADDITION*, *CNT_DOT*, and *ST_CURVE*. When you are finished, the dialog should look like this:

SEPATH - 3627

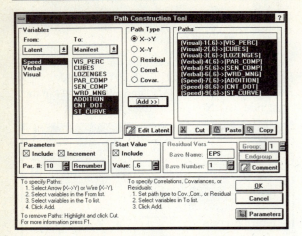

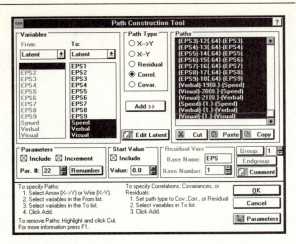

Then create the residual variable paths. Click the *Residual* radio button in the *Path Type* group. Type *.64* in the *Value* field in the *Start Value* box. Highlight all the manifest variables, and click *Add>>*.

Your model is fully specified. Click *OK,* and the model will appear in the *New.cmd* window.

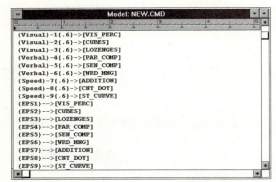

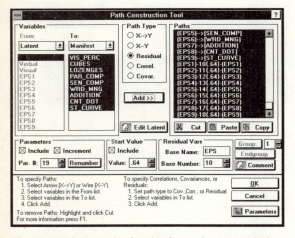

Before continuing, compare your file with the commands in *Monte1.cmd* to verify that they agree.

You are now ready to perform the Monte Carlo experiment. Before setting up and running the Monte Carlo study, we shall choose the appropriate parameters from the *Analysis Parameters* dialog. From the *Startup Panel*, click *Set Parameters* and select *Correlations* under *Data to Analyze*.

Finally, add the paths for the factor intercorrelations. Select *Latent* in the *To:* variables list. Click on the *Correl.* radio button for *Path Type*. Highlight the variables *Speed, Visual,* and *Verbal* in the *To:* list. Type *0.0* as the *Value* in the *Start Value* group. Then click *Add>>*.

11. STRUCTURAL MODELING - MONTE CARLO METHODS

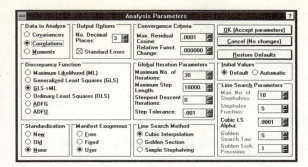

Since most factor analyses are performed on correlation matrices, this mode is preferred over analyzing covariances.

Note that, in the *Global Iteration Parameters* box, the *Maximum Number of Iterations* has a default value of *30*. In most Monte Carlo studies, it is a good idea to increase the maximum number of iterations, to avoid the somewhat ambiguous results that are generated when iteration is prematurely terminated because more than 30 iterations are required. Increase the maximum number of iterations to *150* by typing this value into the edit field. Now click *OK*.

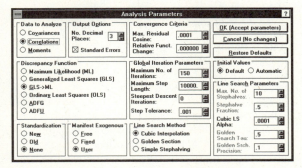

From the *Startup Panel*, click *Setup/Run Analysis* in the *Monte Carlo* group to open the *Monte Carlo Setup* dialog.

In this case, you want to save extra information about parameter estimates, standard errors, and fit indices, so make sure all three boxes are checked in the *Store Extra Information* group.

Because this is an illustrative example, we will run only *50* replications, and use the default seed value of *1234567*.

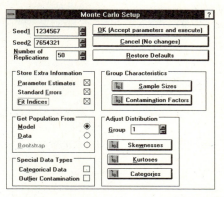

Set these values, then click *OK* to return to the *Startup Panel*.

We wish to examine the long run behavior of the estimation procedure when the sample size is 75, i.e., roughly equivalent to that actually used in the study discussed by Lawley. In order to set the sample size for the Monte Carlo simulation, click on *Sample Sizes* to bring up the *Set Monte Carlo Sample Sizes* dialog.

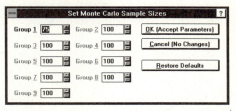

Enter *75* in the first edit field, then click *OK*.

You are now ready to begin the Monte Carlo Study. Click *OK (Accept Parameters and Execute)*. You will be asked if you are sure you want to continue.

11. STRUCTURAL MODELING - MONTE CARLO METHODS

Click *Yes*, and the study will proceed. *SEPATH* will now automatically generate 50 simulated samples taken from a population corresponding to the model you created. Each simulated sample will be analyzed according to the parameters you selected.

Progress of your Monte Carlo study will be described in the *Status Bar* message area at the bottom of the screen.

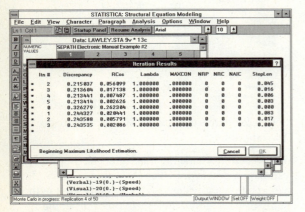

After a few Monte Carlo replications, you can gauge the approximate amount of time it will require to complete the Monte Carlo analysis. (Typically, a 100 Mhz Pentium-based computer will require about five minutes to complete this Monte Carlo analysis.) If you are running the analysis in multitasking mode, you will be able to run other applications while the Monte Carlo study is proceeding. Moreover, you can interrupt the analysis at any time with a *Cancel* command. (Use the ESC key at any time. You may also (1) click *Cancel* on the *Interation Window* if iteration is in progress, or (2) click *Cancel* on the status bar during Monte Carlo data generation.) At that point, all completed results will be available, and may be saved.

When your Monte Carlo results are complete, the *Monte Carlo Results* dialog will come up automatically.

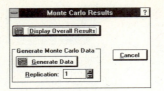

Click *Display Overall Results*, and the following Scrollsheet will appear:

Your first step after completing a Monte Carlo experiment should be to save the overall results, so they will not be lost if there is some problem that causes you computer to malfunction. You can save this file as a *STATISTICA* data file, Scrollsheet, or both. Click on *Save As Data* from the *File* menu, and save the Scrollsheet as a data file called *Monte1.sta*.

Now that you have saved your data, you can proceed to analyze the Monte Carlo results immediately, using the commands invoked by using the Right-mouse-button. Click on the scrollbar on the bottom of the screen until variable number 11 (*BOUNDARY*) is in view. This variable is the number of "boundary cases" or active inequality constraints at the conclusion of iteration, found on each replication. Click on the variable name (*BOUNDARY*), to highlight the entire column.

SEPATH - 3630

Then click the Right-mouse-button. This action will bring up the menu of quick Scrollsheet graphic and statistics capabilities. Click *Block Stats/Columns,* then click *Graphs,* followed by *Histograms.*

You will see the graph shown below. As usual, you can customize the appearance of the resulting graph in countless ways to suit your own preferences. For example, the fitted normal curve, which appears by default, is superfluous in this case, and can be eliminated by double-clicking on the area outside the graph.

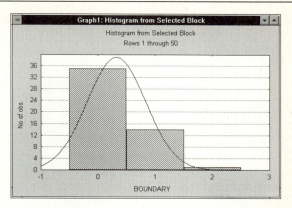

This will call up the *General Layout: 2D Graphs* dialog. Select *Plot Layout,* then select *Fit Off* to remove the normal curve.

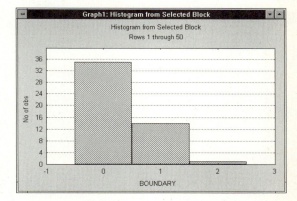

This histogram shows that 35 of the 50 replications resulted in no boundary cases, but 14 replications yielded one boundary case, and 1 replication actually yielded two boundary cases.

Overall, then, 30% (15 of 50) of the replications resulted in at least one boundary value. For example, the very first Monte Carlo replication produced a boundary value. If you scroll across the parameter values in the data file, you will discover that the variable *PAR_11*, the unique variance for the manifest variable *CUBES*, had a value of *0.000* at convergence. Chances are, this value would have iterated to a negative value except that *SEPATH* constrains all variance estimates to be non-negative during iteration. This boundary case, known as a "Heywood case" to factor analysts, often will result in distorted values for some of the other parameters.

Clearly, it is an undesirable state of affairs to have a prior probability of approximately .30 of obtaining a Heywood case, when in fact none of the unique variances in the population is less than .64, and the factor model fits perfectly in the population.

The question is, would a larger sample size have avoided this problem? Why not find out? When time permits, feel free to experiment with the Monte Carlo module. Below is a scatterplot graph (with exponential fit) representing the results of a brief study. This study examined percentage of boundary cases at sample sizes of 75, 150, and 225.

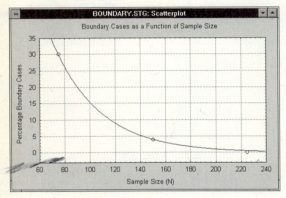

Information such as that obtained in the study just described could be extremely valuable to someone planning a confirmatory factor analytic study, because it allows you to examine, before gathering data, how the estimation procedure will behave "if everything goes according to plan." Needless to say, if the estimation procedure has problems under such ideal conditions, you need to try to eliminate the source of the problem.

The kind of "prior" Monte Carlo analysis we just conducted can furnish additional valuable information. One of the most useful areas of information is an estimate of *precision of estimation*

you are likely to achieve with a given level of sample size.

Suppose, for example, you gather a sample of size 75. What kind of precision of estimation will you achieve in your confirmatory factor analysis if everything goes well?

The Monte Carlo study has given you a considerable amount of relevant information. Simply examine the parameter estimates generated for each parameter across the 50 replications. For example, highlight the variable *PAR_1* by clicking its name with the Right-mouse-button. Then use the Right-mouse-button to call up the *Block Stats/Columns* menu, select *Graphs* followed by *Histograms*.

	PAR_1	PAR_2	PAR_3	PAR_4	PAR_5
1	.51448299	.80974169	.51137167	.35148100	.42389729
2	.79372692	.68888490	.77293879	.64463334	.64747135
3	.51825344	.41146489	.75947628	.63449612	.73910031
4	.32746057	.47143814	1.00000000	.84937693	.58434643
5	1.00000000	.20352020	.24798416	.82283506	.52800326
6	.84675006	.43988311	.64467730	.38256863	.55712860
7	.68208190	.59842003	.55971032	.68439866	.52580149
8	.86785290	.48406250	.49722416	.64594900	.66780036
9	.45962077	.92730496	.59724576	.49992651	.61599378
10	.69815218	.77834528	.62306461	.74258565	.64016606
11	.45614554	1.00000000	.26450451	.54494038	.59726459
12	.66672141	.66621212	.67673482	.58450166	.52872108
13	.58032076	.60749960	.91696407	.70162316	.56807808
14	.80608219	.45276198	.50049697	.69510502	.72129449
15	.52244080	.54190373	.55442766	.53524510	.67821862
16	1.00000000	.40534680	.32695433	.43364046	.85156111
17	.73063176	.55292093	.49042044	.43846875	.85994215
18	.52408893	.58347996	.80721192	.67283733	.76198336
19	.47931986	.67149724	.51364513	.56970551	.54282355
20	.80378000	.57307153	.61129766	.58557649	.65042449
21	.62345610	.46018161	.60405072	.52622462	.64635239

This action will generate a histogram of the values generated for parameter number 1. This histogram furnishes you with a bootstrapped approximation to the sampling distribution of parameter number 1. The standard deviation of this bootstrapped distribution provides an estimate of the standard error of the parameter estimate.

11. STRUCTURAL MODELING - MONTE CARLO METHODS

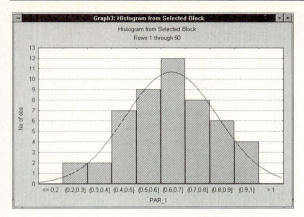

To obtain this standard deviation, click the Right-mouse-button again, and select *All* from the *Stats/Columns* menu. This will generate basic statistics for the 50 Monte Carlo replications. Scroll down to beyond row 50 in the Scrollsheet to examine the statistics that have been calculated.

	PAR_1	PAR_2	PAR_3	PAR_4	PAR_
39	.50055492	.57171417	.52324523	.44330936	.72078
40	.52280982	.83005025	.39657722	.63000478	.54408
41	.78167656	.53327497	.51911438	.59528015	.44832
42	.44100831	.56547374	.81517246	.79905102	.69006
43	.38600531	.49849213	.43090350	.66518523	.73523
44	.68030366	.53705931	.89726903	.79679946	.63161
45	.57888308	.54227379	.60206507	.85919580	.49899
46	.65713102	.67790499	.69643897	.65452632	.62340
47	.67688640	.63317775	.75053494	.54348865	.84320
48	.65286443	.76654854	.48633679	.38578331	.28866
49	.69864074	.51741987	.70029496	.49975066	.62489
50	.41317370	.44727017	1.00000000	.66855443	.69096
MEAN(1-50)	.64342979				
MEDIAN(1-50)	.66192621				
SD(1-50)	.18698244				
VALID_N(1-50)	50.0000000				
SUM(1-50)	32.1714893				
MIN(1-50)	.2049742				
MAX(1-50)	1.0000000				
25th%(1-50)	.5025458				
75th%(1-50)	.7768265				

The standard deviation is *.187*. Since a 95% confidence interval is roughly plus or minus two standard errors, the estimated confidence intervals for each parameter are typically going to be the parameter estimate ±.37. Many experimenters might conclude that this sort of range falls somewhat short of the precision desirable in such an estimate.

In this case, our bootstrapped estimate of the standard error may be biased because we included,

in our calculations, the trials on which Heywood cases occurred. By using the *Block Stats* facilities (available via the *Edit* pull-down menu, or by pressing the right mouse button on the respective Scrollsheet column), or by saving the results Scrollsheet as a data file, and using the *Quick Basic Stats* options, you can examine this notion yourself.

In a similar fashion, you could examine, *a priori,* the kind of precision you will obtain in confidence interval estimates of the non-centrality based fit indices.

In the preceding example, we showed how you can execute a Monte Carlo study, save the results, and analyze them immediately without ever leaving the *SEPATH* module. You have many other analytic options, of course. You can open the Monte Carlo data file you have saved, and use *Quick Stats Graphs*, *Quick Basic Stats* (both available at a click of the Right-mouse-button) to explore the trends further. Alternatively, you can call up another *STATISTICA* module (like *Basic Statistics and Tables*) and pursue the analysis even further. The point is, the Monte Carlo analysis facility is closely integrated with other analytic facilities and tools of the *STATISTICA* system. You can design, execute, analyze, and graph your Monte Carlo results, all without leaving the *STATISTICA* environment. This is in keeping with our philosophy that such analysis, performed *prior* to actual gathering of data, can be crucial to the proper design and execution of research involving structural models.

Performance of GLS and ML Estimation in the Comparison of Correlation Matrices

In the example on comparison of correlation matrices (see page 3622), we found that the *Chi-square* statistic based on the ML estimation procedure produced a rejection of the null hypothesis at the .05 significance level, while the analogous statistic based on GLS estimation failed to reach significance. This contrast raised questions about the relative performance of the two methods. It could be that the difference relates solely to the difference in power between ML and GLS methods. (ML methods are generally somewhat more powerful.) On the other hand, the ML procedure may be *excessive* (i.e., reject a true null hypothesis more frequently than the nominal rate) at the rather small sample sizes used in the Jennrich (1970) example (Jennrich himself questions the adequacy of the sample sizes in his original paper). *SEPATH* provides tools for delving further into these questions. Specifically, you can perform a Monte Carlo study to assess the relative Type I error rates for the GLS and ML estimation procedures.

A full Monte Carlo study to examine the general question of relative performance of the GLS and ML statistics would, of course, require many conditions, varying relative sample sizes and the characteristics of the population correlation matrices being simulated. This section demonstrates how a user could set up one highly relevant condition from such a study.

The simulation examined here assumes that the population correlation matrices are, in fact, the same, and that sample sizes are 90 and 40, roughly equivalent to those in the Jennrich example. This simulation is rather time-consuming. Typically, each replication will require approximately 3 minutes on a computer with a 486DX-33 processor. A set of 100 replications would thus require about 5 hours on a 486-33 (about 50 minutes on a 100 Mhz Pentium-based computer). Depending on your computer's microprocessor, you may have to schedule this example for a time (possibly overnight) when the computer will not be required for other purposes.

The Monte Carlo study will require two runs. The first run will collect data for the Maximum Likelihood discrepancy function, the second run will gather data for the GLS discrepancy function.

A problem is specifying reasonable choices for the hypothetical population correlation matrices for the Monte Carlo study. A choice that seems natural and eminently reasonable in this case is to use, as the hypothetical population, the estimates for the population parameters obtained when testing the hypothesis of equal correlation matrices with a particular estimation procedure. When treated as population values, these estimates will, of necessity, specify a situation where the null hypothesis is true.

SEPATH makes it particularly easy to use the obtained estimates as population parameters. Here is how you do it.

Test the hypothesis of equal correlation matrices, using GLS estimation. The model estimates will appear in the file *New.txt*. After estimation has concluded, call up the *New.txt* window, and copy the parameter estimates (these are *PATH1* statements with the parameter estimates and standard errors in braces after each parameter number) to the Clipboard using CTRL+C. Then return to the *Startup Panel* and open a new model file. Paste the contents of the Clipboard into this new file. (You will have to manually remove any page breaks that were inserted.) The contents of this file should be identical to a file called *Monte2GL.cmd* in the examples directory. This file will serve as the population model for the Monte Carlo study.

Analysis Parameters must be chosen appropriately. After starting *SEPATH*, open file *Jennrich.sta*.

	1 VAR1	2 VAR2	3 VAR3	4 VAR4	5 VAR5	6 VAR6	7 VAR7	8 VAR8	9 VAR9
VAR1	1.000								
VAR2	.543	1.000							
VAR3	.560	.684	1.000						
VAR4	.532	.414	.484	1.000					
VAR5	.526	.485	.503	.870	1.000				
VAR6	.498	.602	.614	.755	.763	1.000			
VAR7	.444	.354	.521	.500	.572	.576	1.000		
VAR8	.336	.323	.264	.372	.423	.433	.611	1.000	
VAR9	.399	.427	.514	.493	.601	.612	.793	.542	1.000
VAR10	.398	.648	.491	.560	.662	.680	.466	.288	.549
VAR11	.568	.527	.546	.619	.672	.650	.503	.520	.646
VAR12	.654	.674	.712	.527	.616	.651	.570	.508	.663
Means									
Std.Dev.									
No.Cases	40.000								

Open the *Monte2GL.cmd* model file.

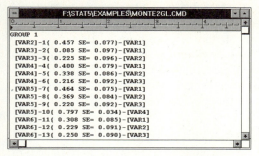

In the *Analysis Parameters* menu, select *Correlations* as the *Data to Analyze*, and adjust the *Maximum No. of Iterations* parameter to *150*. Select *Maximum Likelihood* as the *Discrepancy Function*, and select the *Automatic* option for *Initial Values*.

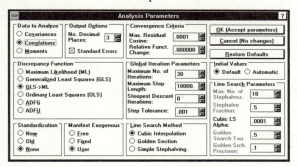

Click *OK* to return to the *Startup Panel*. Then, from the *Startup Panel*, click *Setup/Run Analysis* in the *Monte Carlo* group.

In the *Monte Carlo Setup* dialog, make sure all 3 boxes in the *Store Extra Information* group are checked. Click *Sample Sizes* in the *Group Characteristics* group, and adjust the sample sizes to *40* for group 1, and *90* for group 2.

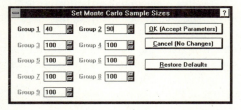

You are now ready to begin the Monte Carlo study. Set the number of Monte Carlo *Replications* to *100*.

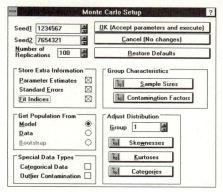

The seed value controls the data generated for each Monte Carlo replication. If you begin two Monte Carlo experiments with the same sample sizes with the same seeds, they will process the same data. Consequently, rerunning the experiment with the same seed, but using ML as the estimation procedure, will create an analogous data set for the ML estimation method. Since these two data sets will have been based on identical sample data, they are more directly comparable than if two different seeds were used for the two runs.

In case your computer is an older model, or your work schedule will not permit running this lengthy experiment, we have run the Monte Carlo study for you. Results from 100 replications of the ML estimation procedure are in the file *Monte2ML.sta*. Each of these runs was generated with a starting seed of *1234567*. Results on the same data for the GLS estimation procedure are in the file *Monte2GL.sta*.

Examination of these Monte Carlo results, using the data analytic facilities of *STATISTICA*, will provide us with some information that is relevant to the questions we addressed earlier.

Open the file *Monte2ML.sta*. This file contains output from 100 replications of the maximum likelihood estimation procedure.

11. STRUCTURAL MODELING - MONTE CARLO METHODS

Case	1 SEED1	2 TERMCODE	3 DISCREP	4 RCOS	5 GRADIENT
1	1234567	0	.72479738	.00006089	.00004351
2	761169069	0	.53489040	.00004182	.00002997
3	936766388	0	.59476701	.00005998	.00007873
4	978650720	0	.53322413	.00003694	.00006415
5	479677633	0	.66429975	.00006793	.00005848
6	572043145	0	.41417452	.00002784	.00003865
7	1364604053	0	.61272338	.00005112	.00008231
8	1488156296	0	.59310173	.00003889	.00005632
9	1858718599	0	.53955804	.00005038	.00007402
10	699581522	0	.45442290	.00004406	.00005592
11	1274321521	0	.43681523	.00002721	.00003300
12	126227091	0	.46226421	.00005814	.00010157
13	970892945	0	.60190351	.00004029	.00005052

Begin by examining the *TERMCODE* variable. Ideally, all termination codes should be zero, indicating that normal convergence occured on that replication. In this case, all 100 replications converged normally.

As a secondary check on the iterative process, construct a histogram of the number of iterations required for the 100 replications, and look for outliers. Click on the variable name *NUM_ITER*, then click with the Right-mouse-button to bring up the *Quick Stats Graphs* menu. Open this menu, and click on *Histogram of NUM_ITER*. Click the *Regular* option to generate the following histogram.

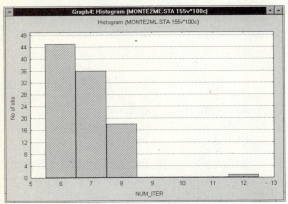

In this case, we find that most cases required between 7 and 9 iterations. This narrow range of low values suggests that the iterative algorithms performed in a consistent and rather stable manner on these data.

Next, examine the performance of the test statistic itself. Recall that the example was constructed to simulate a situation where the null hypothesis is, in fact, true. In this case, the test statistic has an asymptotic distribution that is central *Chi-square*, with 66 degrees of freedom. The question is, "How well is the asymptotic distribution approximated by the actual distribution of the test statistic in this case?"

We shall use *STATISTICA*'s data analytic and graphical capabilities to analyze the performance of the test statistic. First, calculate quick descriptive statistics for the *CHI_SQR* variable by (1) clicking on the variable name to highlight it, (2) clicking the Right-mouse-button, then clicking on *Quick Basic Stats,* (3) clicking on *Descriptives of CHI_SQR*.

You should see the following data for the mean, 95% confidence interval endpoints, and standard deviation.

Variable	Valid N	Mean	Confid. -95.000%	Confid. +95.000%
CHI_SQR	100	74.19509	71.44703	76.94314

Recall that a *Chi-square* variable with 66 degrees of freedom should have a mean of 66, and a variance of 132, corresponding to a standard deviation of 11.489. Moreover, a *Chi-square* with 66 degrees of freedom has a distribution that is very close to normal in shape. Consequently, the standard (normal theory) confidence interval on the population mean should be quite accurate in this case. *STATISTICA* calculates the interval as ranging from *71.44* to *76.94*.

The fact that the interval excludes 66 by a wide margin indicates that the actual distribution of the *Chi-square* statistic clearly departs from the asymptotic distribution, and tends to have values that are too high. Moreover, the standard deviation, *13.84*, is somewhat higher than the theoretical value of 11.489. These distributional characteristics would tend to lead to too frequent rejections in practice. To get some idea of the true Type I error rate for the test statistic when the nominal rate is .05, we delve further into the data by constructing a frequency distribution of the data for the variable *PLEVEL*. *PLEVEL* represents the probability level for the *Chi-square* test statistic, under the assumption that it has a central χ^2_{66} distribution. To tell the actual proportion of Type I errors when tests are run at the α significance level, simply examine what proportion of the time the variable *PLEVEL* has a value less than α.

In this case, we will require the services of the *Basic Statistics and Tables* module. Open this module, and from the *Startup Panel* select *Frequency Tables* and click *OK*. Then, in the *Frequency Tables* dialog, under *Categorization Method for tables and graphs*, select *Step Size* of *.01, Starting at 0.0*. Make sure the box for starting at *Minimum* is not checked. Then click on *Frequency Tables*. The top section of the resulting Scrollsheet should contain the following data. Since the total *N* is 100 replications, counts are identical to percentages in this case. Note that 20% of the *PLEVEL* values were less than .05, meaning that the actual Type I error rate was about .20 when the nominal error rate was .05.

PLEVEL (monte2ml.sta)				
Category	Count	Cumul. Count	Percent	Cumul. Percent
0.00000<=x<.010000	8	8	8.000000	8.0000
.010000<=x<.020000	4	12	4.000000	12.0000
.020000<=x<.030000	5	17	5.000000	17.0000
.030000<=x<.040000	2	19	2.000000	19.0000
.040000<=x<.050000	1	20	1.000000	20.0000
.050000<=x<.060000	4	24	4.000000	24.0000
.060000<=x<.070000	2	26	2.000000	26.0000
.070000<=x<.080000	2	28	2.000000	28.0000
.080000<=x<.090000	0	28	0.000000	28.0000
.090000<=x<.100000	0	28	0.000000	28.0000
.100000<=x<.110000	0	28	0.000000	28.0000
.110000<=x<.120000	0	28	0.000000	28.0000
.120000<=x<.130000	1	29	1.000000	29.0000
.130000<=x<.140000	0	29	0.000000	29.0000
.140000<=x<.150000	3	32	3.000000	32.0000

These data suggest that, to be considered "significant at the .05 level," values of the test statistic should have a nominal significance level less than .01. (This is based on the fact that 8% of the cases had *p-level* values less than .01). After performing this Monte Carlo examination, the sophisticated user might well hesitate to declare the two population correlation matrices significantly different at the .05 level.

Examination of similar data for the GLS test statistic show a rather different trend. You can examine these data by opening the file *Monte2GL.sta*. The descriptive statistics show, in this case, a mean and standard deviation that are subtantially lower than they should be.

Descriptive Statistics (monte2gl.sta)				
Variable	Valid N	Mean	Confid. -95.000%	Confid. +95.000%
CHI_SQR	100	55.13747	53.45488	56.82007

Moreover, the frequency distribution for the variable *PLEVEL* shows that there were no rejections at the .05 level. Hence, while the ML statistic is *excessive*, the GLS statistic is far too *conservative*, i.e., it rejects far too infrequently. Its conservative nature is almost certainly accompanied by very low power in this case.

It appears that the sample size was simply too low to allow for adequate precision for the attempted analysis. Ideally, in structural modeling the number of observations should be at least 10-20 times the number of variables, and the sample sizes of 40 and 89 were just too low.

There are probably dozens of examples of published papers employing structural equation modeling which report test statistics based on questionable sample sizes. Using *SEPATH*, you can reconstruct some of these analyses, run Monte Carlo experiments, and find out for yourself how the test statistics tend to perform with particular combinations of model and sample size.

Prior analysis of system performance, using Monte Carlo methods, can inform you in advance about adequacy of sample sizes, potential convergence problems, etc. We urge you to make use of this information when using structural equation modeling in your research.

In this case, you could employ the Mont Carlo procedure to determine appropriate levels of sample size in the experiment.

Bootstrapping

Bootstrapping (Efron, 1982) is a general technique for estimating sampling distributions. If N independent observations from a population are available, bootstrapping simulates the sampling distribution of any statistic by treating the observed data as if it were the entire (discrete) statistical population under study. Suppose, for example, you wished to estimate the sampling distribution of the correlation coefficient, in order to set a confidence interval for the correlation between two variables. You only have a sample of size 200 from the relevant population. If the population variables can be reasonably assumed to have a bivariate normal distribution, you could use the well-known procedures available in virtually any textbook. However, suppose your data depart seriously from a bivariate normal distribution, and, moreover, you do not have access to more general results (e.g., Steiger and Hakstian, 1982) on the distribution of the correlation coefficient with non-normal data. In this case, setting up a confidence interval using the standard "normal theory" procedures might yield serious errors. How might you proceed in this case?

The way bootstrapping works is as follows. On each replication, a random sample of size N is selected, *with replacement*, from the available data. The statistic of interest is calculated on this "bootstrapped subsample," and recorded. The process is repeated for some reasonable number of replications. Finally, the distribution of all the bootstrapped statistics is tabulated. This distribution furnishes an approximation to the actual sampling distribution of the statistic. Note again that, in effect, bootstrapping assumes that the population distribution can be approximated by a discrete distribution identical to that manifested in your sample.

SEPATH's Monte Carlo module has a bootstrapping facility built into it. This facility takes random samples (with replacement) of size N from the current data file, fits the current model to that "bootstrapped subsample," and stores the results. You can use this facility to estimate the sampling distribution of model parameters, or to perform Monte Carlo simulations of sampling from discrete multivariate distributions.

In this case, bootstrapping will be employed to estimate the standard error of a correlation coefficient from a non-normal multivariate population.

The data file *Bootsamp.sta* contains a sample of 200 observations on two variables. Open this file to begin the example.

	VAR1	VAR2
1	.011	.657
2	1.599	.648
3	-1.471	-1.919
4	-4.209	-.488
5	-.121	-.056
6	-.253	-4.386
7	-2.405	-.768
8	.269	.248
9	-.198	.172
10	-.144	-1.571
11	.256	.722
12	.197	-.505
13	-.094	-.341
14	-1.374	-.693
15	-.059	.344

Next, open the model file *2var.cmd*. This "fully saturated" model simply estimates the population correlation matrix.

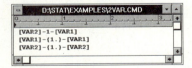

We first fit this model to the observed data. In the *Analysis Parameters* dialog, select *Correlations* in the *Data to Analyze* group. Make sure the default *Discrepancy Function, GLS->ML* is selected, then click *OK*.

11. STRUCTURAL MODELING - MONTE CARLO METHODS

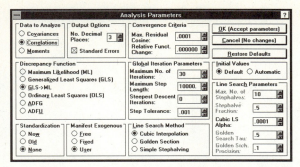

When you return to the *Startup Panel,* begin estimation by clicking *OK (Execute Current Model).* Since the model is fully saturated, there are no degrees of freedom, and you will quickly converge to a discrepancy function that is essentially zero. Click *Model Summary* in the *Results Dialog.* You will see the correlation parameter estimated at *.499,* and its standard error at *.053.* Can this latter estimate be trusted?

Model Estimates (bootsamp.sta)				
	Parameter Estimate	Standard Error	T Statistic	Prob. Level
[VAR2]-1-[VAR1]	.499	.053	9.385	.000
[VAR2]-(1.)-[VAR1]	....	....	....	....
[VAR2]-(1.)-[VAR2]	....	....	....	....

There is strong evidence that the population distribution is not multivariate normal in this case. The variables have high kurtosis. (See *Evaluating Multivariate Normality*, page 3643, for an analysis of distributional characteristics of these data.) Consequently, the estimated standard error obtained under the assumption of multivariate normality may be seriously biased.

In this case, there are two options open to the user. One is to employ *Asymptotically Distribution Free* (*ADF*) estimation procedures (see page 3575). Another is to employ bootstrapping.

Next we shall bootstrap an estimate of the sampling variability of the correlation coefficient, using the *Monte Carlo* module's bootstrapping function. Open *Bootsamp.sta* and *2var.cmd*, and choose *Correlations* as the *Data to Analyze* in the *Analysis Parameters* dialog. Click *Setup/Run Analysis* in the

Monte Carlo group in the *Startup Panel.* This will bring up the *Monte Carlo Setup* dialog. Choose *Bootstrap* as the option in the *Get Population From* group, and set the *Sample Size* to *200* in the resulting dialog.

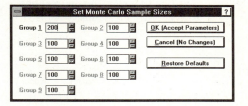

Enter *200* as the *Number of Replications*, and make sure the boxes for *Parameter Estimates* and *Standard Errors* are checked off under *Store Extra Information.*

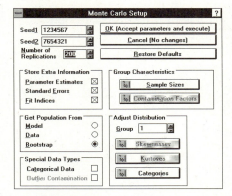

Click *OK*, and start the Monte Carlo study. When the bootstrap data have been gathered, click *Display Overall Results* in the *Monte Carlo Results* dialog. In a few seconds you will see a Scrollsheet containing the results of the Monte Carlo analysis. While the Scrollsheet is the active window, select *Save as Data* from the *File* menu. Enter *Boot1.sta* as the file name, and save the file.

The bootstrap data have now been saved as a *STATISTICA* data file. To construct an empirical estimate of the characteristics of the sampling distribution of the correlation coefficient, call up the *Basic Statistics and Tables* module, and open the *Boot1.sta* file.

SEPATH - 3639

11. STRUCTURAL MODELING - MONTE CARLO METHODS

	1 SEED1	2 TERMCODE	3 DISCREP	4 RCOS	5 GRADIENT
Case					
1	1234567	0	0.00000000	.33244017	0.00000000
2	491320317	0	.00000000	.30192377	0.00000000
3	1944537011	0	0.00000000	.21101889	0.00000000
4	1965193582	0	0.00000000	.16560112	0.00000000
5	884670728	0	0.00000000	.31306769	0.00000000
6	1126761052	0	0.00000000	.31605577	0.00000000

Begin by constructing a histogram of the variable *PAR_1*, which contains the estimated correlation coefficients. To construct the histogram, click *Frequency Tables* from the *Startup Panel*. In the *Frequency Tables* dialog, select *PAR_1* as the variable, and *Step Size .025 Starting at Minimum* under *Categorization Method for Tables and Graphs*.

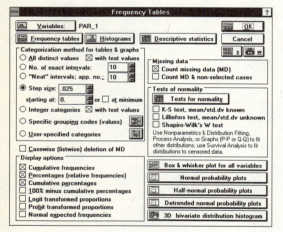

Then click *Histograms* to produce the graph shown below. This histogram provides an estimate of the sampling distribution of the correlation coefficient, based on your data. As you can see, the distribution of the sample correlation coefficient follows a normal distribution very closely.

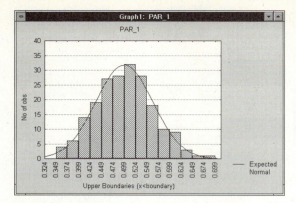

Next, calculate the standard deviation of the parameter estimates to obtain an estimate of the standard error of the correlation coefficient. From the *Startup Panel*, click *Descriptive Statistics* and compute *Detailed Descriptive Statistics* for variable *PAR_1*. You should obtain the following:

The standard deviation of the boostrap value estimates the standard error of the correlation coefficient as *.0627*, substantially larger than the estimate provided by the maximum likelihood estimation procedure.

We can compare this estimate with one obtained via ADF estimation. Return to *SEPATH* and open *Bootsamp.sta* and *2Var.cmd* again; then estimate the model with *ADFU* selected as the *Discrepancy Function*, and *Correlations* as the *Data to Analyze*.

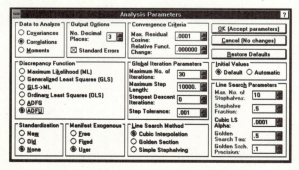

You will obtain an estimated standard error of *.059* for the correlation parameter.

	Model Estimates [bootsamp.sta]			
	Parameter Estimate	Standard Error	T Statistic	Prob. Level
[VAR2]-1-[VAR1]	.499	.059	8.513	.000
[VAR1]-(1.)-[VAR1]	....	....	....	....
[VAR2]-(1.)-[VAR2]	....	....	....	....

Actually, the data in *Bootsamp.sta* were generated by the *SEPATH* Monte Carlo module. The population distribution is multivariate nonnormal. The variables were transformed using the Vale-Maurelli procedure to have population skewnesses of *0*, but population kurtoses of *15* each. By running a large Monte Carlo study (using *SEPATH*), we determined the *actual* standard error of the correlation coefficient to be *.069* in this case. You can see that the bootstrap procedure provided the most accurate estimate of the standard error, the ADF a somewhat less accurate estimate, and the maximum likelihood procedure the worst estimate.

The ADF and bootstrapping estimates might have been more accurate, but in this case the sample kurtoses for the data in *Bootsamp.sta* were rather substantial underestimates of the actual population values of 15.

EVALUATING MULTIVARIATE NORMALITY

Introduction

SEPATH uses procedures based on asymptotic distribution theory. The default (maximum likelihood) procedure, used in about 99% of all published studies, assumes a Wishart distribution of the sample covariance matrix **S**. If this assumption is violated, the *Chi-square* test statistic may be seriously in error. Departures from a Wishart distribution are quite likely to occur if the population distribution is not multivariate normal, or if individual variates depart from normality. In this section, we examine some artificial data that depart from normality, demonstrating how the analytic capabilities of *STATISTICA* would enable you to detect this departure.

Constructing Histograms

STATISTICA provides the user with an unparalled array of powerful and convenient graphical analyses. Unfortunately, there has not been much emphasis on graphical analysis (or assessment of assumptions) in the literature on structural modeling. Part of the reason for this lack of emphasis is that most texts and program manuals concentrate on classic examples from the literature that never presented raw data. Without raw data, the opportunity for analysis of distributional characteristics is extremely limited.

When applying structural modeling techniques to your own data, or to data that are available in raw form, your first step should be to examine the distributions of individual variables, to assess departures from multivariate normality.

Bring up the *Basic Statistics and Tables* module and open the data file *Bootsamp.sta*. This file contains data for two variables. A first step to analyzing these data would be to assess their normality. Begin by constructing a histogram and fitting a normal curve to it.

Click *Frequency Tables,* and select *20* as the *No. of Exact Intervals* in the *Categorization Method for Tables & Graphs* group.

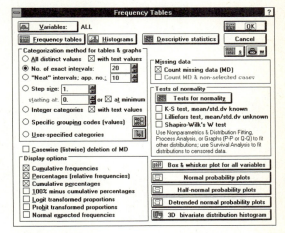

Click *Histograms* to produce the two graphs shown below. The graphs will show a bump in the center that rises far above the plot line for the expected frequency in a normal distribution. This is because these variables are *leptokurtic*, i.e., have high kurtosis (see page 3682).

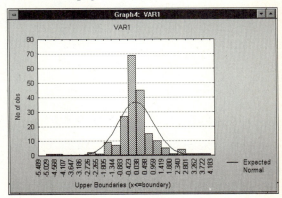

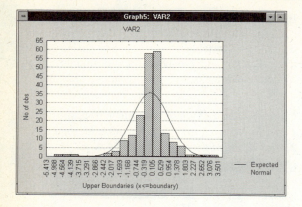

Assessing Departures from Normality

Maximum likelihood procedures for analysis of covariance structures are especially sensitive to departures from zero kurtosis. *SEPATH* provides statistics for assessing univariate kurtosis, as well as multivariate kurtosis. In this section, we examine some of these statistics briefly.

Bring up *SEPATH*, and open the file *2Var.cmd*. In the *Analysis Parameters* dialog, select *Correlations* as the *Data To Analyze*.

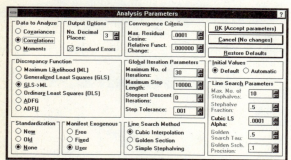

Click *OK*, and begin execution of the model estimation procedure.

Once the model has been analyzed, you can call up a number of diagnostic statistics from the *Results* dialog.

Click on *Univariate Kurtoses* to display the following results:

Univariate Skewness Indices (bootsamp.sta)			
	Skewness	Corrected Skewness	Normalized Skewness
VAR1	-.164	-.166	-.949
VAR2	-1.124	-1.132	-6.488

The *Kurtosis* value is the standard sample kurtosis estimate, while the *Corrected Kurtosis* is an estimate that is unbiased if the population distribution is normal. These values should be close to zero if the distributions of *VAR1* and *VAR2* are normal. The *Normalized Kurtosis* estimate, based on asymptotic distribution theory, transforms the *Kurtosis* estimate to a variable that has an approximate standard normal distribution if the population distribution is normal. Thus, for assessing the hypothesis that the distribution is normal, these values can be treated like standard normal test statistics. Absolute values greater than 1.96 can be treated as "significant at the .05 level."

Clearly, these variables have substantial kurtosis, far beyond that which might be expected solely through sampling error. This point is reinforced by the analysis of multivariate kurtosis measures. In particular, the *Mardia Coefficient of Multivariate Kurtosis* and its normalized transformation, the *Normalized Multivariate Kurtosis*, suggest that the data depart widely from multivariate normality. If the data are sampled from a multivariate normal population, the *Normalized Multivariate Kurtosis* has a distribution that is approximately standard normal, and so it can be interpreted like other standard normal test statistics.

Measures of Multivariate Kurtosis (bootsamp.sta)	
	Value
Mardia Coefficient of Multivariate Kurtosis	15.784
Normalized Multivariate Kurtosis	27.903
Mardia-Based Kappa	1.973
Mean Scaled Univariate Kurtosis	2.071
Adjusted Mean Scaled Univariate Kurtosis	2.071
Relative Multivariate Kurtosis	2.973

In this case, the *Normalized Multivariate Kurtosis* is 27.9, significant beyond the .0001 level. Clearly these data are not multivariate normal!

SOLVING ITERATION PROBLEMS

Introduction

Structural modeling programs generally must obtain their parameter estimates by using iterative techniques. These techniques are special cases of *nonlinear optimization* procedures for minimizing a function of n unknowns. Nonlinear optimization is an extremely challenging area of numerical analysis, and the problems and issues discussed in the chapter on *Nonlinear Estimation* (Volume III) remain relevant here. Any textbook on nonlinear optimization (e.g., Dennis and Schnabel, 1983) will quickly warn the reader that it is an art as well as a science, and that no optimization procedure works "best" for all problems. In general, the more unknowns, and the more nonlinear the function to be minimized, the more difficult the problem becomes. Problems with more than 100 unknowns are, in general, extremely difficult to solve unless you can start the iteration process rather close to the actual solution point.

Ironically, most textbooks on structural modeling completely conceal this fact from the reader, presenting only examples that, with respect to the iteration process, are trivially easy and well-behaved. Armed only with an education from such texts, the beginner to structural modeling can be confused and frustrated by problems encountered during iteration.

If you analyze a significant number of problems with *SEPATH*, you can expect to encounter iteration problems at some point. Here are some guidelines for dealing with these problems. Before continuing, we suggest that you attempt to read the discussion of technical aspects of iterative methods (page 3664) in the section on *Technical Aspects of SEPATH*.

How Iteration Procedures Work

When iteration begins, each parameter in your model has an *initial value,* or *start value*. These values are "plugged in" to your model equations and used to generate an estimated covariance matrix. This estimated covariance matrix is compared to the actual sample covariance matrix, and the value of the *discrepancy function* is calculated.

The program then has to decide how to alter the parameter values to improve the discrepancy function value, i.e., make it smaller. The program calculates a *step direction*, using an approximation procedure called the Gauss-Newton method. A *step increment* vector is calculated, and added to the original parameter estimates to produce a new set of values.

The program takes this new set of values and computes a new value of the discrepancy function. If the discrepancy function has improved sufficiently, the program goes on to the next iteration. If you are anywhere near the correct solution, this is generally what will happen. The process will then continue smoothly until you reach the minimum, usually in 20 iterations or less.

When Iteration Procedure "Hangs Up"

If you are far away from the solution point, several things can go wrong. Besides calculating the direction to step the parameters, the program also has to decide how far to step. As the parameters near the solution point, the discrepancy function graph is usually rather flat and smooth, and the program can gauge very accurately how far to step. However, when the parameters are far from the correct solution, both the step direction and step size may be inaccurate. In that case, the discrepancy function may actually *increase* with the first new set of parameter values the program tries. Usually both

the step direction and step length are in error, but if the program takes a smaller step in the same direction, the discrepancy function will decrease. Consequently, the program starts trying out smaller steps in the same direction. At this stage, the program is performing a *line search*. There are several *line search* algorithms available in *SEPATH* (see page 3665 for a technical discussion). In any particular situation, one may work where others fail.

The way line search algorithms work is they multiply the step increment vector by a constant (usually less than *1*), use this altered vector to recompute new parameter values, and recompute the discrepancy function. If the discrepancy function has not improved, they try again, etc. Hence, when the initial step does not work well, you will see a slowdown in iteration and values of *Lambda* in the *Iteration Results* dialog that are less than *1*. In some cases, iteration may seem to "hang up" briefly. Ultimately, iteration will cease, with a very small value of *Lambda* and *Steplen*.

If the program cannot find a multiple of the step increment vector that reduces the discrepancy function, it means the program has chosen a very bad step direction, or possibly that the program has iterated to a "saddle point." The program may try reversing the step direction, or may terminate iteration. This kind of situation will generally occur only when the initial values are very far from the solution point, or when the discrepancy function surface is extremely irregular for the particular problem being analyzed.

There are other telltale signs that indicate that you have iterated into a "bad" region of parameter space. In computing the step direction, *SEPATH* computes an approximation to the covariance matrix of the parameters. If the program iterates to a particularly bad region, this matrix may become singular. In some cases, the program can "iterate out" of this problem area by partialling out the parameters that are causing the problem, and continuing the iteration

by altering only the other parameters (Jennrich and Sampson, 1968).

An indication that this has happened is that the *NRP* value will change during iteration from zero to some other value, indicating that the information matrix has changed from nonsingular to singular. (The *NRP* value will also be non-zero, generally from the start of iteration, if the model has redundant parameters, i.e., is not identified properly.)

When the program ends up in such an unfortunate situation, there is little you can do but to try again.

For the maximum likelihood discrepancy function to be computable, both the sample covariance and the covariance matrix reproduced by your model and parameter estimates must be nonsingular. Sometimes during iteration, the program steps to values that yield a *singular* estimated covariance matrix. If this happens, iteration will terminate immediately. Generally, this problem will occur only in the early stages of iteration.

If iteration hangs due to any of the above conditions, there are several things to try:

1) If iteration failed using arbitrary initial values (the default), try the *Automatic* option for *Initial Values* in the *Analysis Parameters* dialog (page 3576). These values may get closer to a solution than arbitrary starting values, especially when the covariance matrix is ill-scaled, i.e., has variables that vary widely in variance.

2) Often, a large step is taken on the first few iterations. Sometimes, this step will carry the parameters into a region from which the iterative process cannot recover. Consequently, one thing to try is to reduce the step size the program is allowed to take. Try reducing the step size to a small value (like *.1*) and see what happens. Keep in mind that iteration to a solution will take longer under these conditions. During the early phases of iteration, the program will indicate that the largest allowable step size has been taken by

placing an asterisk next to the *StepLen* value printed in the *Iteration* dialog window.

3) Some times, iteration will fail with one line search algorithm, but succeed with another. Try an exact (*Golden Search*) line search by selecting this option in the *Line Search Method* group in the *Analysis Parameters* dialog.

4) When iteration encounters problems early, it is often because the approximate Hessian is too inaccurate far from the solution to be of use in calculating a proper step direction. In such cases, inserting *steepest descent* iterations at the beginning of iteration will eliminate the problem. Steepest descent iterations use only first derivative information in selecting a step direction, and do not employ the approximate Hessian calculated by the Gauss-Newton procedure.

5) If you are testing your model on the covariance matrix, try testing it on the correlation matrix instead. In many situations, models that fail to iterate properly with an ill-scaled covariance matrix will work fine when tested on the correlation matrix.

6) Keep in mind that you may need to try several of the above options in combination to produce successful iteration. Feel free to experiment.

7) Remember that a badly mis-specified model may fail to fit properly. One of the common mistakes beginners make is to forget to add disturbance terms to endogenous latent variables. The *Structural Modeling Wizard* will add these for you automatically, but if you are copying a model from a path diagram, remember that some authors consider disturbance terms to be "implicitly obvious," and don't bother to put them in the diagram itself.

An Example

Open the data file *Hh.sta* and the model file *Hh.cmd*. In the *Analysis Parameters* dialog, restore all defaults by clicking on the *Restore Defaults* button, then choose *Correlations* as the *Data to Analyze* and *Generalized Least Squares (GLS)* as the *Discrepancy Function*.

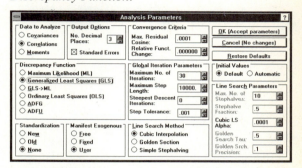

Click *OK*, and start the iteration by clicking on *OK (Execute Current Model)* from the *Startup Panel*.

Observe carefully what happens as the iteration "hangs up." After iteration terminates (as the number of iterations reaches 30), click the *Iteration History* button to produce a Scrollsheet with the information printed during iteration. The iteration history shown below demonstrates some of the phenomena discussed in the preceding section. On the first iteration, the program takes a relatively modest step (*.265*) resulting in a substantial drop in the discrepancy function (from *18.08* to *3.27*). Unfortunately, the approximate Hessian (and associated information matrix) has become singular at the new parameter values, resulting in a *NRP* value of *2*. The program continues iteration, but immediately iterates into an improper region, and has to constrain one or more parameters on the boundary, the number being indicated by *NAIC* values other than zero.

The program takes a large step on the 4th iteration, and immediately jumps into a region (again) where the Hessian is singular. Unfortunately, the program never finds its way out of this region, and iteration stalls out, with step sizes getting smaller and smaller, but no real progress being made.

ITN	DISC	RCOS	LAMBDA	MAXCON	NRP	NRC	NAIC	STEPLN
0	18.08	.788	1.000	1.250	0	0	0	0.000
1	3.27	.720	1.000	.527	2	0	0	.265
2	2.95	.707	1.000	.496	0	0	1	.026
3	1.23	.478	1.000	.293	0	0	2	.272
4	1.18	.599	.250	.336	0	0	3	5.432
5	.90	.522	1.000	.199	3	1	1	.239
6	.78	.541	.250	.238	2	1	2	2.134
7	.67	.538	1.000	.424	1	0	2	.522
8	.64	.525	.125	.392	0	0	0	2.326
9	.60	.481	1.000	.347	1	0	0	.057
10	.44	.089	1.000	.041	1	0	0	.090
11	.43	.085	1.000	.036	1	0	0	.109
12	.42	.063	1.000	.012	1	0	0	.058
13	.42	.057	1.000	.005	1	0	0	.038
14	.42	.054	1.000	.002	1	0	0	.022
15	.42	.054	1.000	.001	1	0	0	.014
16	.42	.055	1.000	.000	1	0	0	.008
17	.42	.056	1.000	.000	1	0	0	.005
18	.42	.056	1.000	.000	1	0	0	.003
19	.42	.056	1.000	.000	1	0	0	.002
20	.42	.056	1.000	.000	1	0	1	.001
21	.42	.056	1.000	.000	1	0	1	.001
22	.42	.057	1.000	.000	1	0	1	.000
23	.42	.057	1.000	.000	1	0	1	.000
24	.42	.057	1.000	.000	1	0	1	.000
25	.42	.057	1.000	.000	1	0	1	.000
26	.42	.057	1.000	.000	1	0	1	.000
27	.42	.057	1.000	.000	1	0	1	.000
28	.42	.057	1.000	.000	1	0	1	.000
29	.42	.057	1.000	.000	1	0	1	.000
30	.42	.057	1.000	.000	1	0	1	.000

One approach to solving the problem is to use automatic starting values. From the *Startup Panel*, click *Set Parameters*, then select *Automatic* as the option in the *Initial Values* group, *Correlations* as the *Data to Analyze,* and *Generalized Least Squares (GLS)* as the *Discrepancy Function*.

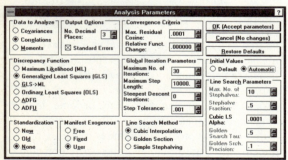

Click *OK*, then begin execution again. Notice how, from the new initial values, iteration proceeds smoothly and effectively to the minimum.

ITN	DISC	RCOS	LAMBDA	MAXCON	NRP	NRC	NAIC	STEPLN
0	13.480	.6864	1.0000	.9723	0	0	0	0.0000
1	1.529	.5919	1.0000	.4561	0	0	0	.2373
2	.456	.2951	1.0000	.1070	0	0	0	.1364
3	.373	.3349	1.0000	.1057	0	0	0	.1043
4	.318	.0328	1.0000	.0193	0	0	0	.0325
5	.318	.0071	1.0000	.0003	0	0	0	.0058
6	.318	.0031	1.0000	.0000	0	0	0	.0023
7	.318	.0016	1.0000	.0000	0	0	0	.0010
8	.318	.0013	1.0000	.0000	0	0	0	.0007
9	.318	.0007	1.0000	.0000	0	0	0	.0004
10	.318	.0005	1.0000	.0000	0	0	0	.0003
11	.318	.0003	1.0000	.0000	0	0	0	.0002
12	.318	.0002	1.0000	.0000	0	0	0	.0001
13	.318	.0001	1.0000	.0000	0	0	0	.0001
14	.318	.0001	1.0000	.0000	0	0	0	.0000

Another approach will also work in this case. From the *Startup Panel* click *Set Parameters*, then select *Default* for *Initial Values, Correlations* as the *Data to Analyze,* and *Generalized Least Squares (GLS)* as the *Discrepancy Function*. These are the conditions that resulted in iteration "stalling out" before.

This time, we will control the step size so that the program can only take small steps. The hope is that this will keep the iteration from jumping into a problem region during the early going, when step direction is gauged less accurately.

For *Maximum Step Length* in the *Global Iteration Parameters* group, enter a value of *0.1*. (The default value of *10000* leaves step size, in effect, completely unconstrained.) Because the steps will be small, you may need more iterations before convergence occurs, so set *Maximum No. of Iterations* to *300* in the *Global Iteration Parameters* window.

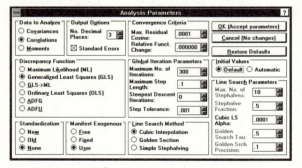

Click *OK* and restart the iteration by clicking *OK (Execute Current Model)* from the *Startup Panel*.

Unfortunately, the iteration fails to converge to a minimum, encountering difficulties similar to what occurred previously. Notice that the problems started on iteration number one, where the program stepped into a region where the Hessian was singular. Now let's try inserting some *steepest descent* iterations, to remove the approximate Hessian from the calculation of step direction. Go back to the *Analysis Parameters* dialog and set *No. of Steepest Descent Iterations* to *10* in the *Global Iteration Parameters* group.

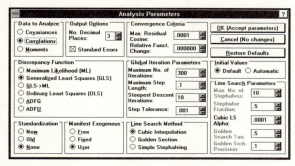

Then click *OK* and restart iteration by clicking *OK (Execute Current Model)* from the *Startup Panel*.

Although the iteration runs into difficulties at several points (by now you are probably learning to recognize when these occur by reading the output parameters during iteration), convergence eventually occurs after more than *50* iterations.

This example illustrates how you must be prepared to be both analytical and artistic when you attempt to solve problems encountered during iteration. Read the *Iteration History* results carefully, take appropriate action, and above all, don't be afraid to experiment.

TECHNICAL ASPECTS OF *SEPATH*

This section presents some of the more important technical and theoretical details on the models, methods, and techniques underlying *SEPATH*. It is not *absolutely* necessary to understand any of this material to use the program. However, prospective users of the program should at least browse through this section.

Two aspects of *SEPATH* are of particular theoretical interest. First, the structural model equation system employed by *SEPATH* is novel. Second, *SEPATH* provides theory for, and a practical implementation of, asymptotic maximum likelihood statistical estimation and confidence intervals for measures of model fit.

Models and Methods

Structural equation models have achieved increasing popularity in the social sciences. Much of the credit for this popularity can be attributed to the flexibility and power of the methods themselves. Equally important has been the availability of computer software for performing the modeling process.

An enormous amount of material has been written on structural modeling. There are now numerous textbooks and monographs for the beginner. Users in need of such materials are referred to books by Long (1983a, b), James, Mulaik and Brett (1982), Kenny (1979), Everitt (1984), Loehlin (1987), Bollen (1989), and Hayduk (1987) among others. All of these books have significant virtues. The reader with a serious interest in the subject should probably at least browse through all of these books.

For a very interesting debate on the value of structural models in the social sciences, the *Summer 1987* issue of the *Journal of Educational Statistics* is strongly recommended. This issue contains a critique of path analysis by D.A. Freedman, and responses to that critique by a number of writers.

Discussion of the deeper apects of the theoretical connections between causal inference and statistical modeling is beyond the scope of this chapter. For important and illuminating accounts of this topic, see the books and papers by Scheines (1994), Spirtes, Glymour, and Scheines (1993), and Steyer (1992, 1994).

The LISREL Model

In the following sections, all variables will be assumed to be in deviation score form (i.e., have zero means) unless explicitly stated otherwise.

The statistical model behind *SEPATH* is best understood in historical context, and so this section begins with a review of several important models for the analysis of covariance structures.

In his 1986 review of developments in structural modeling, Bentler described 3 general approaches to covariance structure representations. The first and most familiar involved integration of the psychometric factor analytic (FA) tradition with the econometric simultaneous equations model (SEM). This approach, originated by a number of authors including Keesling, Wiley, and Jöreskog was described by Bentler with the neutral acronym FASEM. The well-known LISREL model is of course the best known example of this approach.

The LISREL model can be written in three interlocking equations. Perhaps the key equation is the structural equation model, which relates latent variables.

$$\eta = B\eta + \Gamma\xi + \zeta \qquad (1)$$

The endogenous, or "dependent" latent variables are collected in the vector η, while the exogenous, or "independent" latent variables are in ξ. B and Γ are coefficient matrices, while ζ is a random vector of

residuals, sometimes called "errors in equations" or "disturbance terms." The elements of **B** and **Γ** would be path coefficients for directed relationships among latent variables. It is assumed in general that **ζ** and **ξ** are uncorrelated, and that **I − B** is of full rank.

Because usually **η** and **ξ** are not observed without error, there are also factor model (or "measurement model") equations to account for measurement of these latent variables through manifest variables. The measurement models for the two sets of latent variables are

$$y = \Lambda_y \eta + \varepsilon \quad (2)$$

and

$$x = \Lambda_x \xi + \delta \quad (3)$$

With

$$\Sigma = \begin{bmatrix} \Sigma_{yy} & \Sigma_{yx} \\ \Sigma_{xy} & \Sigma_{xx} \end{bmatrix} \quad (4)$$

the LISREL model is that

$$\Sigma_{yy} = \Lambda_y (I - B)^{-1} (\Gamma \Phi \Gamma' + \Psi)(I - B')^{-1} \Lambda_y' + \Theta_\varepsilon \quad (5)$$

$$\Sigma_{xx} = \Lambda_x \Phi \Lambda_x' + \Theta_\delta \quad (6)$$

$$\Sigma_{xy} = \Lambda_x \Phi \Gamma' (I - B')^{-1} \Lambda_y' \quad (7)$$

Φ, Ψ, Θ_ε, and **Θ_δ** are the covariance matrices for **ξ, ζ, ε,** and **δ** respectively. There seems to be considerable confusion in the literature about the precise assumptions required for Equations 5 through 7 to hold. Jöreskog and Sörbom (1989) state the assumptions that (1) **ζ** is uncorrelated with **ξ**, (2) **ε** is uncorrelated with **η**, (3) **δ** is uncorrelated with **ξ**, (4) **ζ, ε,** and **δ** are mutually uncorrelated, and (5) **I − B** is of full rank. However, it appears that Equation 7 also requires an assumption not stated by Jöreskog and Sörbom (1989), i.e., that **ε** and **ξ** are uncorrelated.

This model reduces to a number of well-known special cases. For example, if there are no *y*-variables, then the model reduces to the common factor model, as can be seen from Equation 6.

An important aspect of the LISREL approach is that, in using it, variables must be arranged according to type. Manifest and latent, "exogenous" and "endogenous" variables are used in different places in different equations. Moreover, LISREL's typology for manifest variables is somewhat different from that used by other models. Specifically, in LISREL a manifest variable is designated as x or y on the basis of the type (exogenous or endogenous) of latent variable it loads on.

It is, of course, possible to translate models from a path diagram representation of a model to a LISREL model. However, this is not always easy. In some well known cases special strategies must be used to "trick" the LISREL model into analyzing a path diagram representation. For example, LISREL does not allow direct representation of a path in which an arrow goes from a manifest exogenous variable to a latent endogenous variable. Consequently a dummy latent exogenous variable (identical to the manifest variable) must be created in such cases.

In his review, Bentler (1986) referred to the models of McArdle (1978) and Bentler and Weeks (1979) as "generic" approaches, in that their emphasis was on the distinction between independent (exogenous) and dependent (endogenous) variables, rather than manifest and latent variables.

McArdle (1978) proposed an approach that was considerably simpler than the LISREL model. This approach, in essence, did not require any partitioning of variables into types. One could represent all paths in only two matrices, one representing directed relationships among variables, the other undirected relationships. McArdle's approach, which he called the RAM model, could be tested easily as a special case of McDonald's COSAN model.

SEPATH - 3652

Copyright © StatSoft, 1995

McArdle's specification was innovative, and offered substantial benefits. It allowed path models to be grasped and fully specified in their simplest form — as linear equations among manifest and latent variables. Instead of 18 model matrices, and a plethora of different variable types, one only needed 3 matrices! After reading some of McArdle's early papers, the present author was motivated to seek an automated approach to structural modeling. So *SEPATH,* and its precursor EzPATH, owe a special debt to McArdle.

Ironically, it took some time for McArdle's work to receive the attention it deserved. The work initially met with a lukewarm reception from journal editors and rather harsh opposition from some reviewers. It took 4 years for a detailed algebraic treatment (McArdle & McDonald, 1984) to pass through the review process and achieve publication. By then, unfortunately, the full credit due to McArdle had been diluted.

The COSAN Model

This section begins with a brief description of the McDonald's COSAN model. Let Σ be a population variance-covariance matrix for a set of manifest variables. The COSAN model (McDonald, 1978) holds if Σ may be expressed as

$$\Sigma = \mathbf{F}_1\mathbf{F}_2...\mathbf{F}_k\mathbf{P}\mathbf{F}_k'...\mathbf{F}_2'\mathbf{F}_1' \qquad (8)$$

where $\mathbf{P}$ is symmetric and Gramian, and any of the elements of any $\mathbf{F}$ matrix or $\mathbf{P}$ may be constrained under the model to be a function of the others, or to be specified numerical values. As a powerful additional option, any square $\mathbf{F}$ matrix may be specified to be the *inverse* of a patterned matrix. This "*patterned inverse*" option is critical for applications to path analysis. A COSAN model with k $\mathbf{F}$ matrices is referred to as "a COSAN model of order k."

Obvious special cases are: Orthogonal and oblique common factor models, confirmatory factor models, and patterned covariance matrices.

McDonald's COSAN model is a powerful and original approach which offers many benefits to the prospective tester of covariance structure models. Testing and estimation for the model were implemented in a computer program called, aptly enough, COSAN (See Fraser and McDonald, 1988 for details on a recent version of this program, which has been available since 1978).

In 1978, J. J. McArdle proposed some simple rules for translating any path diagram directly to a structural model. In collaboration with McDonald, he proposed an approach which yielded a model directly testable with the COSAN computer program.

McArdle's RAM Model

McArdle's approach is based on the following covariance structure model, which he has termed the RAM model:

Let $\mathbf{v}$ be a $(p + n) \times 1$ random vector of p manifest variables and n latent variables in the path model, possibly partitioned into manifest and latent variables subsets in $\mathbf{m}$ and $\mathbf{l}$, respectively, in which case

$$\mathbf{v} = \begin{bmatrix}\mathbf{m}':\mathbf{l}'\end{bmatrix}' \qquad (9)$$

(This partitioning is somewhat convenient, but not necessary.) For simplicity assume all variables have zero means. Let $\mathbf{F}$ be a matrix of multiple regression weights for predicting each variable in $\mathbf{v}$ from the $p + n - 1$ other variables in $\mathbf{v}$. $\mathbf{F}$ will have all diagonal elements equal to zero. In general, some elements of $\mathbf{F}$ may be constrained by hypothesis to be equal to each other, or to specified numerical values (often zero). Let $\mathbf{r}$ be a vector of latent exogenous variables, including residuals. The path model may then be written

$$\mathbf{v} = \mathbf{F}\mathbf{v} + \mathbf{r} \qquad (10)$$

In path models, all endogenous variables are perfectly predicted through the arrows leading to

them. (Recall our original simple example on page 3557. Since endogenous variables are dependent variables in one or more linear equations, their variances and covariances can be determined from the variances and covariances of the variables with arrows pointing to them. Ultimately, the variances and covariances of all endogenous variables are explained by a knowledge of the linear equation set up and the variances and covariances of exogenous variables in the system.) Consequently, elements of $\mathbf{r}$ corresponding to endogenous variables in $\mathbf{v}$ will be null. The matrix $\mathbf{F}$ contains the regression coefficients normally placed along the arrows in a path diagram. F_{ij} is the path coefficient from v_j to v_i. If a variable v_i is exogenous, i.e., has no arrow pointing to it, then row i of $\mathbf{F}$ will be null, and $r_i = v_i$. Hence, the non-null elements of the variance covariance matrix of $\mathbf{r}$ will be the coefficients in the "undirected" relationships in the path diagram.

Define $\mathbf{P} = E(\mathbf{rr}')$. Furthermore, let $\mathbf{W} = E(\mathbf{vv}')$, and $\mathbf{\Sigma} = E(\mathbf{mm}')$. The implications of Equation 10 for the structure of $\mathbf{\Sigma}$, the variance-covariance matrix of the manifest variables, can now be derived. Regardless of whether the manifest and latent variables were partitioned into distinct subsets in $\mathbf{v}$, it is easy to construct a "filter matrix" $\mathbf{J}$ which carries $\mathbf{v}$ into $\mathbf{m}$. If the variables in $\mathbf{v}$ are partitioned into manifest and latent variables, one obtains

$$\mathbf{J} = [\mathbf{I} : \mathbf{0}] \qquad (11)$$

$$\mathbf{m} = \mathbf{Jv} \qquad (12)$$

and consequently

$$\mathbf{\Sigma} = E(\mathbf{mm}') = \mathbf{J}\, E(\mathbf{vv}')\, \mathbf{J}' = \mathbf{JWJ}' \qquad (13)$$

Since (assuming $\mathbf{I} - \mathbf{F}$ is nonsingular) Equation 13 may be rewritten in the form

$$\mathbf{v} = (\mathbf{I} - \mathbf{F})^{-1}\mathbf{r} \qquad (14)$$

one obtains

$$\mathbf{W} = (\mathbf{I} - \mathbf{F})^{-1}\mathbf{P}(\mathbf{I} - \mathbf{F})^{-1'} \qquad (15)$$

Equations 13 and 15 imply

$$\mathbf{\Sigma} = \mathbf{J}(\mathbf{I} - \mathbf{F})^{-1}\mathbf{P}(\mathbf{I} - \mathbf{F})^{-1'}\mathbf{J}' \qquad (16)$$

This shows that any path model may be written in the form

$$\mathbf{\Sigma} = \mathbf{F}_1\mathbf{F}_2\mathbf{P}\mathbf{F}_2'\mathbf{F}_1' \qquad (17)$$

as a COSAN model of order 2, where

$$\mathbf{F}_1 = \mathbf{J} = [\mathbf{I} : \mathbf{0}] \qquad (18)$$

and

$$\mathbf{F}_2 = (\mathbf{F} - \mathbf{I})^{-1} = \mathbf{B}^{-1} \qquad (19)$$

McArdle's formulation may thus be characterized as follows:

(1) For convenience, order the manifest variables in the vector $\mathbf{m}$, and the latent variables in the vector $\mathbf{l}$. The path model is then tested as a COSAN model of order 2, in which

(2) $\mathbf{F}_1 = [\mathbf{I} : \mathbf{0}]$, where $\mathbf{I}$ is of order $p \times p$ and $\mathbf{0}$ is $p \times n$.

(3) $\mathbf{F}_2$ is the *inverse* of a square matrix $\mathbf{B}$ of "directed relationships." $\mathbf{B}$ is constructed from the path diagram as follows. Set all diagonal entries of $\mathbf{B}$ to -1. Examine the path diagram for arrows. For each arrow pointing from v_j to v_i, record its path coefficient in position b_{ij} matrix $\mathbf{B}$.

(4) $\mathbf{P}$ contains coefficients for "undirected" paths between variable v_i and v_j recorded in positions p_{ij} and p_{ji}.

Obviously, *SEPATH* could have been written around the elegant and straightforward RAM model. The approach would require simply creating a list of manifest and latent variables, ordering them, and filling the matrices $\mathbf{B}$ and $\mathbf{P}$ with coefficients obtained by parsing PATH1 model statements.

The Bentler-Weeks Model

The RAM model is somewhat wasteful in terms of the size of some of its matrices. Bentler and Weeks (1979) produced an alternative model which is somewhat more efficient in the size of its matrices. Specifically, the $\mathbf{F}_2$ and $\mathbf{P}$ matrices are quite large in the RAM model, and have a large number of zero elements. Bentler and Weeks showed how, in situations where there are no manifest exogenous variables (i.e., all manifest variables have at least one arrow pointing to them), the McArdle-McDonald approach may be modified to reduce the size of the model matrices.

Partition $\mathbf{v}$ in the form

$$\mathbf{v} = \begin{bmatrix} \mathbf{m} \\ \mathbf{l}_n \\ \mathbf{l}_x \end{bmatrix} \tag{20}$$

where the subscripts x and n refer to "exogenous" and "endogenous," respectively.

Then one may write $\mathbf{v} = \mathbf{F}\mathbf{v} + \mathbf{r}$ in a partitioned form as

$$\mathbf{v} = \begin{bmatrix} \mathbf{m} \\ \mathbf{l}_n \\ \mathbf{l}_x \end{bmatrix} = \begin{bmatrix} \mathbf{F}_1 & \mathbf{F}_2 & \mathbf{F}_3 \\ \mathbf{F}_4 & \mathbf{F}_5 & \mathbf{F}_6 \\ \mathbf{0} & \mathbf{0} & \mathbf{0} \end{bmatrix} \begin{bmatrix} \mathbf{m} \\ \mathbf{l}_n \\ \mathbf{l}_x \end{bmatrix} + \begin{bmatrix} \mathbf{0} \\ \mathbf{0} \\ \mathbf{l}_x \end{bmatrix} \tag{21}$$

Now define $\mathbf{n}$ as a vector containing all the endogenous, or "dependent" variables. We may partition $\mathbf{m}$ into manifest exogenous and endogenous variables, i.e.,

$$\mathbf{m} = \begin{bmatrix} \mathbf{m}_x \\ \mathbf{m}_n \end{bmatrix} \tag{22}$$

Then

$$\mathbf{n} = \begin{bmatrix} \mathbf{m}_n \\ \mathbf{l}_n \end{bmatrix} \tag{23}$$

One may then write

$$\mathbf{n} = \mathbf{F}_0 \mathbf{n} + \Gamma \mathbf{l}_x \tag{24}$$

where

$$\mathbf{F}_0 = \begin{bmatrix} \mathbf{F}_1 & \mathbf{F}_2 \\ \mathbf{F}_4 & \mathbf{F}_5 \end{bmatrix}, \text{ and } \Gamma = \begin{bmatrix} \mathbf{F}_3 \\ \mathbf{F}_6 \end{bmatrix} \tag{25}$$

The derivation now proceeds with an algebraic development similar to the RAM-COSAN equations. Rearranging Equation 24, one obtains

$$(\mathbf{I} - \mathbf{F}_0)\mathbf{n} = \Gamma \mathbf{l}_x \tag{26}$$

$$\mathbf{n} = (\mathbf{I} - \mathbf{F}_0)^{-1} \Gamma \mathbf{l}_x \tag{27}$$

$$\mathbf{m}_n = [\mathbf{I} \mid \mathbf{0}](\mathbf{I} - \mathbf{F}_0)^{-1} \Gamma \mathbf{l}_x = \mathbf{J}(\mathbf{I} - \mathbf{F}_0)^{-1} \Gamma \mathbf{l}_x \tag{28}$$

whence, letting

$$\mathbf{G} = [\mathbf{I} \mid \mathbf{0}] \tag{29}$$

$$\mathbf{F}_2 = (\mathbf{I} - \mathbf{F}_0)^{-1} \tag{30}$$

$$\mathbf{F}_3 = \Gamma \tag{31}$$

and

$$\mathbf{P} = E(\mathbf{l}_x \mathbf{l}_x') \tag{32}$$

we have

$$\Sigma = \mathbf{G}\mathbf{F}_2\mathbf{F}_3\mathbf{P}\mathbf{F}_3'\mathbf{F}_2'\mathbf{G}' \tag{33}$$

$\mathbf{G}$ is a filter matrix similar to $\mathbf{J}$ in the McArdle-McDonald specification. $\mathbf{F}_2 = \mathbf{B}_2^{-1}$, where $\mathbf{B}_2$ is a matrix containing path coefficients for directed relationships *among endogenous variables only*, and having -1 as each diagonal element. $\mathbf{F}_3$ contains path coefficients *from exogenous variables to endogenous variables only*, and $\mathbf{P}$ contains coefficients for undirected relationships, i.e., the variance-covariance parameters for the latent exogenous variables.

This clever algebraic refinement allowed some of the virtues of the McArdle approach to be retained, while expressing the essential relationships in smaller matrices. (Notice how several of the null

submatrices are eliminated.) However, this model also had some minor drawbacks. It required partitioning variables into exogenous and endogenous types, and it did not allow direct expression of manifest exogenous variables.

An alternative model allows us to treat manifest exogenous variables explicitly. If you add a vector of manifest variables to each of the two variable lists in the Bentler-Weeks (1979) model, and modify the regression coefficient matrices accordingly, you arrive at the model used in *SEPATH*. In this model, which is similar to one given by Bentler and Weeks (1980), variables are partitioned into two groups.

The SEPATH Model

Let $\mathbf{m}_x$ be a vector of manifest exogenous variables. Partition the variables into vectors $\mathbf{s}_1$ and $\mathbf{s}_2$ as follows:

$$\mathbf{s}_1 = \begin{bmatrix} \mathbf{m}_x \\ \mathbf{m}_n \\ \mathbf{l}_n \end{bmatrix} \quad (34)$$

and

$$\mathbf{s}_2 = \begin{bmatrix} \mathbf{m}_x \\ \mathbf{l}_x \end{bmatrix} \quad (35)$$

Then one may write

$$\mathbf{s}_1 = \mathbf{B}\mathbf{s}_1 + \mathbf{\Gamma}\mathbf{s}_2 \quad (36)$$

where

$$\mathbf{B} = \begin{bmatrix} \mathbf{0} & \mathbf{0} & \mathbf{0} \\ \mathbf{0} & \mathbf{F}_1 & \mathbf{F}_2 \\ \mathbf{0} & \mathbf{F}_4 & \mathbf{F}_5 \end{bmatrix} \quad (37)$$

and

$$\mathbf{\Gamma} = \begin{bmatrix} \mathbf{I} & \mathbf{0} \\ \mathbf{F}_7 & \mathbf{F}_3 \\ \mathbf{F}_8 & \mathbf{F}_6 \end{bmatrix} \quad (38)$$

Assuming a nonsingular $\mathbf{I} - \mathbf{B}$, Equation 23 may be rewritten as

$$\mathbf{s}_1 = (\mathbf{I} - \mathbf{B})^{-1}\mathbf{\Gamma}\mathbf{s}_2. \quad (39)$$

Let $\mathbf{G}$ be a filter matrix which extracts the manifest variables from $\mathbf{s}_1$, and let $\Xi = E(\mathbf{s}_2\mathbf{s}_2')$ be the covariance matrix for $\mathbf{s}_2$.

Then

$$\mathbf{m} = \begin{bmatrix} \mathbf{m}_x \\ \mathbf{m}_n \end{bmatrix} = \mathbf{G}\mathbf{s}_1 = \mathbf{G}(\mathbf{I} - \mathbf{B})^{-1}\mathbf{\Gamma}\mathbf{s}_2 \quad (40)$$

and one obtains the following model for covariance structure:

$$\Sigma = \mathbf{G}(\mathbf{B} - \mathbf{I})^{-1}\mathbf{\Gamma}\Xi\mathbf{\Gamma}'(\mathbf{B}' - \mathbf{I})^{-1}\mathbf{G}' \quad (41)$$

The covariance matrix $\text{Cov}(\mathbf{s}_1) = \Psi$ for manifest exogenous, manifest endogenous, and latent endogenous variables may be computed as

$$\Psi = (\mathbf{B} - \mathbf{I})^{-1}\mathbf{\Gamma}\Xi\mathbf{\Gamma}'(\mathbf{B}' - \mathbf{I})^{-1} \quad (42)$$

The model of Equation 41 allows direct correspondence between all permissible PATH1 statements and the algebraic model. There is no need to concoct dummy latent variables. All possible types of relationships among manifest and latent variables are accounted for. After a model is complete, all variables can immediately be assigned to one of the 4 vectors $\mathbf{m}_n$, $\mathbf{m}_x$, $\mathbf{l}_n$, or $\mathbf{l}_x$. All coefficients (for arrows) are then assigned to the matrices $\mathbf{F}_1$ through $\mathbf{F}_8$. The column index for a variable (in any of these 8 matrices) represents the variable from which the arrow points, the row index the variable to which the arrow points. Coefficients for wires are represented in a similar manner in the matrix Ξ.

The model of Equation 41 sacrifices some of the simplicity of the RAM model, because variables must be assigned to 4 types before the location of model coefficients can be determined. However, in our typology and with the *SEPATH* diagramming rules the typing of each variable into one of 4

categories can be determined by looking *only at that variable in the path diagram*. Because two headed arrows are eliminated, a variable is endogenous if and only if it has an arrowhead directed toward it. A variable is latent if and only if it appears in an oval or circle. (If it is not already obvious, let us note that with two headed arrows one must look away from the variable of interest to determine if the variable is endogenous, because an arrowhead attached to the variable and pointing to it might be two-headed! *Not only is the SEPATH system less cluttered, but it is also visually more efficient.*)

Two final points should be emphasized. First, it is not clear which of the above models is, in any overall sense, "superior" to the others. The *SEPATH* model of Equation 41 was chosen primarily because it offered a good trade-off between certain conceptual and computational advantages. However, there are also definite advantages, both conceptual and computational, in each of the other model formulations.

Second, it is possible to express some of the models as special cases of the others. For example, the LISREL model can be written easily as a COSAN model. To see why, suppose that the manifest and latent variables were ordered in the **v** of Equation 10 so that

$$\mathbf{v} = \begin{bmatrix} \mathbf{y} \\ \mathbf{x} \\ \mathbf{\eta} \\ \mathbf{\xi} \end{bmatrix} \quad (43)$$

Then it follows immediately that one may write $\mathbf{v} = \mathbf{F}^*\mathbf{v} + \mathbf{r}^*$ where

$$\mathbf{F}^* = \begin{bmatrix} \mathbf{0} & \mathbf{0} & \mathbf{\Lambda}_y & \mathbf{0} \\ \mathbf{0} & \mathbf{0} & \mathbf{0} & \mathbf{\Lambda}_x \\ \mathbf{0} & \mathbf{0} & \mathbf{B} & \mathbf{\Gamma} \\ \mathbf{0} & \mathbf{0} & \mathbf{0} & \mathbf{0} \end{bmatrix} \quad (44)$$

and

$$\mathbf{r}^* = \begin{bmatrix} \mathbf{\epsilon} \\ \mathbf{\delta} \\ \mathbf{\zeta} \\ \mathbf{\xi} \end{bmatrix} \quad (45)$$

If $\mathbf{P}^*$ is defined as the covariance matrix of $\mathbf{r}^*$, then clearly one can test any LISREL model as a COSAN model of the form

$$\mathbf{\Sigma} = \mathbf{G}\,(\mathbf{F}^* - \mathbf{I})^{-1}\,\mathbf{P}\,(\mathbf{F}'^* - \mathbf{I})^{-1}\mathbf{G}' \quad (46)$$

where **G** is a matrix which filters **x** and **y** from **v**.

Statistical Estimation

The preceding section outlined the statistical model for Σ which *SEPATH* attempts to fit to the sample data. If the model fits perfectly in the population, then Equation 41 holds. Such perfection is, of course, extremely unlikely to happen.

It is, in fact, virtually certain that Equation 41 does *not* hold exactly for your statistical population, and that in fact an additional error term $\mathbf{E}_{pop}$ should be added to the right side of the equation. The size of the elements of this error matrix would reflect how badly a particular model fits in the population. You could find out what $\mathbf{E}_{pop}$ was if you somehow knew Σ. (You would simply input Σ to *SEPATH* and fit your model to it.) If you did, you would be faced with a difficult problem of exactly how to quantify the information in $\mathbf{E}_{pop}$.

There is an additional complication. In practice, you do not know Σ. You only have **S**, an estimate of Σ from sample data. It is this estimate, usually the ordinary sample covariance matrix based on *N* independent observations, which one attempts to fit with *SEPATH*.

Consequently, in practice one attempts to fit **S** rather than Σ with the model of Equation 41, and one obtains, as a result of this model fitting procedure, a *sample* matrix of residuals $\mathbf{E}_{samp}$. In general the object of the estimation process is to make the elements of $\mathbf{E}_{samp}$ as "small as possible" in some sense. This notion of "smallness" is quantified in a "discrepancy function."

General Properties of Discrepancy Functions

Define θ as the current vector of free parameter values. Let $\Sigma(\theta)$ represent a function which models Σ as a function of the *t* free parameter values in θ. The traditional approach to statistical estimation states as a model that

$$H_0: \Sigma = \Sigma(\theta) \tag{47}$$

$\Sigma(\theta)$ is assumed in our general discussion to be any twice differentiable function of θ. In practice, it is usually restricted to the particular form of the general model supported by the covariance structure software used for fitting the model to data. For example, when fitting covariance matrices, *SEPATH* is restricted to the model of Equation 41. In this case, assuming $\mathbf{B}$, Γ, and Ξ have elements which are either fixed numerical values or elements of θ, one may write

$$\Sigma(\theta) = \mathbf{G}(\mathbf{B} - \mathbf{I})^{-1} \Gamma \Xi \Gamma' (\mathbf{B}' - \mathbf{I})^{-1} \mathbf{G}' \tag{48}$$

The discrepancy function $F(\mathbf{S}, \Sigma(\theta))$ is a measure on **S** and $\Sigma(\theta)$. In general, if a model is identified (see *Model Identification* later in this section), minimization of a discrepancy function satisfying the following three restrictions will lead to consistent estimates for the elements of θ:

$F(\mathbf{S}, \Sigma(\theta)) \geq 0$

$F(\mathbf{S}, \Sigma(\theta)) = 0$ if and only if $\mathbf{S} = \Sigma(\theta)$.

$F(\mathbf{S}, \Sigma(\theta))$ is continuous in **S** and $\Sigma(\theta)$.

The above notation, which is employed in many books and papers on structural equation modeling, can be quite confusing in practice, because θ may stand for different quantities in different situations. For example, when we refer above to the discrepancy function $F(\mathbf{S}, \Sigma(\theta))$, we are referring to *any* permissible set of numbers employed as parameters in a model. In other contexts, the values in θ may acquire a more specific meaning. For example, when we are referring to the outcome of a maximum likelihood minimization process in which the maximum likelihood discrepancy function has been minimized as a function of θ, the elements of θ are now "maximum likelihood estimates."

Besides the *sample discrepancy function* $F(\mathbf{S}, \Sigma(\theta))$, we may also discuss the *population discrepancy function* $F(\Sigma, \Sigma(\theta))$, which we would obtain if we somehow knew Σ, the population covariance matrix, and used our estimation algorithm to fit the structural model to Σ rather than **S**. We may write

$$\Sigma = \Sigma(\theta) + \mathbf{E}_{pop} \tag{49}$$

Thus, the null hypothesis in Equation 47 may be expressed in several equivalent forms. For example,

$$H_0: \ F(\Sigma, \Sigma(\theta)) = 0 \tag{50}$$

or

$$H_0: \ \mathbf{E}_{pop} = \mathbf{0} \tag{51}$$

As a simple consequence of the preceding definitions, we can see that, when θ is *identified* (see *Model Identification*, page 3662), and the null hypothesis is true, θ is uniquely defined for *any* discrepany function. However, suppose the null hypothesis is *not* true, which under most conditions is the reasonable assumption. In this case, we might define the "population parameters" as those we would obtain if we somehow knew Σ, and fit a model to Σ by minimizing a discrepancy function. The parameters in θ would then be defined as those that "fit best in the population." The subtle problem here is that different discrepancy functions will usually produce different θ values. Hence, although the point is hardly ever discussed in the literature, θ is, in practice, hardly ever uniquely defined, unless you choose a *particular* discrepancy function (say, maximum likelihood) as your criterion for choosing θ "in the population." The problem is that discrepancy functions have been chosen primarily on the basis of their optimality properties for fitting **S** to a model, *not* for fitting Σ. The reader should keep that subtle point in mind when reading the following discussion of discrepancy functions.

OLS Estimation

One simple measure of how badly $\Sigma(\theta)$ fits **S** is to examine the sum of squared elements of $\mathbf{E}_{samp} = \mathbf{S} - \Sigma(\theta)$. The function is known as the Ordinary Least Squares (OLS) discrepancy function, and may be written

$$F_{OLS}(\mathbf{S}, \Sigma(\theta)) = \tfrac{1}{2} \operatorname{Tr}(\mathbf{S} - \Sigma(\theta))^2$$
$$= \tfrac{1}{2} \operatorname{Tr}(\mathbf{E}_{samp})^2 \quad (52)$$

where Tr() denotes the trace operator. The values in θ which, for a given **S**, minimize F_{OLS} are called *Ordinary Least Squares (OLS)* estimates. That is

$$\hat{\theta}_{OLS} = \arg\min_{\theta}\left(F_{OLS}(\mathbf{S}, \Sigma(\theta))\right) \quad (53)$$

The OLS discrepancy function has a number of difficulties, summarized nicely by Everitt (1984). In particular, it is not *scale free* — different scalings of the manifest variables can produce different discrepancy function values. Moreover, when calculated on sample discrepancies, simple sums of squares may be inappropriate from a statistical standpoint, because the elements of **S** are not independent random variables, and because they usually have different sampling variances.

GLS Estimation

The Generalized Least Squares (GLS) discrepancy function compensates for these problems by, in effect, standardizing each element of $\mathbf{E}_{samp}$. The resulting discrepancy function is

$$F_{GLS}(\mathbf{S}, \Sigma(\theta)) = \tfrac{1}{2} \operatorname{Tr}\left[(\mathbf{S} - \Sigma(\theta))\mathbf{S}^{-1}\right]^2 \quad (54)$$

and the estimates resulting from minimizing the function are called *Generalized Least Squares (GLS)* estimates, denoted

$$\hat{\theta}_{GLS} = \arg\min_{\theta}\left(F_{GLS}(\mathbf{S}, \Sigma(\theta))\right) \quad (55)$$

Maximum Wishart Likelihood Estimation

A more complex function is the Maximum Wishart Likelihood (ML) discrepancy function. This function may be written

$$F_{ML}(\mathbf{S}, \Sigma(\theta)) = \ln|\Sigma(\theta)| - \ln|\mathbf{S}|$$
$$+ \operatorname{Tr}(\mathbf{S}\Sigma(\theta)^{-1}) - p \quad (56)$$

where Tr( ) denotes the trace operator, |**S**| the determinant of **S**, the *unbiased* sample covariance matrix, and p is the number of manifest variables. If **S** has a Wishart distribution (a somewhat less restrictive assumption than the requirement that the observed variables follow a multivariate normal distribution), minimizing the ML discrepancy function produces *Maximum Wishart Likelihood Estimates*, i.e.,

$$\hat{\theta}_{ML} = \arg\min_{\theta}\left(F_{ML}(\mathbf{S}, \Sigma(\theta))\right) \quad (57)$$

Iteratively Reweighted GLS Estimation

Bentler (1989) discussed the properties of an "iteratively reweighted" generalized least squares (IRGLS) discrepancy function

$$F_{IRGLS}(\mathbf{S}, \Sigma(\theta)) = \tfrac{1}{2} \text{Tr}\left[(\mathbf{S} - \Sigma(\theta))\Sigma(\theta)^{-1}\right]^2 \quad (58)$$

with associated estimates

$$\hat{\theta}_{IRGLS} = \arg\min_{\theta}\left(F_{IRGLS}(\mathbf{S}, \Sigma(\theta))\right) \quad (59)$$

Citing Lee and Jennrich (1979), Bentler (1989) stated that $\hat{\theta}_{IRGLS} = \hat{\theta}_{ML}$.

Chi-square Test Statistics

If $\mathbf{S}$ has a Wishart distribution, the model is identified, and θ has t free parameters, then under fairly general conditions $(N-1)F_{ML}(\mathbf{S}, \Sigma(\theta))$, $(N-1)F_{GLS}(\mathbf{S}, \Sigma(\theta))$, and $(N-1)F_{IRGLS}(\mathbf{S}, \Sigma(\theta))$, all have an asymptotic Chi-square distribution with $p(p+1)/2 - t$ degrees of freedom.

Such a *Chi-square* statistic, often described as a "goodness-of-fit" statistic (but perhaps more accurately called a "badness-of-fit" statistic) allows us to test statistically whether a particular model fits Σ *perfectly* in the population (i.e., whether $\Sigma = \Sigma(\theta)$). There is a long tradition of performing such a test, although it is becoming increasingly clear that the procedure is seldom appropriate.

Browne (1974) showed that, under typical assumptions for maximum likelihood estimation, the two statistics $(N-1)F_{ML}$ and $(N-1)F^*_{ML}$, where

$$F^*_{ML} = \tfrac{1}{2}\text{Tr}\left[(\mathbf{S}-\Sigma(\theta))\left\{\Sigma(\hat{\theta}_{ML})\right\}^{-1}\right]^2 \quad (60)$$

will converge stochastically, and will both be distributed as *Chi-square* variates as $N \to \infty$. Moreover, as $N \to \infty$, the probability that the two discrepancy functions F_{ML} and F^*_{ML} will be minimized by different θ vectors converges to zero.

In practice, then, there tend to be only trivial differences, if any, between estimates which minimize F_{ML} and those which minimize F^*_{ML}. This suggests that there will seldom be differences in practice between θ which minimize F_{ML} and those which minimize F_{IRGLS}. As mentioned above, Bentler (1989) has stated that the F_{IRGLS} and F_{ML} are equivalent methods, i.e., lead to the same θ.

ADF Estimation

The *Chi-square* statistics based on ML and GLS estimation procedures assume multivariate normality of the data. When this assumption is violated, the resulting test statistic will, in general, no longer have a χ^2 distribution, and can therefore mislead during the fit evaluation process. Browne (1982, 1984) proposed procedures which will lead to a correct *Chi-square* statistic under much more general conditions.

Suppose the covariance structure model is expressed in the form

$$\mathbf{S} = \Sigma(\theta) + \mathbf{E}_{samp} \quad (61)$$

For any $p \times p$ symmetric matrix $\mathbf{A}$, let $\mathbf{vecs}(\mathbf{A})$ be the vector composed of the $p(p+1)/2$ nonduplicated elements of $\mathbf{A}$. Defining

$$\mathbf{s} = \mathbf{vecs}(\mathbf{S}) \quad (62)$$

$$\sigma(\theta) = \mathbf{vecs}(\Sigma(\theta)) \quad (63)$$

$$\varepsilon = \mathbf{vecs}(\mathbf{E}_{samp}) \quad (64)$$

one can express the model of Equation 61 alternatively as

$$\mathbf{s} = \sigma(\theta) + \varepsilon \quad (65)$$

Browne (1984) showed that

$$\varepsilon^* = (N-1)^{1/2}\varepsilon \quad (66)$$

has, under a true null hypothesis, an asymptotic distribution which is multivariate normal with null mean vector and variance-covariance matrix Ψ. Moreover, the generalized least squares discrepancy function of the form

$$F(\mathbf{s}, \sigma(\theta)) = (\mathbf{s} - \sigma(\theta))' \mathbf{U}^{-1} (\mathbf{s} - \sigma(\theta)) \quad (67)$$
$$= \varepsilon' \mathbf{U}^{-1} \varepsilon$$

has the property that if $\mathbf{U}$ is selected to be a consistent estimator of Ψ, then

$$\chi^2 = (N-1) F(\mathbf{s}, \sigma(\theta)) \quad (68)$$

will have an asymptotic *Chi-square* distribution under very general distributional assumptions. Both the IRGLS and GLS discrepancy functions discussed above can be expressed in the form of Equation 67. However, they incorporate a $\mathbf{U}$ matrix which is generally a consistent estimator of Ψ only under the assumption of multivariate normality. Modifying $\mathbf{U}$ so that it is a consistent estimator of Ψ under more general assumptions will allow the assumption of multivariate normality to be dispensed with, thus leading to "asymptotically distribution free" (ADF) procedures for the analysis of covariance structures.

ADF procedures have seldom been used in practice, although it seems that the assumption of multivariate normality is frequently contestable with data in the behavioral sciences. One reason for the lack of popularity of ADF procedures is that they were not implemented in widely available computer software like LISREL VI, EzPATH 1.0, or COSAN.

There are other serious practical problems with ADF estimation procedures. First, $\mathbf{U}$ can, in practice, be a very large matrix, thus imposing practical limits on the size of the problem which can be processed. Second, the elements of $\mathbf{U}$ require estimates of second and fourth-order moments of the manifest variables. Such estimates have large sampling variability at small to moderate sample sizes, so, in general, one might expect the *Chi-square* test statistic (and associated estimates) based on ADF estimation to converge somewhat more slowly to the asymptotic behavior than comparable normal theory estimation procedures.

However, there seems little justification for performing normal theory estimation and testing procedures on data which are clearly non-normal, especially when the sample size is large and the numbers of variables and unknowns moderate. The current version of *SEPATH* therefore supports two versions of ADF estimation.

Browne (1984) showed that, with the typical definitions for the first, second, and fourth-order sample moments about the mean, i.e.,

$$\overline{x}_{\bullet j} = N^{-1} \sum_{i=1}^{N} x_{ij} \quad (69)$$

$$t_{ij} = (x_{ij} - \overline{x}_{\bullet j}) \quad (70)$$

$$w_{jk} = N^{-1} \sum_{i=1}^{N} t_{ij} t_{ik} \quad (71)$$

$$w_{jkhm} = N^{-1} \sum_{i=1}^{N} t_{ij} t_{ik} t_{ih} t_{im} \quad (72)$$

a typical element of $\mathbf{U}_g$, a consistent and Gramian (but not unbiased) estimator of Ψ, is given by

$$u_{jk,hm} = w_{jkhm} - w_{jk} w_{hm} \quad (73)$$

The ADFG option in *SEPATH* minimizes the discrepancy function in Equation 67 with $\mathbf{U} = \mathbf{U}_g$.

Browne (1984) also showed how to obtain unbiased estimates of the elements of Ψ. A matrix $\mathbf{U}_{un}$ containing these unbiased estimates has typical element

$$u_{jk,hm} = \frac{N}{(N-2)(N-3)} K, \text{ with}$$

$$K = (N-1)(w_{jkhm} - w_{jk} w_{hm}) \quad (74)$$
$$- \left(w_{jh} w_{km} + w_{jm} w_{kh} - \frac{2}{N-1} w_{jk} w_{hm} \right)$$

The "ADF Unbiased" option in *SEPATH* minimizes the discrepancy function in Equation 67 with $\mathbf{U} = \mathbf{U}_{un}$.

Model Identification

For practical purposes it is usually not enough to have a particular model which, when expressed in the framework of Equation 41, reproduces Σ. For a model to be of much conceptual or practical value, its parameters must be identified. That is, there must exist *only one* parameter vector θ for which $\Sigma = \Sigma(\theta)$.

Perhaps the simplest example of a covariance structure model which is not identified is a common factor analysis model with two manifest variables and one common factor. In this case (assuming the common factor has a variance of 1) the covariance structure model becomes

$$\Sigma = \mathbf{ff}' + \mathbf{U}^2 \qquad (75)$$

In this case the parameter vector θ has 4 elements, the two elements in $\mathbf{f}$ and the two diagonal elements of $\mathbf{U}^2$

Suppose

$$\Sigma = \begin{bmatrix} 1 & .5 \\ .5 & 1 \end{bmatrix} \qquad (76)$$

If

$$\mathbf{f} = \begin{bmatrix} .7071 \\ .7071 \end{bmatrix}, \text{ and } \mathbf{U}^2 = \begin{bmatrix} .5 & 0 \\ 0 & .5 \end{bmatrix}, \qquad (77)$$

$\Sigma = \mathbf{ff}' + \mathbf{U}^2$, and the model fits perfectly. In this case

$$\theta' = \begin{bmatrix} .7071 & .7071 & .5 & .5 \end{bmatrix} \qquad (78)$$

But there are other values of θ which will reproduce Σ equally well. In fact there are infinitely many such values.

For example, let

$$\theta' = \begin{bmatrix} .9000 & .5556 & .1900 & .6914 \end{bmatrix} \qquad (79)$$

If $\mathbf{U}^2$ is restricted to be positive definite, clearly any two values for the first two elements of θ which have a product of .5, and are both less than one in absolute value will produce a discrepancy function value of zero. The diagonal elements of $\mathbf{U}^2$ are then obtained by subtracting the square of the corresponding element of $\mathbf{f}$ from 1.0.

Note that this is *not* a problem of the well known "rotational indeterminacy" in factor analysis. (With only one factor, there is no rotation.) Rather it is an example of a lesser known phenomenon, namely, that the elements of $\mathbf{U}^2$ may not be identified in the common factor model. If $\mathbf{U}^2$ is not identified, then there may exist common factor patterns which reproduce Σ equally well, but which are not obtainable from each other by rotation.

Even in the relatively comfortable confines of the common factor model, the phenomena of model identification are not well understood. Some of the most significant textbooks on factor analysis have failed to ever mention the problem. Moreover, several authoritative figures in the history of psychometrics have produced "results" on model identification in factor analysis which they have later had to retract or correct.

In general, necessary and sufficient conditions for identification are not available. However, it is often possible to determine that a model is *not* identified by showing that a necessary condition is violated.

There are some results available on when $\mathbf{U}^2$ in the factor model is definitely *not* identified. One of the best-known was given by Anderson and Rubin (1956). They showed that if, in unrestricted factor analysis, *under any orthogonal or oblique rotation*, there existed a factor pattern with only 2 non-zero elements in any column, then $\mathbf{U}^2$ is not identified. Clearly then, if such a situation exists (see Everitt, 1984, pages 45-49 for an example), additional constraints will have to be imposed to yield an identified solution.

The Anderson-Rubin result has an important implication which is often overlooked in discussions of the identification issue. Namely, *it may not be possible to prove identification in the population* without knowing Σ! In other words the same model may be identified for one Σ, but not for another. One cannot prove identification merely by counting equations and unknowns.

For some (relatively simple) models, it may be possible to prove identification by deriving unique equations, showing each parameter as a function of the elements of Σ. Unfortunately this approach is often impractical, and so checking for identification usually involves two stages.

First, very obvious sources of lack of identification should be removed. The most obvious source of underidentification in path models occurs when the measurement scale of an exogenous latent variable is left indeterminate. Consider the oblique common factor model, which can be written

$$\Sigma = \mathbf{FWF}' + \mathbf{U}^2 \qquad (80)$$

The variances of the common factors are found on the diagonal of $\mathbf{W}$. The factor loading coefficient for manifest variable i on factor j is found in element $\mathbf{F}_{ij}$. It is easy to show that unless restrictions are imposed on this model, the variance for factor j and the loadings on this factor are jointly indeterminate. To see why, suppose you were to multiply all the factor variances by 2. If you were to multiply all the columns of $\mathbf{F}$ by .7071, you would have exactly the same Σ. More generally, if we were to scale the diagonal of $\mathbf{W}$ with a diagonal scaling matrix $\mathbf{D}$, we could compensate by scaling the columns of $\mathbf{F}$ with $\mathbf{D}^{-1}$. In other words, for positive definite $\mathbf{D}$,

$$\begin{aligned}\Sigma &= \mathbf{FWF}' + \mathbf{U}^2 \\ &= (\mathbf{FD}^{-1})(\mathbf{DWD})(\mathbf{D}^{-1}\mathbf{F}') + \mathbf{U}^2 \qquad (81) \\ &= \mathbf{F}^*\mathbf{W}^*\mathbf{F}'^* + \mathbf{U}^2\end{aligned}$$

so that for any $\mathbf{F}$ and $\mathbf{W}$ there are infinitely many $\mathbf{F}$ and $\mathbf{W}$ which reproduce Σ equally well.

There are several ways of eliminating the lack of identification problem in practice. One way is to fix the variances of the exogenous latent variables at 1. (This fix may not be sufficient in all cases.) Another approach is to apply some constraint to the factor loading coefficients themselves. This approach is popular in structural models where the main interest is in the relations between latent variables. In this case, identification is often obtained by fixing one of the coefficients on a particular variable to 1.

When *unstandardized* latent variable models are fit, usually the variance of endogenous latent variables will not be identified. In such situations, the traditional "fix" has been to set one of the coefficients from the latent variable to 1. However, when the "Standardization New" option is used, this is not necessary, as *SEPATH* imposes internal constraints on the estimation process which result in all endogenous latent variables having unit variance.

Once obvious sources of non-identification have been eliminated, it is productive to examine whether either of the following easily tested conditions is violated.

(1) The number of degrees of freedom for the model must be nonnegative. That is $p(p+1)/2 \geq t$, where p is the order of Σ, and t is the number of free parameters in the model.

(2) The Hessian (the matrix of second derivatives of the discrepancy function with respect to the parameters) must be positive definite.

Violation of either of these conditions usually indicates an identification problem (for exceptions, see Shapiro & Browne, 1983), and *SEPATH* warns the user if they are violated.

Unconstrained Minimization Techniques

SEPATH produces its estimates for the elements of θ by minimizing a discrepancy function under choice of θ. The problem of finding the θ which minimizes the discrepancy function is certainly a difficult one. It is, in fact, non-trivial even when θ has only one element!

"Unconstrained minimization," i.e., the minimization of a function of θ, is a major area in the field of numerical analysis. The interested reader is urged to read an especially clear treatment of this area given by Dennis and Schnabel (1983).

The discussion begins with a non-technical overview. The discrepancy function is minimized by an iterative process. The iteration starts with initial estimates (often referred to as "starting values") for the elements of θ. On each iteration, the current value of the function is calculated, and the program estimates, using derivatives, which direction of change for θ will produce a further decrease in the discrepancy function. θ is changed in that direction by an initial amount (called the "step length"), and the function is recalculated. It may be that, according to certain criteria, the initial step went either too far, or not far enough. In that case the step length may go through several adjustments during an iteration. Once the "best" step length is estimated, the program moves on to the next iteration. When the discrepancy function stops improving, the algorithm terminates.

Now for some technical details. *SEPATH* uses a minor variation of the Gauss-Newton type algorithm discussed by Mels (1989) and Browne (1982). Let $d(\theta)$ be a vector of first partial derivatives (of the elements of $\Sigma(\theta)$) with respect to the elements of θ. That is,

$$d(\theta) = \partial \operatorname{Vec} \Sigma(\theta)/\partial \theta \tag{82}$$

In the Gauss-Newton approach H_k is an approximate Hessian given by

$$H_k = 2d'(\theta_k)d(\theta_k) \tag{83}$$

and g_k is the negative gradient of the discrepancy function with respect to θ, i.e,

$$g_k = -\partial F(S, \Sigma(\theta))/\partial \theta \big|_{\theta=\theta_k} \tag{84}$$

In the Gauss-Newton algorithm, the estimate on the kth iteration is related to the estimate on the next iteration by the formula

$$\begin{aligned}\hat{\theta}_{k+1} &= \hat{\theta}_k + \lambda_k H_k^{-1} g_k \\ &= \hat{\theta}_k + \lambda_k \delta_k\end{aligned} \tag{85}$$

The vector δ_k establishes the direction of change for the parameters on the kth iteration, while the scalar parameter λ_k establishes, jointly with δ_k, the length of the step vector.

Especially during the early phases of iteration, the program may attempt to take extremely large steps. Often, this causes no problem, and in fact hastens the progress toward a correct solution. However, sometimes large steps result in a set of parameter estimates that cause iteration to "blow up," either because the estimates yield a singular estimated covariance matrix, or because the estimates end up in a region from which recovery is impossible.

SEPATH allows the user to control the length of steps the program takes, in the following manner. There is a *Maximum Step Length* parameter β in the *Global Iteration Parameters* box in the *Analysis Parameters* dialog (see page 3576). If the vector δ_k has a length greater than β, it is rescaled by a positive constant so that its length becomes exactly β.

When δ_k is first calculated, it is compared to β and rescaled if necessary. Then λ_k is set to 1, θ_{k+1} is calculated via equation 85, and $F(S, \Sigma(\theta_{k+1}))$ is calculated.

If the new function value is less than the value on the preceding iteration by a "reasonable" amount, the algorithm proceeds to the next iteration. The interpretation of "reasonable" is controlled by the line search parameter α. When α is small (for example, near the default value of .0001), virtually any reduction in the discrepancy function is acceptable. When α is larger, a greater change is required. (For a full technical discussion of the line search paramater α and how it is employed, consult Dennis and Schnabel, 1983, especially eq. 6.3.3a.)

Usually, an acceptable change occurs immediately. However, on some occasions $F(\mathbf{S}, \Sigma(\boldsymbol{\theta}_{k+1}))$ may actually exceed $F(\mathbf{S}, \Sigma(\boldsymbol{\theta}_k))$, or be much closer to it than expected. In such a case, λ_k is adjusted by a "line search" algorithm until an acceptable value of the discrepancy function is found. Note that, in effect, the minimization in terms of the t unknown elements of $\boldsymbol{\theta}$ is temporarily reduced to a minimization problem in one unknown, namely λ_k. Usually, a good value of the discrepancy function can be found for $0 < \lambda_k \leq 1$. Some algorithms assume, in effect, this will always occur by constraining λ_k to within those limits. However, in some circumstances this will not occur — either a λ_k greater than 1 is required to achieve improvement in the discrepancy function, or the step direction is wrong (in effect a negative λ_k is necessary to reduce the discrepancy function). In such cases algorithms (such as the simple "stephalving" approach advocated by Mels, 1989, and Bentler, 1989) requiring $0 < \lambda_k \leq 1$ may exhibit erratic behavior and/or fail to converge.

Steepest Descent Iterations

When initial values for $\boldsymbol{\theta}$ are far from the ultimate minimum, the approximate Hessian $\mathbf{H}_k$ may fail to yield a proper step direction during iteration. In this case, the program may iterate into a region of the parameter space from which recovery (i.e., successful iteration to the true minimum point) is not possible. In such cases, SEPATH offers several options to help control the iteration process. One option is to precede the Gauss-Newton procedure with a few iterations utilizing the "method of steepest descent." In the steepest descent approach, values of $\boldsymbol{\theta}$ on each iteration are obtained as

$$\hat{\boldsymbol{\theta}}_{k+1} = \hat{\boldsymbol{\theta}}_k + \lambda_k \mathbf{g}_k \qquad (86)$$

Line Search Options

SEPATH offers several line search options:

(1) A simple stephalving approach (but with user adjustable parameters). In this approach, λ_k is set at 1 at the beginning of each iteration. If an acceptable function value is not found, λ_k is multiplied by a "stephalving fraction" between zero and 1, and the discrepancy function is recalculated. The process continues until an acceptable discrepancy function is found, or a maximum number of "stephalves" has occurred. In SEPATH both the stephalving fraction and the maximum number of stephalving operations is user selectable. For further details see the SEPATH command reference.

(2) A cubic interpolation procedure (the default). This procedure uses the previous function evaluation information during a particular iteration to attempt to pinpoint, quickly, a reasonably close approximation to the "best" λ_k. SEPATH uses the algorithm A6.3.1 in the Appendix of Dennis & Schnabel (1983), modified in several ways to make it more robust than the version used in EzPATH 1.0.

(3) An "exact" line search procedure, with user selectable parameters, which allows the user to pinpoint the "best" λ_k to a very high degree of accuracy on a particular iteration. The algorithm used in SEPATH is the Golden Search algorithm described in Kennedy and Gentle (1980, page 432).

Each of the line search options in *SEPATH* has its merits and drawbacks. The stephalving approach is inexpensive and usually works, but occasionally fails with difficult problems. The cubic interpolation approach is quite reliable and not too expensive. Consequently, it is used as the default in *SEPATH*. The Golden Section approach can be useful when a discrepancy function is an extremely nonlinear function of λ_k on a particular iteration. Remember that, with the ability to break out of the iterative process, change the line search parameters, and restart iteration, the highly sophisticated user can manipulate the iterative process to help achieve a difficult solution with difficult problems.

Boundary Constraints

During iteration, it is not uncommon for the approximate Hessian matrix $\mathbf{H}_k$ to become singular, especially when the parameter estimates are still far from the desired local or global minimum. Jennrich and Sampson (1968) introduced a stepwise regression procedure which deals with this problem quite effectively in practice. During iteration, the Jennrich-Sampson procedure checks whether parameter estimates are within prescribed bounds, and constrains the estimates to remain within these bounds. This facility is quite useful in the context of structural modeling, where variances, if estimated as unconstrained free parameters, may frequently take on negative values. Browne and DuToit (1982) describe a FORTRAN implementation of the Jennrich-Sampson procedure which is especially adapted to facilitate the constrained estimation procedure (described below) for standardizing the variance of endogenous latent variables during iteration. *SEPATH* now uses the Jennrich-Sampson procedure to calculate δ_k on each iteration.

Convergence Criteria

Once a new θ is found which has reduced the discrepancy function by a reasonable amount, the whole cycle is repeated until at least one of several *convergence criteria* are met:

(1) The discrepancy function is extremely close to zero.

(2) An iteration fails to reduce the discrepancy function by more than a very small percentage. In *SEPATH*, a "relative function change" criterion is checked on each iteration. The criterion on the k'th iteration is

$$F_{crit} = \left| \frac{F_k - F_{k-1}}{1 + |F_k|} \right| \qquad (87)$$

(3) The "residual cosine" criterion of Browne (1982, eq. 1.9.5) falls below a specified tolerance.

One generally finds with path models that the more elements there are in θ, the more difficult it is to find the actual θ which minimizes the discrepancy function. Thus, all other things being equal, you might expect to need more iterations to converge to a solution for a large problem than for a small one. Unfortunately, the larger the problem, the longer each iteration tends to take. Consequently, good initial estimates can be very important for large problems. Indeed, without good initial estimates, even the best "state of the art" non-linear optimization routine will fail to find solutions for some problems.

Noncentrality-Based Indices of Fit

Besides the *Chi-square* value and its probability level, *SEPATH* prints a number of indices of fit. which can be used to interpret how well a model fits the data. The indices printed here are all single model indices, i.e., values which can be computed from a single model tested on one data set.

General Theoretical Orientation

When attempting to assess how well a model fits a particular data set, one must realize at the outset that the classic hypothesis-testing approach is inappropriate. Consider common factor analysis. When maximum likelihood estimation became a practical reality, the *Chi-square* "goodness-of-fit" statistic was originally employed in a sequential testing strategy. According to this strategy, one first picked a small number of factors, and tested the null hypothesis that this factor model fit the population Σ perfectly. If this hypothesis was rejected, the model was assumed to be too simple (i.e., to have too few common factors) to fit the data. The number of common factors was increased by one, and the preceding procedure repeated. The sequence continued until the hypothesis test failed to reject the hypothesis of perfect fit.

Steiger and Lind (1980) pointed out that this logic was essentially flawed, because, for any population Σ (other than one constructed as a numerical example directly from the common factor model!) the *a priori* probability is essentially 1 that the common factor model will not fit perfectly so long as degrees of freedom for the *Chi-square* statistic were positive.

In essence, then, population fit for a covariance structure model with positive degrees of freedom is never really perfect. Testing whether it is perfect makes little sense. It is what statisticians sometimes call an "accept-support" hypothesis test, because accepting the null hypothesis supports what is generally the experimenter's point of view, i.e., that the model does fit.

Accept-support hypothesis tests are subject to a host of problems. In particular, of course, the traditional priorities between Type I and Type II error are reversed. If the proponent of a model simply performs the *Chi-square* test with low enough power, the model can be supported. As a natural consequence of this, hypothesis testing approaches to the assessment of model fit *should* make some attempt at power evaluation. Steiger and Lind (1980) demonstrated that performance of statistical tests in common factor analysis could be predicted from a noncentral *Chi-square* approximation. A number of papers dealing with the theory and practice of power evaluation in covariance structure analysis have been published (Matsueda & Bielby, 1986; Satorra and Saris, 1985; Steiger, Shapiro, & Browne, 1985). Unfortunately, power estimation in the analysis of a multivariate model is a difficult, somewhat arbitrary procedure, and such power estimates have not, in general, been reported in published studies.

The main reason for evaluating power is to gain some understanding of precision of estimation in a particular situation, to guard against the possibility that a model is "accepted" simply because of insufficient power. An alternative (and actually more direct) approach to the evaluation of precision is to *construct a confidence interval on the population noncentrality parameter* (or some particularly useful function of it). This approach, first suggested in the context of covariance structure analysis by Steiger and Lind (1980) offers two worthwhile pieces of information at the same time. It allows one, for a particular model and data set to express (1) how bad fit is in the population, and (2) how precisely the *population* badness-of-fit has been determined from the *sample* data.

Noncentrality-Based Parameter Estimates and Confidence Intervals

Let **S** be the sample covariance matrix based on N observations, and for notational convenience, define $n = N - 1$. $\Sigma(\theta)$ is the attempt to reproduce **S** with a particular model and a particular parameter vector θ. $\Sigma(\theta_{ML})$ is the corresponding matrix constructed from the vector of maximum likelihood estimates θ_{ML} obtained by minimizing the discrepancy function of Equation 56. Equations 56 and 58 give two alternative discrepancy functions which lead to *Chi-square* statistics for testing structural models. Suppose one has obtained maximum likelihood estimates. Then under conditions (i.e., the "population drift" conditions in Steiger, Shapiro, and Browne, 1985) designed to simulate the situation where the model fits well but not perfectly, $nF_{ML}(S, \Sigma(\theta))$ has an asymptotic noncentral *Chi-square* distribution with $p(p+1)/2 - t$ degrees of freedom, where t is the number of free parameters in the model, and p is the order of S. The noncentrality parameter is $nF^* = nF_{ML}(\Sigma, \Sigma(\theta))$. F^* is the value of the statistic in Equation 56 obtained if S is replaced by the population covariance matrix Σ, and maximum likelihood estimation is performed on Σ instead of S. Hence, for the quadratic form statistic, the noncentrality parameter is in effect the "population badness-of-fit statistic."

Interestingly, if one divides by the noncentrality paramber by n, one obtains a measure of population badness-of-fit which depends only on the model, Σ, and the method of estimation.

If one has a single observation from a noncentral *Chi-square* distribution, it is very easy to obtain an unbiased estimate of the noncentrality parameter. By well known theory, if noncentral *Chi-square* variate X has noncentrality parameter λ and degrees of freedom ν, the expected value of X is given by

$$E(X) = \nu + \lambda \qquad (88)$$

whence it immediately follows that an unbiased estimate of λ is simply $X - \nu$. Consequently a large sample "biased corrected" estimate of F^* is $(X - \nu)/n$. Since F can never be negative, the simple unbiased estimator is generally modified in practice by converting negative values to zero. The estimate

$$F^+ = \max\{(X - \nu)/n, 0\} \qquad (89)$$

is the result.

It is also possible, by a variety of methods, to obtain, from a single observation from a non-central *Chi-square* distribution with ν degrees of freedom, a maximum likelihood estimate of the noncentrality parameter λ, and confidence intervals for λ as well. (See, e.g., Saxena and Alam, 1982; Spruill, 1986.)

Before continuing, recall some very basic statistical principles.

(1) Under very general conditions, if $\hat{\theta}$ is a maximum likelihood estimator for a parameter θ, then for any monotonic strictly increasing function $f(\ )$, $f(\hat{\theta})$ is a maximum likelihood estimator of $f(\theta)$.

(2) Moreover, if x_{low} and x_{high} are valid limits of a $100(1-\alpha)\%$ confidence interval for θ, $f(x_{low})$ and $f(x_{high})$ are valid limits of a $100(1-\alpha)\%$ confidence interval for $f(\theta)$.

These principles immediately imply that, since one can obtain a maximum likelihood estimate and confidence interval for nF^*, one can obtain a confidence interval and maximum likelihood estimate for F^* by dividing by n.

SEPATH obtains a point estimate and confidence interval for nF^* by iterative methods. The $100(1-\alpha)\%$ confidence limits for the noncentrality parameter λ of a $\chi^2_{\nu,\lambda}$ distribution are obtained by finding (via quasi-Newton iteration) the values of λ which place the observed value of the *Chi-square* statistic at the $100(\alpha/2)$ and $100(1-\alpha/2)$ percentile points of a $\chi^2_{\nu,\lambda}$ distribution. The "point estimate" of the "population noncentrality index" printed by *SEPATH* is the simple bias-corrected estimate F^+ (see Equation 89) recommended by McDonald (1988).

When the IRGLS or GLS estimation methods are employed, the population noncentrality index is a quadratic form, and as such is a weighted sum of squares of the residuals. Suppose you were to place the non-redundant elements of Σ in a vector $\sigma =$

vecs (Σ). Recalling the result of Equation 67, the discrepancy function is of the form $\varepsilon'W\varepsilon$, i.e., a weighted sum of squared discrepancies.

Steiger-Lind RMSEA Index

The Population Noncentrality Index F^* (PNI) offers some significant virtues as a measure of badness-of-fit (see, e.g., Steiger & Lind, 1980; McDonald, 1989). First, it is a weighted sum of discrepancies. Second, unlike the Akaike information criterion, for example, it is relatively unaffected by sample size.

However, there are two obvious problems with using the population noncentrality index as an index of population badness-of-fit.

The PNI is not in the metric of the original standardized parameters, because the quadratic form squares the weighted residuals.

The PNI fails to compensate for model complexity. In general, for a given Σ, the more complex the model the better it fits. A method for assessing population fit which fails to compensate for this will inevitably lead to choosing the most complex models, even when much simpler models fit the data nearly as well. The PNI fails to compensate for the size or complexity of a model. Hence it has limited utility as a device for comparing models.

The RMS index, first proposed by Steiger and Lind (1980), takes a relatively simplistic (but not altogether unreasonable) approach to solving these problems. Since model complexity is reflected directly in the number of free parameters, and inversely in the number of degrees of freedom, the PNI is divided by degrees of freedom, then the square root is taken to return the index to the same metric as the original standardized parameters.

Hence

$$R^* = \sqrt{\frac{F^*}{\nu}} \qquad (90)$$

The RMS index R^* can be thought of roughly as a root mean square standardized residual. Values above .10 indicate an inadequate fit, values below .05 a very good fit. Point estimates below .01 indicate an outstanding fit, and are seldom obtained.

In practice, point and interval estimates of the population RMS index are calculated as follows. First, we obtain point and interval estimates of the PNI. (Negative point estimates are replaced by zero.) Since all these are non-negative, and R^* is a monotonic transform of the PNI, point estimates and a confidence interval for R^* are obtained by inserting the corresponding values for F^* in Equation 90.

It may be shown easily that a bound on the point estimate of R^* implies a corresponding bound on the ratio of the *Chi-square* statistic to its degrees of freedom. Specifically, suppose, for example, you have decided that, for your purposes, the point estimate of the RMS index should be less than some value c. Manipulating the interval, we have

$$R^* < c$$

Letting $\chi^2 = nF$, the expression becomes

$$\sqrt{\frac{\frac{\chi^2 - \nu}{n}}{\nu}} < c$$

This in turn implies that

$$\frac{\chi^2}{\nu} < 1 + nc^2 \qquad (91)$$

So, for example, the rule of thumb that, for "close fit," RMS should be less than .05 translates into a rule that

$$\frac{\chi^2}{\nu} < 1 + \frac{n}{400} \qquad (92)$$

With this criterion, if $n = 400$, the ratio of the *Chi-square* to its degrees of freedom should be less than 2. Note that this rule implies a *less stringent* criterion for the ratio χ^2 / ν as sample size increases.

Rules of thumb that cite a single value for a critical ratio of χ^2 / ν ignore the point that the *Chi-square* statistic has an expected value that is a function of degrees of freedom, population badness of fit, *and* N. Hence, for a fixed level of population badness of fit, the expected value of the *Chi-square* statistic will increase as sample size increases. The rule of Equation 91 compensates for this, and hence it may be useful as a quick and easy criterion for assessing fit.

To avoid misinterpretation, we should emphasize at this point that our primary emphasis is on a confidence interval based approach, rather than one based on point estimates. The confidence interval approach incorporates information about precision of estimate into the assessment of population badness of fit. Simple rules of thumb (such as that of Equation 91) based on point estimates ignore these finer statistical considerations.

Population *Gamma* Index

Tanaka and Huba (1985, 1989) have provided a general framework for conceptualizing certain fit indices in covariance structure analysis. In their first paper, Tanaka and Huba (1985, their Equation 19) gave a general form for the *sample* fit index for covariance structure models under arbitrary generalized least squares estimation.

In the Tanaka-Huba treatment, it is assumed that a covariance structure model has been fit by minimizing an arbitrary GLS discrepancy function of the form

$$F(\mathbf{S}, \Sigma(\theta)|\mathbf{V}) = \tfrac{1}{2} \text{Tr}\left[(\mathbf{S} - \Sigma(\theta))\mathbf{V}\right]^2 \qquad (93)$$

or, equivalently (see Browne, 1974)

$$F(\mathbf{s}, \sigma|\mathbf{W}) = (\mathbf{s} - \sigma)' \mathbf{W}(\mathbf{s} - \sigma) \qquad (94)$$

where $\mathbf{s} = \text{vecs}(\mathbf{S})$, and $\sigma = \text{vecs}(\Sigma(\theta))$. $\mathbf{V}$ in Equation 93 and $\mathbf{W}$ in Equation 94 are arbitrary matrices. Appropriate choice of $\mathbf{V}$ or $\mathbf{W}$ can yield GLS or IRGLS estimators. For example, minimization of Equation 93 with $\mathbf{V} = \mathbf{S}^{-1}$ if $\mathbf{S}$ has a Wishart distribution yields the well-known GLS estimators (Browne, 1974). Setting $\mathbf{V} = [\Sigma(\theta)]^{-1}$ yields IRGLS estimators. Bentler (1989, page 216), citing Lee & Jennrich, 1979, states that IRGLS estimators are equivalent to ML estimators. Setting

$$\mathbf{V} = \left[\Sigma(\hat{\theta}_{ML})\right]^{-1} \qquad (95)$$

yields a discrepancy function which, according to Browne (1974), is usually minimized by the same θ which minimizes the maximum likelihood discrepancy function.

The Tanaka-Huba fit index can be written as

$$\gamma = 1 - [\varepsilon' \mathbf{W} \varepsilon / \mathbf{s}' \mathbf{W} \mathbf{s}] \qquad (96)$$

In their more recent paper, Tanaka and Huba (1989) demonstrate a deceptively simple, but important result which holds for models which are invariant under a constant scaling function (ICSF). A covariance structure model is ICSF if multiplication of any covariance matrix which fits the model by a positive scalar yields another covariance matrix which also satisfies the model exactly (though possibly with different free parameter values).

If a model which is ICSF has been estimated by minimizing a discrepancy function of the form given in Equations 93 and 94, then

$$\varepsilon' \mathbf{W} \sigma = 0 \qquad (97)$$

i.e., ε and σ are orthogonal "in the metric of $\mathbf{W}$," and, consequently,

$$\mathbf{s}' \mathbf{W} \mathbf{s} = \sigma' \mathbf{W} \sigma + \varepsilon' \mathbf{W} \varepsilon \qquad (98)$$

If Equation 98 holds, then γ may be written

$$\gamma = \sigma' \mathbf{W} \sigma / \mathbf{s}' \mathbf{W} \mathbf{s} = 1 - \varepsilon' \mathbf{W} \varepsilon / \mathbf{s}' \mathbf{W} \mathbf{s} \qquad (99)$$

where, as in Equation 64, $\varepsilon = s - \sigma$. In this form, γ defines a weighted coefficient of determination.

Under the conditions of Equation 98, with maximum likelihood estimation, one immediately obtains

$$F(S, \Sigma(\theta)|\Sigma(\hat{\theta}_{ML})) = \varepsilon' W \varepsilon$$
$$= \tfrac{1}{2} \text{Tr}\left[(S - \Sigma(\hat{\theta}_{ML}))\{\Sigma(\hat{\theta}_{ML})\}^{-1}\right]^2 \quad (100)$$

$$s'Ws = \tfrac{1}{2} \text{Tr}\left[S\{\Sigma(\hat{\theta}_{ML})\}^{-1}\right]^2 \quad (101)$$

whence

$$\gamma_{ML} = 1 - \left\{ \frac{\text{Tr}\left[S\{\Sigma(\hat{\theta}_{ML})\}^{-1} - I\right]^2}{\text{Tr}\left[S\{\Sigma(\hat{\theta}_{ML})\}^{-1}\right]^2} \right\} \quad (102)$$

which is equivalent to the Jöreskog and Sörbom (1984) GFI index.

Moreover, if the model is ICSF, then, under maximum Wishart likelihood estimation, there is the simplifying result (Browne, 1974, Proposition 8)

$$\text{Tr}\left[S\{\Sigma(\hat{\theta}_{ML})\}^{-1}\right] = p \quad (103)$$

Substituting in Equation 100, one finds

$$F(S, \Sigma(\theta)|\Sigma(\hat{\theta}_{ML})) =$$
$$\tfrac{1}{2}\left(\text{Tr}\left[S\{\Sigma(\hat{\theta}_{ML})\}^{-1}\right]^2 - p\right) \quad (104)$$

and so

$$\gamma_{ML} = \frac{p}{2s'Ws} = \frac{p}{\text{Tr}\left[S\{\Sigma(\hat{\theta}_{ML})\}^{-1}\right]^2} \quad (105)$$

Tanaka and Huba (1985, 1989) based their derivation of γ on sample quantities. However, in principle one is interested in a sample index primarily as a vehicle for estimating the corresponding *population* index. The corresponding population quantities are obtained by substituting Σ for S, and $\Sigma(\theta)$ for $\Sigma(\theta_{ML})$ in Equations 104 and 105.

One obtains

$$\Gamma_1 = \frac{p}{\text{Tr}\left[\Sigma\{\Sigma(\theta)\}^{-1}\right]^2} \quad (106)$$

Γ_1 can be thought of as a *weighted population coefficient of determination for the multivariate (ICSF) model*. (It may also be thought of as the population equivalent of the Jöreskog-Sörbom GFI index.)

An accurate point estimate for Γ_1 will provide useful information about the extent to which a model reproduces the information in Σ. A confidence interval, however, provides even more useful information, because it conveys not only the size of Γ_1, but also the precision of our estimate.

Let F^* be the Population Noncentrality Index $F(\Sigma, \Sigma(\theta_{ML}))$. From Equations 104 and 105, it is easy to see that

$$\Gamma_1 = \frac{p}{2F^* + p} \quad (107)$$

Equation 107 demonstrates that, under maximum likelihood estimation with ICSF models, Γ_1 can be expressed solely as a function of the Population Noncentrality Index and p, the number of manifest variables. Any consistent estimate of F^* will yield a consistent estimate of Γ_1 when substituted in Equation 107.

Equation 107 implies that an asymptotic maximum likelihood estimate (AMLE) for the population noncentrality index F^* may be converted readily to

an AMLE for Γ_1, simply by substituting the AMLE for F* in Equation 107.

Similarly, substitution of the endpoints of the confidence interval for F* in Equation 107 will generate a confidence interval for Γ_1.

Equation 107 and the accompanying derivation were first presented in Steiger (1989). Maiti and Mukherjee (1990), working completely independently of Steiger (1989), produced the identical result (their Equation 17). Steiger (1989) had suggested that the sample GFI was a biased estimator of the population value. Maiti and Mukerjee quantified the bias with the following (their Equation 16) approximate expression (for a *Chi-square* statistic with ν degrees of freedom).

$$E(\gamma_{ML}) \approx \frac{p}{p + 2F^* + \frac{2}{N}\nu} \qquad (108)$$

This can be rewritten in perhaps a more revealing form as

$$E(\gamma_{ML}) \approx \Gamma_1 \left(\frac{p}{p + (2\gamma\nu/N)} \right) \qquad (109)$$

Adjusted Population *Gamma* Index

Γ_1, like F*, fails to compensate for the effect of model complexity. Consider a sequence of *nested* models, where the models with more degrees of freedom are special cases of those with fewer degrees of freedom. (See Steiger, Shapiro, and Browne, 1985, for a discussion of the statistical properties of *Chi-square* tests with nested models.) For a nested sequence of models, the more complex models (i.e., those with more free parameters and fewer degrees of freedom) will always have Γ_1 coefficients as low or lower than those which are less complex.

Goodness of fit, as measured by Γ_1, improves more or less inevitably as more parameters are added. The adjusted population *gamma* index Γ_2 attempts to compensate for this tendency.

Just as Γ_1 is computed by subtracting a ratio of sums of squares from 1, Γ_2 is obtained by subtracting a corresponding ratio of mean squares from 1. Let $p^* = p(p+1)/2$. Let σ be a $p^* \times 1$ vector of non-duplicated elements of the population reproduced covariance matrix $\Sigma(\theta)$, as in Equation 99, for a model with ν degrees of freedom, and ε a corresponding vector of residuals. Then Γ_2 is

$$\begin{aligned}\Gamma_2 &= 1 - \frac{\varepsilon' W \varepsilon / \nu}{\sigma' W \sigma / p^*} \\ &= 1 - (p^*/\nu)(1 - \Gamma_1)\end{aligned} \qquad (110)$$

Consistent estimates and confidence intervals for Γ_1 may thus be converted into corresponding quantities for Γ_2 by applying Equation 110.

McDonald's Index of Noncentrality

McDonald proposed this index of noncentrality in a 1989 article in the *Journal of Classification*. The index represents one approach to transforming the population noncentrality index F^* into the range from 0 to 1. The index does not compensate for model parsimony, and the rationale for the exponential transformation it uses is primarily pragmatic.

The index may be expressed as

$$e^{-\frac{F^*}{2}}$$

Good fit is indicated by values above .95.

Extensions to Multiple Group Analysis

When more than one group is analyzed, the *Chi-square* statistic is a weighted sum of the discrepancy functions obtained from the individual groups (see page 3681). If the sample sizes are equal, the noncentrality-based indices discussed above generalize in a way that is completely straightforward. When sample sizes are unequal, this is not so, although *SEPATH* will still compute modified versions of the indices as described below, and these will still be of considerable value in assessing model fit.

With K independent samples, the overall *Chi-square* statistic is of the form

$$F = \sum_{k=1}^{K} c_k F_k \qquad (111)$$

where

$$c_k = \frac{N_k - 1}{N_{total} - K} \qquad (112)$$

and

$$N_{total} = \sum_{k=1}^{K} N_k \qquad (113)$$

The *Chi-square* statistic is then computed as

$$\chi^2 = (N_{total} - K)F \qquad (114)$$

This statistic has, under the assumptions of Steiger, Shapiro, and Browne (1985) a large sample distribution that is approximated by a noncentral *Chi-square* distribution, with ν degrees of freedom, and a noncentrality parameter equal to

$$\lambda = \sum_{k=1}^{K} (N_k - 1) F_k^* \qquad (115)$$

where F_k^* is the population discrepancy function for the kth group.

One can estimate this noncentrality parameter and set confidence intervals on it. However, inference to relevant population quantities is less straightforward. Consider, for example, the point estimate analogous to the single sample case. The statistic

$$\frac{\chi^2 - \nu}{N_{total} - K} \qquad (116)$$

has an expected value of approximately

$$\frac{\lambda}{N_{total} - K} = \sum_{k=1}^{K} \frac{N_k - 1}{N_{total} - K} F_k^* \qquad (117)$$

or

$$\sum_{k=1}^{K} c_k F_k^* \qquad (118)$$

where c_k is as defined above. This demonstrates that we can estimate a *weighted average* of the discrepancies for each sample, where the weights sum to 1, and are a function of sample size. If the sample sizes are equal, the weighted average becomes the simple arithmetic average, or mean, and so we can also estimate the unweighted sum of discrepancies.

How one should this information to produce multiple group versions of the RMSEA, and population gamma indices is open to some question when sample sizes are not equal. Perhaps the most natural candidates for the population RMSEA would be an "unweighted" index,

$$RMSEA_{unweighted} = \sqrt{\sum_{k=1}^{K} F_k^* / \nu} \qquad (119)$$

and a "weighted" index

$$RMSEA_{weighted} = \sqrt{\sum_{k=1}^{K} c_k F_k^* / \left(\nu / k\right)} \qquad (120)$$

When sample sizes are equal, both are the same.

Unfortunately, since we can only estimate the *weighted average* of population discrepancies, we must choose the second option when sample sizes are unequal. *SEPATH* currently reports point and

interval estimates for the *weighted* coefficient, which represents the square root of the ratio of a weighted average of discrepancies to an average number of degrees of freedom.

In calculating analogs of the population gamma indices, *SEPATH* substitutes K times the estimate of of the *weighted average* of discrepancies in place of F^* in equations 107 and 110.

Other Indices of Fit

Jöreskog-Sörbom GFI

This sample based index of fit is computed as γ_{ML} in Equation 105.

Jöreskog-Sörbom Adjusted GFI

This sample based index of fit is computed as

$$1 - (p^*/\nu)(1 - \gamma_{ML}) \tag{121}$$

where γ_{ML} is the GFI index, as defined in Equation 105.

Rescaled Akaike Information Criterion

In a number of situations the user must decide among a number of competing *nested* models of differing dimensionality. (The most typical example is the choice of the number of factors in common factor analysis.) Akaike (1973, 1983) proposed a criterion for selecting the dimension of a model. Steiger and Lind (1980) presented an extensive Monte Carlo study of the performance of the Akaike criterion. Here the criterion is rescaled (without affecting the decisions it indicates) so that it remained more stable across differing sample sizes. The rescaled Akaike criterion is as follows.

Let $F_{ML,k}$ be the maximum likelihood discrepancy function and f_k be the number of free parameters for the model M_k. Let N be the sample size

Select the model M_k for which

$$A_k = F_{ML,k} + \frac{2f_k}{N-1} \tag{122}$$

is a minimum.

Schwarz's Bayesian Criterion

This criterion (Schwarz, 1978) is similar in use to Akaike's index, selecting, in a sequence of nested models, the model for which

$$S_k = F_{ML,k} + \frac{f_k \ln(N)}{N-1} \tag{123}$$

is a minimum.

Browne-Cudeck Single Sample Cross-Validation Index

Browne and Cudeck (1989) proposed a single sample cross-validation index as a follow-up to their earlier (Cudeck & Browne, 1983) paper on cross-validation. Cudeck and Browne had proposed a cross-validation index which, for model M_k in a set of competing models is of the form $F_{ML}(\mathbf{S}_v, \mathbf{\Sigma}_k(\boldsymbol{\theta}))$. In this case, F is the maximum likelihood discrepancy function, $\mathbf{S}_v$ is the covariance matrix calculated on a cross-validation sample, and $\mathbf{\Sigma}_k(\boldsymbol{\theta})$ the reproduced covariance matrix obtained by fitting model M_k to the original calibration sample. In general, better models will have smaller cross-validation indices.

The drawback of the original procedure is that it requires two samples, i.e., the calibration sample for fitting the models, and the cross-validation sample. The new measure estimates the original cross-validation index from a single sample.

The measure is

$$C_k = F_{ML}(S_v, \Sigma_k(\theta)) + \frac{2\nu_k}{N-p-2} \quad (124)$$

where ν_k is the number of free parameters in model k, p is the number of manifest variables, and N is the sample size.

Independence Model
Chi-square and df

These are the *Chi-square* goodness-of-fit statistic, and associated degrees of freedom, for the hypothesis that the population covariances are all zero. Under the assumption of multivariate normality, this hypothesis can only be true if the variables are all independent. The "Independence Model" is used as the "Null Model" in several comparative fit indices.

Bentler-Bonett Normed Fit Index

One of the most historically important and original fit indices, the Bentler-Bonett index measures the *relative decrease* in the discrepancy function caused by switching from a "Null Model" or baseline model, to a more complex model. It is defined as:

$$B_k = \frac{F_0 - F_k}{F_0}, \text{ where} \quad (125)$$

F_0 = the discrepancy function for the "Null Model"

F_k = the discrepancy function for the k'th model

This index approaches 1 in value as fit becomes perfect. However, it does not compensate for model parsimony.

Bentler-Bonett Non-Normed Fit Index

This comparative index takes into account model parsimony. It may be written as

$$BNN_k = \frac{\frac{\chi_0^2}{\nu_0} - \frac{\chi_k^2}{\nu_k}}{\frac{\chi_0^2}{\nu_0} - 1} = \frac{\frac{F_0}{\nu_0} - \frac{F_k}{\nu_k}}{\frac{F_0}{\nu_0} - \frac{1}{N-1}}, \quad (126)$$

χ_0^2 = Chi-square for the "Null Model"

χ_k^2 = Chi-square for the k'th model

ν_0 = degrees of freedom for the "Null Model"

ν_k = degrees of freedom for the k'th model

Bentler Comparative Fit Index

This comparative index estimates the relative decrease in population noncentrality obtained by changing from the "Null Model" to the k'th model. The index may be computed as:

$$1 - \frac{\hat{\tau}_k}{\hat{\tau}_0} \quad (127)$$

$\hat{\tau}_k$ = estimated non-centrality parameter for the k'th model

$\hat{\tau}_0$ = estimated non-centrality parameter for the "Null Model"

James-Mulaik-Brett Parsimonious Fit Index

This index was one of the earliest (along with the Steiger-Lind index) to compensate for model parsimony. Basically, it operates by rescaling the Bentler-Bonnet Normed fit index to compensate for model parsimony. The formula for the index is:

$$\pi_k = \frac{\nu_k}{\nu_0} B_k \quad (128)$$

ν_0 = degrees of freedom for the "Null Model"

ν_k = degrees of freedom for the k'th model

B_k = Bentler-Bonnet normed fit index

Bollen's *Rho*

This comparative fit index computes the relative reduction in the discrepancy function per degree of freedom when moving from the "Null Model" to the k'th model. It is computed as:

$$\rho_k = \frac{\dfrac{F_0}{\nu_0} - \dfrac{F_k}{\nu_k}}{\dfrac{F_0}{\nu_0}} \qquad (129)$$

ν_0 = degrees of freedom for the "Null Model"

ν_k = degrees of freedom for the k'th model

F_0 = the discrepancy function for the "Null Model"

F_k = the discrepancy function for the k'th model

Comparing Equations 129 and 126, we see that, for even moderate *N*, there is bound to be virtually no difference between Bollen's *Rho* and the Bentler-Bonnet Non-normed fit index.

Bollen's *Delta*

This index is also similar in form to the Bentler-Bonnet index, but rewards simpler models (those with higher degrees of freedom). It is computed as:

$$\Delta_k = \frac{F_0 - F_k}{F_0 - \dfrac{\nu_k}{N}} \qquad (130)$$

ν_k = degrees of freedom for the k'th model.

F_0 = the discrepancy function for the "Null Model"

F_k = the discrepancy function for the k'th model

N = sample size

New Method for Standardizing Endogenous Latent Variables

For interpreting linear structural relationships, it is often desirable to have structural parameters *standardized*, i.e., constrained so that all latent variables have unit variance. It is rather easy, in traditional computational methods of analysis of covariance structures, to constrain the variances of *exogenous* latent variables to unity, since these variances appear as parameters in the standard model specification. One simply sets these parameters equal to a fixed value of 1. This approach was not available with *endogenous* latent variables, because their variances could not be specified directly. Consequently, "standardized" solutions were generated in EzPATH 1.0 (as in, say, EQS 3.0, LISREL VI, and CALIS) by first computing the unstandardized solution, then computing (non-iteratively) the values of standardized coefficients after the fact, using standard regression algebra. There are, in practice, some problems with such solutions. First, standard errors are not available. Second some equality constraints specified in the model coefficients, which are satisfied in the unstandardized solution, may not be achieved in the standardized version.

SEPATH offers an option (the *New* option in the *Standardization* group in the *Analysis Parameters* window), which produces a standardized solution by constraining the variances of endogenous latent variables during iteration. This method, described by Browne and DuToit (1987), and Mels (1989), is a constrained Fisher Scoring algorithm. The algorithm works as follows. Describe the *r* constraints on the endogenous latent variable variances in the form

$$\mathbf{c}(\boldsymbol{\theta}) = \mathbf{0} \qquad (131)$$

where $\mathbf{c}(\boldsymbol{\theta})$ is a differentiable, continuous function of the parameter vector $\boldsymbol{\theta}$. Let $\mathbf{L}$ be the Jacobian matrix of $\mathbf{c}(\boldsymbol{\theta})$, i.e.,

$$L = L(\theta) = \frac{\partial c(\theta)}{\partial \theta'} \quad (132)$$

During minimization, approximate the constraint function with its first-order Taylor expansion, i.e.,

$$c(\theta) \approx c(\hat{\theta}_k) + L(\hat{\theta}_k)(\theta - \hat{\theta}_k) \quad (133)$$

The nonlinear constraints required to establish unit variances for the endogenous latent variables (i.e., those of Equation 131) can thus be approximated by the linear constraints

$$L(\hat{\theta}_k)\delta_k = -c(\hat{\theta}_k)$$

On each iteration, the increment vector δ_k is calculated by solving the linear equation system

$$\begin{bmatrix} H_k & L'_k \\ L_k & 0 \end{bmatrix} \begin{bmatrix} \delta_k \\ \lambda_k \end{bmatrix} = \begin{bmatrix} g_k \\ -c_k \end{bmatrix} \quad (134)$$

(where g_k is the negative gradient, $c_k = c(\hat{\theta}_k)$, and $L_k = L(\theta_k)$), using the Jennrich-Sampson (1968) stepwise regression approach. The vector λ_k consists of r Lagrange multipliers corresponding to the r constraints. If there are t free parameters in the model, the degrees of freedom for the *Chi-square* statistic is

$$\nu = p(p+1)/2 - t - r \quad (135)$$

The above approach can be used as a general method to minimize a discrepancy function subject to constraints on the parameters. Here some very specific constraints are of interest. Specifically, suppose there are r endogenous latent variables in a model. The last r diagonal elements of the matrix Ψ (see Equation 42) must be constrained to be equal to unity. Hence, in this special case, a typical element of $c(\hat{\theta}_k)$ is given by

$$[c(\hat{\theta}_k)]_i = \Psi_{p+i,p+i} - 1 \quad (136)$$

where p is the number of manifest variables. Mels (1989, page 35) shows how to calculate a typical element of the Jacobian matrix L_k.

During iteration, progress is monitored so that the augmented discrepancy functions satisfy the inequality

$$F(S, \Sigma(\hat{\theta}_{k+1})) + 2\sum_{i=1}^{r}[\lambda_k]_i[c_{k+1}]_i$$
$$< F(S, \Sigma(\hat{\theta}_k)) + 2\sum_{i=1}^{r}[\lambda_k]_i[c_k]_i \quad (137)$$

Once the algorithm has converged, an estimate of the covariance matrix of the elements of θ may be obtained by dividing the first $t \times t$ principal submatrix of the *inverse* of the augmented Hessian

$$H_{aug} = \begin{bmatrix} H_k & L'_k \\ L_k & 0 \end{bmatrix} \quad (138)$$

by $N-1$. Further details, including detailed formulae for calculating the necessary derivatives for implementing the constrained estimation procedure, are provided in clear and compact form by Mels (1989).

Analyzing Correlation Matrices

SEPATH implements a procedure, pioneered by Mels (1989), for correctly analyzing correlation matrices. Traditional models and procedures for analysis of covariance structures are based on the assumption that the *sample covariance matrix* is being analyzed. This assumption is often inconvenient in practice. In many situations, the sample covariance matrix is ill-scaled. In addition, variables standardized to the same scale (i.e., unit variance) are generally easier to interpret. Moreover, in some situations involving reanalysis of older studies, only the correlation matrix is available.

The above considerations have led many researchers to input sample *correlation* matrices to covariance structure analysis programs as though they were covariance matrices. Cudeck (1989) points out that this often can lead to incorrect results. In particular, unless the model is invariant under diagonal rescaling, the calculated standard errors will almost certainly be incorrect, and the observed test statistic may also be incorrect.

SEPATH implements, via the *Correlations* option in the *Data to Analyze* group in the *Analysis Parameters* dialog, a completely transparent system for *correctly* analyzing the sample correlation matrix. If this command is given, *SEPATH* computes and analyzes the sample correlation matrix, regardless of whether the input file is a covariance matrix, correlation matrix, or rectangular data file. Thus, for models which *SEPATH* can analyze, the problems detailed by Cudeck (1989) are completely eliminated.

SEPATH solves the problem of analyzing correlations by utilizing an augmented version of the *SEPATH* model, and following the general analytical strategy advocated by Cudeck (1989) and Mels (1989). This strategy begins by converting the structural model into one which is scale free. This conversion is done as follows.

(1) Start with the path diagram as it would be if covariances were analyzed.

(2) Replace each manifest variable with a "dummy" latent variable, with a single arrow (having a free "scaling" parameter) pointing to the original manifest variable. If the original manifest variable is exogenous, its dummy latent variable will be exogenous. If the original manifest variable is endogenous, its dummy latent variable will be endogenous.

The resulting model must be invariant under changes in scale, because any effect of change in the scale of the observed variables can be absorbed by the scaling parameters. Moreover, the path coefficients can be estimated with all the dummy latent variables constrained to have unit variances. Simply set the variance of the exogenous dummy latent variables to a fixed value of 1, and use the new constrained estimation procedure to fix the variances of the endogenous dummy latent variables to 1.

In the revised model,

$$\mathbf{s}_1 = \begin{bmatrix} \mathbf{l}_x^* \\ \hline \mathbf{l}_n^* \\ \hline \mathbf{l}_n \end{bmatrix} \quad (139)$$

and

$$\mathbf{s}_2 = \begin{bmatrix} \mathbf{l}_x^* \\ \hline \mathbf{l}_x \end{bmatrix} \quad (140)$$

where $\mathbf{l}_x^*$ and $\mathbf{l}_n^*$ are the dummy latent variables. As mentioned above, each dummy latent variable is connected, via a single arrow, to a manifest variable. Hence, the model equations are the same as before, except for the additional relationships, which may be written

$$\mathbf{m}_x = \mathbf{D}_x \mathbf{l}_x^* \quad (141)$$

and

$$\mathbf{m}_n = \mathbf{D}_n \mathbf{l}_n^* \quad (142)$$

Letting

$$\mathbf{D}_s = \begin{bmatrix} \mathbf{D}_x & \vline & \mathbf{0} \\ \hline \mathbf{0} & \vline & \mathbf{D}_n \end{bmatrix} \quad (143)$$

the revised equation for Σ can be written, in the notation of Equation 41 as

$$\begin{aligned} \Sigma &= \mathbf{D}_s \mathbf{P} \mathbf{D}_s \\ &= \mathbf{D}_s \mathbf{G}(\mathbf{B} - \mathbf{I})^{-1} \Gamma \Xi \Gamma'(\mathbf{B}' - \mathbf{I})^{-1} \mathbf{G}' \mathbf{D}_s \end{aligned} \quad (144)$$

where $\mathbf{D}_s$ is a matrix of scaling factors, and $\mathbf{P}$ is the covariance matrix of the dummy latent variables. These variables, as mentioned above, may be constrained to have unit variance. Since they are related to the manifest variables by scaling constants, they may be thought of as the standardized unit variance equivalents of the manifest variables, so the matrix **P** is, in fact, a correlation structure model for the manifest variables.

Since the above model is invariant under rescaling of Σ, the identical discrepancy function will be produced, regardless of how the sample covariance matrix is scaled. Any changes in scaling will simply be absorbed in the nuisance scaling parameters in $\mathbf{D}_s$. Consequently, when the *Analyze Correlations* option is in effect, the sample covariance matrix is automatically rescaled into a correlation matrix, which is then analyzed. This rescaling tends, in practice, to improve performance of the iterative algorithms, especially in cases where **S**, the sample covariance matrix, has variances that differ by orders of magnitude.

Fully Standardized Path Models

By combining the *New Standardization* and *Analyze Correlations* options, a completely standardized path analysis can be performed, and standard errors can be estimated correctly for all path coefficients.

Analyzing Invariance Properties

Introduction

Some important recent papers (Browne & Shapiro, 1989, 1991; Dijkstra, 1990) have investigated invariance properties of covariance structures. This topic is important for several reasons. First, some of the statistics calculated (i.e, the Γ_1 and Γ_2 coefficients) depend on a certain scale invariance property. Second, some analyses discussed in the preceding section depend on invariance properties of the covariance structure. Finally, and perhaps most important, to fully understand the relationship between data and model, it is crucial to understand what aspects of the model are affected by rescalings of the data, and what aspects (if any) are unaffected.

Shapiro and Browne (1989) established a number of key invariance properties for covariance structures. Their paper concerned situations where a model fitting a covariance matrix Σ of a random vector **x** would continue to fit if the manifest variables in **x** were linearly transformed, i.e.,

$$\Sigma \to \mathbf{A}\Sigma\mathbf{A}', \quad \mathbf{x} \to \mathbf{A}\mathbf{x} \qquad (145)$$

Shapiro and Browne (1989) studied how covariance structures remained invariant when Σ is allowed to be transformed by **A** matrices of various types. They used the following definition. Consider a multiplicative group G of nonsingular $p \times p$ matrices. That is, if $\mathbf{A} \in G$ and $\mathbf{B} \in G$, then $\mathbf{AB}^{-1} \in G$. Under certain side conditions met by the groups under consideration, G constitutes a Lie group with matrix multiplication as the group operation. Associated with G is a corresponding Lie group G^* of transformations defined on the set of symmetric positive definite matrices by $\Sigma \to \mathbf{A}\Sigma\mathbf{A}'$. A *covariance structure* is a symmetric matrix valued function $\Sigma(\theta)$ which relates a parameter vector θ from a subset $\mathcal{P}_\theta$ of $\Re^q$ to Σ.

Definition. A covariance structure $\Sigma(\theta)$ is said to be invariant under the group G^* if for every $\theta \in \mathcal{P}_\theta$ and $\mathbf{A} \in G$ there exists a $\theta^* \in \mathcal{P}_\theta$ such that $\Sigma(\theta^*) = \mathbf{A}\Sigma(\theta)\mathbf{A}'$.

This means that for any $\mathbf{A} \in G$, the set of positive definite matrices corresponding to the given model remains invariant under the transformation $\Sigma \to \mathbf{A}\Sigma\mathbf{A}'$.

Browne and Shapiro (1989) studied several types of **A** matrices corresponding to different kinds of invariance.

These included two kinds of invariance which are of particular interest to *SEPATH* users (see the following section).

Types of Invariance

Invariance under a constant scaling factor.
In this case, $\mathbf{A}$ is in the group G_1 where

$$G_1 = \{\mathbf{A} : \mathbf{A} = \tau \mathbf{I}_p, \tau \neq 0\} \quad (146)$$

In essence, a model is invariant under a constant factor if multiplying $\mathbf{\Sigma}$ by a non-zero constant will not change the fact that a model fits $\mathbf{\Sigma}$. Most covariance structures of any practical interest are at least invariant under a constant scaling factor. Some of the non-centrality based fit indices developed by Steiger (1989, 1990) require that the model fitted be invariant under a constant scaling factor (ICSF).

Invariance under changes of scale.
In this case $\mathbf{A}$ is in the group G_2, where

$$G_2 = \{\mathbf{A} : \mathbf{A} = \text{Diag}(\tau), \tau_i \neq 0, i = 1, \ldots, p\} \quad (147)$$

where $\text{Diag}(\tau)$ is a diagonal matrix with diagonal elements given by the p components of the vector τ.

Analyzing Invariance of Fitted Covariance Structures

An important aspect of the Browne and Shapiro (1989) study of invariance was their derivation of methods for studying invariance of fitted moment structures. They defined a *reflector matrix* corresponding to each discrepancy function minimized by *SEPATH*, and showed how this reflector matrix could be analyzed to shed light on the invariance properties of a particular model when fitted with a particular discrepancy function.

Reflector Matrices

In this section the *reflector matrix* corresponding to each type of discrepancy function will be defined, followed by a discussion of how these matrices may be used to analyze invariance properties. The abbreviated notation will use $\hat{\mathbf{\Sigma}}$ to stand for the reproduced covariance matrix $\mathbf{\Sigma}(\hat{\boldsymbol{\theta}})$.

The reflector matrices for the discrepancy functions analyzed here are

$$\hat{\mathbf{\Omega}}_{\text{ML}} = \hat{\mathbf{\Sigma}}^{-1}(\hat{\mathbf{\Sigma}} - \mathbf{S}) \quad (148)$$

$$\hat{\mathbf{\Omega}}_{\text{LS}} = (\hat{\mathbf{\Sigma}} - \mathbf{S})\hat{\mathbf{\Sigma}} \quad (149)$$

$$\hat{\mathbf{\Omega}}_{\text{GLS}} = (\mathbf{S}^{-1}\hat{\mathbf{\Sigma}} - \mathbf{I})\mathbf{S}^{-1}\hat{\mathbf{\Sigma}} \quad (150)$$

$$\hat{\mathbf{\Omega}}_{\text{IRGLS}} = \hat{\mathbf{\Sigma}}^{-1}\mathbf{S}\hat{\mathbf{\Sigma}}^{-1}(\hat{\mathbf{\Sigma}} - \mathbf{S}) \quad (151)$$

$$\hat{\mathbf{\Omega}}_{\text{ADF}} = \mathbf{Q}\hat{\mathbf{\Sigma}} \quad (152)$$

where the elements of the symmetric matrix $\mathbf{Q}$ are

$$[\mathbf{Q}]_{ij} = \tfrac{1}{2}(2 - \delta_{ij})\left[\mathbf{U}^{-1}(\hat{\boldsymbol{\sigma}} - \mathbf{s})\right]_{\text{lex}(i,j)} \quad (153)$$

where $\delta_{i,j}$ is the Kronecker *delta*,

$$\text{lex}(i,j) = \frac{\max(i,j)(\max(i,j)-1)}{2} + \min(i,j) \quad (154)$$

and $\mathbf{U}$ is the weight matrix referred to in Equation 67.

Using Reflector Matrices

Browne and Shapiro (1989) give several corollaries which establish properties of reflector matrices implied by various types of scale invariances. Their results include (Shapiro & Browne, 1989, page 8).

Corollary 1.1 *If* $\Sigma(\theta)$ *is invariant under* G_1^*, *then the sum of the diagonal elements of* $\hat{\mathbf{\Omega}}$ *is zero.*

Corollary 1.2 *If* $\Sigma(\theta)$ *is invariant under* G_2^*, *then all diagonal elements of* $\hat{\mathbf{\Omega}}$ *are zero.*

These corollaries provide convenient devices for falsifying the dual assertion that a minimum has been obtained for a given discrepancy function, *and* that the fitted model possesses an invariance property.

Specifically, for a given discrepancy function, the reflector matrix is computed after convergence. If the trace of the reflector matrix is not zero, then *either a minimum has not been obtained, or the model is not invariant under a constant scaling factor, or both*. Then the individual diagonal elements of the reflector matrix are examined. If they are not all zero, one can conclude that *either a minimum has not been obtained, or the model is not invariant under changes of scale, or both*.

From a practical standpoint, one must remember issues of machine precision. One would only expect the above criteria to be met to an acceptable level of machine precision. Consequently, *SEPATH*, besides printing the reflector matrix, also reports (1) the trace of the reflector matrix, and (2) the largest absolute value on the diagonal of the reflector matrix.

Multiple Sample Models

In the case of *K independent* samples, the *Chi-square* statistic is essentially the sum of the *Chi-squares* computed on independent samples. Consequently, most of what was stated about the single sample estimation process remains true in the multiple sample case.

Specifically, for each sample k having sample size N_k, one computes a discrepancy function F_k in the same manner as with a single sample. These are then combined to produce an overall discrepancy function of the form

$$F = \sum_{k=1}^{K} \frac{(N_k - 1)}{(N_{total} - K)} F_k \quad (155)$$

where

$$N_{total} = \sum_{k=1}^{K} N_k \quad (156)$$

This overall discrepancy function F is minimized during the iterative process. The *Chi-square* statistic is then computed as

$$\chi^2 = (N_{total} - K)F \quad (157)$$

Models with Structured Means or Intercept Variables

Analysis of covariance structures can be extended to handle modeling of the means of variables as well as their variances and covariances. However, some important restrictions apply to these model extensions. In particular, in order to obtain a *Chi-square* statistic, the assumption of multivariate normality of the observed variables is required. The estimates obtained are "maximum normal likelihood" estimates, rather than "maximum Wishart likelihood" estimates, as in standard covariance structure models.

The extended *SEPATH* model has the same form as the model in Equation 36, except the variables $\mathbf{m}_x, \mathbf{m}_n, \mathbf{l}_x$, and $\mathbf{l}_n$ are replaced in $\mathbf{s}_1$ and $\mathbf{s}_2$ by

$$\mathbf{m}_x^* = \mathbf{m}_x + \boldsymbol{\tau}_x \quad (158)$$

$$\mathbf{m}_n^* = \mathbf{m}_n + \boldsymbol{\tau}_n \quad (159)$$

$$\mathbf{l}_x^* = \mathbf{l}_x + \boldsymbol{\kappa}_x \quad (160)$$

$$\mathbf{l}_n^* = \mathbf{l}_n + \boldsymbol{\kappa}_n \quad (161)$$

where $\boldsymbol{\tau}_x, \boldsymbol{\tau}_n, \boldsymbol{\kappa}_x$, and $\boldsymbol{\kappa}_n$ are constants, sometimes referred to as "intercept terms" in the model.

Under these conditions, the observed manifest variables $\mathbf{m}_x^*$ and $\mathbf{m}_n^*$ will no longer have zero means, but rather will have expected values that are a straightforward function of the model parameters in $\mathbf{B}, \boldsymbol{\Gamma}, \boldsymbol{\tau}_x, \boldsymbol{\tau}_n, \boldsymbol{\kappa}_x$, and $\boldsymbol{\kappa}_n$. Define

$$\boldsymbol{\mu}_x = E(\mathbf{m}_x^*) \quad (162)$$

$$\boldsymbol{\mu}_n = E(\mathbf{m}_n^*) \quad (163)$$

and

$$\mu = \begin{bmatrix} \mu_x \\ \mu_n \end{bmatrix} \quad (164)$$

If the population distribution is multivariate normal, then sample means and covariances are statistically independent, and two following discrepancy functions are of particular interest.

The *Maximum Normal Likelihood (ML)* discrepancy function is

$$F_{ML}(S^*, \Sigma(\theta), \mu, \overline{x}) = \ln|\Sigma(\theta)| - \ln|S^*|$$
$$+ \mathrm{Tr}(S^* \Sigma(\theta)^{-1}) - p \quad (165)$$
$$+ (\overline{x} - \mu)' S^{*-1} (\overline{x} - \mu)$$

where $\overline{x}$ is a vector of sample means, and

$$S^* = \frac{1}{N} \sum_{i=1}^{N} (x_i - \overline{x})(x_i - \overline{x})' \quad (166)$$

is the (biased) maximum normal likelihood estimate of the covariance matrix. The *Normal Theory Generalized Least Squares (GLS)* discrepancy function is

$$F_{GLS}(S^*, \Sigma(\theta), \mu, \overline{x}) =$$
$$\tfrac{1}{2} \mathrm{Tr}\big[(S^* - \Sigma(\theta)) S^{*-1}\big]^2 \quad (167)$$
$$+ (\overline{x} - \mu)' S^{*-1} (\overline{x} - \mu)$$

Some straightforward algebra shows how the mathematics developed for covariance structure models can be employed to fit models for mean and covariance structures. Specifically, if one uses the standard ML discrepancy function in Equation 56, but replaces S with the augmented moment matrix M_A, and Σ with

$$\Sigma = \begin{bmatrix} \Sigma + \mu\mu' & \mu' \\ \mu & 1 \end{bmatrix} \quad (168)$$

one obtains a result identical to the discrepancy function in Equation 165. A similar result holds for the GLS discrepancy functions

For a set of p variables, the augmented moment matrix M_A is a $(p+1) \times (p+1)$ square matrix. The first p rows and columns contain the matrix of moments about zero, while the last row and column contain the sample means for the p variables. The matrix is therefore of the form:

$$M_A = \begin{bmatrix} M & \overline{x} \\ \overline{x}' & 1 \end{bmatrix} \quad (169)$$

where M is a matrix with element

$$M_{jk} = \frac{1}{N} \sum_{i=1}^{N} X_{ij} X_{ik} \quad (170)$$

and $\overline{x}$ is a vector with the means of the variables.

Indices of Skewness and Kurtosis

SEPATH computes a number of indices of skewness and kurtosis. The univariate indices assess skewness and kurtosis for each variable individually. The multivariate kurtosis indices evaluate overall kurtosis for the entire set of variables.

Indices of Univariate Skewness

The sample mean and variance for the *jth* variable x_j are defined in the standard fashion as

$$\overline{X}_{\bullet j} = \frac{1}{N} \sum_{i=1}^{N} X_{ij} \quad (171)$$

and

$$s_j^2 = \frac{1}{N} \sum_{i=1}^{N} (X_{ij} - \overline{X}_{\bullet j})^2 \quad (172)$$

Uncorrected univariate skewness. The *uncorrected univariate skewness* for variable x_j is defined as

$$\gamma_{1(j)} = \frac{N \sum_{i=1}^{N} (X_{ij} - \overline{X}_{\bullet j})^3}{\sqrt{N \left[\sum_{i=1}^{N} (X_{ij} - \overline{X}_{\bullet j})^2 \right]^3}} \quad (173)$$

Corrected univariate skewness. The uncorrected estimate is biased. The following *corrected univariate skewness* is an unbiased estimate when the distribution is normal.

$$\gamma_{1(j)}^* = \frac{N}{(N-1)(N-2)} \frac{\sum_{i=1}^{N} (X_{ij} - \overline{X}_{\bullet j})^3}{s_j^3} \quad (174)$$

The asymptotic variance of these measures is $6/N$, which is used to standardize the uncorrected skewness to produce the "normalized" skewness printed by the program.

Indices of Univariate Kurtosis

Uncorrected univariate kurtosis. For variable x_j, this index is defined as

$$\gamma_{2(j)} = \frac{N \sum_{i=1}^{N} (X_{ij} - \overline{X}_{\bullet j})^4}{\left[\sum_{i=1}^{N} (X_{ij} - \overline{X}_{\bullet j})^2 \right]^2} - 3 \quad (175)$$

The uncorrected estimate is biased. The following *corrected univariate kurtosis* is an unbiased estimate when the distribution is normal.

Corrected univariate kurtosis. This index is defined as

$$\gamma_{2(j)}^* = C \frac{\sum_{i=1}^{N} (X_{ij} - \overline{X}_{\bullet j})^4}{s_j^4} - D \quad (176)$$

where

$$C = \frac{N(N+1)}{(N-1)(N-2)(N-3)} \quad (177)$$

and

$$D = \frac{3(N-1)^2}{(N-2)(N-3)} \quad (178)$$

The asymptotic variance of these measures is $24/N$, which is used to standardize the uncorrected kurtosis to produce the "normalized" kurtosis printed by the program.

Indices of Multivariate Kurtosis

Mardia's multivariate kurtosis. This index is defined as

$$\gamma_2 = \frac{1}{N} \sum_{i=1}^{N} \left\{ (\mathbf{x}_i - \overline{\mathbf{x}})' \mathbf{S}^{-1} (\mathbf{x}_i - \overline{\mathbf{x}}) \right\}^2 - p(p+2) \quad (179)$$

where $\mathbf{x}_i$ is the *ith* vector of observations, $\overline{\mathbf{x}}$ is a vector of sample means, p is the number of observed variables, and $\mathbf{S}$ is the sample covariance matrix.

Relative multivariate kurtosis. This index is defined as

$$\gamma_2^* = \frac{\gamma_2 + p(p+2)}{p(p+2)} \quad (180)$$

Normalized multivariate kurtosis. This index is defined as

$$\kappa_0 = \frac{\gamma_2}{\sqrt{8p(p+2)/N}} \quad (181)$$

Mardia-based kappa. This index is defined as

$$\kappa_1 = \frac{\gamma_2}{p(p+2)} \qquad (182)$$

Mean scaled univariate kurtosis. This index is defined as

$$\kappa_2 = \frac{1}{3p} \sum_{j=1}^{p} \gamma_{2(j)} \qquad (183)$$

Adjusted mean scaled univariate kurtosis. This index is defined as

$$\kappa_3 = \frac{1}{3p} \sum_{j=1}^{p} g_{2(j)} \qquad (184)$$

with

$$g_{2(j)} = \begin{cases} \gamma_{2(j)} & \text{if } \gamma_{2(j)} > \frac{-6}{p+2} \\ \frac{-6}{p+2} & \text{otherwise} \end{cases} \qquad (185)$$

Monte Carlo Data Generation Techniques

SEPATH generates Monte Carlo data using pseudo-random number generation techniques. All random number procedures begin with generation of uniform random variates. The uniform random numbers are produced by a standard linear congruential generator. The generator creates the ith random integer via the formula

$$x_i = 742938285 x_{i-1} (\bmod 2^{31} - 1) \qquad (186)$$

The integer can be transformed to a floating point number between zero and 1 by dividing by 2^{31}.

Uniform numbers are converted to simulated normal random numbers using Marsaglia's (1962) polar method. Independent normal random numbers are converted to multivariate normal numbers having a desired covariance structure by multiplying the p-variate vector by a Choleski factor of the desired covariance matrix Σ.

Multivariate normal random variates are transformed into variates with desired (marginal) skewness and kurtosis using the technique of Vale and Maurelli (1983).

INDEX

A

Adding paths, 3585
ADF estimation, 3660
Adjusted mean scaled univariate
  kurtosis, 3684
Augmented moment matrix, 3682

B

Bentler-Bonnet non-normed fit index,
  3675
Bentler-Bonnet normed fit index, 3675
Bentler-Weeks model, 3655
Bollen's *delta*, 3676
Bollen's *rho*, 3676
Bootstrapping, 3638
Boundary constraints in structural
  modeling, 3666

C

Causal modeling, 3545
Confirmatory factor analysis, 3545
  introductory example, 3551
Confirmatory factor analysis wizard
  correlated factors, 3554
  final dialog, 3554
  step 1, 3553
congeneric tests, 3614
Convergence criteria
  maximum residual cosine, 3576
  relative function change, 3576
Corrected univariate kurtosis, 3683
Correlation structure models, 3545
COSAN
  Computer program, 3653
COSAN model, 3653
Covariance structure models, 3545

D

Discrepancy function
  ADFG, 3575
  ADFU, 3575
  asymptotically distribution free
    (ADF), 3660
  generalized least squares, 3574

Discrepancy function (continued)
  generalized least squares(GLS),
    3659
  IRGLS, 3659
  maximum likelihood, 3574
  maximum Wishart likelihood
    (ML), 3659
  ordinary least squares, 3575
  ordinary least squares (OLS),
    3659
Discrepancy functions
  general properties, 3658

G

GLS estimation, 3659
Goodness of fit indices
  adjusted population *gamma*
    index, 3672
  Bentler comparative index, 3675
  Bentler-Bonnet non-normed
    index, 3675
  Bentler-Bonnet normed index,
    3675
  Bollen's *delta*, 3676
  Bollen's rho, 3676
  Browne-Cudeck single sample
    cross-validation index, 3674
  general theory, 3667
  Jöreskog-Sörbom adjusted GFI,
    3674
  Jöreskog-Sörbom GFI, 3674
  McDonald's index of
    noncentrality, 3672
  noncentrality-based, 3667
  Parsimonious fit index, 3675
  population *gamma* index, 3670
  rescaled Akaike information
    criterion, 3674
  Schwarz's Bayesian criterion,
    3674
  Steiger-Lind RMSEA index, 3669
  structural equation modeling,
    3666

H

Heywood cases, 3626

I

Identification
  in structural equation modeling,
    3662

Invariance
  under a constant scaling factor
    (ICSF), 3680
  under changes of scale (ICS),
    3680
Invariance properties
  structural equation models, 3679
Iteration dialog, 3590
Iteration problems
  solving, 3645
Iterative techniques
  in structural equation modeling,
    3664

K

Kappa
  Mardia-based, 3683
Kurtosis
  adjusted mean scaled univariate,
    3684
  corrected univariate, 3683
  indices of multivariate, 3683
  indices of univariate, 3683
  mean scaled univariate, 3684
  normalized multivariate, 3683
  relative multivariate, 3683
  uncorrected univariate, 3683

L

LaGrange multiplier statistics, 3594
Line search
  cubic interpolation, 3577
  golden section, 3577
  parameters, 3577
  simple stephalving, 3577
Line search methods, 3577
LISREL model, 3651

M

Manifest exogenous variables, 3575
Mardia's multivariate kurtosis
  coefficient, 3595
Mardia's multivariate kurtosis index,
  3683
Mardia-based kappa, 3683
Matrix
  augmented moment, 3682
Maximum residual cosine, 3576
Mean scaled univariate kurtosis, 3684
Minimization
  Unconstrained, 3664

11. STRUCTURAL MODELING - INDEX

Model summary scrollsheet, 3593
Monte Carlo methods, 3625
Multiple group matrix format, 3573
Multiple sample models
    in structural equation modeling, 3681
Multivariate kurtosis
    index of, 3683
Multivariate normality
    assumption of, 3575, 3595

N

Noncentrality-based indices, 3594

O

OLS estimation, 3659

P

Path construction tool, 3568, 3583
Paths
    adding, 3585
    editing, 3585
    removing, 3585
    renumbering, 3586

R

RAM model, 3653
Reflector matrix, 3596, 3680
Regression models, 3545
Relative function change criterion, 3576
Relative multivariate kurtosis, 3683
Renumbering paths, 3586
Rules
    for path diagrams, 3559
    for the PATH1 language, 3561

S

Second order factor analysis, 3545
Seed values
    Monte Carlo analysis, 3587
SEPATH (Structural Equation Modeling and Path Analysis), 3539, 3545
SEPATH examples
    a multiple regression model, 3612
    comparing correlation matrices, 3622

SEPATH examples (continued)
    comparing dependent variances, 3616
    comparing factor structure in two groups, 3608
    confirmatory factor analysis, 3599
    confirmatory factor analysis with identifying constraints, 3601
    effect of peer influence on ambition, 3602
    factor analysis with an intercept variable, 3607
    home environment and math achievement, 3613
    longitudinal factor model, 3619
    multi-trait, multi-method model, 3617
    personality and drug use, 3620
    standardized solution for the effects of peer influence, 3604
    test for compound symmetry, 3621
    test theory models for sets of congeneric tests, 3614
    testing for circumplex structure, 3609
    testing for stability of a correlation matrix over time, 3611
SEPATH model, 3656
Skewness
    corrected univariate, 3683
    indices of univariate, 3682
    uncorrected univariate, 3683
Standardized models
    in path analysis, 3679
Statistical estimation in structural modeling, 3657
Steepest descent iterations, 3665
Structural equation modeling
    algebraic models, 3651
    analysis parameters, 3555, 3573
    analysis results, 3591
    analyzing correlations, 3574
    analyzing moments, 3574
    basic idea behind, 3546
    Bentler-Weeks model, 3655
    bootstrapping, 3638
    boundary constraints, 3666
    chi square goodness-of-fit statistics, 3660
    chi square statistic, 3593
    conceptual overview, 3545

Structural equation modeling (cont.)
    conditions for model identification, 3663
    confirmatory factor analysis example, 3551
    convergence criteria for iteration, 3666
    corrrelation structures, 3677
    COSAN model, 3653
    data to analyze, 3573
    define endogenous factors, 3566
    evaluating multivariate normality, 3643
    goodness of fit indices, 3666
    goodness-of-fit statistics, 3594
    identification of a model, 3662
    indices of kurtosis, 3682
    indices of skewness, 3682
    initial values, 3576
    inputting path diagrams, 3561
    introductory examples, 3551
    iteration dialog, 3555, 3590
    iteration problems, 3645
    line search methods, 3577, 3665
    LISREL model, 3651
    measurement model, 3565
    measures of skewness, 3595
    ML estimation, 3659
    model identification, 3662
    model invariance properties, 3679
    models with intercept variables, 3681
    Monte Carlo results, 3596
    Monte Carlo setup, 3587
    multiple group analysis, 3572
    multiple group matrix format, 3573
    multiple sample models, 3681
    multivariate kurtosis measures, 3595
    normalized residuals, 3596
    path construction tool, 3568
    path diagrams, 3557
    PATH1 language rules, 3561
    program overview, 3547
    RAM model, 3653
    reflector matrices, 3680
    resolving ambiguities in path diagrams, 3560
    results dialog, 3555
    results text window, 3592
    rules for path diagrams, 3559
    SEPATH model, 3656
    specifying groups, 3572

Structural equation modeling (cont.)
- specifying structural paths, 3566
- standardization methods, 3676
- standardized residuals, 3596
- standardized solution, 3575
- start values, 3576
- startup panel, 3552, 3571
- statistical estimation theory, 3657
- structured means models, 3681
- technical aspects, 3651
- tests of assumptions, 3595
- unconstrained minimization techniques, 3664
- univariate kurtosis measures, 3595

Structural equation modeling
- results text window, 3592

U

Unconstrained minimization techniques, 3664
Uncorrected univariate kurtosis, 3683

W

Wizard
- confirmatory factor analysis, 3553, 3578
- structural equation modeling, 3563
- structural modeling, 3580

11. STRUCTURAL MODELING - INDEX

Chapter 12:
MEGAFILE MANAGER

Table of Contents

Introductory Overview	3695
Dialogs and Options	3697
Select Columns	3697
Select a Single Column	3698
Select Rows	3698
View Column Values	3698
Find Value	3699
Replace Value	3699
Global Width	3700
Create New *Megafile Manager* Data	3700
Open *Megafile Manager* Data	3702
Save *Megafile Manager* File	3702
Save As *Megafile Manager* File	3702
Importing Files in *Megafile Manager*	3702
Append *STATISTICA* File	3711
Exporting Files in *Megafile Manager*	3712
Convert to a *STATISTICA* File	3717
Print MFM Data	3717
Merge Files	3724
Merge Columns	3725
Merge Rows	3726
Overlay Merge	3727
Subset	3727
Sort Data	3728
Optimize Size	3728
Verify Column Names	3729
Incorrect Variable Name	3730
Descriptive Statistics	3730
Correlations	3731
Frequencies	3733
Index	3737

The *Detailed Table of Contents* follows on the next page.

12. MEGAFILE MANAGER - CONTENTS

Detailed
Table of Contents

INTRODUCTORY OVERVIEW 3695
    General Overview 3695
    Megafile Manager and other *STATISTICA* Facilities 3695
    Unique Features of *Megafile Manager* 3696

DIALOGS AND OPTIONS 3697
    Select Columns 3697
        Selecting Columns 3697
        De-selecting Previously Selected Columns 3697
        Next >> and *Previous* << Buttons 3697
        Select All 3697
        Spread/Shrink 3697
        Zoom 3698
    Select a Single Column 3698
        Selecting Columns 3698
        Next >> and *Previous* << Buttons 3698
        Spread/Shrink 3698
        Zoom 3698
    Select Rows 3698
    View Column Values 3698
        Column Values 3699
    Find Value 3699
        Find What 3699
        Direction 3699
        Match Whole Word Only 3699
        Match Case 3699
        Find Next 3699
    Replace Value 3699
        Find What 3700
        Replace With 3700
        Direction 3700
        Match Whole Word Only 3700
        Match Case 3700
        Find Next 3700
        Replace All 3700
    Global Width 3700
    Create New *Megafile Manager* Data 3700
        New File Name 3701
        Number of Columns 3701

12. MEGAFILE MANAGER - CONTENTS

Number of Rows	3701
Column Name Prefix	3701
Format Statement	3701
One-Line File Header	3702
Open *Megafile Manager* Data	3702
Save *Megafile Manager* File	3702
Save As *Megafile Manager* File	3702
Importing Files in *Megafile Manager*	3702
Options	3703
Import *STATISTICA* Data File	3703
Case Names	3704
Variables with Text Values	3704
Import Excel File	3704
Range	3704
Column Names	3704
Import Lotus File	3705
Range	3705
Review Ranges	3705
Column Names	3706
Import Quattro File	3706
Range	3706
Column Names	3706
Import SPSS File	3707
Import Value Labels	3707
Import dBASE File	3707
Range	3707
Import ASCII File	3708
Import ASCII Fixed File	3708
File Size	3708
Format Statement	3708
Import ASCII Free File	3709
File Size	3709
Format Statement	3709
Separators	3710
Additional	3710
Append *STATISTICA* File	3711
Export Options	3712
Export to *STATISTICA* Data File	3712
Export Text Columns	3713
Export Case Names	3713
Auto Numeric/Text Match	3713
Range	3714
Export File to Excel	3714
Export Column Names	3714

MM - 3691

Copyright © StatSoft, 1995

12. MEGAFILE MANAGER - CONTENTS

Export File to Lotus	3714
Export Column Names	3714
Define Sheets	3714
Export File to Quattro	3715
Export Column Names	3715
Export File to dBASE	3715
Export File to SPSS	3715
Export File to ASCII	3716
Export File to ASCII Free Format	3716
Line Length	3716
Separators	3716
Export File to ASCII Fixed Format	3717
Line Length	3717
Convert to a *STATISTICA* File	3717
Print MFM Data	3717
Columns	3717
Rows	3717
Print Data Values	3717
Print Column Specifications	3717
Print/Eject Pages after this Printout	3718
Add Columns	3718
Number of Columns to Add	3718
Insert After Column	3718
Prefix	3718
Type	3718
Length	3719
MD	3719
Display Format	3719
Move Columns	3719
From Column	3720
To Column	3720
Insert After	3720
Copy Columns	3720
From Column	3720
To Column	3720
Insert After	3720
Delete Columns	3720
From Column	3721
To Column	3721
Edit Column Specs	3721
Name	3721
Type	3721
Length	3722
MD	3722

12. MEGAFILE MANAGER - CONTENTS

Display Format	3722
Long Name	3722
Values	3722
Add Rows	3722
Number of Rows to Add	3722
Insert After Row	3723
Move Rows	3723
From Row	3723
To Row	3723
Insert After	3723
Copy Rows	3723
From Row	3723
To Row	3723
Insert After	3723
Delete Rows	3724
From Row	3724
To Row	3724
File Header	3724
One Line Header	3724
File Information/Notes	3724
Merge Files	3724
Merge	3724
2nd File	3725
Merge Columns	3725
Mode	3725
Relational Merge Options	3726
Unmatched Cases	3726
Merge Rows	3726
Overlay Merge	3727
Update Columns	3727
Update Rows	3727
Subset	3727
Columns	3727
Rows	3727
Sort Data	3728
Column	3728
Sorting Order	3728
More Keys	3728
Optimize Size	3728
Verify Column Names	3729
Column Names	3729
Formats	3730
Incorrect Variable Name	3730
Variable Name	3730

MM - 3693

Copyright © StatSoft, 1995

12. MEGAFILE MANAGER - CONTENTS

    Ignore 3730
    Accept 3730
    Break 3730
    Undo All 3730
Descriptive Statistics 3730
    Columns 3731
    Rows 3731
    Descriptive Statistics 3731
Correlations 3731
    Columns 3731
    Rows 3731
    Pairwise Missing Data Deletion 3731
    Display n 3732
    Display p-levels 3732
Frequencies 3733
    Columns 3733
    Rows 3733
    Mode 3733
    Warning Level 3734
Megafile Manager Language (*MML*) 3734

INDEX 3737

12. MEGAFILE MANAGER - INTRODUCTORY OVERVIEW

Chapter 12:
MEGAFILE MANAGER

INTRODUCTORY OVERVIEW

General Overview

Megafile Manager is a specialized data base management system accessible from the *Data Management* module of *STATISTICA*. Its unique feature is that it can manage and directly process types of data which need to be transformed, aggregated, extracted, or cleaned before they can be directly accessed by statistical or graphics procedures (e.g., data organized into extremely long records or data embedded inside very long text values).

Megafile Manager can process extremely large records of data (*rows*): up to 32,000 columns with up to 255 characters each (up to 8 megabytes per row). Data organized in such long records can, for example, be produced by some automated quality control measurement devices or other data acquisition or monitoring equipment. Also, such files are sometimes useful in maintaining integrated, large archival data banks consisting of numerous merged or concatenated files.

Megafile Manager and other STATISTICA Facilities

Most likely, the majority of common data processing needs can be easily addressed with the standard procedures available in every module of *STATISTICA*. Therefore, *Megafile Manager* will typically be used only when there is a need for facilities to handle very unusual data importing, management, and pre-processing needs, for example, in order to maintain very large data bases, or perform pre-processing of imported long text values, etc. Note that the standard procedures (offered in every module of *STATISTICA*) can also handle large size and very complex tasks; for example:

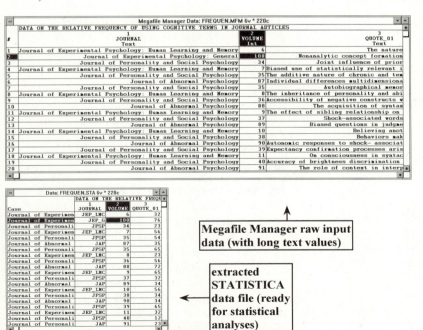

- to compute descriptive statistics or correlation matrices for very many variables (e.g., over 1,000) in a single run, use the *Quick Basic Stats* facilities available in every module via the *Analysis* pull-down menu or the toolbar button (see Volume I for details);

- to perform complex recoding and aggregation of data, extensive operations on text

MM - 3695

12. MEGAFILE MANAGER - INTRODUCTORY OVERVIEW

values, or complex cleaning and verification of data, use *STATISTICA BASIC*, available in every module of *STATISTICA* from the *Analysis* pull-down menu (see Volume I for a brief description, and Volume V for a detailed description of *STATISTICA BASIC*);

- to import large files from other applications (e.g., Excel, Lotus, Quattro Pro, Paradox, etc.), use the *Import* options available from the pull-down menu *File* in every module of *STATISTICA* (see Volume I);

- to perform sequential, hierarchical, or relational merging of files, use the *Merge* option from the *Analysis* pull-down menu in the *Data Management* module (see Volume I), or use the functions available in *STATISTICA BASIC* for accessing external files (and then use the available powerful *STATISTICA BASIC* functions to merge data via simple or complex user-defined algorithms; refer to Volume V for a detailed description of *STATISTICA BASIC*).

Unique Features of *Megafile Manager*

Maintaining large, archival data banks; hierarchical relations between data bases. *Megafile Manager* offers options for aggregating data sets from other applications and setting up very large (e.g., 8 megabytes per record), efficient archival data bases. It also supports links between related (and hierarchically organized) data sets. Subsets of columns from such archival data banks can be extracted and used with other applications (such as *STATISTICA,* Excel, or Paradox).

Preprocessing large records of raw data. Another unique application of *Megafile Manager* is at the stage of analysis when raw data need to be aggregated or preprocessed before meaningful indices are obtained for use in data analysis. Such raw data sets (e.g., from automated quality control measurement devices or other data acquisition equipment) may feature records that are too long to fit into any standard application (e.g., 32,000 measures per row). *Megafile Manager* can be used to access such data sets, convert them into meaningful indices, and transfer to another application (such as *STATISTICA* or Excel) for further analysis. Such raw data often need to be cleaned and verified before they can be preprocessed. Custom-designed data verification (and interactive correction) can be performed in *Megafile Manager* using its integrated programming language which features specialized functions for interactive data editing and verification.

Data processing, analysis. Thus, *Megafile Manager* not only offers facilities to aggregate, store, and maintain long-record files, but it can also efficiently process them. Its integrated programming language (*MML*) features a variety of data analytic options and a library of functions. *Megafile Manager* also includes basic statistics facilities that can process data regardless of the record size. For example, it can tabulate data, compute descriptive statistics, or generate correlation matrices of practically unlimited size (the size of correlation matrices that could be generated by *Megafile Manager* exceeds the capacity of any existing storage device).

Long text values. Another specific feature of *Megafile Manager* is its ability to process very long text values. Also, its integrated programming language (*MML*) offers a comprehensive selection of functions to manipulate text data.

Exchanging data with *STATISTICA* data files. *Megafile Manager* uses a specialized file format optimized for its specific applications (e.g., maintaining data types from a variety of programs). However, easy to use and flexible facilities are provided in the *Data Management* module to move data in and out between *STATISTICA* data files and archival *Megafile Manager* data bases. For more information on these facilities, see *Data Management*, Volume I.

MM - 3696

Copyright © StatSoft, 1995

12. MEGAFILE MANAGER - DIALOGS AND OPTIONS

DIALOGS AND OPTIONS

Select Columns

This dialog opens whenever you are requested to select columns from a single list.

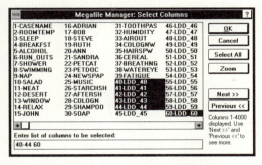

If the list of columns is too large for the display window, then the list will be scrollable.

Selecting Columns

Select columns in this dialog either by highlighting them or entering the column numbers in the edit window (e.g., *1 25-50 56*) and then clicking on the *OK* button. If you choose to select the columns by highlighting them in the list, you can select a continuous block of columns by holding down the left-mouse-button and dragging the cursor over the columns that you want to highlight.

You can also highlight a continuous block of columns by clicking on the first column name that you want to select, and while holding down the SHIFT key, clicking on the last column name that you want to select. This method will select (highlight) the first and last columns and all columns in-between. You can select discontinuous blocks of columns by holding down the CTRL key and then clicking on each column that you want to select.

De-selecting Previously Selected Columns

You can de-select discontinuous or continuous blocks of previously selected columns by holding down the CTRL key and then clicking on each column that you want to de-select. Alternatively, you can de-select columns by deleting the column number(s) from the list of columns in the edit window.

Next >> and *Previous* << Buttons

Up to 4,000 columns can be displayed in the scrollable window (i.e., page) in this dialog. If the *Megafile Manager* file contains more than 4,000 columns, then the display of the columns in this window will be "broken down" into pages, and *Next* and *Previous* buttons will appear allowing you to move forward or backward among the "pages" of column listings. Selections made in one page of the column listings will be retained in the edit window while making selections in other pages. In other words, highlighting column names in the window will affect only that part of the numeric list of columns in the edit box which apply to the current "page" of column names.

Select All

Click this button to automatically select all of the columns in the list. Alternatively, you can enter an * (asterisk) in the edit window to select the entire list of columns.

Spread/Shrink

When you click on this button, any long names associated with each column will be displayed. Click on this button again to remove the long names

MM - 3697

Copyright © StatSoft, 1995

12. MEGAFILE MANAGER - DIALOGS AND OPTIONS

from the list. Note that this option may not be available when the list of columns is very long.

Zoom

Clicking this button will open the *Column Values Window* for the first of the highlighted columns (see below) in which you can browse through a scrollable sorted list of the data values.

Select a Single Column

This dialog opens whenever you are requested to select one column from a list of columns.

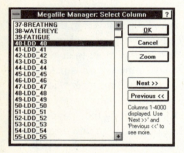

If the list of columns is too large for the display window, then the list will be scrollable.

Selecting Columns

Select a column in this dialog either by double-clicking on the column name or by highlighting it (clicking once on the column name) and then clicking on the *OK* button.

Next >> and *Previous* << Buttons

Up to 4,000 columns can be displayed in the scrollable window (i.e., page) in this dialog. If the *Megafile Manager* file contains more than 4,000 columns, then the display of the columns in this window will be "broken down" into pages and *Next* and *Previous* buttons will appear allowing you to move forward or backward among the "pages" of column listings.

Spread/Shrink

When you click on this button, the long names (if there are any) associated with each column in the list will be displayed. Click on this button again to remove the long names from the list. Note that this option may not be available when the list of columns is very long.

Zoom

Clicking this button will open the *Column Values Window* (see below) in which you can browse through a scrollable, sorted list of the data values for the highlighted column.

Select Rows

The *Select Rows* dialog opens whenever you are requested to specify a range of rows.

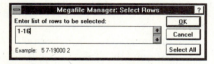

Enter either a continuous (e.g., *1-280*) or discontinuous (e.g., *1 25-50 235-240*) range of row numbers in this dialog.

Optionally, you can click on the *Select All* button in order to automatically select all rows in the current *Megafile Manager* file.

View Column Values

This window allows you to examine the values of a column, as well as its name, missing data value, format and long name (if there is one).

12. MEGAFILE MANAGER - DIALOGS AND OPTIONS

Column Values

The column values are presented in a scrollable, sorted list through which you can browse. This window is accessed by pressing the *Zoom* button in a dialog, and you can return to that dialog by clicking *OK* in this window or by selecting the *Column Values* option from the right-mouse-button flying menu.

Find Value

The *Find Value* dialog (accessible by selecting the *Find* option in the pull-down menu *Edit*) will enable you to search for a specific data value (text or numeric) in the currently highlighted column of the *Megafile Manager* file.

Find What

Specify here the column value (text or numeric) that you want to search for in the *Megafile Manager* file.

Direction

Find will search either above (select the *Up* option button) or below (select the *Down* option button) the currently highlighted cell, but will stop the search when it reaches the top or bottom of the file.

Match Whole Word Only

Select this option when you want to search for a whole word (e.g., *male*) and not places where the word may be a subset of a larger word (e.g., fe*male*).

Match Case

Select this option when you want to match the uppercase and lowercase formatting of the word exactly (e.g., *Females* instead of *females*).

Find Next

After you have specified the above options, click on this button to begin the search. When there is more than one match, *Find* will pause so that you can either exit (*Cancel* button) the dialog or click *Find Next* to search for the next match.

Shortcut. You can repeat the search by pressing CTRL+F.

Replace Value

The *Replace Value* dialog (accessible by selecting the *Replace* option in the pull-down menu *Edit*) will enable you to search for a specific data value (text or numeric) in the currently highlighted column and replace it with a different value.

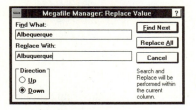

MM - 3699

Copyright © StatSoft, 1995

12. MEGAFILE MANAGER - DIALOGS AND OPTIONS

Find What

Specify the column value (text or numeric) that you want to search for and replace.

Replace With

Specify the value (text or numeric) here that will be used to replace the value or text that you are searching for.

Direction

Replace will search either above (select the *Up* option button) or below (select the *Down* option button) the currently highlighted cell or text, but will stop the search when it reaches the top or bottom of the file.

Match Whole Word Only

Select this option when you want to search for a whole word (e.g., *male*) and not places where the word may be a subset of a larger word (e.g., fe*male*).

Match Case

Select this option when you want to match the uppercase and lowercase formatting of the word exactly (e.g., *Females* instead of *females*).

Find Next

After you have specified the above options, click on this button to begin the search. When *Megafile Manager* has found the specified text or value, you will be given the following choices to replace it:

Yes. Click this button to replace the text or value with your specified replacement.

No. Click this button to skip over this value (it will not be replaced).

Yes to all. Click this button to replace all of the text or values that match without asking for confirmation.

Cancel. Click this button to quit the search and leave the dialog.

Replace All

Click this button instead of the *Find Next* button to replace all values or text that match. You will not be prompted to accept or reject each replacement.

Global Width

The value entered in this dialog (minimum = 1, maximum = 63) determines the width of all selected columns (*From* column, *To* column) in the *Megafile Manager* file.

This is a global operation, in that the value which you specify here will be applied to the specified range (inclusive) of columns in the file. Note that text values longer than 63 characters (up to 255 characters) can be reviewed by scrolling within their respective fields.

Create New Megafile Manager Data

You can create new *Megafile Manager* files with the *Create New File* dialog.

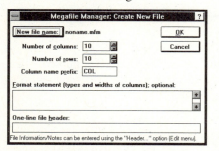

12. MEGAFILE MANAGER - DIALOGS AND OPTIONS

This dialog is accessible from the *New MFM Data* option in the *Megafile Manager* window *File* pull-down menu or from the *MFM: Create New Data* option from the *Megafile Manager* window *Analysis* pull-down menu or the *Data Management* startup panel.

New File Name

Click on this button to open the *New Data: Specify File Name* dialog in which you can designate the file name for the new *Megafile Manager* file.

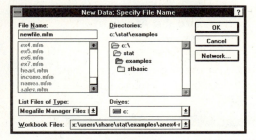

Number of Columns

Specify the number of columns to be included in the new data file. Up to 32,000 columns may be created and maintained in *Megafile Manager*.

Number of Rows

Enter the number of rows to be included when creating the data file. *STATISTICA* imposes practically no limit on the number of rows in one file (up to 2.14 billion rows may be specified).

Column Name Prefix

Specify the column name prefix here (all column names in this data file will have this prefix). The column name prefix which is defined here will have a sequential number added to the end (e.g., *COL1*, *COL2*, *COL3*, etc.).

Format Statement

The (optional) format statement is a set of instructions for interpreting the structure of the new *Megafile Manager* file (optionally, you could double-click on a column title to open the *Column Specs* dialog in which you can specify the format for that particular column). This option explicitly defines for *STATISTICA* the exact contents of the *Megafile Manager* file. The format statement is entered as a list of formats in the form nXm, where n is an integer multiplier indicating the number of times the format is to be repeated (no multiplier = 1); X is the column type; m is the column width [e.g., *10A8* means 10 text value (alphanumeric) fields, each 8 characters long].

The following column type specifiers are supported.

A — Text (Alphanumeric)

F — Float (also *R* - Real, equivalent to *Real 4B* in the *Column Specs* dialog, page 3721)

D — Double Float (also *DR* - Double Real, equivalent to *Real 8B* in the *Column Specs* dialog, page 3721)

I — Integer (also *NI* - Normal Integer, equivalent to *Int 2B* in the *Column Specs* dialog, page 3721)

S — Short Integer (also *SI*, equivalent to *Int 1B* in the *Column Specs* dialog, page 3721)

LI — Long Integer (also *J*, equivalent to *Int 4B* in the *Column Specs* dialog, page 3721)

L — Logical

For example, the format statement *2a15 i3 f5* defines four columns: the first two columns contain text (symbol *a*), 15 characters in display width; the third column contains *integer* values (symbol *i*) of display

12. MEGAFILE MANAGER - DIALOGS AND OPTIONS

width 3; the fourth *float* column (symbol *f*) contains values 5 characters in display width. For more information on column types and widths in *Megafile Manager*, see *Edit Column Specs*, page 3721.

One-Line File Header

Enter one line (up to 77 characters) of comment about the file here. Note that up to one page of text can be entered into the *File Header and Info Page* dialog (see page 3724).

Open *Megafile Manager* Data

Selecting this option will bring up the *Open Megafile Manager File* dialog from which you can select a *Megafile Manager* file (file with the file name extension *.mfm*) to open.

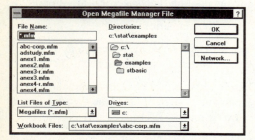

Megafile Manager files contain data values and also labels and formats (see *Column Specs*, page 3721), headers and notes (see *Header*, page 3724) and other information. In order to open a non-*Megafile Manager* format file (e.g., Excel, Lotus, Quattro, dBASE, ASCII, and others), use the *Import* option in the *File* pull-down menu (see below). For additional details concerning the *Workbook* facilities, refer to Volume I.

Save *Megafile Manager* File

Save a *Megafile Manager* data file with this option. If the file has not been saved before, then the *Save File As* dialog will open (see below) in which you can specify the file name and disk location for it to be saved. Once the file name and location have been specified, whenever you select *Save*, the spreadsheet will automatically be saved, overwriting the previous copy of the file. If you want to change the name or location of the file, select the *Save As* option.

Save As *Megafile Manager* File

Every time you select this option, the *Save File As* dialog will open in which you can save the *Megafile Manager* file.

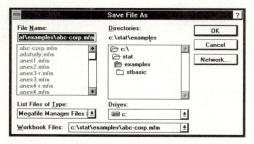

By default, the current file name and disk location will be displayed in this dialog. You can save the *Megafile Manager* file under a new file name or edited default name or specify a different disk location for the *Megafile Manager* file.

Importing Files in *Megafile Manager*

Importing data to *Megafile Manager* by converting foreign data files (e.g., worksheets or data bases) is one of several ways (offered in *STATISTICA*) of accessing data from other applications. Other methods available in *Megafile Manager* include using the Clipboard (which in *STATISTICA* recognizes special Clipboard formatting conventions used by other Windows applications such as Excel or Lotus) and the *Append STATISTICA File* option from the *Megafile Manager* window *File* pull-down menu (page 3711).

Importing is a very simple operation; in the *STATISTICA Megafile Manager* window, select the *Import* option from the *File* pull-down menu (or the startup panel), then follow the steps outlined below.

Step 1. Select the file name in the *Select File to Import* dialog.

Step 2. Confirm or change the type of file to be imported in the *Import Data From* dialog (see below).

Step 3. Click the *OK* button in the latter dialog to import and store the data in the *Megafile Manager* file format.

Each of the options below represents a group of formats (e.g., *Lotus* can import files from Lotus version 1A, 2, 3, Symphony, etc.). When importing files, *STATISTICA* will use the column/field/variable names of the source file. (If an imported name is longer than eight characters, the first eight characters will be used as the *Megafile Manager* column name.) For more detailed information on each of the import formats, see the respective topic.

- Import *STATISTICA* File (see below)
- Import Excel File (page 3704)
- Import Lotus File (page 3705)
- Import Quattro File (page 3706)
- Import dBASE File (page 3707)
- Import SPSS File (page 3707)
- Import ASCII Free (page 3709)
- Import ASCII Fixed (page 3708)

Options

Click on this button to open a dialog that will contain importing options specific to the import format that you choose.

Import *STATISTICA* Data File

The quickest way to convert a *STATISTICA* data file to *Megafile Manager* format is to open the data file in the *Data Management* module, and then to choose option *Convert to MFM Data* from the *File* pull-down menu (see Volume I for details). You can also use the *Append* option (see page 3711).

The *Import* facility described here offers additional options. After you select a *STATISTICA* file in the *Select File to Import* dialog, the *Import Data From* dialog will open (see above). Click on the *OK* button in this dialog to import the *STATISTICA* data file that you previously selected, or click on the *Options* button to open the *STATISTICA Input Options* dialog.

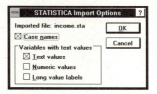

During import, *STATISTICA* variable long names (if they exist) become *Megafile Manager* column long names, and the *STATISTICA* one-line file header and comment are imported as the *Megafile Manager* one-line file header and comment.

Case Names

When you select this option, the *STATISTICA* case names will be imported as standard *Megafile Manager* text values and placed in the first column of the *Megafile Manager* file.

Variables with Text Values

STATISTICA's double notation of variables (text/numeric) may be expanded via these options to become up to three *Megafile Manager* columns with names related to the *STATISTICA* variable (representing the numeric values, text values, and value labels of the *STATISTICA* variable). The settings of the following options will determine the creation of these columns:

Text values. When you select this option, a column containing the text values will be created. If the *STATISTICA* variable name is shorter than 8 characters, then the *Megafile Manager* text column name will have a *$* added to the end; if the name is 8 characters long, then the second character will be changed to a *$*. This is done to avoid later conflicts regarding duplicated column names with the requirements of *MML* and other programs to which the data may be exported, and so that the automatic matching of text and numeric columns will be facilitated, if you later export the file to *STATISTICA* (or *Convert to STATISTICA*, see page 3717).

Numeric values. Select this option in order to create a column with the numeric equivalents to the text values.

Long value labels. When you select this option, then a column with value labels will be created (if the *STATISTICA* variable name is shorter than 8 characters, then the *Megafile Manager* text column name will have an *@* added to the end; if the name is 8 characters long, then the second character will be changed to an *@*).

Import Excel File

When importing an Excel file, column types are dynamically determined according to the Excel spreadsheet contents: If a spreadsheet column contains both labels and numeric values, then column type *Text* is used; if it contains both integer and real values, then a *Float* format will be used.

Please note that *Megafile Manager* will automatically convert the Excel spreadsheet into a "data base" at the point of import; therefore, it is not necessary to save the Excel file as a "data base" before importing it.

Range

By default, the entire active area of the spreadsheet is imported. This option permits selection of a specific range of rows and/or columns to be imported.

Columns. Specify the range of columns to be imported (*From* which column, *To* which column, inclusive).

Rows. Specify the range of rows to be imported (*From* which row, *To* which row, inclusive).

Column Names

This option specifies the source in Excel for *Megafile Manager* column names.

From spreadsheet. When you select this option, *Megafile Manager* will use the Excel column designators (*A*, *B*, *C*, etc.) as *Megafile Manager* column names.

From first row of selected range. Select this option to use the labels from the first row of the specified import range (see above). If any label is longer than eight characters, it will be truncated and only the first eight characters will be used as the *Megafile Manager* Column name.

Create (COL1, COL2, . . .). This option will result in the use of a combination of the prefix *COL* and a sequential number (e.g., *COL1*, *COL2*, etc.) as column names in the imported file.

Note that in addition to the file import operation, Excel data can be accessed by *STATISTICA* via the Clipboard (*STATISTICA* recognizes the Excel Clipboard format and will properly interpret "formatted" values such *$100* or *2,000*).

Import Lotus File

Megafile Manager will import worksheets from Lotus 1-2-3 Release 1A (Lotus/1), Release 2.01/2.2 (Lotus/2), Release 3 (Lotus/3), and Lotus for Windows, as well as Symphony. Due to their 3-dimensional structure, 1-2-3/3 and Lotus for Windows files are imported with each page placed in a successive set of columns in the *Megafile Manager* file. Column types are dynamically determined according to the worksheet contents: If a spreadsheet column contains both labels and numeric values, then column type *Text* is used; if it contains both integer and real values, then a *Float* format will be used.

After you select a Lotus file from the *Select File to Import* dialog, the *Import Data From* dialog will open (by default, *Lotus* will be selected; see page 3702). In this dialog, click on the *Options* button to bring up the *Lotus Import Options* dialog (the same dialog is used for all versions of Lotus and Symphony files).

Range

By default, the entire active area of the worksheet is imported. A portion of the worksheet can be selected by defining the range (see below).

Columns. Specify the range of columns to be imported (*From* which column, *To* which column, inclusive).

Rows. Specify the range of rows to be imported (*From* which row, *To* which row, inclusive).

Sheets. This option is only available if a Lotus 1-2-3 Release 3 or Lotus/w file is being imported (otherwise the option is dimmed). By default, all sheets are imported; however, you can select specific sheets to be imported using this option (*From* which sheet, *To* which sheet, inclusive).

Review Ranges

In Lotus, you can define "named ranges" of data in order to define distinctive sections of the worksheet. Clicking the *Review Ranges* button will bring up a list of defined ranges for the Lotus worksheet that you want to import.

Now, instead of importing the whole Lotus worksheet (which might include column headings and other text), you can import only a subset of data by selecting the named worksheet range (highlight the range name and click the *OK* button).

Column Names

This option specifies the source for names of columns.

From spreadsheet. Use Lotus 1-2-3 column designators (*A*, *B*, *C*, etc.) as *Megafile Manager* file column names (Lotus 1-2-3 Release 3 or Lotus/w uses names *A:A*, *A:B*, *B:A*, *B:B*, etc.; they will appear in *Megafile Manager* as *A_A*, *A_B*, *B_A*, *B_B*) when you select this option. This option is especially useful in working with Release 3 or Lotus/w files which will eventually be exported once again (using Lotus' column designators will preserve the identification of columns originating in different sheets).

From first row of selected range. *Megafile Manager* will use labels from the first row of the specified range when you select this option. If any label is longer than eight characters, only the first eight characters of the label will be used as the *Megafile Manager* column name.

Create (COL1, COL2, . . .). This option will result in the use of a combination of the prefix *COL* and a sequential number (e.g., *COL1*, *COL2*, etc.) as column names in the imported file.

Note that in addition to the file import operation, Lotus data can be accessed by *STATISTICA* via Clipboard (*STATISTICA* recognizes the Lotus Clipboard format and will properly interpret "formatted" values such *$100* or *2,000*).

Import Quattro File

Megafile Manager can import worksheets from Quattro Pro (the DOS and Windows versions; see the note about importing Quattro for Windows, below).

Range

By default, the entire active area of the worksheet is imported. A portion of the worksheet can be selected by defining the range (see below).

Columns. Specify the range of columns to be imported (*From* which column, *To* which column, inclusive).

Rows. Specify the range of rows to be imported (*From* which row, *To* which row, inclusive).

Column Names

This option specifies the source for *Megafile Manager* column names.

From spreadsheet. Use Quattro column designators (*A*, *B*, *C*, etc.) as *Megafile Manager* column names.

From first row of selected range. *Megafile Manager* will use labels from the first row of the specified range when you select this option. If any label is longer than eight characters, only the first eight characters of the label will be used as the *Megafile Manager* column name.

Create (COL1, COL2, . . .). This option will result in the use of a combination of the prefix COL and a sequential number (e.g., *COL1*, *COL2*, etc.) as column names in the imported file.

Note that in addition to the file import operation, Quattro for Windows data can be accessed by *STATISTICA* via Clipboard (*STATISTICA* recognizes the Quattro for Windows Clipboard format and will properly interpret "formatted" values such as *$100* or *2,000*).

Import SPSS File

Import SPSS Portable files, as well as all other files which follow the PFF (*Portable File Format*) convention, with this option.

SPSS column names will become *Megafile Manager* column names, and SPSS variable long names (up to 80 characters long) will become *Megafile Manager* column long names.

Import Value Labels

When you select this option, *Megafile Manager* will import the SPSS value labels (up to 40 characters). In the case of numeric SPSS columns, the program will also create text equivalents for the numeric values out of the first 8 characters of the value label.

Note that missing data in SPSS columns may be defined in different ways:

- If a single MD value has been assigned in a column, then that MD value will be used as the MD code in the *STATISTICA* variable.

- If a range of values have been defined as MD for a column in SPSS, then *Megafile Manager* imports the actual data values (to maximize the information extracted from the source file), and assigns the first value from the list of MD values as the MD code in the *Megafile Manager* file.

Import dBASE File

STATISTICA can import data files from dBASE versions III and IV. dBASE files contain a flag within each record indicating whether the record is to be read or not; *STATISTICA* will ignore all records marked as having been deleted. After you select a dBASE file from the *Select File to Import* dialog, the *Import Data From* dialog will open.

Now, in this dialog, click on the *Options* button to bring up the *dBASE Import Options* dialog (the same dialog is used for either version of dBASE that you are importing).

Range

By default, the entire active area of the data file is imported. A portion of the data file can be selected by defining the range (see below).

Columns. Specify the range of columns to be imported (*From* which column, *To* which column, inclusive).

Rows. Specify the range of rows to be imported (*From* which row, *To* which row, inclusive).

Import ASCII File

Megafile Manager can import both ASCII fixed (undelimited) and ASCII free (delimited) files. Each type of ASCII file has its own dialog of options. For more information on these options, see the following two topics.

Import ASCII Fixed File

After you select an ASCII fixed (fixed format) file from the *Select File to Import* dialog, the *Import Data From* dialog will open. By default, ASCII will be selected, and if the file name extension is *.fix*, then *Fixed* will automatically be checked as the ASCII version. If *Megafile Manager* is unable to recognize (from the file name extension) that the file is an ASCII *Fixed* file, then the *Free* ASCII version will be selected and you will need to change the type to *Fixed*.

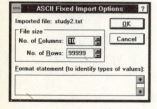

File Size

As in *ASCII Free Format Files*, in order for *Megafile Manager* to properly interpret the file, the number of columns and rows must be accurately specified.

Number of columns. Specify the number of columns (variables, fields) in the ASCII file here.

Number of rows. Specify the number of rows in the ASCII file here. If you are uncertain about the exact number of rows, you may overestimate (*Megafile Manager* will detect the actual length of the file during import). Note that each line in the ASCII source file may be up to 4,000 characters in length. This limit only applies to the individual line length and not the total length of a "row" of data: Each imported row may be represented by many lines of data in the source file.

Format Statement

This option explicitly defines for *Megafile Manager* the exact contents of the input ASCII file. (This is an important distinction: The format statement is a set of instructions for interpreting the structure of the input file, not a definition of the *Megafile Manager* columns to be created.)

The format statement is entered as a list of formats in the form *nXm*, where *n* is an integer multiplier indicating the number of times the format is to be repeated (no multiplier = 1); *X* is the column type; *m* is the column width [e.g., *10A8* means 10 text value (alphanumeric) fields, each 8 characters long].

The following column type specifiers are supported:

A - Text (Alphanumeric)

F - Float (also *R* - Real, equivalent to *Real 4B* in the *Column Specs* dialog, page 3721)

D - Double Float (also *DR* - Double Real, equivalent to *Real 8B* in the *Column Specs* dialog, page 3721)

I - Integer (also *NI* - Normal Integer, equivalent to *Int 2B* in the *Column Specs* dialog, page 3721)

S - Short Integer (also *SI*, equivalent to *Int 1B* in the *Column Specs* dialog, page 3721)

LI - Long Integer (also *J*, equivalent to *Int 4B* in the *Column Specs* dialog, page 3721)

L - Logical

The format specified as *logical* is expected to contain text designators of *true* and *false*. The following three conventions will be recognized by *Megafile Manager* when the data are imported:

TRUE or *FALSE* if the field length is 5 or more.

YES or *NO* if it is 3 or 4 characters long.

Y, *N*, or *T*, *F* respectively, if it is 1 character long.

Values of *TRUE* or *YES* will be imported as *1*, *FALSE* or *NO* will be imported as *0*.

The additional specifier *Tx* is available to instruct *STATISTICA* to jump to character number *x* in the current line in the input file (i.e., to skip to that position). This may be used to jump over unwanted characters or to return to an earlier character (if *x* specifies a character position which has already been read).

The slash character (/) may be used to indicate that the remainder of the line should be ignored (i.e., skip to the next line in the input file). Example: *10I3 A8 T55 3I5 /* specifies the import of 14 values: ten 3-digit integers, followed by an 8-character text string, then skip to the 55th character in the record, import three 5-digit integers and ignore the rest of the record. If the multiplier precedes a list of formats enclosed in parentheses, the list will be repeated as many times as specified in the multiplier [e.g., *10(2A3,I5)*].

Import ASCII Free File

After you select an ASCII *Free* file from the *Select File to Import* dialog, the *Import Data From* dialog will open (by default, ASCII *Free* will be selected). Now, in this dialog, click on the *Options* button to bring up the *ASCII Import Options* dialog.

File Size

As in *ASCII Fixed Format Files* (see above), the number of columns must be accurately specified.

Number of columns. Specify the number of columns (variables, fields) in the ASCII file.

Number of rows. Specify the number of rows in the ASCII file. If you are uncertain about the exact number of rows, you may overestimate (*Megafile Manager* will detect the actual length of the file during import). Note that each line in the ASCII source file may be up to 4,000 characters in length. This limit only applies to the individual line length and not the total length of a "row" of data: Each imported row may be represented by many lines of data in the source file.

Format Statement

This option explicitly defines for *Megafile Manager* the exact contents of the input ASCII file. (This is an important distinction: the format statement is a set of instructions for interpreting the structure of the input file, not a definition of the *Megafile Manager* columns to be created.) The format statement is entered as a list of formats in the form *nX* where *n* is an integer multiplier indicating the number of times the format is to be repeated (no multiplier = 1) and *X* is the column type (e.g., *40F* means 40 fields containing numeric (here float) values).

The following column type specifiers are supported:

12. MEGAFILE MANAGER - DIALOGS AND OPTIONS

A   - Text (Alphanumeric).

F   - Float (also *R* - Real, equivalent to *Real 4B* in the *Column Specs* dialog, page 3721).

D   - Double Float (also *DR* - Double Real, equivalent to *Real 8B* in the *Column Specs* dialog, page 3721).

I   - Integer (also *NI* - Normal Integer, equivalent to *Int 2B* in the *Column Specs* dialog, page 3721).

S   - Short Integer (also *SI*, equivalent to *Int 1B* in the *Column Specs* dialog, page 3721).

LI   - Long Integer (also *J*, equivalent to *Int 4B* in the *Column Specs* dialog, page 3721).

L   - Logical

The format specified as *logical* is expected to contain text designators of *true* and *false*. The following three conventions will be recognized by *Megafile Manager* when the data are imported:

TRUE or *FALSE* if the field length is 5 or more.

YES or *NO* if it is 3 or 4 characters long.

Y, *N*, or *T*, *F* respectively, if it is 1 character long.

Values of *TRUE* or *YES* will be imported as *1*, *FALSE* or *NO* will be imported as *0*.

Text fields (type *A*) must also include a length value from 1 to 255 immediately following the letter *A*, indicating the maximum possible length of text in this field.

The slash character (/) may be used to indicate that the remainder of the current line in the input file should be ignored (i.e., skip to the next line in the input file).

If the multiplier precedes a list of formats enclosed in parentheses, then the list of formats within the parentheses will be repeated the number of times specified by the multiplier.

For example, *2(2L a5)* is equivalent to *L L A5 L L A5* and specifies two *Logical* columns followed by a *Text* column that can hold up to five characters, then two more *Logical* columns and a final *Text* column (again up to five characters).

Separators

Within this option, you will be able to define the characters used in the input file as delimiters. (The final list of separators to be used will be the combination of the set of selected *Basic* and any *Additional* separators.)

Basic. Select the type of delimiter used in the input file from four predefined sets of separators (*CR* stands for carriage return, *LF* stands for line feed, and *FF* stands for form feed).

Standard set. *, ; <space> <tab> <CR\LF>*

Undefined. Do not use a predefined set (see *Additional* button, below)

Blank characters. *<space> <tab> <FF> <LF> <CR\LF>*

Non-numeric. (all but: *0 - 9 . - +*)

Additional

Click on this button to bring up the *ASCII Free Separators* dialog from which you can select an undefined character as the delimiter.

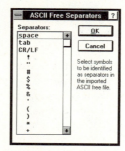

This dialog contains a list of characters that you can choose from. The characters or symbols listed in this dialog are as follows (*CR/LF* stands for carriage return/line feed):

<*space*> <*tab*> <*CR/LF*>

! " # $ % & ') * (, [\ + - .

0 1 2 3 4 5 6 7 8 9

] ^ _ / : ; < = > ? @ ` { | } ~

Upper case alphabet (i.e., *A B C . . . Z*)

Lower case alphabet (i.e., *a b c . . . z*)

Note that if you want to import numeric values, you should not specify any of the numbers (i.e., 0 through 9) or the "+", "-", or "." as separators.

Treat multiple separators as MD. When you select this option, *Megafile Manager* will interpret each pair of adjacent separators as an occurrence of missing data (an absent value) and will place the default missing data value (-9999) in the position between the adjacent separators. If this option is de-selected, then multiple separator characters are treated as one separator, and missing data must be explicitly coded into the ASCII file as a unique value (for instance *-9999*).

This option is particularly useful if, for instance, individual values in the data file are separated by spaces, with a variable number of spaces between values. If *spaces* are used as separators, then each pair of *spaces* would be seen as an occurrence of missing data, and the resulting file would be full of missing values.

Use quotation marks as text boundaries. Select this option if double (") or single (') quotation marks were used as text boundaries and the specified separator characters appear within the values of text columns in the input file (e.g., *"John Jones, Ph.D."*, uses the comma both as part of the text and as a separator after it). In this case, when *STATISTICA* imports the data, the quotation marks will be recognized only as boundaries around the text values, keeping the text values and the embedded separator character together (the quotation marks

will not be included as part of the imported text values).

Note that if a text string is to contain quotation marks as part of the string itself (e.g., as in the titles of books such as *"Moby Dick"*), then two methods may be used to import them:

- Select this option and enclose the entire text string within the alternate quotation mark (opposite of the embedded one); e.g., '*William Shakespeare, "King Lear", Act 2, Scene 1*' will be imported as *William Shakespeare, "King Lear", Act 2, Scene 1*.

- Select this option and then double the quotation mark wherever it is to be preserved; e.g., *"William Shakespeare's ""King Lear"", ""Macbeth"" and ""Hamlet"""* will be imported as *William Shakespeare's "King Lear", "Macbeth" and "Hamlet"*.

If this option is de-selected, then the character will be interpreted as a separator and not as a part of the text value, and the quotation marks will be imported as part of the text; e.g., '*William Shakespeare, "King Lear", Act 2, Scene 1*' will be imported as '*William Shakespeare* in one column, "*King Lear*" in another column, *Act 2* in different column and *Scene 1*' in another column.

Append *STATISTICA* File

When you select this option, the *Select STATISTICA File to Append* dialog will open in which you can select the file that will be appended to the *Megafile Manager* file. The selected *STATISTICA* file will be appended to the current *Megafile Manager* file after the last column in the *Megafile Manager* file. In order to adjust the order of columns or insert *STATISTICA* columns in a particular location in the *Megafile Manager* data set, use the *Move Columns* option (page 3719) after the *STATISTICA* data are appended. If one file has fewer rows than the other

12. MEGAFILE MANAGER - DIALOGS AND OPTIONS

file, then *STATISTICA* will fill the empty cells with missing values. For example, consider the following files:

Megafile Manager file (2 rows):

```
Col1    Col2    Col3
----    ----    ----
 34      21     102
 45      10      98
```

STATISTICA data file (5 cases):

```
Var1    Var2
----    ----
 78      90
 64     125
 24      36
 36      12
```

When the *STATISTICA* file is appended to the *Megafile Manager* file, the resulting *Megafile Manager* file will appear as follows (where the empty cells in the first three columns contain missing data):

```
Col1    Col2    Col3    Var1    Var2
----    ----    ----    -----   -----
 34      21     102      78      90
 45      10      98      64     125
                         24      36
                         36      12
```

Export Options

Megafile Manager files can easily be exported to other applications. When you choose the *Export* option in *Megafile Manager*, the *Export Data To* dialog will open in which you can select the export program.

Exporting is a very simple operation; in the *STATISTICA Megafile Manager* window, select the *Export* option from the *File* pull-down menu (or the startup panel), then follow the steps outlined below.

Step 1. Select the *Export Data* option from the *File* pull-down menu to open the *Export Data To* dialog.

Step 2. Select the type of file into which the data will be exported. You may click on the *Options* button to select from several exporting options (described under the specific export option, see below).

Step 3. Click the *OK* button in the latter dialog to export and store (you will need to select a file name) the file in the desired file format.

For more information on specific export options, see the respective export option (listed below).

- *STATISTICA* (see below; see also *Convert* for a quick method, page 3717)
- Excel (page 3714)
- Lotus, Symphony, and 3D Lotus files (page 3714)
- Quattro (page 3715)
- dBASE III and IV (page 3715)
- SPSS Portable File Format (page 3715)
- ASCII (free and fixed format, pages 3716 and 3717, respectively)

Note that because many applications cannot accommodate files as large as those that can be managed in *Megafile Manager* (32,000 columns, unlimited rows), it may be necessary to create a subset of the data file prior to exporting it (see page 3727).

Export to *STATISTICA* Data File

The quickest method to convert *Megafile Manager* data (or its subset) to the *STATISTICA* format is using the Clipboard (to paste a respective block of *Megafile Manager* data into a *STATISTICA* spreadsheet) or using the *Convert* button on the *Megafile Manager* window toolbar (see page 3717). Missing data codes from *Megafile Manager* columns will be used in creating the *STATISTICA* file.

STATISTICA files use a "double notation" format for variables: Each variable may contain paired numeric and text (up to 8 characters) values. Each numeric/text pair may also have a value label up to 40 characters in length. When importing a *STATISTICA* data file (see *Import STATISTICA file*, page 3703), "double variables" are automatically "split" (into two or three columns, depending on whether they are labeled) to become sequential *Megafile Manager* columns, which can be automatically put together in the process of exporting them back to *STATISTICA*. To create these "double variables" when exporting to the *STATISTICA* format, *Megafile Manager* columns are combined into pairs or triplets under the following conditions: A numeric column can be matched with one or two immediately following text columns which will become the text values and value labels for the *STATISTICA* variable. The first of the text columns may be up to 8 characters long and will become the text values for the *STATISTICA* variable; its column name must be the same as the numeric column name (with a *$* added to the end if the numeric column name is less than 8 characters, or if the name is 8 characters long, the second character should be replaced by a *$*). (Note that if the text values in this *$* column are longer than 8 characters, then only the first 8 characters will be exported.) The second of the text columns will become the value labels; its column name must be the same as the numeric column name with a *@* added to the end (if the numeric column name is less than 8 characters -- if the name is 8 characters, the second character should be replaced by *@*). A one-to-one correspondence between the numeric and text values in these columns must exist: Each numeric value must be paired with only one unique text string in the text column(s). This correspondence will be checked by *Megafile Manager* during the process of creating the *STATISTICA* file.

Megafile Manager also exports column long names as *STATISTICA* variable long names, case names (from a designated *Megafile Manager* column), and the one-line file header and the notes/comments (see *Header*, page 3724).

Export Text Columns

When you select this option, *Megafile Manager* will export unpaired text columns (columns to which no numeric column has been matched) into *STATISTICA* "double variables" by assigning a numeric code (consecutive integer) for each text value. If this option is de-selected, then no text variables will be created from the unpaired text columns. If this option is de-selected and the *Auto numeric/text match* option is also de-selected (see below), then all text columns will be excluded from the export file.

Export Case Names

Select this option in order to create *STATISTICA* case names from the first data column of the current *Megafile Manager* file.

Auto Numeric-/Text Match

Selecting this option will match *Megafile Manager* columns to become *STATISTICA* "double variables" based on pairing a numeric column with one or two text columns with related names (see above). De-selecting this option will cause no pairing of columns. If this option is de-selected and the *Export text columns* option is selected (see above), and then all text columns are automatically converted into new "double variables" with numeric (consecutive integers) and text values (up to 8 characters long).

Range

By default, the entire *Megafile Manager* file is imported. A portion of the file can be selected by defining the range (see below).

Columns. Specify the range of columns to be imported (*From* which column, *To* which column, inclusive).

Rows. Specify the range of rows to be imported (*From* which row, *To* which row, inclusive).

Export File to Excel

Excel files may contain up to 256 columns and 16384 rows. If the current *Megafile Manager* file is larger than this, then you can create *Subsets* of the file (see page 3727) and export each of those subsets separately. When you select the *Excel* option and click on the *Options* button in the *Export Data To* dialog, click on the *Options* button to open the *Excel Export Options* dialog.

Export Column Names

When this option is selected, *STATISTICA* will export the *Megafile Manager* column names into the first row of the Excel spreadsheet. If this option is de-selected, then the column names will not be included in the Excel file.

Export File to Lotus

Megafile Manager can export to Lotus 1-2-3 Release 1A, 2.01/2.2, Release 3, Lotus for Windows, and Symphony format files. Lotus 1-2-3 Release 2.01/2.2, and Release 3 files may contain up to 256 columns and 8,192 rows.

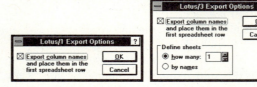

If the current *Megafile Manager* file is larger than this, then you can create *Subsets* of the file (see page 3727) and export each of those subsets separately.

Export Column Names

When this option is selected, *Megafile Manager* will export the column names into the first row of the Lotus spreadsheet. If this option is de-selected, then column names will not be included in the Lotus file.

Define Sheets

These options are available only if the Lotus 1-2-3 Release 3 or Lotus for Windows format is selected from the *Export Data To* dialog (see the *Lotus Export Options* dialog on the right, above).

How many. By default, all columns in the *Megafile Manager* file are exported as one worksheet. If desired, *Megafile Manager* columns may be equally divided among a specified number of worksheets. You can specify that number here.

By names. Alternatively, the *Megafile Manager* columns may be assigned to specific worksheets *by names*: Columns named *A_A* through *A_IV* will be placed in worksheet *A*, columns *B_A* through *B_IV* in worksheet *B*, etc. (This will be the case if the option *From Spreadsheet* was selected when originally importing column names from a Release 3 or Lotus for Windows worksheet; see page 3705.)

Export File to Quattro

You may export *Megafile Manager* files to Quattro. Quattro files may contain up to 256 columns and 8,192 rows. When you select the *Quattro* option from the *Export Data To* dialog, click on the *Options* button to open the *Quattro Export Options* dialog.

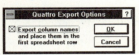

Export Column Names

When this option is selected, *Megafile Manager* will export the column names into the first row of the Quattro spreadsheet. If this option is de-selected, then column names will not be included in the Quattro file.

Export File to dBASE

Megafile Manager exports to dBASE III+ and IV file formats. dBASE III+ files are limited to 128 fields and dBASE IV to 255 fields. If the current *Megafile Manager* file is larger than this, then you can create *Subsets* of the file and export each of those subsets separately. Field names in dBASE must begin with a letter A to Z and may contain only uppercase letters (during export, lowercase letters in *Megafile Manager* column names are converted to uppercase), as well as numerals and the underscore (_). The *Verify* option from the *Megafile Manager Analysis* pull-down menu may be used to check *Megafile Manager* column names for agreement with these conventions and to correct any which do not comply (see page 3729).

Export File to SPSS

This *Megafile Manager* option exports to the SPSS Portable File Format (also compatible with SAS). When importing an SPSS Portable File (see page 3707), data values and value labels are automatically "split" to become sequential *Megafile Manager* columns, which can be automatically put together in the process of exporting them back to SPSS. A numeric (or text column up to 8 characters long) may be paired with a long text column with a related name to become an SPSS Portable File column with value labels. If the name of the data column is less than 8 characters, the name of the related column has the same name with either an @ or a $ added to the end (in the same manner as data imported from *STATISTICA*, page 3703); if the name is 8 characters long, then the second character is replaced with an @ or $. Each numeric or text value in the data column must correspond to a unique text value in the matching text column. Unpaired text columns will become text variables without value labels. *Megafile Manager* files with up to 2,000 columns may be exported to an SPSS Portable File; however, on some systems, SPSS imposes lower file size limits (check the SPSS manual for the particular system). If the current *Megafile Manager* file is larger than 2,000 columns, then you can create *Subsets* of the file (page 3727) and export each of those subsets separately.

SPSS column names may be up to 8 characters long, must begin either with an uppercase letter (A to Z; lowercase letters in *Megafile Manager* column names are automatically converted to uppercase during export), or the characters $ or @, and may also contain digits, periods, #, and underscore (_). Certain names are reserved as keywords and may not be used as column names, including *ALL*, *AND*, *BY*, *EQ*, *GE*, *GT*, *LE*, *LT*, *NE*, *NOT*, *OR*, *TO*, and *WITH*.

The *Verify* option from the *Megafile Manager Analysis* pull-down menu may be used to check *Megafile Manager* columns names for agreement with these conventions and to correct any which do not comply (see page 3729).

Export File to ASCII

Megafile Manager can export to both ASCII fixed (undelimited) and ASCII free (delimited) files. Each type of ASCII file has its own dialog of options. For more information on these options, see the appropriate section (below and on page 3717).

Export File to ASCII Free Format

Megafile Manager can export to ASCII free format files. If you choose instead to export to *ASCII fixed format files* (page 3717), then values which are shorter than the format are padded with blanks so that the absolute position of each value is "fixed" within the line in the output file. Free format (delimited) ASCII files are created without padding values to fill out column formats since user-specified characters are used to separate or delimit successive values. When you select the *ASCII Free* option from the *Export Data To* dialog, click on the *Options* button to open the *ASCII Free Export Options* dialog.

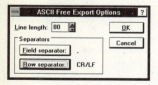

Line Length

The number of columns per line in the output file is determined by the specified line length. Enter the length of the line in the ASCII file here (values can be from 80 to 4,000). If the specified line length is insufficient to contain all columns, then the line will be broken after the last complete column (i.e., a column will not be divided across lines). If unbroken records are desired (i.e., each record occupies only one line in the output file), specify a line length long enough to accommodate the row length.

Separators

Field separator. Click on this button to bring up the *ASCII Free Separator* dialog from which you can select a single character to be used to separate successive values in the output file (by default, a comma will be used as the field separator).

The numeric characters *0* through *9*, period (decimal point), "+" and "-" are not permitted as delimiters.

Row separator. The *Row Separator* serves to indicate the end of each of the original rows of data (one row of data may extend across several lines of the output file). Click on this button to bring up the *ASCII Free Separator* dialog (see above) from which you can select a single character to be used to indicate the completion of each row of data in the output file (by default, a comma will be used as the row separator). The numeric characters *0* through *9*, period (decimal point), "+" and "-" are not allowed as row separators. Note that if the length of the row in the *Megafile Manager* file is longer than the specified output *Line Length* (see above), then the row will extend across one or more lines in the output file.

12. MEGAFILE MANAGER - DIALOGS AND OPTIONS

Export File to ASCII Fixed Format

Unlike *ASCII free format file* (see above), fixed format (undelimited) ASCII files will be created using the current column formats (values which are shorter than the format are padded with blanks so that the absolute position of each value is "fixed" within the line in the output file). When you select the *ASCII Fixed* option from the *Export Data To* dialog, click on the *Options* button to open the *ASCII Fixed Export Options* dialog.

Line Length

The number of columns per line in the output file is determined by the specified line length. Enter the length of the line in the ASCII file here (values can be from 80 to 4,000). If the specified line length is insufficient to contain all columns, then the line will be broken after the last complete column (i.e., a column will not be divided across lines). If unbroken records are desired (i.e., each record occupies only one line in the output file), specify a line length no less than the sum of all column formats.

Convert to a STATISTICA File

This option will allow you to quickly open the *STATISTICA Export Options* dialog in order to convert (i.e., export) a *Megafile Manager* file to a *STATISTICA* data file. For more information on these export options, see *STATISTICA Export Options*, page 3712.

Print MFM Data

When you select this option, the *Megafile Manager: Print* dialog will open in which you can choose from several printing options (described below).

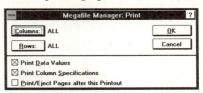

Shortcut. Click on the *Print* Toolbar button (see below) to bypass the *Print* dialog (i.e., the previously selected settings or the default settings will be used) and print the *Megafile Manager* file. Note that if a block is selected in the file, then that block will be printed; otherwise, the previously specified columns and rows will be printed.

Columns

Click on this button to open the *Megafile Manager: Select Columns* dialog (page 3697) in which you can select the columns to be printed.

Rows

Click on this button to open the *Megafile Manager: Select Rows* dialog (page 3698) in which you can select the rows to be printed.

Print Data Values

Select this option to print the *Megafile Manager* file data values for the selected rows and columns.

Print Column Specifications

Select this option to print the *Megafile Manager* file column specifications for the selected columns.

MM - 3717

Copyright © StatSoft, 1995

12. MEGAFILE MANAGER - DIALOGS AND OPTIONS

Print/Eject Pages after this Printout

When selected, this option will cause the requested printout to automatically be sent to the printer or output file when you click OK in this dialog or when you press the *Print* button on the toolbar (any other text previously left in the internal buffer will also be printed). Thus, each requested printout (table) will start on a new page. If this option is de-selected, then output will be stored in the internal buffer and sent to the printer depending on the current setting of the check box *Automatically Eject Each Filled Page* (either whenever there is enough to fill one page or all at once later, e.g., when you exit the module or when you empty the buffer by using the option *Print/Eject Current Pages*). Note that this option will only apply locally (temporarily) to the current *STATISTICA* module and the current session, whereas, selecting the *Print/Eject Pages after Each Printout* option in the *Page/Output Setup* dialog (see Volume I) will apply globally (permanently).

Add Columns

Add columns to the *Megafile Manager* file by designating the number of new columns to add, as well as where to add them (after which column) in the *Add Columns* dialog.

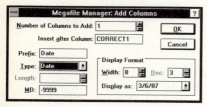

You can click on the column edit field in the this dialog to bring up the list of columns in the current *Megafile Manager* file and select a column that the new column(s) will be placed after (see *Megafile Manager: Select Column*, page 3698).

Number of Columns to Add

Choose the desired number of columns to add to the current *Megafile Manager* file. You may include up to 32,000 columns (with up to 255 characters per column) in *Megafile Manager*.

Insert After Column

Designate the location (column) in the *Megafile Manager* file that the new column(s) will be added after. The new columns will be added to the right of the column specified here.

How to insert new columns before column 1. In order to insert a new column or block of columns before column 1, first insert the column(s) after column number 1 and then move (page 3719) column number 1 to the right of the inserted column(s) (you can also use the *Copy Columns* option (page 3720) to first copy the new columns and then move them).

Prefix

You can specify the prefix for the new column names which will be followed by consecutive numbers (e.g., *NEW1*, *NEW2*, *NEW3*, ...). In order to change the names of individual columns, double-click on the column name in the *Megafile Manager* spreadsheet to open the *Column Specs* dialog (page 3721).

Type

Specify here the type of data values for the new columns. Each type is described below:

Type	Type of Values	Examples	Bytes
text	any char.	John Brown	1-255
short integer (Int 1B)	±127	-98	1
integer (Int 2B)	±32767	9834	2
long integer (Int 4B)	±2147483647	12345678	4
float (Real 4B)	±3.4*10^{+38}	1234.5678	4
double float (Real 8B)	±1.7*10^{+308}	123456.78974	8
logical	yes/no/ true/false	YES	1
date	dates from Jan.1,1900	02/08/55	8

Length

This option will become available when you select *Text* as the column type (see above). Enter here the desired length (up to 255 characters) of the text values.

MD

Enter a specific value for the missing data code (value of the column that indicates data points for which no data are available) or accept the default code of -9999 (the default MD code for all *text*, *date*, and *logical* values in *Megafile Manager* is fixed and cannot be changed). Missing data values are displayed as blanks in the cells of the *Megafile Manager* file (the value can be viewed in the *Megafile Manager* window toolbar *Show Window* when a blank cell is highlighted) and are ignored in all data analyses.

Display Format

The format determines the length of the field used to display (or print in reports) values from the respective column; the format does not affect the contents or precision of values stored in *Megafile Manager* data files and can be adjusted at any point.

For *Real* (*4B* or *8B*) type columns, the format also determines the number of decimal places displayed; if a numeric value cannot fit in its format, *Megafile Manager* uses "compressed scientific notation" (e.g., a value of *30,000* will be displayed in a 3-character format as *3E5*, i.e., $3*10^5$); a string of asterisks (e.g., *****) is displayed when a value cannot be displayed even in compressed notation.

The formats that you specify here will be applied to each column added to the *Megafile Manager* file. In order to change the format for an individual column, double-click on the column name in the *Megafile Manager* spreadsheet to open the *Column Specs* dialog (page 3721).

Width. Edit the width of the values of the columns to be added here. The default format width is 8 characters.

Decimal. Specify the number of decimal places for the added columns. The default format is 3 decimal places.

Display as. This option will become available when you select either *Dates* or *Logical* as the column type (see above). You can click on the combo box to select the desired *Date* or *Logical* display for those values.

Move Columns

Move one or more columns in the *Megafile Manager* file by designating the range (inclusive) of columns to be moved and the location (column to insert after) in the *Move Columns* dialog.

You can easily replace one column name with another by clicking on the column edit field in the *Move Columns* dialog to bring up the list of columns

12. MEGAFILE MANAGER - DIALOGS AND OPTIONS

in the current *Megafile Manager* file from which you can select the desired column.

From Column

Designate the first column (inclusive) in the block of columns to be moved. (Note that if you are moving only one column, then designate the same column in both the *From Column* and *To Column* lists.)

To Column

Designate the last column (inclusive) in the block of columns to be moved.

Insert After

Designate the location (column) in the *Megafile Manager* file that the column(s) will be moved after. (you cannot insert a block of columns *after* a column that is included in the range of columns to be moved). The block of columns will be "cut" from their current file location and moved to the right of the column specified here.

How to insert before column 1. In order to insert a column or block of columns before column 1, first move the column(s) after column number 1 and then move column number 1 to the right of the moved column(s). You can also use the *Copy Columns* option (see below) to first copy the columns and then move them.

Copy Columns

Place a copy of the specified column(s) at a designated location by specifying the range (inclusive) of columns to be copied and the *Megafile Manager* file location (insert after which column) in the *Copy Columns* dialog.

The copy will include not only the data but also the format, long name, display specifications, etc.

You can easily replace one column name with another by clicking on the column edit field in the *Copy Columns* dialog to bring up the list of columns in the current *Megafile Manager* file from which you can select the desired column.

From Column

Designate the first column (inclusive) in the block of columns to be copied. (Note that if you are copying only one column, then designate the same column in both the *From Column* and *To Column* lists.)

To Column

Designate the last column (inclusive) in the block of columns to be copied.

Insert After

Designate the location (column) in the spreadsheet that the column(s) will be placed after copying (you cannot insert a block of columns *after* a column that is included in the range of columns to be copied). The copied columns will be located to the right of the column specified here.

How to insert before column 1. In order to insert a column or block of columns before column 1, first copy the column(s) after column number 1 and then move column number 1 to the right of the copied column(s) using the *Move Columns* option (see above).

Delete Columns

One or more columns can be deleted from the current *Megafile Manager* file when you designate the range (inclusive) of columns to be deleted in the *Delete Columns* dialog.

You can easily replace one column name with another by clicking on the column edit field in the

MM - 3720

Copyright © StatSoft, 1995

Delete Columns dialog to bring up the list of columns in the current *Megafile Manager* file from which you can select the desired column.

Unlike deleting (clearing) the contents of a highlighted range or block of values in the *Megafile Manager* file, this option will remove both the contents of the column(s) and the column(s) itself; thus the subsequent columns will be moved to the left and the size of the file will decrease.

From Column

Designate the first column (inclusive) in the block of columns to be deleted. (Note that if you are deleting only one column, then designate the same column in both the *From Column* and *To Column* lists.)

To Column

Designate the last column (inclusive) in the block of columns to be deleted.

Edit Column Specs

Specifications of the currently highlighted column (i.e., column name and format, missing data value, and label) can be edited in the *Column Specs* dialog.

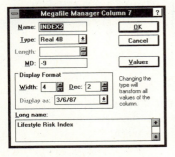

Shortcut. Double-click on a column name in order to open the *Column Specs* dialog for that column.

Name

Edit the column name here. Column names can be up to 8 characters long and may contain all keyboard characters, including upper and lower case letters, numbers, spaces, punctuation, etc.). In addition to their names, columns can also be referred to by their consecutive numbers, (e.g., in all cases where a column or a list of columns is to be entered).

Type

Specify here the type of data values for the new columns. Each type is described below:

```
               Type of
Type           Values           Examples       Bytes
----           ----------       ----------     -----
text           any char.        John Brown     1-255
short
integer
(Int 1B)       ±127             -98            1
integer
(Int 2B)       ±32767           9834           2
long
integer
(Int 4B)       ±2147483647      12345678       4
float
(Real 4B)      ±3.4*10+38       1234.5678      4
double
float
(Real 8B)      ±1.7*10+308      123456.78974   8
logical        yes/no/          YES            1
               true/false
date           dates from       02/08/55       8
               Jan.1,1900
```

The general type of data value (e.g., *Integer* for any integer type) that you choose here is displayed in the third line of the column header in the *Megafile Manager* spreadsheet.

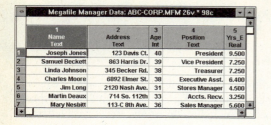

Length

This option will become available when you select *Text* as the column type (see above). Enter here the desired length (up to 255 characters) of the text values.

MD

Enter a specific value for the missing data code (value of the column that indicates data points for which no data are available) or accept the default code of -9999 [the default MD code for all *text*, *date*, and *logical* values in *Megafile Manager* is fixed (set to a so-called "null character") and cannot be changed]. Missing data values are displayed as blanks in the cells of the *Megafile Manager* file (the value can be viewed in the *Megafile Manager* window toolbar *Show Window* when a blank cell is highlighted) and are ignored in all data analyses.

Display Format

The format determines the length of the field used to display (or print in reports) values from the respective column; the format does not affect the contents or precision of values stored in *Megafile Manager* data files and can be adjusted at any point.

For *Real* (*4B* or *8B*) type columns, the format also determines the number of decimal places displayed; if a numeric value cannot fit in its format, *Megafile Manager* uses "compressed scientific notation" (e.g., a value of *30,000* will be displayed in a 3-character format as *3E5*, i.e., $3*10^5$); a string of asterisks (e.g., ***) is displayed when a value cannot be displayed even in compressed notation.

Width

Edit the width of the values of the columns to be added here. The default format width is 8 characters.

Decimal

Specify the number of decimal places for the added columns. The default format is 3 decimal places.

Display as

This option will become available when you select either *Dates* or *Logical* as the column type (see above). You can click on the combo box to select the desired *Date* or *Logical* display for those values.

Long Name

For each column, you can assign a column long name up to 128 characters long. The label can contain notes, including any printable characters.

Values

Click on this button to open the *Column Values* window (page 3698) in which you can view the values of the current column.

Add Rows

Add rows to the *Megafile Manager* file by designating the number of new rows to add, as well as where to add them (after which row) in the *Add Rows* dialog.

Number of Rows to Add

Choose the desired number of rows to add to the current *Megafile Manager* file. The number of rows that you can add to a file is unlimited.

Insert After Row

Designate the row number after which the new row(s) will be added. The added rows will be located after the row specified here.

How to insert before row 1. In order to insert a row or block of rows before row 1, first insert the row(s) after row number 1 and then move row number 1 below the inserted row(s) (see below); you can also use the *Copy Rows* option (page 3723) to first copy the rows and then move them.

Move Rows

Move one or more rows in the *Megafile Manager* file by designating the range (inclusive) of rows to be moved and the location (row to insert after) in the *Move Rows* dialog.

From Row

Designate the first row (inclusive) in the block of row to be moved. (Note that if you are moving only one row, then designate the same row in both the *From Row* and *To Row* lists.)

To Row

Designate the last row (inclusive) in the block of rows to be moved.

Insert After

Designate the location (row) in the *Megafile Manager* file that the row(s) will be moved after. The moved rows will be located after the row specified here.

How to insert before row 1. In order to insert a row or block of rows before row 1, first move the row(s) after row number 1 and then move row number 1 below the moved row(s) [you can also use the *Copy Rows* option (see below) to first copy the rows and then move them].

Copy Rows

Place a copy of the specified row(s) at a designated location by specifying the range (inclusive) of rows to be copied and the *Megafile Manager* file location (insert after which row) in the *Copy Rows* dialog.

From Row

Designate the first row (inclusive) in the block of rows to be copied. (Note that if you are copying only one row, then designate the same row in both the *From Row* and *To Row* lists.)

To Row

Designate the last row (inclusive) in the block of rows to be copied.

Insert After

Designate the location (row) in the *Megafile Manager* file that the row(s) will be placed after copying. The copied rows will be located after the row specified here.

How to insert before row 1. In order to insert a row or block of rows before row 1, first copy the row(s) after row number 1 and then move row number 1 below the copied row(s) using the *Move Rows* option (see above).

Delete Rows

One or more rows can be deleted from the current *Megafile Manager* file when you designate the range (inclusive) of rows to be deleted in the *Delete Rows* dialog.

Unlike deleting (clearing) the contents of a highlighted range or block of values in the *Megafile Manager* file, this option will remove both the contents of the row(s) and the row(s) itself, thus the subsequent rows will be moved up and the size of the file will decrease.

From Row

Designate the first row (inclusive) in the block of rows to be deleted. (Note that if you are deleting only one row, then designate the same row in both the *From Row* and *To Row* lists.)

To Row

Designate the last row (inclusive) in the block of rows to be deleted.

File Header

Make notes or comments about the current *Megafile Manager* file in the *Megafile Manager File Header and Info Page* dialog.

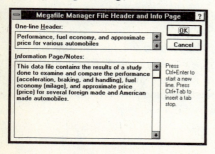

You can include either a one line file header, several lines of notes, or both.

One Line Header

Enter a line (up to 77 characters) of comment about the file here. The *One Line Header* is saved as part of the file and is printed along with the data when the *Print Column Specifications* option is selected in the *Megafile Manager Print* dialog (see page 3717).

File Information/Notes

Make a comment about the *Megafile Manager* file here. These notes can be up to 1,970 characters in length and can include any printable characters. The *File Information/Notes* are printed along with the data when the *Print Column Specifications* option is selected in the *Megafile Manager Print* dialog (see page 3717).

Merge Files

Merge operations combine data from two files, either by pasting files together (*column* and *row* merge) or by replacing data in a segment of the first file with data from a second file (*overlay* merge).

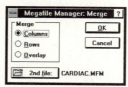

After selecting the type of merge and entering the name of the second *Megafile Manager* file (which will be merged to the current file), the respective merge dialog will open (see below).

Merge

Select one of three merging options.

12. **MEGAFILE MANAGER** - DIALOGS AND OPTIONS

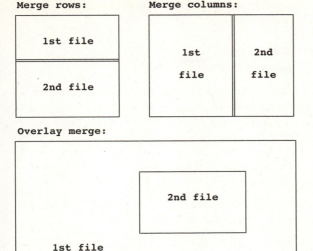

For a more detailed description and example of a specific merging option, see the respective section, below.

- Merge Columns (see below)
- Merge Rows (page 3726)
- Overlay Merge (page 3727)

2nd File

Click on this button to open the *Second File* dialog from which you can select a file to merge with the current file.

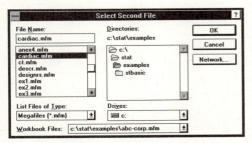

Please note that you cannot merge a file to itself (i.e., you cannot enter the current file as the second file name). If you wish to "double" the file, use the *Save As* option from the *File* pull-down menu to create a second copy of the file. For additional about the options in this dialog (e.g., the *Workbook*

Merge Columns

When you select the *Merge Columns* option from the *Merge Files* dialog (see above), *Megafile Manager* will add the second file's columns after the columns of the first file.

The following options (available from the *Merge Columns* dialog) are used to align (match) the rows of the second file with those of the first.

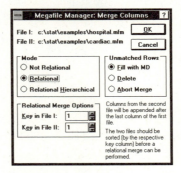

Mode

If the number and order of rows in the two *Megafile Manager* files are not precisely the same (i.e., if at least one row is missing in the second file or there is one row that has no equivalent in the first file), then all subsequent data would be merged with the wrong rows.

In order to avoid this problem, you will need to specify a mode of merging here. When relationally merging rows from two files, you need to specify a *key* (row identifier) column in each file; then for each row, the program will check the values of this key in both files and merge the rows only if their respective keys match.

MM - 3725

Copyright © StatSoft, 1995

12. MEGAFILE MANAGER - DIALOGS AND OPTIONS

Not relational. Columns of the second file are simply added alongside those of the first when this mode is selected (this is the default mode).

Relational. When you select this mode, the rows from the second file will be matched with those of the first file, based on the values of a specified *key* (see below).

Relational hierarchical. This mode differs from the simple relational mode (see above) in the handling of multiple records with the same key value in either the primary or secondary file.

In the standard relational mode (see above), successive records with identical key values will be merged. If there are uneven numbers of records with identical key values in the two files, missing data are added to "pad" the file with the lesser number of records.

In contrast, in the *Relational hierarchical* mode the file is padded with the values found in the last identical key record that was matched. For example:

Standard *Relational Merge*, unmatched cases filled with missing data:

```
File 1          File 2          Merged File
---------       ---------       --------------------
Key1 Var1       Key2 Var2       Key1 Var1   Key2 Var2
----  ----      ----  ----      ----  ----  ----  ----
 1    1          1    1          1    1      1    1
 2    4          1    2          Miss Miss   1    2
 2    5          1    3          Miss Miss   1    3
 3    2          2    4          2    4      2    4
 3    3          2    5          2    5      2    5
                                 3    2      Miss Miss
                                 3    3      Miss Miss
```

Hierarchical Relational Merge, unmatched cases filled with missing data:

```
                                Merged File
                                --------------------
                                Key1 Var1   Key2 Var2
                                ----  ----  ----  ----
                                 1    1      1    1
                                 1    1      1    2
                                 1    1      1    3
                                 2    4      2    4
                                 2    5      2    5
                                 3    2      Miss Miss
                                 3    3      Miss Miss
```

Note that the sequence of rows is not changed during merging; therefore, each file should be sorted (see *Sort Data*, page 3728) in order of its key column before merging.

Relational Merge Options

When you choose either the *Relational Merge* or *Relational Hierarchical Merge* option, you will need to specify a key column for each file in order to merge the two files. *Megafile Manager* will expect the two files to be sorted (see *Sort Data*, page 3728) by their respective key column before a relational merge can be performed.

Key in file 1. Enter here the number of the column to be used as the key in the first file.

Key in file 2. Enter here the number of the column to be used as the key in the second file.

Unmatched Cases

Select one of three ways of dealing with unmatched rows when the two files are merged. Unmatched rows may result from unequal numbers of rows in the merged files or because some of the rows do not meet the *Relational merge* criteria (see above).

Fill with MD. Unfilled columns in unmatched rows are padded with missing data when you select this option.

Delete. Rows from either file which cannot be matched will be removed from the merged file if you select this option.

Abort merge. The presence of unmatched rows in either file will cause an error message to be displayed and the merge procedure to be abandoned.

Merge Rows

Select this option from the *Merge Files* dialog (page 3724) when you want to merge rows from the second file after the rows of the first file. (Note that all column types in the second file must match those in the first file.)

Overlay Merge

This merging option will replace data in the first file with data from the second file starting in the user-defined file location (column/row). When you select this option in the *Merge Files* dialog (page 3724), the *Overlay Merge* dialog will open.

Update Columns

Start with column. The column number that you specify here will determine the starting position in which the data from the second file will occupy within the first file by matching column numbers.

Based on names. Alternatively, you can select this option in order to determine the starting position that the data from the second file will occupy within the first file by matching column names. When you select this option, *Megafile Manager* will search for identical column names and will begin the overlay merge at the first identical column name that it encounters.

Update Rows

Start with row. When you select this option, you can specify the starting row in the first file to receive data from the second file.

Based on keys (relational). Rows from the second file will be matched to rows in the first file based on values of *key columns* (see below) when you select this option. *Megafile Manager* will expect the two files to be sorted (see *Sort*, page 3728) by their respective key column before an overlay merge can be performed.

Key in file I. Specify the key column from the first file.

Key in file II. Specify the key column from the second file.

Subset

Use this *Megafile Manager* option to extract a section (subset) of the current file and store the subset under a new *Megafile Manager* name. When you select this option from the *Analysis* pull-down menu, the *Subset* dialog will open.

After you have created this subset file, you will need to save it under a different file name using the *Save As* option from the *File* pull-down menu.

Columns

Click on this button to bring up the *Select Columns* dialog (page 3697) from which you may select the columns to be included in the subsetted file.

Rows

Click on this button to open the *Select Rows* dialog (page 3698) from which you may select a subset of rows to be included in the subsetted file.

Sort Data

Megafile Manager performs hierarchical (nested) sorting on the basis of user-specified key columns (i.e., if two rows have identical values in a given key, the values of the next key are used to determine ranking). When you select the *Sort* option from the *Analysis* pull-down menu, the *Sort Options* dialog will open.

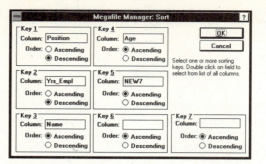

You can choose up to three key columns by which to sort in this dialog (up to seven if you press the *More Keys* ... button). Each key column has the same set of options (described below).

Column

Type in the key column name or double-click on this window to open the *Select Column* dialog (page 3697). The *Select Column* dialog will display a list of columns from which you can select the key column.

Sorting Order

Choose to use the *Ascending* or *Descending* sort order.

More Keys

When this dialog first opens, only three keys are displayed (see above).

If you need to enter more keys than this, click on the *More Keys* button to increase the number of displayed keys to seven (the maximum).

Optimize Size

Select this option in order to change both column type and format for best fit to the selected columns. When you select this option, the *Optimize Size* dialog will open in which you can select the columns to be optimized.

This option is useful in reducing wasted space in files (e.g., a column type of *Text-255* characters which contains less than 80 characters of data). Some imported files (such as from the SPSS Portable File Format) may contain large amounts of wasted space, as all numeric columns are specified as *Double Float* (i.e., *Real 8B*); optimization compacts the column formats for the most efficient use of disk space. For example, if data in a column specified as *Double Float* (*Real 8B*; format F16.6) are only 2 digit integers, optimization will change the column type to *Short Integer* (i.e., *Int 1B*), saving seven bytes per record (87.5% savings).

Note. Values in *Double Float* columns which are within the range of normal *Float* values ($\pm 3.4*10^{38}$) will be converted to *Float* values during optimization and will be rounded to 7 significant digits. To preserve values with more than 7 significant digits, do not include them in the list of columns to be optimized. Also note that MD codes are considered in the determination of optimal formats and they are preserved: For example, if an *Integer* column contains data values which fall within the range of *Short Integers* (±127), but contains a MD value which is outside this range (e.g., -9999), then the column will not be converted to *Short Integer*.

Verify Column Names

This verification facility may be used to check for (and adjust) the "fit" of column *Formats* (see below) to the data, to check for duplicated column names, or to check their compliance with selected standards (e.g., *STATISTICA*, dBASE, SPSS). When you select this option from the *Analysis* pull-down menu, the *Verify* dialog will open.

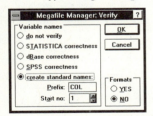

Column Names

This option will check the correctness of the column names according to the requirements of *STATISTICA*, dBASE, or SPSS, or it will create standard names. Select one of the options described below. If an invalid or duplicated name is found, an appropriate *Name Verification* dialog will open (see below), allowing you to correct the name (either of the two names can be edited if a duplicate is found).

Do not verify. Select this option if you are interested in only verifying column formats.

***STATISTICA* correctness.** This option is used to check all column names for duplications, either for compliance with the requirements of *STATISTICA BASIC* (see Volume V), *MML*, or prior to exporting the file.

dBASE correctness. When selected, this option checks for compliance with the requirements of dBASE. All columns must have names which may contain the letters *A-Z*, *a-z*, the digits *0-9*, and the underscore (_). Each column name must begin with a letter, and all column names must be unique (which may be checked by using the *STATISTICA Correctness* option above).

SPSS correctness. When selected, this option checks for agreement with the requirements of SPSS. All columns must have names which may contain the letters *A-Z*, *a-z*, the digits *0-9*, underscore (_), period (.), @, #, and $. Each column name must begin with either a letter, $, or @, and cannot be one of the reserved SPSS keywords (*ALL, AND, BY, EQ, GE, GT, LE, LT, NE, NOT, OR, TO,* and *WITH*).

Create standard names. When selected, this option assigns "standard" names (a prefix plus a sequential number) to all columns through the following options.

- *Prefix*: Specify a five-letter prefix for standard column names.

- *Start Number*: Select the starting value for the numeric portion of the names (useful when you intend to merge the file with another *Megafile Manager* file).

Formats

When you select *Yes*, *STATISTICA* will examine all specified columns to determine whether the column formats are an appropriate "fit" to the data values. For example, the value 1234.5 entered in a column whose format is *4.3* would result in the column format being expanded to *8.3* in order to accommodate the data (represented as 1234.500). Also, if the format is longer than any of the values in the column (e.g., the value 12.34 in a column with a *15.7* format), the format length will be adjusted to the actual length of the longest value (however, the number of decimal places remains unchanged).

Incorrect Variable Name

If incorrect names are encountered while verifying the correctness of variable names in the *Verify Names/Format* dialog (see above), then the *Name Verification* dialog will open allowing the incorrect names to be edited.

Variable Name

The incorrect variable name will be listed in an edit field next to its variable number. You can edit the name here.

Ignore

Click this button if you want to leave the variable name as it appears (uncorrected).

Accept

Click this button to accept the edited variable name and proceed to the next incorrect name.

Break

Clicking this button will stop the verification process (all changes to names up to this point will be saved).

Undo All

Click this button to stop the verification procedure and undo all changes made during verification.

Descriptive Statistics

Select this option in order to compute the mean, standard deviation, standard error, variance, skewness, and kurtosis for numeric and logical columns of data in *Megafile Manager* files. The results of these analyses are saved as *Megafile Manager* files which may then be used for further processing (e.g., editing, exporting, merging, etc.).

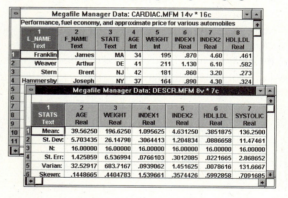

The following options are available when you select *Descriptives* from the *Analysis* pull-down menu.

The descriptive statistics output is saved to a *Megafile Manager* file (you will be prompted to select the file name). Text variables are automatically excluded from the analysis. The first column of the output file always contains names of the descriptive statistics (*Mean*, *Std.Dev*, etc.). All other columns are *Double Float* and they store the values of the respective statistics with 15-digit precision.

Columns

Click on this button to open the *Select Columns* dialog (page 3697) in which you can select the columns for the analyses.

Rows

Click on this button to open the *Select Rows* dialog (page 3698) in which you can select the rows for the analyses.

Descriptive Statistics

Select the desired descriptive statistics here.

Mean. Set this check box to compute the means of the selected columns.

Standard deviation. Set this check box to compute the standard deviations. The standard deviation is computed as the square root of the sum of squared deviations (from the mean) divided by $n-1$.

Valid n. Set this check box to compute the number of valid rows for each selected column.

Standard error of the mean. The standard error of the mean is computed as the standard deviation divided by the square root of n.

Variance. Set this check box to compute the variances. The variance of a column is computed as the sum of squared deviations (from the mean) divided by $n-1$.

Skewness. The skewness is a measure of the symmetry of the distribution of values. If the distribution is symmetrical, then the skewness is equal to zero (see, *Basic Statistics and Tables*, Volume I).

Kurtosis. The kurtosis is a measure of "peakedness" of the distribution of values. If the distribution follows the standard normal distribution, then the value of the kurtosis is zero (see, *Basic Statistics and Tables*, Volume I).

Correlations

This option computes the Pearson product-moment correlations between selected columns.

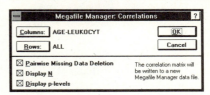

The following options are used to select the range of data to be analyzed and the output format.

Columns

Click on this button to open the *Select Columns* dialog (page 3697) in which you can select the columns of the current *Megafile Manager file* to be included in the correlation analysis.

Rows

Click on this button to open the *Select Rows* dialog (page 3698) in which you can select the rows of the current *Megafile Manager file* to be included in the correlation analysis.

Pairwise Missing Data Deletion

If this check box is set, then each correlation will be computed from all rows that have complete (valid) data for the respective pair of columns (i.e., *pairwise deletion of missing data*), possibly resulting in unequal *n* for different correlations. If this check box is not set (i.e., *casewise deletion of missing data*), then only the rows that have complete (valid) data for *all* columns that are currently selected for the analysis will be included in the computation of the correlations.

Display *n*

If this check box is set, then clicking *OK* in this dialog will produce an expanded format correlation matrix (see below); each cell in the matrix will not only contain the respective correlation coefficient but also the valid *n* (if the *Pairwise deletion of missing data* check box is set).

Display *p*-levels

If this check box is set, then clicking *OK* in this dialog will produce an expanded format correlation matrix (see below); each cell in the matrix will not only contain the respective correlation coefficient but also the *p*-level (if the *Pairwise missing data deletion* check box is set).

Note. The correlation matrix output is saved to a *Megafile Manager* file (you will be prompted to select the file name). The format of the output varies according to the requested statistics. The first column always contains the column names (and optionally the labels for the *n*'s and *p*-levels). If the *Display n* and/or *Display p*-levels options are selected, then all columns of the output file will be created as *text* columns (to permit the inclusion of text strings such as *p<*), and the correlations will be displayed with six digits of precision.

If neither the *Display n* and/or *Display p*-levels options are selected, then the correlation matrix will be stored in *double float* columns (and the coefficients may be displayed with up to 15 digits of precision).

Therefore, if you are planning to process the output correlation matrix with *STATISTICA* (see *Convert to STATISTICA*, page 3717, and *Export to STATISTICA*, page 3712) or another application (see *Export*, page 3712), do not request to display the *n*'s and *p*-levels.

Note also that in order to use the *Megafile Manager* correlation matrix as input for analyses in *STATISTICA* modules, (e.g., *Multiple Regression*, *Canonical Correlation*, *Reliability and Item Analysis*, *Cluster Analysis*, *Multidimensional Scaling*, *Factor Analysis*, *etc.*), you will need to add four rows to the end of the resulting *STATISTICA* file and specific case names so that *STATISTICA* will recognize it as a matrix format file (for a detailed description of the matrix file format, refer to the *Data Management* chapter, Volume I). Large correlation matrices for single and multiple groups can also be computed and saved in matrix file format via the *Quick Basic Stats* options (see Volume I for details).

Frequencies

This option allows you to compute frequency tables for the selected variables based on the specified mode of categorization (see below).

After clicking *OK* in this dialog, you will be requested to specify a file name under which the resulting frequencies can be saved.

Columns

Click on this button to open the *Select Columns* dialog (page 3697) in which you can select the columns for the analyses.

Rows

Click on this button to open the *Select Rows* dialog (page 3698) in which you can select the rows for the analyses.

Mode

The settings in this box will determine how the currently selected columns will be categorized or tabulated for the frequency tables.

Integer values. If this radio button is set, then the frequency tables will be based on integer category (interval) boundaries and step sizes, starting with the smallest integer value found in the respective column.

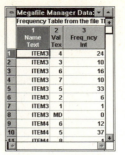

All non-integer values will be rounded up or down to integer values in order to place them in integer categories.

Distinct values. Select this option if you want the frequencies to be based on all of the distinct values for each of the selected columns. They can be either numeric

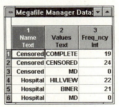

or text (or logical) values.

# of intervals. If this radio button is set, then the entire range of values for each numeric column will be divided into the user-specified number of intervals.

12. MEGAFILE MANAGER - DIALOGS AND OPTIONS

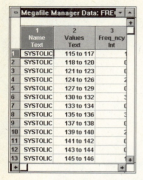

Text values are always tabulated by *Distinct values* (see above).

Warning Level

Specify here the limit on the number of categories for *integer* and *distinct values* tabulations. If the limit is exceeded, then a warning message *Specified number of distinct values exceeded. Continue? (Yes/No/Cancel)* will be issued, permitting the analysis to be completed or terminated. This option is always used when tabulating text values, but has no effect when tabulating numeric values using the *No. of intervals* mode of categorization (see above).

Note. The frequency output is saved to a *Megafile Manager* file (you will be prompted to select the file name). The first column of the output file contains the name of each tabulated column (repeated as many times as there are distinctive values or ranges for that column). The second column (also text) contains the names of consecutive specific values or ranges. The third column (labeled *Freq_ncy*) contains the actual frequency counts (stored in *LongInteger* format) for each value or range. If desired, the output file may be reorganized by performing a hierarchical (nested) sort (see page 3728) as follows: select the first column (*Name*, i.e., the column names) as the first key, then select as the second key either the second column (*Values*, if each set of frequencies is to be sorted by values) or the third column (*Freq_ncy*, if each set of values is to be sorted by frequency).

Megafile Manager Language (*MML*)

MML (*Megafile Manager Language*) is a specialized data base management and data transformation programming language integrated into the *Megafile Manager* window (accessible from the *Data Management* module). *MML* is a simple but very powerful programming language, similar to Pascal in terms of its general syntax conventions. *MML* offers extensions designed to aid transformations of data and statistical data base management.

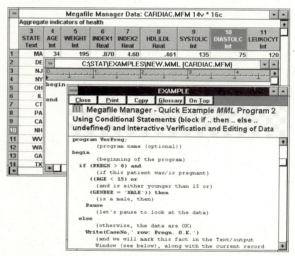

Selecting the *Language (MML)* option from the *Megafile Manager Analysis* pull-down menu will open an *MML* program window for the current *Megafile Manager* data file (i.e., the one which is active when *MML* is invoked).

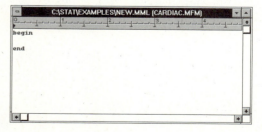

MM - 3734

Copyright © StatSoft, 1995

Each *MML* program window is associated with the *Megafile Manager* data file window from which it was opened, and the name of that data file is displayed in the caption bar of that *MML* program window (e.g., the *MML* window shown above is associated with the *Megafile Manager* data file *300.mfm*).

In order to process a different *Megafile Manager* data file using a program that is already in a *MML* window not associated with that data file, open a new *MML* program window for that data set [i.e., select the *Language (MML)* option when that data set window is active] and copy the program to that window.

For more specific information about *MML*, refer to Volume V of the manual, or review the syntax reference and examples in the *Electronic Manual*.

INDEX

A

Append STATISTICA file, 3711
ASCII fixed
    export files to, 3717
    importing files from, 3708
ASCII free
    export files to, 3716
    importing files from, 3709
    separators, 3710

C

Column specs, editing in Megafile Manager, 3721
Convert to a STATISTICA file, 3717
Correlations, 3731
Create new MFM file, 3700

D

Data Management
    archival data banks, 3696
    exchanging data with Megafile Manager, 3696
    large data files, 3696
    long records, 3696
    long text values, 3696
    Megafile Manager, 3696
    preprocessing of large records, 3696
    text values, 3696
Data processing, analysis, 3696
dBASE
    export files to, 3715
    importing files from, 3707
    verify column names for correctness, 3729
Descriptive statistics, 3730

E

Edit column specs, 3721
Excel
    export files to, 3714
    importing files from, 3704
Exchanging data with STATISTICA data files, 3696

Export
    ASCII fixed files, 3717
    ASCII free files, 3716
    dBASE files, 3715
    Excel files, 3714
    Lotus files, 3714
    Quattro files, 3715
    SPSS files, 3715
    STATISTICA files, 3712

F

File header, 3724
Frequencies, 3733

H

Hierarchical relations between data bases in Megafile Manager, 3696

I

Import
    ASCII fixed files, 3708
    ASCII free files, 3709
    ASCII free separators, 3710
    dBASE files, 3707
    Excel files, 3704
    Lotus files, 3705
    Lotus review ranges, 3705
    Quattro files, 3706
    SPSS files, 3707
    STATISTICA files, 3703
Importing data, 3702
Incorrect variable name, 3730

K

Kurtosis, 3731

L

Lotus
    export files to, 3714
    importing files from, 3705
    review ranges, 3705

M

Maintaining large, archival data banks in Megafile Manager, 3696
Mean, 3731

Megafile Manager
    add columns, 3718
    add rows, 3722
    append STATISTICA file, 3711
    ASCII free separators, 3710
    convert to a STATISTICA file, 3717
    copy columns, 3720
    copy rows, 3723
    correlations, 3731
    create new MFM file, 3700
    data processing, analysis, 3696
    delete columns, 3720
    delete rows, 3724
    descriptive statistics, 3730
    edit column specs, 3721
    exchanging data with STATISTICA data files, 3696
    export to ASCII fixed, 3717
    export to ASCII free, 3716
    export to dBASE, 3715
    export to Excel, 3714
    export to Lotus, 3714
    export to Quattro, 3715
    export to SPSS, 3715
    export to STATISTICA, 3712
    file header, 3724
    find value, 3699
    frequencies, 3733
    global width, 3700
    hierarchical relations between data bases, 3696
    import ASCII fixed files, 3708
    import ASCII free files, 3709
    import dBASE files, 3707
    import Excel files, 3704
    import Lotus files, 3705
    import Quattro files, 3706
    import SPSS files, 3707
    import STATISTICA files, 3703
    importing files, 3702
    incorrect variable name, 3730
    long text values, 3696
    maintaining large, archival data banks, 3696
    merge columns, 3725
    merge files, 3724
    merge rows, 3726
    MML, 3734
    move columns, 3719
    move rows, 3723
    not relational merge, 3726
    open MFM file, 3702

Megafile Manager (continued)
    optimize size, 3728
    overlay merge, 3727
    overview, 3695
    preprocessing large records of raw data, 3696
    print MFM data, 3717
    relational hierarchical merge, 3726
    relational merge, 3726
    replace value, 3699
    save MFM file, 3702
    select columns, 3697
    select rows, 3698
    sort data, 3728
    subset, 3727
    text values, 3696
    verify column names, 3729
    view column values, 3698
Merge files
    merge columns, 3725
    merge rows, 3726
    not relational merge, 3726
    overlay merge, 3727
    relational hierarchical merge, 3726
    relational merge, 3726
Missing data deletion
    casewise deletion of missing data, 3731
MML programming language, 3695, 3734

O

Optimize size, 3728

P

Pairwise deletion of missing data, 3731
Print MFM data, 3717

Q

Quattro
    export files to, 3715
    importing files from, 3706

R

Relational hierarchical merge, 3726
Relational merge, 3726

S

Skewness, 3731
Sort data, 3728
SPSS
    export files to, 3715
    importing files from, 3707
    verify column names for correctness, 3729
Standard deviation, 3731
Standard error of the mean, 3731
Subset Megafile Manager data files, 3727

V

Valid n, 3731
Variance, 3731
Verify column names, 3729

REFERENCES

Abraham, B., & Ledolter, J. (1983). *Statistical methods for forecasting*. New York: Wiley.

Adorno, T. W., Frenkel-Brunswik, E., Levinson, D. J., & Sanford, R. N. (1950). *The authoritarian personality*. New York: Harper.

Akaike, H. (1973). Information theory and an extension of the maximum likelihood principle. In B. N. Petrov and F. Csaki (Eds.), *Second International Symposium on Information Theory*. Budapest: Akademiai Kiado.

Akaike, H. (1983). Information measures and model selection. *Bulletin of the International Statistical Institute: Proceedings of the 44th Session, Volume 1*. Pages 277-290.

Aldrich, J. H., & Nelson, F. D. (1984). *Linear probability, logit, and probit models*. Beverly Hills, CA: Sage Publications.

Almon, S. (1965). The distributed lag between capital appropriations and expenditures. *Econometrica*, *33*, 178-196.

American Supplier Institute (1984-1988). *Proceedings of Supplier Symposia on Taguchi Methods*. (April, 1984; November, 1984; October, 1985; October, 1986; October, 1987; October, 1988), Dearborn, MI: American Supplier Institute.

Anderson, O. D. (1976). *Time series analysis and forecasting*. London: Butterworths.

Anderson, S. B., & Maier, M. H. (1963). 34,000 pupils and how they grew. *Journal of Teacher Education*, *14*, 212-216.

Anderson, T. W. (1958). *An introduction to multivariate statistical analysis*. New York: Wiley.

Anderson, T. W., & Rubin, H. (1956). Statistical inference in factor analysis. *Proceedings of the Third Berkeley Symposium on Mathematical Statistics and Probability.* Berkeley: The University of California Press.

Andrews, D. F. (1972). Plots of high-dimensional data. *Biometrics*, *28*, 125-136.

AT&T (1956). *Statistical quality control handbook, Select code 700-444*. Indianapolis, AT&T Technologies.

Auble, D. (1953). Extended tables for the Mann-Whitney statistic. *Bulletin of the Institute of Educational Research, Indiana University*, *1*, No. 2.

Bails, D. G., & Peppers, L. C. (1982). *Business fluctuations: Forecasting techniques and applications*. Englewood Cliffs, NJ: Prentice-Hall.

Bain, L. J. (1978). *Statistical analysis of reliability and life-testing models*. New York: Decker.

Baird, J. C. (1970). *Psychophysical analysis of visual space*. New York: Pergamon Press.

Baird, J. C., & Noma, E. (1978). *Fundamentals of scaling and psychophysics*. New York: Wiley.

Barcikowski, R., & Stevens, J. P. (1975). A Monte Carlo study of the stability of canonical correlations, canonical weights, and canonical variate-variable correlations. *Multivariate Behavioral Research*, *10*, 353-364.

Barker, T. B. (1986). Quality engineering by design: Taguchi's philosophy. *Quality Progress*, *19*, 32-42.

Barlow, R. E., & Proschan, F. (1975). *Statistical theory of reliability and life testing*. New York: Holt, Rinehart, & Winston.

Barnard, G. A. (1959). Control charts and stochastic processes. *Journal of the Royal Statistical Society*, Ser. B, *21*, 239.

Bartholomew, D. J. (1984). The foundations of factor analysis. *Biometrika*, *71*, 221-232.

REFERENCES

Bates, D. M., & Watts, D. G. (1988). *Nonlinear regression analysis and its applications.* New York: Wiley.

Bayne, C. K., & Rubin, I. B. (1986). *Practical experimental designs and optimization methods for chemists.* Deerfield Beach, FL: VCH Publishers.

Becker, R. A., Denby, L., McGill, R., & Wilks, A. R. (1986). Datacryptanalysis: A case study. *Proceedings of the Section on Statistical Graphics, American Statistical Association*, 92-97.

Bendat, J. S. (1990). *Nonlinear system analysis and identification from random data.* New York: Wiley.

Bentler, P. M, & Bonett, D. G. (1980). Significance tests and goodness of fit in the analysis of covariance structures. *Psychological Bulletin, 88,* 588-606.

Bentler, P. M. (1986). Structural modeling and Psychometrika: A historical perspective on growth and achievements. *Psychometrika, 51,* 35-51.

Bentler, P. M. (1989). *EQS Structural equations program manual.* Los Angeles, CA: BMDP Statistical Software.

Bentler, P. M., & Weeks, D. G. (1979). Interrelations among models for the analysis of moment structures. *Multivariate Behavioral Research, 14,* 169-185.

Bentler, P. M., & Weeks, D. G. (1980). Linear structural equations with latent variables. *Psychometrika, 45,* 289-308.

Berkson, J., & Gage, R. R. (1950). The calculation of survival rates for cancer. *Proceedings of Staff Meetings, Mayo Clinic, 25,* 250.

Bhote, K. R. (1988). *World class quality.* New York: AMA Membership Publications.

Binns, B., & Clark, N. (1986). The graphic designer's use of visual syntax. *Proceedings of the Section on Statistical Graphics, American Statistical Association,* 36-41.

Birnbaum, Z. W. (1952). Numerical tabulation of the distribution of Kolmogorov's statistic for finite sample values. *Journal of the American Statistical Association, 47,* 425-441.

Birnbaum, Z. W. (1953). Distribution-free tests of fit for continuous distribution functions. *Annals of Mathematical Statistics, 24,* 1-8.

Bishop, Y. M. M., Fienberg, S. E., & Holland, P. W. (1975). *Discrete multivariate analysis.* Cambridge, MA: MIT Press.

Bjorck, A. (1967). Solving linear least squares problems by Gram-Schmidt orthonormalization. *Bit, 7,* 1-21.

Blackman, R. B., & Tukey, J. (1958). *The measurement of power spectral from the point of view of communication engineering.* New York: Dover.

Blackwelder, R. A. (1966). *Taxonomy: A text and reference book.* New York: Wiley.

Blalock, H. M. (1972). *Social statistics* (2nd ed.). New York: McGraw-Hill.

Bloomfield, P. (1976). *Fourier analysis of time series: An introduction.* New York: Wiley.

Bock, R. D. (1963). Programming univariate and multivariate analysis of variance. *Technometrics, 5,* 95-117.

Bock, R. D. (1975). *Multivariate statistical methods in behavioral research.* New York: McGraw-Hill.

Bolch, B.W., & Huang, C. J. (1974). *Multivariate statistical methods for business and economics.* Englewood Cliffs, NJ: Prentice-Hall.

Bollen, K. A. (1989). *Structural equations with latent variables.* New York: John Wiley & Sons.

Borg, I., & Lingoes, J. (1987). *Multidimensional similarity structure analysis*. New York: Springer.

Borg, I., & Shye, S. (in press). *Facet Theory*. Newbury Park: Sage.

Bowker, A. G. (1948). A test for symmetry in contingency tables. *Journal of the American Statistical Association*, *43*, 572-574.

Box, G. E. P. (1954a). Some theorems on quadratic forms applied in the study of analysis of variance problems: I. Effect of inequality of variances in the one-way classification. *Annals of Mathematical Statistics*, *25*, 290-302.

Box, G. E. P. (1954b). Some theorems on quadratic forms applied in the study of analysis of variance problems: II. Effect of inequality of variances and of correlation of errors in the two-way classification. *Annals of Mathematical Statistics*, *25*, 484-498.

Box, G. E. P., & Anderson, S. L. (1955). Permutation theory in the derivation of robust criteria and the study of departures from assumptions. *Journal of the Royal Statistical Society*, *17*, 1-34.

Box, G. E. P., & Behnken, D. W. (1960). Some new three level designs for the study of quantitative variables. *Technometrics*, *2*, 455-475.

Box, G. E. P., & Cox, D. R. (1964). An analysis of transformations. *Journal of the Royal Statistical Society*, *26*, 211-253.

Box, G. E. P., & Cox, D. R. (1964). An analysis of transformations. *Journal of the Royal Statistical Society, B26*, 211-234.

Box, G. E. P., & Draper, N. R. (1987). *Empirical model-building and response surfaces*. New York: Wiley.

Box, G. E. P., & Jenkins, G. M. (1970). *Time series analysis*. San Francisco: Holden Day.

Box, G. E. P., & Jenkins, G. M. (1976). *Time series analysis: Forecasting and control*. San Francisco: Holden-Day.

Box, G. E. P., & Tidwell, P. W. (1962). Transformation of the independent variables. *Technometrics*, *4*, 531-550.

Box, G. E. P., Hunter, W. G., & Hunter, S. J. (1978). *Statistics for experimenters: An introduction to design, data analysis, and model building*. New York: Wiley.

Brenner, J. L., et al. (1968). Difference equations in forecasting formulas. *Management Science*, *14*, 141-159.

Brent, R. F. (1973). *Algorithms for minimization without derivatives*. Englewood Cliffs, NJ: Prentice-Hall.

Breslow, N. E. (1970). A generalized Kruskal-Wallis test for comparing *K* samples subject to unequal pattern of censorship. *Biometrika*, *57*, 579-594.

Breslow, N. E. (1974). Covariance analysis of censored survival data. *Biometrics*, *30*, 89-99.

Brigham, E. O. (1974). *The fast Fourier transform*. Englewood Cliffs, NJ: Prentice-Hall.

Brillinger, D. R. (1975). *Time series: Data analysis and theory*. New York: Holt, Rinehart. & Winston.

Brown, D. T. (1959). A note on approximations to discrete probability distributions. *Information and Control*, *2*, 386-392.

Brown, R. G. (1959). *Statistical forecasting for inventory control*. New York: McGraw-Hill.

Browne, M. W. (1968). A comparison of factor analytic techniques. *Psychometrika*, *33*, 267-334.

Browne, M. W. (1974). Generalized least squares estimators in the analysis of covariance structures. *South African Statistical Journal*, *8*, 1-24.

Browne, M. W. (1982). Covariance Structures. In D. M. Hawkins (Ed.) *Topics in Applied Multivariate Analysis*. Cambridge, MA: Cambridge University Press.

REFERENCES

Browne, M. W. (1984). Asymptotically distribution free methods for the analysis of covariance structures. *British Journal of Mathematical and Statistical Psychology, 37*, 62-83.

Browne, M. W., & Cudeck, R. (1990). Single sample cross-validation indices for covariance structures. *Multivariate Behavioral Research, 24*, 445-455.

Browne, M. W., & Cudeck, R. (1992). Alternative ways of assessing model fit. In K. A. Bollen and J. S. Long (Eds.), *Testing structural equation models.* Beverly Hills, CA: Sage.

Browne, M. W., & DuToit, S. H. C. (1982). *AUFIT* (Version 1). A computer programme for the automated fitting of nonstandard models for means and covariances. Research Finding WS-27. Pretoria, South Africa: Human Sciences Research Council.

Browne, M. W., & DuToit, S. H. C. (1987). *Automated fitting of nonstandard models*. Report WS-39. Pretoria, South Africa: Human Sciences Research Council.

Browne, M. W., & DuToit, S. H. C. (1992). Automated fitting of nonstandard models. *Multivariate Behavioral Research, 27*, 269-300.

Browne, M. W., & Mels, G. (1992). *RAMONA User's Guide.* The Ohio State University: Department of Psychology.

Browne, M. W., & Shapiro, A. (1989). *Invariance of covariance structures under groups of transformations*. Research Report 89/4. Pretoria, South Africa: University of South Africa Department of Statistics.

Browne, M. W., & Shapiro, A. (1991). Invariance of covariance structures under groups of transformations. *Metrika, 38*, 335-345.

Buffa, E. S. (1972). *Operations management: Problems and models* (3rd. ed.). New York: Wiley.

Buja, A., & Tukey, P. A. (Eds.) (1991). *Computing and Graphics in Statistics.* New York: Springer-Verlag.

Buja, A., Fowlkes, E. B., Keramidas, E. M., Kettenring, J. R., Lee, J. C., Swayne, D. F., & Tukey, P. A. (1986). Discovering features of multivariate data through statistical graphics. *Proceedings of the Section on Statistical Graphics, American Statistical Association*, 98-103.

Burman, J. P. (1979). Seasonal adjustment - a survey. *Forecasting, Studies in Management Science, 12*, 45-57.

Burns, L. S., & Harman, A. J. (1966). *The complex metropolis, Part V of profile of the Los Angeles metropolis: Its people and its homes.* Los Angeles: University of Chicago Press.

Campbell D. T., & Fiske, D. W. (1959). Convergent and discriminant validation by the multitrait-multimethod matrix. *Psychological Bulletin, 56*, 81-105

Carmines, E. G., & Zeller, R. A. (1980). *Reliability and validity assessment.* Beverly Hills, CA: Sage Publications.

Carroll, J. D., & Wish, M. (1974). Multidimensional perceptual models and measurement methods. In E. C. Carterette and M. P. Friedman (Eds.), *Handbook of perception.* (Vol. 2, pp. 391-447). New York: Academic Press.

Cattell, R. B. (1966). The scree test for the number of factors. *Multivariate Behavioral Research, 1*, 245-276.

Cattell, R. B., & Jaspers, J. A. (1967). A general plasmode for factor analytic exercises and research. *Multivariate Behavioral Research Monographs.*

Chambers, J. M., Cleveland, W. S., Kleiner, B., & Tukey, P. A. (1983). *Graphical methods for data analysis.* Bellmont, CA: Wadsworth.

Chan, L. K., Cheng, S. W., & Spiring, F. (1988). A new measure of process capability: C_{pm}. *Journal of Quality Technology, 20,* 162-175.

Chernoff, H. (1973). The use of faces to represent points in k-dimensional space graphically. *Journal of American Statistical Association, 68,* 361-368.

Christ, C. (1966). *Econometric models and methods.* New York: Wiley.

Clarke, G. M., & Cooke, D. (1978). *A basic course in statistics.* London: Edward Arnold.

Cleveland, W. S. (1979). Robust locally weighted regression and smoothing scatterplots. *Journal of the American Statistical Association, 74,* 829-836.

Cleveland, W. S. (1984). Graphs in scientific publications. *The American Statistician, 38,* 270-280.

Cleveland, W. S. (1985). *The elements of graphing data.* Monterey, CA: Wadsworth.

Cleveland, W. S., Harris, C. S., & McGill, R. (1982). Judgements of circle sizes on statistical maps. *Journal of the American Statistical Association, 77,* 541-547.

Cliff, N. (1983). Some cautions concerning the application of causal modeling methods. *Multivariate Behavioral Research, 18,* 115-126.

Cochran, W. G. (1950). The comparison of percentages in matched samples. *Biometrika, 37,* 256-266.

Conover, W. J. (1974). Some reasons for not using the Yates continuity correction on 2 x 2 contingency tables. *Journal of the American Statistical Association, 69,* 374-376.

Cook, R. D. (1977). Detection of influential observations in linear regression. *Technometrics, 19,* 15-18.

Cook, R. D., & Nachtsheim, C. J. (1980). A comparison of algorithms for constructing exact D-optimal designs. *Technometrics, 22,* 315-324.

Cooke, D., Craven, A. H., & Clarke, G. M. (1982). *Basic statistical computing.* London: Edward Arnold.

Cooley, J. W., & Tukey, J. W. (1965). An algorithm for the machine computation of complex Fourier series. *Mathematics of Computation, 19,* 297-301.

Cooley, W. W., & Lohnes, P. R. (1971). *Multivariate data analysis.* New York: Wiley.

Cooley, W. W., & Lohnes, P. R. (1976). *Evaluation research in education.* New York: Wiley.

Coombs, C. H. (1950). Psychological scaling without a unit of measurement. *Psychological Review, 57,* 145-158.

Coombs, C. H. (1964). *A theory of data.* New York: Wiley.

Corballis, M. C., & Traub, R. E. (1970). Longitudinal factor analysis. *Psychometrika, 35,* 79-98.

Cormack, R. M. (1971). A review of classification. *Journal of the Royal Statistical Society, 134,* 321-367.

Cornell, J. A. (1990a). *How to run mixture experiments for product quality.* Milwaukee, WI: ASQC.

Cornell, J. A. (1990b). *Experiments with mixtures: designs, models, and the analysis of mixture data* (2nd ed.). New York: Wiley.

Cox, D. R. (1957). Note on grouping. *Journal of the American Statistical Association, 52,* 543-547.

Cox, D. R. (1959). The analysis of exponentially distributed life-times with two types of failures. *Journal of the Royal Statistical Society, 21,* 411-421.

Cox, D. R. (1964). Some applications of exponential ordered scores. *Journal of the Royal Statistical Society, 26,* 103-110.

Cox, D. R. (1970). *The analysis of binary data.* New York: Halsted Press.

REFERENCES

Cox, D. R. (1972). Regression models and life tables. *Journal of the Royal Statistical Society, 34,* 187-220.

Cox, D. R., & Oakes, D. (1984). *Analysis of survival data.* New York: Chapman & Hall.

Cramer, H. (1946). *Mathematical methods in statistics.* Princeton, NJ: Princeton University Press.

Crowley, J., & Hu, M. (1977). Covariance analysis of heart transplant survival data. *Journal of the American Statistical Association, 72,* 27-36.

Cudeck, R. (1989). Analysis of correlation matrices using covariance structure models. *Psychological Bulletin, 105,* 317-327.

Cudeck, R., & Browne, M. W. (1983). Cross-validation of covariance structures. *Multivariate Behavioral Research, 18,* 147-167.

Cutler, S. J., & Ederer, F. (1958). Maximum utilization of the life table method in analyzing survival. *Journal of Chronic Diseases, 8,* 699-712.

Dahlquist, G., & Bjorck, A. (1974). *Numerical Methods.* Englewood Cliffs, NJ: Prentice-Hall.

Daniel, C. (1976). *Applications of statistics to industrial experimentation.* New York: Wiley.

Daniell, P. J. (1946). Discussion on symposium on autocorrelation in time series. *Journal of the Royal Statistical Society, Suppl. 8,* 88-90.

Darlington, R. B. (1990). *Regression and linear models.* New York: McGraw-Hill.

Darlington, R. B., Weinberg, S., & Walberg, H. (1973). Canonical variate analysis and related techniques. *Review of Educational Research, 43,* 433-454.

Davies, P. M., & Coxon, A. P. M. (1982). *Key texts in multidimensional scaling.* Exeter, NH: Heinemann Educational Books.

De Boor, C. (1978). *A practical guide to splines.* New York: Springer-Verlag.

De Gruitjer, P. N. M., & Van Der Kamp, L. J. T. (Eds.). (1976). *Advances in psychological and educational measurement.* New York: Wiley.

Deming, S. N., & Morgan, S. L. (1993). *Experimental design: A chemometric approach.* (2nd ed.). Amsterdam, The Netherlands: Elsevier Science Publishers B.V.

Deming, W. E., & Stephan, F. F. (1940). The sampling procedure of the 1940 population census. *Journal of the American Statistical Association, 35,* 615-630.

Dempster, A. P. (1969). *Elements of Continuous Multivariate Analysis.* San Francisco: Addison-Wesley.

Dempster, A. P., Laird, N. M., & Rubin, D. B. (1977). Maximum likelihood from incomplete data via the EM algorithm. *Journal of the Royal Statistical Society, 39,* 1-38.

Dennis, J. E., & Schnabel, R. B. (1983). *Numerical methods for unconstrained optimization and nonlinear equations.* Englewood Cliffs, NJ: Prentice Hall.

Diamond, W. J. (1981). *Practical experimental design.* Belmont, CA: Wadsworth.

Dijkstra, T. K. (1990). Some properties of estimated scale invariant covariance structures. *Psychometrika, 55,* 327-336.

Dixon, W. J. (1954). Power under normality of several non-parametric tests. *Annals of Mathematical Statistics, 25,* 610-614.

Dixon, W. J., & Massey, F. J. (1983). *Introduction to statistical analysis* (4th ed.). New York: McGraw-Hill.

Dodd, B. (1979). Lip reading in infants: Attention to speech presented in- and out-of-synchrony. *Cognitive Psychology, 11,* 478-484.

Dodge, Y. (1985). *Analysis of experiments with missing data.* New York: Wiley.

Dodge, Y., Fedorov, V. V., & Wynn, H. P. (1988). *Optimal design and analysis of experiments.* New York: North-Holland.

Duncan, A. J. (1974). *Quality control and industrial statistics.* Homewood, IL: Richard D. Irwin.

Duncan, O. D., Haller, A. O., & Portes, A. (1968). Peer influence on aspiration: A reinterpretation. *American Journal of Sociology, 74,* 119-137.

Durbin, J. (1970). Testing for serial correlation in least-squares regression when some of the regressors are lagged dependent variables. *Econometrica, 38,* 410-421.

Durbin, J., & Watson, G. S. (1951). Testing for serial correlations in least squares regression. II. *Biometrika, 38,* 159-178.

Dykstra, O. Jr. (1971). The augmentation of experimental data to maximize |X'X|. *Technometrics, 13,* 682-688.

Eason, E. D., & Fenton, R. G. (1974). A comparison of numerical optimization methods for engineering design. *ASME Paper 73-DET-17.*

Efron, B. (1982). *The jackknife, the bootstrap, and other resampling plans.* Philadelphia, PA: Society for Industrial and Applied Mathematics.

Elandt-Johnson, R. C., & Johnson, N. L. (1980). *Survival models and data analysis.* New York: Wiley.

Elliott, D. F., & Rao, K. R. (1982). *Fast transforms: Algorithms, analyses, applications.* New York: Academic Press.

Enslein, K., Ralston, A., & Wilf, H. S. (1977). *Statistical methods for digital computers.* New York: Wiley.

Everitt, B. S. (1977). *The analysis of contingency tables.* London: Chapman & Hall.

Everitt, B. S. (1984). *An introduction to latent variable models.* London: Chapman and Hall.

Ewan, W. D. (1963). When and how to use Cu-sum charts. *Technometrics, 5,* 1-32.

Feigl, P., & Zelen, M. (1965). Estimation of exponential survival probabilities with concomitant information. *Biometrics, 21,* 826-838.

Fetter, R. B. (1967). *The quality control system.* Homewood, IL: Richard D. Irwin.

Fienberg, S. E. (1977). *The analysis of cross-classified categorical data.* Cambridge, MA: MIT Press.

Finn, J. D. (1974). *A general model for multivariate analysis.* New York: Holt, Rinehart & Winston.

Finn, J. D. (1977). Multivariate analysis of variance and covariance. In K. Enslein, A. Ralston, and H. S. Wilf (Eds.), *Statistical methods for digital computers. Vol. III.* (pp. 203-264). New York: Wiley.

Finney, D. J. (1971). *Probit analysis.* Cambridge, MA: Cambridge University Press.

Fisher, R. A. (1936). The use of multiple measurements in taxonomic problems. *Annals of Eugenics, 7,* 179-188.

Fletcher, R. (1969). *Optimization.* New York: Academic Press.

Fletcher, R., & Powell, M. J. D. (1963). A rapidly convergent descent method for minimization. *Computer Journal, 6,* 163-168.

Fletcher, R., & Reeves, C. M. (1964). Function minimization by conjugate gradients. *Computer Journal, 7,* 149-154.

Fomby, T.B., Hill, R.C., & Johnson, S.R. (1984). *Advanced econometric methods.* New York: Springer-Verlag.

Fraser, C., & McDonald, R. P. (1988). COSAN: Covariance structure analysis. *Multivariate Behavioral Research, 23,* 263-265.

Friedman, M. (1937). The use of ranks to avoid the assumption of normality implicit in the analysis of variance. *Journal of the American Statistical Association, 32,* 675-701.

REFERENCES

Friedman, M. (1940). A comparison of alternative tests of significance for the problem of *m* rankings. *Annals of Mathematical Statistics, 11,* 86-92.

Fries, A., & Hunter, W. G. (1980). Minimum aberration $2^{(k-p)}$ designs. *Technometrics, 22,* 601-608.

Frost, P. A. (1975). Some properties of the Almon lag technique when one searches for degree of polynomial and lag. *Journal of the American Statistical Association, 70,* 606-612.

Fuller, W. A. (1976). *Introduction to statistical time series.* New York: Wiley.

Gale, N., & Halperin, W. C. (1982). A case for better graphics: The unclassed choropleth map. *The American Statistician, 36,* 330-336.

Galil, Z., & Kiefer, J. (1980). Time- and space-saving computer methods, related to Mitchell's DETMAX, for finding D-optimum designs. *Technometrics, 22,* 301-313.

Gara, M. A., & Rosenberg, S. (1979). The identification of persons as supersets and subsets in free-response personality descriptions. *Journal of Personality and Social Psychology, 37,* 2161-2170.

Gara, M. A., & Rosenberg, S. (1981). Linguistic factors in implicit personality theory. *Journal of Personality and Social Psychology, 41,* 450-457.

Gardner, E. S., Jr. (1985). Exponential smoothing: The state of the art. *Journal of Forecasting, 4,* 1-28.

Garvin, D. A. (1987). Competing on the eight dimensions of quality. *Harvard Business Review,* November/December, 101-109.

Gbur, E., Lynch, M., & Weidman, L. (1986). An analysis of nine rating criteria on 329 U. S. metropolitan areas. *Proceedings of the Section on Statistical Graphics, American Statistical Association,* 104-109.

Gedye, R. (1968). *A manager's guide to quality and reliability.* New York: Wiley.

Gehan, E. A. (1965a). A generalized Wilcoxon test for comparing arbitrarily singly-censored samples. *Biometrika, 52,* 203-223.

Gehan, E. A. (1965b). A generalized two-sample Wilcoxon test for doubly-censored data. *Biometrika, 52,* 650-653.

Gehan, E. A., & Siddiqui, M. M. (1973). Simple regression methods for survival time studies. *Journal of the American Statistical Association, 68,* 848-856.

Gehan, E. A., & Thomas, D. G. (1969). The performance of some two sample tests in small samples with and without censoring. *Biometrika, 56,* 127-132.

Gerald, C. F., & Wheatley, P. O. (1989). *Applied numerical analysis* (4th ed.). Reading, MA: Addison Wesley.

Gibbons, J. D. (1976). *Nonparametric methods for quantitative analysis.* New York: Holt, Rinehart, & Winston.

Gibbons, J. D. (1985). *Nonparametric statistical inference* (2nd ed.). New York: Marcel Dekker.

Gifi, A. (1990). *Nonlinear multivariate analysis.* New York: Wiley.

Gill, P. E., & Murray, W. (1972). Quasi-Newton methods for unconstrained optimization. *Journal of the Institute of Mathematics and its Applications, 9,* 91-108.

Gill, P. E., & Murray, W. (1974). *Numerical methods for constrained optimization.* New York: Academic Press.

Glass, G. V., & Stanley, J. (1970). *Statistical methods in education and Psychology.* Englewood Cliffs, NJ: Prentice-Hall.

Glasser, M. (1967). Exponential survival with covariance. *Journal of the American Statistical Association, 62,* 561-568.

Gnanadesikan, R., Roy, S., & Srivastava, J. (1971). *Analysis and design of certain quantitative multiresponse experiments*. Oxford: Pergamon Press, Ltd.

Golub, G. H., & Van Loan, C. F. (1983). *Matrix computations*. Baltimore: Johns Hopkins University Press.

Goodman, L .A., & Kruskal, W. H. (1972). Measures of association for cross-classifications IV: Simplification of asymptotic variances. *Journal of the American Statistical Association, 67*, 415-421.

Goodman, L. A. (1954). Kolmogorov-Smirnov tests for psychological research. *Psychological Bulletin, 51*, 160-168.

Goodman, L. A. (1971). The analysis of multidimensional contingency tables: Stepwise procedures and direct estimation methods for models building for multiple classification. *Technometrics, 13*, 33-61.

Goodman, L. A., & Kruskal, W. H. (1954). Measures of association for cross-classifications. *Journal of the American Statistical Association, 49*, 732-764.

Goodman, L. A., & Kruskal, W. H. (1959). Measures of association for cross-classifications II: Further discussion and references. *Journal of the American Statistical Association, 54*, 123-163.

Goodman, L. A., & Kruskal, W. H. (1963). Measures of association for cross-classifications III: Approximate sampling theory. *Journal of the American Statistical Association, 58*, 310-364.

Grant, E. L., & Leavenworth, R. S. (1980). *Statistical quality control* (5th ed.). New York: McGraw-Hill.

Green, P. E., & Carmone, F. J. (1970). *Multidimensional scaling and related techniques in marketing analysis*. Boston: Allyn & Bacon.

Greenhouse, S. W., & Geisser, S. (1958). Extension of Box's results on the use of the F distribution in multivariate analysis. *Annals of Mathematical Statistics, 29*, 95-112.

Greenhouse, S. W., & Geisser, S. (1959). On methods in the analysis of profile data. *Psychometrika, 24*, 95-112.

Gross, A. J., & Clark, V. A. (1975). *Survival distributions: Reliability applications in the medical sciences*. New York: Wiley.

Gruvaeus, G., & Wainer, H. (1972). Two additions to hierarchical cluster analysis. *The British Journal of Mathematical and Statistical Psychology, 25*, 200-206.

Guttman, L. (1954). A new approach to factor analysis: the radex. In P. F. Lazarsfeld (Ed.), *Mathematical thinking in the social sciences*. New York: Columbia University Press.

Guttman, L. (1968). A general nonmetric technique for finding the smallest coordinate space for a configuration of points. *Pyrometrical, 33*, 469-506.

Haberman, S. J. (1972). Loglinear fit for contingency tables. *Applied Statistics, 21*, 218-225.

Haberman, S. J. (1974). *The analysis of frequency data*. Chicago: University of Chicago Press.

Hahn, G. J., & Shapiro, S. S. (1967). *Statistical models in engineering*. New York: Wiley.

Hakstian, A. R., Rogers, W. D., & Cattell, R. B. (1982). The behavior of numbers of factors rules with simulated data. *Multivariate Behavioral Research, 17*, 193-219.

Hald, A. (1952). *Statistical theory with engineering applications*. New York: Wiley.

Harman, H. H. (1967). *Modern factor analysis*. Chicago: University of Chicago Press.

Harris, R. J. (1976). The invalidity of partitioned U tests in canonical correlation and multivariate analysis of variance. *Multivariate Behavioral Research, 11*, 353-365.

REFERENCES

Hart, K. M., & Hart, R. F. (1989). *Quantitative methods for quality improvement.* Milwaukee, WI: ASQC Quality Press.

Hartigan, J. A. (1975). *Clustering algorithms.* New York: Wiley.

Hartley, H. O. (1959). Smallest composite designs for quadratic response surfaces. *Biometrics, 15,* 611-624.

Haviland, R. P. (1964). *Engineering reliability and long life design.* Princeton, NJ: Van Nostrand.

Hayduk, L. A. (1987). *Structural equation modelling with LISREL: Essentials and advances.* Baltimore: The Johns Hopkins University Press.

Hays, W. L. (1981). *Statistics* (3rd ed.). New York: CBS College Publishing.

Hays, W. L. (1988). *Statistics* (4th ed.). New York: CBS College Publishing.

Heiberger, R. M. (1989). *Computation for the analysis of designed experiments.* New York: Wiley.

Henley, E. J., & Kumamoto, H. (1980). *Reliability engineering and risk assessment.* New York: Prentice-Hall.

Hettmansperger, T. P. (1984). *Statistical inference based on ranks.* New York: Wiley.

Hibbs, D. (1974). Problems of statistical estimation and causal inference in dynamic time series models. In H. Costner (Ed.), *Sociological Methodology 1973/1974* (pp. 252-308). San Francisco: Jossey-Bass.

Hilton, T. L. (1969). *Growth study annotated bibliography.* Princeton, NJ: Educational Testing Service Progress Report 69-11.

Hochberg, J., & Krantz, D. H. (1986). Perceptual properties of statistical graphs. *Proceedings of the Section on Statistical Graphics, American Statistical Association,* 29-35.

Hocking, R. R., & Speed, F. M. (1975). A full rank analysis of some linear model problems. *Journal of the American Statistical Association, 70,* 707-712.

Hoerl, A. E. (1962). Application of ridge analysis to regression problems. *Chemical Engineering Progress, 58,* 54-59.

Hoff, J. C. (1983). *A practical guide to Box-Jenkins forecasting.* London: Lifetime Learning Publications.

Hogg, R. V., & Craig, A. T. (1970). *Introduction to mathematical statistics.* New York: Macmillan.

Holzinger, K. J., & Swineford, F. (1939). *A study in factor analysis: The stability of a bi-factor solution.* University of Chicago: Supplementary Educational Monographs, No. 48.

Hooke, R., & Jeeves, T. A. (1961). Direct search solution of numerical and statistical problems. *Journal of the Association for Computing Machinery, 8,* 212-229.

Hotelling, H. (1947). Multivariate quality control. In Eisenhart, Hastay, and Wallis (Eds.), *Techniques of Statistical Analysis.* New York: McGraw-Hill.

Hotelling, H., & Pabst, M. R. (1936). Rank correlation and tests of significance involving no assumption of normality. *Annals of Mathematical Statistics, 7,* 29-43.

Hsu, P. L. (1938). Contributions to the theory of Student's t test as applied to the problem of two samples. *Statistical Research Memoirs, 2,* 1-24.

Huba, G. J., & Harlow, L. L. (1987). Robust structural equation models: implications for developmental psychology. *Child Development, 58,* 147-166.

Huberty, C. J. (1975). Discriminant analysis. *Review of Educational Research, 45,* 543-598.

Huynh, H., & Feldt, L. S. (1970). Conditions under which mean square ratios in repeated measures designs have exact F-distributions. *Journal of the American Statistical Association, 65,* 1582-1589.

Ireland, C. T., & Kullback, S. (1968). Contingency tables with given marginals. *Biometrika, 55*, 179-188.

Jaccard, J., Weber, J., & Lundmark, J. (1975). A multitrait-multimethod factor analysis of four attitude assessment procedures. *Journal of Experimental Social Psychology, 11*, 149-154.

Jacobs, D. A. H. (Ed.). (1977). *The state of the art in numerical analysis.* London: Academic Press.

Jacoby, S. L. S., Kowalik, J. S., & Pizzo, J. T. (1972). *Iterative methods for nonlinear optimization problems.* Englewood Cliffs, NJ: Prentice-Hall.

James, L. R., Mulaik, S. A., & Brett, J. M. (1982). *Causal analysis. Assumptions, models, and data.* Beverly Hills, CA: Sage Publications.

Jardine, N., & Sibson, R. (1971). *Mathematical taxonomy.* New York: Wiley.

Jastrow, J. (1892). On the judgment of angles and position of lines. *American Journal of Psychology, 5*, 214-248.

Jenkins, G. M., & Watts, D. G. (1968). *Spectral analysis and its applications.* San Francisco: Holden-Day.

Jennrich, R. I, & Sampson, P. F. (1968). Application of stepwise regression to non-linear estimation. *Technometrics, 10*, 63-72.

Jennrich, R. I. (1970). An asymptotic χ^2 test for the equality of two correlation matrices. *Journal of the American Statistical Association, 65*, 904-912.

Jennrich, R. I. (1977). Stepwise regression. In K. Enslein, A. Ralston, & H.S. Wilf (Eds.), *Statistical methods for digital computers.* New York: Wiley.

Jennrich, R. I. (1977). Stepwise discriminant analysis. In K. Enslein, A. Ralston, & H.S. Wilf (Eds.), *Statistical methods for digital computers.* New York: Wiley.

Jennrich, R. I., & Moore, R. H. (1975). Maximum likelihood estimation by means of nonlinear least squares. *Proceedings of the Statistical Computing Section*, American Statistical Association, 57-65.

Johnson, L. W., & Ries, R. D. (1982). *Numerical Analysis* (2nd ed.). Reading, MA: Addison Wesley.

Johnson, N. L. (1961). A simple theoretical approach to cumulative sum control charts. *Journal of the American Statistical Association, 56*, 83-92.

Johnson, N. L., & Leone, F. C. (1962). Cumulative sum control charts - mathematical principles applied to their construction and use. *Industrial Quality Control, 18*, 15-21.

Johnson, R. A. & Wichern, D. W. (1988). *Applied multivariate statistical analysis.* Englewood Cliffs, NJ: Prentice Hall.

Johnson, S. C. (1967). Hierarchical clustering schemes. *Psychometrika, 32*, 241-254.

Johnston, J. (1972). *Econometric methods.* New York: McGraw-Hill.

Jöreskog, K. G. (1973). A general model for estimating a linear structural equation system. In A. S. Goldberger and O. D. Duncan (Eds.), *Structural Equation Models in the Social Sciences.* New York: Seminar Press.

Jöreskog, K. G. (1974). Analyzing psychological data by structural analysis of covariance matrices. In D. H. Krantz, R. C. Atkinson, R. D. Luce, and P. Suppes (Eds.), *Contemporary Developments in Mathematical Psychology, Vol. II.* New York: W. H. Freeman and Company.

Jöreskog, K. G. (1978). Structural analysis of covariance and correlation matrices. *Psychometrika, 43*, 443-477.

Jöreskog, K. G., & Lawley, D. N. (1968). New methods in maximum likelihood factor analysis. *British Journal of Mathematical and Statistical Psychology, 21*, 85-96.

REFERENCES

Jöreskog, K. G., & Sörbom, D. (1979). *Advances in factor analysis and structural equation models.* Cambridge, MA: Abt Books.

Jöreskog, K. G., & Sörbom, D. (1982). Recent developments in structural equation modeling. *Journal of Marketing Research, 19*, 404-416.

Jöreskog, K. G., & Sörbom, D. (1984). *Lisrel VI. Analysis of linear structural relationships by maximum likelihood, instrumental variables, and least squares methods.* Mooresville, Indiana: Scientific Software.

Jöreskog, K. G., & Sörbom, D. (1989). *Lisrel 7. A guide to the program and applications.* Chicago: SPSS Inc.

Judge, G. G., Griffith, W. E., Hill, R. C., Luetkepohl, H., & Lee, T. S. (1985). *The theory and practice of econometrics.* New York: Wiley.

Juran, J. M. (1960). Pareto, Lorenz, Cournot, Bernnouli, Juran and others. *Industrial Quality Control, 17*, 25.

Juran, J. M. (1962). *Quality control handbook.* New York: McGraw-Hill.

Juran, J. M., & Gryna, F. M. (1970). *Quality planning and analysis.* New York: McGraw-Hill.

Juran, J. M., & Gryna, F. M. (1980). *Quality planning and analysis* (2nd ed.). New York: McGraw-Hill.

Juran, J. M., & Gryna, F. M. (1988). *Juran's quality control handbook* (4th ed.). New York: McGraw-Hill.

Kachigan, S. K. (1986). *Statistical analysis: An interdisciplinary introduction to univariate & multivariate methods.* New York: Redius Press.

Kackar, R. M. (1985). Off-line quality control, parameter design, and the Taguchi method. *Journal of Quality Technology, 17*, 176-188.

Kackar, R. M. (1986). Taguchi's quality philosophy: Analysis and commentary. *Quality Progress, 19*, 21-29.

Kaiser, H. F. (1958). The varimax criterion for analytic rotation in factor analysis. *Pyrometrical, 23*, 187-200.

Kaiser, H. F. (1960). The application of electronic computers to factor analysis. *Educational and Psychological Measurement, 20*, 141-151.

Kalbfleisch, J. D., & Prentice, R. L. (1980). *The statistical analysis of failure time data.* New York: Wiley.

Kane, V. E. (1986). Process capability indices. *Journal of Quality Technology, 18*, 41-52.

Kaplan, E. L., & Meier, P. (1958). Nonparametric estimation from incomplete observations. *Journal of the American Statistical Association, 53*, 457-481.

Karsten, K. G., (1925). *Charts and graphs.* New York: Prentice-Hall.

Keeves, J. P. (1972). *Educational environment and student achievement.* Melbourne: Australian Council for Educational Research.

Kendall, M. G. (1948). *Rank correlation methods.* (1st ed.). London: Griffin.

Kendall, M. G. (1975). *Rank correlation methods* (4th ed.). London: Griffin.

Kendall, M. G. (1984). *Time Series.* New York: Oxford University Press.

Kendall, M., & Ord, J. K. (1990). *Time series* (3rd ed.). London: Griffin.

Kendall, M., & Stuart, A. (1979). *The advanced theory of statistics* (Vol. 2). New York: Hafner.

Kennedy, A. D., & Gehan, E. A. (1971). Computerized simple regression methods for survival time studies. *Computer Programs in Biomedicine, 1*, 235-244.

Kennedy, W. J., & Gentle, J. E. (1980). *Statistical computing.* New York: Marcel Dekker, Inc.

Kenny, D. A. (1979). *Correlation and causality.* New York: Wiley.

Keppel, G. (1973). *Design and analysis: A researcher's handbook*. Engelwood Cliffs, NJ: Prentice-Hall.

Keppel, G. (1982). *Design and analysis: A researcher's handbook* (2nd ed.). Engelwood Cliffs, NJ: Prentice-Hall.

Keselman, H. J., Rogan, J. C., Mendoza, J. L., & Breen, L. L. (1980). Testing the validity conditions for repeated measures F tests. *Psychological Bulletin, 87*, 479-481.

Khuri, A. I., & Cornell, J. A. (1987). *Response surfaces: Designs and analyses*. New York: Marcel Dekker, Inc.

Kiefer, J., & Wolfowitz, J. (1960). The equivalence of two extremum problems. *Canadian Journal of Mathematics, 12*, 363-366.

Kim, J. O., & Mueller, C. W. (1978a). *Factor analysis: Statistical methods and practical issues*. Beverly Hills, CA: Sage Publications.

Kim, J. O., & Mueller, C. W. (1978b). *Introduction to factor analysis: What it is and how to do it*. Beverly Hills, CA: Sage Publications.

Kirk, D. B. (1973). On the numerical approximation of the bivariate normal (tetrachoric) correlation coefficient. *Psychometrika, 38*, 259-268.

Kirk, R. E. (1968). *Experimental design: Procedures for the behavioral sciences*. (1st ed.). Monterey, CA: Brooks/Cole.

Kirk, R. E. (1982). *Experimental design: Procedures for the behavioral sciences*. (2nd ed.). Monterey, CA: Brooks/Cole.

Kivenson, G. (1971). *Durability and reliability in engineering design*. New York: Hayden.

Klecka, W. R. (1980). *Discriminant analysis*. Beverly Hills, CA: Sage.

Klein, L. R. (1974). *A textbook of econometrics*. Englewood Cliffs, NJ: Prentice-Hall.

Kline, P. (1979). *Psychometrics and psychology*. London: Academic Press.

Kline, P. (1986). *A handbook of test construction*. New York: Methuen.

Kmenta, J. (1971). *Elements of econometrics*. New York: Macmillan.

Kolata, G. (1984). The proper display of data. *Science, 226*, 156-157.

Kolmogorov, A. (1941). Confidence limits for an unknown distribution function. *Annals of Mathematical Statistics, 12*, 461-463.

Korin, B. P. (1969). On testing the equality of k covariance matrices. *Biometrika, 56*, 216-218.

Kruskal, J. B. (1964). Nonmetric multidimensional scaling: A numerical method. *Pyrometrical, 29*, 1-27, 115-129.

Kruskal, J. B., & Wish, M. (1978). *Multidimensional scaling*. Beverly Hills, CA: Sage Publications.

Kruskal, W. H. (1952). A nonparametric test for the several sample problem. *Annals of Mathematical Statistics, 23*, 525-540.

Kruskal, W. H. (1975). Visions of maps and graphs. In J. Kavaliunas (Ed.), *Auto-carto II, proceedings of the international symposium on computer assisted cartography*. Washington, DC: U. S. Bureau of the Census and American Congress on Survey and Mapping.

Kruskal, W. H., & Wallis, W. A. (1952). Use of ranks in one-criterion variance analysis. *Journal of the American Statistical Association, 47*, 583-621.

Ku, H. H., & Kullback, S. (1968). Interaction in multidimensional contingency tables: An information theoretic approach. *J. Res. Nat. Bur. Standards Sect. B, 72*, 159-199.

Ku, H. H., Varner, R. N., & Kullback, S. (1971). Analysis of multidimensional contingency tables. *Journal of the American Statistical Association, 66*, 55-64.

Kullback, S. (1959). *Information theory and statistics*. New York: Wiley.

Kvålseth, T. O. (1985). Cautionary note about R^2. *The American Statistician, 39*, 279-285.

REFERENCES

Lagakos, S. W., & Kuhns, M. H. (1978). Maximum likelihood estimation for censored exponential survival data with covariates. *Applied Statistics, 27*, 190-197.

Lance, G. N., & Williams, W. T. (1966). A general theory of classificatory sorting strategies. *Computer Journal, 9*, 373.

Lance, G. N., & Williams, W. T. (1966). Computer programs for hierarchical polythetic classification ("symmetry analysis"). *Computer Journal, 9*, 60.

Larsen, W. A., & McCleary, S. J. (1972). The use of partial residual plots in regression analysis. *Technometrics, 14*, 781-790.

Lawless, J. F. (1982). *Statistical models and methods for lifetime data*. New York: Wiley.

Lawley, D. N., & Maxwell, A. E. (1971). *Factor analysis as a statistical method.* New York: American Elsevier.

Lawley, D. N., & Maxwell, A. E. (1971). *Factor analysis as a statistical method* (2nd. ed.). London: Butterworth & Company.

Lee, E. T. (1980). *Statistical methods for survival data analysis.* Belmont, CA: Lifetime Learning.

Lee, E. T., & Desu, M. M. (1972). A computer program for comparing *K* samples with right-censored data. *Computer Programs in Biomedicine, 2*, 315-321.

Lee, E. T., Desu, M. M., & Gehan, E. A. (1975). A Monte-Carlo study of the power of some two-sample tests. *Biometrika, 62*, 425-532.

Lee, S., & Hershberger, S. (1990). A simple rule for generating equivalent models in covariance structure modeling. *Multivariate Behavioral Research, 25*, 313-334.

Lehmann, E. L. (1975). *Nonparametrics: Statistical methods based on ranks.* San Francisco: Holden-Day.

Lilliefors, H. W. (1967). On the Kolmogorov-Smirnov test for normality with mean and variance unknown. *Journal of the American Statistical Association, 64*, 399-402.

Lindeman, R. H., Merenda, P. F., & Gold, R. (1980). *Introduction to bivariate and multivariate analysis*. New York: Scott, Foresman, & Co.

Lindman, H. R. (1974). *Analysis of variance in complex experimental designs.* San Francisco: W. H. Freeman & Co.

Linfoot, E. H. (1957). An informational measure of correlation. *Information and Control, 1*, 50-55.

Linn, R. L. (1968). A Monte Carlo approach to the number of factors problem. *Psychometrika, 33*, 37-71.

Lipson, C., & Sheth, N. C. (1973). *Statistical design and analysis of engineering experiments*. New York: McGraw-Hill.

Lloyd, D. K., & Lipow, M. (1977). *Reliability: Management, methods, and mathematics*. New York: McGraw-Hill.

Loehlin, J. C. (1987). *Latent variable models: An introduction to latent, path, and structural analysis.* Hillsdale, NJ: Erlbaum.

Long, J. S. (1983a). *Confirmatory factor analysis.* Beverly Hills: Sage.

Long, J. S. (1983b). *Covariance structure models: An introduction to LISREL.* Beverly Hills: Sage.

Longley, J. W. (1967). An appraisal of least squares programs for the electronic computer from the point of view of the user. *Journal of the American Statistical Association, 62*, 819-831.

Longley, J. W. (1984). *Least squares computations using orthogonalization methods.* New York: Marcel Dekker.

Lord, F. M. (1957). A significance test for the hypothesis that two variables measure the same trait except for errors of measurement. *Psychometrika, 22*, 207-220.

Lorenz, M. O. (1904). Methods of measuring the concentration of wealth. *American Statistical Association Publication*, *9*, 209-219.

Lucas, J. M. (1976). The design and use of cumulative sum quality control schemes. *Journal of Quality Technology*, *8*, 45-70.

Lucas, J. M. (1982). Combined Shewhart-CUSUM quality control schemes. *Journal of Quality Technology*, *14*, 89-93.

Maddala, G. S. (1977) *Econometrics*. New York: McGraw-Hill.

Maiti, S. S., & Mukherjee, B. N. (1990). A note on the distributional properties of the Jöreskog-Sörbom fit indices. *Psychometrika*, *55*, 721-726.

Makridakis, S. G. (1983). Empirical evidence versus personal experience. *Journal of Forecasting*, *2*, 295-306.

Makridakis, S. G. (1990). *Forecasting, planning, and strategy for the 21st century.* London: Free Press.

Makridakis, S. G., & Wheelwright, S. C. (1978). *Interactive forecasting: Univariate and multivariate methods* (2nd ed.). San Francisco, CA: Holden-Day.

Makridakis, S. G., & Wheelwright, S. C. (1989). *Forecasting methods for management* (5th ed.). New York: Wiley.

Makridakis, S. G., Wheelwright, S. C., & McGee, V. E. (1983). *Forecasting: Methods and applications* (2nd ed.). New York: Wiley.

Makridakis, S., Andersen, A., Carbone, R., Fildes, R., Hibon, M., Lewandowski, R., Newton, J., Parzen, R., & Winkler, R. (1982). The accuracy of extrapolation (time series) methods: Results of a forecasting competition. *Journal of Forecasting*, *1*, 11-153.

Malinvaud, E. (1970). *Statistical methods of econometrics*. Amsterdam: North-Holland Publishing Co.

Mandel, B. J. (1969). The regression control chart. *Journal of Quality Technology*, *1*, 3-10.

Mann, H. B., & Whitney, D. R. (1947). On a test of whether one of two random variables is stochastically larger than the other. *Annals of Mathematical Statistics*, *18*, 50-60.

Mann, N. R., Schafer, R. E., & Singpurwalla, N. D. (1974). *Methods for statistical analysis of reliability and life data*. New York: Wiley.

Mantel, N. (1966). Evaluation of survival data and two new rank order statistics arising in its consideration. *Cancer Chemotherapy Reports*, *50*, 163-170.

Mantel, N. (1967). Ranking procedures for arbitrarily restricted observations. *Biometrics*, *23*, 65-78.

Mantel, N. (1974). Comment and suggestion on the Yates continuity correction. *Journal of the American Statistical Association*, *69*, 378-380.

Mantel, N., & Haenszel, W. (1959). Statistical aspects of the analysis of data from retrospective studies of disease. *Journal of the National Cancer Institute*, *22*, 719-748.

Marascuilo, L. A., & McSweeney, M. (1977). *Nonparametric and distribution free methods for the social sciences.* Monterey, CA: Brooks/Cole.

Marple, S. L., Jr. (1987). *Digital spectral analysis.* Englewood Cliffs, NJ: Prentice-Hall.

Mardia, K. V., Kent, J. T., and Bibby, J. M. (1979). *Multivariate analysis*. New York: Academic Press.

Marsaglia, G. (1962). Random variables and computers. In J. Kozenik (Ed.), *Information theory, statistical decision functions, random processes: Transactions of the Third Prague Conference*. Prague: Czechoslovak Academy of Sciences.

Mason, R. L., Gunst, R. F., & Hess, J. L. (1989). *Statistical design and analysis of experiments with applications to engineering and science.* New York: Wiley.

REFERENCES

Massey, F. J., Jr. (1951). The Kolmogorov-Smirnov test for goodness of fit. *Journal of the American Statistical Association, 46*, 68-78.

Matsueda, R. L., & Bielby, W. T. (1986). Statistical power in covariance structure models. In N. B. Tuma (Ed.), *Sociological methodology.* Washington, DC: American Sociological Association.

McArdle, J. J. (1978). A structural view of structural models. Paper presented at the *Winter Workshop on Latent Structure Models Applied to Developmental Data, University of Denver, December, 1978.*

McArdle, J. J., & McDonald, R. P. (1984). Some algebraic properties of the Reticular Action Model for moment structures. *British Journal of Mathematical and Statistical Psychology, 37*, 234-251.

McCleary, R., & Hay, R. A. (1980). *Applied time series analysis for the social sciences.* Beverly Hills, CA: Sage Publications.

McDonald, R. P. (1980). A simple comprehensive model for the analysis of covariance structures. *British Journal of Mathematical and Statistical Psychology, 31*, 59-72.

McDonald, R. P. (1989). An index of goodness-of-fit based on noncentrality. *Journal of Classification, 6*, 97-103.

McDonald, R. P., & Hartmann, W. M. (1992). A procedure for obtaining initial value estimates in the RAM model. *Multivariate Behavioral Research, 27*, 57-76.

McDonald, R. P., & Mulaik, S. A. (1979). Determinacy of common factors: A nontechnical review. *Psychological Bulletin, 86*, 297-306.

McDowall, D., McCleary, R., Meidinger, E. E., & Hay, R. A. (1980). *Interrupted time series analysis.* Beverly Hills, CA: Sage Publications.

McKenzie, E. (1984). General exponential smoothing and the equivalent ARMA process. *Journal of Forecasting, 3*, 333-344.

McKenzie, E. (1985). Comments on 'Exponential smoothing: The state of the art' by E. S. Gardner, Jr. *Journal of Forecasting, 4*, 32-36.

McLain, D. H. (1974). Drawing contours from arbitrary data points. *The Computer Journal, 17*, 318-324.

McLean, R. A., & Anderson, V. L. (1984). *Applied factorial and fractional designs.* New York: Marcel Dekker.

McLeod, A. I., & Sales, P. R. H. (1983). An algorithm for approximate likelihood calculation of ARMA and seasonal ARMA models. *Applied Statistics,* 211-223 (Algorithm AS).

McNemar, Q. (1947). Note on the sampling error of the difference between correlated proportions or percentages. *Psychometrika, 12*, 153-157.

McNemar, Q. (1969). *Psychological statistics* (4th ed.). New York: Wiley.

Mels, G. (1989). *A general system for path analysis with latent variables.* M. S. Thesis: Department of Statistics, University of South Africa.

Mendoza, J. L., Markos, V. H., & Gonter, R. (1978). A new perspective on sequential testing procedures in canonical analysis: A Monte Carlo evaluation. *Multivariate Behavioral Research, 13*, 371-382.

Meredith, W. (1964). Canonical correlation with fallible data. *Psychometrika, 29*, 55-65.

Miller, R. (1981). *Survival analysis.* New York: Wiley.

Milligan, G. W. (1980). An examination of the effect of six types of error perturbation on fifteen clustering algorithms. *Psychometrika, 45*, 325-342.

Milliken, G. A., & Johnson, D. E. (1984). *Analysis of messy data: Vol. I. Designed experiments.* New York: Van Nostrand Reinhold, Co.

REFERENCES

Mitchell, T. J. (1974a). Computer construction of "D-optimal" first-order designs. *Technometrics, 16*, 211-220.

Mitchell, T. J. (1974b). An algorithm for the construction of "D-optimal" experimental designs. *Technometrics, 16*, 203-210.

Monro, D. M. (1975). Complex discrete fast Fourier transform. *Applied Statistics, 24*, 153-160.

Monro, D. M., & Branch, J. L. (1976). The chirp discrete Fourier transform of general length. *Applied Statistics, 26*, 351-361.

Montgomery, D. C. (1976). *Design and analysis of experiments*. New York: Wiley.

Montgomery, D. C. (1985). *Statistical quality control*. New York: Wiley.

Montgomery, D. C. (1991) *Design and analysis of experiments* (3rd ed.). New York: Wiley.

Montgomery, D. C., & Wadsworth, H. M. (1972). Some techniques for multivariate quality control applications. *Technical Conference Transactions*. Washington, DC: American Society for Quality Control.

Montgomery, D. C., Johnson, L. A., & Gardiner, J. S. (1990). *Forecasting and time series analysis* (2nd ed.). New York: McGraw-Hill.

Mood, A. M. (1954). *Introduction to the theory of statistics*. New York: McGraw Hill.

Morris, M., & Thisted, R. A. (1986). Sources of error in graphical perception: A critique and an experiment. *Proceedings of the Section on Statistical Graphics, American Statistical Association*, 43-48.

Morrison, A. S., Black, M. M., Lowe, C. R., MacMahon, B., & Yuasa, S. (1973). Some international differences in histology and survival in breast cancer. *International Journal of Cancer, 11*, 261-267.

Morrison, D. (1967). *Multivariate statistical methods*. New York: McGraw-Hill.

Moses, L. E. (1952). Non-parametric statistics for psychological research. *Psychological Bulletin, 49*, 122-143.

Mulaik, S. A. (1972). *The foundations of factor analysis*. New York: McGraw Hill.

Muth, J. F. (1960). Optimal properties of exponentially weighted forecasts. *Journal of the American Statistical Association, 55*, 299-306.

Nachtsheim, C. J. (1979). *Contributions to optimal experimental design*. Ph.D. thesis, Department of Applied Statistics, University of Minnesota.

Nachtsheim, C. J. (1987). Tools for computer-aided design of experiments. *Journal of Quality Technology, 19*, 132-160.

Nelder, J. A., & Mead, R. (1965). A Simplex method for function minimization. *Computer Journal, 7*, 308-313.

Nelson, L. (1984). The Shewhart control chart - tests for special causes. *Journal of Quality Technology, 15*, 237-239.

Nelson, L. (1985). Interpreting Shewhart X-bar control charts. *Journal of Quality Technology, 17*, 114-116.

Nelson, W. (1982). *Applied life data analysis*. New York: Wiley.

Neter, J., Wasserman, W., & Kutner, M. H. (1985). *Applied linear statistical models: Regression, analysis of variance, and experimental designs*. Homewood, IL: Irwin.

Neter, J., Wasserman, W., & Kutner, M. H. (1989). *Applied linear regression models* (2nd ed.). Homewood, IL: Irwin.

Nisbett, R. E., Fong, G. F., Lehman, D. R., & Cheng, P. W. (1987). Teaching reasoning. *Science, 238*, 625-631.

Noori, H. (1989). The Taguchi methods: Achieving design and output quality. *The Academy of Management Executive, 3*, 322-326.

Nunally, J. C. (1970). *Introduction to psychological measurement*. New York: McGraw-Hill.

REFERENCES

Nunnally, J. C. (1978). *Psychometric theory*. New York: McGraw-Hill.

Nussbaumer, H. J. (1982). *Fast Fourier transforms and convolution algorithms* (2nd ed.). New York: Springer-Verlag.

O'Brien, R. G., & Kaiser, M. K. (1985). MANOVA method for analyzing repeated measures designs: An extensive primer. *Psychological Bulletin, 97*, 316-333.

O'Neill, R. (1971). Function minimization using a Simplex procedure. *Applied Statistics, 3*, 79-88.

Okunade, A. A., Chang, C. F., & Evans, R. D. (1993). Comparative analysis of regression output summary statistics in common statistical packages. *The American Statistician, 47*, 298-303.

Olds, E. G. (1949). The 5% significance levels for sums of squares of rank differences and a correction. *Annals of Mathematical Statistics, 20*, 117-118.

Olson, C. L. (1976). On choosing a test statistic in multivariate analysis of variance. *Psychological Bulletin, 83*, 579-586.

Ostle, B., & Malone, L. C. (1988). *Statistics in research: Basic concepts and techniques for research workers* (4th ed.). Ames, IA: Iowa State Press.

Ostrom, C. W. (1978). *Time series analysis: Regression techniques*. Beverly Hills, CA: Sage Publications.

Overall, J. E., & Speigel, D. K. (1969). Concerning least squares analysis of experimental data. *Psychological Bulletin, 83*, 579-586.

Page, E. S. (1954). Continuous inspection schemes. *Biometrics, 41*, 100-114.

Page, E. S. (1961). Cumulative sum charts. *Technometrics, 3*, 1-9.

Palumbo, F. A., & Strugala, E. S. (1945). Fraction defective of battery adapter used in handie-talkie. *Industrial Quality Control, November*, 6-8.

Pankratz, A. (1983). *Forecasting with univariate Box-Jenkins models: Concepts and cases*. New York: Wiley.

Parzen, E. (1961). Mathematical considerations in the estimation of spectra: Comments on the discussion of Messers, Tukey, and Goodman. *Technometrics, 3*, 167-190; 232-234.

Patil, K. D. (1975). Cochran's Q test: Exact distribution. *Journal of the American Statistical Association, 70*, 186-189.

Peace, G. S. (1993). *Taguchi methods: A hands-on approach*. Milwaukee, WI: ASQC.

Pearson, K., (Ed.). (1968). *Tables of incomplete beta functions* (2nd ed.). Cambridge, MA: Cambridge University Press.

Pedhazur, E. J. (1973). *Multiple regression in behavioral research*. New York: Holt, Rinehart, & Winston.

Pedhazur, E. J. (1982). *Multiple regression in behavioral research* (2nd ed.). New York: Holt, Rinehart, & Winston.

Peressini, A. L., Sullivan, F. E., & Uhl, J. J., Jr. (1988). *The mathematics of nonlinear programming*. New York: Springer.

Peto, R., & Peto, J. (1972). Asymptotically efficient rank invariant procedures. *Journal of the Royal Statistical Society, 135*, 185-207.

Phadke, M. S. (1989). *Quality engineering using robust design*. Englewood Cliffs, NJ: Prentice-Hall.

Piepel, G. F. (1988). Programs for generating extreme vertices and centroids of linearly constrained experimental regions. *Journal of Quality Technology, 20*, 125-139.

Pike, M. C. (1966). A method of analysis of certain class of experiments in carcinogenesis. *Biometrics, 22*, 142-161.

Pillai, K. C. S. (1965). On the distribution of the largest characteristic root of a matrix in multivariate analysis. *Biometrika, 52*, 405-414.

Plackett, R. L., & Burman, J. P. (1946). The design of optimum multifactorial experiments. *Biometrika, 34*, 255-272.

Porebski, O. R. (1966). Discriminatory and canonical analysis of technical college data. *British Journal of Mathematical and Statistical Psychology, 19*, 215-236.

Powell, M. J. D. (1964). An efficient method for finding the minimum of a function of several variables without calculating derivatives. *Computer Journal, 7*, 155-162.

Prentice, R. (1973). Exponential survivals with censoring and explanatory variables. *Biometrika, 60*, 279-288.

Priestley, M. B. (1981). *Spectral analysis and time series*. New York: Academic Press.

Pyzdek, T. (1989). *What every engineer should know about quality control*. New York: Marcel Dekker.

Ralston, A., & Wilf, H.S. (Eds.). (1960). *Mathematical methods for digital computers*. New York: Wiley.

Ralston, A., & Wilf, H.S. (Eds.). (1967). *Mathematical methods for digital computers* (Vol. II). New York: Wiley.

Randles, R. H., & Wolfe, D. A. (1979). *Introduction to the theory of nonparametric statistics*. New York: Wiley.

Rao, C. R. (1951). An asymptotic expansion of the distribution of Wilks' criterion. *Bulletin of the International Statistical Institute, 33*, 177-181.

Rao, C. R. (1965). *Linear statistical inference and its applications*. New York: Wiley.

Rhoades, H. M., & Overall, J. E. (1982). A sample size correction for Pearson chi-square in 2 x 2 contingency tables. *Psychological Bulletin, 91*, 418-423.

Ripley, B. D. (1981). *Spacial statistics*. New York: Wiley.

Rogan, J. C., Keselman, J. J., & Mendoza, J. L. (1979). Analysis of repeated measurements. *British Journal of Mathematical and Statistical Psychology, 32*, 269-286.

Rosenberg, S. (1977). New approaches to the analysis of personal constructs in person perception. In A. Landfield (Ed.), *Nebraska symposium on motivation* (Vol. 24). Lincoln, NE: University of Nebraska Press.

Rosenberg, S., & Sedlak, A. (1972). Structural representations of implicit personality theory. In L. Berkowitz (Ed.), *Advances in experimental social psychology* (Vol. 6). New York: Academic Press.

Roskam, E. E., & Lingoes, J. C. (1970). *MINISSA-I*: A Fortran IV program for the smallest space analysis of square symmetric matrices. *Behavioral Science, 15*, 204-205.

Ross, P. J. (1988). *Taguchi techniques for quality engineering: Loss function, orthogonal experiments, parameter, and tolerance design*. Milwaukee, WI: ASQC.

Roy, J. (1958). Step-down procedure in multivariate analysis. *Annals of Mathematical Statistics, 29*, 1177-1187.

Roy, J. (1967). *Some aspects of multivariate analysis*. New York: Wiley.

Roy, R. (1990). *A primer on the Taguchi method*. Milwaukee, WI: ASQC.

Royston, J. P. (1982). An extension of Shapiro and Wilk's W test for normality to large samples. *Applied Statistics, 31*, 115-124.

Rozeboom, W. W. (1979). Ridge regression: Bonanza or beguilement? *Psychological Bulletin, 86*, 242-249.

Rozeboom, W. W. (1988). Factor indeterminacy: the saga continues. *British Journal of Mathematical and Statistical Psychology, 41*, 209-226.

REFERENCES

Runyon, R. P., & Haber, A. (1976). *Fundamentals of behavioral statistics*. Reading, MA: Addison-Wesley.

Ryan, T. P. (1989). *Statistical methods for quality improvement.* New York: Wiley.

Sandler, G. H. (1963). *System reliability engineering*. Englewood Cliffs, NJ: Prentice-Hall.

SAS Institute, Inc. (1982). *SAS user's guide: Statistics, 1982 Edition.* Cary, NC: SAS Institute, Inc.

Satorra, A., & Saris, W. E. (1985). Power of the likelihood ratio test in covariance structure analysis. *Psychometrika, 50,* 83-90.

Saxena, K. M. L., & Alam, K. (1982). Estimation of the noncentrality parameter of a chi squared distribution. *Annals of Statistics, 10,* 1012-1016.

Scheffé, H. (1953). A method for judging all possible contrasts in the analysis of variance. *Biometrica, 40,* 87-104.

Scheffé, H. (1959). *The analysis of variance.* New York: Wiley.

Scheffé, H. (1963). The simplex-centroid design for experiments with mixtures. *Journal of the Royal Statistical Society, B25,* 235-263.

Scheffé, H., & Tukey, J. W. (1944). A formula for sample sizes for population tolerance limits. *Annals of Mathematical Statistics, 15,* 217.

Scheines, R. (1994). Causation, indistinguishability, and regression. In F. Faulbaum, (Ed.), *SoftStat '93. Advances in statistical software 4.* Stuttgart: Gustav Fischer Verlag.

Schiffman, S. S., Reynolds, M. L., & Young, F. W. (1981). *Introduction to multidimensional scaling: Theory, methods, and applications*. New York: Academic Press.

Schmidt, P., & Muller, E. N. (1978). The problem of multicollinearity in a multistage causal alienation model: A comparison of ordinary least squares, maximum-likelihood and ridge estimators. *Quality and Quantity, 12,* 267-297.

Schmidt, P., & Sickles, R. (1975). On the efficiency of the Almon lag technique. *International Economic Review, 16,* 792-795.

Schmidt, P., & Waud, R. N. (1973). The Almon lag technique and the monetary versus fiscal policy debate. *Journal of the American Statistical Association, 68,* 11-19.

Schnabel, R. B., Koontz, J. E., and Weiss, B. E. (1985). A modular system of algorithms for unconstrained minimization. *ACM Transactions on Mathematical Software, 11,* 419-440.

Schneider, H. (1986). *Truncated and censored samples from normal distributions*. New York: Marcel Dekker.

Schönemann, P. H., & Steiger, J. H. (1976). Regression component analysis. *British Journal of Mathematical and Statistical Psychology, 29,* 175-189.

Schrock, E. M. (1957). *Quality control and statistical methods*. New York: Reinhold Publishing.

Schwarz, G. (1978). Estimating the dimension of a model. *Annals of Statistics, 6,* 461-464.

Scott, D. W. (1979). On optimal and data-based histograms. *Biometrika, 66,* 605-610.

Searle, S. R. (1987). *Linear models for unbalanced data*. New York: Wiley.

Searle, S. R., Casella, G., & McCullock, C. E. (1992). *Variance components*. New York: Wiley.

Seber, G. A. F., & Wild, C. J. (1989). *Nonlinear regression*. New York: Wiley.

Sebestyen, G. S. (1962). *Decision making processes in pattern recognition*. New York: Macmillan.

Sen, P. K., & Puri, M. L. (1968). On a class of multivariate multisample rank order tests, II: Test for homogeneity of dispersion matrices. *Sankhya, 30*, 1-22.

Shapiro, A., & Browne, M. W. (1983). On the investigation of local identifiability: A counter example. *Psychometrika, 48*, 303-304.

Shapiro, S. S., Wilk, M. B., & Chen, H. J. (1968). A comparative study of various tests of normality. *Journal of the American Statistical Association, 63*, 1343-1372.

Shewhart, W. A. (1931). *Economic control of quality of manufactured product.* New York: D. Van Nostrand.

Shewhart, W. A. (1939). *Statistical method from the viewpoint of quality.* Washington, DC: The Graduate School Department of Agriculture.

Shirland, L. E. (1993). *Statistical quality control with microcomputer applications.* New York: Wiley.

Shiskin, J., Young, A. H., & Musgrave, J. C. (1967). *The X-11 variant of the census method II seasonal adjustment program.* Technical paper no. 15. Washington, DC: Bureau of the Census.

Shumway, R. H. (1988). *Applied statistical time series analysis.* Englewood Cliffs, NJ: Prentice Hall.

Siegel, A. E. (1956). Film-mediated fantasy aggression and strength of aggressive drive. *Child Development, 27*, 365-378.

Siegel, S. (1956). *Nonparametric statistics for the behavioral sciences.* New York: McGraw-Hill.

Siegel, S., & Castellan, N. J. (1988). *Nonparametric statistics for the behavioral sciences* (2nd ed.) New York: McGraw-Hill.

Simkin, D., & Hastie, R. (1986). Towards an information processing view of graph perception. *Proceedings of the Section on Statistical Graphics, American Statistical Association*, 11-20.

Sinha, S. K., & Kale, B. K. (1980). *Life testing and reliability estimation.* New York: Halstead.

Smirnov, N. V. (1948). Table for estimating the goodness of fit of empirical distributions. *Annals of Mathematical Statistics, 19*, 279-281.

Smith, D. J. (1972). *Reliability engineering.* New York: Barnes & Noble.

Smith, K. (1953). Distribution-free statistical methods and the concept of power efficiency. In L. Festinger and D. Katz (Eds.), *Research methods in the behavioral sciences* (pp. 536-577). New York: Dryden.

Sneath, P. H. A., & Sokal, R. R. (1973). *Numerical taxonomy.* San Francisco: W. H. Freeman & Co.

Snee, R. D. (1975). Experimental designs for quadratic models in constrained mixture spaces. *Technometrics, 17*, 149-159.

Snee, R. D. (1979). Experimental designs for mixture systems with multi-component constraints. *Communications in Statistics - Theory and Methods, A8(4)*, 303-326.

Snee, R. D. (1985). Computer-aided design of experiments - some practical experiences. *Journal of Quality Technology, 17*, 222-236.

Snee, R. D. (1986). An alternative approach to fitting models when re-expression of the response is useful. *Journal of Quality Technology, 18*, 211-225.

Sokal, R. R., & Mitchener, C. D. (1958). A statistical method for evaluating systematic relationships. *University of Kansas Science Bulletin, 38*, 1409.

Sokal, R. R., & Sneath, P. H. A. (1963). *Principles of numerical taxonomy.* San Francisco: W. H. Freeman & Co.

Spirtes, P., Glymour, C., & Scheines, R. (1993). *Causation, prediction, and search.* Lecture Notes in Statistics, V. 81. New York: Springer-Verlag.

REFERENCES

Spjotvoll, E., & Stoline, M. R. (1973). An extension of the *T*-method of multiple comparison to include the cases with unequal sample sizes. *Journal of the American Statistical Association*, *68*, 976-978.

Springer, M. D. (1979). *The algebra of random variables*. New York: Wiley.

Spruill, M. C. (1986). Computation of the maximum likelihood estimate of a noncentrality parameter. *Journal of Multivariate Analysis*, *18*, 216-224.

Steiger, J. H. (1979). Factor indeterminacy in the 1930's and in the 1970's; some interesting parallels. *Psychometrika*, *44*, 157-167.

Steiger, J. H. (1980a). Tests for comparing elements of a correlation matrix. *Psychological Bulletin*, *87*, 245-251.

Steiger, J. H. (1980b). Testing pattern hypotheses on correlation matrices: Alternative statistics and some empirical results. *Multivariate Behavioral Research*, *15*, 335-352.

Steiger, J. H. (1988). Aspects of person-machine communication in structural modeling of correlations and covariances. *Multivariate Behavioral Research*, *23*, 281-290.

Steiger, J. H. (1989). *EzPATH: A supplementary module for SYSTAT and SYGRAPH*. Evanston, IL: SYSTAT, Inc.

Steiger, J. H. (1990). Some additional thoughts on components and factors. *Multivariate Behavioral Research*, *25*, 41-45.

Steiger, J. H., & Browne, M. W. (1984). The comparison of interdependent correlations between optimal linear composites. *Psychometrika*, *49*, 11-24.

Steiger, J. H., & Hakstian, A. R. (1982). The asymptotic distribution of elements of a correlation matrix: Theory and application. *British Journal of Mathematical and Statistical Psychology*, *35*, 208-215.

Steiger, J. H., & Lind, J. C. (1980). Statistically-based tests for the number of common factors. Paper presented at the annual Spring Meeting of the Psychometric Society in Iowa City, IA. May 30, 1980.

Steiger, J. H., & Schönemann, P. H. (1978). A history of factor indeterminacy. In S. Shye, (Ed.), *Theory Construction and Data Analysis in the Social Sciences*. San Francisco: Jossey-Bass.

Steiger, J. H., Shapiro, A., & Browne, M. W. (1985). On the multivariate asymptotic distribution of sequential chi-square statistics. *Psychometrika*, *50*, 253-264.

Stelzl, I. (1986). Changing causal relationships without changing the fit: Some rules for generating equivalent LISREL models. *Multivariate Behavioral Research*, *21*, 309-331.

Stenger, F. (1973). Integration formula based on the trapezoid formula. *Journal of the Institute of Mathematics and Applications*, *12*, 103-114.

Stevens, J. (1986). *Applied multivariate statistics for the social sciences*. Hillsdale, NJ: Erlbaum.

Stevens, W. L. (1939). Distribution of groups in a sequence of alternatives. *Annals of Eugenics*, *9*, 10-17.

Stewart, D. K., & Love, W. A. (1968). A general canonical correlation index. *Psychological Bulletin*, *70*, 160-163.

Steyer, R. (1992). *Theorie causale regressionsmodelle* [Theory of causal regression models]. Stuttgart: Gustav Fischer Verlag.

Steyer, R. (1994). Principles of causal modeling: a summary of its mathematical foundations and practical steps. In F. Faulbaum, (Ed.), *SoftStat '93. Advances in statistical software 4*. Stuttgart: Gustav Fischer Verlag.

Taguchi, G. (1987). *Jikken keikakuho* (3rd ed., Vol I & II). Tokyo: Maruzen. English translation edited by D. Clausing. *System of experimental design*. New York: UNIPUB/Kraus International

Tanaka, J. S., & Huba, G. J. (1985). A fit index for covariance structure models under arbitrary GLS estimation. *British Journal of Mathematical and Statistical Psychology, 38*, 197-201.

Tanaka, J. S., & Huba, G. J. (1989). A general coefficient of determination for covariance structure models under arbitrary GLS estimation. *British Journal of Mathematical and Statistical Psychology, 42*, 233-239.

Tatsuoka, M. M. (1970). *Discriminant analysis*. Champaign, IL: Institute for Personality and Ability Testing.

Tatsuoka, M. M. (1971). *Multivariate analysis*. New York: Wiley.

Tatsuoka, M. M. (1976). Discriminant analysis. In P. M. Bentler, D. J. Lettieri, and G. A. Austin (Eds.), *Data analysis strategies and designs for substance abuse research*. Washington, DC: U.S. Government Printing Office.

Thorndyke, R. L., & Hagen, E. P. (1977). *Measurement and evaluation in psychology and education*. New York: Wiley.

Thurstone, L. L. (1947). *Multiple factor analysis*. Chicago: University of Chicago Press.

Timm, N. H. (1975). *Multivariate analysis with applications in education and psychology*. Monterey, CA: Brooks/Cole.

Timm, N. H., & Carlson, J. (1973). *Multivariate analysis of non-orthogonal experimental designs using a multivariate full rank model*. Paper presented at the American Statistical Association Meeting, New York.

Timm, N. H., & Carlson, J. (1975). Analysis of variance through full rank models. *Multivariate behavioral research monographs*, No. 75-1.

Tribus, M., & Sconyi, G. (1989). An alternative view of the Taguchi approach. *Quality Progress, 22*, 46-48.

Trivedi, P. K., & Pagan, A. R. (1979). Polynomial distributed lags: A unified treatment. *Economic Studies Quarterly, 30*, 37-49.

Tucker, L. R., Koopman, R. F., & Linn, R. L. (1969). Evaluation of factor analytic research procedures by means of simulated correlation matrices. *Psychometrika, 34*, 421-459.

Tufte, E. R. (1983). *The visual display of quantitative information*. Cheshire, CT: Graphics Press.

Tufte, E. R. (1990). *Envisioning information*. Cheshire, CT: Graphics Press.

Tukey, J. W. (1953). *The problem of multiple comparisons*. Unpublished manuscript, Princeton University.

Tukey, J. W. (1967). An introduction to the calculations of numerical spectrum analysis. In B. Harris (Ed.), *Spectral analysis of time series*. New York: Wiley.

Tukey, J. W. (1977). *Exploratory data analysis*. Reading, MA: Addison-Wesley.

Tukey, J. W. (1984). *The collected works of John W. Tukey*. Monterey, CA: Wadsworth.

Tukey, P. A. (1986). A data analyst's view of statistical plots. *Proceedings of the Section on Statistical Graphics, American Statistical Association*, 21-28.

Tukey, P. A., & Tukey, J. W. (1981). Graphical display of data sets in 3 or more dimensions. In V. Barnett (Ed.), *Interpreting multivariate data*. Chichester, U.K.: Wiley.

Vale, C. D., & Maurelli, V. A. (1983). Simulating multivariate nonnormal distributions. *Psychometrika, 48*, 465-471.

Vandaele, W. (1983). *Applied time series and Box-Jenkins models*. New York: Academic Press.

Vaughn, R. C. (1974). *Quality control*. Ames, IA: Iowa State Press.

REFERENCES

Velicer, W. F., & Jackson, D. N. (1990). Component analysis vs. factor analysis: some issues in selecting an appropriate procedure. *Multivariate Behavioral Research*, *25*, 1-28.

Velleman, P. F., & Hoaglin, D. C. (1981). *Applications, basics, and computing of exploratory data analysis*. Belmont, CA: Duxbury Press.

Wainer, H. (1995). Visual revelations. *Chance*, *8*, 48-54.

Wald, A. (1947). *Sequential analysis*. New York: Wiley.

Walker, J. S. (1991). *Fast Fourier transforms*. Boca Raton, FL: CRC Press.

Wallis, K. F. (1974). Seasonal adjustment and relations between variables. *Journal of the American Statistical Association*, *69*, 18-31.

Wang, C. M., & Gugel, H. W. (1986). High-performance graphics for exploring multivariate data. *Proceedings of the Section on Statistical Graphics, American Statistical Association*, 60-65.

Ward, J. H. (1963). Hierarchical grouping to optimize an objective function. *Journal of the American Statistical Association*, *58*, 236.

Wei, W. W. (1989). *Time series analysis: Univariate and multivariate methods*. New York: Addison-Wesley.

Welstead, S. T. (1994). *Neural network and fuzzy logic applications in C/C++*. New York: Wiley.

Wescott, M. E. (1947). Attribute charts in quality control. *Conference Papers, First Annual Convention of the American Society for Quality Control*. Chicago: John S. Swift Co.

Wheaton, B., Múthen, B., Alwin, D., & Summers G. (1977). Assessing reliability and stability in panel models. In D. R. Heise (Ed.), *Sociological Methodology*. New York: Wiley.

Wheeler, D. J., & Chambers, D.S. (1986). *Understanding statistical process control*. Knoxville, TN: Statistical Process Controls, Inc.

Wherry, R. J. (1984). *Contributions to correlational analysis*. New York: Academic Press.

Whitney, D. R. (1948). *A comparison of the power of non-parametric tests and tests based on the normal distribution under non-normal alternatives*. Unpublished doctoral dissertation, Ohio State University.

Whitney, D. R. (1951). A bivariate extension of the U statistic. *Annals of Mathematical Statistics*, *22*, 274-282.

Wiggins, J. S., Steiger, J. H., and Gaelick, L. (1981). Evaluating circumplexity in models of personality. *Multivariate Behavioral Research*, *16*, 263-289.

Wilcoxon, F. (1945). Individual comparisons by ranking methods. *Biometrica Bulletin*, *1*, 80-83.

Wilcoxon, F. (1947). Probability tables for individual comparisons by ranking methods. *Biometrics*, *3*, 119-122.

Wilcoxon, F. (1949). *Some rapid approximate statistical procedures*. Stamford, CT: American Cyanamid Co.

Wilde, D. J., & Beightler, C. S. (1967). *Foundations of optimization*. Englewood Cliffs, NJ: Prentice-Hall.

Wilks, S. S. (1946). *Mathematical statistics*. Princeton, NJ: Princeton University Press.

Williams, W. T., Lance, G. N., Dale, M. B., & Clifford, H. T. (1971). Controversy concerning the criteria for taxonometric strategies. *Computer Journal*, *14*, 162.

Wilson, G. A., & Martin, S. A. (1983). An empirical comparison of two methods of testing the significance of a correlation matrix. *Educational and Psychological Measurement*, *43*, 11-14.

Winer, B. J. (1962). *Statistical principles in experimental design*. New York: McGraw-Hill.

Winer, B. J. (1971). *Statistical principles in experimental design* (2nd ed.). New York: McGraw-Hill.

Wolynetz, M. S. (1979a). Maximum likelihood estimation from confined and censored normal data. *Applied Statistics*, *28*, 185-195.

Wolynetz, M. S. (1979b). Maximum likelihood estimation in a linear model from confined and censored normal data. *Applied Statistics*, *28*, 195-206.

Wonnacott, R. J., & Wonnacot, T. H. (1970). *Econometrics*. New York: Wiley.

Woodward, J. A., Bonett, D. G., & Brecht, M. L. (1990). *Introduction to linear models and experimental design*. New York: Harcourt, Brace, Jovanovich.

Woodward, J. A., & Overall, J. E. (1975). Multivariate analysis of variance by multiple regression methods. *Psychological Bulletin*, *82*, 21-32.

Woodward, J. A., & Overall, J. E. (1976). Calculation of power of the *F* test. *Educational and Psychological Measurement*, *36*, 165-168.

Woodward, J. A., Douglas, G. B., & Brecht, M. L. (1990). *Introduction to linear models and experimental design*. New York: Academic Press.

Yokoyama, Y., & Taguchi, G. (1975). *Business data analysis: Experimental regression analysis*. Tokyo: Maruzen.

Youden, W. J., & Zimmerman, P. W. (1936). Field trials with fiber pots. *Contributions from Boyce Thompson Institute*, *8*, 317-331.

Young, F. W, & Hamer, R. M. (1987). *Multidimensional scaling: History, theory, and applications*. Hillsdale, NJ: Erlbaum

Young, F. W., Kent, D. P., & Kuhfeld, W. F. (1986). Visuals: Software for dynamic hyper-dimensional graphics. *Proceedings of the Section on Statistical Graphics, American Statistical Association*, 69-74.

Younger, M. S. (1985). *A first course in linear regression* (2nd ed.). Boston: Duxbury Press.

Yuen, C. K., & Fraser, D. (1979). *Digital spectral analysis*. Melbourne: CSIRO/Pitman.

Zippin, C., & Armitage, P. (1966). Use of concomitant variables and incomplete survival information in the estimation of an exponential survival parameter. *Biometrics*, *22*, 665-672.

Zupan, J. (1982). *Clustering of large data sets*. New York: Research Studies Press.

Zwick, W. R., & Velicer, W. F. (1986). Comparison of five rules for determining the number of components to retain. *Psychological Bulletin*, *99*, 432-442.

REFERENCES

INDEX - A

A

A priori
  classification probabilities, 3074, 3088, 3099, 3101
  predictions, 3073
  versus post hoc classification, 3089
Absolute value of x, 3058
ACF (autocorrelation function), 3272, 3275, 3331
Active work area, in Time Series, 3313
Actuarial life table, 3499
Adding paths, in SEPATH, 3585
Additive error, 3009
Additive seasonality, 3285, 3288
ADF estimation, 3660
Adjusted mean scaled univariate kurtosis, 3684
Aggregated life table, 3500
Almon distributed lag, 3299, 3420
Alpha level for highlighting, 3224
Amalgamation or linkage rules
  complete linkage (furthest neighbor), 3171
  joining (tree clustering), 3178, 3185
  overview, 3171, 3175, 3178, 3185
  single linkage (nearest neighbor), 3171
  unweighted pair-group average, 3171
  unweighted pair-group centroid, 3172
  Ward's method, 3172
  weighted pair-group average, 3172
  weighted pair-group centroid, 3172
Amalgamation schedule, 3180, 3190
Analysis of variance, 3067
Animated stratification, review of 3D box plot, 3193
Append STATISTICA file, 3711
AR process, 3274
Arc sine of x, 3058
ARIMA, 3310
  abrupt-permanent impact, 3278, 3364
  abrupt-temporary impact, 3365
  ACF (autocorrelation function), 3275

ARIMA (continued)
  approximate max. likelihood method, 3277, 3358
  asymptotic standard errors, 3277
  autocorrelation, 3275, 3343, 3348, 3360
  autocorrelation of residuals, 3369
  autoregressive moving average model, 3274
  autoregressive parameters, 3358
  autoregressive process, 3274
  backcasting, 3359
  constant, 3274, 3275
  convergence criterion, 3362
  d parameter, 3274
  delta, 3279, 3364
  dialogs, options, statistics, 3357
  differencing, 3275, 3343, 3358
  ds parameter, 3276
  estimation, 3275, 3276, 3277, 3358, 3361, 3363
  exact max. likelihood method, 3277, 3358
  examples, 3341
  forecasting, 3275, 3277, 3346, 3354, 3367
  gradual-permanent impact, 3278, 3364
  identification, 3275, 3341, 3349
  impact analysis, 3363
  impact patterns, 3278, 3364
  interrupted time series, 3278, 3310, 3349, 3353, 3363
  intervention analysis, 3310, 3363
  intervention types, 3353, 3364
  invertibility, 3274
  iterations, 3362
  log transformation, 3342, 3358
  long seasonality, 3277
  MA process, 3274
  McLeod & Sales method, 3277, 3358
  Melard method, 3277, 3358
  missing data, 3314
  model, 3274
  model identification, 3275, 3341, 3349
  model parameters, 3357
  moving average parameters, 3358
  moving average process, 3274
  multiplicative seasonality, 3270, 3276, 3342
  normal probability plot, 3347
  omega, 3278, 3364

ARIMA (continued)
  overview, 3273
  p parameter, 3274, 3276
  PACF (partial autocorrelation function), 3275
  parameter covariances, 3367
  parameter estimates, 3366
  parameter estimation, 3276, 3345, 3361
  parameter standard errors, 3277
  parameter start values, 3362
  parameters, 3345
  partial autocorrelation, 3275, 3360
  partial autocorrelation of residuals, 3369
  penalty value, 3277
  permanent impact, 3278
  power transformation, 3358
  program overview, 3310
  ps parameter, 3276
  q parameter, 3274, 3276
  qs parameter, 3276
  quasi-Newton method, 3276, 3361
  random walk model, 3325
  residuals, 3278, 3347, 3354, 3368
  results, 3277, 3310, 3346, 3365
  seasonal autoregressive parameters, 3358
  seasonal differencing, 3344
  seasonal moving average parameters, 3358
  seasonality, 3276, 3342
  specifying the model, 3345, 3352
  standard error of autocorrelation, 3360
  standard errors, 3277
  start values, 3362
  stationary series, 3274, 3275
  temporary impact, 3279
  types of interventions, 3364
  white noise standard errors, 3360
ASCII fixed (see also Volume I)
  export files to, 3717
  importing files from, 3708
ASCII free (see also Volume I)
  export files to, 3716
  importing files from, 3709
  separators, 3710
Associations between design variables, 3453

INDEX - A-C

Assumptions and effects of violating assumptions
  correlations between means and variances, 3072
  homogeneity of variances/covariances, 3071
  matrix ill-conditioning, 3072, 3092
  normality, 3071, 3144
  of the Cox proportional hazard model, 3490, 3533
  sample size, 3144
  tolerance values, 3072
Asymptotic standard errors, 3018
Attenuation correction, 3113, 3123, 3128, 3129
Augmented moment matrix, 3682
Autocorrelation, 3271, 3275, 3309, 3325, 3326, 3331, 3343, 3360
Automatic model fitting, 3451, 3468
Autoregressive moving average model, 3274
Autoregressive parameters, 3358
Autoregressive process, 3274

B

Backcasting, 3359
Bartlett window, 3304, 3338
BASIC (STATISTICA) programs, useful for nonlinear regression, 3024, 3060
Batch processing/printing, 3126
  in Cluster Analysis, 3186, 3188
  in X-11 seasonal decomposition, 3406, 3407, 3412
Bentler-Bonnet non-normed fit index, 3675
Bentler-Bonnet normed fit index, 3675
Bentler-Weeks model, 3655
Binary logarithm of x (base 2), 3058
Biquartimax normalized, 3228
Biquartimax raw, 3228
Bollen's delta, 3676
Bollen's rho, 3676
Bootstrapping, 3638
Boundary constraints in structural modeling, 3666
Box and whisker plots
  by group, 3095

Box and whisker plots (continued)
  for all variables, 3080, 3094, 3121, 3127, 3129, 3145, 3147, 3156, 3220
Box plot, 3D (of reordered data matrix), 3193
Box-Cox and Box-Tidwell transformations, 3060
Box-Jenkins ARIMA, 3273, 3341
Box-Ljung Q, 3331
Breakpoint regression, 3012, 3036, 3045

C

Canonical correlation
  assumptions, 3144
  canonical R, 3149
  canonical roots, 3140, 3150
  canonical scores, 3142, 3151
  canonical variates, 3140
  canonical weights (coefficients), 3142, 3158
  clustering of cases, 3152
  correlations, 3141, 3159
  descriptive statistics, 3155
  discriminant function analysis, 3069, 3086, 3097, 3100
  eigenvalues, 3141, 3158, 3159
  example, 3147
  factor loadings, 3142, 3151
  factor structure, 3142
  factor structure vs. canonical weights, 3142
  factor structures and redundancy, 3150, 3158
  input data, 3145
  matrix file input, 3145
  matrix ill-conditioning, 3144
  missing data, 3155
  outliers, 3144
  overview, 3139
  redundancy, 3143, 3149, 3150
  saving canonical scores, 3145, 3159
  sequential significance test, 3141, 3150, 3158
  specifying the canonical analysis, 3149
  startup panel, 3155
  structure coefficients, 3151
  sum scores, 3139
  variance extracted, 3143, 3150
Canonical R, 3149

Canonical roots, 3140
Canonical scores, 3142, 3151
Canonical scores for each case, 3101
Canonical variates, 3140
Canonical weights (coefficients), 3142, 3158
Casewise deletion of missing data, 3186, 3217, 3526
Categorical variables, in Cox regression models, 3492
Causal modeling, 3545
Censored observations, 3485
Census method I, 3391, 3395
  additive seasonality, 3285, 3395
  additive trend-cycle component, 3285
  centered moving averages, 3395
  components, 3284, 3396
  computations, 3286
  dialogs, options, statistics, 3395
  example, 3391
  introductory overview, 3284
  irregular component, 3287, 3396
  missing data, 3314
  multiplicative seasonality, 3395
  multiplicative trend-cycle component, 3285
  program overview, 3311
  random component, 3287, 3396
  seasonality, 3391, 3395
  seasonally adjusted series, 3286, 3396
  trend-cycle component, 3284, 3396
Census method II (see also X-11), 3287, 3289, 3401
Centroid method, 3218
Centroids, 3074, 3089
Chebychev distance, 3171
Chi-square tests
  exponential regression model, 3535
  goodness of fit, 3218, 3225, 3451, 3487, 3493, 3495, 3501, 3507
  marginal association Chi-square, 3460
  maximum likelihood ratio Chi-square, 3451
  multiple sample test, survival analysis, 3489, 3511
  normal regression model, 3535
  partial association Chi-square, 3460

INDEX - C

Chi-square tests (continued)
    Pearson Chi-square, 3451
    proportional hazard model, 3533
    stratified analysis, regression
        models, 3494
    successive roots, 3086, 3100
    test of successive roots, 3158
City-block (Manhattan) distance, 3170
Classification
    a priori classification
        probabilities, 3074, 3088,
        3099, 3101
    a priori versus post hoc
        classification, 3073, 3089
    classification functions, 3073,
        3088, 3097
    classification matrix, 3089, 3098
    classification of cases, 3073,
        3089, 3098
    classification scores, 3073
    Mahalanobis distances, 3074,
        3089
    posterior classification
        probabilities, 3074, 3089,
        3098
Classification functions, 3073, 3088, 3097
Classification of cases, 3073, 3089, 3098
    results, 3052
Classification scores, 3073
Classifications and distances, 3192
Cluster analysis
    amalgamation rules, 3171, 3175,
        3178, 3185
    amalgamation schedule, 3180,
        3190
    analysis of variance, 3181, 3190
    animated stratification, review of
        3D box plot, 3193
    areas of application, 3169
    batch printing, 3186, 3188
    Chebychev distance, 3171
    city-block (Manhattan) distance, 3170
    cluster cases, 3185, 3187
    cluster distances, 3183
    cluster variables, 3185, 3187
    complete linkage (furthest
        neighbor), 3171
    descriptive statistics for each
        cluster, 3182
    diagonal branches, 3189

Cluster analysis (continued)
    distance matrix, 3186, 3190
    distance measures, 3178, 3186
    Euclidean distance, 3170
    example, 3177
    graph of amalgamation schedule,
        3180, 3190
    hierarchical tree plot, 3170, 3179,
        3189
    horizontal hierarchical tree plot,
        3179, 3189
    icicle plot, 3170, 3189
    identification of clusters, 3180,
        3182
    initial cluster centers, for k-means
        clustering, 3181, 3187
    joining (tree clustering), 3169,
        3177, 3185, 3188
    k-means clustering, 3173, 3181,
        3186, 3190
    linkage rules, 3171, 3175, 3178,
        3185
    Manhattan (city-block) distance,
        3170
    maximum number of iterations,
        3187
    number of clusters, 3187
    overview, 3169
    percent disagreement, 3171
    power distances, 3171, 3186
    rectangular branches, 3189
    scale tree to dlink/dmax*100,
        3189
    single linkage (nearest neighbor),
        3171
    squared Euclidean distance, 3170
    startup panel, 3185
    statistical significance testing,
        3169
    threshold value, 3188
    tree diagram, 3170, 3179, 3189
    two-way joining (block
        clustering), 3172, 3188,
        3192
    unweighted pair-group average,
        3171
    unweighted pair-group centroid,
        3172
    vertical icicle tree plot, 3179,
        3189
    vs. discriminant function analysis,
        3176
    vs. factor analysis, 3175

Cluster analysis (continued)
    vs. multidimensional scaling
        (MDS), 3175
    Ward's method, 3172
    weighted pair-group average,
        3172
    weighted pair-group centroid,
        3172
    X-ray review tool, for 3D box
        plot, 3193
Cluster distances, 3183
Clustering of cases, 3152
Coefficient of alienation, 3250, 3255
Coefficients for canonical variables, 3100
Coherency, 3307
Column specs, editing in Megafile Manager, 3721
Combining two variables into a single factor, 3202
Common logarithm of x (base 10), 3058
Communalities, 3205, 3216, 3224
    computed as multiple R-square,
        3218
Comparing groups, via Nonlinear Estimation, 3013
Comparing survival in two or more groups, 3504
Complete linkage (furthest neighbor), 3171
Complex numbers, 3301
Confirmatory factor analysis, 3201, 3208, 3545
    introductory example, 3551
Confirmatory factor analysis wizard
    correlated factors, 3554
    final dialog, 3554
    step 1, 3553
Congeneric tests, 3614
Constant, in ARIMA, 3274, 3275
Constraining parameters, 3016, 3059
Continuous response functions, 3010
Contour plot, 3193
Convergence criteria
    maximum residual cosine, 3576
    Nonlinear Estimation, 3016
    relative function change, 3576
Convert to a STATISTICA file, 3717
Corrected univariate kurtosis, 3683
Correction for attenuation, 3113, 3123, 3129

INDEX - C-D

Correlation matrix, 3119, 3125, 3127, 3128
   input file, 3211, 3217
   matrix ill-conditioning and the modified correlation
   reproduced and residual, 3215, 3225
   reviewing, 3219
   saving, 3220

Correlation structure models, 3545

Correlations, computed in Megafile Manager, 3731

Correlations between means and variances, 3072

Correlogram, 3272

COSAN Computer program, 3653

Cosine of x, 3058

Covariance matrix, factor analysis of, 3231

Covariance structure models, 3545

Cox F test, 3488, 3518

Cox proportional hazard model, 3527, 3532
   assumptions, 3490, 3533
   categorical variables (factors), 3492
   editing expressions for time-dependent covariates, 3524
   equal coefficients, different baseline h0, 3523
   experimental designs, 3492
   likelihood function, 3533
   overview, 3490
   parameter estimation, 3492, 3533
   retrieving expressions, 3525
   reviewing results, 3509
   saving expressions, 3525
   segmented time-dependent covariates, 3492
   separate baseline hazards in different groups, 3494, 3523
   specifying time-dependent covariates, 3509, 3523
   stratification, 3491
   syntax for specifying time-dependent covariates, 3524, 3533
   technical notes, 3533
   testing the proportionality assumption, 3491
   time-dependent covariates, 3490
   time-dependent covariates example, 3508, 3524
   Wald statistic, 3526, 3535

Cox-Mantel test, 3489, 3518
Create new MFM file, 3700
Cronbach's alpha, 3113, 3131
Cross-amplitude, 3307
Cross-correlation function, 3332
Cross-density, 3307
Crossproduct ratios, for logit/probit regression results, 3052
Cross-spectrum analysis
   Bartlett window, 3439
   coherency, 3307, 3438
   cosine coefficients, 3437
   cospectral density, 3438
   cross-amplitude, 3307, 3438
   cross-density, 3307
   cross-periodogram values, 3438
   Daniell window, 3439
   detrending, 3430
   frequency, 3437
   gain, 3438
   Hamming window, 3439
   introductory overview, 3300, 3306
   n largest values, 3439
   padding, 3430
   Parzen window, 3439
   period, 3437
   periodogram, 3440
   phase spectrum, 3307, 3438
   plots, 3440
   program overview, 3311
   quadrature spectrum, 3307, 3438
   results, 3311, 3437
   sine coefficients, 3438
   subtract mean, 3430
   tapering, 3337
   transformations, 3337
   Tukey window, 3439

Crosstabulation of data, 3449
Cumulative proportion surviving, 3486

D

Damped trend, 3380
Daniell window, 3303, 3338
Data Management (see also Volume I)
   archival data banks, 3696
   exchanging data with Megafile Manager, 3696
   large data files, 3696
   long records, 3696
   long text values, 3696

Data Management (continued)
   Megafile Manager, 3696
   preprocessing of large records, 3696
   text values, 3696

Data reduction method, 3201
Data windows, in spectrum analysis, 3303
dBASE data files (see also Volume I)
   export files to, 3715
   importing files from, 3707
   verify column names for correctness, 3729

Deleting items from a scale, 3115, 3122
Delta
   in ARIMA, 3279
   in exponential smoothing, 3283

Descriptive statistics, 3211, 3219, 3730
Descriptive statistics for each cluster, 3182
Design variables, 3449, 3453
Designing a reliable scale, 3114
D-hats, 3250, 3251, 3255
Differencing, 3273, 3275, 3337, 3343
Discontinuous regression models
   breakpoint regression, 3012, 3036
   comparing groups, 3013
   piecewise linear regression, 3012

Discrepancy function
   ADFG, 3575
   ADFU, 3575
   asymptotically distribution free (ADF), 3660
   generalized least squares (GLS), 3574, 3659
   general properties, 3658
   IRGLS, 3659
   maximum likelihood, 3574
   maximum Wishart likelihood (ML), 3659
   ordinary least squares (OLS), 3575, 3659

Discriminant analysis
   assumptions, 3071
   backward stepwise analysis, 3068
   canonical analysis, 3069, 3086, 3097
   canonical scores for each case, 3101
   centroids, 3074, 3089
   classification, 3072, 3088

INDEX - D-E

Discriminant analysis (continued)
   coefficients for canonical variables, 3100
   computational approach, 3067
   descriptive statistics, 3080
   discriminant functions for multiple groups, 3069
   eigenvalues, 3087
   example, 3079
   F to enter, F to remove, 3068, 3083, 3091
   factor structure coefficients, 3070, 3087, 3100
   Fisher linear discriminant analysis, 3069
   forward stepwise analysis, 3068
   group centroids, 3074, 3089
   homogeneity of variances/covariances, 3071
   Mahalanobis distances, 3074, 3089
   means of canonical variables, 3087, 3100
   multiple variables, 3068
   overview, 3067
   partial Wilks' lambda, 3084, 3103
   raw discriminant function coefficients, 3100
   reviewing results, 3084
   R-square, 3085, 3097
   specifying the analysis, 3079
   specifying the discriminant analysis, 3082
   squared Mahalanobis distances, 3089, 3097, 3098
   standardized discriminant function coefficients, 3100
   startup panel, 3091
   stepwise discriminant analysis, 3068
   stop rules, 3083
   tolerance values, 3083, 3085, 3092, 3096
   two-group discriminant function, 3069
   Wilks' lambda, 3084, 3103
Discriminant function coefficients, 3086
Discriminant functions for multiple groups, 3069
Distance matrix, 3186, 3240, 3251
Distance measures
   Chebychev distance, 3171

Distance measures (continued)
   city-block (Manhattan) distance, 3170
   Euclidean distance, 3170
   Manhattan (city-block) distance, 3170
   overview, 3170
   percent disagreement, 3171
   power distances, 3171
   squared Euclidean distance, 3170
Distributed lags analysis, 3298
   Almon distributed lag, 3299, 3420
   dialogs, options, statistics, 3421
   example, 3419
   independent variable, 3421
   introductory overview, 3298
   lag length, 3421
   method, 3421
   misspecification of lags, 3300
   model, 3299
   polynomial order, 3421
   program overview, 3311
   results, 3420
   specifying the analysis, 3419
Distributions (see also Volumes I, II)
   exponential distribution, 3531
   fitting, 3486, 3499, 3501, 3514, 3531
   Gompertz distribution, 3531
   linear exponential distribution, 3531
   Weibull distribution, 3502, 3531
Distributions and their integrals, in user-defined regression models, 3058
Double exponential smoothing, 3378
Drug responsiveness and half-maximal response, 3012
D-stars, 3250, 3251, 3255

E

Edit column specs, 3721
Editing expressions for time-dependent covariates, 3524
Eigenvalues, 3087, 3141, 3158, 3159, 3203, 3212, 3224, 3231
Endogenous variable, 3299
Epsilon, 3250
Equamax normalized, 3228
Equamax raw, 3228
Error
   additive, 3009

Error (continued)
   multiplicative, 3009
Estimated function and loss function, 3046
Estimating linear models, 3007
Estimating nonlinear models, 3007
Eta for finite difference approximation, 3048
Euclidean distance, 3170
Ex post MSE, 3281
Excel (see also Volume I)
   export files to, 3714
   import files from, 3704
Exchanging data with STATISTICA data files, 3696
Exogenous variable, 3299
Exponential distribution, 3531
Exponential order statistic, 3493, 3528, 3535
Exponential regression, 3031, 3044, 3493, 3528, 3535
Exponential smoothing, 3279, 3310, 3371
   additive seasonality, 3283, 3377
   alpha, 3280, 3372, 3382
   automatic parameter search, 3281, 3282, 3375, 3384, 3387
   damped trend, 3284, 3380
   delta, 3283, 3382
   dialogs, options, statistics, 3377
   double exponential smoothing, 3378
   error, 3281
   ex post MSE, 3281
   example, 3371
   exponential trend, 3379
   forecasting, 3279, 3280, 3371
   gamma, 3284, 3383
   grid search, 3374, 3384, 3387
   Holt's method, 3282, 3373, 3378
   initial value, 3282, 3383
   lack of fit indicators, 3281, 3388
   linear trend, 3373
   MAE (mean absolute error), 3281, 3388
   MAPE (mean absolute percentage error), 3282, 3389
   mean error, 3281
   mean percentage error, 3282
   mean square error, 3388
   missing data, 3314
   models, 3371, 3377
   MPE (mean percentage error), 3282

Exponential smoothing (continued)
  MSE (mean square error), 3281, 3388
  multiplicative seasonality, 3283, 3378
  parameters, 3382
  percentage error, 3282
  phi, 3284, 3383
  program overview, 3310
  results, 3310, 3376
  seasonality, 3282, 3374
  simple, 3280, 3372
  SSE (sum of squared error), 3281
  summary plot, 3384
  trend, 3282, 3284, 3373
  triple exponential smoothing, 3374, 3379
  unconstrained parameter estimation, 3388
  Winters' method, 3282, 3374, 3379
Export (see also Volume I)
  ASCII fixed files, 3717
  ASCII free files, 3716
  dBASE files, 3715
  Excel files, 3714
  Lotus files, 3714
  Quattro files, 3715
  SPSS files, 3715
  STATISTICA files, 3712
Extracting principal components, 3202
Extraction methods
  centroid method, 3218
  iterated communalities (MINRES method), 3218
  maximum likelihood factors, 3218
  maximum number of factors, 3218
  maximum number of iterations, 3219
  minimum change in communality, 3219
  minimum eigenvalue, 3219
  multiple R-square, computing communalities as, 3218
  principal axis method, 3218
  principal components, 3218
Extreme values, in X-11 decomposition, 3289

F

F to enter, F to remove, 3068, 3083, 3091
Factor analysis
  2D plot of loadings, 3226, 3230
  3D plot of loadings, 3226, 3229
  alternative procedures, 3210
  assumptions, 3219
  centroid method, 3218
  classification method, 3205
  combining two variables into a single factor, 3202
  communalities, 3205, 3216, 3224
  communalities computed as R-square, 3218
  confirmatory factor analysis, 3201, 3208, 3551
  covariance matrix, factor analysis of, 3231
  data reduction method, 3201
  deciding on the number of factors, 3213
  descriptive statistics, 3219
  eigenvalues, 3203, 3212, 3224, 3231
  example, 3211
  extraction method, 3212, 3218
  factor extraction, 3217
  factor loadings, 3206, 3213, 3225
  factor rotation, 3225
  factor score coefficients, 3216, 3226
  factor scores, 3208, 3216, 3226
  generalizing to the case of multiple variables, 3202
  hierarchical analysis of oblique factors, 3207, 3226
  how many factors to extract, 3203
  interpreting the factor structure, 3207
  iterated communalities, 3218
  Kaiser criterion, 3204
  matrix ill-conditioning, 3208, 3231
  matrix input, 3209
  maximum likelihood factors, 3218
  MINRES method, 3205, 3218
  moment matrix, factor analysis of, 3231
  multiple orthogonal factors, 3203
  multiple regression, 3222
  oblique factors, 3207, 3226, 3228

Factor analysis (continued)
  partial correlation, 3223
  principal axis method, 3218
  principal factors analysis, 3204
  principal factors vs. principal components, 3205
  reproduced and residual correlations, 3208, 3215, 3225
  rotating the factor structure, 3206, 3214
  rotational strategies, 3206, 3227
  scree test, 3213, 3225
  startup panel, 3217
  tolerance, 3223
Factor extraction, 3217
Factor loadings, 3151, 3206, 3213, 3225
Factor rotation, 3225
  biquartimax normalized, 3228
  biquartimax raw, 3228
  equamax normalized, 3228
  equamax raw, 3228
  quartimax normalized, 3228
  quartimax raw, 3228
  varimax normalized, 3228
  varimax raw, 3227
Factor score coefficients, 3216, 3226
Factor scores, 3208, 3216, 3226
Factor structure
  and redundancy, 3150, 3158
  coefficients, 3070, 3087, 3100
  vs. canonical weights, 3142
Fast Fourier transform, 3305, 3338
Finite difference approximation of standard errors, 3020
Fisher linear discriminant analysis, 3069
Forecasting, 3277, 3279, 3280, 3367, 3371
  in ARIMA, 3275, 3346
Forty two-fifty three (4253) H filter, 3324, 3336
Fourier analysis (see also spectrum analysis), 3300, 3338, 3425, 3429
Frequencies, tabulation in Megafile Manager, 3733
Frequency, in spectrum analysis, 3300
Function minimization algorithms, 3015

G

Gain, 3307
Gamma, in exponential smoothing, 3284
Gaussian quadrature, computation of tetrachoric correlations, 3117, 3125
Gauss-Jordan algorithm, matrix inversion, 3161
Gehan's Wilcoxon test, 3518
General growth model, 3010
General logistic regression model, 3011
General syntax conventions
    loss functions, 3059
    regression equations, 3057
GLS estimation, 3659
Gompertz distribution, 3531
Goodness of fit, 3487, 3493, 3495, 3501
Goodness of fit Chi-square, 3018, 3218, 3225
Goodness of fit indices
    adjusted population gamma index, 3672
    Bentler comparative index, 3675
    Bentler-Bonnet non-normed index, 3675
    Bentler-Bonnet normed index, 3675
    Bollen's delta, 3676
    Bollen's rho, 3676
    Browne-Cudeck single sample cross-validation index, 3674
    general theory, 3667
    Log-Linear Analysis, 3451, 3461
    Jöreskog-Sörbom adjusted GFI, 3674
    Jöreskog-Sörbom GFI, 3674
    McDonald's index of noncentrality, 3672
    noncentrality-based, 3667
    Parsimonious fit index, 3675
    population gamma index, 3670
    rescaled Akaike information criterion, 3674
    Schwarz's Bayesian criterion, 3674
    Steiger-Lind RMSEA index, 3669
    structural equation modeling, 3666

Graph of amalgamation schedule, 3180, 3190
Grid search, in exponential smoothing, 3374, 3384, 3387
Group centroids, 3074, 3089
Guttman split-half reliability, 3124, 3131
Guttman-Lingoes method, 3249

H

Half-maximal response, 3012
Hamming window, 3304, 3338
Hazard rate, 3486, 3502, 3515
Henderson curve moving average, 3409
Hessian matrix, 3018
Heywood cases, 3626
Hierarchical factor analysis, 3207, 3226
Hierarchical relations between data bases in Megafile Manager, 3696
Hierarchical tests of alternative models, 3462
Hierarchical tree plot, 3170, 3179, 3189
Histograms (see also Volume II)
    2D histograms, 3026, 3028, 3080, 3120, 3148, 3221, 3467
    3D bivariate histograms, 3221
    3D histograms, 3467
    categorized histograms, 3081, 3095
    combined histogram for all groups, 3094, 3101
    histogram for selected groups, 3094
    histogram of canonical scores for selected group, 3101
Holt's exponential smoothing model, 3282, 3373, 3378
Homogeneity of variances/covariances, 3071
Hooke-Jeeves pattern moves, 3017
Horizontal hierarchical tree plot, 3189
Householder reduced matrix, 3161
Hyperbolic sine of x, 3058

I

Icicle plot, 3170, 3189
Icon plot, 3098

Identification
    in structural equation modeling, 3662
    of clusters, in Cluster Analysis, 3180, 3182
Imaginary numbers, 3301
Impact analysis, 3278, 3310
Import (see also Volume I)
    ASCII fixed files, 3708
    ASCII free files, 3709
    ASCII free separators, 3710
    dBASE files, 3707
    Excel files, 3704
    Lotus files, 3705
    Lotus review ranges, 3705
    Quattro files, 3706
    SPSS files, 3707
    STATISTICA files, 3703
Incremental fit, 3029
Index of reliability, 3112
Initial cluster centers, for k-means clustering, 3181, 3187
Initial value, in exponential smoothing, 3282
Integrating a time series, 3337
Interaction effects, 3450
Internal-consistency reliability, 3113
Interrupted time series, 3278, 3310, 3349
Intervention analysis, 3278, 3310, 3349
Intrinsically linear regression models, 3008
Intrinsically nonlinear regression models, 3009
Invariance properties, 3679
    under a constant scaling factor (ICSF), 3680
    under changes of scale (ICS), 3680
Inverse Fourier transform, 3339
Invertibility, 3274
Irregular component, 3287
Item difficulty, 3115
Item-total statistics, 3122, 3128
Iterated communalities, 3218
Iteration dialog, 3590
Iteration problems
    solving, 3645
Iterative proportional fitting, 3451, 3475
Iterative techniques
    in structural equation modeling, 3664

J

Joining (tree clustering), 3169, 3177, 3185, 3188

K

Kaiser criterion, 3204
Kaplan-Meier product-limit estimator, 3487, 3503, 3511, 3515, 3516
Kappa, Mardia-based, 3683
K-factor interactions, 3452, 3459
K-means clustering, 3173, 3181, 3186, 3190
Kuder-Richardson-20 formula, 3113
Kurtosis (see also Volume I), 3731
    adjusted mean scaled univariate, 3684
    corrected univariate, 3683
    indices of multivariate, 3683
    indices of univariate, 3683
    mean scaled univariate, 3684
    normalized multivariate, 3683
    relative multivariate, 3683
    uncorrected univariate, 3683

L

LaGrange multiplier statistics, 3594
Leakage, 3303, 3337, 3429
Least squares estimation, 3013
Life table analysis, 3485, 3495, 3499, 3500, 3511, 3512, 3513, 3514, 3531
Likelihood function, 3533
Line search
    cubic interpolation, 3577
    golden section, 3577
    methods, 3577
    parameters, 3577
    simple stephalving, 3577
Linear exponential distribution, 3531
Linkage rules, 3171, 3175, 3178, 3185
LISREL model, 3651
Local minima, 3016
Logistic transformation, 3011
Logit and probit models, 3010, 3015, 3025, 3043
Log-linear analysis
    automatic model fitting, 3451, 3468
    convergence criterion, 3469

Log-linear analysis (continued)
    crosstabulation of data, 3449
    delta, 3468
    design variables, 3449, 3453
    example, 3457
    Freeman-Tukey deviates, 3472
    goal of the analysis, 3457
    goodness of fit, 3451, 3461
    hierarchical tests of alternative models, 3462
    interaction effects, 3450
    iterative proportional fitting, 3451, 3475
    k-factor interactions, 3452, 3459
    log-linear model, 3451
    Mantel-Haenszel test, 3455
    marginal associations, 3452, 3460, 3468
    marginal frequencies, 3449
    marginal tables, 3453, 3472
    maximum likelihood ratio Chi-square, 3472
    maximum number of iterations, 3468
    model fitting approach, 3450
    multi-way frequency tables, 3450
    partial associations, 3452, 3460, 3468
    residual frequencies, 3451
    response variables, 3449, 3453
    rules for specifying a model, 3461
    startup panel, 3465
    stepwise model selection, 3463, 3470
    structural zeros, 3454, 3468
    two-way frequency tables, 3449
Log-normal regression, 3493, 3529, 3535
Log-rank test, 3518
Loss functions, 3013, 3023, 3032, 3039, 3059
Lotus 1-2-3 (see also Volume I)
    export files to, 3714
    importing files from, 3705
    review ranges, 3705

M

MA process, 3274
MAE (mean absolute error), 3281, 3388
Mahalanobis distances, 3074, 3089

Maintaining large, archival data banks in Megafile Manager, 3696
Manhattan (city-block) distance, 3170
Manifest exogenous variables, 3575
Mantel-Haenszel test, 3455, 3489, 3496, 3532
MAPE (mean absolute percentage error), 3282, 3389
Mardia's multivariate kurtosis coefficient, 3595, 3683
Marginal associations, 3452, 3460, 3468
Marginal frequencies, 3449
Marginal tables, 3453, 3472
Math errors, 3059
Matrix
    augmented moment, 3682
    file input, 3145
    matrix ill-conditioning, 3072, 3092, 3144, 3208, 3231
    matrix plots (see also Volume II), 3049
Maximum likelihood estimation, 3014, 3015
Maximum likelihood factors, 3218, 3551
Maximum residual cosine, 3576
MCD (month for cyclical dominance), 3290
MDS (see multidimensional scaling), 3237
Mean, 3731
Mean absolute error (MAE), 3281
Mean absolute percentage error (MAPE), 3282
Mean error (ME), 3281
Mean scaled univariate kurtosis, 3684
Mean substitution of missing data, 3217, 3526
Means of canonical variables, 3087, 3100
Median smoothing, 3271, 3335
Median survival time, 3486
Megafile Manager
    add columns, 3718
    add rows, 3722
    append STATISTICA file, 3711
    ASCII free separators, 3710
    convert to a STATISTICA file, 3717
    copy columns, 3720

INDEX - M

Megafile Manager (continued)
   copy rows, 3723
   correlations, 3731
   create new MFM file, 3700
   data processing, analysis, 3696
   delete columns, 3720
   delete rows, 3724
   descriptive statistics, 3730
   edit column specs, 3721
   exchanging data with STATISTICA data files, 3696
   export to ASCII fixed, 3717
   export to ASCII free, 3716
   export to dBASE, 3715
   export to Excel, 3714
   export to Lotus, 3714
   export to Quattro, 3715
   export to SPSS, 3715
   export to STATISTICA, 3712
   file header, 3724
   find value, 3699
   frequencies, 3733
   global width, 3700
   hierarchical relations between data bases, 3696
   import ASCII fixed files, 3708
   import ASCII free files, 3709
   import dBASE files, 3707
   import Excel files, 3704
   import Lotus files, 3705
   import Quattro files, 3706
   import SPSS files, 3707
   import STATISTICA files, 3703
   importing files, 3702
   incorrect variable name, 3730
   long text values, 3696
   maintaining large, archival data banks, 3696
   merge columns, 3725
   merge files, 3724
   merge rows, 3726
   MML, 3734
   move columns, 3719
   move rows, 3723
   not relational merge, 3726
   open MFM file, 3702
   optimize size, 3728
   overlay merge, 3727
   overview, 3695
   preprocessing large records of raw data, 3696
   print MFM data, 3717

Megafile Manager (continued)
   relational hierarchical merge, 3726
   relational merge, 3726
   replace value, 3699
   save MFM file, 3702
   select columns, 3697
   select rows, 3698
   sort data, 3728
   subset, 3727
   text values, 3696
   verify column names, 3729
   view column values, 3698
Melard's exact maximum likelihood method, 3277
Merge files
   merge columns, 3725
   merge rows, 3726
   not relational merge, 3726
   overlay merge, 3727
   relational hierarchical merge, 3726
   relational merge, 3726
Method of rotating coordinates, 3017
Minimization, unconstrained, 3664
MINRES method, 3205, 3218
Misfit (sources of), 3238
Missing data deletion, 3080, 3117, 3125, 3155
   casewise deletion of missing data, 3186, 3217, 3526, 3731
   mean substitution of missing data, 3217, 3526
   pairwise deletion of missing data, 3217, 3220
   substitution by means, 3186
Missing data replacement, in Time Series
   interpolation from adjacent points, 3314
   mean of n adjacent points, 3315, 3321
   median of n adjacent points, 3315, 3321
   overall mean, 3314, 3321
   predicted values from linear trend regression, 3315, 3321
MML programming language, 3695, 3734
Model identification, ARIMA, 3275
Models for binary responses probit and logit, 3010
Moment matrix, factor analysis of, 3231, 3455

Monotone regression transformation, 3250, 3251, 3255
Monte Carlo methods, 3625
Month for cyclical dominance, 3290
Moving average parameters, 3274, 3358
Moving average smoothing, 3271, 3335
MPE (mean percentage error), 3282
Multidimensional scaling
   coefficient of alienation, 3250
   computational approach, 3237
   D-hats, 3250, 3251, 3255
   distance matrix, 3251
   D-stars, 3250, 3251, 3255
   epsilon, 3250
   estimation, 3250
   example, 3245
   graph D-hat vs. distances, 3253
   graph D-star vs. distances, 3253
   how many dimensions to specify, 3238
   interpreting the dimensions, 3239, 3247
   logic of MDS, 3237
   MDS and factor analysis, 3241
   misfit (sources of), 3238
   monotone regression transformation, 3250, 3251, 3255
   orientation of axes in MDS, 3237
   overview, 3237
   rank-image permutation, 3250, 3255
   raw stress, 3246, 3250, 3255
   reproduced and observed distances, 3246
   scree test, 3239, 3248
   Shepard diagram, 3238, 3247, 3253
   standardized stress, 3250
   starting configurations, 3249
   startup panel, 3249
   steepest descent method, 3245, 3255
Multiple group matrix format, 3573
Multiple orthogonal factors, 3203
Multiple regression analysis, 3217, 3222
Multiple regression of items with scale, 3122, 3126
Multiple sample models, in structural equation modeling, 3681
Multiplicative error, 3009

INDEX - M-P

Multiplicative seasonality, 3270, 3276, 3285, 3288, 3342
Multivariate kurtosis, 3683
Multivariate normality, assumption of, 3575, 3595
Multi-way frequency tables, 3450

N

Natural logarithm of x (base Euler), 3058
Newton-Raphson iterations, computation of tetrachoric correlations, 3117, 3125
Noncentrality-based indices, 3594
Nonlinear estimation
    BASIC (STATISTICA) programs, 3024, 3060
    Box-Cox and Box-Tidwell transformations, 3060
    breakpoint regression, 3012, 3036, 3045
    classification of cases, 3052
    comparing groups, 3013
    comparing two learning curves, 3033
    compound operators, 3058
    constants, 3058
    constraining parameters, 3016, 3059
    convergence criteria, 3016
    crossproduct ratios, 3052
    distributions and their integrals, 3058
    estimating linear and nonlinear models, 3007
    estimating the breakpoint, 3037
    estimating two different models, 3035
    eta for finite difference approximation, 3048
    evaluating the fit of the model, 3018
    exponential regression models, 3031, 3044
    finite difference approximation of standard errors, 3020
    function minimization algorithms, 3015
    functions, 3058
    general growth model, 3010
    general logistic regression model, 3011

Nonlinear estimation (continued)
    half-maximal response, 3012, 3038
    Hooke-Jeeves pattern moves, 3017
    incremental fit with more than one variable, 3029
    intrinsically linear regression models, 3008
    intrinsically nonlinear regression models, 3009
    least squares estimation, 3013
    local minima, 3016
    logical expressions, 3057
    logical operators, 3058
    logistic transformation, 3011
    logit and probit models, 3010, 3043
    logit models, 3010, 3025, 3030
    loss functions, 3013, 3032, 3039
    making nonlinear models linear, 3009
    math errors, 3059
    maximum likelihood estimation, 3014, 3028
    maximum likelihood estimation and probit/logit models, 3015
    method of rotating coordinates, 3034
    models that are nonlinear in the parameters, 3008
    normal distribution function, 3058
    odds ratios, 3052
    operators, 3057
    overview, 3007
    parameter estimates, 3052
    penalty functions, 3016, 3059
    piecewise linear regression, 3036, 3038, 3045
    polynomial regression, 3008
    precedence, of computations in user-specified models, 3058
    predicting recovery from injury, 3031
    predicting redemption of coupons, 3029
    predicting success/failure, 3025
    probit models, 3010
    proportion of variance explained, 3018
    quasi-Newton method, 3017

Nonlinear estimation (continued)
    regression in pieces, 3012, 3036, 3045
    Rosenbrock pattern search, 3017, 3034
    simplex procedure, 3017
    standard errors, 3052
    start values, 3016
    STATISTICA BASIC programs, 3024, 3060
    step sizes, 3016
    stepwise analysis, 3023, 3029
    syntax for user-specified regression models, 3057
    two-stage least squares, 3060
    user-defined models, 3046
    weighted least squares, 3014, 3015, 3024, 3060
Nonlinear in the parameters, 3008
Nonlinear in the variables, 3008
Normal distribution, 3071
Normal regression, 3493, 3529, 3535
Number of backups, in Time Series, 3313
Number of cases at risk, 3486
Number of items and reliability, 3112

O

Oblique factors, 3207, 3228
Odds ratios, for logit/probit regression results, 3052
OLS (ordinary least squares) estimation, 3659
Omega, in ARIMA, 3278
One-way analysis of variance, 3067
Optimize size, 3728
Orientation of axes in MDS, 3237
Orthogonal factors, 3203
    and oblique factors, 3207, 3226**

P

PACF (partial autocorrelation function), 3272, 3275, 3332
Pairwise deletion of missing data, 3217, 3220, 3731
Pairwise means, 3220
Pairwise standard deviations, 3220
Parameter estimation, 3051, 3052
Partial associations, 3452, 3460, 3468
Partial autocorrelation, 3272, 3275, 3332, 3360
Partial correlation, 3223

INDEX - P-R

Partial Wilks' lambda, 3084, 3096, 3103
Parzen window, 3304, 3338
Path construction tool, 3568, 3583
Paths, in SEPATH
    adding, 3585
    editing, 3585
    removing, 3585
    renumbering, 3586
PE (percentage error), 3282
Pearson r correlation coefficients, 3125
Penalty functions, 3016, 3059
Percent disagreement, 3171
Percentiles of survival function, 3504, 3516
Period, in spectrum analysis, 3300
Periodogram, 3302, 3426
Peto and Peto's Wilcoxon test, 3519
Phase spectrum, 3307
Phi, in exponential smoothing, 3284
Piecewise linear regression, 3012, 3036, 3038, 3045
Plot of fitted 2D function and observed values, 3032, 3039, 3053
Plot of fitted 3D function and observed values, 3053
Plot of observed vs. predicted values, 3019
Polynomial distributed lags, 3311
Polynomial regression, 3008
Pooled within-groups covariances and correlations, 3093
Post hoc predictions, 3073
Posterior classification probabilities, 3074, 3089, 3098
Power distances, 3171, 3186
Predefined loss functions, 3023
Predicting success/failure, 3025
Principal axis method, 3218
Principal components analysis
    extracting principal components, 3202
    extraction method, 3218
    overview, 3202
    reviewing the results of a principal components analysis, 3203
Principal factors analysis, 3204
Principal factors vs. principal components, 3205
Print MFM data, 3717
Prior daily weights, in X-11, 3407

Probability density function, 3502, 3515
Probability plots (see also Volume II)
    categorized normal probability plot, 3096
    detrended normal probability plots, 3333, 3347
    half-normal probability plots, 3019, 3333
    normal probability plots, 3019, 3028, 3221, 3333, 3347
Probit and logit models, 3011, 3025, 3043
Product-limit estimator, 3487, 3495, 3503, 3511, 3515, 3516
Proportion failing, 3486
Proportion of variability explained, 3140, 3141, 3143, 3151
Proportion surviving, 3486

Q

QCD (quarter for cyclical dominance), 3290
QL algorithm with implicit shifts, 3161
Quadrature spectrum, 3307
Quarter for cyclical dominance, 3290
Quartimax normalized, 3228
Quartimax raw, 3228
Quasi-Newton method, 3017, 3047, 3276, 3361
Quattro (see also Volume I)
    export files to, 3715
    importing files from, 3706

R

RAM model, 3653
Random walk model, 3325
Rank-image permutation, 3250, 3255
Ratio-to-moving-averages method, 3284, 3395
Raw discriminant function coefficients, 3100
Raw stress, 3250, 3255
Real and imaginary numbers, 3301
Redundancy, 3143, 3149, 3150, 3158
Reflector matrix, 3596, 3680
Regression in pieces, 3012, 3036, 3045
Regression models, 3545
Relational (hierarchical) merge, 3726

Relative function change criterion, 3576
Relative multivariate kurtosis, 3683
Reliability and item analysis
    analysis of variance, 3128
    attenuation correction, 3113, 3123, 3129
    classical testing model, 3112
    codes for dichotomized variables, 3126
    correction of attenuation, 3113, 3123, 3129
    Cronbach's alpha, 3113
    deleting items from the scale, 3115, 3122
    descriptive statistics, 3119, 3126
    designing a reliable scale, 3114
    example, 3119
    Gaussian quadrature, computation of tetrachoric
    Guttman split-half reliability, 3124, 3131
    how many more items, 3129
    index of reliability, 3112
    internal-consistency reliability, 3113
    items with zero variances, 3131
    item-total statistics, 3122, 3128
    Kuder-Richardson-20 formula, 3113
    measures of reliability, 3112
    missing data deletion, 3117, 3125
    multiple regression of items with scale, 3122, 3126
    Newton-Raphson iterations, computation of tetrachoric
    number of items and reliability, 3112
    overview, 3111
    reliability, 3112
    Spearman-Brown split-half coefficient, 3113
    split-half reliability, 3113, 3123, 3125, 3128, 3130
    standardized alpha, 3131
    startup panel, 3125
    sum scales, 3112
    tetrachoric correlations, 3115
    tetrachoric r (iterative approx), 3125
    tetrachoric r (quick cos p approx), 3125
    true scores and error, 3112

Reliability and item analysis (cont.)
    what if more items option, 3123, 3129
    what if... analyses, 3122
    zero variances, items with, 3131
Renumbering paths, 3586
Reordered data matrix, 3192
Reproduced and observed distances, 3246
Reproduced and residual correlations, 3208, 3215, 3225
Residual frequencies, 3451
Response variables, 3449, 3453
Result tables, in X-11 decomposition, 3290
Retrieving expressions for time-dependent covariates, 3525
Return the sign of x, 3058
Rosenbrock pattern search, 3017, 3034
Rotational strategies, 3206, 3227
R-square, 3085, 3097
Rules
    for path diagrams, 3559
    for the PATH1 language, 3561

S

Saving canonical scores, 3145, 3159
Saving expressions for time-dependent covariates, 3525
Scale tree to dlink/dmax*100, 3179
Scatterplots (see also Volume II)
    2D scatterplot, 3120, 3222
    3D scatterplot, 3222
    3D scatterplot with fitted quadratic surface, 3222
    categorized scatterplot, 3095
    fitted frequencies vs. components of the maximum
    fitted frequencies vs. Freeman-Tukey deviates, 3473
    fitted frequencies vs. standardized residuals, 3473
    matrix scatterplots, 3049, 3081, 3094, 3120, 3127, 3128, 3148, 3219
    observed vs. expected frequencies, 3472
    residual frequencies, 3473
    scatterplot of canonical correlations, 3159
    scatterplot of canonical scores, 3087, 3101

Scree test, 3204, 3213, 3225, 3239, 3248
Seasonal decomposition, 3284, 3311, 3391, 3395, 3401
    additive trend-cycle component, 3285
    components, 3284
    irregular component, 3287
    multiplicative trend-cycle component, 3285
    random component, 3287
    seasonally adjusted series, 3286
    trend-cycle component, 3284, 3287
    X-11 method, 3287
Seasonal differencing, 3344
Seasonality, 3270, 3276, 3277, 3282, 3288, 3342, 3391
Second order factor analysis, 3545
Seed values, in Monte Carlo analysis, 3587
Segmented time-dependent covariates, 3492
Select cases for classification, 3078
SEPATH (see Structural Equation Modeling and Path Analysis), 3539, 3545
SEPATH examples
    a multiple regression model, 3612
    comparing correlation matrices, 3622
    comparing dependent variances, 3616
    comparing factor structure in two groups, 3608
    confirmatory factor analysis, 3599
    confirmatory factor analysis with identifying constraints, 3601
    effect of peer influence on ambition, 3602
    factor analysis with an intercept variable, 3607
    home environment and math achievement, 3613
    longitudinal factor model, 3619
    multi-trait, multi-method model, 3617
    personality and drug use, 3620
    standardized solution for the effects of peer influence, 3604
    test for compound symmetry, 3621

SEPATH examples (continued)
    test theory models for sets of congeneric tests, 3614
    testing for circumplex structure, 3609
    testing for stability of a correlation matrix over time, 3611
SEPATH model, 3656
Sequential significance test, 3141, 3150, 3158
Serial correlation, 3271
Series G, 3270, 3341, 3371
Shepard diagram, 3247
Significance of discriminant functions, 3071
Significance of roots, 3086
Similarity matrix, 3240
Simple structure, 3206, 3214
Simplex procedure, 3017
Sine of x, 3058
Single linkage (nearest neighbor), 3171
Skewness, 3731
    corrected univariate, 3683
    indices of univariate, 3682
    uncorrected univariate, 3683
Smoothing, 3335, 3337
    median smoothing, 3271
    moving average smoothing, 3271
Sort data, 3728
Spearman-Brown split-half coefficient, 3113
Spectral density, 3426
Spectrum analysis, 3425
    Bartlett window, 3304, 3338, 3435
    coherency, 3307
    complex numbers, 3301
    computation, 3305
    cosine coefficients, 3434
    cross-amplitude, 3307
    cross-density, 3307
    cross-periodogram, 3307
    Daniell window, 3303, 3338, 3435
    data windows, 3303, 3337, 3434
    density estimates, 3426
    detrending, 3430
    dialogs, options, statistics, 3429
    example, 3302, 3425
    fast Fourier transformation algorithm, 3305, 3338
    frequency, 3300, 3434

INDEX - S

Spectrum analysis (continued)
    gain, 3307
    general model, 3301
    Hamming window, 3304, 3338, 3435
    imaginary numbers, 3301
    introductory overview, 3300
    leakage, 3303, 3337, 3429
    missing data, 3314
    model, 3301
    multiple regression estimates, 3301
    n largest values, 3434
    padding, 3303, 3430
    Parzen window, 3304, 3338, 3435
    period, 3300, 3434
    periodogram, 3302, 3426, 3434, 3435
    phase spectrum, 3307
    plots, 3435
    preparing data, 3304
    program overview, 3311
    quadrature-spectrum, 3307
    results, 3311, 3426, 3433
    sine coefficients, 3434
    single series Fourier analysis, 3429
    spectral density estimates, 3426, 3434, 3437
    split-cosine-bell tapering, 3303, 3429
    subtract mean, 3430
    sunspot data, 3425
    tapering, 3303, 3337, 3429
    transformations, 3337
    Tukey window, 3304, 3338, 3435
    two-series Fourier analysis, 3429
    white noise periodogram, 3304
Split-cosine-bell tapering, 3303, 3337, 3429
Split-half reliability, 3113, 3123, 3125, 3128, 3130
SPSS (see also Volume I)
    export files to, 3715
    importing files from, 3707
    verify column names for correctness, 3729
Square root of x, 3058
Squared Euclidean distance, 3170
Squared Mahalanobis distances, 3089, 3097, 3098
Standard deviation, 3731
Standard errors, 3048, 3052, 3731

Standard exponential order statistic, 3493, 3535
Standardized alpha, 3131
Standardized discriminant function coefficients, 3100
Standardized models in path analysis, 3679
Standardized stress, 3250
Start values, in Nonlinear Estimation, 3016, 3050
Starting configuration
    from a STATISTICA file, 3250
    last final, 3249
    standard Guttman-Lingoes, 3249
Stationary series, 3274, 3275
STATISTICA BASIC programs, useful for nonlinear regression, 3024, 3060
Statistical estimation in structural modeling, 3657
Steepest descent method, 3245, 3255, 3665
Step sizes, 3016, 3050
Stepwise analysis, in Nonlinear Estimation, 3023, 3029
Stepwise discriminant analysis
    backward stepwise analysis, 3068, 3092
    capitalizing on chance, 3069
    F to enter, F to remove, 3068, 3091, 3092
    forward stepwise analysis, 3068, 3091
Stepwise model selection, 3463, 3470
Stop rules, 3083
Stratified analyses, 3491, 3493
Stress measure, 3237, 3246, 3248
Structural equation modeling
    algebraic models, 3651
    analysis parameters, 3555, 3573
    analysis results, 3591
    analyzing correlations, 3574
    analyzing moments, 3574
    basic idea behind, 3546
    Bentler-Weeks model, 3655
    bootstrapping, 3638
    boundary constraints, 3666
    Chi-square goodness-of-fit statistics, 3660
    Chi-square statistic, 3593
    conceptual overview, 3545
    conditions for model identification, 3663

Structural equation modeling (cont.)
    confirmatory factor analysis example, 3551
    convergence criteria for iteration, 3666
    corrrelation structures, 3677
    COSAN model, 3653
    data to analyze, 3573
    define endogenous factors, 3566
    evaluating multivariate normality, 3643
    goodness of fit indices, 3594, 3666
    identification of a model, 3662
    indices of kurtosis, 3682
    indices of skewness, 3682
    initial values, 3576
    inputting path diagrams, 3561
    introductory examples, 3551
    iteration dialog, 3555, 3590
    iteration problems, 3645
    line search methods, 3577, 3665
    LISREL model, 3651
    measurement model, 3565
    measures of skewness, 3595
    ML estimation, 3659
    model identification, 3662
    model invariance properties, 3679
    models with intercept variables, 3681
    Monte Carlo results, 3596
    Monte Carlo setup, 3587
    multiple group analysis, 3572
    multiple group matrix format, 3573
    multiple sample models, 3681
    multivariate kurtosis measures, 3595
    normalized residuals, 3596
    path construction tool, 3568
    path diagrams, 3557
    PATH1 language rules, 3561
    program overview, 3547
    RAM model, 3653
    reflector matrices, 3680
    resolving ambiguities in path diagrams, 3560
    results dialog, 3555
    rules for path diagrams, 3559
    SEPATH model, 3656
    specifying groups, 3572
    specifying structural paths, 3566
    standardization methods, 3676
    standardized residuals, 3596

INDEX - S-T

Structural equation modeling (cont.)
  standardized solution, 3575
  start values, 3576
  startup panel, 3552, 3571
  statistical estimation theory, 3657
  structured means models, 3681
  technical aspects, 3651
  tests of assumptions, 3595
  unconstrained minimization techniques, 3664
  univariate kurtosis measures, 3595
Structural zeros, 3454, 3468
Structural equation modeling results text window, 3592
Structure coefficients, 3151
Subset Megafile Manager data files, 3727
Substitution of missing data by means, 3186
Sum scales, 3111, 3112
Summary measures, in X-11 seasonal decomposition, 3406
Surface plot (see also Volume II), 3222
Survival analysis
  aggregated life table, 3500
  censored observations, 3485
  censoring indicator variable, 3512
  codes for complete and censored responses, 3513, 3516, 3517, 3520, 3523
  comparing multiple groups, 3488, 3504, 3511, 3520
  comparing two groups, 3488, 3504, 3505, 3511, 3517
  computation of the standard exponential order statistic, 3535
  correcting intervals without terminations (deaths), 3531
  Cox proportional hazard model, 3490, 3527, 3532
  Cox-Mantel test, 3489, 3518
  cumulative proportion surviving, 3486
  distribution fitting, 3486, 3501, 3514, 3531
  equal coefficients, different baseline h0, 3523
  exponential distribution, 3531
  exponential order statistic (alpha), 3528, 3535

Survival analysis (continued)
  exponential regression, 3493, 3528, 3535
  F test, 3518
  Gehan's Wilcoxon test, 3518
  Gompertz distribution, 3531
  goodness of fit, 3487, 3493, 3495, 3501
  hazard rate, 3486, 3502, 3515
  Kaplan-Meier product-limit estimator, 3487, 3503, 3511, 3515, 3516
  life table analysis, 3485, 3511, 3512, 3513, 3514, 3531
  linear exponential distribution, 3531
  log-normal regression, 3493, 3529, 3535
  log-rank test, 3518
  Mantel-Haenszel test, 3489, 3496, 3532
  median survival time, 3486
  multiple sample test, 3489
  normal regression, 3493, 3529, 3535
  number of cases at risk, 3486
  percentiles of the survival function, 3516
  Peto and Peto's Wilcoxon test, 3519
  plot of survival function, 3501
  probability density function, 3486, 3502, 3515
  proportion failing, 3486
  proportion surviving, 3486
  regression models, 3489, 3506, 3512, 3522, 3526
  saving Workbook files, 3525
  segmented time-dependent covariates, 3492
  separate baseline hazards in different groups, 3494, 3523
  setting parameters for estimation, 3525
  specifying survival times, 3512
  specifying tabulated survival data as input, 3531
  specifying time-dependent covariates, 3509
  standard exponential order statistic, 3493, 3535
  startup panel, 3511
  stratified analysis, 3493, 3527

Survival analysis (continued)
  survival for user-defined values of the independent
  survival function, 3486, 3515
  survival function for mean values of covariates, 3527
  survival times and scores, 3520
  syntax for specifying time-dependent covariates, 3533
  time-dependent covariates, 3490, 3508, 3523
  two-sample test, 3489
  unequal proportions of censored data, 3489
  Weibull distribution, 3502, 3531
  weighted least squares, 3531
Survival distributions, 3501
Survival function, 3486, 3515
Syntax for specifying time-dependent covariates, 3524, 3533
Syntax for user-specified nonlinear regression models, 3057

T

Tangent of x, 3058
Tapering, 3303, 3337, 3429
Testing the proportionality assumption, 3491
Tetrachoric correlations, 3115, 3125
  iterative approximation, 3125
  quick cos p approximation, 3125
Theoretical construct, 3112
Three-dimensional plot of final configuration, 3240, 3252
Threshold value, 3188
Time series
  ACF (autocorrelation function), 3272
  active work area, 3313, 3319, 3323
  additive seasonality, 3285, 3288
  additive trend-cycle component, 3285
  adjustment for trading-day variation, 3294, 3295
  Almon distributed lag, 3299
  alpha, in exponential smoothing, 3280, 3372
  approximate max. likelihood method, 3277
  AR process, 3274
  ARIMA, 3273, 3274, 3275, 3310

INDEX - T

Time series (continued)
- ARIMA estimation, 3276
- ARIMA examples, 3341
- ARIMA forecasting, 3277
- ARIMA parameters, 3345, 3357
- ARIMA residuals, 3278, 3368
- ARIMA results, 3277, 3310, 3346, 3365
- ARIMA standard errors, 3277
- autocorrelation, 3271, 3275, 3309, 3325, 3326, 3331, 3343, 3348
- autoregressive process, 3274
- backcasting, 3359
- backups, 3313, 3320, 3329
- Bartlett window, 3304, 3338
- Box-Jenkins ARIMA, 3273, 3341
- Box-Ljung Q, 3331
- Census method I, 3284
- coherency, 3307
- comparison of parameter estimation methods, 3277
- complex numbers, 3301
- components, 3284
- constant, in ARIMA, 3274, 3275
- conventions, 3319
- correlogram, 3272
- cross-amplitude, 3307
- cross-correlation, 3332
- cross-density, 3307
- cross-spectrum analysis, 3300, 3306
- cross-spectrum results, 3437
- d parameter, 3274
- damped trend, 3284, 3380
- Daniell window, 3303, 3338
- data windows, 3303, 3337
- deleting series, 3314, 3329
- delta, in ARIMA, 3279, 3364
- delta, in exponential smoothing, 3283
- dialog, options, statistics, 3329
- differencing, 3273, 3275, 3337, 3343
- display/plot options, 3315
- distributed lags analysis, 3298, 3311
- endogenous variable, 3299
- ex post MSE, 3281
- exact max. ARIMA likelihood, 3277
- exogenous variable, 3299
- exponential smoothing, 3279, 3310, 3335, 3371, 3377

Time series (continued)
- exponential smoothing parameters, 3382
- exponential trend, 3379
- extreme values, 3289
- fast Fourier transform algorithm, 3305, 3338
- fitting a function, 3271
- forecasting, 3280, 3346, 3354, 3367, 3371
- forty two-fifty three (4253) H filter, 3324, 3336
- Fourier analysis, 3311, 3338
- frequency, 3300
- gain, 3307
- general conventions, 3313, 3319
- general overview, 3269
- grid search, in exponential smoothing, 3374, 3387
- Hamming window, 3304, 3338
- Holt's exponential smoothing, 3282, 3373, 3378
- identification, in ARIMA, 3341
- identifying patterns, 3270
- imaginary numbers, 3301
- integrating a series, 3337
- interrupted time series, 3349, 3353, 3363
- intervention analysis, 3278, 3310, 3349
- inverse Fourier transform, 3339
- invertibility, 3274
- irregular component, 3287
- labeling plots, 3315, 3330
- lagging a series, 3336
- leakage, 3303
- locked series, 3314, 3320, 3325
- long seasonality, 3277
- MA process, 3274
- MAE (mean absolute error), 3281, 3388
- main goals, 3269
- MAPE (mean absolute percentage error), 3282, 3389
- MCD (month for cyclical dominance), 3290
- McLeod & Sales method, 3277
- mean absolute percentage error (MAPE), 3282
- mean error, 3281
- mean percentage error, 3282
- median smoothing, 3271, 3335
- missing data, 3314, 3320

Time series (continued)
- model identification, ARIMA, 3275
- moving average parameters, 3274, 3358
- moving average smoothing, 3271, 3335
- moving median smoothing, 3335
- MPE (mean percentage error), 3282
- MSE (mean square error), 3388
- multiplicative seasonality, 3270, 3276, 3288, 3342
- multiplicative trend-cycle component, 3285
- naming conventions, 3313, 3319
- normal probability plot, 3333
- number of backups, 3313, 3320, 3329
- omega, 3278, 3364
- p parameter, 3274
- PACF (partial autocorrelation function), 3272
- parameter estimation, ARIMA, 3276
- partial autocorrelation, 3272, 3275, 3332, 3360
- Parzen window, 3304, 3338
- percentage error, 3282
- period, 3300
- periodogram, 3302
- phase spectrum, 3307
- plot options, 3315
- plotting time series, 3322, 3329
- prior daily weights, 3407
- program overview, 3309
- ps parameter, 3276
- q parameter, 3274
- QCD (quarter for cyclical dominance), 3290
- qs parameter, 3276
- quadrature spectrum, 3307
- quarterly seasonal adjustment, 3410
- quasi-Newton method, 3276, 3361
- random walk model, 3325
- ratio-to-moving-averages method, 3284
- related procedures, 3312
- removing serial dependency, 3273
- residuals, 3347
- reviewing the time series, 3321, 3322, 3330

INDEX - T-X

Time series (continued)
    saving series in active work area, 3314, 3325, 3329
    scaling of plots, 3316, 3331
    seasonal autoregressive parameters, 3358
    seasonal decomposition, 3284, 3311, 3391
    seasonal differencing, 3344
    seasonal moving average parameters, 3358
    seasonality, 3270, 3276, 3282, 3288, 3342, 3391
    serial correlation, 3271
    Series G, 3270, 3341, 3371
    shifting a series, 3336
    simple exponential smoothing, 3372
    smoothing, 3271, 3335
    spectrum analysis, 3300, 3311, 3336, 3425, 3429
    split-cosine-bell tapering, 3303, 3337, 3429
    standard error of autocorrelation, 3331, 3360
    standard error of cross-correlation, 3333
    standard error of partial autocorrelation, 3332
    stationarity, 3275
    stationary series, 3274
    subset plots, 3330
    summary measures, 3406
    tapering, 3303, 3337, 3429
    trading-day regression, 3289, 3403
    transformations, 3309, 3319, 3323, 3329, 3334
    trend, 3270, 3284
    trend subtraction, 3334
    trend-cycle component, 3284, 3287, 3289
    triple exponential smoothing, 3374
    Tukey window, 3304, 3338
    white noise, 3270
    white noise standard errors, 3360
    Winters' exponential smoothing, 3282, 3379
    X-11 charts, 3297
    X-11 result tables, 3290, 3291, 3415
    X-11 seasonal decomposition, 3287, 3311, 3401

Time-dependent covariates
    categorical variables (factors), 3492
    editing expressions, 3524
    example, 3508, 3524
    likelihood function, 3533
    overview, 3490
    parameter estimation, 3492, 3533
    retrieving expressions, 3525
    reviewing results, 3509
    saving expressions, 3525
    segmented time-dependent covariates, 3492
    specifying, 3509, 3523
    stratification, 3491
    syntax for specifying, 3524, 3533
    technical notes, 3533
    testing the proportionality assumption, 3491
Tolerance values, 3072, 3083, 3085, 3092, 3096, 3223
Trading-day adjustment, 3289, 3403
Transformations, in Time Series, 3319, 3323, 3329, 3334
Tree diagram, 3170, 3179, 3189
    diagonal branches, 3179
    hierarchical tree, 3179
    horizontal hierarchical tree plot, 3179
    icicle plot, 3170, 3189
    rectangular branches, 3179
    scale tree to dlink/dmax*100, 3179
    vertical icicle tree plot, 3179, 3189
Trend analysis, 3270
Trend component, 3270
Trend-cycle component, 3284, 3285, 3287, 3289
Triple exponential smoothing, 3374, 3379
Tukey window, 3304, 3338
Two-dimensional plot of final configuration, 3240, 3252
Two-group discriminant function, 3069
Two-stage least squares, 3060
Two-way frequency tables, 3449
Two-way joining (block clustering), 3172, 3188, 3192
    graph, 3193

U

Unconstrained minimization techniques, 3664
Uncorrected univariate kurtosis, 3683
Unweighted pair-group average, 3171
Unweighted pair-group centroid, 3172
User-specified loss functions, 3023
User-specified regression models, 3023, 3031, 3046, 3057

V

Variance extracted, 3143, 3150
Variance maximizing (varimax) rotation, 3202
Variance/covariance matrix for parameters, 3019, 3020
Varimax normalized, 3228
Varimax raw, 3227
Verify column names, 3729
Vertical icicle tree plot, 3179, 3189

W

Wald statistic, 3526, 3535
Ward's method, 3172
Weibull distribution, 3502, 3531
Weighted least squares, 3014, 3015, 3024, 3039, 3060
    for itting survival distributions, 3531
Weighted pair-group average, 3172
Weighted pair-group centroid (median), 3172
What if more items option, 3123
White noise, 3270
White noise standard errors, 3360
Wilks' lambda, 3084, 3096, 3103
Winters' exponential smoothing model, 3282, 3374
Workbook files (see also Volume I), 3525

X

X-11 method
    additive seasonality, 3288, 3401, 3405
    batch printing, 3407, 3412
    Census method II, 3289

X-11 method (continued)

charts, 3297, 3406, 3411
dates (start of series), 3406, 3411
dialogs, options, statistics, 3405
example, 3401
extreme values, 3289, 3408
final estimation of components (Part D), 3290
final trading-day variation (Part C), 3290
full printout, 3406, 3411
general model, 3288
Henderson moving average, 3409
introductory overview, 3287
length of month allowance, 3407
long printout, 3406, 3411
MCD (month for cyclical dominance), 3290
missing data, 3314
modified series (Part E), 3290
monthly adjustment, 3405
moving averages for seasonal factors, 3408, 3415
multiplicative seasonality, 3288, 3401, 3405
outliers, 3289
output detail, 3291
preliminary estimate of trading day variation (part B), 3290
printout detail, 3406, 3411
prior adjustment (part A), 3290
prior daily weights, 3407
program overview, 3311
QCD (quarter for cyclical dominance), 3290
quarterly adjustment, 3410
result tables, 3290, 3291, 3403, 3415
results, 3311
seasonality, 3288, 3401
sigma limits for graduating extremes, 3408, 3412
standard printout, 3406, 3411
strikes, 3410, 3412
summary measures, 3406, 3411
summary statistics, 3289
trading-day regression, 3289, 3403, 3407
trend-cycle component, 3289, 3409

X-11 seasonal decomposition, 3287
X-ray review tool, for 3D box plot, 3193

INDEX

STATISTICA

StatSoft

INDEX - 3782

Copyright © StatSoft, 1995